Bioprocessing for Biomolecules Production

# Bioprocessing for Biomolecules Production

*Edited by*

*Gustavo Molina*
Laboratory of Food
Biotechnology – Food Engineering
Institute of Science and Technology – UFVJM
Diamantina, Minas Gerais
Brazil

*Vijai Kumar Gupta*
ERA Chair of Green Chemistry
Department of Chemistry and Biotechnology
Tallinn University of Technology
Tallinn
Estonia

*Brahma N. Singh*
CSIR-National Botanical Research Inst
Lucknow, Uttar Pradesh
India

*Nicholas Gathergood*
ERA Chair of Green Chemistry
Tallinn University of Technology
Tallinn
Estonia

WILEY

This edition first published 2020


*Registered Offices*
John Wiley & Sons, Inc., 111 River Street, Hoboken, NJ 07030, USA

John Wiley & Sons Ltd, The Atrium, Southern Gate, Chichester, West Sussex, PO19 8SQ, UK

*Editorial Office*
The Atrium, Southern Gate, Chichester, West Sussex, PO19 8SQ, UK

For details of our global editorial offices, customer services, and more information about Wiley products visit us at www.wiley.com.

Wiley also publishes its books in a variety of electronic formats and by print-on-demand. Some content that appears in standard print versions of this book may not be available in other formats.



*Library of Congress Cataloging-in-Publication Data*

Names: Molina, Gustavo, 1983- editor. | Gupta, Vijai Kumar, editor. | Singh, Brahma N., 1981- editor. | Gathergood, Nicholas, 1972- editor.
Title: Bioprocessing for biomolecules production / Gustavo Molina, Inst. of Science & Technology, UFVJM, Diamantina, Minas Gerais, BR, Vijai K Gupta, ERA Chair of Green Chemistry, Department of Chemistry and Biotechnology, Tallinn University of Technology, Tallinn, EST, Brahma N. Singh, CSIR-National Botanical Research Inst., Lucknow, Uttar Pradesh, IN, Nicholas Gathergood, ERA Chair of Green Chemistry, Tallinn University of Technology, Tallinn, EST.
Description: First edition. | Hoboken : Wiley, 2019. | Includes bibliographical references and index.
Identifiers: LCCN 2019023222 (print) | LCCN 2019023223 (ebook) | ISBN 9781119434320 (cloth) | ISBN 9781119434368 (adobe pdf) | ISBN 9781119434405 (epub)
Subjects: LCSH: Biomolecules. | Biochemical engineering.
Classification: LCC TP247 .B565 2019 (print) | LCC TP247 (ebook) | DDC 660.6/3–dc23
LC record available at https://lccn.loc.gov/2019023222
LC ebook record available at https://lccn.loc.gov/2019023223

Cover Design: Wiley
Cover Image: © FOTOGRIN/Shutterstock

Set in 10/12pt WarnockPro by SPi Global, Chennai, India
Printed and bound in Singapore by Markono Print Media Pte Ltd

10 9 8 7 6 5 4 3 2 1

# Contents

## Contributors

*Rafael Ferraz Alves*
School of Food Engineering
University of Campinas (UNICAMP)
Campinas-SP
Brazil

Novozymes Latin America Ltda.
Araucária-PR
Brazil

*Ashokan Anushya*
Department of Biotechnology,
Kumaraguru College of Technology
Coimbatore
India

*KB Arun*
Rajiv Gahdhi Center for Biotechnology
Thiruvananthapuram
India

*Mayara C.S. Barcelos*
Laboratory of Food
Biotechnology – Food Engineering
Institute of Science and
Technology – UFVJM
Diamantina, Minas Gerais
Brazil

*Sunil Kumar Behera*
Department of Bioscience and
Bioinformatics
Khallikote University
Berhampur, Odisha
India

*Meganathan Bhuvaneswari*
Department of Biotechnology
Sona College of Arts and Science
Salem
India

*Parameswaran Binod*
Microbial Processes and Technology
Division
Council of Scientific and Industrial
Research – National Institute for
Interdisciplinary Science and Technology
(CSIR-NIIST)
Thiruvananthapuram, Kerala
India

*Eduarda Roberta Bordin*
Laboratory of Microbiology and
Bioprocesses
Federal University of Fronteira Sul
Erechim, Rio Grande do Sul
Brazil

*Gustavo Pagotto Borin*
Brazilian BioRenewables National
Laboratory (LNBR)
Brazilian Center for Research in Energy
and Materials (CNPEM)
Campinas-SP
Brazil

Graduate Program in Genetics and Molecular Biology
Institute of Biology, University of Campinas (UNICAMP)
Campinas-SP
Brazil

*Aline Frumi Camargo*
Laboratory of Microbiology and Bioprocesses
Federal University of Fronteira Sul
Erechim, Rio Grande do Sul
Brazil

*Anuj Kumar Chandel*
Department of Biotechnology, Engineering School of Lorena (EEL)
University of São Paulo
Lorena, São Paulo
Brazil

*Vinay Singh Chauhan*
Department of Biotechnology
Bundelkhand University
Jhansi
India

*Gwendoline Christophe*
Université Clermont Auvergne
CNRS, Sigma Clermont
Institut Pascal
Aubière
France

*Fabiano Jares Contesini*
Department of Biochemistry and Tissue Biology
Institute of Biology, University of Campinas – (Unicamp)
Campinas, São Paulo
Brazil

*Laurent Dufossé*
Laboratoire de Chimie des Substances Naturelles et des Sciences des Aliments
LCSNSA
Ecole Supérieure d'Ingénieurs Réunion Océan Indien ESIROI
Département Agroalimentaire
Université de La Réunion
Sainte-Clotilde
France

*Om Prakash Dwivedi*
Department of Botany
S.V.M. Science and Technology P.G. College
Lalganj, Uttar Pradesh
India

*Edina Eminagic*
Department of Bioengineering
Industrial Biotechnology and Systems Biology (IBSB) Research Group
Marmara University
Istanbul
Turkey

*Antônio Djalma Nunes Ferraz Júnior*
Centre for Environmental Policy
Imperial College
London
UK

*Leonardo de Figueiredo Vilela*
Department of Basic Sciences
Federal University of the Jequitinhonha and Mucuri Valleys
Diamantina, Minas Gerais
Brazil

*Juliana F. Floriano*
Department of Obstretrics and Gynecology, Medical School of Botucatu
São Paulo State University (UNESP)
Botucatu
Brazil

*Pierre Fontanille*
Université Clermont Auvergne
CNRS, Sigma Clermont
Institut Pascal
Aubière
France

*Edgard Gnansounou*
Ecole Polytechnique Federale de Llausanne
ENAC GR-GN
Lausanne
Switzerland

*Simone Maria Golunski*
Laboratory of Microbiology and Bioprocesses
Federal University of Fronteira Sul
Erechim, Rio Grande do Sul
Brazil

*Eduardo J. Gudiña*
CEB - Centre of Biological Engineering
Campus de Gualtar, University of Minho
Braga
Portugal

*Vijai Kumar Gupta*
ERA Chair of Green Chemistry
Department of Chemistry of Biotechnology, School of Science
Tallinn University of Technology
Tallinn
Estonia

*Abd El-Latif Hesham*
Genetics Department, Faculty of Agriculture
Assiut University
Assiut
Egypt

Muhammed Yusuf Kandur
Department of Bioengineering
Industrial Biotechnology and Systems Biology (IBSB) Research Group
Marmara University
Istanbul
Turkey

*Ceyda Kasavi*
Department of Bioengineering
Industrial Biotechnology and Systems Biology (IBSB) Research Group
Marmara University
Istanbul
Turkey

*Tanvir Kaur*
Department of Biotechnology
Akal College of Agriculture
Eternal University
Baru Sahib, Sirmour
Himachal Pradesh
India

*Divjot Kour*
Department of Biotechnology, Akal College of Agriculture
Eternal University
Baru Sahib, Sirmour
Himachal Pradesh
India

*Ashok Kumar*
Department of Biotechnology and Bioinformatics
Jaypee University of Information
Waknaghat, Solan
India

*Sunil Kumar*
Department of Biotechnology, Akal College of Agriculture
Eternal University Baru Sahib, Sirmour
Himachal Pradesh
India

***Christian Larroche***
Université Clermont Auvergne
CNRS, Sigma Clermont
Institut Pascal
Aubière
France

***Aravind Madhavan***
Microbial Processes and Technology Division
Council of Scientific and Industrial Research – National Institute for Interdisciplinary Science and Technology (CSIR-NIIST)
Thiruvananthapuram
Kerala, India

***Glaucia Maria***
Laboratory of Food Biotechnology – Food Engineering
Institute of Science and Technology – UFVJM
Diamantina, Minas Gerais
Brazil

***Sabrina Evelin Martiniano***
Department of Biotechnology, Engineering School of Lorena (EEL)
University of São Paulo
Lorena, São Paulo
Brazil

***Shashank Mishra***
Quality Control and Quality Assurance Laboratory
Biotech Park, Sector G, Jankipuram
Lucknow, Uttar Pradesh
India

***Tanuja Mishra***
Department of Biotechnology, Akal College of Agriculture
Eternal University
Baru Sahib, Sirmour
Himachal Pradesh
India

***Tatiani Andressa Modkovski***
Laboratory of Microbiology and Bioprocesses
Federal University of Fronteira Sul
Erechim, Rio Grande do Sul
Brazil

***Gustavo Molina***
Laboratory of Food Biotechnology - Food Engineering
Institute of Science and Technology - UFVJM
Diamantina, Minas Gerais
Brazil

***Altemir José Mossi***
Laboratory of Microbiology and Bioprocesses
Federal University of Fronteira Sul
Erechim, Rio Grande do Sul
Brazil

***Puneet Negi***
Department of Physics, Akal College of Basic Sciences
Eternal University
Baru Sahib, Sirmour
India

***Ebru Toksoy Oner***
Department of Bioengineering
Industrial Biotechnology and Systems Biology (IBSB) Research Group
Marmara University
Istanbul
Turkey

***Emrah Ozcan***
Department of Bioengineering
Industrial Biotechnology and Systems Biology (IBSB) Research Group
Marmara University
Istanbul
Turkey

Department of Bioengineering
Gebze Technical University
Kocaeli
Turkey

***Sandeep Kumar Panda***
School of Biotechnology
KIIT University
Bhubaneswar
India

***Ashok Pandey***
CSIR-Indian Institute of Toxicology
Research (CSIR-IITR)
Lucknow, Uttar Pradesh
India

***Bruno Nicolau Paulino***
Laboratory of Bioflavors and Bioactive
Compounds, Department of Food
Science
School of Food Engineering
University of Campinas
Campinas, São Paulo
Brazil

Faculty of Pharmaceutical Sciences
Federal University of Amazonas
Manaus
Brazil

***Glaucia Maria Pastore***
Laboratory of Bioflavors and Bioactive
Compounds, Department of Food
Science
School of Food Engineering
University of Campinas
Campinas, São Paulo
Brazil

***Franciele M. Pelissari***
Laboratory of Food
Biotechnology – Food Engineering
Institute of Science and
Technology – UFVJM
Diamantina, Minas Gerais
Brazil

***Arthur P.A. Pereira***
Luiz de Queiroz College of Agriculture
University of São Paulo
Piracicaba
Brazil

Hawkesbury Institute for the
Environment
Hawkesbury Campus
Western Sydney University
Penrith
Australia

***Marina Gabriel Pessôa***
Laboratory of Bioflavors and Bioactive
Compounds
Department of Food Science
School of Food Engineering
University of Campinas
Brazil

***Rafael Rodrigues Philippini***
Department of Biotechnology,
Engineering School of Lorena (EEL)
University of São Paulo
Lorena, São Paulo
Brazil

***HL Raghavendra***
Department of Biochemistry,
School of Medicine
Wollega Univeristy, Nekemte University
Nekemte
Ethiopia

Department of Obstetrics and
Gynecology
Medical School of Botucatu
São Paulo State University (UNESP)
Botucatu
Brazil

College of Medical and Health Sciences
Wollega University
Nekemte, Oromia
Ethiopia

***Cintia Lacerda Ramos***
Department of Basic Sciences
Federal University of the Jequitinhonha and Mucuri Valleys
Diamantina, Minas Gerais
Brazil

***Kusam Lata Rana***
Department of Biotechnology, Akal College of Agriculture
Eternal University
Baru Sahib, Sirmour
Himachal Pradesh
India

***Ali A. Rastegari***
Department of Molecular and Cell Biochemistry, Falavarjan Branch
Islamic Azad University
Isfahan
Iran

***Ramesh Chandra Ray***
ICAR-Regional Center of Central Tuber Crops Research Institute
Bhubaneswar
India

***Lígia R. Rodrigues***
CEB - Centre of Biological Engineering
Campus de Gualtar, University of Minho
Braga
Portugal

***Susana Rodriguez-Couto***
IKERBASQUE, Basque Foundation for Science
Bilbao
Spain

***Vanusa Rossetto***
Laboratory of Microbiology and Bioprocesses
Federal University of Fronteira Sul
Erechim, Rio Grande do Sul
Brazil

***Lopamudra Sahu***
Department of Botany, Utkal University
Bhuabaneswar
India

***Júlio César dos Santos***
Department of Biotechnology, Engineering School of Lorena (EEL)
University of São Paulo
Lorena, São Paulo
Brazil

***Thamarys Scapini***
Laboratory of Microbiology and Bioprocesses
Federal University of Fronteira Sul
Erechim, Rio Grande do Sul
Brazil

***Jochen Schmid***
Chemistry of Biogenic Resources
Technical University of Munich
Campus Straubing for Biotechnology and Sustainability
Straubing
Germany

Norwegian University of Science and Technology
Department of Biotechnology and Food Science
Sem Sælands vei, Trondheim
Norway

*Rosane Freitas Schwan*
Department of Biology
Federal University of Lavras
Lavras, Minas Gerais
Brazil

*Sinem Selvin Selvi*
Department of Bioengineering
Industrial Biotechnology and Systems
Biology (IBSB) Research Group
Marmara University
Istanbul
Turkey

*Sushma Sharma*
Department of Biotechnology, Akal
College of Agriculture
Eternal University
Baru Sahib, Sirmour
India

*Volker Sieber*
Chemistry of Biogenic Resources
Technical University of Munich
Campus Straubing for Biotechnology and
Sustainability
Straubing
Germany

*Silvio Silvério da Silva*
Department of Biotechnology,
Engineering School of Lorena (EEL)
University of São Paulo
Lorena, Sao Paulo
Brazil

*Bianca M.P. Silveira*
Laboratory of Food
Biotechnology – Food Engineering
Institute of Science and
Technology – UFVJM
Diamantina, Minas Gerais
Brazil

*Raveendran Sindhu*
Microbial Processes and Technology
Division
Council of Scientific and Industrial
Research – National Institute for
Interdisciplinary Science and Technology
(CSIR-NIIST)
Thiruvananthapuram
Kerala
India

*Bhanumati Singh*
Department of Biotechnology
Bundelkhand University
Jhansi
India

*Saurabh Singh*
Institute of Environment and Sustainable
Development
Banaras Hindu University
Varanasi
India

*Surabhi Singh*
Department of Botany
S.V.M. Science and Technology P.G.
College
Lalganj, Uttar Pradesh
India

*Nallusamy Sivakumar*
Department of Biology,
College of Science
Sultan Qaboos University
Muscat, Sultanate of Oman

*Selvaraju Sivamani*
Department of Biotechnology
Kumaraguru College of Technology
Coimbatore
India

*Victoria Ramos Sodré*
Graduation Program of Functional and Molecular Biology, Institute of Biology
State University of Campinas (Unicamp)
Campinas, São Paulo
Brazil

*Moorthi Swathika*
Department of Biotechnology
Kumaraguru College of Technology
Coimbatore
India

*Sapna Thakur*
Department of Biotechnology, Akal College of Agriculture
Eternal University
Baru Sahib, Sirmour
Himachal Pradesh
India

*Robson Tramontina*
Brazilian Bioethanol Science and Technology Laboratory (CTBE)
Brazilian Center for Research in Energy and Materials (CNPEM)
Campinas, São Paulo
Brazil

Graduation Program of Biosciences and Technology of Bioactive Products
Institute of Biology
State University of Campinas (Unicamp)
Campinas, São Paulo
Brazil

*Helen Treichel*
Laboratory of Microbiology and Bioprocesses
Federal University of Fronteira Sul
Erechim, Rio Grande do Sul
Brazil

*Shivaleela P. Upashe*
College of Medical and Health Sciences
Wollega University
Nekemte, Oromia
Ethiopia

College of Nursing Sciences
School of Health Sciences
Dayananda Sagar University
Shavige Malleshwara Hills
Bangalore, Karnataka
India

*Bruno Venturin*
Laboratory of Microbiology and Bioprocesses
Federal University of Fronteira Sul
Erechim, Rio Grande do Sul
Brazil

*Jay Prakash Verma*
Institute of Environment and Sustainable Development
Banaras Hindu University
Varanasi
India

Hawkesbury Institute for the Environment
Hawkesbury Campus, Western Sydney University
Penrith
Sydney
Australia

*Kele A.C. Vespermann*
Laboratory of Food Biotechnology – Food Engineering
Institute of Science and Technology – UFVJM
Diamantina, Minas Gerais
Brazil

*Natalia Videira*
Graduation Program of Biosciences and Technology of Bioactive Products, Institute of Biology
State University of Campinas (Unicamp)
Campinas, São Paulo
Brazil

Brazilian Biosciences National Laboratory (LNBio)
Brazilian Center for Research in Energy and Materials (CNPEM)
Campinas, São Paulo
Brazil

*Ajar Nath Yadav*
Department of Biotechnology, Akal College of Agriculture
Eternal University
Baru Sahib, Sirmour
Himachal Pradesh
India

*Neelam Yadav*
Gopi Nath P.G. College
Veer Bahadur Singh Purvanchal University
Jaunpur, Uttar Pradesh
India

# Part I

# General Overview of Biotechnology for Industrial Segments: An Industrial Approach

# 1

# An Overview of Biotechnological Processes in the Food Industry

*Bianca M.P. Silveira, Mayara C.S. Barcelos, Kele A.C. Vespermann, Franciele M. Pelissari, and Gustavo Molina*

*Laboratory of Food Biotechnology – Food Engineering, Institute of Science and Technology – UFVJM Diamantina, Minas Gerais, Brazil*

## 1.1 Introduction

The problems related to climate change and the rising costs of supplies and fuel associated with population growth have led to the search of renewable and sustainable technologies in order to provide products that meet market and consumers demand (Zilberman et al. 2013). This technological development is crucial for the implementation of *bioeconomics,* in which biotechnological processes are fundamental for providing economic opportunities for the chemical, food and feed, pulp and paper, textiles, automotives, electronics, and energy sectors (De Buck et al. 2016; Diehl 2017).

Plants and animals have interesting commercial compounds (e.g., flavors, polysaccharide, fatty acids), and research in recent years has focused on extracting those compounds. As a bonus, they have the advantage of being labeled as *natural,* which consumers increasingly desire (Akacha and Gargouri 2015; Nigam and Luke 2016). However, the availability of these resources is insufficient to meet industrial needs, due to seasonality, ecological, social, and political factors, in addition to the low achievable yields that limit the use of several compounds obtained from this source (Felipe et al. 2017).

The limitations encountered in the extraction processes have encouraged the search for alternative methods to obtain commercial products such as chemical synthesis (Longo and Sanromán 2006). Chemical synthesis is characterized as a low-cost process with flexible production and efficiency, and is commonly used in industrial processes (Carroll et al. 2016; Serra et al. 2005). However, the environmental damages resulting from its use, as well as the distaste for products seen as modified or *unnatural,* has resulted in disadvantages for the use of this process (Abu Yazid et al. 2017).

Bioprocesses have thus become a promising alternative for obtaining natural products due to the climatic, seasonal limitations, and the availability of natural sources used in the extraction processes. They are aimed at developing sustainable and efficient industrial processes with the potential to result in commercially applicable products, such as polysaccharides (Arisawa and Watanabe 2017; Barnard et al. 2012).

The fermentation process is one of the earliest methods of using biotechnology, considered a key component of several industrial applications that involve the use of

*Bioprocessing for Biomolecules Production,* First Edition.
Edited by Gustavo Molina, Vijai Kumar Gupta, Brahma N. Singh, and Nicholas Gathergood.

biological material. As an example, it contributes to several food additives, besides the range of other high-value-added products, which mainly serve the food industry (Chotani et al. 2017; Lokko et al. 2018). In 2015, the market for fermentation-derived products was valued at US$ 24.3 billion, with an expected annual growth of 7.7%, reaching US$ 35.1 billion in 2020 (Felipe et al. 2017; PRNewswire 2015).

Through technological development, the options for using these biotechnological processes have expanded, making it possible to obtain natural flavors and dyes; emulsifiers; new vitamins, and improved enzymes. Also, the use of genetic modification (GM) techniques allows for the creation of new products, optimization of processes, and new waste-treatment procedures, from "greener" manufacturing processes (Maryam 2017; Nyambok and Robinson 2016).

Thus, considering the advantages of using biotechnological processes for the food industry, the purpose of this chapter is to present an overview of the industrial landscape of final products and food additives obtained from biotechnological processes, as well as general perspectives and economic data of the sectors.

## 1.2 Biotechnological Process Applied to Food Products

Consumer demand for natural food products has encouraged the use of biotechnological processes and their development, in order to obtain economically viable and optimized commercial products (Gupta et al. 2017). In this way, several companies, such as Novozymes, Danisco, and Nestlé, have pursued expansions in the microorganisms-related market to capitalize on that demand. Mergers between companies have also become an alternative to achieve greater development and space in the sector, as in the case of Bayer-Monsanto, Syngenta-ChemChina, and Dow-DuPont (Messeni Petruzzelli et al. 2015; Novozymes 2016).

The presence of biotechnology in food-processing chain occurs with the use of microorganisms for the preservation and production of a variety of food products or aiming at the production of food additives, such as enzymes, vitamins, organic acids, aminoacids, flavor compounds, microbial lipids, proteins, carbohydrates, and several others (Balciunas et al. 2013; Karihaloo and Perera 2010; Kawasaki and Ueda 2017; Ramachandra Rao and Ravishankar 2002). Hence, the main industrial products obtained from biotechnological processes of the sector will be presented in this section.

### 1.2.1 Organic Acids

Organic acids are considered organic compounds with weak acidic properties that do not completely dissociate in the presence of water. Being commonly produced from the primary metabolism of the microorganisms, they present several applications in the industrial sector of food processing, nutrition and feed, pharmaceuticals, and oil and gas stimulation units (Chen and Nielsen 2016; Panda et al. 2016).

Biotechnological methods are the most used to produce organic acids, with the first processes documented in 1823 and 1913, to obtain acetic acid and citric acid, respectively. The application of DNA recombination techniques and metabolic engineering to obtain organic acids has been used with the intention of increasing the efficiency of these biotechnological processes, as well as to meet the global market demand for these bioproducts (Becker et al. 2015).

These products can support shelf-life extension, improve moisture retention, prevent oxidation of lipid components, accelerate solidification and coagulation, and enhance flavor and ion chelation (Kawasaki and Ueda 2017). Organic acids are thus considered one of the most important additives in the food industry, and are widely used in beverages, food, and in the processing of these products (Quitmann et al. 2013).

The most popular organic acids used in food processing are citric acid, acetic acid, formic acid, and lactic acid (Kawasaki and Ueda 2017). In 2011, citric acid was most used in the market, with a bioproduction around 1.75 million tons (Chen and Nielsen 2016). However, by 2016 and 2017, acetic acid dominated the market, with values around US$ 1.8 billion. Both of these organic acids are of great interest in the food and pharmaceutical sectors (Report Buyer 2019).

Global market of organic acids reached US$ 16.8 billion in 2016, dominated by the citric acid and formic acid of microbial origin (Sahu 2017). With the growing demand for natural products, it is expected the value will grow 7% by 2022, with expected income of US$ 20 billion (Shrivastava 2017).

The main companies representing the organic acid market are BASF SE (Germany), Dow Chemical (United States), BP Plc, BioAmber Inc. (Canada), Tate and Lyle Plc (United Kingdom), Archer Daniels Midland (United States), Corbion NV (The Netherlands), Elekeiroz SA (Brazil), Cargill Inc. (United States), Henan Jindan Lactic Acid Technology Co. Ltd. (China), Myriant Corporation (United States), Jungbunzlauer Suisse AG (Switzerland), Nature Works LLC (United States), Celanese Corporation (United States), and Eastman Chemicals Company (United States) (Transparency Market Research 2017a).

### 1.2.2 Flavors

Flavors can be defined as volatile organic compounds (VOCs), which can be subdivided mainly as alcohols, aldehydes, carboxylic acids, furans, fatty acids, esters, ethers, hydrocarbons, ketones, lactones, pyrazines, and terpenes (Longo and Sanromán 2006; Wylock et al. 2015).

Among the processes used for flavor production, chemical synthesis is most used (Chreptowicz et al. 2016). However, due to the negative image associated with the chemical synthesis processes, mainly related to their classifications as "artificial," several companies have sought alternative methods, such as biotechnological processes (Akacha and Gargouri 2015).

The biotechnological processes offer advantages for the generation of volatile compounds, with the potential to obtain high optical purity and not having seasonal interferences (Felipe et al. 2017). In addition, these processes have the appeal of sustainable production, an attractive factor for industries (Berger 2015).

Flavors are considered the largest segment of the food additives market, besides being one of the main components of the global market, registering in 2016 a value of US$ 12.1 billion, characterized as a highly specialized, technical, and innovative category (Mordor Intelligence 2017).

These compounds are mainly used in dairy products, beverages, confectionery products, nondairy ice cream, bakery products, and nutraceuticals. Such compounds may be marketed as powder, liquid, and paste, of which vanilla, vanillin, and fruity flavors are the products with the greatest prominence in the sector. In relation to vanillin, commercial projections indicate that in 2023 the consumption of this product would be approximately 500 tons (Verma 2016; Market Research Future 2018a).

Companies in the biotech-flavors sector have invested in research and innovation, in order to excel in the market. As an example, Givaudan S.A. (Switzerland) stood out with a market share of 18.7% in 2016 (Goeke et al. 2017; Schroeder and Ruethi 2017). In the same year, Firmenich S.A. (Switzerland) (13.5%), International Flavors and Fragrances Inc. (United States) (12.3%), and Symrise A.G. (Germany) (9.2%) also represented the sector, with market values estimated at US$ 24.7 billion at the end of that period (Leffingwell & Associates Flavor 2017).

### 1.2.3 Polysaccharides

Polysaccharides obtained from biotechnological processes are high-molecular-weight biopolymers, being water-soluble gums with novel and unique physical properties (Bergfeld et al. 2011). With similar application to those polymers extracted from natural sources, biopolymers have stable characteristics, with unlimited availability in the market, and well-controlled production processes on a large scale within a comparatively limited space and production time, thus favoring their production and commercial application (Giavasis 2013).

The industrial application of polysaccharides can take place in several sectors, such as textiles, detergents, adhesives, microbial enhanced oil recovery, wastewater treatment, dredging, brewing, downstream processing, cosmetology, pharmacology, and food additives (Rütering et al. 2016). In the food industry, polysaccharides are mainly used as thickeners, stabilizers, and gelling agents in a wide range of food products (Ramalingam et al. 2014).

The current global hydrocolloids market was estimated at US$ 5.8 billion in 2017, with projected growth of 3.6% over the period 2017–2025, having an expected value of US$ 7.6 billion at the end of that period (Transparency Market Research 2017b). Among the main commercial microbial polysaccharides, hydrocolloids are xanthan gum, dextran, scleroglucan, pullulan, and levan (Ates 2015).

Among those, xanthan gum is the most significant exopolysaccharide on the market, accounting for 6% of the total market value of hydrocolloids, with annual production of about 30 000 tons (Ates and Oner 2017). Produced from *Xanthomonas campestris*, xanthan gum stands out for its rheological properties, pseudoplasticity, thickening property, and heat stability, acid and alkali, and was the first biopolymer produced on an industrial scale (Li et al. 2016). These factors directly reflect its increased demand of 5–10% per year, demonstrating its importance (Habibi and Khosravi-Darani 2017).

The companies that stand out in recent years are DuPont (United States), Cargill Inc. (United States), Darling Ingredients (United States), Kerry Group (Ireland), CP Kelco (United States), Fuerst Day Lawson (United Kingdom), Ingredion Inc. (United States), Ashland Inc. (United States), DSM N.V. (The Netherlands), and Rico Carrageenan (Philippines) (Market Research Reports Search Engine 2017).

### 1.2.4 Amino Acids

Amino acids play a fundamental role as building blocks of proteins (Mitsuhashi 2014). Nine amino acids are considered essential for human and animal metabolic processes:

*L*-valine, *L*-leucine, *L*-isoleucine, L-lysine, *L*-threonine, *L*-methionine, *L*-histidine, *L*-phenylalanine, and *L*-tryptophan (Leuchtenberger et al. 2005).

Amino acids have several industrial applications in the food, cosmetic, pharmaceutical, and animal feed industries (Tonouchi and Ito 2016). Focusing on the food industry, these compounds are used for dietary supplementation, mainly in food products linked to infants' and children's diets (Mitsuhashi 2014).

The amino acids with the highest commercial interest are *L*-glutamate (monosodium), L-lysine (chloride), *L*-tryptophan, methionine, and *L*-phenylalanine, with glutamate being the most required in the industry (IMARC Group 2017; Tonouchi and Ito 2016).

In 2015, the global market of amino acids was estimated at US$ 4.88 billion, with an expected growth of 6.5% between 2017 and 2023 (Research and Markets 2017). The food amino acids market is expected to grow 7.8% by 2022 due to the market's encouragement of healthy foods and constant process innovations, with projected values for the end of this period of US$ 6.82 billion (Markets and Markets 2017a).

Among the main companies in the food amino acid market are Ajinomoto Co. Inc. (Japan), Kyowa Hakko Kirin Group (Japan), Evonik Industries A.G. (Germany), Sigma-Aldrich Co. LLC (United States), Prinova Group LLC. (United States), and Daesang Corporation (South Korea), among others (Markets and Markets 2017a).

### 1.2.5 Enzymes

Enzymes are considered natural reactions catalysts, presenting the ability to accelerate specific chemical reactions, thus reducing the activation energy required for the process. These proteins have great application in the food and beverage industries, being used to control the brewing process (Singh et al. 2016).

The enzyme market is driven by the need to improve flavoring, texture, and quality of food products. It encourages the search for new technologies in large-scale fermentation processing and the expansion of the food and beverage industries (Verma 2017). The essential enzymes have broad application in the food and beverage industry, such as amylase, cellulase, xylanase, pectinase, protease, and lipase, among others, mainly used in baking, juice production, dairy, and brewing (Kumar et al. 2014; Singh et al. 2016).

Among the main commercial enzymes, amylases stands out in the industrial sector, accounting for about 25% of the global enzyme market, followed by proteases (Khemakhem et al. 2018). Projections made for the sector by the Freedonia Group (2016) suggest growth of global demand for enzymes around 4.6% by 2020. This data represents a total global revenue of US$ 7.2 billion, just for the food and beverage sector, projections of US$ 2.3 billion in 2020.

Among the main companies in this sector, Novozymes (Denmark), DSM N.V. (The Netherlands), AB Enzymes (United States), Amano Enzyme Inc. (Japan), DuPont (Danisco) (United States), Soufflet Group (French), Megazyme (United States), Boli Bioproducts (China) can be highlighted (Hegde 2017). Novozymes is considered the market leader for industrial enzymes, with a 48% share in the 2016, followed by DuPont, which had an estimated share of 19% in the same year (Novozymes 2016).

### 1.2.6 Surfactants

Microbial surfactants, also known as biosurfactants, are amphiphilic surface active molecules featuring the ability to alter the surface and interfacial properties of a liquid, resulting in the formation of microemulsions (Campos et al. 2013). As an advantage over conventional surfactants, biosurfactants can be produced with less environmental impact when renewable substrates are used, as well as being biodegradable and presenting lower toxicity (Henkel et al. 2017).

The potential for application of these compounds is mainly as emulsification agents in pharmaceutical, food, dye, cosmetic, and agrochemical industries (Sivapathasekaran and Sen 2017). Among the main commercial products are rhamnolipids, sophorolipids, methyl ester sulfonates, alkyl polyglucosides, sorbitan esters, and sucrose esters (Panjiar et al. 2017; Variant Market Research 2017).

The food-processing sector of the biosurfactants market was 5.8% in 2013, mainly due to the demand of the dairy, confectionery, and bakery sectors, being used as stabilizers, foams, gels, and emulsifiers (Ahuja and Singh 2018a).

The sector of biosurfactants is expected to grow by 4.3% in value between 2014 and 2020, with projection of US$ 2.6 billion by 2023 (Grand View Research 2015). Growth in this period is mainly associated with biosurfactants rhamnolipids (forecast of 8% increase), sophorolipids (forecast of $ 3.3 million), alkyl polyglucosides (forecast of 4.5%), and methyl ester sulfonates (forecast of $ 900 million) (Ahuja and Singh 2018a). There is expected to be a consumption of consumption over 6.5 tons by 2023 by the food-processing industry in the United States alone (Ahuja and Singh 2018b).

Companies such as Mitsubishi Chemical Corporation (Japan), Urumqi Unite Bio-Technology Company Ltd. (China), Jeneil Biotech Inc. (United States), Croda International Plc (United Kingdom), Saraya Co. Ltd. (Japan), Cargill Inc. (United States), and Evonik Industries A.G. (Germany), stand out in the biosurfactants sector (Variant Market Research 2017).

### 1.2.7 Pigments

Considered the first sensory parameter observed by the consumers, pigments influence the perception of the food by the consumer, stimulating the appetite and directly impacting in the consumption of these products (Damant 2011; Solymosi et al. 2015). Industrial pigments are divided into synthetic, also called artificial, and natural pigments. However, due to the consumer risk of carcinogenicity, hyper-allergenicity, and other toxicological problems, many of these artificial pigments have been banned by organizations such as the European Food Standards Authority (EFSA), the World Health Organization (WHO), and the US Food and Drug Administration (FDA) (Tuli et al. 2015; Venil et al. 2013). Due to the risks associated with artificial pigments and the market demand for natural products, the production and commercialization of natural pigments have been prioritized (Rodriguez-Amaya 2016).

The microorganisms present the ability to synthesize natural pigments through an efficient and controlled process, resulting in products with potential benefits to consumer's health, presenting potential medicinal properties such as antioxidant, antimicrobial, anticancer, and immunoregulation (Nigam and Luke 2016).

Technological development and genetic engineering methods allowed the improvement of the biosynthesis process, allowing the use of microbiological pigments in the food industry (Kumar et al. 2015). Due to these factors, it is possible to find pigments on the market obtained from biotechnological processes, with the most common being riboflavin, β-Carotene, lycopene, astaxanthin, and canthaxanthin (Nigam and Luke 2016).

During 2015, for example, the production of pigments reached 2 tons, with a 10% increase expected until 2050. One of the factors involved in this growth is related to their use in food and beverage industries, applicable in bakery and confectionery, beverages, dairy, frozen, meat products, oil and fat, and fruits and vegetables (Grand View Research 2017a). Another factor is related to the risks associated with the consumption of the synthetic pigments, which drive the natural food colors market, having an expected growth in 2025 of US$ 2.5 billion (Grand View Research 2017b).

Main companies in the field of natural dyes include DSM N.V. (The Netherlands), CHR Hansen (Denmark), FMC Corp. (United States), Allied Biotech Corporation (Taiwan), Chenguang BioTech Group Co. Ltd. (China), and Roha Dye Chem Pvt. Ltd. (India), among others (Goldstein Research 2017).

## 1.3 Genetically Modified Organisms (GMO)

A widely adopted GM technique can be applied to yeasts, virus, bacteria, animals, or plants in typically unrelated species (Nsanzabera 2016). As a method of genome manipulation, the GM process is divided according to the material to be modified. Thus, when the process occurs in food, it is known as genetically engineered, genetically modified foods or transgenic, as when the modification occurs in organisms, such as bacteria and yeasts is known as genetically modified organisms (GMOs) when the genetic changes (Dizon et al. 2016).

GMO applications are found in human health, agriculture, environment, food, chemicals, paper, and textile industries (Saxena 2015). Through the genetic changes implemented, it is possible to optimize fermentation processes, thus obtaining higher yields, higher efficiency, and lower process cost (Maryam 2017).

These advantages occur due to improved metabolic potential, greater stability, and specificity of GMOs, which results in better fermentation conditions when compared to the preexisting processes (Han 2004; Mishra et al. 2017). Examples are observed in industrial fermentation processes, such as enzyme production processes, performing genetic alterations in *Bacillus subtilis* to obtain α-amylase and in *Escherichia coli* and *Aspergillus niger* to get rennin, or lipase from *E. coli* with GDSL lipase encoding gene (Memarpoor-Yazdi et al. 2017). In addition, breweries can use GM yeast with foreign gene encoding glucoamylase (Sewalt et al. 2016).

The GM strains can also be used to obtain bioproducts of industrial interest, such as organic acids from *Yarrowia lipolytica* (Liu et al. 2017), *A. niger* (Hu et al. 2017), *Ustilaginaceae* species (Paulino et al. 2017). Food flavors from *Pseudomonas putida* S12 (Groeneveld et al. 2016) and *Thymus albicans* (Filipe et al. 2017) or polysaccharides from *X. campestris* and *Agrobacterium* strains (Schmid et al. 2015) can also be obtained.

Regarding the global market for GMOs, the projections for 2022 are US$ 6.28 billion, indicating a growth of 14.5% during that period (Markets and Markets 2017b). Among the companies associated with this market are Thermo Fisher Scientific Inc. (United States), Horizon Discovery Group Plc. (United Kingdom), New England Biolabs (United States), Transposagen Biopharmaceuticals Inc. (United States), Genscript Biotech Corporation (United States), Merck KGaA (Germany), Lonza Group Ltd. (Switzerland), Inc.

(United States), Origene Technologies, Inc. (United States), and Integrated DNA Technologies, Inc. (United States) (Market Research Future 2018b).

The global food market trend of GMOs is growing, having an expected annual growth of 3.2% by the year 2021 (Research Nester 2019). It is possible to infer that this growth is closely linked to the use of GMOs, a technique well established in the industrial sector. As a characteristic, this technique is involved in several fermentative processes in the food industry, such as microbiological enzyme production processes, or recombination of yeast strains (Saxena 2015).

However, consumer hesitation is still a hindrance, since the consumer believes that these products can cause health hazards. The lack of information on processes and food safety has led to consumer mistrust of GM products (Gwira Baumblatt et al. 2017). Herewith, it is necessary that the target audiences of food companies overcome their fear, leading to a greater acceptance of the products. Presenting the processing would be a factor of extreme importance for the development of the sector and increase the acceptance of these products.

Thus, it is possible to observe the impact on the development and production of the food sector by the use of GMOs. Associated with the fermentation processes, they gained traction and enabled superior yields and efficiency in the industrial processes. This contributed to a greater growth of the biotechnology market, as well as a greater incentive of investment in the sector.

## 1.4 Future Perspectives of Biotechnological Processes in the Food Industry

Biotechnology is a multidisciplinary field present in the industrial sector, which enables the development of new processes and products, with less environmental impact than preexisting products, better quality, and with acceptable economic feasibility (Gupta et al. 2017). With expected growth of 7.4% by the year 2025, the worldwide biotechnology market is expected to reach US$ 727.1 billion by the end of this period (Grand View Research 2017c).

The development and economic aspects of any technology, as in the case of biotechnological processes, is driven by many factors, from market demand and profit generation to the creation of new process and innovation techniques of preexisting techniques (Felipe et al. 2017; Messeni Petruzzelli et al. 2015). In this context, patents have become tools that drive the increase in the demand of this market, allowing investment firms and researchers to benefit from their discoveries and creations. In addition, the contribution of investors for the generation of new products on the market are also important for the development of this market (Lokko et al. 2018).

Investments in research for the production and use of enzymes in the last four years have been significant, with many of these studies associated with genetic alterations and solution of problems in industrial processes (Adrio and Demain 2014; Singh et al. 2016).

Companies like Novozymes, DSM, and DuPont have shown great interest in the sector (Nielsen and Loft 2015). These companies have been granted 160 patents within the last five years covering a wide range of solutions of problems of industrial processes (Cramer et al. 2018; Jenner et al. 2014; Persson and Banke 2013), including proteins production (Stocks et al. 2017), ethanol production (Tsang et al. 2014a,b,c,d,e,f), creation of methodologies for the production of dairy product with enzymes that present lactase activity (Henriksen et al. 2017), enzymes applications at fermented milk products (Eisele et al. 2017; Yu et al. 2016), and production of food and beverage products (mainly in brewing) (Sorensen and Miller 2017).

With the aim of meeting the expectations of consumers seeking healthier foods, there is a greater tendency for research in this field, such as regarding the production of sweeteners from natural methods. GMs were made into *Solanum lycopersicum* and *Yarrowia* strains, for example, to obtain optimum processes conditions for the production of steviol gycoside through fermentation (Boer et al. 2018; Bosch et al. 2018). Another interesting sweetener investigation was related to erythritol in fruit juices. The aim of the study was to use microorganisms capable of metabolizing sugars into sugar erythritol. The process is carried out by microorganisms of the genera *Pichia, Yarrowia, Penicillium, Aspergillus, Candida, Torulopsis, Trigonopsis, Moniliella, Aureobasidium,* and *Trichosponon* spp., with the purpose of producing a fermented drink with low carbohydrate content (Hugenholtz et al. 2015).

Since 2014, there has also been a strong market trend in microorganism GMs. Companies including Ajinomoto® Co. have shown great interest in this sector, producing *L*-amino acids from modified microorganisms belonging to the Enterobacteriaceae family (Cerceau et al. 2016; Doi 2016; Livshits et al. 2016; Mitsuhashi 2014; Rybak et al. 2016; Vasilievich et al. 2016).

Other companies have also shown interest in GMOs, such as Cargill Inc., in order to produce of lactic acid from modified bacteria (Brazeau 2015) and fatty acids by means of modified microorganisms (Hans et al. 2015). In addition, BASF SE studied ways to obtain succinic acid from strains from the family of Enterobacteriaceae, Pasteurellaceae, Bacilli, or Actinobacteria (Hartwig et al. 2017).

The future of biotechnology in the food sector is mainly related to the consumer's perception of the benefits associated with these biotechnologies. Researchers, governments, and the food industry itself all play an important role in educating consumers about the technology, its advantages, and its limitations, and thus demonstrate overall that the food products and additives obtained are safe (Gupta et al. 2017).

## 1.5 Concluding Remarks and Perspectives

Biotechnological products application is diverse, including pharmaceutical, textile, and food industries, among several others. Focusing on the food sector, it is important to highlight the availability of food additives, such as organic acids, flavors, polysaccharides, amino acids, enzymes, surfactants, and pigments, which stand out in the sector and encourage the use of bioprocess routes.

Associated with these processes, several methods of genetic alterations are used in order to optimize preexisting processes as well as to develop new ones. These methods contribute to the development of the food industry, allowing better yields and higher efficiency. Thus, the potential associated with these biotechnological products is interesting for application on an industrial scale, in order to supply the market demand for several commercial food products.

## References

Abu Yazid, N., Barrena, R., Komilis, D., and Sánchez, A. (2017). Solid-state fermentation as a novel paradigm for organic waste valorization: a review. *Sustainability* 9 (2): 224.

Adrio, J. and Demain, A. (2014). Microbial enzymes: tools for biotechnological processes. *Biomolecules* 4 (1): 117–139.

Ahuja, K., and S. Singh. 2018a. Biosurfactants market size by product (sophorolipids, rhamnolipids, alkyl polyglucosides [apg], methyl ethyl sulfonates [mes], sucrose esters, sorbitan esters), by application (household detergents, personal care, industrial cleaners, food processing, oil. Global Market Insights. https://www.gminsights.com/industry-analysis/biosurfactants-market-report.

Ahuja, K., and S. Singh. 2018b. Biosurfactants market will exceed $2.6 Billion by 2023. Global Market Insights. https://www.gminsights.com/pressrelease/biosurfactants-market-size.

Akacha, N.B. and Gargouri, M. (2015). Microbial and enzymatic technologies used for the production of natural aroma compounds: synthesis, recovery modeling, and bioprocesses. *Food Bioprod. Process.* 94 (May): 675–706.

Arisawa, A. and Watanabe, A. (2017). Pursuing the unlimited potential of microorganisms-progress and prospect of a fermentation company. *Biosci. Biotechnol. Biochem.* 81 (1): 43–47.

Ates, O. (2015). Systems biology of microbial exopolysaccharides production. *Front. Bioeng. Biotechnol.* 3: 1–16.

Ates, O. and Oner, E.T. (2017). Microbial xanthan, levan, gellan, and curdlan as food additives. In: *Microbial Functional Foods and Nutraceuticals* (eds. V. Gupta, H. Treichel, V. Shapaval, et al.), 149–173. Chichester, UK: Wiley.

Balciunas, E.M., Martinez, F.A.C., and Todorov, S.D. (2013). Novel biotechnological applications of bacteriocins: a review. *Food Control* 32 (1): 134–142.

Barnard, F., Foltz, J.C., and Yeager, E.A. (2012). *Agribusiness Management*, 4e. New York: Taylor & Francis Group.

Becker, J., Lange, A., Fabarius, J., and Wittmann, C. (2015). Top value platform chemicals: bio-based production of organic acids. *Curr. Opin. Biotechnol.* 36, n. Figure 1: 168–175.

Berger, R.G. (2015). Biotechnology as a source of natural volatile flavours. *Curr. Opin. Food Sci.* 1 (1): 38–43.

Bergfeld, W.F., Belsito, D.V., Hill, R.A. et al. (2011). *Microbial Polysaccharides*, vol. 121, 1–44. Cosmetic Ingredient Review Expert Panel http://www.cir-safety.org/sites/default/files/microb112111SLR_faa_mmf_faa.pdf.

Boer, V.M., Van Leeuwen, J.G.E., Zwartjens, P., and Kolen, C.P.A.M. DSM IP Assets BV, 2018. Udp-glycosyltransferases. U.S. Patent Application 15/558,133.

Bosch, H. J., Beekwilder, M. J., and Boer, V. M. Udp-glycosyltransferases from solanum lycopersicum. U.S. Patent Application 15/560,318, 2018.

Brazeau, B. J. Methods of using fungi to acidify milk, 2015. https://worldwide.espacenet.com/publicationDetails/biblio?CC=US&NR=2018192661A1&KC=A1&FT=D&ND=3&date=20180712&DB=EPODOC&locale=en_EP>

Campos, J.M., Stamford, T.L.M., Sarubbo, L.A. et al. (2013). Microbial biosurfactants as additives for food industries. *Biotechnol. Prog.* 29 (5): 1097–1108.

Carroll, A.L., Desai, S.H., and Atsumi, S. (2016). Microbial production of scent and flavor compounds. *Curr. Opin. Biotechnol.* 37: 8–15.

Cerceau, C.I., Barbosa, L.C., Filomeno, C.A. et al. (2016). An optimized and validated 1H NMR method for the quantification of α-pinene in essentials oils. *Talanta* 150: 97–103.

Chen, Y. and Nielsen, J. (2016). Biobased organic acids production by metabolically engineered microorganisms. *Curr. Opin. Biotechnol.* 37: 165–172.

Chotani, G.K., Dodge, T.C., and Arbige, M.V. (2017). Industrial biotechnology: discovery to delivery. In: *Handbook of Industrial Chemistry and Biotechnology* (eds. J.A. Kent, T.V. Bommaraju and S.D. Barnicki), 1495–1570. Cham: Springer International Publishing.

Chreptowicz, K., Wielechowkia, M., Glówscyk-Zubek, J. et al. (2016). Production of natural 2-phenylethanol: from biotransformation to purified product. *Food Bioprod. Process.* 100: 275–281.

Cramer, J. F., Kellett-Smith, A.H., Jensen, L.B., and Larsen, B. DuPont Nutrition Biosciences APS, 2018. Aldc production methods. U.S. Patent Application 15/572,939. 2018.

Damant, A.P. (2011). Food colourants. In: *Handbook of Textile and Industrial Dyeing*, vol. 2p (ed. M. Clark), 252–305. Elsevier.

De Buck, S., de Oliveira, D., and Van Montagu, M. (2016). Key innovations in plant biotechnology and applications in agriculture, industrial processes, and healthcare. In: *Innovative Farming and Forestry Across the Emerging World: the Role of Genetically Modified Crops and Trees* (eds. S. De Buck, I. Ingelbrecht, M. Heijde and M. Van Montagu), 13–33. International Industrial Biotechnology Network (IIBN).

Diehl, P. (2017). What is biotechnology and how did it make the biotech industry? *Balance*: 1–4.

Dizon, F., Costa, S., Rock, C. et al. (2016). Genetically modified (GM) foods and ethical eating. *J. Food Sci.* 81 (2): R287–R291.

Doi, H. Method for producing L-amino acid from seaweed-derived biomass, 2016.

Eisele, T., Bejder, H. C., and Vargas, E. G. D. DuPont Nutrition Biosciences APS, 2018. Proteases for high protein fermented milk products. U.S. Patent Application 15/552,372. 2017.

Felipe, L.D.O., Oliveira, A.M., and Bicas, J.L. (2017). Bioaromas – perspectives for sustainable development. *Trends Food Sci. Technol.* 62: 141–153.

Filipe, A., Cardoso, J.C., Miguel, G. et al. (2017). Molecular cloning and functional characterization of a monoterpene synthase isolated from the aromatic wild shrub *Thymus albicans*. *J. Plant Physiol.* 218 (July): 35–44.

Giavasis, I. (2013). Production of microbial polysaccharides for use in food. In: *Microbial Production of Food Ingredients, Enzymes and Nutraceuticals* (eds. B. McNeil, D. Archer, I. Giavasis and L. Harvey), 413–468. Woodhead Publishing.

Goeke, A., Kraft, P., and Zou, Y. Organic Compounds, 2017.

Goldstein Research. Natural Food Colors Market Outlook 2024: Global Opportunity and Demand Analysis, Market Forecast, 2016-2024. https://www.goldsteinresearch.com/report/natural-food-colors-market-outlook-2024-global-opportunity-and-demand-analysis-market-forecast-2016-2024. 2017.

Grand View Research. Biosurfactants market analysis by product (rhamnolipids, sophorolipids, MES, APG, sorbitan esters, sucrose esters) and segment forecast to 2020. http://www.grandviewresearch.com/industry-analysis/biosurfactants-industry. 2015.

Grand View Research. Colorants market analysis by application (plastic, textile, food, personal care, printing, paints and coatings), by region (North America, Europe, Asia Pacific, Central and South America, MEA), and segment forecasts, 2014–2025. http://www.grandviewresearch.com/industry-analysis/colorants-market. 2017a.

Grand View Research. Natural food colors market estimates and trend analysis by product (curcumin, carotenoids, anthocyanin, carmine, chlorophyllin), by application (bakery and confectionery, beverages, dairy and frozen products, meat products), and segment forecasts, 2018–2025. http://www.grandviewresearch.com/industry-analysis/natural-food-colors-market. 2017b.

Grand View Research. Biotechnology market worth $ 727.1 billion by 2025. http://www.grandviewresearch.com/press-release/global-biotechnology-market. 2017c.

Groeneveld, M., van Beek, H.L., Duetz, W.A., and Fraaije, M.W. (2016). Identification of a novel oxygenase capable of regiospecific hydroxylation of d-limonene into (+)-trans-carveol. *Tetrahedron* 72 (46): 7263–7267.

Gupta, V., Sengupta, M., Prakash, J., and Tripathy, B.C. (2017). *Basic and Applied Aspects of Biotechnology*. Singapore: Springer Singapore.

Gwira Baumblatt, J.A., Carpenter, L.R., Wiedeman, C. et al. (2017). Population survey of attitudes and beliefs regarding organic, genetically modified, and irradiated foods. *Nutr. Health* 23 (1): 7–11.

Habibi, H. and Khosravi-Darani, K. (2017). Effective variables on production and structure of xanthan gum and its food applications: a review. *Biocatal. Agric. Biotechnol.* 10: 130–140.

Han, L. (2004). Genetically modified microorganisms. In: *The GMO Handbook*, vol. 24 (ed. L. Han), 85–130.

Hans, L. Spindler E., Warner J. R et al. Microorganisms and Methods for the Production of Fatty Acids and Fatty Acid Derived Products, 2015. https://worldwide.espacenet.com/publicationDetails/biblio?DB=EPODOC&II=4&ND=3&adjacent=true&locale=en_EP&FT=D&date=20161124&CC=US&NR=2016340700A1&KC=A1.

Hartwig, S. Haefner, S., von Abendroth, G. et al. Novel Microbial Succinic Acid Producers And Purification of Succinic Acid, 2017. https://worldwide.espacenet.com/publicationDetails/biblio?DB=EPODOC&II=1&ND=3&adjacent=true&locale=en_EP&FT=D&date=20170414&CC=MY&NR=161185A&KC=A.

Hegde, A. Brewing Enzymes Market worth over $ 390 million by 2024: Global Market Insights, Inc. https://globenewswire.com/news-release/2017/08/15/1084664/0/en/Brewing-Enzymes-Market-worth-over-390-million-by-2024-Global-Market-Insights-Inc.html. 2017.

Henkel, M., Geissler, M., Weggenmann, F., and Hausmann, R. (2017). Production of microbial biosurfactants: status quo of rhamnolipid and surfactin towards large-scale production. *Biotechnol. J.* 12 (7): 1600561.

Henriksen, H. V. et al. Method for producing a dairy product, 2017.

Hu, W., Li, W., Chen, H. et al. (2017). Changes in transcript levels of starch hydrolysis genes and raising citric acid production via carbon ion irradiation mutagenesis of *Aspergillus niger*. *PLoS One* 12 (6): e0180120.

Hugenholtz, J., Trachotta, T., Bekker, M. et al. Compositions and methods for reduced carbohydrates and increased erythritol in beverages, 2015. U.S. Patent Application 14/436,588.

IMARC Group. Amino Acids Market: Global Industry Trends, Share, Size, Growth, Opportunity and Forecast 2017-2022. Disponível em: http://www.imarcgroup.com/amino-acid-technical-material-market-report.2017.

Jenner, R. L. B. et al. USE, 2014.

Karihaloo, J.L., and O. Perera Agricultural biotechnologies in developing countries: Options and opportunities in crops, forestry, livestock, fisheries and agro-industry to face the challenges of food insecurity and climate change (ABDC-10). FAO International Technical Conference, Guadalajara, Mexico. March 2010, http://www.fao.org/fileadmin/user_upload/abdc/documents/apaari.pdf.

Kawasaki, H. and Ueda, K. (2017). Microbial innovations in the world of food. *Biosci. Biotechnol. Biochem.* 81 (1): 48–53.

Khemakhem, B., Smaoui, S., El Abed, H. et al. (2018). Improving changes in physical, sensory and texture properties of cake supplemented with purified amylase from fenugreek (*Trigonella foenum graecum*) seeds. *3 Biotech* 8 (3): 174.

Kumar, V., Singh, D., Sangwan, P., and Gill, P.K. (2014). Global market scenario of industrial enzymes. In: *Industrial Enzymes: Trends, Scope and Relevance* (ed. V. Kumar), 173–196. New York: Nova Science Publishers.

Kumar, A., Vishwakarma, H.S., Singh, J. et al. (2015). Microbial pigments: production and their applications in various industries. *Int. J. Pharm. Chem. Biol. Sci.* 5 (1): 203–212.

Leffingwell & Associates. 2013–2017 Flavor & Fragrance Industry Leaders. http://www.leffingwell.com/top_10.htm. 2017.

Leuchtenberger, W., Huthmacher, K., and Drauz, K. (2005). Biotechnological production of amino acids and derivatives: current status and prospects. *Appl. Microbiol. Biotechnol.* 69 (1): 1–8.

Li, P., Li, T., Zeng, Y. et al. (2016). Biosynthesis of xanthan gum by *Xanthomonas campestris* LRELP-1 using kitchen waste as the sole substrate. *Carbohydr. Polym.* 151: 684–691.

Liu, H., Madzak, C., Sun, M.-L. et al. (2017). Engineering *Yarrowia lipolytica* for arachidonic acid production through rapid assembly of metabolic pathway. *Biochem. Eng. J.* 119: 52–58.

Livshits, V. A. Zakataeva N, Aleshin V, et al. Novel Gene and Method for Producing L-Amino Acids, 2016.

Lokko, Y., Heijde, M., Schebesta, K. et al. (2018). Biotechnology and the bioeconomy—towards inclusive and sustainable industrial development. *New Biotechnol.* 40: 5–10.

Longo, M.A. and Sanromán, M.A. (2006). Production of food aroma compounds: microbial and enzymatic methodologies. *Food Technol. Biotechnol.* 44 (3): 335–353.

Market Research Future. Biotech Flavors Market 2017 Global Trend, Segmentation and Opportunities Forecast To 2022. https://www.mordorintelligence.com/industry-reports/global-food-additives-market-industry. 2018a.

Market Research Future. Global Genetic Engineering Market Projected to Grow Radiantly by 2027; Asserts MRFR Unleashing Industry Forecast. https://www.marketresearchfuture.com/press-release/genetic-engineering-market>. 2018b.

Market Research Reports Search Engine. Food hydrocolloids market–global industry analysis size share growth trends and forecast 2017–2025. http://www.digitaljournal.com/pr/3463316. 2017.

Markets and Markets. Food Amino Acids Market by Application, Type, Source, and Region–Global Forecast to 2022. https://www.reportbuyer.com/product/4804571/food-amino-acids-market-by-application-type-source-and-region-global-forecast-to-2022.html. 2017a.

Markets and Markets. Genome Editing/Genome Engineering Market worth 6.28 Billion USD by 2022. https://www.marketsandmarkets.com/PressReleases/genome-editing-engineering.asp. 2017b.

Maryam, B.M. (2017). The role of biotechnology in food production and processing. *Ind. Eng.* 1 (1): 24–35.

Memarpoor-Yazdi, M., Karbalaei-Heidari, H.R., and Khajeh, K. (2017). Production of the renewable extremophile lipase: valuable biocatalyst with potential usage in food industry. *Food Bioprod. Process.* 102: 153–166.

Messeni Petruzzelli, A., Rotolo, D., and Albino, V. (2015). Determinants of patent citations in biotechnology: an analysis of patent influence across the industrial and organizational boundaries. *Technol. Forecasting Social Change* 91: 208–221.

Mishra, S.S., Ray, R.C., Panda, S.K., and Montet, D. (2017). Technological innovations in processing of fermented foods. In: *Fermented Foods – Part II: Technological Interventions* (eds. R.C. Ray and D. Montet), 21–45. Boca Raton, FL: CRC Press.

Mitsuhashi, S. (2014). Current topics in the biotechnological production of essential amino acids, functional amino acids, and dipeptides. *Curr. Opin. Biotechnol.* 26: 38–44.

Mordor Intelligence. Global Food Additives Market - Growth, Trends & Forecasts (2017–2022). https://www.mordorintelligence.com/industry-reports/global-food-additives-market-industry. 2017.

Nielsen, P. H., and Loft, B. D. Strategy: Growth and Priorities for 2020. https://s21.q4cdn.com/655485906/files/doc_downloads/capital_markets_days/Novozymes-CMD-2015_-1_Strategy_and_Priorities-presentation_FINAL-UPDATED.pdf. 2015.

Nigam, P.S. and Luke, J.S. (2016). Food additives: production of microbial pigments and their antioxidant properties. *Curr. Opin. Food Sci.* 7: 93–100.

Novozymes. Macro and industry trends, 2016. https://report2016.novozymes.com/our-business/trends>

Nsanzabera, F. (2016). Modern biotechnology and new food varieties. *Adv. Biochem.* 4 (3): 26.

Nyambok, E. and Robinson, C. (2016). The role of food additives and chemicals in food allergy. *Ann. Food Process. Preserv.* 1 (1): 1–3.

Panda, S.K., Mishra, S.S., Kayitesi, E., and Ray, R.C. (2016). Microbial-processing of fruit and vegetable wastes for production of vital enzymes and organic acids: biotechnology and scopes. *Environ. Res.* 146: 161–172.

Panjiar, N., Sachan, S.G., and Sachan, A. (2017). Biosurfactants: a multifunctional microbial metabolite. In: *Microbial Applications*, vol. 2p (ed. V.C. Kalia), 213–229. Cham: Springer International Publishing.

Paulino, B.N., Pessôa, M.G., Molina, G. et al. (2017). Biotechnological production of value-added compounds by ustilaginomycetous yeasts. *Appl. Microbiol. Biotechnol.* 101 (21): 7789–7809.

Persson, J. M., and Banke, N. Reduction of culture viscosity by manganese addition, 2013.

PRNewswire. World Markets for Fermentation Ingredients - Reportlinker Review. https://www.prnewswire.com/news-releases/world-markets-for-fermentation-ingredients---reportlinker-review-300160023.html. 2015.

Quitmann, H., Fan, R., and Czermak, P. (2013). Acidic organic compounds in beverage, food, and feed production. *Adv. Biochem. Eng. Biotechnol.* 143: 91–141.

Ramachandra Rao, S. and Ravishankar, G.A. (2002). Plant cell cultures: chemical factories of secondary metabolites. *Biotechnol. Adv.* 20 (2): 101–153.

Ramalingam, C., Priya, J., and Mundra, S. (2014). Applications of microbial polysaccharides in food industry. *Int. J. Pharm. Sci. Rev. Res.* 27 (1): 322–324.

Report Buyer. Organic Acids Market - Global Industry Analysis, Size, Share, Growth, Trends, and Forecast 2017 - 2022. https://www.reportbuyer.com/product/5143316/organic-acids-market-global-industry-analysis-size-share-growth-trends-and-forecast-2017-2022.html. 2019.

Research and Markets. Amino acid market: global industry analysis, trends, market size & forecasts to 2023. https://www.researchandmarkets.com/research/xsjjhs/amino_acid. Acesso em: 3 out. 2017.

Research Nester. Global genetically modified (gmo) food market analysis & opportunity outlook 2021. https://www.researchnester.com/reports/global-genetically-modified-gmo-food-market-analysis-opportunity-outlook-2021/117. 2019.

Rodriguez-Amaya, D.B. (2016). Natural food pigments and colorants. *Curr. Opin. Food Sci.* 7: 20–26.

Rütering, M., Schmid, J., Rühmann, B. et al. (2016). Controlled production of polysaccharides–exploiting nutrient supply for levan and heteropolysaccharide formation in *Paenibacillus* sp. *Carbohydr. Polym.* 148: 326–334.

Rybak, K. V., Slivinskaya, E.A., Voroshilova, E.B., and Kozlov, Y.I. Method for producing an l-amino acid using a bacterium of the enterobacteriaceae family having a pathway of glycogen biosynthesis disrupted. U.S. Patent 7,422,880., 2016.

Sahu, Y. Organic Acids Market by Type (Acetic Acid, Citric Acid, Formic Acid, Lactic Acid, Itaconic Acid, Succinic Acid, Gluconic Acid, Ascorbic Acid, Fumaric Acid, and Propionic Acid), Source (Biomass, Molasses, Starch, Chemical Synthesis, Agro- Industrial Residue), End-User (Food & Beverage, Animal Feed, Chemicals & Industrial, Pharmaceuticals, Personal Care, Agriculture) - Global Opportunity Analysis and Industry Forecast, 2017–2023. https://www.alliedmarketresearch.com/organic-acids-market>. 2017.

Saxena, S. (2015). Microbial technology and biotechnology. In: *Applied Microbiology*, 13–18. New Delhi: Springer India.

Schmid, J., Sieber, V., and Rehm, B. (2015). Bacterial exopolysaccharides: biosynthesis pathways and engineering strategies. *Front. Microbiol.* 6 (496): 1–24.

Schroeder, F., and Ruethi, F. Organic Compounds, 2017.

Serra, S., Fuganti, C., and Brenna, E. (2005). Biocatalytic preparation of natural flavours and fragrances. *Trends Biotechnol.* 23 (4): 193–198.

Sewalt, V., Shanahan, D., Gregg, L. et al. (2016). The generally recognized as safe (GRAS) process for industrial microbial enzymes. *Ind. Biotechnol.* 12 (5): 295–302.

Shrivastava, V. Organic acids market–global opportunity analysis and industry forecast (2017–2022). https://www.meticulousresearch.com/product/organic-acids-market-2022. 2017.

Singh, R., Kumar, M., Mittal, A., and Mehta, P.K. (2016). Microbial enzymes: industrial progress in 21st century. *3 Biotech* 6 (2): 174.

Sivapathasekaran, C. and Sen, R. (2017). Origin, properties, production and purification of microbial surfactants as molecules with immense commercial potential. *Tenside Surfactants Deterg.* 54 (2): 92–104.

Solymosi, K., Latruffe, N., Morant-Manceau, A. et al. (2015). Food colour additives of natural origin. In: *Colour Additives for Foods and Beverages* (ed. M.J. Scotter), 3–34. Woodhead Publishing.

Sorensen, J. F., and Miller, L.B. Enzymes, US Patent 9683224 2017.

Stocks, S. M., Lehmann, L., and Albaek, M. O. Direct inoculation, 2017.
The Freedonia Group (2016). Demand and sales forecasts, market share, market size, market leaders. In: *World Enzymes*, 397.
Tonouchi, N. and Ito, H. (2016). Present global situation of amino acids in industry. *Adv. Biochem. Eng. Biotechnol.* 123p: 3–14.
Transparency Market Research. Organic Acids Market (Product Type – Acetic Acid, Formic Acid, Lactic Acid, Citric Acid, Propionic Acid, Ascorbic Acid, Gluconic Acid, Fumaric Acid, Malic Acid; Application – Bakery and Confectionery, Dairy, Beverages, Poultry, Meat and Seafood, Livestock Feed, Animal Feed, Pharmaceuticals, Industrial) – Global Industry Analysis, Size, Share, Growth, Trends, and Forecast 2018–2026. https://www.transparencymarketresearch.com/organic-acids-market.html.2017a.
Transparency Market Research. Food hydrocolloids market – global industry analysis size share growth trends and forecast 2017–2025. https://www.transparencymarketresearch.com/food-hydrocolloids-market.html. 2017b.
Tsang, A., Powlowski, J., and Butler, G. Novel cell wall deconstruction enzymes of thielavia australiensis and uses thereof, U.S. Patent Application 14/405,602. 2014a.
Tsang, A., Powlowski, J., and Butler, G. Patentes cell wall deconstruction enzymes of Pseudocercosporella herpotrichoides and uses thereof, 2014b.
Tsang, A., Powlowski, J., and Butler, G. Cell wall deconstruction enzymes of Myceliophthora fergusii (Corynascus thermophilus) and uses thereof, 2014c.
Tsang, A., Powlowski, J., and Butler, G. Cell wall deconstruction enzymes of Thermoascus aurantiacus and uses thereof, 2014d.
Tsang, A., Powlowski, J., and Butler, G. Cell wall deconstruction enzymes of Paecilomyces byssochlamydoides and uses thereof, 2014e.
Tsang, A., Powlowski, J., and Butler, G. Cell wall deconstruction enzymes of Malbranchea cinnamomea and uses thereof, 2014f.
Tuli, H.S., Chaudhary, P., and Beniwal, V. (2015). Microbial pigments as natural color sources: current trends and future perspectives. *J. Food Sci. Technol.* 52 (8): 4669–4678.
Variant Market Research. Biosurfactants Market (Rhamnolipids, Sophorolipids, Methyl Ether Sulfonates, Alkyl Polyglucosides, Sorbitan Esters, Sucrose Esters; Household Detergents, Personal Care, Industrial & Institutional Cleaners, Food Processing, Oilfield Chemicals, Agricultural. https://www.variantmarketresearch.com/about-us. 2017.
Vasilievich, S. V., Sergeevna, E. N., and Viktorovna, S. N. Method for producing an l-amino acid using a bacterium of the family Enterobacteriaceae having attenuated expression of a phosphate transporter-encoding gene, 2016.
Venil, C.K., Zakaria, Z.A., and Ahmad, W.A. (2013). Bacterial pigments and their applications. *Process Biochem.* 48 (7): 1065–1079.
Verma, V. 2016. Bio vanillin market size by application (food & beverage [ice cream, chocolate, confectionaries & baked goods], fragrances, pharmaceuticals), industry analysis report, regional outlook (U.S., Canada, Mexico, Germany, UK, France, Italy, Spain, China, India, Japan, Australia, South Korea, South Africa, GCC, Brazil), production routes, downstream application potential (methoxyhydroquinone, vanillic acid, vanillyl alcohol), price trend, competitive market share & forecast, 2016–2023. https://www.gminsights.com/industry-analysis/bio-vanillin-market.
Verma, V. 2017. Enzymes market size by product (proteases, lipases, carbohydrases [amylases, xylanases/hemicellulase, cellulase, pectinase], polymerases & nucleases,

phytases), by application (food & beverages, detergents, animal feed, biofuels, textile, pulp & paper, Pe., p. 1–410, Global Market Insights.
Wylock, C., Mballa, E., Heilporn, C. et al. (2015). Review on the potential technologies for aromas recovery from food industry flue gas. *Trends Food Sci. Technol.* 46 (1): 68–74.
Yu, S. et al. Method, 2016.
Zilberman, D., Kim, E., Kirschner, S. et al. (2013). Technology and the future bioeconomy. *Agric. Econ.* 44 (s1): 95–102.

# 2

# Status of Biotechnological Processes in the Pharmaceutical Industry

*Natalia Videira*[1,2], *Robson Tramontina*[1,3], *Victoria Ramos Sodré*[4], *and Fabiano Jares Contesini*[5]

[1] *Graduation Program of Biosciences and Technology of Bioactive Products, Institute of Biology, State University of Campinas (Unicamp), Campinas, São Paulo, Brazil*
[2] *Brazilian Biosciences National Laboratory (LNBio), Brazilian Center for Research in Energy and Materials (CNPEM), 13083-970, Campinas, São Paulo, Brazil*
[3] *Brazilian Bioethanol Science and Technology Laboratory (CTBE), Brazilian Center for Research in Energy and Materials (CNPEM), 13083-970, Campinas, São Paulo, Brazil*
[4] *Graduation Program of Functional and Molecular Biology, Institute of Biology, State University of Campinas (Unicamp), Campinas, São Paulo, Brazil*
[5] *Department of Biochemistry and Tissue Biology, Institute of Biology, University of Campinas – (Unicamp), Campinas, São Paulo, 13083-862, Brazil*

## 2.1 Introduction

Since their major inception in the 1980s, biotech drugs based on natural proteins or produced via biotechnological processes have become the difference between life and death for millions of patients, treating diabetes, cancer, infectious diseases, and many genetic disorders. Especially with the scientific progress concerning recombinant DNA techniques, biotechnology is now present in the pharmaceutical sector. Pharmaceutical biotechnology has been consolidated as a field of study contemplating the use of basic science, such as biology and chemistry, as well the knowledge of applied sciences, which includes engineering and pharmaceutical sciences.

The first step in developing a biotechnological medicament is to genetically modify an organism to introduce a genetic code sequence that produces the chosen protein or induce the production of the desired metabolite by biotechnological process. Therefore, it is a multidisciplinary area that uses industrial and scientific knowledge to produce complex biomolecules (such as antibodies, enzymes, and antibiotics) that are tools for health promotion (Vitolo et al. 2015).

In the past decades, biotechnology provided several contributions to pharmacology and medicine, through the discovery and development of specific and safe medicines for complex and rare diseases, of new diagnosis tools, of new pathways for drug production (semisynthetic drugs) and of potential gene therapies (Lybecker 2016). The traditional synthetic routes for drug synthesis, which established medicines as

*Bioprocessing for Biomolecules Production*, First Edition.
Edited by Gustavo Molina, Vijai Kumar Gupta, Brahma N. Singh, and Nicholas Gathergood.

Paracetamol (analgesic), Ketamine (anesthetic), Acyclovir (antiviral), Azidothymidine (AZT, antiviral), have started to show signs of lower productivity and is losing ground to molecular biology and biotechnology, which are now the major source of innovation in the pharmaceutical industry (Lybecker 2016).

The history of polypeptides, proteins, or glycoproteins being used as therapeutic agents started with Pasteur in 1885, when he injected inactivated rabies virus in a child bitten by a rabid dog (Cardoso 2013). In 1904, L-asparaginase, probably the main enzyme used as a drug, was first observed by Lang and Uber (1904) and is the first therapeutic enzyme with antineoplastic properties (Szymanska et al. 2012).

Alexander Fleming reported the first and probably most famous example of antibiotic discovery in 1929. Fleming accidentally found an antibacterial compound produced by a *Penicillium* species that inhibited the growth of some *Staphylococcus* colonies. That compound was named penicillin, and initially it was thought to be produced by a *Penicillium chrysogenum* strain. After full sequencing of the genome of that strain the fungus was reidentified as *Penicillium rubens* (Houbraken et al. 2011). The discovery of antibiotics was a groundbreaking innovation in the twentieth century and helped save a large number of lives in wars.

In the twentieth century, Herbert Boyer and others succeeded at genetically manipulating plasmids of *Escherichia coli* bacteria to produce insulin with the same amino acid sequence as seen in humans. In 1982, insulin was the first genetically engineered drug to be approved for marketing in the United States (Cardoso 2013). Moreover, the first recombinant Hepatitis B vaccine and the first monoclonal antibody (mAb) therapy against liver-transplant rejection were launched in 1986 (Ecker et al. 2015). Fomivirsen was the first oligonucleotide developed for therapeutically purposes, and was approved in 1998 (Orr 2001).

The increasing knowledge of the diseases' mechanisms at a molecular level has enabled the pharmaceutical industry to model and synthesize more and more specific molecules (*National Research Council* 2004). These advances were only possible due to the biotechnological revolution, accomplished with -omic studies (e.g. genomics, functional genomics, proteomics, metabolomics, and cytometry) that provided information about the functional and/or structural identification of tissues and cells; about gene expression patterns, screening of DNA mutations and polymorphisms, and metabolic profiles (Kayser and Müller 2004; Vitolo et al. 2015).

With this information, it is now possible to identify physiological and/or metabolic alterations induced by a disease, and even evaluate the pharmaceutical effect of a medicine on a given organism. Thus, greater understanding of the genetic causes of diseases allows for early detection and treatment, and the new field of gene therapy may enable treatment and even total recovery of a disease (Kayser and Müller 2004).

In terms of the current scenario, biotechnology has already made significant strides for human health. New drugs have been created, especially for rare or still-untreated diseases (Gould Rothberg et al. 2008). The advances of biotechnology enabled the development of biopharmaceuticals, semisynthetic drugs, enzymes, more powerful antibiotics, gene therapies and diagnosis tools (Kayser and Müller 2004; Vitolo et al. 2015).

Of these aforementioned biotechnology advances, biopharmaceuticals (also called biologicals), contribute to a great share in the income of pharmaceutical industries. Biopharmaceuticals can be proteins (including antibodies, enzymes, blood factors) or nucleic acids (DNA, RNA, or antisense oligonucleotides) used for therapeutic or *in vivo* diagnostic purposes and are produced by means other than direct extraction from a native (nonengineered) biological source (Vitolo et al. 2015). While traditional synthetic drugs are composed of small molecules typically defined as between 100 and 1000 atoms; biopharmaceuticals are large, complex proteins, consisting of thousands of atoms, and being mostly chemically and biochemically unstable. This molecular complexity of biopharmaceuticals and their biosynthesis in living cells make the final product very sensitive to variations in production conditions. Nevertheless, these biotechnology production methods provide safer versions of existing treatments and potentially in unlimited quantities (Geigert 2013).

The *first-generation biopharmaceuticals* are usually made of recombinant proteins with amino acids sequences identical to the natural proteins (Vitolo et al. 2015). The *second-generation* ones, however, have a planned alteration in the amino acid sequence, with the purpose of increasing or decreasing the peak of the product biological activity, increase the half-life or decrease the immunogenicity of the product (Geigert 2013). However, compared to synthetic-route drugs, there are no identical biopharmaceuticals among two different batches or suppliers, because this type of medicine is produced by independent cell lines (Geigert 2013). To overcome this limitation, biopharmaceuticals are submitted to numerous and rigorous tests to attest to their safety, stability, and function.

Not only peptides and nucleotides can be produced through biotechnological process for health promotion. Nowadays, brand-new antibiotics are industrially produced from organisms, such as fungi. In addition, several molecules can be chemically modified using specific enzymes to create more potent analogues of their natural counterparts. For instance, microbial lipases have been applied to kinetic resolution of racemic mixtures in order to produce enantiopure drugs (Carvalho et al. 2006; Łukowska-Chojnacka et al. 2017).

Accordingly, this chapter describes biotech products that are on the market as well as a summary of their production process. An overview of the economic aspects, intellectual property scenario, and perspectives about therapeutically antibodies, biologicals, and enzymes are also discussed.

## 2.2 Main Biotechnological Products in the Pharmaceutical Industry

### 2.2.1 Antibiotics in the Pharmaceutical Industry

Antibiotics are secondary metabolism compounds with low molecular weight (<1500 Da) usually produced during the late-growth phase by microorganisms, including bacteria and fungi. These molecules are one of the most important classes of drugs used in several infectious diseases. They inhibit growth of microorganisms even at low concentrations.

Antibiotics account for a significant share of the pharmaceutical industry economic income. As an example, in 2009 the market of antibiotics corresponded to 46% of sales of anti-infective drugs that also include antiviral drugs and vaccines, as well as 5% of the global pharmaceutical market generating sales of US$42 billion (Hamad 2010; IMS Health. IMS MIDAS 2009).

Biotechnological production of antibiotics is focused on fermentation processes of some microorganisms, including *Penicillium* spp. and bacteria like *Streptomyces* spp. In addition, there are semisynthetic antibiotics that are first produced by fermentation, purified and, then, followed by chemical alterations through organic synthesis tools. Finally, some antibiotics are completely chemically synthesized.

There are different criteria for classification of antibiotics and different applications. With regards to human health, β-lactam antibiotics are probably the most relevant group of antibiotics. Marketed β-lactam antibiotics include penicillin-derived, like ampicillin and floxacillin; cephalosporins, such as cefalothin and cefadroxyl, among other relevant antibiotics. These broad-spectrum antibiotics contain a β-lactam ring in their molecular structures. This ring is composed of one nitrogen and three carbon atoms. Their action mechanism is based on the inhibition of the last stage of bacterial cell wall synthesis (Suárez and Gudiol 2009). In addition, they have been produced by some fungi, such as *P. chrysogenum* and *Acremonium chrysogenum* (Elander 2003).

Apart from production of antibiotics by fungi, several relevant antibiotics are produced by actinomycetes, mainly *Streptomyces* spp. These Gram-positive bacteria produce several medically relevant antibiotics including tetracyclin (*Streptomyces aureofaciens*) and streptomycin (*Streptomyces griseus*).

For antibiotic production, the main biosynthetic pathways are those involved with formation of peptides, polyketides, isoprenes, oligosaccharides, aromatic compounds, and β-lactam rings (Sánchez et al. 2010). These pathways involve the action of individual enzymes or multifunctional polypeptides, including different synthases and synthetases (Demain 1998). Most genes involved with such pathways are usually chromosomal. In only a few cases genes have plasmidial origin. Additionally, genes involved with antibiotic synthesis are usually clustered, but not necessarily as single operons (Sánchez et al. 2010).

The classical method for screening of antibiotics is based on the cultivation of strains in different growth conditions, followed by the extraction of metabolites and evaluation of the bioactivity. If positive results are acquired, more detailed analysis are performed, including the identification of the compounds using chromatography and mass spectrometry. Although these approaches are still important, the availability of genome sequence of different microorganisms allowed the identification of bioactive molecules that were previously not possible. In addition, metabolic engineering techniques based on genome editing, more specifically on gene deletion and overexpression, result in improvements never seen before.

Examples of metabolic engineering of microbial strains to overproduce antibiotics include the overexpression of the malonyl-CoA synthesizing complex (acetyl-CoA carboxylase complex), leading to an increase in actinorhodin production in *Streptomyces coelicolor* (Ryu et al. 2006). Another example is the improvement of oxytetracycline production in *Streptomyces rimosus* M4018 by metabolic engineering of the G6PDH gene in the pentose phosphate pathway (Tang et al. 2011).

### 2.2.2 Enzymes in the Pharmaceutical Industry

Enzymes are biocatalysts with remarkable biotechnological applications. Most enzymes are proteins, except some cases in which they are RNA molecules. They have varied functions and hence several biotechnological uses. Naturally, they act intra, extracellularly, or on the surface of a cell membrane (Maurer 2001). These proteins are capable of catalyzing reactions under mild reactions conditions regarding pH and temperature. They present high selectivity for the substrate, and therefore, produce high levels of added-value compounds with a low number of toxic byproducts.

The pharmaceutical industry is one of the most important areas of enzyme applications. In this field, some enzymes can be used as drugs, because of their capability to treat diseases, including some types of cancer. Others, due to their major selectivity for substrates, can be used in organic synthesis to obtain enantiopure drugs. Enzymes can also be used as biomarkers for diagnosis purposes. Regarding their production, recombinant enzymes can be produced through homologous or heterologous production using efficient hosts (Yang et al. 2011; Ahmad et al. 2014; Contesini et al. 2017b).

#### 2.2.2.1 Enzymes as Pharmaceuticals

Enzymes have been studied for use as drugs for decades (Vellard 2003). They are of great interest as pharmaceuticals because they frequently bind to their targets with great affinity and specificity, converting target molecules into the desired products. There are many therapeutic enzymes and, in this section, we will describe some representative examples.

L-asparaginase is probably the most important enzyme used as a drug. This enzyme catalyzes the degradation of asparagine, an essential amino acid for leukemic cells, into ammonia and aspartate. This enzyme was first observed by Lang and Uber (1904) and is the first therapeutic enzyme with antineoplastic properties. This enzyme is applied to the treatment of acute lymphoblastic leukemia in combination with other drugs (vincristine and a glucocorticoid) (Szymanska et al. 2012). In addition, L-asparaginase is used in the treatment of other diseases, such as Hodgkin's disease, acute myelocytic leukemia, acute myelomonocytic leukemia, chronic lymphocytic leukemia, lymphosarcoma, reticulosarcoma, and melanosarcoma (Broome 1961; Kidd 1953). Since L-asparagine is an essential amino acid for many tumor cells, its conversion into aspartic acid and ammonia is the basis for its medical use. Therefore, malignant cells are deprived of important growth factors, often resulting in cell death. L-asparaginase is produced by several microorganisms, including bacteria and fungi. However, only L-asparaginases from *E. coli* and *Erwinia chrysanthemi* have been approved to be used as part of a multiagent chemotherapy to treat acute lymphoblastic leukemia (Verma et al. 2012). This enzyme is commercially available as *Crasnitin, Crisantas, Pasum, Kidrolase, Elspar, Erwinaze PEG-asparaginase,* and *Pegasparagasum.*

α-Galactosidase A is an important example of enzyme as pharmaceutical due to its use in cases of Fabry disease. Patients with this disease present an X-linked inborn error of glycolipid metabolism. The main cause for this is the absence of the lysosomal α-galactosidaseA that hydrolyzes terminal α-galactoside linkages in different glycolipids. In males, this leads to early death due to occlusive disease of the heart, kidney, and brain. As an alternative, efforts have been made to replace the defective enzyme with normal enzyme obtained from different human and recombinant DNA sources (Mapes et al. 1970; Brady et al. 1973; Desnick et al. 1979). Therefore, some α-galactosidase A is commercialized under different names, including Replagal™ and Fabrazyme®.

Other therapeutic enzymes are Laronidase, Rasburicase, Sacrosidase, Pegaspargase, Pegademase bovine, Dornase $\alpha$, among others (Vellard 2003).

#### 2.2.2.2 Enzymes Used to Obtain Pharmaceuticals

Many enzymes present high selectivity for the substrate allowing the obtainment of specific compounds that frequently are difficult to produce using chemical catalysts. This is the case for enantiopure drugs, as single enantiomers. Since several drugs are optically active due to the presence of one or more chiral centers, if these drugs are chemically synthesized in the absence of a source of stereoinduction, the resulting drug will be a racemate. The problem for the pharmaceutical industries often relies on the fact that, in many cases, enantiomers often show important differences in their activity and pharmacokinetics (Calvey 1995). Furthermore, the separation or resolution of the racemic mixture by chemical methods is highly difficult due to the similar physicochemical properties of both enantiomers.

Enantiopure drugs can be produced by different approaches, including (i) asymmetric catalysis that requires the synthesis of chiral chemical catalysts, or (ii) the use of enzymes capable of recognizing one of the enantiomers from a racemic mixture. These enzymes catalyze enantioselective reactions, facilitating the separation of the enantiomers, in a process called kinetic resolution. This step can be applied to pharmaceuticals or to precursors used in the chemical synthesis of the drug. The process can be carried out by extracellular enzymes or whole cells as the biocatalysts. Among the extracellular enzymes, hydrolases, mainly lipases, are the most used enzymes for obtaining enantiomerically pure compounds. On the other hand, oxidases and reductases are frequently used within the cell that are applied to the synthesis of enantiomerically pure drugs since these enzymes often requires intracellular cofactors such as adenosine triphosphate (ATP), nicotinamide adenine dinucleotide phosphate (NAD(P)H), and flavin adenine dinucleotide (FADH) (Jeon et al. 2017; Sikora et al. 2017; Tramontina et al. 2017).

Lipases naturally hydrolyze oils and fat. However, in the presence of hydrophobic organic solvents, such as hexane, these enzymes are capable of catalyzing esterification reactions. Therefore, lipases from different organisms can catalyze hydrolytic and synthetic enantioselective reactions. For example, a lipolytic extract from *Aspergillus niger* has been studied for the enantioselective esterification of the racemic nonsteroidal anti-inflammatory drugs ibuprofen and ketoprofen (Carvalho et al. 2006). Studies have also been focused on the enantioselective hydrolysis of naproxen methyl ester by *Candida rugosa* immobilized-lipase showing remarkable values of enantioselectivity (Sayin et al. 2014).

Whole cells are also utilized in the production of enantiopure drugs, mainly using oxidases/reductases. In this case, although it is based on enzymatic reactions, the biocatalyst is considered the whole cell. One interesting example is the production of enantiopure (+)- and (−)-pinoresinol by kinetic resolution. (+)-Pinoresinol showed to be effective against different diseases, including cancer (Sepporta et al. 2013). In the work of Ricklefs et al. (2016), an *E. coli*-based whole-cell system was developed based on the heterologous expression of three enzymes, a reductase from *Forsythia intermedia*, a vanillyl-alcohol oxidase from *Penicillium simplicissimum* and a bacterial laccase from *Corynebacterium glutamicum*, showing interesting results.

#### 2.2.2.3 Enzymes Used for Diagnostics Purposes

Several diseases associated with many components of the enzyme metabolism systems are now widely applied in clinical exams as biomarkers for diseases. For example, the presence of some of the proteins in the serum indicates that tissue or cellular damage has occurred resulting in the release of intracellular components into the blood (Raja et al. 2011). For example, widely assayed enzymes are the amino transferases: alanine transaminase, ALT and aspartate aminotransferase, AST; lactate dehydrogenase, LDH; creatine kinase, CK; gamma-glutamyl transpeptidases, GGTs (Raja et al. 2011).

Beyond that, enzymes can be used to detect other metabolites, such as blood glucose levels using glucose oxidase method (Bergmeyer et al. 1974).

Another example of enzyme application is the use of DNA polymerase in polymerase chain reaction (PCR) technique for the diagnoses of cancer mutations, celiac disease, among other genetic disorders (Barra et al. 2011). It is also possible to perform paternity test and forensic analysis. *Thermo Fisher Scientific* is the multinational players in this field, developing interesting PCR application, including PCR reactions to be performed in space without complex PCR machineries (Boguraev et al. 2017). However, emerging companies are growing in many countries, such as *Ecra Biotech* located in Brazil, which are producing reliable PCR kits for human diagnostics purposes.

#### 2.2.2.4 Enzyme Production

All types of organisms that can be considered good sources for gene prospection. However, when it comes to enzyme production, microorganisms are considered the best tools, for recombinant or nonrecombinant enzymes, because these organisms grow faster and easier than mammalian and plant cells. Additionally, several microorganisms present remarkable capacity of enzyme secretion. This is advantageous, since this simplifies downstream processing, thereby reducing costs further (Tufvesson et al. 2011; Abood et al. 2015).

One of the main bottlenecks for enzyme application is the large-scale production, mainly due to the costs involved in growing media that reach 50% of the expenses (Burkert 2004). Therefore, for a feasible application of industrial enzymes the use of engineered hosts for homologous or heterologous protein production is often required.

The selection of an efficient host for recombinant enzyme production depends on different aspects; however, for extracellular enzymes the use of an efficient protein secretor is important. Good candidates are filamentous fungi, such as *Aspergillus* spp. (Rubio et al. 2015; Contesini et al. 2017a). Several fungi have their secretory capacity engineered in order to produce considerable amount of proteins per liter of cultivation media. Apart from the selection of good secretors, it is important to select an effective expression system. Since each enzyme production must be optimized case by case, in general, the genetic engineering aspects to be addressed are the use of strong promoters, optimization of the number of gene copies in the genome, as well as codon optimization (Wang and Xia 2011; Favaro et al. 2012; Zhao et al. 2012).

After the selection of the best enzyme producer, it is necessary to optimize fermentation parameters, including the selection of low-cost substrates. Good examples are agro-industrial residues, such as sugarcane molasses as carbon sources or corn steep liquor as nitrogen source (Santos et al. 2016). Other important parameters are aeration, pH, and temperature (Kawaguti et al. 2006; Oriente et al. 2015; Tramontina et al. 2016; Contesini et al. 2017a).

### 2.2.3 Antibodies in the Pharmaceutical Industry

Immunoglobulins – antibodies able to neutralize pathogens and toxins – are among the best-studied proteins in biology: from the first observations concerning cowpox vaccination in the late XVII century, to recent automated peptide synthesis techniques, antibody molecules represent a fast-growing class of biopharmaceuticals and one of the best biological platforms to treat a variety of diseases. According to USA National Institute for Health Clinical Trial Database (clinicaltrials.gov), over 5800 clinical trials around the world have used immunoglobulins as the main therapeutic intervention to treat more than 1600 different medical conditions. Approximately half of those studies are currently underway, and up to 104 studies are on the fourth and final phase of development.

According to a 2017 Business Communication Company Inc. Research Report, therapeutic mAbs represented an 80.8 billion dollars market in 2016 and should grow over 6% per year at least until 2022, when it is expected to reach over US$114 billion. Indeed, according to the Food and Drug Administration (FDA), a United States federal regulatory agency, over half of the top 10 biological drugs sold in 2015 represented mAbs, such as *Remira® (infliximab; Johnson & Johnson), Rituxan® (rituximab; Roche), Avastin® (bevacizumab), Herceptin® (trastuzumab; Roche), Lucentis® (ranibizumab; Roche, Novartis)* and *Humira® (adalimumab; AbbVie)*. In fact, *Humira* was not only the best-selling biological drug in 2015, but also the overall best-selling drug worldwide, regardless of class.

The market success can be attributed to the unique array of biochemical and structural properties of antibodies, which allow these molecules to bind virtually any possible antigen target, while maintaining relatively high specificity and stability. Generally, all antibodies are formed by two pairs of polypeptide chains, called heavy (H) and light (L) chains, linked together by disulfide bonds. Each H and L chain presents a constant region, which is conserved among the individuals of a certain species, and a variable region, which differs between individuals and also contains a hypervariable portion responsible for antigen recognition and selectivity. In general, antigens contain several epitopes, short sequences containing five to six amino acids, which can be recognized by antibodies; therefore, a single antigen may bind to more than one type of antibody (Abbas Abul et al. 2014).

In humans, immunoglobulins are secreted by mature B-lymphocytes present in the blood, namely plasma cells, and can be further classified in IgG, IgM, IgA, IgD, and IgE, according to the heavy-chain amino acid sequence and biological function. Immunoglobulins of IgG class are the most abundant in the serum and also present an extended half-life (15–20 days) in comparison to other classes. For that reason, IgG molecules, in particular $IgG_1$, $IgG_2$, and $IgG_4$, are extensively used for treatment of chronic conditions and other diseases. An enzyme extracted from the papaya fruit, Papain, is capable of cleaving the IgG molecule in the "hinge" positions, rendering two major fractions. One, called Fab (Antigen Binding Fragment), contains the light chains and retains the antigen binding capacity. The other, called Fc, is composed of the heavy chains and, while incapable of antigen recognition, is easily crystallized. Similarly, Pepsin is also capable of cleaving the IgG in a slightly lower position than Papain, which maintains the "hinge" structure and results in a Fab fraction containing two antigen recognition sites, named $Fab_2$ antibody (Abbas Abul et al. 2014).

The advance of immunoglobulin purification techniques allowed the assembly of fractions containing different proportions of specific Fab, $Fab_2$, and Fc fragments. As seen previously, a target antigen may span different antibody-recognition sites, reducing specificity. For therapeutic applications, it is crucial to produce mAbs, which are prevenient from a single B cell lineage and recognize only one unique epitope. In this section, we will cover the main classes of current mAbs used in the pharmaceutical industry – mouse, chimeric, humanized, and human – and explore representative examples to delineate the main development process and pharmacological aspects of each.

#### 2.2.3.1 Mouse mAbs

In 1975, Köhler and Milstein published in *Nature* a method for production of the first mAb-secreting cell lines: the hybridoma technology. The innovation brought together the specificity of antibodies secreted from B cells, obtained from immunized mice, to the immortality of human myeloma cells, allowing stable mAb production in laboratory cultures. The resulting hybrid contains genetic material from both parental lineages and must be stabilized by successive rounds of cloning. In 1984, Köhler and Milstein were awarded the Nobel Prize in Medicine and Physiology for their groundbreaking work.

In 1986, the first FDA-approved mAb obtained from hybridoma cell lines for clinical use in humans: muromonab CD-3 (*Orthoclone OKT3*®) was made available for use in kidney transplant patients with acute allograft rejection. Muromonab CD-3 was designed to bind the epsilon-chain of T-cell receptor-CD3 complex (TcR-CD3) complex, a glycoprotein directly involved in the activation and differentiation of T-cells during this process. The immediate effect is steric occlusion of the ligation between T-cell's membrane receptor TcR and antigen presenting cells major histocompatibility complex class II (MHC II) receptor, leading to dysfunction of affected T-cells and their subsequent opsonization and removal by circulating macrophages. This ensues a drastic reduction in $CD3^+$ T-cells levels, largely diminishing transplant rejection; moreover, surviving $CD3^+$ T-cells undergo internalization, rendering them immunologically inactive (Caillat-Zucman et al. 1990).

The hybridoma is obtained by immunizing mice with appropriated antigen complexes (mixture of antigens and adjuvants that stimulate immune response), such as isolated TcR-CD3 molecules or whole human T-cells. Following an initial screening, responsive cells are selected, isolated, and fused with murine myeloma cell lines, resulting in antibody-secreting hybridomas. The latter are tested for both survival and specificity, and positive candidates are expanded. Finally, the desired $IgG_2$ mAbs are purified and microbiologically tested (Becker 2007).

Most patients exhibit a variety of side effects upon the administration of the first few doses of muromonab CD-3, which has been designed as Cytokine Release Syndrome (CRS). CRS symptoms range from more usual mild "flu-like" fever, joint/muscle pains and weakness, to rarer shock-like allergic reactions, but generally cease after four to six hours. It is believed that the CRS mechanism of action is ensued by initial activation of T-cells upon muromonab CD-3 injection, leading to a transient increase in the secretion of pro-inflammatory cytokines before T-cell function is depleted. Indeed, CRS resolution usually follows continuous administration of the drug and, therefore, interruption of the treatment is not necessary nor advised in most cases (Becker 2007).

Nevertheless, side effects such as CRS evidenced the undesired immunogenic reactions triggered by murine mAbs; indeed, although several drugs have been developed, only a few have been approved by the FDA, such as muromonab itself, *Zevalin® (Ibritumomab tiuxetan; Biogen Idec)* and a few diagnostic drugs (*Leukoscan*® and *CEA-scan*®, *Immunomedics*). The immunogenic effect of murine mAbs, namely human-against-mouse antibody (HAMA) response has propelled the need to design novel antibody molecules with as little nonhuman sequences as possible.

#### 2.2.3.2 Chimeric Monoclonal Antibodies

Evidence that administration of murine Abs ensues an inflammatory response and production of HAMAs, with considerable side effects for the patients, led to extensive effort directed to minimize these immunogenic reactions. Importantly, the basic science around the antibody structure and immunogenicity determinant regions, together with the advent of recombinant DNA technology in the late 1980s, made it possible to design tailor-made "chimeric" antibodies, effectively substituting nonessential murine amino acid sequences with corresponding sequences obtained from human immunoglobulin. Depending on the length of substitution, these fusion proteins can be classified as chimeric or humanized antibodies. Generally, a chimeric antibody refers to a fusion protein presenting a human immunoglobulin backbone, and variable regions of the heavy and light chains – which are responsible for antigen binding – from mouse mAbs (Ho and Gibaldi 2013).

Abciximab was the first chimeric mAb to be approved (by the FDA, in 1994, and by the European Medicines Agency-EMA, the following year) and is still a viable option for treatment of high-risk angioplasty. The drug consists of the Fab fragment of chimeric murine/human mAb 7E3, obtained by papain digestion, which is perfused intravenously. Abciximab recognizes the glycoprotein IIa/IIIb present in the surface of platelets; therefore, preventing the aggregation of coagulation factors such as fibronectin, von Willebrand factor, among others. Since it binds to platelets, which present a very short turnover time, abciximab plasma half-life is approximately 10 minutes; nevertheless, its effects on overall coagulation can be observed up to 48 hours after the treatment has ceased (Usta et al. 2016).

Even though chimeric mAbs still elicit immunogenic responses (albeit considerably less than murine mAbs), the intensity of the side effects varies greatly, and are usually less than for murine mAbs. Since the first drug release in 1994, several FDA-approved chimeric antibodies are currently in use for a wide array of conditions, from arthritis and psoriasis (Infliximab) to several types of cancer (Retuximab, Cetuximab) and coagulation issues (Abciximab) (actip.org, accessed on December 2017).

#### 2.2.3.3 Humanized Monoclonal Antibodies

The humanization of mAbs, also known as complementarity determining region (CDR) grafting, further refines the binding affinities and immunogenic characteristics of chimeric mAbs. Humanized mAbs contain only the animal sequences responsible for the minimal antigen-binding ability from mice or other animals, called CDRs, as opposed to the whole animal-derived variable region observed in chimeric antibodies (Ho and Gibaldi 2013).

As chimeric mAbs, humanized mAbs also explore the possibilities of relatively straightforward recombinant DNA techniques; however, the rational design of humanized mAbs is no trivial task. A simplified workflow for designing such antibodies consists mainly of subsequential steps for selection of chimeric/murine mAbs donor sequences, search for suitable human frameworks, identification of putative mutations and possible redesign, construction and purification of the final humanized mAbs and, finally, clinical testing of the candidates.

The first step is selecting the appropriate CDRs among the six highly variable sequences present in the variable portion of heavy (H) and light (L) chains of the donor mAbs. Since the CDRs play an important role in antigen binding, it is essential that they are not only retained in the humanized antibody, but that the main sequence and structure of its canonical residues are maintained as well. Two methods are currently used for selecting suitable CDRs: choice by sequence, according to Kabat et al. (1984), or choice by structure, according to Chothia et al. (1989), whereas the latter is considered less immunogenic due to the possibility of choosing only the minimal residues responsible for structure maintenance. Today, it is also possible to use online tools to identify the canonical structure and important residues of both CDRs and frameworks alike (www.bioinf.org.uk/abs/chothia.html).

The choice of an adequate human framework directly affects the specificity, affinity, and immunogenicity of the final humanized antibody. The frameworks can be chosen according to preestablished models (fixed frameworks) or by similarity with the donor sequence (homology search), and each option retains advantages and disadvantages. Fixed frameworks, albeit not necessarily ideal for the specific application of interest, are curated under databases based on experience and may, therefore, be more robust. That was the case for some of the first humanized antibodies to enter the market, such as *Campath-1*® (Riechmann et al. 1988) and *Actemra*® (Sato et al. 1993). On the other hand, the choice of homologue matches could lead to more specific and less immunogenic humanized antibodies. An interesting study comparing those two approaches was conducted by Graziano et al. (1995), describing the humanization process of antibody M22 and in that case, homologue matches seemed to be the more advantageous choice. Finally, it is possible to mix and match the available options using 3D modeling tools and recombinant DNA techniques to identify the optimal framework structure.

The use of 3D-computer modeling of the donor antibody structure, human framework and, if possible, of antigen docking is a powerful tool to compare the final humanized antibody with the source mAbs, especially if crystal structures have not been defined. Indeed, the superposition of initial (donor) and final (humanized) structures is important to check the preservation of determinant characteristics such as charge, size, hydrophobicity, and presence/absence of hydrogen bonds. The modification of such features may be related to the presence of backmutations, which can be identified and analyzed. It is also possible to automatically build models using softwares such as protein data bank (PDB) Swiss Prot, although double-checking of the process by the antibody designer is indispensable.

An illustrative example of humanized mAb development is anti-IgE *Omalizumab (rhuMab-E25; Xolair*®), an anti IgE mAb especially designed for the treatment of moderate to severe allergic asthma, which has been approved for use in several countries. *Omalizumab* therapeutic effects are due to its capacity of specific binding to free IgE antibodies in the circulation, which are deeply involved in allergic manifestations and, since up to two-thirds of asthma patients are allergic, could be involved in the disease's pathophysiology (Kroegel and Foerster 2008).

IgE molecules contain a ε-heavy chain constant region, which is capable of binding to low (FcεRII) and high-affinity (FcεRI) receptors present in cells such as mast cells, eosinophils, and dendritic cells. The binding of antigen/allergen molecules to the variable portion of IgE, concomitant to the recognition of the Fc region by cell receptors, induces a cross-linking mediated cell response, which culminates in the activation of effector pathways. Ultimately, it induces the release of chemical mediators, such as histamine, leukotrienes, and prostaglandins, as well as proinflammatory cytokines by responsive cells.

Due to IgE's pathophysiological effects being triggered upon cross-linking with cell receptors, it was important that the anti-IgE mAb binds very specifically only to free IgE molecules. Thus, homology scanning mutagenesis studies were used to analyze the IgG constant sequence homologous to IgE's Fcε chain. Six key amino acids (R408, S411, K414, E452, R465, M469) thought to mediate the binding to high-affinity FcεRI receptors were identified, and available mAbs capable of specifically binding to this epitope were evaluated (Nissim et al. 1993; Presta et al. 1994). A murine mAb, MAE11, capable of binding to free IgE simultaneously as the free FcεRI receptor and therefore competing for the binding, was selected and used as template for humanization (Saban et al. 1994; Shields et al. 1994).

Using a humanized $IgG_1$ framework, dozens of variants were generated and the best-fit, named rhuMab-E25, retained the IgE binding affinity and biological activity of the donor murine mAb MAE11. The final humanized antibody contained only 5% of non-human sequences, which corresponded to the CDRs residues necessary for binding. Early *in vitro* assays demonstrated that rhuMab-E25 prevented IgE-mediated activation of FcεRI bearing cells, while preclinical *in vivo* studies in cynomolgus monkeys showed that, even in doses as high as $50\,mg\,kg^{-1}$ or by direct subepidermal injection of 1 μg, Omalizumab did not trigger anaphylaxis and was hereinafter considered safe. In terms of comparison, injection of as little as 1 ng of a cross-linking murine mAb was able to trigger a wheal and flare reaction, demonstrating the deeply beneficial effects obtained by the humanization process (Saban et al. 1994; Shields et al. 1994).

There are currently several humanized mAbs in the market that represent impressive sales numbers, such as *Lemtrada® (alemtuzumab; Sanofi), Kadcylan® (ado-trastuzumab emtansine; Roche), Perjeta® (pertuzumab; Roche), Actemra (tocilizumab; Roche), Cimziag® (certolizumab pegol, UCB), Soliris® (eculizumab; Alexion Pharmaceuticals), Lucentiso® (ranibizumab; Roche, Novartis), Tysabri® (natalizumab; Biogen Idec), Avastin (bevacizumab, Roche),* and many others. Together, the above-mentioned drugs accounted for a global sales revenue of more than US$16.5 billion in 2013 alone (Ecker et al. 2015).

#### 2.2.3.4 Human Monoclonal Antibodies

As previously seen, murine mAbs were the first immunoglobulins to enter the clinical market due to its facilitated production process; however, several immunogenicity issues raised concern regarding the safety of animal-derived proteins. To circumvent these problems, chimeric and humanized antibodies were developed during the 1980s and constituted the majority of clinical trials during the 1990s; indeed, in 2010, up to two-thirds of the commercially available mAb based therapies were either chimeric or humanized forms. Human mAbs, on the other hand, lagged behind during the last decades of the twentieth century, which comprehended the early stages of DNA recombinant technologies (Nelson et al. 2010).

Nevertheless, in terms of immunogenicity and specificity, human mAbs constituted the "golden standard" of therapeutic immunoglobulins and attempts to implement serial production of these molecules started as early as the beginning of 1980s. At the time, the main strategies were analogous to the production process of murine mAbs and involved *in vitro* immunization of mononuclear peripheral blood cells, followed by establishment of human hybridomas obtained by fusing human lymphocytes and myeloma cell lines.

Another popular approach was the artificial immortalization of human lymphocytes lineages with the Epstein-Barr virus. However, several issues arose associated mainly with cell line instability, poor secretion of antibodies and contamination of cell cultures. In addition, as opposed to murine mAbs, it is unethical to immunize human patients with specific antigens and subsequently collect the responsive lymphocytes for establishment of hybridomas. For those reasons, 15 out of 16 early human mAbs candidates were ruled out during the clinical tests. In the late 1990s, the advent of more sophisticated recombinant DNA methods allowed the production of human mAbs by transgenic mice lineages or phage display approaches (Nelson et al. 2010).

The use of transgenic, also called transchromosomic, mice relies on the disruption of endogenous *loci* encoding for the heavy and light chains of murine immunoglobulin; instead, transgenes encoding for human Ig components are introduced and expressed. This technique allows for the introduction of different genes encoding for the variable region, and thus increases the overall potential array of antigens recognized by the mAbs. Analogously, the introduction of genes for different heavy-chain isotypes ensues the expression of immunoglobulins capable of diverse biological functions. The use of transgenic mice also overcomes the HAMA issue observed with murine mAbs, although the strains used for the former are not always capable of triggering an immune response as effective as the latter (Yamashita et al. 2007).

The selection of antigen-specific variable region gene variants for tailor-made mAbs design is largely done by the phage display method. In this technique, a combinatorial library comprising thousands of single chain variable fragments (i.e., the random combination of gene segments encoding heavy- and light-chain variable regions) is introduced in bacteriophages to be assembled. The recombinant proteins are displayed on the surface of the phage and are challenged with an antigen of interest, allowing selection of those capable of interacting with higher affinity. The use of refined gene and protein engineering allows even further optimization of the process by improving the affinity and specificity of the single chain variable fragments for antigens (Yamashita et al. 2007).

The advance of molecular biology and protein engineering techniques, in addition to improvements in the area of cell culture and manipulation of organisms, have propelled the rapid expansion of human mAbs in the market. In 2002, *adalimumab (Humira, Abbott)*, an anti-TNF$\alpha$ antibody used in the treatment of rheumatoid arthritis, was the first human mAb to be approved by the FDA. In the following eight years, up to 2010, another six human mAbs were developed and approved by the FDA and EMA for therapeutic use. In contrast, since 2010, 19 additional novel human mAbs have been approved and introduced to the market (actip.org, accessed in December 2017).

## 2.3 Prospects for Area Development

### 2.3.1 Patent Generation

Considered a revolution in modern medicine, the biotechnology trend in the global pharmaceutical market brings innovation for the treatment of various complex and high incidence diseases. During the years of 2016 and 2017, the FDA had granted its approval of 40 biological products. Currently, it is estimated that one-fifth of the total number of drugs sold each year comes from biotechnological processes (US FDA 2017).

Around 300 million patients worldwide are already benefiting from these new drugs for the treatment of chronic diseases such as hepatitis C (interferon alpha), kidney cancer (interleukin), hemophilia (factor VIII), diabetes (human insulin), multiple sclerosis (interferon beta), and many others (DiMasi et al. 2016). However, such revolutionary drugs also come with large price tags. It takes several million dollars to discover, develop, and gain regulatory approval for a new medicine. US-based biopharmaceutical companies invested about $75 billion in R&D in 2015 (DiMasi et al. 2016). In fact, according to the American National Science Foundation, the sector accounts for the single largest share of all US business R&D, representing 1 out of every 6 dollars (17%) spent on domestic R&D by US businesses (TEConomy Partners 2017).

Therefore, patent protections are very important for pharmaceutical companies, as it allows the appropriation of the benefits from innovation due to the expensive pharmaceutical R&D process. In the absence of patent protection, or some equivalent barrier, imitators could "free ride" on the innovator's FDA approval and duplicate the compound for a small fraction of the originator's costs (Eccleston-Turner 2016).

One of the reasons R&D is so costly for pharmaceuticals is that most new drug candidates fail to reach the market, even with several preclinical experiments, some drug candidates fail at the clinical trials due to the complexity of the human organism that cannot be predicted with *in vitro* and *in vivo* tests. Furthermore, the full R&D process from synthesis to FDA approval involves undertaking successive trials of increasing size and complexity (DiMasi et al. 2016). As part of the pharmaceutical industry, biopharmaceuticals are supported and protected by the patent system. Thus, when it expires, there is the possibility of producing the so-called biosimilars, which are the very similar copies of the original biopharmaceuticals (Geigert 2013). A biosimilar is not regarded as a generic of a biological medicine. This is mostly because the natural variability and more complex manufacturing of biological medicines do not allow an exact replication of the molecular micro heterogeneity (European Medicines Agency 2017).

While currently a growing source of revenue, the market exclusivity enjoyed by many of these products will erode as patents expire and competing products enter the market. In 2016, biologics, with US sales totaling $60 billion, lost patent protection in the US market (Lybecker 2016; Morton and Boller 2017). The expiration of patents from products developed in the last few decades create an opportunity for the market of biosimilars. In 2006, the EMA approved its first biosimilar: the growth hormone somatropin. With the increasing price of new biologicals and continuing pressure on health-care budgets, biosimilars are expected to make up an increasing share of the biologicals market (European Medicines Agency 2017). Although the European Union (EU) previously defined a period of 10 years for data exclusivity, this was revised at the latest review of pharmaceutical EU legislation to the following: 10 years if the reference product is centrally approved or application to the centralized procedure has been made before 20 November 2005; or 8 years data exclusivity +2 years market exclusivity +1 year possible extension if a full dossier is submitted on or after 30 October 2005 via a national procedure or after 20 November 2005 via the centralized procedure (GaBI Journal Editor 2015). There are also more than 200 new biotechnology products in the pipeline (phase II to be registered), all of which could be future targets for biosimilars.

In Table 2.1, the estimated patent expiry dates for several of the best-selling biological molecules are listed. As shown, biologicals are entering in an industry competition in the near future driven by the insight that the reference biological patents are expiring as well as other intellectual property rights considerations.

Accordingly, governments encourage the production of biosimilars as the patents expire to combat high prices of reference biopharmaceuticals, specially aiming to promote high-quality public health programs (GaBI Journal Editor 2015).

Over the next five years, it is expected that the biosimilars market will rapidly expand over 20-fold (Eccleston-Turner 2016). Some of the main reasons of this expansion are related to the strategic collaborations between government and big pharma and the rise of new diseases. Moreover, emerging economies in Asia and Latin America are also trialing the adoption of biosimilars initiatives. For example, *BioNovis* located in Brazil are producing and developing different biopharmaceuticals (Dalgaard et al. 2013).

### 2.3.2 Perspectives for Biotechnology in the Pharmaceutical Sector

Biotechnology will continue to provide further advances in clinical research in the coming years, and will also continue to offer alternatives or improvements to the current conventional treatments available. The perspective is that in the next few years, tools and discoveries that are still in the research field will be applied to the clinic, bringing more innovation and profit to the pharmaceutical industries.

Regarding biopharmaceuticals, new drugs can be developed for new diseases or processes can be optimized for a more efficient therapy. There are biopharmaceuticals undergoing clinical trials with the expectation that some reach the market in the next years. According to Drugs.com, an accurate database information on drugs, some example are: Cancer therapeutic – larotrectinib (Phase 1 Loxo Oncology); Episodic migraine – erenumab (Phase 3 – Amgen); Zika virus vaccine (Phase 1 – Takeda Pharmaceutical Company); Schizophrenia treatment – lumateperone (FDA Fast Track designation – Intra-Cellular Therapies); severe postpartum depression – brexanolone (Phase 3 – Sage Therapeutics).

As previously described, siltuximab, bevacizumab, and remicade will soon end their patent protection. After the end of a product's license, other companies race to launch a biosimilar and to get part of a biopharmaceutical's lucrative market. The production of a biosimilar is complex, and undergoes more rigorous tests than the production of a biopharmaceutical de novo. Although a biosimilar and the reference biopharmaceutical are the same protein, they are produced in different batch of cells, with some minor modifications in the DNA coding sequencing. US FDA has strict guidelines to approve commercialization of a biosimilar. Therefore, biosimilars must undergo several structural, stability, and functional tests to show that they have the same properties of the reference. Nevertheless, all the tests required for the approval of a biosimilar are only a minor amount of the costs of developing a new biopharmaceutical, and the profit from the sales of a blockbuster drug will compensate the expenses with the approval tests.

In the coming years, the selection by the physician of the right biopharmaceutical treatment for a certain disease will be guided by the particularities of the patient genome. Great efforts in the field of Precision Medicine are being done in the last years, with the creation of several biobanks (Olson et al. 2014) and initiatives such as Precision Medicine Initiative, 2015 U.S., collecting genetic data, biopsies samples, and health

**Table 2.1** Biologicals patent expiration dates.

| | Biological | Approval date | Patent expiration (US) | Use |
|---|---|---|---|---|
| Not humanized antibodies | *Adcetris (brentuximab vedotin)* | 19/09/2011 | 2015–2031 | Refractory Hodgkin lymphoma (HL) and systemic anaplastic large cell lymphoma (ALCL) |
| | *Aranesp (darbepoetin alfa)* | 17/09/2001 | 15/05/2024 | Anemia associated with chronic renal failure (CRF) |
| | *Avonex/Rebif (interferon beta-1b)* | 07/02/2003 | 2015 | Multiple sclerosis |
| | *Betaferon/Betaseron (Interferon beta-1b)* | 23/06/1993 | 07/05/2013 | Multiple sclerosis |
| | *Bexxar (tositumomab)* | 27/07/2003 | 01/01/2029 | Non-Hodgkin's lymphomas |
| | *Prostascint (capromab)* | 28/10/1996 | 01/11/2036 | Prostate cancer |
| | *Enbrel (etanercept)* | 02/11/1998 | 22/11/2028 | Rheumatoid arthritis, plaque psoriasis, ankylosing spondylitis, and polyarticular juvenile idiopathic arthritis |
| | *Erbitux (cetuximab)* | 12/02/2004 | 13/02/2016 | Cancers of the colon and rectum |
| | *Orthoclone OKT3 (muromonab-CD3)* | 14/09/1992 | 01/01/2021 | Immunosuppressant |
| | *Remicade (infliximab)* | 24/08/1998 | 08/09/2018 | Autoimmune diseases |
| | *Rituxan (rituximab)* | 26/11/1997 | 22/09/2016 | Follicular CD20-positive non-Hodgkin's lymphoma |
| | *Reopro (abciximab)* | 16/12/1993 | 01/06/2016 | Antiplatelet agent |
| | *Simulect (basiliximab)* | 12/05/1998 | | Immunosuppressant |
| | *Sylvany (siltuximab)* | 23/04/2014 | 01/08/2017 | Multicentric Castleman's Disease (MCD) |
| | *Zevalin (ibritumomab)* | 19/02/2002 | 19/02/2016 | Non-Hodgkin's lymphoma |
| Humanized antibodies | *Abthrax (raxibacumab)* | 14/12/2012 | 01/01/2016 | Prophylaxis and treatment of inhaled anthrax |
| | *Actemra (tocilizumab)* | 08/01/2010 | 22/12/2015 | Rheumatoid arthritis |
| | *Avastin (bevacizumab)* | 26/02/2004 | 04/06/2019 | Cancer treatment |
| | *Arzera (ofatumumab)* | 26/10/2009 | 01/01/2013 | Chronic lymphocytic leukemia (CLL) |

| | | | | |
|---|---|---|---|---|
| | *Benlysta (belimumab)* | 09/03/2011 | 01/01/2013 | Systemic lupus erythematosus |
| | *Campath (alemtuzumab)* | 07/05/2001 | 06/07/2021 | Chronic lymphocytic leukemia (CLL), cutaneous T-cell lymphoma (CTCL) and T-cell lymphoma |
| | *Cinzia (certolizumab)* | 22/04/2008 | 13/02/2024 | Crohn's disease and certain types of arthritis |
| | *Humira (adalimumab)* | 31/12/2002 | 25/11/2017 | Rheumatoid arthritis, chronic plaque psoriasis, Crohn's disease, ankylosing spondylitis and psoriatic arthritis |
| | *Lucentis (ranibizumab)* | 30/06/2006 | 30/06/2020 | Age-related macular degeneration |
| | *Soliris (eculizumab)* | 16/03/2007 | 16/03/2021 | Atypical hemolytic uremic syndrome (aHUS) and paroxysmal nocturnal hemoglobinuria (PNH) |
| | *Tysabri (natalizumab)* | 23/11/2004 | 01/03/2015 | Multiple sclerosis |
| | *Vecttibix (panitumumab)* | 27/09/2006 | 08/04/2020 | Wild-type RAS metastatic colorectal cancer |
| | *Xolair (omalizumab)* | 20/06/2003 | 20/06/2017 | Allergic asthma |
| | *Yervoy (ipilimumab)* | 25/03/2011 | 01/01/2023 | Metastatic melanoma |
| | *Zonapax (daclizumab)* | 10/12/1997 | 01/03/2013 | Multiple sclerosis |
| Not antibodies | *Epogen (epoetin alfa)* | 20/08/1989 | 20/08/2013 | Anemia |
| | *Eylea (aflibercept)* | 18/11/2011 | 21/06/2027 | Age-related macular degeneration |
| | *Forteo (teriparatide)* | 26/11/2002 | 08/12/2012 | Osteoporosis |
| | *Kinepret (anakinra)* | 14/11/2001 | 21/06/2027 | Rheumatoid arthritis |
| | *Nulojix (belatacept)* | 15/06/2011 | 01/01/2023 | Immunosuppressants |
| | *Neupogen (filgrastim)* | 20/02/1991 | 03/12/2013 | Neutropenia |
| | *Neulasta (pegfilgrastim)* | 31/01/2002 | 20/10/2015 | Neutropenia |
| | *NeutripinAq (somatropin)* | 16/10/1986 | Expired | Growth failure |
| | *Novorapid (insulin aspart)* | 07/07/2000 | 07/12/2014 | Diabetes |
| | *Oncaspar (pegaspargase)* | 01/01/1994 | 10/06/2027 | Acute lymphocytic leukemia (ALL) |
| | *Orencia (abatacept)* | 23/12/2005 | 01/10/2019 | Rheumatoid arthritis |
| | *Lantus (insulin glargine)* | 24/04/2000 | 01/01/2014 | Diabetes |
| | *Lovenox (enoxaparin)* | 29/03/1993 | Expired | Anticoagulant |

records for identification of biomarkers. In the future, these biomarkers will guide the personalized treatment with specific drugs for each individual (Collins et al. 2016). This is already a reality in some cases of cancer treatments, for example, trastuzumab. This well-known personalized medicine consists of a humanized IgG1 mAb for the treatment of breast cancer in patients whose tumors overexpress the oncogene HER2 (Jackson and Chester 2015).

Biotechnology techniques are also making advances in the field of biosensors that are soon to enter the market. Biosensors are devices based on a physical transducer (voltammetric, amperometric, conductometric, spectrophotometric, etc.) (Gil and de Melo 2010). Glucose biosensors are already widely used in clinical applications for diagnosis of diabetes mellitus and at home to monitor blood-glucose levels (Mehrotra 2016). In the near future, biosensors will have a great share at the pharmaceutical market. For example, they can be used to diagnose infectious diseases, to identify end-stage heart failure patients, or they can be a biochip for a quick and accurate detection of multiple cancer markers and for diagnostics of Parkinson's disease (Mehrotra 2016).

New gene therapy technologies are now in the research phase or have recently been approved for clinical use. In 2017, the FDA approved two therapies using genetically engineered immune cells collected from patients, CAR T-cell therapy: tisagenlecleucel (*Kymriah*™) – for children and adults with advanced leukemia from Novartis; and axicabtagene ciloleucel (*Yescarta*™) – for patients with large-B-cell lymphoma from Kite pharma (Ronson et al. 2016). The treatment with CAR T-cell therapy is customized for each patient using the patient's T cells, but should be used as a last resort; until now the infusion of CAR T cells can lead to inflammatory side effects, like CRS. Also, in 2017, a novel cancer vaccine, called mRNA-4157 from Moderna Therapeuticals began the tests for Clinical Phase I (www.clinicaltrials.gov 2018). This new treatment uses engineered mRNA targeting solid tumors, and represents a huge step forward, not only for cancer treatment, but also for personalized medicine, since each vaccine is tailored to a specific patient and their tumor.

Genome-editing technologies using clustered regularly interspersed palindromic repeats (CRISPR) in combination with CRISPR-associated systems (Cas) have revolutionized the research field in this decade (Lindsay-Mosher and Su, 2016). Accordingly, in 2017, two clinical trials for cancer treatment were announced in China and the United States using CRISPR-Cas to engineer patient T cells *in vitro* to destroy cancer cells (Lindsay-Mosher and Su 2016).

The patent on many of the medicines launched in the 1990s will expire over the next few years, leaving many companies very exposed. To avoid reduced profits after patents expire, pharmaceuticals industries will have to focus on specialized therapies for diseases with currently no effective treatments. The companies can also invest in methodologies, such as human-on-a-chip and virtual patients, aiming to decrease time and costs involved in researching and developing such medicines to ensure more affordable treatments for society and governments.

## 2.4 Conclusion

Biotechnology advances allow a better understanding of the mechanism of several diseases. Consequently, biotechnology allows designing and producing new drugs, with greater efficacy and specificity and, consequently, fewer side effects compared to the traditional treatment. Biotechnology offers targeted products for specific diseases using innovative technologies – in particular, genetics approaches. More than 350 biological drugs are on the market for the treatment of more than 150 diseases (Kepplinger, 2015), including treatments for rare diseases and some cancers.

Biotechnology offers greater control over the manufacturing process, allowing a significant reduction of the risks of contamination by infectious agents. One good example is the blood products used to treat hemophilia (Almeida et al. 2011).

Some products that are not naturally created in sufficient amounts for therapeutic purposes can be produced in large-scale using biotechnological tools. Heterologous expression allows the production of enzymes, proteins, and antibodies used for treatment or diagnosis; and overexpression in microorganisms allows the purification of compounds such as antibiotics. Biotechnological products can not only be used as drugs but also as means to produce the final medicine – for example, enzymes that can be also used for the production of enantiopure compounds.

Although promising, these technologies aiming to develop biopharmaceuticals still show some limitations. For example, (i) the requirement of better analytical methods to interpret and correlate the large amount of data generated; (ii) a better way to rationally and effectively apply these data on drug development, and (iii) better development of administration techniques and more knowledge on absorption, distribution, metabolism, and excretion of these medicines (Jenkins and Parker, 2015; Lopes, 2015).

The biotech sector is now the most important source of innovation in pharmaceutical industries, where more than US$75 billion were spent in R&D, and thousands of patents were created in the last 20 years. Beyond great economical value, this market segment is also a strategic sector from social point of view, creating knowledge and new jobs.

Biotechnology will continue to bring innovation and profit for the pharmaceutical industry. This is due to improvements in current treatments and development of new products using recent discoveries from the research fields, such as gene therapies and biosensors.

## References

Abbas Abul, K., Lichtman, A.H., and Pillai, S. (2014). *Cellular and Molecular Immunology*. Elsevier.

Abood, A., Al-Fahad, A., Scott, A. et al. (2015). Kinetic characterisation of the FAD dependent monooxygenase TropB and investigation of its biotransformation potential. *RSC Adv.* 5: 49987–49995. https://doi.org/10.1039/C5RA06693J.

Actip.org. 2017. Monoclonal Antibodies Approved by the EMA and FDA for Therapeutic Use – ACTIP. http://www.actip.org/products/monoclonal-antibodies-approved-by-the-ema-and-fda-for-therapeutic-use. (Accessed 10 January 2018).

Ahmad, M., Hirz, M., Pichler, H. et al. (2014). Protein expression in *Pichia pastoris*: recent achievements and perspectives for heterologous protein production. *Appl. Microbiol. Biotechnol.* 98 (12): 5301–5317. https://doi.org/10.1007/s00253-014-5732-5.

Almeida, H., Amaral, M.H., and Lobão, P. (2011). Drugs obtained by biotechnology processing. *Braz. J. Pharm. Sci.* 47 (2): 199–207. https://doi.org/10.1590/S1984-82502011000200002.

Barra, G.B., Caixeta, M.C.S.S., Costa, P.G.G. et al. (2011). Diagn{ó}stico molecular – passado, presente e futuro (Molecular diagnosis – past, present and future). *Rev. Bras. Anal. Clin* 43 (3): 254–260.

Becker, H. 2007. 'Muromonab-CD3 (Orthoclone OKT3)', in Handbook of Therapeutic Antibodies, pp. 905-940. ISBN 978-3-527-31453-9.

Bergmeyer, H.U., Gawehn, K., and Grassl, M. (1974). *Methods of Enzymatic Analysis*, 2e. New York: Academic Press, Inc.. Available at: http://www.sigmaaldrich.com/technical-documents/protocols/biology/enzymatic-assay-of-glucose-oxidase.html#sthash.JPNV9u3i.dpuf.

Boguraev, A.-S., Christensen, H.C., Bonneau, A.R. et al. (2017). Successful amplification of DNA aboard the International Space Station. *npj Microgravity* 3 (1): 26. https://doi.org/10.1038/s41526-017-0033-9.

Brady, R.O., Tallman, J.F., Johnson, W.G. et al. (1973). Replacement therapy for inherited enzyme deficiency. *N. Engl. J. Med.* 289 (1): 9–14. https://doi.org/10.1056/NEJM197307052890103.

Broome, J.D. (1961). Evidence that the L-asparaginase activity of guinea pig serum is responsible for its antilymphoma effects. *Nature* 191 (4793): 1114–1115. https://doi.org/10.1038/1911114a0.

Burkert, J. (2004). Optimization of extracellular lipase production by *Geotrichum* sp. using factorial design. *Bioresour. Technol.* 91 (1): 77–84. https://doi.org/10.1016/S0960-8524(03)00152-4.

Caillat-Zucman, S., Blumenfeld, N., Legendre, C. et al. (1990). The OKT3 immunosuppressive effect. In situ antigenic modulation of human graft-infiltrating T cells. *Transplantation* 49 (1): 156–160.

Calvey, T.N. (1995). Isomerism and anaesthetic drugs. *Acta Anaesthesiol. Scand.* 39: 83–90. https://doi.org/10.1111/j.1399-6576.1995.tb04316.x.

Cardoso, F. (2013). *Medicamentos Biológicos na Prática Médica: A importância dos medicamentos biológicos*, 1e. São Paulo: AMB.

Carvalho, P.d.O., Contesini, F.J., Bizaco, R. et al. (2006). Optimization of enantioselective resolution of racemic ibuprofen by native lipase from *Aspergillus niger*. *J. Ind. Microbiol. Biotechnol.* 33 (8): 713–718. https://doi.org/10.1007/s10295-006-0138-8.

Chothia, C., Lesk, A.M., Tramontano, A. et al. (1989). Conformations of immunoglobulin hypervariable regions. *Nature* 342 (6252): 877–883. https://doi.org/10.1038/342877a0.

Clinical trials 2018. Safety, Tolerability, and Immunogenicity of mRNA-4157 Alone in Subjects With Resected Solid Tumors and in Combination With Pembrolizumab in Subjects With Unrespectable Solid Tumors. https://www.clinicaltrials.gov/ct2/show/NCT03313778?term=Moderna&rank=3. (Accessed 10 January 2018).

Collins, H., Calvo, S., Greenberg, K. et al. (2016). Information needs in the precision medicine era: how genetics home reference can help. *Interact. J. Med. Res.* 5 (2): e13. https://doi.org/10.2196/ijmr.5199.

Contesini, F.J., Liberato, M.V., Rubio, M.V. et al. (2017a). Structural and functional characterization of a highly secreted α-l-arabinofuranosidase (GH62) from *Aspergillus nidulans* grown on sugarcane bagasse. *Biochim. Biophys. Acta, Proteins Proteomics* 1865 (12): 1758–1769. https://doi.org/10.1016/j.bbapap.2017.09.001.

Contesini, F.J., de Melo, R.R., and Sato, H.H. (2017b). An overview of *Bacillus* proteases: from production to application. *Crit. Rev. Biotechnol.* 1–14. https://doi.org/10.1080/07388551.2017.1354354.

Dalgaard, K., Evers, M., and da Silva, J. S. 2013. Biosimilars seven years on: Where are we and what 's next?', *McKinsey & Company*, p. 9.
Demain, A.L. (1998). Induction of microbial secondary metabolism. *Int. Microbiol.* 1 (4): 259–264. https://doi.org/10.2436/im.v1i4.26.
Desnick, R.J., Dean, K.J., Grabowski, G. et al. (1979). Enzyme therapy in Fabry disease: differential in vivo plasma clearance and metabolic effectiveness of plasma and splenic alpha-galactosidase A isozymes. *Proc. Natl. Acad. Sci. U.S.A.* 5326–5330. https://doi.org/10.1073/pnas.76.10.5326.
DiMasi, J.A., Grabowski, H.G., and Hansen, R.W. (2016). Innovation in the pharmaceutical industry: new estimates of R&D costs. *J. Health Econ.* 47: 20–33. https://doi.org/10.1016/j.jhealeco.2016.01.012.
Eccleston-Turner, M. (2016). The economic theory of patent protection and pandemic influenza vaccines. *Am. J. Law Med.* 42 (2–3): 572–597. https://doi.org/10.1177/0098858816658280.
Ecker, D.M., Jones, S.D., and Levine, H.L. (2015). The therapeutic monoclonal antibody market. *mAbs* 7 (1): 9–14. https://doi.org/10.4161/19420862.2015.989042.
Elander, R.P. (2003). Industrial production of β-lactam antibiotics. *Appl. Microbiol. Biotechnol.* 61 (5–6): 385–392. https://doi.org/10.1007/s00253-003-1274-y.
European Medicines Agency: Biosimilars in the EU, Information guide for healthcare professionals. 2017. http://www.ema.europa.eu/docs/en_GB/document_library/Leaflet/2017/05/WC500226648.pdf.
Favaro, L., Jooste, T., Basaglia, M. et al. (2012). Codon-optimized glucoamylase sGAI of *Aspergillus awamori* improves starch utilization in an industrial yeast. *Appl. Microbiol. Biotechnol.* 95 (4): 957–968. https://doi.org/10.1007/s00253-012-4001-8.
Generics and Biosimilars Initiative Journal (GaBI Journal Editor) (2015). Patent expiry dates for best-selling biologicals. *GaBI Journal* 4 (4): 178–179. https://doi.org/10.5639/gabij.2015.0404.040.
Geigert, J. (2013). Biologics are not chemical drugs. In: *The Challenge of CMC Regulatory Compliance for Biopharmaceuticals and Other Biologics*, 21–33. New York: Springer New York https://doi.org/10.1007/978-1-4614-6916-2_2.
Gil, E.d.S. and de Melo, G.R. (2010). Electrochemical biosensors in pharmaceutical analysis. *Braz. J. Pharm. Sci.* 46 (3): 375–391. https://doi.org/10.1590/S1984-82502010000300002.
Gould Rothberg, B.E., Pea, C.E.A., and Rothberg, J.M. (2008). A systems biology approach to target identification and validation for human chronic disease drug discovery. In: *Modern Biopharmaceuticals*, 99–125. Weinheim, Germany: Wiley-VCH Verlag GmbH https://doi.org/10.1002/9783527620982.ch4.
Graziano, R.F., Tempest, P.R., White, P. et al. (1995). Construction and characterization of a humanized anti-gamma-Ig receptor type I (Fc gamma RI) monoclonal antibody. *J. Immunol.* 155 (10): 4996–5002. Available at: http://www.ncbi.nlm.nih.gov/pubmed/7594506.
Hamad, B. (2010). The antibiotics market. *Nat. Rev. Drug Discovery* 9 (9): 675–676. https://doi.org/10.1038/nrd3267.
Ho, R.J.Y. and Gibaldi, M. (2013). *Biotechnology and Biopharmaceuticals*. Hoboken, NJ: Wiley https://doi.org/10.1002/9781118660485.
Houbraken, J., Frisvad, J.C., and Samson', R.A. (2011). Fleming's penicillin producing strain is not *Penicillium chrysogenum* but *P. rubens*. *IMA Fungus* 2 (1): 87–95. https://doi.org/10.5598/imafungus.2011.02.01.12.

IMS Health. IMS MIDAS 2009. http://www.imshealth.com/portal/site/ims.

Jackson, S.E. and Chester, J.D. (2015). Personalised cancer medicine. *Int. J. Cancer* 137 (2): 262–266. https://doi.org/10.1002/ijc.28940.

Jenkins, S.M. and Parker, D.D. (2015). Diagnosing biopharmaceutical limitations. In: *Discovering and Developing Molecules with Optimal Drug-Like Properties*, AAPS Advances in the Pharmaceutical Sciences Series, vol. 15 (eds. A. Templeton, S. Byrn, R. Haskell and T. Prisinzano), 141–175. New York: Springer https://doi.org/10.1007/978-1-4939-1399-2_4.

Jeon, H., Yoon, S., Ahsan, M. et al. (2017). The kinetic resolution of racemic amines using a whole-cell biocatalyst co-expressing amine dehydrogenase and NADH oxidase. *Catalysts* 7 (9): –251. https://doi.org/10.3390/catal7090251.

Kabat, E.A., Wu, T., Bilofsky, H. et al. (1984). Sequences of proteins of immunological interest. *Anal. Biochem.* 265. https://doi.org/10.1016/0003-2697(84)90805-4.

Kawaguti, H.Y., Manrich, E., and Sato, H.H. (2006). Production of isomaltulose using *Erwinia* sp. D12 cells: culture medium optimization and cell immobilization in alginate. *Biochem. Eng. J.* 29 (3): 270–277. https://doi.org/10.1016/j.bej.2006.01.006.

Kayser, O. and Müller, R.H. (eds.) (2004). *Pharmaceutical Biotechnology*. Weinheim, FRG: Wiley-VCH Verlag GmbH & Co. KGaA https://doi.org/10.1002/3527602410.

Kepplinger, E.E. (2015). FDA's expedited approval mechanisms for new drug products. *Biotechnol. Law Rep.* 34 (1): 15–37. https://doi.org/10.1089/blr.2015.9999.

Kidd, J.G. (1953). Regression of transplanted lymphomas induced in vivo by means of normal guinea pig serum. I. Course of transplanted cancers of various kinds in mice and rats given guinea pig serum, horse serum, or rabbit serum. *J. Exp. Med.* 98 (6): 565–582. Available at: http://www.ncbi.nlm.nih.gov/pubmed/13109110.

Kroegel, C. and Foerster, M. (2008). Omalizumab (Xolair): anti-immunoglobulin E treatment in allergic diseases. In: *Handbook of Therapeutic Antibodies*, 951–997. https://doi.org/10.1002/9783527619740.ch37.

Lang, S. and Uber (1904). Desaminierung im Tierkörper. In: *Beitraege zur Chemischen Physiologie und Pathologie*, 321–345.

Lindsay-Mosher, N. and Su, C. (2016). Cancer gene therapy: innovations in therapeutic delivery of CRISPR-Cas9. *Drug Discovery Today Dis. Models* 21: 17–21. https://doi.org/10.1016/j.ddmod.2017.02.009.

Lopes, A.G. (2015). Single-use in the biopharmaceutical industry: a review of current technology impact, challenges and limitations. *Food Bioprod. Process.* 93: 98–114. https://doi.org/10.1016/j.fbp.2013.12.002.

Łukowska-Chojnacka, E., Kowalkowska, A., and Napiórkowska, A. (2017). Lipase-catalyzed kinetic resolution of novel antitubercular benzoxazole derivatives. *Chirality* 30 (4): 457–468. https://doi.org/10.1002/chir.22806.

Lybecker, K. (2016). *The Biologics Revolution in the Production of Drugs*, 1e. Vancouver: Fraser Institute Available at: www.fraserinstitute.org.

Mapes, C.A., Anderson, R.L., Sweeley, C.C. et al. (1970). Enzyme replacement in Fabry's disease, an inborn error of metabolism. *Science* 169 (3949): 987–989.

Maurer, H.R. (2001). Bromelain: biochemistry, pharmacology and medical use. *Cell. Mol. Life Sci.* 58 (9): 1234–1245. https://doi.org/10.1007/PL00000936.

Mehrotra, P. (2016). Biosensors and their applications – a review. *J. Oral Biol. Craniofac. Res.* 6 (2): 153–159. https://doi.org/10.1016/j.jobcr.2015.12.002.

Morton, F. S., and Boller, L. T. 2017. Enabling competition in pharmaceutical markets. *Hutchins Center Working Paper*. Available at: https://www.brookings.edu/wp-content/uploads/2017/05/wp30_scottmorton_competitioninpharma1.pdf.

National Research Council (2004). *Health and Medicine: Challenges for the Chemical Sciences in the 21st Century*. Washington, D.C.: National Academies Press https://doi.org/10.17226/10889.

Nelson, A.L., Dhimolea, E., and Reichert, J.M. (2010). Development trends for human monoclonal antibody therapeutics. *Nat. Rev. Drug Discovery* 9 (10): 767–774. https://doi.org/10.1038/nrd3229.

Nissim, A., Schwarzbaum, S., Siraganian, R., and Eshhar, Z. (1993). Fine specificity of the IgE interaction with the low and high affinity Fc receptor. *J. Immunol.* 150 (4): 1365–1374. Available at: http://www.ncbi.nlm.nih.gov/pubmed/7679426.

Olson, J.E., Bielinski, S.J., Ryu, E. et al. (2014). Biobanks and personalized medicine. *Clin. Genet.* 86 (1): 50–55. https://doi.org/10.1111/cge.12370.

Oriente, A., de Andrades, D., Henn, C. et al. (2015). Characterization of a novel *Aspergillus niger* beta-glucosidase tolerant to saccharification of lignocellulosic biomass products and fermentation inhibitors. *Chem. Pap.* 69 (8): https://doi.org/10.1515/chempap-2015-0111.

Orr, R.M. (2001). Technology evaluation: fomivirsen, Isis Pharmaceuticals Inc/CIBA vision. *Curr. Opin. Mol. Ther.* 3 (3): 288–294. Available at: http://www.ncbi.nlm.nih.gov/pubmed/11497353.

Presta, L., Shields, R., O'Connell, L. et al. (1994). The binding site on human immunoglobulin E for its high affinity receptor. *J. Biol. Chem.* 269 (42): 26368–26373. Available at: http://www.ncbi.nlm.nih.gov/pubmed/7929356.

Raja, M.M.M., Raja, A., Imran, M.M. et al. (2011). Enzymes application in diagnostic prospects. *Biotechnology* 10 (1): 51–59. https://doi.org/10.3923/biotech.2011.51.59.

Ricklefs, E., Girhard, M., and Urlacher, V.B. (2016). Three-steps in one-pot: whole-cell biocatalytic synthesis of enantiopure (+)- and (−)-pinoresinol via kinetic resolution. *Microb. Cell Fact.* 15 (1): 78. https://doi.org/10.1186/s12934-016-0472-0.

Riechmann, L., Clark, M., Waldmann, H., and Winter, G. (1988). Reshaping human antibodies for therapy. *Nature* 332 (6162): 323–327. https://doi.org/10.1038/332323a0.

Ronson, A., Tvito, A., and Rowe, J.M. (2016). Treatment of relapsed/refractory acute lymphoblastic leukemia in adults. *Curr. Oncol. Rep.* 18 (6): 39. https://doi.org/10.1007/s11912-016-0519-8.

Rubio, M.V., Tramontina, R., Gonçalves, T.A. et al. (2015). Heterologous expression of carbohydrate-active enzymes in filamentous fungi. In: *Mycology: Current and Future Developments*, vol. 1 (eds. M.V. Rubio, R. Tramontina, T. Augusto, et al.), 148–201. Bentham Science Publishers https://doi.org/10.2174/9781681080741115010009.

Ryu, Y.-G., Butler, M.J., Chater, K.F. et al. (2006). Engineering of primary carbohydrate metabolism for increased production of actinorhodin in *Streptomyces coelicolor*. *Appl. Environ. Microbiol.* 72 (11): 7132–7139. https://doi.org/10.1128/AEM.01308-06.

Saban, R., Haak-Frendscho, M., Zine, M. et al. (1994). Human FcERI-IgG and humanized anti-IgE monoclonal antibody MaE11 block passive sensitization of human and rhesus monkey lung. *J. Allergy Clin. Immunol.* 94 (5): 836–843. Available at: http://www.ncbi.nlm.nih.gov/pubmed/7525678.

Sánchez, S., Chavez, A., Forero, A. et al. (2010). Carbon source regulation of antibiotic production. *J. Antibiot.* 63 (8): 442–459. https://doi.org/10.1038/ja.2010.78.

Santos, S.C., de Sousa, A.S., Dionísio, S.R. et al. (2016). Bioethanol production by recycled *Scheffersomyces stipitis* in sequential batch fermentations with high cell density using xylose and glucose mixture. *Bioresour. Technol.* 219: https://doi.org/10.1016/j.biortech.2016.07.102.

Sato, K., Tsuchiya, M., Saldanha, J. et al. (1993). Reshaping a human antibody to inhibit the interleukin 6-dependent tumor cell growth. *Cancer Res.* 53 (4): 851–856. Available at: http://www.ncbi.nlm.nih.gov/pubmed/8428365.

Sayin, S., Akoz, E., and Yilmaz, M. (2014). Enhanced catalysis and enantioselective resolution of racemic naproxen methyl ester by lipase encapsulated within iron oxide nanoparticles coated with calix[8]arene valeric acid complexes. *Org. Biomol. Chem.* 12 (34): 6634–6642. https://doi.org/10.1039/C4OB01048E.

Sepporta, M.V., Mazza, T., Morozzi, G. et al. (2013). Pinoresinol inhibits proliferation and induces differentiation on human HL60 leukemia cells. *Nutr. Cancer* 65 (8): 1208–1218. https://doi.org/10.1080/01635581.2013.828089.

Shields, R.L., Whether, W.R., Zioncheck, K. et al. (1994). Inhibition of allergic reactions with antibodies to IgE. *Int. Arch. Allergy Immunol.* 107 (1–3): 308–312. Available at: http://www.ncbi.nlm.nih.gov/pubmed/7613156.

Sikora, A., Sroka, W.D., Siódmiak, T. et al. (2017). Kinetic resolution of (R,S)-atenolol with the use of lipases in various organic solvents. *Curr. Org. Synth.* 14 (5): https://doi.org/10.2174/1570179414666161230120414.

Suárez, C. and Gudiol, F. (2009). Antibióticos betalactámicos. *Enferm. Infecc. Microbiol. Clin.* 27 (2): 116–129. https://doi.org/10.1016/j.eimc.2008.12.001.

Szymanska, B., Wilczynska-Kalak, U., Kang, M.H. et al. (2012). Pharmacokinetic modeling of an induction regimen for in vivo combined testing of novel drugs against pediatric acute lymphoblastic leukemia xenografts. *PLoS One*. Edited by F. Prosper 7 (3): e33894. https://doi.org/10.1371/journal.pone.0033894.

Tang, Z., Xiao, C., Zhuang, Y. et al. (2011). Improved oxytetracycline production in *Streptomyces rimosus* M4018 by metabolic engineering of the G6PDH gene in the pentose phosphate pathway. *Enzym. Microb. Technol.* 49 (1): 17–24. https://doi.org/10.1016/j.enzmictec.2011.04.002.

TEConomy Partners; for PhRMA. The Biopharmaceutical Industry's Role in Fueling the Economy and Global Competitiveness. Columbus, OH: TEConomy Partners; (2017, July). http://phrmadocs.phrma.org/industryprofile/pdfs/2017IndustryProfile_TheBiopharmaceuticalIndustrysRole.pdf.

Tramontina, R., Robl, D., Maitan-Alfenas, G.P. et al. (2016). Cooperation of *Aspergillus nidulans* enzymes increases plant polysaccharide saccharification. *Biotechnol. J.* 11 (7): https://doi.org/10.1002/biot.201500116.

Tramontina, R., Cairo, J.P.L.F., Liberato, M.V. et al. (2017). The Coptotermes gestroi aldo-keto reductase: a multipurpose enzyme for biorefinery applications. *Biotechnol. Biofuels* 10 (1): https://doi.org/10.1186/s13068-016-0688-6.

Tufvesson, P., Lima-Ramos, J., Nordblad, M. et al. (2011). Guidelines and cost analysis for catalyst production in biocatalytic processes. *Org. Process Res. Dev.* 15 (1): 266–274. https://doi.org/10.1021/op1002165.

Usta, C., Turgut, N.T., and Bedel, A. (2016). How abciximab might be clinically useful. *Int. J. Cardiol.* 222: 1074–1078. https://doi.org/10.1016/j.ijcard.2016.07.213.

Vellard, M. (2003). The enzyme as drug: application of enzymes as pharmaceuticals. *Curr. Opin. Biotechnol.* 14 (4): 444–450. https://doi.org/10.1016/S0958-1669(03)00092-2.

Verma, N., Bansal, M., and Kumar, S. (2012). Whole cell based miniaturized fiber optic biosensor to monitor L-asparagine. *J. Appl. Sci. Res.* 3: 809–814.

Vitolo, M., Pessoa, A. Jr.,, de Souza, G.M. et al. (2015). *Biotecnologia Farmacêutica: aspectos sobre aplicação industrial*, 1e. São Paulo: Edgard Blücher.

Wang, B. and Xia, L. (2011). High efficient expression of cellobiase gene from *Aspergillus niger* in the cells of *Trichoderma reesei*. *Bioresour. Technol.* 102 (6): 4568–4572. https://doi.org/10.1016/j.biortech.2010.12.099.

Yamashita, M., Katakura, Y., and Shirahata, S. (2007). Recent advances in the generation of human monoclonal antibody. *Cytotechnology* 55 (2–3): 55–60. https://doi.org/10.1007/s10616-007-9072-5.

Yang, H., Liu, L., Li, J. et al. (2011). Heterologous expression, biochemical characterization, and overproduction of alkaline α-amylase from *Bacillus alcalophilus* in *Bacillus subtilis*. *Microb. Cell Fact.* 10 (1): 77. https://doi.org/10.1186/1475-2859-10-77.

Zhao, G., Hou, L., Yao, Y. et al. (2012). Comparative proteome analysis of *Aspergillus oryzae* 3.042 and *A. oryzae* 100-8 strains: towards the production of different soy sauce flavors. *J. Proteomics* 75 (13): 3914–3924. https://doi.org/10.1016/j.jprot.2012.04.056.

# 3

# Current Status of Biotechnological Processes in the Biofuel Industries

*Gustavo Pagotto Borin*[1,2], *Rafael Ferraz Alves*[3,4], *and Antônio Djalma Nunes Ferraz Júnior*[5]

[1] *Brazilian BioRenewables National Laboratory (LNBR), Brazilian Center for Research in Energy and Materials (CNPEM), Campinas-SP, Brazil*
[2] *Graduate Program in Genetics and Molecular Biology, Institute of Biology, University of Campinas (UNICAMP), Campinas-SP, Brazil*
[3] *School of Food Engineering, University of Campinas (UNICAMP), Campinas-SP, Brazil*
[4] *Novozymes Latin America Ltda., Araucária-PR, Brazil*
[5] *Centre for Environmental Policy, Imperial College, London, UK*

## 3.1 Introduction

Biofuels have gained increasing attention worldwide, driven by the necessity of replacing fossil fuels and developing new energy sources. Several factors have contributed to the growth of the biofuels market, including overall concern about global warming, reliance on oil and other related nonsustainable commodities, and the possibility of depleting fossil fuels (Jönsson and Martín 2016).

Biofuels are the most promising alternative to fossil fuels due to their capacity for reducing $CO_2$ emissions while stimulating socioeconomic development. They can be generated by abundant and inexpensive feedstocks, such as the lignocellulose present in vegetal biomass (Dahiya et al. 2018; Jönsson and Martín 2016). Moreover, there is a trend of turning the current fossil fuels-based economy into a bioeconomy supported by renewable sources and biobased products that are produced in an integrated and environmentally friendly way. International agreements, such as COP21-Paris, involving almost 200 countries, have also encouraged nations to adopt governmental actions toward the reduction of greenhouse gases (GHGs), which has triggered the development of biofuels (Chen et al. 2016; Lynd et al. 2017). According to the United Nations, it is expected that the second generation (2G) biofuels consumed by the road transportation sector, for example, will achieve 193.41 billion liters in 2020, a 58% increase compared to the consumption in 2013 (122.3 billion liters) (United Unions 2016).

In the context of replacing fossil fuels and minimizing GHG emissions, several international directives have emerged and are driving important public policies in order to stimulate biofuel production and development. One of them is the Renewable Energy Directive (RED) agreed with the European Union countries in 2016. It has established that the European transport sector will have to use 10% of its energy demanding matrix with renewable biofuels until 2020 (European Commission 2009). After a revision of

*Bioprocessing for Biomolecules Production*, First Edition.
Edited by Gustavo Molina, Vijai Kumar Gupta, Brahma N. Singh, and Nicholas Gathergood.

the international agreement in 2016, it is expected that 27% of the energy consumption should come from renewable sources until 2030 (Carneiro et al. 2017; European Commission 2016). Another important milestone for the biofuels sector was the Renewable Fuel Standard (RES) agreement signed by the United States, one of the main producers and consumers of fossil fuels (responsible for 14% and 21% of the worldwide production and consumption share, respectively) (Renewable Fuels Association 2017a). It defines that the United States must increasingly blend renewable biofuels into transportation fuel, and the expectation is to achieve 136 billion liters (36 billion gallons) of renewable fuel consumption by 2022 (UNICA 2012). The production of cellulosic ethanol, made from lignocellulosic biomass, is expected to rise from 5.5 billion gallons in 2017 to 16 billion gallons by 2022 (an increase of 190%) (Carneiro et al. 2017; EPA 2013). Along with these governmental programs, the Brazilian RenovaBio approved in the Senate and by the ex-president Michel Temer (law n° 13.576/17) is another crucial program to boost and strengthen Brazil's sugar-energy sector (MME 2017). Among other aims, the RenovaBio program aims by 2030: (i) to decrease the GHGs emission in 43% and (ii) to increase the participation of renewable sources in the energetic matrix to 45% (Addington 2017; Brazilian Ministry of Mines and Energy 2017).

All these public policies indicate the real concern of the main players to make decisions and create initiatives that can help to protect the renewable energy sector, increase biofuels production, and also mitigate GHG emissions. For all the reasons and data pointed out, biofuels are undoubtedly a promising market that must be more developed to decrease fossil fuel consumption and encourage more sustainable technologies.

Even though biofuels have emerged as one the main alternatives to replace fossil fuels, several challenges must be overcome to make them economically feasible and attractive. In this context, biotechnology has arisen as a valuable platform to overtake obstacles faced in industrial processes and hence improve the productivity of bioproducts (Chen 2014). Biotechnology allows industries and companies to develop new strains able to use a metabolic pathway of interest and to produce new metabolites of high value-added. It provides various tools with different applications that can be used in synthetic biology and genetic engineering, aiming to optimize a particular process, for example. However, although biotechnological improvements have been made through the past years, several issues still hamper the production of biofuels in industries, including high demand of energy, biological contamination, necessity of microorganisms able to ferment pentose and hexose sugars, high loading of enzyme mixtures for breaking down the feedstock, expensive treatments for effluents having toxic compounds, and high capital investment on equipment and qualified workforce (Chen and Jiang 2018; Chen 2014; Dashtban et al. 2009). These challenges make the biofuel production still not economically competitive, especially when compared to fossil fuels production while oil prices are falling.

In this regard, some issues need to be addressed and explored, including (i) which is the best pretreatment and condition for each biomass, considering the operational and equipment costs, and formation of inhibitory fermentation products; (ii) how to use genetic improvement to develop plants having higher-quantity cellulosic fibers or cell wall structure more accessible to polysaccharide-degrading enzymes; (iii) and how to increase the efficiency of enzymatic cocktails and lower the enzyme loading for lignocellulose degradation. All these issues must be investigated in detail. Biotechnology has thus been proved to be essential for the advance of biofuels (Tan et al. 2016).

This chapter presents an overview of some economic and biotechnological aspects of the main important biofuels: bioethanol, biodiesel, biobutanol, and biogas. The third generation (3G) biofuel production based on algal biomass will also be briefly covered. In the next chapters, their industrial processes, main producer companies, patents, and also industrial challenges will be discussed. The aim is to give a whole picture of the industrial scenario of these biofuels and to show how the biotechnology is present and must be used to overcome the bottlenecks that impair the progress of these sustainable fuels. Finally, future perspectives on the development of biofuels will be discussed.

## 3.2 Biofuels and an Overview of the Industrial Processes

### 3.2.1 Bioethanol

For many years, ethanol has been widely used to fuel vehicles and it is considered one of the most important fuels to replace oil, coal, and natural gas. As a fuel additive, ethanol has several advantages compared to gasoline, including (i) higher octane number (106–110 versus 91–96), which shortens the burning time and lower engine knock; (ii) higher heat of vaporization (840 kJ $kg^{-1}$ versus 305 kJ $kg^{-1}$), which promotes an enhancement of power output; (iii) safer alternative to the additive MTBE (methyl tertiary butyl ether), which might contaminate underground water; and (iv) production from renewable sources (Jeuland et al. 2004; Yao et al. 2009; Zabed et al. 2017).

In relation to ethanol production, the principal producers are the United States, Brazil, European Union, and China, being responsible for 57%, 27%, 5%, and 3% of global production, respectively (Renewable Fuels Association 2017b). Comparing worldwide ethanol production, there was an increase of 13.7% (23.4–26.6 billion gallons) in the years 2013–2016, reaching a market value of USD 55 billion in 2016 (Araújo 2016). According to this positive trend, it is proposed that biofuels will continue to expand in order to meet future energy needs (Araújo 2016).

Ethanol can be produced from several different feedstocks, including sugar crops, such as corn, sugarcane, beet, wheat, and rice (first-generation [1G] ethanol); nonfood lignocellulosic biomass, such as municipal waste, crop residues, and food processing waste (2G ethanol; or cellulosic ethanol); and algae (3G ethanol). The 1G ethanol technology has already been consolidated, mainly because of the expertise acquired and large investment programs made over the last decades, such as the Brazilian ProAlcool program launched in the 1970s (Araújo 2016).

The 2G ethanol produced from lignocellulosic materials is already a reality, and biorefineries across the globe produce this fuel on a large scale. However, as previously stated, some bottlenecks must be addressed to make this ethanol production economically viable. Research on 3G ethanol from algae is in the early stages, mainly conducted at a laboratory scale. Thus, it is paramount that 3G ethanol technology be investigated more comprehensively – in particular, it is worth exploring algae as an attractive feedstock and the possibilities of scaling up 3G ethanol (Guo et al. 2015; Zabed et al. 2017). As this chapter will focus on only some aspects of the bioethanol industry, the term *bioethanol* will hereafter refer to 2G ethanol, excluding 1G and 3G ethanol.

In the face of the food versus energy debate - "does increasing production of energy from food crops lead to an increase in the cost of food?" - development of 2G biofuels from lignocellulose, such as bioethanol, became essential. Associated feedstocks for 2G biofuels do not compete for land used for food production and generally are low-cost and abundant residues. The lignocellulose present in the plant cell wall is the most abundant organic material on Earth, with a production estimated in $150–170 \times 10^9$ tons per year (Hadar 2013).

Lignocellulose is normally composed of cellulose (35–50%), hemicellulose (20–35%), and lignin (10–25%) (Isikgor and Becer 2015). Cellulose consists of glucose units bound by β-1,4 links and organized in crystalline and amorphous regions. Hemicellulose is a more heterogeneous polymer in terms of sugar monomers: it is composed of pentose (xylose and arabinose), hexose (glucose, mannose and galactose), and acid sugars (glucuronic and galacturonic acid, for instance). The most common hemicelluloses are xylan (grasses and angiosperms), mannan (gymnosperms), and xyloglucan (angiosperms). Lignin is a high-energy polyphenol macromolecule formed primarily by three alcohols: *p*-coumaryl, coniferyl, and sinapyl alcohol. Lignin confers recalcitrance to lignocellulose, as it covers cellulose and hemicellulose in a type of impermeable matrix, hampering the access of (hemi)cellulolytic enzymes that catalyze cell wall degradation (Adsul et al. 2011; Horn et al. 2012; Tan et al. 2016). It has also been demonstrated that cellulase enzymes that bind to lignin cannot effectively act on cellulose. All these lignocellulose components must be broken down in order to release simple sugars that are used by industrial engineered yeasts in the fermentation process (Kumar and Sharma 2017).

Therefore, 2G ethanol technology requires fundamentally three steps: (i) pretreatment to reduce the cell wall recalcitrance and allow the access of hydrolytic and oxidative enzymes to the lignocellulose core; (ii) enzymatic hydrolysis to convert long sugar polymer chains in simple sugars; and (iii) fermentation of the simple sugars released (pentoses and hexoses) (Silveira et al. 2015). Pretreatment and enzymatic hydrolysis are the major limitations for the economic feasibility of 2G ethanol, and a cost-effective process is required to make this biofuel economically more attractive and competitive (Bornscheuer et al. 2014; Gupta and Verma 2015; Junqueira et al. 2016a).

A broad range of pretreatments can be employed to reduce the recalcitrance of lignocellulose and release fermentable sugars, including physical, chemical, physicochemical, and biological methods. Representative examples for each pretreatment include irradiation and mechanical grinding for physical pretreatments; alkaline, dilute and concentrated acids, autohydrolysis (or liquid hot water), $CO_2$, $SO_2$, organosolv, ionic liquids (ILs), supercritical solvents (SCFs), wet oxidation and peroxidation for chemical pretreatment; steam explosion and ammonia fiber extraction (AFEX) for physico-chemical pretreatments; and lignocellulose biodegradation performed by white, brown, and soft rot fungi for biological pretreatments (Hassan et al. 2018).

No pretreatment meets the requirements for degrading all types of biomass. However, among the wide variety of alternatives, steam explosion is generally used in physicochemical pretreatments, and alkaline and dilute acid are used in chemical pretreatments. The steam explosion pretreatment is cost-effective, eco-friendly, and has already been used on an industrial scale, although it causes the formation of enzyme inhibitors. Alkaline pretreatment uses sodium, potassium, and ammonium hydroxides as reagents, and it is an advantageous method due to the efficient removal of lignin and hemicellulose with only small carbohydrate losses. Dilute sulfuric acid is also a commonly used pretreatment for lignocellulose breakdown. Dilute acid pretreatment causes less corrosion, lower inhibitor formation, and shows versatility when used to process a variety of feedstocks compared to concentrated acid pretreatments (Auxenfans et al. 2017; Guo et al. 2015; Gupta and Verma 2015; Silveira et al. 2015; Thomas et al. 2017).

Enzymatic hydrolysis is the following step in the 2G ethanol production process. After pretreatment, the complex polysaccharides that were not broken down in the previous step are more accessible to be converted to simple sugars by the action of different enzymes, such as cellulases (cellobiohydrolase, endoglucanase, and β-glucosidase) and hemicellulases (xylanase, β-xylosidase, α-arabinofuranosidase, acetylxylan esterase, ferulic acid esterase, xyloglucanase, and α-glucuronidase) (Barr et al. 2012; Dondelinger et al. 2016; Gupta and Verma 2015). More recently, it has also been demonstrated that oxidative enzymes known as lytic polysaccharide monooxigenases (LPMOs) can boost the action of hydrolytic enzymes yielding a higher (hemi)cellulose degradation (Cragg et al. 2015). Of note, the current commercial cocktails Cellic® CTec2 and Ctec3 (Novozymes, Denmark) have already added oxidative enzymes, along with canonical cellulases and hemicellulases, thus contributing to a decrease in the costs of cellulase preparations (Johansen 2016).

Naturally, several microorganisms are able to produce and/or secrete these enzymes, including the fungi *Trichoderma reesei, Trichoderma harzianum, Aspergillus niger, Penicillium echinulatum, Phanerochaete chrysosporium, Neurospora crassa,* and also bacteria, such as *Ruminococcus* sp., *Bacillus pumilus, Clostridium thermocellum, Pseudomonas* sp., *Pedobacter* sp., *Mucilaginibacter* sp., and actinobacteria (Guerriero et al. 2016; Koeck et al. 2014; Lynd et al. 2017; Salihu et al. 2015). Generally, their cellulolytic system might be found in two ways: (i) enzymes, which are produced inside the cell and are forwarded to the extracellular environment, or (ii) various proteins gathered in a complex structure termed cellulosome. The first system is typical of aerobic microorganisms and the second one is found in most anaerobic bacteria (Ravachol et al. 2016). *T. reesei* is the main fungal source of cellulase and hemicellulase for enzymatic cocktail, and it has been studied for decades as a model for lignocellulose degradation and host for heterologous expression (Borin et al. 2017; dos Santos et al. 2016; Schmoll et al. 2016). This vast diversity of lignocellulose-degrading microorganisms offers a wide range of proteins, metabolic pathways, and biological systems that might be exploited at an industrial scale.

Next, the sugars released in the previous steps are converted into ethanol by industrial yeasts. These microorganisms must be robust enough to overcome the challenging conditions of the fermentation tanks, such as high ethanol concentration, high temperature, an acidic environment, presence of fermentation inhibitors, and competition with microbial contaminants. Moreover, the yeasts must be capable of fermenting xylose as well as glucose, which has proved to be a difficult task as the most commonly used yeast *Saccharomyces cerevisiae* does not consume xylose. In addition, yeast propagation and the ethanol productivity are other key factors that must be considered. First, the carbon source from sugars must be used to produce cell mass for the inoculum, and afterwards it must be converted into ethanol. In this step, a yeast strain able to rapidly ferment pentose and glucose sugars in the presence of inhibitors, and yielding high ethanol titer, is highly desirable. Various works have been carried out in order to overcome the obstacles of lignocellulosic hydrolysate fermentation, including metabolic engineering, mutagenesis, and evolutionary adaptation using culture media having increasing xylose concentration and/or presence of fermentation inhibitors, such as 5-hydroxymethylfurfural (HMF) and acetic acid. Although *S. cerevisiae* is the

**Table 3.1** Main 2G ethanol biorefineries using lignocellulose residues (Araújo 2016; dos Santos et al. 2016; Guo et al. 2015; Lynd et al. 2017; Renewable Fuels Association 2017a).

| Company | Pilot plant localization | Installation year | Ethanol capacity (million gallons/year) | Feedstock | Installation cost (millions) |
|---|---|---|---|---|---|
| DuPont | Nevada (IA), US | 2015 | 30 | Corn stover | USD 225 |
| Abengoa | Hugoton (KS), US | 2014 | 25 | Corn stover, wheat straw, grasses | USD 504 |
| GranBio | São Miguel dos Campos (AL), Brazil | 2014 | 21.6 | Sugarcane bagasse and straw | USD 195 |
| POET-DSM | Emmetsburg (IA), US | 2014 | 20 | Corn stover and cobs | USD 250 |
| Beta Renewables | Crescentino, Italy | 2013 | 13.4 | Wheat straw | € 90 |
| Raízen | Piracicaba (SP), Brazil | 2015 | 10.6 | Sugarcane bagasse and straw | USD 100 |

industrial yeast still preferred in most ethanol refineries, due to its robustness and high ethanol productivity, the unconventional yeasts *Scheffersomyces stipitis, Candida shehatae, Spathaspora passalidarum*, and *Kluyveromyces marxianus* have also been considered alternative yeasts by many researches aiming to produce ethanol and other bioproducts by xylose fermentation (Coutouné et al. 2017; Demeke et al. 2013; dos Santos et al. 2016; Ehsaan et al. 2016; Katahira et al. 2017; Ko et al. 2016; Löbs et al. 2017; Narayanan et al. 2017; Sharma et al. 2017). Therefore, the development of genetically modified microorganisms (GMOs) to meet the needs of lignocellulosic hydrolysate fermentation is crucial to promote 2G ethanol. In Brazil, for example, five GMOs developed for biofuel production have been registered and approved by the Brazilian responsible regulatory agency since 2009. Finally, sugar fermentation from lignocellulosic hydrolysate is another key step in 2G ethanol production, and it has also been critically evaluated for the economic viability of this biofuel (dos Santos et al. 2016).

The largest producer companies of 2G ethanol are, in decreasing order of volume: DuPont (US), Abengoa (US), Granbio (Brazil), POET-DSM (US), Beta Renewables (Italy) and Raízen (Brazil). These companies are responsible for the largest share of cellulosic ethanol produced worldwide, and their main feedstocks are corn, wheat, and sugarcane-based residues (Table 3.1).

The first commercial-scale plant of 2G ethanol was built by the Italian company Beta Renewables in 2013. This facility has an ethanol productivity of 13.4 million gallons per year, using wheat straw as feedstock. The company has also developed a technology named PROESA™, which consists of separating the cellulose and hemicellulose from lignin and breaking down the polysaccharides using enzymes from Novozymes (Denmark). PROESA technology has already been licensed and sold to companies of the biofuels sector. After starting operation in 2014, the Brazilian company Granbio, for instance, obtained license to use PROESA technology to process sugarcane bagasse and straw, main byproducts generated from sugarcane industry (Pereira et al. 2015).

Over the years, more biorefineries have initiated their operations, stimulated by public incentive policies of promoting biofuels development. Despite of the current low oil cost and unstable geopolitical scenario, it is clear that 2G ethanol will be one of the main players regarding mitigation of GHG emissions and development of new green technologies. Evidence of this is the increasing participation of biofuels into the liquid fuel blends, the need for developing new alternatives to fossil fuels in order to meet the global energy demand in the future, and international agreements signed for establishing public policies on biofuels (Guo et al. 2015; Zabed et al. 2017).

### 3.2.2 Biodiesel

Similarly to other biofuels, biodiesel production is expected to increase steadily due to the overall growing energy demand along with public incentive policies for the development of biofuels. Global biodiesel production rose from almost 20 billion liters to 35 billion liters between 2010 and 2016 (an increase of 75%). This production represented one quarter of all biofuels produced worldwide in 2015. The largest biodiesel producers were the European Union (43%), United States (15%), Brazil (13%), Argentina (7%), and Indonesia (4%) (Naylor and Higgins 2017; USDA 2017). Furthermore, the incentive policies of increasing quantities of biodiesel into petrol diesel (fuel blends) over the years have also strongly promoted this biofuel (da Silva et al. 2017).

Biodiesel is a promising fuel substitute for petrol diesel, and it might be produced by esterification and/or transesterification of fatty acids. These fatty acids can be extracted from various feedstocks, including vegetable oil (camelina, canola, castor bean, coconut, jatropha, palm, peanut, rapeseed, soybean, sunflower, and tung), animal fat, algal lipids, and waste greases (used cooking oil and sewage sludge, for instance). Globally, the main feedstocks used for biodiesel production are rapeseed oil (European Union), soybean oil (United States, Brazil, and Argentina) and palm oil (Indonesia), which, taken together, represent more than 50% of the biodiesel feedstock supply (Guo et al. 2015; Naylor and Higgins 2017).

The most common method of biodiesel production is a chemical process based on the transesterification of vegetable oils using methanol or ethanol as solvents, and potassium hydroxide or sodium hydroxide as catalysts. After a heating step in a reactor using mechanical stirring, fatty acid esters (biodiesel) and glycerol are formed from the transesterification reaction. Next, some impurities are removed from the biodiesel by washing and drying or by membrane refining. The glycerol formed as a byproduct also undergoes a refining process, and it can be reused for different applications after being removed from the biodiesel production process (Kirubakaran and Arul Mozhi Selvan 2018; de Melo et al. 2016; Vassilev et al. 2017).

The chemical process for biodiesel production presents high productivity and is economically viable. However, the use of large amounts of chemicals is not sustainable, and significant volumes of liquid waste are generated and require adequate treatment. Moreover, depending on what feedstock is used, the chemical process catalyzed by alkali, such as sodium hydroxide, can form soap (i.e. saponification) and other undesirable byproducts that are not easy to separate and must be removed to achieve the required biodiesel purity (de Melo et al. 2016; Taher and Al-Zuhair 2017).

Besides the conventional chemical process, two other transesterification methods have been proposed: (i) supercritical alcohols and (ii) lipase catalysis. The first method does not require a catalyst, and it is proposed to be more suitable for vegetable oils. This is because the alcohol, in supercritical conditions, i.e. when the temperature (~350 °C) and pressure (45 MPa) applied are above the critical point, can tolerate more water, making the transesterification reaction faster and yielding a high biodiesel concentration. However, transesterification by supercritical alcohols is a high-energy-demanding process and has additional costs associated with performing the extreme conditions of temperature and pressure safely (González Prieto et al. 2017).

On the other hand, the second method of transesterification performed by lipases appears to be a promising approach for biodiesel production from various different types of vegetable oils and other feedstocks. These enzymes (EC 3.1.1.3) have optimum activity in mild conditions of temperature (between 20 and 50 °C) without soap formation, and they are easily recovered in the process. Also, lipase-catalyzed transesterification is eco-friendly and requires minimal waste treatment. The main sources of commercial lipases for biodiesel production include *Pseudomonas fluorescens, Pseudomonas cepacia, Candida rugosa, Candida antarctica, Rhizomucor miehei, Rhizopus oryzae, Chromobacterium viscosum,* and *Thermomyces lanuginosus* (Dias Ribeiro et al. 2011; Ivana et al. 2016; Jegannathan et al. 2008; da Silva and Oliveira 2014; Taher and Al-Zuhair 2017).

Nevertheless, the high cost of enzymes hinders the economic feasibility of producing biodiesel by this enzymatic route compared with the traditional chemical process (Dahiya et al. 2018; Dias Ribeiro et al. 2011; da Silva and Oliveira 2014). Immobilization of different lipases and matrices has been developed to overcome this economic obstacle, since this strategy facilitates the reusability and increases the stability of the enzymes, thus decreasing the costs of biodiesel production (Adlercreutz 2013; Li et al. 2017; Tan et al. 2010). Beyond the cost, there are other limitations that must be investigated according to the type of feedstock and enzyme used. These include enzyme denaturation by high temperatures, deactivation by glycerol or methanol, and the optimum dosage of enzymes to be used (da Silva and Oliveira 2014; Taher and Al-Zuhair 2017). Furthermore, taking into account the biodiesel market and the projections for this biofuel, it is of paramount importance that new enzymes and immobilization supports be explored to tackle these limitations and make the biodiesel production economically competitive in the short term. Utilization of low-cost feedstocks, such as waste municipal oils, is another alternative that must receive more attention, since the feedstock cost could represent up to 70–95% of the final biodiesel price (Shahzad et al. 2017).

### 3.2.3 Biobutanol

Butanol is a four-carbon alcohol and is considered a superior biofuel compared to ethanol, regarding such properties: (i) high miscibility with gasoline or diesel; (ii) higher energy density; (iii) low octane value; (iv) lower volatility; and (v) lower vapor pressure (Green 2011; Khanna et al. 2013). Additionally, butanol and its derivatives may be used in other industrial applications, such as surface coatings, plasticizers, and diluents, confirming the versatility and the commercial interest in this compound (Lee et al. 2008). The renewable butanol (biobutanol) can be produced by biochemical route, using the solventogenic *Clostridia* ssp. to ferment different substrates in order to produce solvents. This process is known as acetone–butanol–ethanol (ABE) fermentation, since the major chemical products are acetone, butanol, and ethanol (Sauer 2016).

The global market of butanol was estimated around 3.8 million tons in 2012 (USD 7 billion). The perspective of growth from 2013 to 2018 is 4.6%, and it is expected that the market will achieve USD 9.9 billion in 2020 (Jiang et al. 2015). The global demand for butanol is basically concentrated in North America, Western Europe, and China, the latter being the key consumer (34.8% in 2012) of butanol in the world (Bharathiraja et al. 2017).

Early industrial production of butanol was based on biochemical route, in which starch and molasses were fermented by the bacterium *Clostridium acetobutylicum* to produce mainly butanol and acetone (Jones and Woods 1986). However, with the increasing price of carbon sources because of the high demand for cattle feed and the rapid growth of the petrochemical industry, the biochemical process declined rapidly (Sauer 2016). As a result, many ABE fermentation facilities were closed, causing a significant decline of research on ABE fermentation (Moon et al. 2016).

Most butanol produced today is based on the oxo process that consists in the catalytic hydroformylation of propylene followed by hydrogenation of aldehydes formed to produce a mixture of n- and isobutanol. The main players are: BASF SE (Germany), The Dow Chemical Company (US), BASF-YPC Ltd. (China), OXO Corporation (US), Sasol Ltd. (South Africa), Formosa Plastics Corporation (Taiwan), Eastman Chemical Company (US), Oxichimie SAS (France), KH Neochem Co. Ltd. (Japan), and CNPC (China). Nevertheless, the need for developing a bioeconomy based on renewable sources to replace the use and production of fossil fuels has led several companies around the world to invest in R&D for biobutanol production (Jin et al. 2011) (Table 3.2).

Usually, traditional ABE fermentation uses corn starch or molasses as a substrate, and it is well known that the feedstock chosen for fermentation process is an important parameter for the biofuel economic feasibility (Gapes 2000). In a conventional biobutanol plant using corn starch as a substrate, the feedstock cost accounts for around 70% of the whole process (Pfromm et al. 2010). According to the current situation of butanol industries in China, for instance, the petrochemical process is still more economically advantageous (USD 1.52/kg butanol) than the traditional ABE fermentation process (USD 1.87/kg butanol) (Jiang et al. 2015). One of the strategies to decrease the cost associated to biobutanol production is the use of renewable feedstock, such as lignocellulosic materials (e.g. agricultural waste, paper waste, wood chips) (He et al. 2017; Nanda et al. 2014). Similar to bioethanol, the production of biobutanol using lignocellulosic biomass needs to follow an integrated process involving three steps already mentioned: pretreatment, hydrolysis, and fermentation. The great difference is during the fermentation process, which requires sugar fermentation by *Clostridium* spp., including *C. acetobutylicum*, *Clostridium beijerinkii*, and *Clostridium saccharoperbutylacetonicum* (Cheng et al. 2012).

The ABE fermentation is characterized by two distinct stages: acidogenic and solventogenic. In the acidogenic phase, the organic acids (acetic and butyric) are produced, followed by a solventogenic phase, when acids are re-assimilated and converted into solvents (acetone, butanol, and ethanol) (Guo et al. 2013). One of the main problems regarding the use of lignocellulosic material for biobutanol fermentation is that the inhibitors generated during the process can inhibit cell growth and fermentation (Baral and Shah 2014). Another important challenge is the downstream separation techniques for in situ continuous recovery (Bharathiraja et al. 2017). Although the distillation operation unit has a major role in enabling solvent recovery, it has a high impact in the

**Table 3.2** Production description of biobutanol in the main industries.

| Companies | Production description |
|---|---|
| Gevo | US-based company. A modified strain of *E.coli* is used to ferment different feedstocks (corn, sugar, and beets) for butanol production. Gevo currently operates its first fermentation facility in Luverne, Minnesota (US), where it produces isobutanol, ethanol, and related products like protein-rich animal feed (Gevo 2018). |
| Cobalt Technologies | California-based biofuel company. They produce butanol via continuous fermentation process using non-modified *Clostridium*. The feedstocks include biomass of wood pulp and sugar beets (Bharathiraja et al. 2017). |
| Butamax Advanced Biofuels | The company is a joint venture of BP and Dupont. A genetically modified engineered yeast is used to ferment corn hydrolysate and produce butanol. They have recently announced a step forward the commercialization of bio-isobutanol with the acquisition of an ethanol facility in Kansas, US (Butamax and DuPont Industrial Biosciences 2017). |
| Butyl Fuel, LLC | US-based company. Production of n-butanol using a modified *Clostridium*. According to the company, their technologies increase butanol production by over 30% compared with the alternative process, besides reducing acetone production (38%) and virtually eliminate ethanol production (Bharathiraja et al. 2017). |
| Green Biologics, GBL | UK-based company. The company has developed its microorganisms (clostridia and *Geobacillus*) to produce butanol efficiently. Recently they announced the start of shipments of biobased n-butanol and acetone from its manufacturing facility in Little Falls, Minnesota, EUA (BiofuelsDigest 2016). |
| Metabolic Explorer | Based in Clermont-Ferrand, France. Process relies on the fermentation of biomass for biobutanol production using high performance microorganisms (Bharathiraja et al. 2017). |

economic feasibility of ABE fermentation industry, being responsible for 20% of total cost (Mariano and Filho 2012). Alternative methods are reported, focused on reducing the costs and promoting efficient separation (Kujawska et al. 2015). Besides the high operation costs already mentioned, low titer, yield, productivity, and solvent toxicity are all challenges that must be overcome to make biobutanol economically feasible (Table 3.3).

In order to meet industrial competitiveness, several researchers have investigated the combination of modification in fermentation process and strain development using metabolic engineering (Moon et al. 2016). Recent studies using different approaches (e.g., strain development, fermentation, and downstream process) have led to significant advances in the ABE fermentation and biobutanol production. And in the short term, it is expected that this biofuel will receive more attention as the bottlenecks are overcome.

### 3.2.4 Biogas

The biogas market is growing rapidly due to the (i) flexibility of its production, which includes waste from agricultural, industrial, and urban activities; (ii) use in different productive scales and in decentralized systems; (iii) diversity of final application that meets the energy system needs (i.e. energy, industry, housing, and transport); and (iv) environmental, social, and economic sustainability (Raboni and Urbini 2014).

**Table 3.3** Major bottlenecks of clostridia biobutanol fermentation and approaches to address them (Green 2011; Moon et al. 2016).

| Issues | Approaches |
|---|---|
| Raw material cost | Transition toward cheaper (and more sustainable) feedstocks, such as wastes and agricultural residues. |
| Low titer, yield and productivity | Development and screening of microorganisms with enhanced biobutanol productivity and robustness to toxic compounds. |
| Solvent recovery using conventional distillation is energy intensive and expensive | To develop low energy methods for solvent recovery and purification. |
| Poor genetic engineering tools | To develop more plasmid vectors and more efficient transformation methods; more efficient gene knockout methods (e.g. CRISPR-Cas system); inducible expression system with stringent control. |
| Poor understanding on physiology | Detailed characterization of acidogenic and solventogenic phases and metabolism using metabolic flux. To understand better the complex regulatory circuits and their interaction with the microbial metabolism. |

Widespread in Europe, biogas production in terms of primary energy is equivalent to 10,086 ktoe year$^{-1}$, being up to 50.2% produced in Germany (5,068 ktoe year$^{-1}$). The United Kingdom (1,765 ktoe year$^{-1}$) is the second largest producer in Europe, followed by Italy (1,096 ktoe year$^{-1}$) and France (350 ktoe year$^{-1}$) (Raboni and Urbini 2014).

Interestingly, Brazil has the potential to produce 52 billion m$^3$ year$^{-1}$ of biogas, according to the Brazilian Association of Biogas and Biomethane (Abiogás 2017). This is equivalent to 115 000 GWh year$^{-1}$ or 9,888.2 ktoe year$^{-1}$, equivalent to more than all the energy generated in a year at the Itaipu hydroelectric plant (Brazil/Paraguay, 14,000 MW) that it was the largest plant on the planet, in terms of energy produced by 2015. Yet, in the association's accounts, 9 billion m$^3$ would be produced with agricultural wastes and waste from animals such as pigs, cattle, and poultry; the largest part, 39 billion m$^3$, would come from waste from sugarcane used in the production of alcohol (Abiogás 2017).

The sugar-energy industry is recovering, along with the Brazilian economic sector, and recently it received a remarkable incentive from the federal government by the program RenovaBio. The aim of this program is to expand biofuels by 2030, by strategic guidelines approved by the National Energy Policy Council (CNPE) in 2017. Furthermore, the program foresees expansion of the capacity and investments in new industrial units and technology to also production of the 2G ethanol, which may result in an increase in bioethanol production from 28 to 50 billion liters per year, almost double the current production.

In parallel, slow but important steps have been taken in relation to biogas in Brazil. Initially, biomethane specifications (purity ≥ 96.5%) were established in the Technical Regulation of the National Agency of Petroleum, Natural Gas and Biofuels (ANP No. 1/2015; ANP 2017). Then, there was an unprecedented event in the Brazilian history, in which a large-scale biogas facility won a power generation auction (Auction A-5) promoted by the National Electric Energy Agency (Aneel).[1]

Regardless of the speed and stage of development of sugarcane biorefineries and biogas public policies in Brazil, it has been noted that both are receiving incentives and following the same direction. However, the investment made by the sugar-energy industry on biogas plants have been conservative and the technology is far from optimal compared to modern configurations of bioreactors commercially available. In fact, the biogas plants have been inspired in the "technology" widely used in European farms, especially in Germany, the United Kingdom, Italy, and France. Basically, it consists of a pond or tank covered with PVC (polyvinyl chloride) canvas to retain the biogas produced during the biodigestion process from animal waste (i.e., residues with high solids content and continuous production).

The advantages of the technology of covered lagoons are undeniable compared to other bioreactor configurations. These include ease of construction, operation, and maintenance, as well as their respective reduced costs. In contrast, there are high construction area requirements to meet the low organic loading rate (OLR) supported by the system (0.25–1 kg $COD/m^3$ d), resulting in a high hydraulic retention time (HRT) between three and six days (SABESP 2009). These values are huge when compared to those obtained in lab-scale reactors under high-rate operation. Ferraz Júnior et al. (2016) and Fuess et al. (2017) optimized a biodigestion system of vinasse (OLR of 25–30 kg $COD/m^3$ d) and HRT of less than one day, reporting an average methane yield of 90% of the maximum theoretical value.

In Brazil, biogas production has largely been limited to covered lagoons of sugarcane plants. Perhaps, the insistence on this "technology" by the mills is justified by the mentality and convenience acquired over the years due to the lack of public environmental policies, as well as a culture of company-university partnerships.[2] However, this feeling of having solved the problem has created more than a 25-year[3] gap in scientific/applied researches related to biogas production, and leads to difficulties in the application and expansion of anaerobic digestion in the sugar and alcohol industry.

In this context, vinasse, one of the main byproducts of ethanol distillation in sugarcane mills, could be a valuable residue for anaerobic digestion. Vinasse is generated after ethanol separation from fermented cane juice in the proportion of 10–15 liters for each liter of ethanol produced. Due to the high content of organic matter expressed as chemical oxygen demand (COD; 10–65 g $O_2$ $L^{-1}$) and nutrients (nitrogen, phosphorus, and especially potassium), the most common destination of vinasse is application to the soil as fertilizer for the sugarcane crop (i.e. fertigation).

From an economic point of view, and still respecting the Brazilian environmental legislation, the soil disposal of vinasse represents the least expensive and simplest solution for this voluminous liquid stream. In fact, the fertigation with vinasse is necessary to return water to soil, considering its water content of 93%; and reduce the mineral input

1 The Raízen project, the winner of the auction, traded energy from Biogas Bonfim, 20.8 MW at R$ 251/MWh (Canal Jornal da Bioenergia 2016).

2 The vinasse soil disposal is based only on the maximum potassium content of 400 kg $K_2O$/ha year, as recommended in CETESB (1985, 2006).

3 Souza et al. (1992) was one of the first studies to demonstrate the anaerobic digestion of Brazilian vinasse, in a pilot scale. High values of methane yield (310 NmL $CH_4$ $g^{-1}$ $COD_{removed}$) and COD removal (above 70%) at OLR of 26.5 kg $COD/m^3$ d were reported.

**Table 3.4** Energy comparison of common reactions in anaerobic degradation (Decker et al. 1970; McCarty and Mosey 1991; Thauer et al. 1977).

| Step | Reaction | ΔG° (kJ/reaction) |
|---|---|---|
| Acidogenesis | $C_6H_{12}O_6 + 2H_2O \rightarrow 2CH_3COO^- + 2CO_2 + 2H^+ + 4H_2$ | −206 |
| | $C_6H_{12}O_6 + 2H_2 \rightarrow 2CH_3CH_2COO^- + 2H_2O + 2H^+$ | −358 |
| | $C_6H_{12}O_6 \rightarrow CH_3CH_2CH_2COO^- + 2CO_2 + H^+ + 2H_2$ | −255 |
| Acetogenesis | $CH_3CH_2COO^- + 3H_2O \rightarrow CH_3COO^- + HCO_3^- + H^+ + 3H_2$ | +76.1 |
| | $CH_3CH_2COO^- + 2HCO_3^- \rightarrow CH_3COO^- + H^+ + 3HCOO^-$ | +72.2 |
| | $CH_3CH_2CH_2COO^- + 2H_2O \rightarrow 2CH_3COO^- + H^+ + 2H_2$ | +48.1 |
| Methanogenesis | $CH_3COO^- + H_2O \rightarrow CH_4 + HCO_3^- + 2H_2$ | −31.0 |
| | $H_2 + 0.25HCO_3^- + 0.25H^+ \rightarrow 0.25CH_4 + 0.75H_2O$ | −33.9 |
| | $HCOO^- + 0.25H_2O + 0.25H^+ \rightarrow 0.25CH_4 + 0.75HCO_3^-$ | −32.6 |

in the sugarcane crop. However, this practice can represent a loss of energy for the sugar-energy plant and trigger a series of environmental problems, ranging from soil leaching to GHG emissions and alteration of soil resistome dynamics (Braga et al. 2017).

Anaerobic digestion stands out and has become a requirement for biorefineries not only for alleviating environmental concerns but also for converting the residual liquid and solid streams into a valuable renewable fuel product, biogas. Furthermore, anaerobic digestion can integrate and consolidate the following aspects: (i) system of power and steam generation; (ii) system of the water use (abstraction, treatment, disposal); and (iii) final waste disposal.

Briefly, anaerobic digestion is a worldwide technology based on the fundamentals of acidogenesis, acetogenesis, and methanogenesis processes. Fermentative microorganisms are the first to play their role, being the ones that benefit most energetically. The hydrolyzed products from complex organic material are converted in reduced end-products (mainly acetate, butyrate, propionate), few solvents (ethanol, butanol, and acetone), hydrogen, and carbon dioxide by acidogenic bacteria. In view of that, those reactions are thermodynamically favorable (Table 3.4), have a low generation time (~0.5 hour) and have the highest growth rates of microbial consortium when using acidogenic bacteria. Thus, the acidogenic step will only limit the whole process if the material to be degraded is not easily hydrolyzed (de Aquino and Chernicharo 2005).

On the other hand, acetogenic reactions are thermodynamically unfavorable under standard conditions, but occur naturally in anaerobic reactors due to the syntrophy between methanogenic and acetogenic microorganisms. To overcome the thermodynamic limitations, acetogenesis products must be kept at low levels, which occurs through the activity of hydrogenotrophic methanogenic microorganisms. The hydrogen produced in the acetogenesis step must be continuously removed to ensure that the acetate production is not disrupted or decreased dramatically, since methanogenesis strongly depends on the availability of acetate, the precursor of about 70% of methane produced (Foresti 1994; Leitão et al. 2006). In addition, the anaerobic process may occur in a single-stage system, in which the final product is a methane-enriched biogas (60–70%), or in a two-stage system. In such instances, the biogas generated in the first stage is hydrogen-enriched (30–40%) while the biogas generated in the second stage is methane-enriched (60–70%) (Aryal et al. 2018; Mehariya et al. 2018).

Generally, the analysis of reactors comprises the configuration selection and operational routines most appropriate to the residual liquid stream treatment in question, based on quality parameters, mainly, biogas production and yield; and organic matter removal efficiency. Some studies have reported the potential of using sugarcane vinasse as a substrate for production of hydrogen (Ferraz Júnior et al. 2014) and methane (Ferraz Júnior et al. 2016; Fuess et al. 2017; Kiyuna et al. 2017; Mota et al. 2013; Souza et al. 1992). These studies covered the selection of the most appropriate types of reactors, operational requirements (support material, pH, temperature, HRT, OLR organic matter/sulfate ratio), and identification of the microorganisms involved in vinasse biodigestion.

Regarding the evaluation and technological feasibility of vinasse biodigestion, some milestones were established:

- In terms of anaerobic digestion process and biogas production from vinasse, the adoption of a two-stage thermophilic system (acidogenic and methanogenic) increases the organic compounds degradability by 21%; increases biodigestion stability and energy yield by 25.7%; and reduces the alkalinization contribution in the methanogenic stage by 50% (Ferraz Júnior 2013; Ferraz Júnior et al. 2016).
- In energy terms, the two-stage thermophilic system (acidogenic and methanogenic) presented a 25.7% higher energy production compared to the single-stage system (potential energy of 1 L of vinasse: 45.5 W vs. 36.2 W, respectively). It is also important to note that only 1.5 W (acidogenic reactor) came from hydrogen production (Ferraz Júnior et al. 2016).
- In terms of opportunities of biogas use,[4] biogas (biomethane) can replace 40% of the annual supply of diesel in the agricultural operations of a sugarcane-based biorefinery and still provides approximately 14 MWh annually from cogeneration, considering a national scenario (Moraes et al. 2014).
- In economic terms, the implementation of a biodigestion plant of vinasse with biomethane injection in the grid was feasible in a brownfield plant type (i.e. ready-to-build plant), resulting in a net present value (NPV) and internal rate of return (TIR) of USD 23.3 million and 22.2% per year, respectively (Junqueira et al. 2016b).
- In environmental terms, biodigested vinasse in two-stage thermophilic system (acidogenic and methanogenic) presented a pollutant load four times lower than its initial one (Ferraz Júnior et al. 2016).
- In terms of microbiology, the role of approximately 16 genera was described and associations of syntrophy between bacteria and methanogenic archaea were suggested (data not published).

Sugarcane vinasse is a typical Brazilian effluent that presents specific needs of management and treatment, mainly in relation to operational conditions of biodigestion

4 The use of biomethane in passenger vehicles is already a reality in the International Center for Renewable Energy (CI Biogas), Foz do Iguaçu, Parana, Brazil. Biomethane is produced continuously in reactors fed with pruning remnants, organic waste from the company's refectory and sewage, characterizing a co-digestion process (i.e. biodigestion of more than one substrate). The use of biogas in 70 passenger vehicles represents a monthly savings of R$ 40,000 for CI Biogas.

process. These conditions vary from vinasse generation (i.e., intermittent generation, due to the sugarcane off-season) to its compositional variation throughout the sugarcane harvest season. Thus, it is necessary to adapt existing biodigestion technologies to the "residues" of sugarcane distillation in Brazil that includes sugarcane bagasse and filter cake (solid residues).

Large companies (Raízen, GranBio, and São Martinho Group) have shown great interest in biogas production from vinasse. In parallel, the evaluation of its potential has been made by small and medium-sized companies, incubated in the plants of those large companies (mentioned a priori). However, the convenience of Brazilian industries intensified by the political instability and economic crisis of the country are leading to changes in the investment schedule, including the targets for the implementation of 2G technology that would, theoretically, correspond to a 50% increase in the total ethanol production that means positive impact on mill revenues.

It is mandatory to unlock both public and private funds for continuous research, development, and demonstration in sugar-energy industry and public policy changes to encourage the production and use of biogas in Brazil, which, in terms of electricity, only harnesses 0.06% of the potential energy from that source.

### 3.2.5 Microalgal Biomass for Biofuels Production

Microalgal biomass has recently received more attention as a potential platform to be used as biological resource for 3G biofuel production, namely bioethanol, biodiesel, and biogas. Microalgae cells have a high content of carbohydrate and/or lipid, and a higher photosynthetic efficiency and faster growth rate when compared with other crops. They are autotrophs and can assimilate $CO_2$ by photosynthesis reaction, mitigating GHG emissions in the environment. In addition, algal cultivation does not compete with land dedicated to food crops, alleviating concerns about the "food versus biofuels" issue. Low economic value lands, such as arid soil, could be used to set up microalgae biorefinery, generating various valuable co-products (pigments and proteins, for instance) with different industrial applications. Although microalgae require a large amount of water to grow (in general, it is estimated 1,000 kg of water per kg of microalgal biomass), they can grow on urban and industrial wastewater and recycled spent water. This process can reduce the operational costs and is more environmentally friendly. However, 3G biofuel production still has not made much progress mainly because of the low productivity and high costs associated with microalgae harvesting (this step represents up to 20–30% of the total cost of biofuel production by microalgae), lipid extraction from algal biomass, and other operational activities (Behera et al. 2014; Huy et al. 2018; Klein et al. 2018).

Several works have been conducted showing the technological, economic, and commercial aspects of producing biofuels from microalgae. Behera et al. (2014) published a study summarizing the main aspects of biodiesel, biogas, biohydrogen, and syngas production from algal biomass. A comparison of various studies using algae or terrestrial crops was also presented. Another interesting review is from Zhou et al. (2017). Although the review is not focused only on biofuels, it brings valuable information about microalgae cultivation methods, carbon metabolism, and the value-added products that might be produced. This review still shows the energy and cost requirements for biodiesel and biogas production in different scenarios.

Likewise, Klein et al. (2018) presented an overview of the main features of microalgae and pointed out that is possible to integrate 1G refineries with microalgae cultivation to produce biodiesel and biogas. In Brazil, for example, sugarcane mills could feed microalgae cultures with the vinasse digested for biogas production after ethanol distillation. Thus, the infrastructure of 1G biorefineries could be maximized for the generation of new products (higher flexibility) and reduction of costs. Following this trend of integrating different production systems, various companies are already cultivating microalgae for production of cosmetics (TerraVia Holdings, BASF), animal feed (Cellana Inc.), the pigment astaxanthin (Cyanotech Co., BASF), betaine surfactant (BASF) and, evidently, biofuels (Cellana Inc., J-Power, Euglena Co. Ltd., IHI Co.) (Maeda et al. 2018).

Unfortunately, the full-scale implementation of microalgal biomass for production of biofuels is still economically unfeasible, and the perspectives point to the commercialization of this green technology only within a few decades (Kumar et al. 2017; Passos et al. 2014). Studies evaluating the potential of microalgae to produce 3G biofuels have been carried out mostly at a laboratory scale, delaying the advance of these sustainable fuels. In this context, collaborative efforts from academia and companies should be made to overcome the technological and economic challenges of using biomass from microalgae. The drawbacks for production of 3G biofuels include improvements in genetic engineering of high-performance microalgae strains; optimization of microalgal cultivation (culture system, nutrient recycling) and processing (harvesting, pretreatment, drying) technologies that are high energy-consuming; and integration of microalgal biomass production with the generation of co-products of high value-added (Gouveia et al. 2017; Klein et al. 2018; Zhou et al. 2017). The combination of these elements could reduce the costs associated with microalgal biorefinery, making 3G biofuels a competitive resource (Behera et al. 2014; Kiesenhofer and Fluch 2018).

## 3.3 Conclusion

In this chapter, an overview was presented covering some key points of biofuel production from different feedstocks and technologies. It is evident that the biofuels presented here will be of fundamental importance for the global energy scenario in the next years, mainly due to the growth of population and the industrial activities, public incentive policies, and the necessity of increasingly developing renewable alternatives to fossil fuels. Bioethanol and biobutanol from lignocellulosic residues and biodiesel from vegetable oils have proven to be feasible to commercialize, even though there are some limitations that must be overcome. Regarding biogas and production of biofuels from algae, more efforts from the academic and industrial sides are required to make them economically competitive. Integration of different biofuels production systems and the use of low-cost feedstocks could help to decrease the costs in the facilities and attract more investments. Finally, there is a diverse range of biofuels to be explored, and in the short term more attention must be given to this sector for meeting the global energy needs.

## References

Abiogás, 2017. Abiogás. URL www.abiogas.org.br/setor-no-brasil

Addington, B., 2017. RenovaBio: a paradigm shift for biofuels in Brazil. https://www.udop.com.br/download/noticias/2017/24_07_17_artigo_renovabio.pdf

Adlercreutz, P. (2013). Immobilisation and application of lipases in organic media. *Chem. Soc. Rev.* 42: 6406.

Adsul, M.G., Singhvi, M.S., Gaikaiwari, S.A., and Gokhale, D.V. (2011). Development of biocatalysts for production of commodity chemicals from lignocellulosic biomass. *Bioresour. Technol.* 102: 4304–4312.

ANP, 2017. Agência Nacional de Petróleo Gás Natural e Biocombustíveis (ANP n° 1/2015). URL www.anp.gov.br/wwwanp/biocombustiveis/biometano (accessed 7.1.17).

de Aquino, S.F. and Chernicharo, C.A.L. (2005). Acúmulo de ácidos graxos voláteis (AGVs) em reatores anaeróbios sob estresse: causas e estratégias de controle. *Eng. Sanit. e Ambient.* 10: 152–161.

Araújo, W.A. (2016). Ethanol industry: surpassing uncertainties and looking forward. In: *Global Bioethanol*, 1–33. Elsevier.

Aryal, N., Kvist, T., Ammam, F. et al. (2018). An overview of microbial biogas enrichment. *Bioresour. Technol.* 264: 359–369.

Auxenfans, T., Crônier, D., Chabbert, B., and Paës, G. (2017). Understanding the structural and chemical changes of plant biomass following steam explosion pretreatment. *Biotechnol. Biofuels* 10: 36.

Baral, N.R. and Shah, A. (2014). Microbial inhibitors: formation and effects on acetone-butanol-ethanol fermentation of lignocellulosic biomass. *Appl. Microbiol. Biotechnol.* 98: 9151–9172.

Barr, C.J., Mertens, J.A., and Schall, C.A. (2012). Critical cellulase and hemicellulase activities for hydrolysis of ionic liquid pretreated biomass. *Bioresour. Technol.* 104: 480–485.

Behera, S., Singh, R., Arora, R. et al. (2014). Scope of algae as third generation biofuels. *Front. Bioeng. Biotechnol.* 2: 90.

Bharathiraja, B., Jayamuthunagai, J., Sudharsanaa, T. et al. (2017). Biobutanol – an impending biofuel for future: a review on upstream and downstream processing tecniques. *Renewable Sustainable Energy Rev.* 68: 788–807.

BiofuelsDigest, 2016. Green Biologics begins shipments from 1st commercial plant. URL http://www.biofuelsdigest.com/bdigest/2016/12/13/green-biologics-begins-shipments-from-1st-commercial-plant

Borin, G.P., Sanchez, C.C., de Santana, E.S. et al. (2017). Comparative transcriptome analysis reveals different strategies for degradation of steam exploded sugarcane bagasse by *Aspergillus niger* and *Trichoderma reesei*. *BMC Genomics* 18: 501.

Bornscheuer, U., Buchholz, K., and Seibel, J. (2014). Enzymatic degradation of (ligno)cellulose. *Angew. Chem. Int. Ed.* 53: 10876–10893.

Braga, L.P.P., Alves, R.F., Dellias, M.T.F. et al. (2017). Vinasse fertirrigation alters soil resistome dynamics: an analysis based on metagenomic profiles. *BioData Min.* 10: 1–7.

Brazilian Ministry of Mines and Energy, 2017. RenovaBio - Diretrizes Estratégicas para Bicombustíveis. URL www.mme.gov.br/documents/10584/7948692/RenovaBio+2030++-+Texto+para+Consulta+Pública.pdf/8cfdfb6a-4c38-4827-92b5-ac71f45cecba?version=1.0 (accessed 12.19.17).

Butamax, BP, DuPont Industrial Biosciences, 2017. BP and DuPont Joint Venture, Butamax®, announces next step in commercialization of bio-isobutanol with acquisition of ethanol facility in Kansas. URL: http://www.butamax.com/Portals/0/pdf/2017_04_03_ib_butamax_announce - FINAL.pdf

Canal Jornal da Bioenergia, 2016. Pela primeira vez um projeto de biogás venceu um leilão de energia. URL www.canalbioenergia.com.br/pela-primeira-vez-um-projeto-de-biogas-venceu-um-leilao-de-energia (accessed 7.1.17).

Carneiro, M.L.N.M., Pradelle, F., Braga, S.L. et al. (2017). Potential of biofuels from algae: comparison with fossil fuels, ethanol and biodiesel in Europe and Brazil through life cycle assessment (LCA). *Renewable Sustainable Energy Rev.* 73: 632–653.

CETESB, 1985. Notes on Control Technologies: Fabrication of Sugar and Alcohol - Technical Report CETESP. São Paulo.

CETESB, 2006. Norma Técnica CETESB P 4.231 - Vinhaça - Critérios e procedimentos para aplicação no solo agrícola. URL: https://cetesb.sp.gov.br/wp-content/uploads/2013/11/NTC-P4.231_Vinhaça_-Critérios-e-procedimentos-para-aplicação-no-solo-agrícola-3ª-Ed-2ª-VERSÃO.pdf

Chen, H. (2014). *Biotechnology of Lignocellulose.* Dordrecht: Springer Netherlands.

Chen, G.-Q. and Jiang, X.-R. (2018). Next generation industrial biotechnology based on extremophilic bacteria. *Curr. Opin. Biotechnol.* 50: 94–100.

Chen, M., Smith, P.M., and Wolcott, M.P. (2016). U.S. biofuels industry: a critical review of opportunities and challenges. *Bioprod. Bus.* 1: 42–59.

Cheng, C.L., Che, P.Y., Chen, B.Y. et al. (2012). Biobutanol production from agricultural waste by an acclimated mixed bacterial microflora. *Appl. Energy* 100: 3–9.

Coutouné, N., Mulato, A.T.N., Riaño-Pachón, D.M., and Oliveira, J.V. (2017). Draft genome sequence of *Saccharomyces cerevisiae* Barra Grande (BG-1), a Brazilian industrial bioethanol-producing strain. *Genome Announc.* 5: https://doi.org/10.1128/genomeA.00111-17.

Cragg, S.M., Beckham, G.T., Bruce, N.C. et al. (2015). Lignocellulose degradation mechanisms across the tree of life. *Curr. Opin. Chem. Biol.* 29: 108–119.

Dahiya, S., Kumar, A.N., Shanthi Sravan, J. et al. (2018). Food waste biorefinery: sustainable strategy for circular bioeconomy. *Bioresour. Technol.* 248: 2–12.

Dashtban, M., Schraft, H., and Qin, W. (2009). Fungal bioconversion of lignocellulosic residues; opportunities & perspectives. *Int. J. Biol. Sci.* 5: 578–595.

Decker, K., Jungermann, K., and Thauer, R.K. (1970). Energy production in anaerobic organisms. *Angew. Chem. Int. Ed.* 9: 138–158.

Demeke, M.M., Dietz, H., Li, Y. et al. (2013). Development of a D-xylose fermenting and inhibitor tolerant industrial *Saccharomyces cerevisiae* strain with high performance in lignocellulose hydrolysates using metabolic and evolutionary engineering. *Biotechnol. Biofuels* 6: 89.

Dondelinger, E., Aubry, N., Ben Chaabane, F. et al. (2016). Contrasted enzymatic cocktails reveal the importance of cellulases and hemicellulases activity ratios for the hydrolysis of cellulose in presence of xylans. *AMB Express* 6: 24.

dos Santos, L.V., de Barros Grassi, M.C., Gallardo, J.C.M. et al. (2016). Second-generation ethanol: the need is becoming a reality. *Ind. Biotechnol.* 12: 40–57.

Ehsaan, M., Kuit, W., Zhang, Y. et al. (2016). Mutant generation by allelic exchange and genome resequencing of the biobutanol organism *Clostridium acetobutylicum* ATCC 824. *Biotechnol. Biofuels* 9: 4.

EPA, 2013. The Public Health and Welfare: Chapter 85 - Air pollution prevention and control, Subchapter II - emission standards for moving sources, Part A - Motor vehicle emission and fuel standars, Section 7545 - Regulation of fuels. URL: http://uscode.house

.gov/view.xhtml?req=granuleid:USC-prelim-title42-section7545&num=0&edition=prelim

European Commission, 2009. Directive 2009/28/EC of the European Parliament and of the Council of 23 April 2009 on the promotion of the use of energy from renewable sources and amending and subsequently repealing Directives 2001/77/EC and 2003/30/EC (Text with EEA relevance). URL: https://eur-lex.europa.eu/legal-content/EN/ALL/?uri=CELEX:32009L0028

European Commission, 2016. Proposal for a directive of the European parliament and of the council on the promotion of the use of energy from renewable sources (recast). URL: https://eur-lex.europa.eu/legal-content/EN/TXT/?uri=CELEX:52016PC0767

Ferraz Júnior, A.D.N., 2013. Digestão anaeróbia da vinhaça da cana de açúcar em reator acidogênico de leito fixo seguido de reator metanogênico de manta de lodo. Universidade de São Paulo.

Ferraz Júnior, A.D.N., Wenzel, J., Etchebehere, C., and Zaiat, M. (2014). Effect of organic loading rate on hydrogen production from sugarcane vinasse in thermophilic acidogenic packed bed reactors. *Int. J. Hydrog. Energy* 39: 16852–16862.

Ferraz Júnior, A.D.N., Koyama, M.H., de Araújo Júnior, M.M., and Zaiat, M. (2016). Thermophilic anaerobic digestion of raw sugarcane vinasse. *Renewable Energy* 89: 245–252.

Foresti, E., 1994. Fundamentos do processo de digestão anaeróbia. In: Proceedings of III Taller y Seminario Latinoamericano of Tratamiento Anaeróbio de Aguas Residuales. s.n, Montevideo, Uruguay, pp. 97–110.

Fuess, L.T., Kiyuna, L.S.M., Ferraz, A.D.N. et al. (2017). Thermophilic two-phase anaerobic digestion using an innovative fixed-bed reactor for enhanced organic matter removal and bioenergy recovery from sugarcane vinasse. *Appl. Energy* 189: 480–491.

Gapes, J.R. (2000). The economics of acetone-butanol fermentation: theoretical and market considerations further reading. *J. Mol. Microbiol. Biotechnol.* 2: 27–32.

Gevo, 2018. Gevo. URL http://www.gevo.com.

González Prieto, M., Fortunatti Montoya, M., Hegel, P.E., and Pereda, S. (2017). Supercritical reactors for the production of advanced bio-fuels: a review. *J. Supercrit. Fluids* 134: 106–113.

Gouveia, L., Oliveira, A.C., Congestri, R. et al. (2017). Biodiesel from microalgae. In: *Microalgae-Based Biofuels and Bioproducts*, 235–258. Elsevier.

Green, E.M. (2011). Fermentative production of butanol-the industrial perspective. *Curr. Opin. Biotechnol.* 22: 337–343.

Guerriero, G., Hausman, J.-F., Strauss, J. et al. (2016). Lignocellulosic biomass: biosynthesis, degradation, and industrial utilization. *Eng. Life Sci.* 16: 1–16.

Guo, T., He, A.Y., Du, T.F. et al. (2013). Butanol production from hemicellulosic hydrolysate of corn fiber by a *Clostridium beijerinckii* mutant with high inhibitor-tolerance. *Bioresour. Technol.* 135: 379–385.

Guo, M., Song, W., and Buhain, J. (2015). Bioenergy and biofuels: history, status, and perspective. *Renewable Sustainable Energy Rev.* 42: 712–725.

Gupta, A. and Verma, J.P. (2015). Sustainable bio-ethanol production from agro-residues: a review. *Renewable Sustainable Energy Rev.* 41: 550–567.

Hassan, S.S.; Williams, G.A.; Jaiswal, A.K. (2018). Emerging technologies for the pretreatment of lignocellulosic biomass. *Bioresource Technology*, v. 262, p. 310–318, 1ago.

Hadar, Y. (2013). Sources for lignocellulosic raw materials for the production of ethanol. In: *Lignocellulose Conversion*, 21–38. Berlin, Heidelberg: Springer Berlin Heidelberg.

He, C.R., Kuo, Y.Y., and Li, S.Y. (2017). Lignocellulosic butanol production from Napier grass using semi-simultaneous saccharification fermentation. *Bioresour. Technol.* 231: 101–108.

Horn, S., Vaaje-Kolstad, G., Westereng, B., and Eijsink, V.G. (2012). Novel enzymes for the degradation of cellulose. *Biotechnol. Biofuels* 5: 45.

Huy, M., Kumar, G., Kim, H.-W., and Kim, S.-H. (2018). Photoautotrophic cultivation of mixed microalgae consortia using various organic waste streams towards remediation and resource recovery. *Bioresour. Technol.* 247: 576–581.

Isikgor, F.H. and Becer, C.R. (2015). Lignocellulosic biomass: a sustainable platform for the production of bio-based chemicals and polymers. *Polym. Chem.* 6: 4497–4559.

Ivana, L., Kesic, Z., Zdujić, M., and Skala, D. (2016). Vegetable oil as a feedstock for biodiesel synthesis. In: *Vegetable Oils – Properties, Uses and Benefits* (ed. B. Holt), 83–128. Nova Publishers.

Jegannathan, K.R., Abang, S., Poncelet, D. et al. (2008). Production of biodiesel using immobilized lipase – a critical review. *Crit. Rev. Biotechnol.* 28: 253–264.

Jeuland, N., Montagne, X., and Gautrot, X. (2004). Potentiality of ethanol as a fuel for dedicated engine. *Oil Gas Sci. Technol.* 59: 559–570.

Jiang, Y., Liu, J., Jiang, W. et al. (2015). Current status and prospects of industrial bio-production of n-butanol in China. *Biotechnol. Adv.* 33: 1493–1501.

Jin, C., Yao, M., Liu, H. et al. (2011). Progress in the production and application of n-butanol as a biofuel. *Renewable Sustainable Energy Rev.* 15: 4080–4106.

Johansen, K.S. (2016). Lytic polysaccharide monooxygenases: the microbial power tool for lignocellulose degradation. *Trends Plant Sci.* 21: 926–936.

Jones, D.T. and Woods, D.R. (1986). Acetone-butanol fermentation revisited. *Microbiol. Rev.* 50: 484–524.

Jönsson, L.J. and Martín, C. (2016). Pretreatment of lignocellulose: formation of inhibitory by-products and strategies for minimizing their effects. *Bioresour. Technol.* 199: 103–112.

Junqueira, T.L., Cavalett, O., and Bonomi, A. (2016a). The virtual sugarcane biorefinery – a simulation tool to support public policies formulation in bioenergy. *Ind. Biotechnol.* 12: 62–67.

Junqueira, T.L., Moraes, B., Gouveia, V.L.R. et al. (2016b). Use of VSB to plan research programs and public policies. In: *Virtual Biorefinery – An Optimization Strategy for Renewable Carbon Valorization* (eds. A. Bonomi, O. Cavalett, M. Pereira da Cunha and M.A.P. Lima), 257–282. Cham: Springer.

Katahira, S., Muramoto, N., Moriya, S. et al. (2017). Screening and evolution of a novel protist xylose isomerase from the termite *Reticulitermes speratus* for efficient xylose fermentation in *Saccharomyces cerevisiae*. *Biotechnol. Biofuels* 10: 203.

Khanna, S., Goyal, A., and Moholkar, V.S. (2013). Production of n-butanol from biodiesel derived crude glycerol using *Clostridium pasteurianum* immobilized on Amberlite. *Fuel* 112: 557–561.

Kiesenhofer, D.P. and Fluch, S. (2018). The promises of microalgae – still a long way to go. *FEMS Microbiol. Lett.* 365: 3.

Kirubakaran, M. and Arul Mozhi Selvan, V. (2018). A comprehensive review of low cost biodiesel production from waste chicken fat. *Renewable Sustainable Energy Rev.* 82: 390–401.

Kiyuna, L.S.M., Fuess, L.T., and Zaiat, M. (2017). Unraveling the influence of the COD/sulfate ratio on organic matter removal and methane production from the biodigestion of sugarcane vinasse. *Bioresour. Technol.* 232: 103–112.

Klein, B.C., Bonomi, A., and Maciel Filho, R. (2018). Integration of microalgae production with industrial biofuel facilities: a critical review. *Renewable Sustainable Energy Rev.* 82: 1376–1392.

Ko, J.K., Um, Y., Woo, H.M. et al. (2016). Ethanol production from lignocellulosic hydrolysates using engineered *Saccharomyces cerevisiae* harboring xylose isomerase-based pathway. *Bioresour. Technol.* 209: 290–296.

Koeck, D.E., Pechtl, A., Zverlov, V.V., and Schwarz, W.H. (2014). Genomics of cellulolytic bacteria. *Curr. Opin. Biotechnol.* 29: 171–183.

Kujawska, A., Kujawski, J., Bryjak, M., and Kujawski, W. (2015). ABE fermentation products recovery methods – a review. *Renewable Sustainable Energy Rev.* 48: 648–661.

Kumar, A.K. and Sharma, S. (2017). Recent updates on different methods of pretreatment of lignocellulosic feedstocks: a review. *Bioresour. Bioprocess.* 4: 7.

Kumar, G., Shobana, S., Chen, W.-H. et al. (2017). A review of thermochemical conversion of microalgal biomass for biofuels: chemistry and processes. *Green Chem.* 19: 44–67.

Lee, S.Y., Park, J.H., Jang, S.H. et al. (2008). Fermentative butanol production by clostridia. *Biotechnol. Bioeng.* 101: 209–228.

Leitão, R.C., van Haandel, A.C., Zeeman, G., and Lettinga, G. (2006). The effects of operational and environmental variations on anaerobic wastewater treatment systems: a review. *Bioresour. Technol.* 97: 1105–1118.

Li, K., Fan, Y., He, Y. et al. (2017). *Burkholderia cepacia* lipase immobilized on heterofunctional magnetic nanoparticles and its application in biodiesel synthesis. *Sci. Rep.* 7: 16473.

Löbs, A.-K., Schwartz, C., and Wheeldon, I. (2017). Genome and metabolic engineering in non-conventional yeasts: current advances and applications. *Synth. Syst. Biotechnol.* 2: 198–207.

Lynd, L.R., Liang, X., Biddy, M.J. et al. (2017). Cellulosic ethanol: status and innovation. *Curr. Opin. Biotechnol.* 45: 202–211.

Maeda, Y., Yoshino, T., Matsunaga, T. et al. (2018). Marine microalgae for production of biofuels and chemicals. *Curr. Opin. Biotechnol.* 50: 111–120.

Mariano, A.P. and Filho, R.M. (2012). Improvements in biobutanol fermentation and their impacts on distillation energy consumption and wastewater generation. *Bioenergy Res.* 5: 504–514.

McCarty, P.L. and Mosey, F.E. (1991). Modelling of anaerobic digestion processes (a discussion of concepts). *Water Sci. Technol.* 24: 17–33.

Mehariya, S., Kumar Patel, A., Obulisamy, K. et al. (2018). Co-digestion of food waste and sewage sludge for methane production: current status and perspective. *Bioresour. Technol.* 265: 519–531.

de Melo, R.R., Borin, G.P., Zanini, G.K. et al. (2016). An overview on vegetable oils and biocatalysis. In: *Vegetable Oil: Properties, Uses and Benefits* (ed. B. Holt), 1–24. Nova Publishers.

MME, 2017. Brasil institui Política Nacional dos Biocombustíveis (RenovaBio). URL: http://www.mme.gov.br/

Moon, H.G., Jang, Y., Cho, C. et al. (2016). One hundred years of clostridial butanol fermentation. *FEMS Microbiol. Lett.* 363: 15.

Moraes, B.S., Junqueira, T.L., Pavanello, L.G. et al. (2014). Anaerobic digestion of vinasse from sugarcane biorefineries in Brazil from energy, environmental, and economic perspectives: profit or expense? *Appl. Energy* 113: 825–835.

Mota, V.T., Santos, F.S., and Amaral, M.C.S. (2013). Two-stage anaerobic membrane bioreactor for the treatment of sugarcane vinasse: assessment on biological activity and filtration performance. *Bioresour. Technol.* 146: 494–503.

Nanda, S., Dalai, A.K., and Kozinski, J.A. (2014). Butanol and ethanol production from lignocellulosic feedstock: biomass pretreatment and bioconversion. *Energy Sci. Eng.* 2: 138–148.

Narayanan, V., Schelin, J., Gorwa-Grauslund, M. et al. (2017). Increased lignocellulosic inhibitor tolerance of *Saccharomyces cerevisiae* cell populations in early stationary phase. *Biotechnol. Biofuels* 10: 114.

Naylor, R.L. and Higgins, M.M. (2017). The rise in global biodiesel production: implications for food security. *Glob. Food Sec.* 16: 75–84.

Passos, F., Uggetti, E., Carrère, H., and Ferrer, I. (2014). Pretreatment of microalgae to improve biogas production: a review. *Bioresour. Technol.* 172: 403–412.

Pereira, S., Maehara, L., Machado, C., and Farinas, C. (2015). 2G ethanol from the whole sugarcane lignocellulosic biomass. *Biotechnol. Biofuels* 8: 44.

Pfromm, P.H., Amanor-Boadu, V., Nelson, R. et al. (2010). Bio-butanol vs. bio-ethanol: a technical and economic assessment for corn and switchgrass fermented by yeast or *Clostridium acetobutylicum*. *Biomass Bioenergy* 34: 515–524.

Raboni, M. and Urbini, G. (2014). Production and use of biogas in Europe: a survey of current status and perspectives. *Ambient. Agua* 9: 191–202.

Ravachol, J., de Philip, P., Borne, R. et al. (2016). Mechanisms involved in xyloglucan catabolism by the cellulosome-producing bacterium *Ruminiclostridium cellulolyticum*. *Sci. Rep.* 6: 22770.

Renewable Fuels Association, 2017a. 2017 Ethanol industry outlook - Building partnerships and growing markets. Renewable Fuels Association. URL: https://ethanolrfa.org/wp-content/uploads/2017/02/Ethanol-Industry-Outlook-2017.pdf

Renewable Fuels Association, 2017b. Industry Statistics - World Fuel Ethanol Production. URL http://www.ethanolrfa.org/resources/industry/statistics/#1454099103927-61e598f7-7643 (accessed 12.19.17).

Ribeiro, B.D., Machado De Castro, A., Alice, M. et al. (2011). Production and use of lipases in bioenergy: a review from the feedstocks to biodiesel production. *Enzyme Res.* 2011: 1–16.

SABESP, 2009. Norma técnica SABESP NTS 230 - Projeto de lagoas de estabilização e seu tratamento complementar para esgoto sanitário. São Paulo. URL: http://www2.sabesp.com.br/normas/nts/nts230.pdf

Salihu, A., Abbas, O., Sallau, A.B., and Alam, M.Z. (2015). Agricultural residues for cellulolytic enzyme production by *Aspergillus niger*: effects of pretreatment. *3 Biotech* 5: 1101–1106.

Sauer, M. (2016). Industrial production of acetone and butanol by fermentation – 100 years later. *FEMS Microbiol. Lett.* 363: fnw134.

Schmoll, M., Dattenböck, C., Carreras-Villaseñor, N. et al. (2016). The genomes of three uneven siblings: footprints of the lifestyles of three *Trichoderma* species. *Microbiol. Mol. Biol. Rev.* 80: 205–327.

Shahzad, K., Nizami, A.S., Sagir, M. et al. (2017). Biodiesel production potential from fat fraction of municipal waste in Makkah. *PLoS One* 12: e0171297.

Sharma, N.K., Behera, S., Arora, R., and Kumar, S. (2017). Evolutionary adaptation of *Kluyveromyces marxianus* NIRE-K3 for enhanced xylose utilization. *Front. Energy Res.* 5: 32.

da Silva, C. and Oliveira, J.V. (2014). Biodiesel production through non-catalytic supercritical transesterification: current state and perspectives. *Braz. J. Chem. Eng.* 31: 271–285.

da Silva, A.S., de Sá, L.R.V., Aguieiras, E.C.G. et al. (2017). Productive chain of biofuels and industrial biocatalysis. In: *Biotechnology of Microbial Enzymes*, 545–581. Elsevier.

Silveira, M.H.L., Morais, A.R.C., da Costa Lopes, A.M. et al. (2015). Current pretreatment technologies for the development of cellulosic ethanol and biorefineries. *ChemSusChem* 8: 3366–3390.

Souza, M.E., Fuzaro, G., and Polegato, A.R. (1992). Thermophilic anaerobic digestion of vinasse in pilot plant UASB reactor. *Water Sci. Technol.* 25: 213–222.

Taher, H. and Al-Zuhair, S. (2017). The use of alternative solvents in enzymatic biodiesel production: a review. *Biofuels Bioprod. Biorefin.* 11: 168–194.

Tan, T., Lu, J., Nie, K. et al. (2010). Biodiesel production with immobilized lipase: a review. *Biotechnol. Adv.* 28: 628–634.

Tan, H.-T., Corbin, K.R., and Fincher, G.B. (2016). Emerging technologies for the production of renewable liquid transport fuels from biomass sources enriched in plant cell walls. *Front. Plant Sci.* 7: 1854.

Thauer, R.K., Jungermann, K., and Decker, K. (1977). Energy conservation in chemotrophic anaerobic bacteria. *Bacteriol. Rev.* 41: 100–180.

Thomas, V.A., Donohoe, B.S., Li, M. et al. (2017). Adding tetrahydrofuran to dilute acid pretreatment provides new insights into substrate changes that greatly enhance biomass deconstruction by *Clostridium thermocellum* and fungal enzymes. *Biotechnol. Biofuels* 10: 13.

UNICA, 2012. Strengthening U.S. - Brazil partnership: path to a global ethanol market. URL www.unica.com.br/news/2271118592034872045/strengthening-us-brazil-partnership-por-cento3A-path-to-a-global-ethanol-market

United Unions, 2016. Second-generation biofuel markets: state of play, trade and developing country perspectives. URL: https://unctad.org/en/pages/PublicationWebflyer.aspx?publicationid=1455

USDA, 2017. USDA agricultural projections to 2026. URL: https://www.usda.gov/oce/commodity/projections/USDA_Agricultural_Projections_to_2026.pdf

Vassilev, N., Malusa, E., Reyes Requena, A. et al. (2017). Potential application of glycerol in the production of plant beneficial microorganisms. *J. Ind. Microbiol. Biotechnol.* 44: 735–743.

Yao, C., Yang, X., Roy Raine, R. et al. (2009). The effects of MTBE/ethanol additives on toxic species concentration in gasoline flame. *Energy Fuels* 23: 3543–3548.

Zabed, H., Sahu, J.N., Suely, A. et al. (2017). Bioethanol production from renewable sources: current perspectives and technological progress. *Renewable Sustainable Energy Rev.* 71: 475–501.

Zhou, W., Wang, J., Chen, P. et al. (2017). Bio-mitigation of carbon dioxide using microalgal systems: advances and perspectives. *Renewable Sustainable Energy Rev.* 76: 1163–1175.

# Part II

# Biotechnological Research and Production of Food Ingredients

# 4

# Research, Development, and Production of Microalgal and Microbial Biocolorants

*Laurent Dufossé*

*Laboratoire de Chimie des Substances Naturelles et des Sciences des Aliments LCSNSA, Ecole Supérieure d'Ingénieurs Réunion Océan Indien ESIROI, Département Agroalimentaire, Université de La Réunion, F-97490 Sainte-Clotilde, France*

## 4.1 Introduction

Ingredients derived from microbial fermentation or extracted from algae are steadily gaining ground in the food, feed, and other industries. Thickening or gelling agents (e.g., polysaccharides such as xanthan, curdlan, gellan), flavor enhancers (yeast hydrolysate, monosodium glutamate), polyunsaturated fatty acids (PUFAs), flavor compounds (gamma-decalactone, diacetyl, methyl-ketones), vitamins, essential amino acids, and acidulants (lactic acid, citric acid) are some examples illustrating this trend. Efforts have been made and continue in order to reduce the production costs of pigments produced by algal ponds and microbial fermentation (Mannazzu et al. 2015), since synthetic pigments or those extracted from natural plant sources can often be produced more economically (Dufossé et al. 2005). The successful marketing of natural pigments such as β-carotene, lutein, and astaxanthin derived from algae (i.e. nonconventional sources) (Jaswir et al. 2011) or extracted from flowering plants (conventional sources), both as food colorants and nutritional supplements (Zhou et al. 2015), reflect the presence and importance of niche markets in which consumers are willing to pay a premium for "natural healthy ingredients." Among other nonconventional sources, filamentous fungi are known to produce an extraordinary range of pigments that include several chemical classes such as carotenoids, melanins, azaphilones, flavins, phenazines, quinones, and more specifically, monascins, violacein, and indigo (Dufossé 2006). The success of any class of pigment produced by fermentation depends on its acceptance by the consumers, regulatory approval, and the capital investment required bringing the product onto the market (Furubayashi et al. 2015; Heider et al. 2014a,b). Twenty years ago, influential representatives from the food industry expressed doubts about the successful commercialization of algae-derived and fermented food grade pigments due to the high investment required for open ponds, photo-bioreactors and fermentation facilities, and the extensive and lengthy toxicity studies required by the regulatory authorities. Poor public perception of fungal-derived products for food use had also to be taken into account. Nowadays, some food grade pigments obtained by fermentation are present on the market worldwide. Among them, fungal *Monascus* pigments, Arpink red™ (now Natural Red™) produced by *Penicillium oxalicum*, microalgal phycocyanin

*Bioprocessing for Biomolecules Production*, First Edition.
Edited by Gustavo Molina, Vijai Kumar Gupta, Brahma N. Singh, and Nicholas Gathergood.

from *Arthrospira (Spirulina) platensis*, riboflavin from the mold fungus *Ashbya gossypii*, lycopene and β-carotene from the tropical mold *Blakeslea trispora* (Chen et al. 2016), β-carotene from the microalgae *Dunaliella salina*, and astaxanthin from the bacterium *Paracoccus carotinifaciens* and microalgae *Haematococcus pluvialis* (Grassi et al. 2016), respectively. As an example, the production yield of β-carotene may be as high as $17\,g\,l^{-1}$L of the *B. trispora* culture medium (Dufossé et al. 2014).

Thus, the present chapter emphasizes the crucial role that microorganisms and microalgae are currently playing and are likely to continue to play in future as microbial cell factories for the production of food grade pigments for the industry. This is due to the versatility in their pigment color and chemical profile (Figure 4.1), amenability for easy large-scale cultivation, and a long history of production by well-investigated production strains (Figure 4.2).

In this chapter, attention of both academia and the food industry is drawn toward some stimulating findings in the research of microbial and algal pigments considering various approaches from simple screening (Lebeau et al. 2017; Fouillaud et al. 2017) up to molecular biology.

## 4.2 Carotenoids

### 4.2.1 Lutein and Zeaxanthin

Lutein, being a member of the carotenoid class, namely the xanthophylls subclass, is a yellow pigment occurring in higher plants and other photoautotrophic organisms like algae (Berman et al. 2014). Lutein is used as a feed additive to brighten the colors of bird feathers and standardize the yellow of egg yolk and broilers (Silva et al. 2016). Lutein is also applied to add color to foods. Additionally, lutein, being an effective functional nutrient, may add benefit to human health regarding cardiovascular diseases, cancers, cognitive function, and age-related macular degeneration (AMD). The worldwide market for lutein is expected to grow to US$308 million by 2018 (Lin et al. 2014). Marigold (*Tagetes erecta* L.) flowers are a conventional lutein source; however, they must be harvested periodically, and extraction is a seasonal and labor-intensive process. Therefore, several microalgae have been proposed as potential alternative sources of lutein (Lin et al. 2014), due to their high lutein content. Other advantages of lutein production from microalgae include the following:

1. Microalgae are a cheap and effective bioresource that can be used to produce value-added compounds, including fine chemicals, vitamins, carotenoids, and polysaccharides.
2. Growth rates of microalgae are about 5–10 times higher than of higher plants.
3. Microalgae, which can be cultivated in seawater or brackish water and on nonarable land, do not compete for resources with conventional agricultural crops.
4. Microalgal biomasses can be harvested all year round in many parts of the world (Lin et al. 2014).

The following lutein contents converted to $g\,kg^{-1}$ biomass have been reported for various microalgal species (Lin et al. 2014): *Chlorococcum citroforme* (7.4), *Tetracystis aplanosporum* (5.9), *Coelastrum proboscideum* (3.4–5.0), *Chlorella fusca* (4.2–4.7),

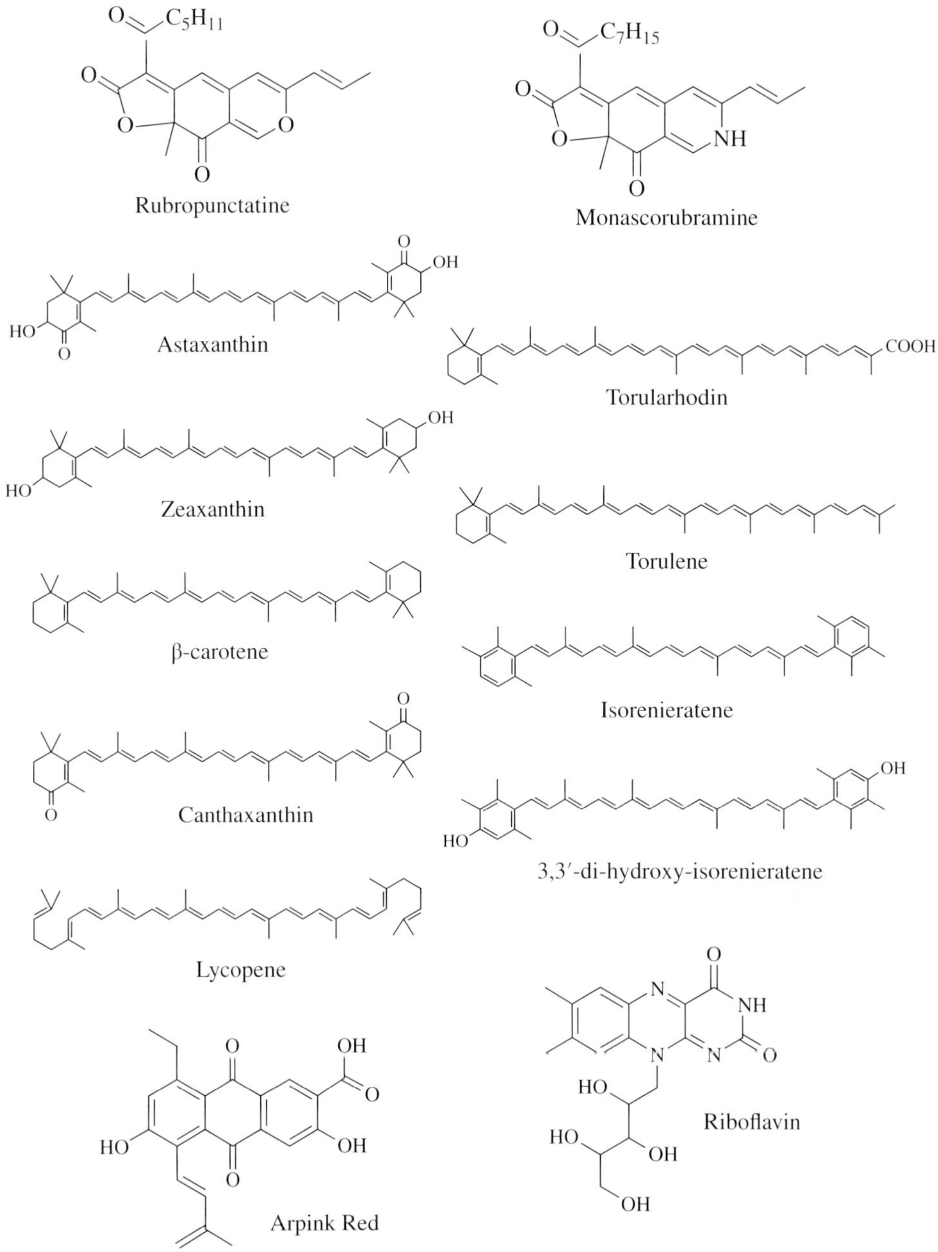


**Figure 4.1** Chemical formulae of some microbial food-grade pigments.

*Neospondiococcum gelatinosum* (4.4), *Tetracystis tetrasporum* (4.4), *Tetracystis intermedium* (3.5), *Chlorella zofingiensis* (2.4–2.8), *Muriella aurantiaca* (2.6), *Muriella decolor* (0.5). Further species yielding high lutein contents have been identified such as *Chlorella protothecoides, Chlorella vulgaris, Scenedesmus obliquus* and other species, as well as *Desmodesmus* sp. A reasonable estimate for lutein yields in microalgal biomass is

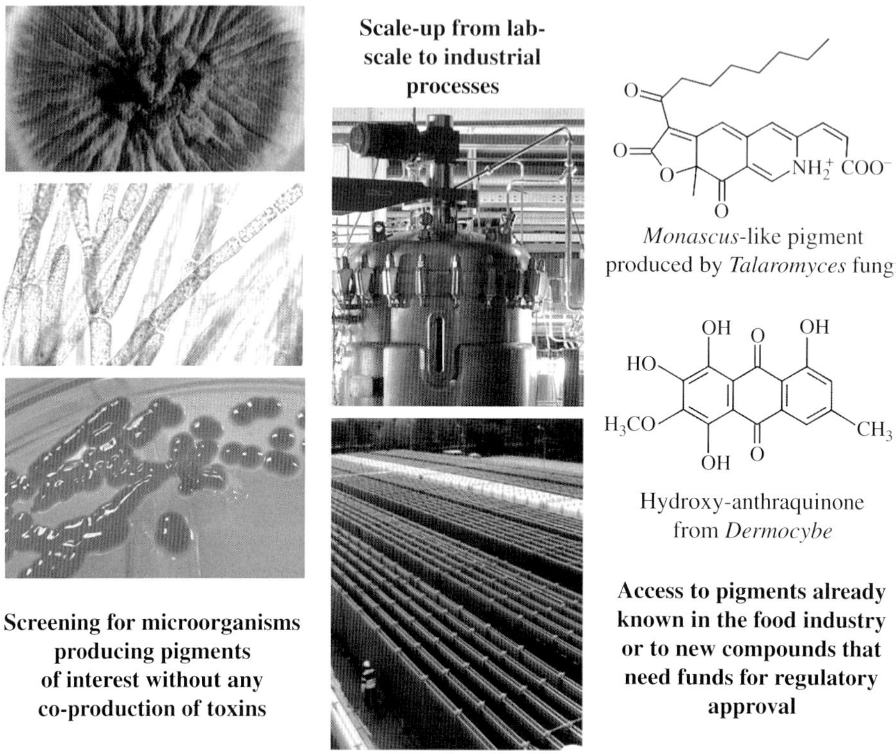


**Figure 4.2** Using microalgae, fungi, yeasts, and bacteria as cell factories for the production of natural food pigments.

$5\,g\,kg^{-1}$, mostly in the free form, while, in plants, lutein and zeaxanthin concentrations from marigold flowers amount to 15 and $2.5\,g\,kg^{-1}$ biomass, respectively, mostly in the ester form (Guedes et al. 2011; Lin et al. 2014).

The second xanthophyll of interest, the yellow pigment known as zeaxanthin or 3,3′-dihydroxy-β-carotene, can be used, for example, as an additive to poultry feed to intensify the yellow color of the broiler and accentuate the color of the egg yolk. In addition to its application as a feed additive or food colorant, this xanthophyll is also suited for the prevention of AMD, e.g., in nutraceuticals and as dietary supplements in the food industry.

In the mid-1960s, scientists at Hoffmann-La Roche isolated several marine bacteria that produced zeaxanthin. *Flavobacterium* sp. (ATCC 21588, classified under the accepted taxonomic standards of that time) when cultured in a nutrient medium containing glucose or sucrose, sulfur-containing amino-acids (such as methionine, cystine, or cysteine), pyridoxine, and bivalent metal ions selected from the group consisting of $Fe^{2+}$, $Co^{2+}$, $Mo^{2+}$, or $Mn^{2+}$ were able to produce up to 190 mg of zeaxanthin per liter, with a specific concentration of $16\,mg\,g^{-1}$ dried cellular mass (US Patent 5,827,652). *Flavobacterium multivorum* (ATCC 55238) is currently under investigation (Thawornwiriyanun et al. 2012).

According to taxonomic reclassification, the zeaxanthin-producing *Flavobacterium* has been grouped into the new genus *Paracoccus*, namely *Paracoccus zeaxanthinifaciens*.

This may become an industrial product in the future, as previously observed for astaxanthin Panaferd-AX™ from *P. carotinifaciens,* which has been developed by JX Nippon Oil and Energy for imparting color to animal feed (US Patent 8,853,460 and US Patent application 20130012594). A second strain was isolated in a mat from an atoll of French Polynesia (Raguenes et al. 2004). This strain is peculiar, as it also produces exopolysaccharides.

Another zeaxanthin-producing bacterium, *Sphingobacterium multivorum,* was recently shown to utilize the deoxyxylulose-5-phosphate (DXP) pathway. For overproduction of zeaxanthin in large quantities, a strain was constructed (US Patent 5,308,759). Since more bacteria are recently being examined for the two different pathways of isopentenyl pyrophosphate (IPP) biosynthesis (Lange et al. 2000; Calegari-Santos et al. 2016), the distribution of the mevalonate and DXP pathways will be better understood. Furthermore, identification of genes involved in IPP biosynthesis lends itself to metabolic engineering of microorganisms having improved productivity of commercially important isoprenoid compounds, including carotenoids.

### 4.2.2 Aryl Carotenoids (Orange Colors and Highly Active Antioxidants) Are Specific to Some Microorganisms

Aryl carotenoids such as isorenieratene and its hydroxyl derivatives (Figure 4.3) are produced by a very small number of microorganisms. Some are anaerobic and photosynthetic (Chlorobiaceae, Chromatiaceae). Being easier to cultivate, nonphotosynthetic bacteria containing aryl carotenoids such as *Brevibacterium linens* (Figure 4.4), *Streptomyces mediolani*, and *Mycobacterium aurum* are of high interest. All mentioned examples belong to the Actinomycetales, an order of Actinobacteria (Dufossé and de Echanove 2005).

The main interest in *B. linens* is that this bacterium is part of the microbiota of smear-ripened cheeses. This may facilitate overcoming regulatory obstacles of using

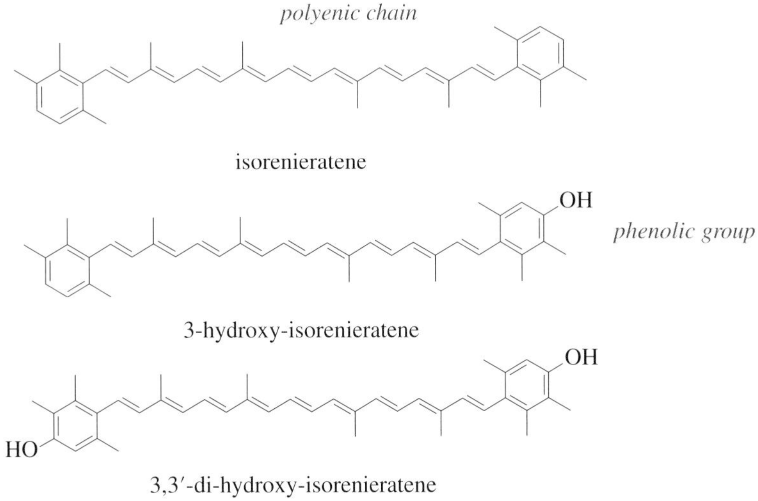


**Figure 4.3** Aryl carotenoids described in *Brevibacterium linens.*

**Figure 4.4** Production of aryl carotenoids by *Brevibacterium linens* on (a) a Petri dish; and (b) on Livarot cheeses.

the carotenoids derived from their bacterial biomass as food ingredients. However, up to now, levels of production in liquid media have been limited to 0.50–1.0 $mg\,l^{-1}$. This concentration of aryl carotenoids is too low to envisage their industrial production (Guyomarc'h et al. 2000).

When comparing the antioxidant potential of carotenoids including astaxanthin, β-cryptoxanthin, zeaxanthin, lycopene, lutein, and the microbial 3,3′-dihydroxyisorenieratene (DHIR), the latter proved to be superior, ranking "top of the list" regarding the radical scavenging and singlet oxygen quenching activity (Wagener et al. 2012). The experimental data obtained based on various antioxidant assays suggest that DHIR acts as a bifunctional radical scavenger attributed to the polyenic and phenolic substructures (Martin et al. 2009).

Aryl carotenoids have been studied for some decades, starting with the first patent in 1971 (US patent 3,598,702). Using *B. linens* as a food-grade bacterium for microbial production, the concentration of aryl carotenoids in the culture medium is not sufficient to allow their commercialization at a competitive price. Consequently, for the short and mid-term, chemical synthesis may be a required alternative route. The German company BASF as well as academic researchers have now developed novel syntheses (Valla et al. 2003); however, aryl carotenoids such as isorenieratene or DHIR are not yet available for large-scale applications. Genetical modifications of producing organisms could also be an alternative to obtain affordable polyenic-phenolic carotenoids (US patent 7 186 523), which proved superior antioxidant activity, thus being of interest as bioactive food ingredients (Patent application WO 2008009675).

### 4.2.3 $C_{50}$ Carotenoids (Sarcinaxanthin, Decaprenoxanthin)

In nature, $C_{50}$ carotenoids are exclusively biosynthesized by bacteria belonging to the *Actinomycetales*. Until now, only three different $C_{50}$ carotenoid biosynthetic pathways have been described:

1. β-cyclic $C_{50}$ carotenoid C.p. 450 pathway in *Dietzia* sp. CQ4; C.p. letters are used, as this carotenoid was first isolated in *Corynebacterium* ***p****oinsettiae*.
2. γ-cyclic $C_{50}$ carotenoid sarcinaxanthin pathway in *Micrococcus luteus* NCTC2665.

3. ε-cyclic $C_{50}$ carotenoid decaprenoxanthin pathway in *Corynebacterium glutamicum*.

More than 95% of all natural carotenoids originate from the symmetric $C_{40}$ phytoene backbone, and only a small number of $C_{30}$ and even fewer $C_{50}$ carotenoids have been discovered in nature. All known $C_{50}$ carotenoids have multiple conjugated double bonds, at least containing one hydroxyl group; both of these features contribute to their strong antioxidant properties. Therefore, these carotenoids, which are only produced by microorganisms, may be suitable for food coloring as well as for nutraceutical and pharmaceutical applications (Heider et al. 2014a,b).

### 4.2.4 Techniques for the Production of Novel Carotenoids with Improved Color Strength/Stability/Antioxidant Properties

In the field of pigment research, the work on carotenoids is most advanced, and cutting-edge techniques have been employed to increase the level of production and to produce novel new carotenoids with improved properties (Castelblanco-Matiz et al. 2015; Chang et al. 2015; Chi et al. 2015).

As an example, directed evolution involves the use of rapid molecular manipulations to mutate the target DNA fragment, followed by a selection or screening process to isolate desirable mutants. In this manner, several carotenogenic enzymes have been improved or optimized for a specific condition using various directed evolution protocols. Directed evolution has been applied to geranylgeranyl diphosphate GGPP synthase from *Archaeoglobus fulgidus*, the rate-controlling enzyme, to enhance the production of carotenoids in metabolically engineered *Escherichia coli*. The library of mutated genes that has been created using these tools was transformed into an *E. coli* strain containing the reconstructed isoprenoid pathway (Jin et al. 2015), and screened for color intensity to optimize the conversion of glucose into astaxanthin or lycopene. In summary, the study included more than 10 000 colonies with eight mutants identified and obtained. The production of lycopene was almost doubled. In addition to the increased production of carotenoids, the directed evolution approach allowed the identification of hot regions that are important for either protein expression or specific activity (Schmidt-Dannert et al. 2000).

A second example refers to the membrane-associated carotenoid synthase that appears to be the major point of control over product diversity. The carotenoid synthase was engineered to accept longer diphosphate substrates and produced synthase variants that can synthesize previously unknown $C_{45}$ and $C_{50}$ carotenoid backbones from the appropriate $C_{20}$ and $C_{25}$ isoprenyl diphosphate precursors. In a previous work, a $C_{35}$ carotenoid backbone was also biosynthesized, and various downstream enzymes (desaturases and cyclases) from the $C_{30}$ and $C_{40}$ carotenoid pathways were functional on this non-natural substrate, which led to the production of a series of novel carotenoids. Thus, it appears that once a carotenoid backbone structure is created, downstream enzymes, either natural or engineered, can accept the new substrate, and a whole series of novel carotenoids can be produced. With the action of carotenoid-modifying enzymes, including desaturases, cyclases, hydroxylases, and cleaving enzymes, on these new backbones, it should be possible to double or even triple the diversity of the carotenoid class comprising about 750 structures up to now.

A different approach to expand the recombinant production of known carotenoids and to synthesize completely novel structures is to engineer novel pathways by combining available biosynthetic genes and evolving new enzyme functions through random mutagenesis, recombination (DNA-shuffling), and selection (Wang et al. 2007). Prerequisites for this approach are that carotenogenic Crt enzymes from different species can function cooperatively in a heterologous host and display enough promiscuity regarding the structure of their substrates (Dufossé 2009). With a few exceptions, such as, for example, zeaxanthin C(5,6) epoxidase, almost all cloned carotenoid biosynthetic genes are functionally expressed in *E. coli*. The success of functional color expression in transgenic *E. coli* for the cloning of a number of carotenoid biosynthesizing genes demonstrates that enzymes from phylogenetically distant species can assemble into a functional membrane-bound multi-enzyme complex at which carotenoid biosynthesis takes place. Both phytoene desaturase (crtI) and lycopene cyclase (crtY) genes have been targeted for *in vitro* evolution to achieve synthesis of novel carotenoids in *E. coli* (Song et al. 2013). A variant enzyme, a desaturase chimera, efficiently catalyzed the extended desaturation of the linear C40 carotenoid pathway, introducing six rather than four double bonds into phytoene, thus enabling the production of the fully conjugated carotenoid, 3,4,3′,4′-tetradehydrolycopene (Song et al. 2013).

A related strategy bypasses the shuffling step. Instead, carotenogenic genes from different bacteria are combined employing unique pathways that maintain altered product specificities in a simple *E. coli* host expressing the biosynthetic machinery for phytoene production. In conjunction with the four-step phytoene desaturase that yields lycopene, a five-step desaturase was used to produce 3,4-didehydrolycopene (Zhang et al. 2012). Further diversification of the $C_{40}$ skeleton using a 1,2-hydratase and a C-3,4 desaturase yielded a range of carotenoids, including acyclic hydroxyl-containing products (1-hydroxy-3,4,3′,4′-tetrahydrolycopene; 1,1′-dihydroxy-3,4-didehydrolycopene; 1′,3-dihydroxy-γ-carotene) with significant antioxidant activity.

Recent achievements in metabolic engineering of *E. coli* (Kim et al. 2016; Li et al. 2015; Takemura et al. 2015) and yeast strains resulted in a significant increase in the productivity of isoprenoid compounds such as carotenoids, by increasing the metabolic flux toward the isoprenoid precursors. However, decades after the beginning of research in this field (i.e., carotenogenic gene cluster from the bacteria *Erwinia herbicola* described in 1991), neither industry nor consumers observed a surge or even the appearance of "engineered" carotenoids on the market, which was disappointing after the considerable investment in time and funding.

## 4.3 Azaphilones

### 4.3.1 Toward Mycotoxin-Free *Monascus* Red

*Monascus* has been used to produce natural colorants and food supplements for more than 1000 years in Asia, and around 1 billion Asian people consume *Monascus*-fermented products as part of their daily diet. The first known source reporting the use of these red colorants was a recipe for the preparation of red pot-roast lamb, in which meat was simmered with hong qu (red rice koji, made with *Monascus purpureus*), as handed down in the Qing Yilu in 965 CE. *Monascus* species are known to produce six major azaphilone pigments, namely the yellow monascin and ankaflavin, the orange monascorubrin and rubropunctatin, and the red monascorubramine and rubropunctamine. To date, more than 50 different chemical structures have been identified (Yang et al. 2015), as azaphilones easily combine with nucleophilic nitrogen-containing compounds.

Using next-generation sequencing and optical mapping approaches, a 24.1-Mb complete genome of a *M. purpureus* YY-1 industrial strain has been described for the first time, and this will allow major improvements in the process in the coming years (Yang et al. 2015). It consists of eight chromosomes and 7491 genes. *M. purpureus* belongs to the Aspergillaceae, mainly comprising the genera *Monascus, Penicillium,* and *Aspergillus.* Phylogenetic analysis at the genome level provides the first comprehensive prediction of the biosynthetic pathway for *Monascus* pigments. Comparative genomic analyses demonstrated that the genome of *M. purpureus* is 13.6–40% smaller than that of closely related filamentous fungi and has undergone significant gene losses, most of which likely occurred during its specialized adaptation to starch-based foods. Some polyketide synthases (PKS) are expressed at high levels under high pigment-yielding conditions. The citrinin PKS C6.123 has also been found in the genome (Yang et al. 2015), paving the way for research aiming at non-mycotoxin producing strains, if suppression of the citrinin gene does not change the ability of the strain to produce pigments, which seems to be feasible, as described by Fu et al. (2007). The latter group has shown that monascorubrin and citrinin are synthesized by two separate pathways, because, when the PKS gene responsible for synthesis of citrinin was disrupted, red pigment production from the fungus was not affected. Comparative transcriptome analysis revealed that carbon starvation stress, resulting from the use of relatively low-quality carbon sources, contributed to the high yield of pigments by suppressing central carbon metabolism and augmenting the acetyl-CoA pool. As for other pigments produced by biotechnology, the problem is to have enough carbon oriented in the correct pathway, i.e. the pigment pathway.

Woo et al. (2014) investigated another filamentous fungus, *Penicillium marneffei,* for production of azaphilones exhibiting black, yellow, and red hues. The polyketide gene cluster and biosynthetic pathway were reported for monascorubrin in this red pigment-producing, thermal dimorphic fungus, taking advantage of available genome sequence and faster growth rate compared to *Monascus* species (Woo et al. 2014). The red pigment of *P. marneffei* consists of a mixture of more than 16 chemical compounds, which are amino acid conjugates of monascorubrin and rubropunctatin. This occurs as amino acids can be conjugated under specific conditions without enzymatic catalysis, i.e., by Schiff base formation (Figure 4.5) (Woo et al. 2014).

The aforementioned polyketide gene cluster and pathway have been shown to also be responsible for the biosynthesis of ankaflavin and citrinin, the latter being a mycotoxin exerting nephrotoxic activity in mammals (Kumar et al. 2014). Twenty-three putative PKS genes and two putative PKS-non-ribosomal peptide synthase hybrid genes were identified in the *P. marneffei* genome (Woo et al. 2014). Woo et al. (2014) systematically knocked out all 25 PKS genes of *P. marneffei.* They also knocked out genes located up and downstream of the PKS gene responsible for red pigment production and characterized the pathway for biosynthesis of the red pigment. However, it is still questionable whether it will be possible to produce mevinolin/lovastatin-free (a cholesterol-lowering drug that is undesired in normal foods) and citrinin-free red pigments from *P. marneffei,* as the latter, a mycotoxin, appears to be an early byproduct of the biosynthetic pathway.

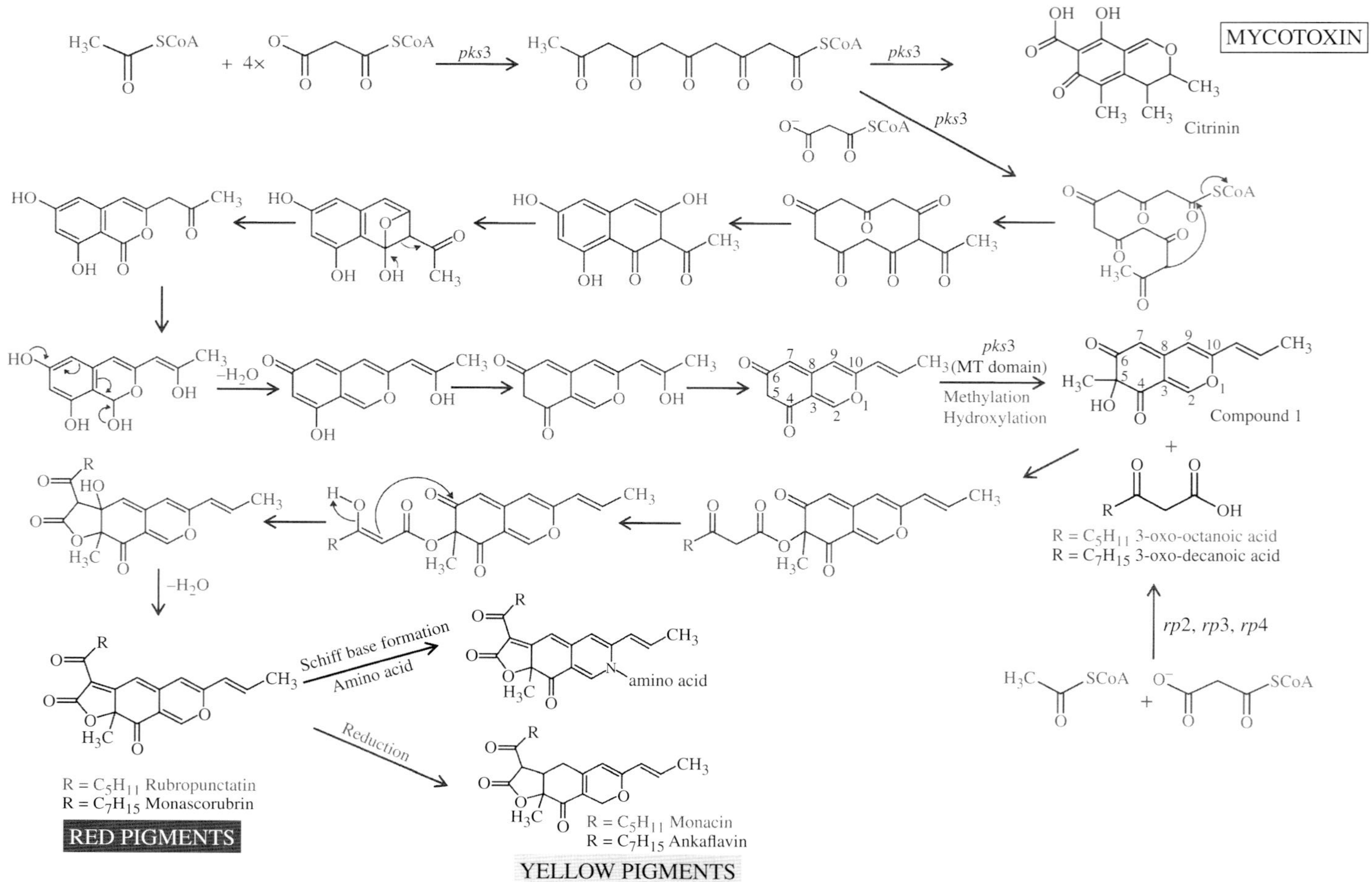


**Figure 4.5** Hypothetical pathway of monascorubrin, ankaflavin, and citrinin biosynthesis in *P. marneffei*. Source: Adapted from Woo et al. (2014).

### 4.3.2 *Monascus*-Like Pigments from Nontoxigenic Fungal Strains

Some species of *Talaromyces* secrete large amounts of red pigments. In literature, this biosynthetic potential has been linked to species such as *Talaromyces purpurogenus, Talaromyces marneffei,* and *Talaromyces minioluteus,* often known under their previous *Penicillium* names. However, since some of them do not exert sufficient stability for pigment production, then such species should be avoided for scale-up production.

Isolates identified as *T. purpurogenus* have been reported to be of industrial interest, as they can produce extracellular enzymes and red pigments. Some of these isolates may also produce unwanted mycotoxins such as rubratoxin A and B and luteoskyrin, in addition to other undesirable extrolites that may be toxic following intraperitoneal (spiculisporic acid) and intravenal (rugulovasine A and B) injections in cats (Frisvad et al. 2004). Consequently, mycotoxin production may limit the use of isolates of a particular species in biotechnology, and Frisvad et al. (2013) concluded that *T. purpurogenus* may thus not be recommended for industrial production of red pigments. *Talaromyces atroroseus* sp. nov., described by the same group, produces the azaphilone biosynthetic families mitorubrins and *Monascus* pigments without being accompanied by mycotoxin synthesis (patent applications WO2012022765, US 20110250656) (Figure 4.6). *Talaromyces albobiverticillius* strain 30548 isolated from Réunion Island lagoon, Indian Ocean, is also under investigation by our group from the University of Réunion Island based on this tropical volcano, located between Madagascar and Mauritius. As it has been found for *Monascus,* these azaphilone pigments may react with amino groups containing compounds, to which reaction they owe their name, providing intense dark red colors (Mapari et al. 2010; Gao et al. 2013).

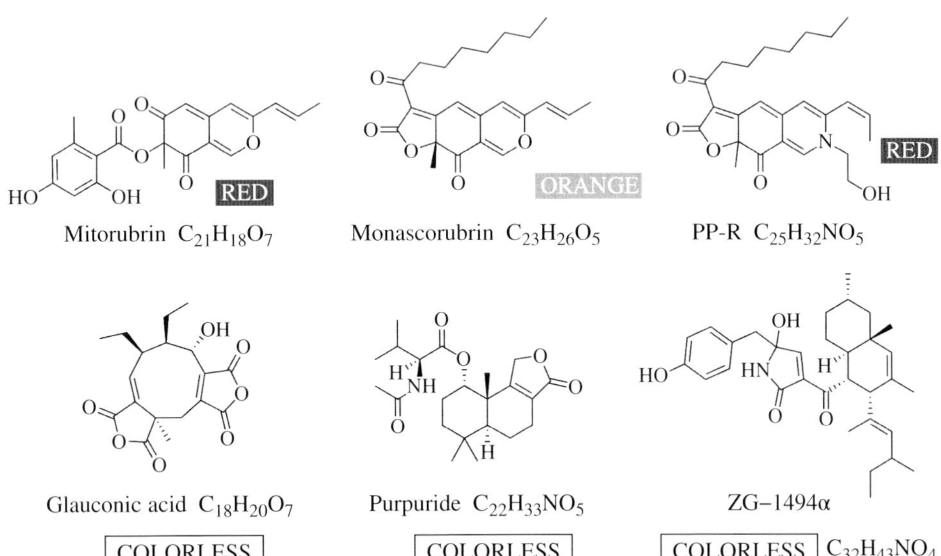


**Figure 4.6** Structures of some of the most characteristic compounds produced by *Talaromyces atroroseus*. Source: Adapted from Frisvad et al. (2013).

## 4.4 Anthraquinones

### 4.4.1 Fungal Natural Red

As the first fungal commercial product claimed as a member of the anthraquinone pigment class, the natural red food colorant Arpink red (now Natural Red) has been manufactured by two Czech companies. Initially by Ascolor Biotech s.r.o., then followed by Natural Red, who both produced the pigment by fermentation and bioprocess engineering using the fungal strain *P. oxalicum* var. Armeniaca CCM 8242, a soil isolate. Numerous patents have been filed by Ascolor, e.g., WO 9,950,434; CZ 285,721; EP 1,070,136; US 6,340,586 cited in Sardaryan et al. (2004).

The cultivation of the fungus in liquid broth requires carbohydrates (such as sucrose and molasses), nitrogen (corn extract, yeast autolysate or extract), zinc sulfate, and magnesium sulfate. The optimum conditions for performing the microbiological synthesis are pH value in the range of 5.6–6.2, and temperature 27–29 °C. On the second day of incubation, the red colorant is released into the broth, increasing up to 1.5–2.0 $g\,l^{-1}$ of broth after 3–4 days (WO 9,950,434; CZ 285,721; EP 1,070,136; US 6,340,586 cited in Sardaryan et al. 2004).

After biosynthesis of the red colorant is completed, the liquid phase is filtered or centrifuged, and then separated from the biomass. The liquid is acidified to pH 3.0–2.5 to precipitate the colorant. The precipitate is dissolved in ethyl alcohol, and filtered. Following removal of alcohol from the filtrate, the microcrystalline form of the colorant is obtained as a dark red powder.

The colorant produces a raspberry-red color in aqueous solution, stable at $pH > 3.5$ (4–60 °C). Solutions at pH 7.0 are even stable after 30 minutes of boiling.

Toxicological data are widely available about this red pigment. This includes patents containing information about acute oral toxicity in mice 90-day subchronical toxicological study, acute dermal irritation, acute eye irritation, anti-tumor activity, micronucleus test in mice, AMES test (*Salmonella typhimurium* reverse mutation assay), estimation of antibiotic activity, and test results of estimation of five mycotoxins (WO 9,950,434; CZ 285,721; EP 1,070,136; US 6,340,586 cited in Sardaryan et al. (2004)).

After evaluating all the documents provided by the company, the Codex Alimentarius Commission made the following statement on the occasion of its Rotterdam meeting on March 11–15, 2002: "…there will not be any objections to use the red colouring matter Arpink Red" in:

- Meat products and meat product analogues in the amount up to 100 $mg\,kg^{-1}$
- Nonalcoholic drinks in the amount up to 100 $mg\,kg^{-1}$
- Alcoholic drinks in the amount up to 200 $mg\,kg^{-1}$
- Milk products and ice creams in the amount up to 150 $mg\,kg^{-1}$
- Confectionery in the amount up to 300 $mg\,kg^{-1}$

Subsequently, this biotechnologically produced anthraquinone was sold and used in Czech Republic for several years. The joint FAO/WHO Expert Committee on Food Additives (JECFA) evaluation process made some progress, and the legal situation concerning Arpink Red was discussed during the 63rd Annual JECFA meeting in Geneva, June 8–17, 2004. Additional data were requested; however, the company Ascolor appeared to stop its activities and a new company, named Natural Red, was established in 2012.

Pros and cons are quite difficult to judge in this case, as private companies using a fungal strain that is not publicly available have conducted the whole development. No academic paper has been published, and much information, in particular, confirmation of genus/species of the fungal strain, chemical structure of the anthraquinone pigment(s), and absence of mycotoxins (e.g. secalonic acid D) is lacking.

### 4.4.2 Other Fungal Anthraquinones

Anthraquinones are widely spread throughout the kingdom of fungi (For example, in *Aspergillus* sp., *Eurotium* sp., *Fusarium* sp., *Dreschlera* sp., *Penicillium* sp., *Emericella purpurea*, *Curvularia lunata*, *Mycosphaerella rubella*, *Microsporum* sp., etc.), and thus might serve as alternative sources independent of agroclimatic conditions. (Caro et al. 2012; Gessler et al. 2013). This is in contrast to plant- and animal-derived sources.

Anthraquinones exhibit a broad range of biological activities, including bacteriostatic, fungicidal, antiviral, herbicidal, and insecticidal effects (Gessler et al. 2013). Presumably, in fungi, these compounds are involved in interspecific interactions. For example, anthraquinones synthesized by endophytic fungi protect the host plant from insects or other microorganisms (Gessler et al. 2013).

The current picture of fungal anthraquinones is quite complex, with a great variety of chemical structures (Figure 4.7) and a huge number of factors or parameters that could affect the composition of quinoidal pigments biosynthesized by a particular species. Among them, e.g., habitat, light, pH, temperature, $O_2$ transfer, liquid/solid media, culture medium, C and N sources, C : N ratio, presence of organic acids, mineral salts, and inoculum have been considered (Caro et al. 2012).

Today, research priority is focused on a small number of fungal anthraquinone-producing species meeting the following profile of requirements established by Mapari et al. (2009) during the identification of potentially safe fungal cell factories for the production of polyketide natural food colorants using chemotaxonomic rationale:

- Fungus shall be nonpathogenic to humans.
- Fungus shall be nontoxigenic under a broad range of production conditions.
- Fungus shall be able to produce in liquid media.

## 4.5 Phycobiliproteins

Phycobiliproteins (PBPs) belong to a group of light-harvesting proteins, which are covalently bound with bilin of the tetrapyrrole prosthetic groups. PBPs are water-soluble chromoproteins present in microalgae belonging to the Rhodophyceae or Cryptophyceae and Cyanophyceae (cyanobacteria). According to their light absorption and types of bilins, PBPs are commonly divided into four subclasses: phycoerythrins (PE, pink-purple, $\lambda$max = 540–570 nm), phycocyanins (PC, blue, $\lambda$max = 610–620 nm), phycoerythrocyanins (PEC, orange, $\lambda$max = 560–600 nm), and allophycocyanins (APC, bluish-green, $\lambda$max = 650–655 nm) (Kuddus et al. 2013). PBPs assemble to form supramolecular complexes called phycobilisomes, contributing to light harvesting and energy transfer. The pigments serve as solar energy collectors in the range beyond chlorophyll *a* (430 nm) absorption. PBPs are generally formed of two

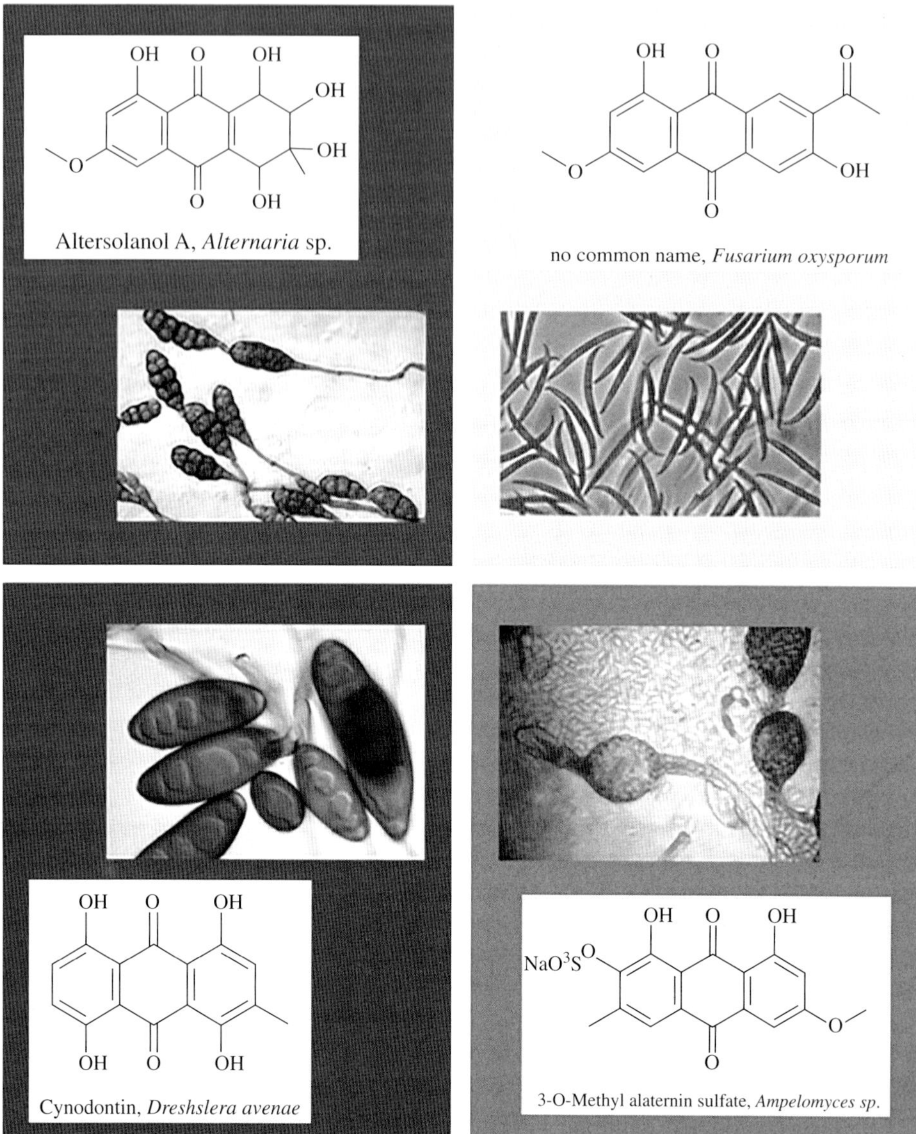


**Figure 4.7** Some anthraquinones from fungal origin (color of the box reflects color of the main pigment produced by the fungus).

chromophore-linked subunits, α and β, and found in trimeric (αβ)*3 (MW ~ 120 kDa) or hexameric units (αβ)*6 (~240 kDa) (Eriksen 2008).

Some authors suggested that phycoerythrin PE may be used as a natural food colorant, by analogy to phycocyanin, exerting "remarkable and extraordinary" performance due to its fluorescent properties – imagine a PE-colored cocktail at night in a bar with fluorescent light! Nowadays, commercial PE is obtained from mesophile rhodophyta belonging to the genus *Porphyridium*. However, PBPs from mesophile organisms are sensitive to heat, thus displaying low stability at high temperatures. Consequently, Pumas et al. (2012) explored hot spring cyanobacteria such as the red-violet cyanobacterium *Leptolyngbya* sp. producing a PE that degrades less than 20% after being incubated at 60 °C for 30 minutes.

## 4.6 Conclusion

As the trend in the food and beverage markets is toward more natural, organic, and clean label products, the demand for a wide choice of natural ingredients is increasing. The trend toward the formulation of recipes containing natural colors has steadily increased. Microbial and microalgal colorants are constituents of commercial products available for the food industry as a natural choice available from many sources. They are either based on alternative production techniques of well-known pigments (e.g. β-carotene and lycopene) or specific molecules so far not biosynthesized by other organisms such as higher plants (e.g., aryl carotenoids, *Monascus* and *Monascus*-like azaphilones) (Table 4.1).

Use of natural colors in functional, beverage, food, and crossover applications requires an understanding of a variety of attributes and concepts, including heat and light stability, in addition to being able to provide exciting color hues. Microbial

**Table 4.1** Microbial and algal production of pigments (already in use as natural food colorants or with high potential in this field).

| Molecule | Color | Microorganism | Status[a)] |
|---|---|---|---|
| Ankaflavin (azaphilone) | Yellow | *Monascus* sp. (fungus)<br>(many companies based in Asia) | IP |
| Anthraquinones | Red and other hues | *Penicillium oxalicum* (and many other fungi)<br>(company Natural Red, chemical structure and strain should be confirmed) | IP |
| Astaxanthin | Pink-red | *Haematococcus pluvialis* (microalgae)<br>(company AstaReal, among many others) | IP |
| Astaxanthin | Pink-red | *Xanthophyllomyces dendrorhous* (yeast), formerly *Phaffia rhodozyma*<br>(company Ajinomoto, among many others) | DS |
| Astaxanthin | Pink-red | *Agrobacterium aurantiacum* (bacteria) | RP |
| Astaxanthin | Pink-red | *Paracoccus carotinifaciens* (bacteria)<br>(company JX Nippon Oil & Energy) | IP |
| Azaphilones | Red | *Talaromyces atroroseus* (fungus)<br>(patents from Denmark Technical University, in close relationship with Chr. Hansen) | DS |
| Azaphilones | Red | *Penicillium purpurogenum* (fungus) | DS |
| Azulenes | Blue | *Lactarius* sp. (fungus) | RP |

**Table 4.1** (Continued)

| Molecule | Color | Microorganism | Status[a)] |
|---|---|---|---|
| β-carotene | Yellow-orange | *Dunaliella salina* (microalgae) (company Henkel-Cognis Australia) | IP |
| β-carotene | Yellow-orange | *Blakeslea trispora* (fungus) (company DSM Nutritional Products) | IP |
| β-carotene | Yellow-orange | *Fusarium sporotrichioides* (fungus) | RP |
| β-carotene | Yellow-orange | *Mucor circinelloides* (fungus) | DS |
| β-carotene | Yellow-orange | *Neurospora crassa* (fungus) | RP |
| β-carotene | Yellow-orange | *Phycomyces blakesleeanus* (fungus) | RP |
| Canthaxanthin | Dark red | *Bradyrhizobium* sp. (bacteria) | RP |
| Chlorophylls | Green | Many microalgae | DS |
| Indigoidine | Blue | *Erwinia chrysanthemi, Streptomyces lavendulae* (bacteria) | RP |
| Isorenieratene and OH derivatives (aryl carotenoids) | Orange | *Brevibacterium linens* (bacteria) | DS |
| Lutein | Yellow | *Chlorella* and many other microalgae | IP |
| Lycopene | Red | *Blakeslea trispora* (fungus) (company DSM Nutritional Products) | IP |
| Lycopene | Red | *Fusarium sporotrichioides* (fungus) | RP |
| Melanin | Black | *Cryptococcus neoformans* var. *nigricans* (yeast) | RP |
| Monascorubramin (azaphilone) | Red | *Monascus* sp. (fungus) | IP |
| Naphtoquinone | Deep blood-red | *Cordyceps unilateralis* (fungus) | RP |
| Phenazines | Broad range of hues | *Pseudomonas* and *Streptomyces* sp. (bacteria) | RP |
| Phycocyanin | Blue | *Arthrospira* sp. *(formerly Spirulina* sp.*)* and many other microalgae and cyanobacteria | IP |
| Phycoerythrin | Red | *Porphyridium cruentum* and many other microalgae and cyanobacteria | DS |
| Prodigiosins | Red | *Serratia marcescens* (and many other bacteria) | DS |
| Riboflavin | Yellow | *Ashbya gossypi* (fungus), *Candida famata* (yeast), *Bacillus subtilis* (bacteria) | IP |
| Rubrolone | Red | *Streptomyces echinoruber* (bacteria) | DS |
| Rubropunctatin (azaphilone) | Orange | *Monascus* sp. (fungus) | IP |
| Torularhodin | Orange-red | *Rhodotorula* sp. (yeast) | DS |
| Violacein | Deep violet | *Chromobacterium violaceum, Janthinobacterium lividum* (bacteria) | DS |
| Zeaxanthin | Yellow | *Flavobacterium* sp. (bacteria) | DS |
| Zeaxanthin | Yellow | *Paracoccus zeaxanthinifaciens* (bacteria) | RP |
| Unknown | Red | *Paecilomyces sinclairii* (fungus) | RP |

a) Industrial production (IP), development stage (DS), research project (RP).

and microalgal colorants have already been proved to exert such properties. Novel compounds from microbial and algal biomass, produced by applying biotechnological processes, should render possible applicable and optimized solutions to the food and feed industry (Sandmann 2015; Nupur et al. 2016; Zhu and Jackson 2015).

## References

Berman, J., Zorrilla-López, U., Farré, G. et al. (2014). Nutritionally important carotenoids as consumer products. *Phytochem. Rev.* 14 (5): 727–743.

Calegari-Santos, R., Diogo, R.A., Fontana, J.D., and Bonfim, T.M.B. (2016). Carotenoid production by halophilic archaea under different culture conditions. *Curr. Microbiol.* 72 (5): 641–651.

Caro, Y., Anamale, L., Fouillaud, M. et al. (2012). Natural hydroxyanthraquinoid pigments as potent food grade colorants: an overview. *Nat. Prod. Bioprospect.* 2: 174–193.

Castelblanco-Matiz, L.M., Barbachano-Torres, A., Ponce-Noyola, T. et al. (2015). Carotenoid production and gene expression in an astaxanthin-overproducing *Xanthophyllomyces dendrorhous* mutant strain. *Arch. Microbiol.* 197 (10): 1129–1139.

Chang, J.-J., Thia, C., Lin, H.-Y. et al. (2015). Integrating an algal β-carotene hydroxylase gene into a designed carotenoid-biosynthesis pathway increases carotenoid production in yeast. *Bioresour. Technol.* 184: 2–8.

Chen, Y., Xiao, W., Wang, Y. et al. (2016). Lycopene overproduction in *Saccharomyces cerevisiae* through combining pathway engineering with host engineering. *Microb. Cell Fact.* 15 (1): 113.

Chi, S., He, Y., Ren, J. et al. (2015). Overexpression of a bifunctional enzyme, CrtS, enhances astaxanthin synthesis through two pathways in *Phaffia rhodozyma*. *Microb. Cell Fact.* 14 (1): 90.

Dufossé, L. (2006). Microbial production of food grade pigments. *Food Technol. Biotechnol.* 44: 313–321.

Dufossé, L. (2009). Pigments, microbial. In: *Encyclopedia of Microbiology*, 3e (ed. M. Schaechter), 457–471. New-York: Elsevier/Academic Press.

Dufossé, L. and de Echanove, C. (2005). The last step in the biosynthesis of aryl carotenoids in the cheese ripening bacteria *Brevibacterium linens* ATCC 9175 (*Brevibacterium aurantiacum* sp. nov.) involves a cytochrome P450-dependent monooxygenase. *Food Res. Int.* 38: 967–973.

Dufossé, L., Galaup, P., Yaron, A. et al. (2005). Microorganisms and microalgae as sources of pigments for food use: a scientific oddity or an industrial reality? *Trends Food Sci. Technol.* 16: 389–406.

Dufossé, L., Fouillaud, M., Caro, Y. et al. (2014). Filamentous fungi are large-scale producers of pigments and colorants for the food industry. *Curr. Opin. Biotechnol.* 26: 56–61.

Eriksen, N.T. (2008). Production of phycocyanin – a pigment with applications in biology, biotechnology, foods and medicine. *Appl. Microbiol. Biotechnol.* 80: 1–14.

Fouillaud, M., Venkatachalam, M., Llorente, M. et al. (2017). Biodiversity of pigmented fungi isolated from marine environment in Reunion Island, Indian Ocean: new resources for colored metabolites. *J. Fungi* 3 (3): 36, (22 pages), DOI:https://doi.org/10.3390/jof3030036.

Frisvad, J.C., Smedsgaard, J., Larsen, T.O., and Samson, R.A. (2004). Mycotoxins, drugs and other extrolites produced by species in *Penicillium* subgenus *Penicillium*. *Stud. Mycol.* 49: 201–241.

Frisvad, J.C., Yilmaz, N., Thrane, U. et al. (2013). *Talaromyces atroroseus*, a new species efficiently producing industrially relevant red pigments. *PLoS One* 8 (12): e84102.

Fu, G., Xu, Y., Li, Y., and Tan, W. (2007). Construction of a replacement vector to disrupt pksCT gene for the mycotoxin citrinin biosynthesis in *Monascus aurantiacus* and maintain food red pigment production. *Asia Pacific J. Clin. Nutr.* 16 (Suppl. 1): 137–142.

Furubayashi, M., Ikezumi, M., Takaichi, S. et al. (2015). A highly selective biosynthetic pathway to non-natural C 50 carotenoids assembled from moderately selective enzymes. *Nat. Commun.* 6: 7534.

Gao, J.M., Yang, S.X., and Qin, J.C. (2013). Azaphilones: chemistry and biology. *Chem. Rev.* 113: 4755–4811.

Gessler, N.N., Egorova, A.S., and Belozerskaya, T.A. (2013). Fungal anthraquinones. *Appl. Biochem. Microbiol.* 49: 85–99.

Grassi, T.L.M., do Espírito Santo, E.F., de Siqueira Marcos, M.T. et al. (2016). Bacterial pigment for Nile tilapia feeding. *Aquacult. Int.* 24 (2): 647–660.

Guedes, A.C., Amaro, H.M., and Malcata, F.X. (2011). Microalgae as sources of carotenoids. *Mar. Drugs* 9: 625–644.

Guyomarc'h, F., Binet, A., and Dufossé, L. (2000). Production of carotenoids by *Brevibacterium linens*: variation among strains, kinetic aspects and HPLC profiles. *J. Ind. Microbiol. Biotechnol.* 24: 64–70.

Heider, S.A., Peters-Wendisch, P., Wendisch, V.F. et al. (2014a). Metabolic engineering for the microbial production of carotenoids and related products with a focus on the rare C50 carotenoids. *Appl. Microbiol. Biotechnol.* 98: 4355–4368.

Heider, S.A.E., Peters-Wendisch, P., Netzer, R. et al. (2014b). Production and glucosylation of C50 and C40 carotenoids by metabolically engineered *Corynebacterium glutamicum*. *Appl. Microbiol. Biotechnol.* 98 (3): 1223–1235.

Jaswir, I., Noviendri, D., Hasrini, R.F., and Octavianti, F. (2011). Carotenoids: sources, medicinal properties and their application in food and nutraceutical industry. *J. Med. Plant Res.* 5 (33): 7119–7131.

Jin, W., Xu, X., Jiang, L. et al. (2015). Putative carotenoid genes expressed under the regulation of Shine–Dalgarno regions in *Escherichia coli* for efficient lycopene production. *Biotechnol. Lett.* 37 (11): 2303–2310.

Kim, S.H., Kim, M.S., Lee, B.Y., and Lee, P.C. (2016). Generation of structurally novel short carotenoids and study of their biological activity. *Sci. Rep.* 6: 21987.

Kuddus, M., Singh, P., Thomas, G., and Al-Hazimi, A. (2013). Recent developments in production and biotechnological applications of C-phycocyanin. *BioMed Res.Int.* https://doi.org/10.1155/2013/742859.

Kumar, M., Dwivedi, P., Sharma, A.K. et al. (2014). Apoptosis and lipid peroxidation in ochratoxin A- and citrinin-induced nephrotoxicity in rabbits. *Toxicol. Ind. Health* 30: 90–98.

Lange, B.M., Rujan, T., Martin, W., and Croteau, R. (2000). Isoprenoid biosynthesis: the evolution of two ancient and distinct pathways across genomes. *Proc. Natl. Acad. Sci. U.S.A.* 97: 13172–13177.

Lebeau, J., Venkatachalam, M., Fouillaud, M. et al. (2017). Production and new extraction method of polyketide red pigments produced by ascomycetous fungi from terrestrial and marine habitats. *J. Fungi* 3 (3): 34, (21 pages), DOI:https://doi.org/10.3390/jof3030034.

Li, X.-R., Tian, G.-Q., Shen, H.-J., and Liu, J.-Z. (2015). Metabolic engineering of *Escherichia coli* to produce zeaxanthin. *J. Ind. Microbiol. Biotechnol.* 42 (4): 627–636.

Lin, J.-H., Lee, D.-J., and Chang, J.S. (2014). Lutein production from biomass: marigold flowers versus microalgae. *Bioresour. Technol.* 184: 421–428.

Mannazzu, I., Landolfo, S., da Silva, T.L., and Buzzini, P. (2015). Red yeasts and carotenoid production: outlining a future for non-conventional yeasts of biotechnological interest. *World J. Microbiol. Biotechnol.* 31 (11): 1665–1673.

Mapari, S.A.S., Meyer, A.S., Thrane, U., and Frisvad, J.C. (2009). Identification of potentially safe promising fungal cell factories for the production of polyketide natural food colorants using chemotaxonomic rationale. *Microb. Cell Fact.* 8: 24.

Mapari, S.A., Thrane, U., and Meyer, A.S. (2010). Fungal polyketide azaphilone pigments as future natural food colorants? *Trends Biotechnol.* 28: 300–307.

Martin, H.D., Kock, S., Scherrers, R. et al. (2009). 3,3′-dihydroxyisorenieratene, a natural carotenoid with superior antioxidant and photoprotective properties. *Angew. Chem. Int. Ed.* 48: 400–403.

Nupur, L.N.U., Vats, A., Dhanda, S.K. et al. (2016). ProCarDB: a database of bacterial carotenoids. *BMC Microbiol.* 16 (1): 96.

Pumas, C., Peerapornpisal, Y., Vacharapiyasophon, P. et al. (2012). Purification and characterization of a thermostable phycoerythrin from hot spring cyanobacterium *Leptolyngbya* sp. KC45. *Int. J. Agric. Biol.* 14: 121–125.

Raguenes, G., Moppert, X., Richert, L. et al. (2004). A novel exopolymer-producing bacterium, *Paracoccus zeaxanthinifaciens* subsp *payriae*, isolated from a "kopara" mat located in Rangiroa, an atoll of French Polynesia. *Curr. Microbiol.* 49: 145–151.

Sandmann, G. (2015). Carotenoids of biotechnological importance. *Adv. Biochem. Eng. Biotechnol.* 148: 449–467.

Sardaryan, E., Zihlova, H., Strnad, R., and Cermakova, Z. (2004). Arpink Red – meet a new natural red food colorant of microbial origin. In: *Pigments in Food, More than Colours…* (ed. L. Dufossé), 207–208. Quimper, France: Université de Bretagne Occidentale Publ.

Schmidt-Dannert, C., Umeno, D., and Arnold, F.H. (2000). Molecular breeding of carotenoid biosynthetic pathways. *Nat. Biotechnol.* 18: 750–753.

Silva, T.P., Paixão, S.M., and Alves, L. (2016). Ability of *Gordonia alkanivorans* strain 1B for high added value carotenoids production. *RSC Adv.* 6 (63): 58055–58063.

Song, G.H., Kim, S.H., Choi, B.H. et al. (2013). Heterologous carotenoid-biosynthetic enzymes: functional complementation and effects on carotenoid profiles in *Escherichia coli*. *Appl. Environ. Microbiol.* 79: 610–618.

Takemura, M., Maoka, T., Osawa, A. et al. (2015). (6E) and (6Z)-9′-Aporhodoxanthinone, novel carotenoids produced in zeaxanthin-synthesizing-*Escherichia coli* by redox stress. *Tetrahedron Lett.* 56 (44): 6063–6065.

Thawornwiriyanun, P., Tanasupawat, S., Dechsakulwatana, C. et al. (2012). Identification of newly zeaxanthin-producing bacteria isolated from sponges in the Gulf of Thailand and their zeaxanthin production. *Appl. Biochem. Biotechnol.* 167: 2357–2368.

Valla, A., Cartier, D., Valla, B. et al. (2003). New syntheses of natural carotene isorenieratene (ϕ,ϕ-carotene) and its 3,3′-dimethoxy analog. *Helv. Chim. Acta* 86: 3314–3319.

Wagener, S., Völker, T., De Spirt, S. et al. (2012). 3,3′-dihydroxyisorenieratene and isorenieratene prevent UV-induced DNA damage in human skin fibroblasts. *Free Radical Biol. Med.* 53: 457–463.

Wang, F., Jiang, J.G., and Chen, Q. (2007). Progress on molecular breeding and metabolic engineering of biosynthesis pathways of C-30, C-35, C-40, C-45, C-50 carotenoids. *Biotechnol. Adv.* 25: 211–222.

Woo, P.C., Lam, C.W., Tam, E.W. et al. (2014). The biosynthetic pathway for a thousand-year-old natural food colorant and citrinin in *Penicillium marneffei*. *Sci. Rep.* 4: 6728. https://doi.org/10.1038/srep06728.

Yang, Y., Liu, B., Du, X. et al. (2015). Complete genome sequence and transcriptomics analyses reveal pigment biosynthesis and regulatory mechanisms in an industrial strain, *Monascus purpureus* YY-1. *Sci. Rep.* 5: 8331. https://doi.org/10.1038/srep08331.

Zhang, J.H., Lu, L.L., Yin, L.J. et al. (2012). Carotenogenesis gene cluster and phytoene desaturase catalyzing both three- and four-step desaturations from *Rhodobacter azotoformans*. *FEMS Microbiol. Lett.* 333: 138–145.

Zhou, P., Ye, L., Xie, W. et al. (2015). Highly efficient biosynthesis of astaxanthin in *Saccharomyces cerevisiae* by integration and tuning of algal crtZ and bkt. *Appl. Microbiol. Biotechnol.* 99 (20): 8419–8428.

Zhu, Q. and Jackson, E.N. (2015). Metabolic engineering of *Yarrowia lipolytica* for industrial applications. *Curr. Opin. Biotechnol.* 36: 65–72.

# 5

# Prospective Research and Current Technologies for Bioflavor Production

*Marina Gabriel Pessôa*[1], *Bruno Nicolau Paulino*[1,2], *Gustavo Molina*[1,3], *and Glaucia Maria Pastore*[1]

[1] *Laboratory of Bioflavors and Bioactive Compounds, Department of Food Science, School of Food Engineering, University of Campinas, Brazil*
[2] *Faculty of Pharmaceutical Sciences, Federal University of Amazonas, Manaus, Brazil*
[3] *Laboratory of Food Biotechnology, Institute of Science and Technology - UFVJM, Diamantina, Brazil*

## 5.1 Introduction

Flavors are volatile organic compounds added to food, cosmetics, and pharmaceutical products in order to provide a characteristic taste, maintain the flavor after processing, modify an already-existing flavor, or to reduce some undesirable aroma, leading to an increase in consumer acceptance and promoting their market success (Berger 2009; Tandon 2017). More than 8000 volatile compounds have been reported; therefore, a great variety of combinations can be obtained, resulting in a specific aroma description for each product (Fanaro et al. 2016). Generally, flavor compounds have low detection thresholds and are usually present as minor components in the final product. However, they represents up to 50% of the total product cost (Fanaro et al. 2016).

In 2017, sales of aromas and fragrances grew about 7% from 2016 and achieved US$26.3 billion (Leffingwell and Associates 2018). This market is larger in North America, Asia-Pacific, and Western Europe, being that in 2013, the European Union sales of flavorings represented a third of the global market, followed by the United States and Asia (UBIC 2014). The flavor industry is a highly competitive sector. Four industries (Givaudan, Firmenich, IFF, and Symrise) were responsible for more than 50% of the market share in 2017 (Figure 5.1). These results are possible due to their vast product portfolio, wide distribution network, and research and development (R&D) investments (Tandon 2017).

Aroma compounds comprise several organic functions, such as lactones, hydrocarbons, alcohols, ketones, vanillin, terpenes, aldehydes, and esters (Table 5.1) (Molina and Fanaro 2016; Bhowmik and Patil 2018). They also differ in solubility, volatility, and temperature and pH stability (Berger 2015, 2007).

Esters are known by their fruity aroma; therefore, they are commonly used as additives in beverages, candies, jams, baked goods, and dairy products. Acetate esters (ethyl acetate, hexyl acetate, isoamyl acetate, and 2-phenylethyl acetate) provide the characteristic flavor in wine and other alcoholic beverages derived from grapes, and they can be produced by non-*Saccharomyces* yeasts, such as *Hanseniaspora guilliermondii*

*Bioprocessing for Biomolecules Production*, First Edition.
Edited by Gustavo Molina, Vijai Kumar Gupta, Brahma N. Singh, and Nicholas Gathergood.

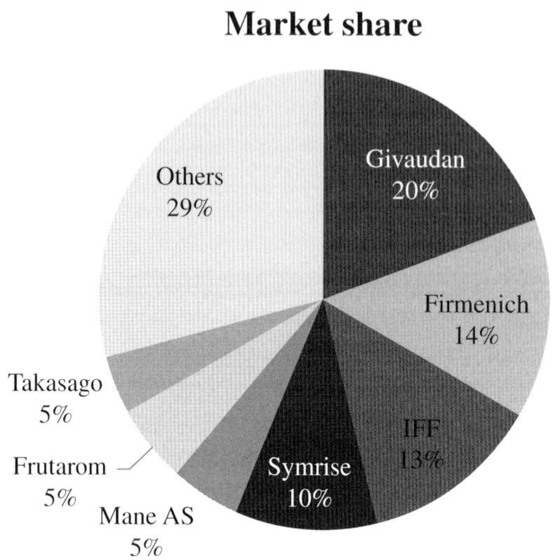


**Figure 5.1** Global market share of the main flavor and fragrance industry leaders, in 2017. Source: Leffingwell and Associates 2018.

and *Pichia anomala*. In cheese, ethyl or methyl esters provide the taste and aroma (Rojas et al. 2001; Shaaban et al. 2016).

Lactones are cyclic esters of γ- and δ-hydroxyl acids, and they contribute taste and flavors like fruity, coconut, creamy, nutty, sweet, and buttery (Burdock 2010; Bhowmik and Patil 2018). Some examples of lactones widely used in the food industry are γ-octalactone and γ-nonalactone, with coconut flavor, 6-pentyl-2-pyrone, that can be produced using *Trichoderma* sp. strains and γ-decalactone and δ-dodecalactone produced from ricinoleic acid and linoleic acid, being commonly used for peach, apricot, and strawberry aromas (Dionísio et al. 2012; Bhowmik and Patil 2018). The natural γ-decalactone extracted from plants can be commercialized with a price of US$6000 $kg^{-1}$, while the chemical synthetized compound costs only US$150 $kg^{-1}$ (Dubal et al. 2008).

In the alcohols group, monoterpenes alcohols are very important in the flavor industry. Menthol is one of the most economically attractive compounds, with a consumption of about 90 metric tons in United States. It is mainly used in cigarettes, cosmetics, toothpastes, chewing gum, candies, and medicines. The isomer (−)-menthol is most often found in nature, and it is appreciated for its mint-like flavor and cooling sensation when in contact with the skin (Koroch et al. 2007; Burdock 2010).

Carveol is another monoterpene alcohol with a spearmint-like odor usually employed in chewing gum, beverages, and candies. It can be extracted from caraway seeds, spearmint, orange juice, mango, and eucalyptus oil in low amounts or produced by oxidation of limonene (Burdock 2010; Dionísio et al. 2012). α-Terpineol and verbenol are mainly used in soaps and cosmetic formulations due to their characteristic aroma. They can be extracted from essential oils or can be obtained from oxidation of pinenes (Vespermann et al. 2017). Besides monoterpenes, 2-phenylethanol (2-PE) is of great interest due to its rose-like smell. Nowadays, it is produced chemically from toluene, benzene, and styrene, but it can be obtained naturally by the bioconversion of 2-phenylalanine (Zhang et al. 2014).

Terpenes are plant-secondary metabolites produced for protection against infection of pathogens and are the main constituents of essential oils (mono and sesquiterpenes).

**Table 5.1** Examples of flavor molecules produced through biotechnology.

| Molecule | Formula | Aroma | Main microorganisms or enzymes | Reference |
|---|---|---|---|---|
| *Esters* | | | | |
| Ethyl acetate | 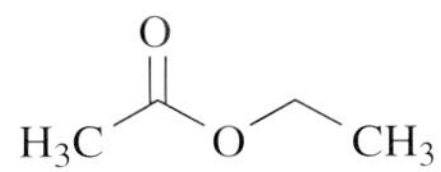 | Fruity, balsamic | *Kluyveromyces lactis, Saccharomyces kluyveri* Y708; *Saccharomyxes cerevisiae; Kloeckera apiculata; Hansenula subpelliculosa, Kluyveromyces marxianus* | Löser et al. 2011, 2014; Møller et al. 2002; Plata et al. 2003; Saerens et al. 2008 |
| Butyl acetate | 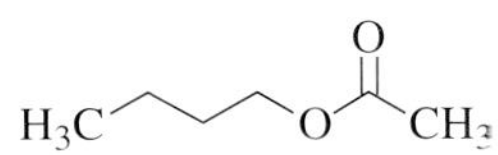 | Fruity, pineapple | *Kluyveromyces marxianus* | Medeiros et al. 2000; Medeiros et al. 2001. |
| Isoamyl acetate | 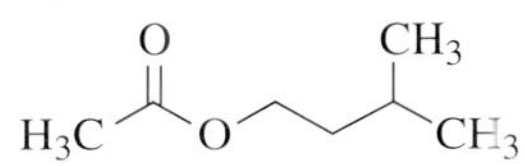 | Fruity, banana, pear | *Kloeckera apiculata; Hansenula subpelliculosa, Kluyveromyces marxianus*; *Saccharomyces kudriavzevii; Saccharomyces uvarum, Saccharomyces cerevisiae*; Immobilized *Mucor miehei* lipase | Plata et al. 2003; Krishna et al. 2000; Stribny et al. 2015; Kuis et al. 2017. |
| Ethyl butyrate | 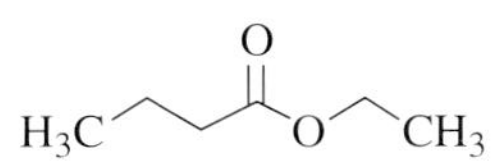 | Fruity, reminiscent of pineapples | Immobilized lipases of *Candida rugosa, Rhizopus oryzae* and *Thermomyces lanuginosus* | Thakar and Madamwar 2005; Grosso et al. 2013; Paludo et al. 2015. |
| Ethyl hexanoate | 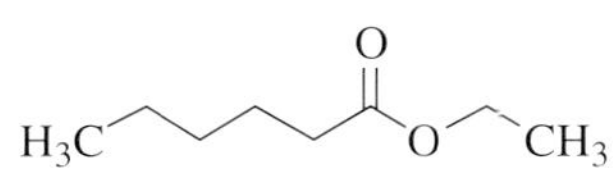 | Apple, fruity, banana, aniseed, strawberry, green apple | *Saccharomyces cerevisiae; Neurospora* sp.; *Bacillus* sp. | Saerens et al. 2008; Han et al. 2009; Berger 2015; Zhao et al. 2017a,b. |
| Benzyl acetate | 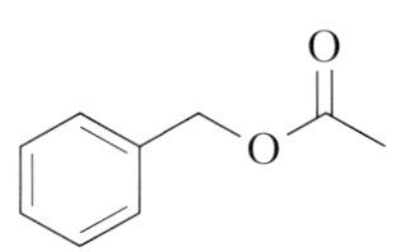 | Jasmine, green apple, strawberry, banana, orchid | Immobilized *Pseudomonas fluorescens* lipase | Badgujar et al. 2015 |
| *Lactones* | | | | |
| γ-decalactone | 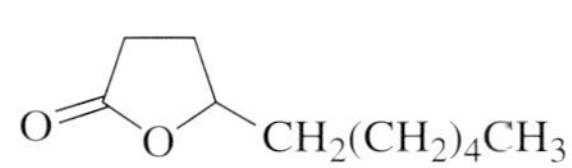 | Peach, coconut, fruity | *Waltomyces lipofer; Yarrowia lipolytica; Lindnera saturnus; Sporobolomyces odorus; Tyromyces sambuceus; Moniliella suaveolens Sporidiobolus salmonicolor* | An and Oh 2013; Moradi et al. 2013; Braga and Belo 2015; Braga and Belo 2016; Prasad et al. 2014; Soares et al. 2017. |

**Table 5.1** (Continued)

| Molecule | Formula | Aroma | Main microorganisms or enzymes | Reference |
|---|---|---|---|---|
| γ-octalactone | 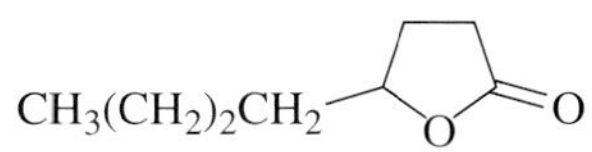 | Fruity, coconut | *Polyporus durus* | Dionísio et al. 2012 |
| 6-pentyl-2-pyrone | 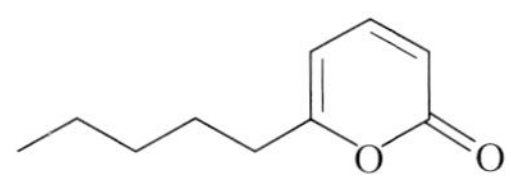 | Coconut | *Trichoderma viridae*<br>*Trichomerma harzianum* | Ramos et al. 2008; Gupta et al. 2015 |
| *Alcohol* | | | | |
| 2-Phenylethanol | 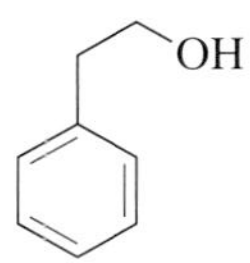 | Rose, peach | *Kluyveromyces marxianus; Metschnikowia pulcherrima; Saccharomyces cerevisiae; Proteus mirabilis* | Hua and Xu 2011; Kim et al. 2014a,b; Chantasuban et al. 2018; Červeňanský et al. 2017; Yin et al. 2015; Liu et al. 2018 |
| Isoamyl alcohol | 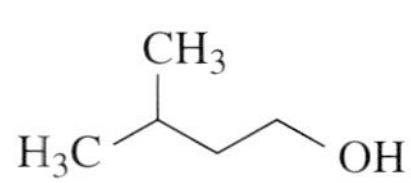 | Fusel oil, whisky, pungent odor and repulsive taste | Engineered strain of *Saccharomyces cerevisiae* | Abe and Horikoshi 2005; Yuan et al. 2017a,b |
| Citronellol | 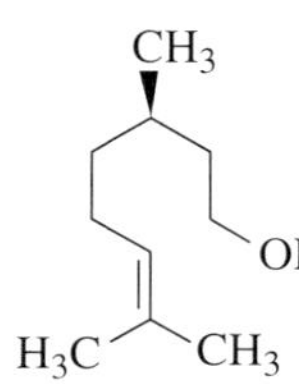 | Rose-like | *Hansenula saturnus; Ceratocysis moniliformis, Kluyveromyces lactis*; *Polyporus brumalis*; Lipase from *Mucor miehei* | Oda et al. 1995; Shaaban et al. 2016; Lee et al. 2015 |
| Carveol | 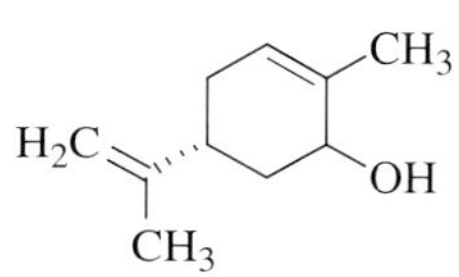 | Spearmint-like odor | *Rhodococcus opacus; Pseudomonas putida* | Duetz et al. 2001; de Carvalho and de Fonseca 2003; Groeneveld et al. 2016 |

| | | | | |
|---|---|---|---|---|
| Verbenol | 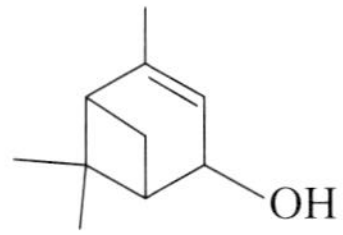 | Fresh pine, ozone odor; camphor and mint notes | *Penicillium* sp.; *Aspergillus* sp.; *Pleorotus sapidus; Chrysosporium pannorum* | Agrawal et al. 1999; Agrawal and Joseph 2000; Rozenbaum et al. 2006; Krings et al. 2009; Trytek et al. 2014 |
| (−) menthol | 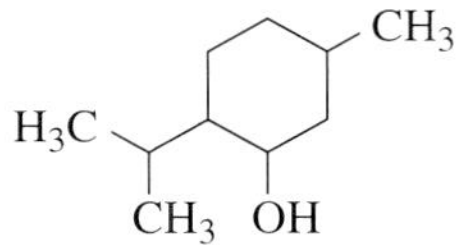 | Mint | Esterase from *Bacillus subtilis;* engineered *Escherichia coli* | Zheng et al. 2009; Zheng et al. 2010; Toogood et al. 2015 |
| *Aldehydes* | | | | |
| Benzaldehyde | 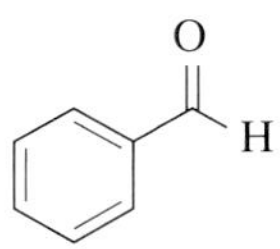 | Almond, floral, spice | *Pichia pastoris; Escherichia coli; Photorhabdus temperate; Gluconobacter oxydans* | Jain et al. 2010; Craig and Daugulis 2013; Craig and Daugulis 2014; Kunjapur et al. 2014; Ullah et al. 2015; Wu et al. 2010 |
| *Ketones* | | | | |
| Carvone | 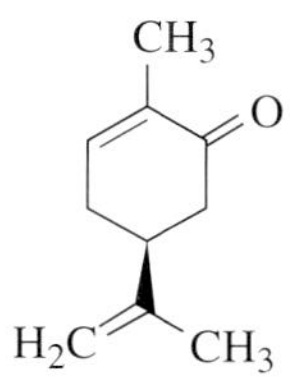 | Herbs (S); spearmint oil (R) | *Bacillus pallidus; Rhodococcus opacus; Phomopsis sp.* | Savithriry et al. 1998; de Carvalho and de Fonseca 2003; Bier et al. 2017 |
| Nootkatone | 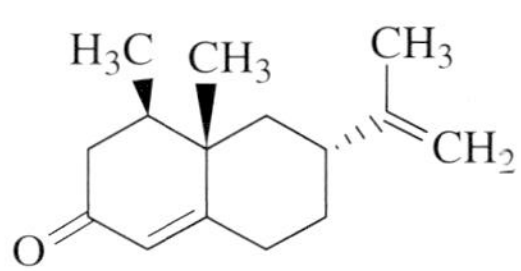 | Grapefruit, citrus, orange and bitter | *Pleurotus sapidus*<br>*Yarrowia lipolytica* | Fraatz et al. 2009a,b; Krügener et al. 2010; Guo et al. 2018a,b |

They are employed as fragrances or ingredients in foods and cosmetics, since they have interesting organoleptic properties and a low sensory threshold (Molina et al. 2014; Aissou et al. 2017). The pathways for some terpenes production have already been identified in microorganisms, being an opportunity for the development of engineered strains with an improved performance for natural flavor production (Schewe et al. 2011; Shaaban et al. 2016). Terpenes can also be microbial or enzymatic oxidized to produce a similar compound with some organic function, such as alcohol or ketone (Maróstica and Pastore 2007; Bicas et al. 2008a,b; Molina et al. 2014).

Pyrazines are heterocyclic, nitrogen-containing compounds with a nutty and roasted flavor. They are naturally produced through a Maillard reaction during conventional cooking or roasting; however, it must be added to microwave cooking foods (Shaaban et al. 2016; Bhowmik and Patil 2018). One example is tetramethylpyrazine, a compound with fermented, coffee aroma widely used in beverages, baked goods, and meat products. It can be produced from amino acids using the strain *Corynebacterium glutamicum* (Burdock 2010; Dionísio et al. 2012).

Methyl ketones are primarily responsible for the characteristic flavor of dairy products. These compounds were first observed in the spores of mold-ripened cheese *Penicillium* and 2-heptanone, 2-nonanone, and 2-undecanone, and are the largest contributors to stale flavor in UHT milk (Longo and Sanromán 2006; Bhowmik and Patil 2018). They have various lengths and can be produced by microorganisms through fatty acid β-oxidation pathway (Carrol et al. 2016).

Vanillin (4-hydroxy-3-methoxybenzaldehyde) is the most consumed flavor compound, employed in foods, beverages, perfumes, pharmaceuticals, and medical industries. Global consumption is estimated at 15 kton/year of vanillin. Structurally, vanillin is a phenol substituted with an aldehyde and methoxy group at specific positions and can be obtained from ferulic acid, phenolic compounds, guaiacol, lignin, eugenol, and isoeugenol (Burdock 2010; Gallage and Møller 2015). It is naturally found in vanilla orchid pods, mainly *Vanilla planifolia*, in a concentration of 1–2%; however, less than 1% of the vanillin global market is covered by the naturally extracted compound, due to the fact that the production of 1 kg of vanillin requires approximately 500 kg of vanilla pods, corresponding to the pollination of approximately 40 000 vanilla orchid flowers (Gallage and Møller 2015). In addition, vanillin extracted from vanilla pods has a market price varying from around US\$1200 to US\$4000 $kg^{-1}$ compared to the chemically produced vanillin with a price of US\$15 $kg^{-1}$ (Gallage and Møller 2015; Felipe et al. 2017). Nowadays, most vanillin is obtained from chemical synthesis, but several biotechnological processes have been developed to produce this compound (Shaaban et al. 2016; Bhowmik and Patil 2018; Gallage and Møller 2018).

Among vanillin, benzaldehyde is another commonly used flavoring agent, with a specific almond-like aroma. Its structure consists of a benzene ring with a formyl substituent, being the simplest aromatic aldehyde, although it is extremely industrially useful (Burdock 2010). The world consumption of this compound is about 7 kton/year, and it is currently produced by fermentation, using cinnamaldehyde or phenylalanine as substrates. The main issue of obtaining this compound is its toxicity for the microorganisms used as biocatalysts; therefore, only low concentrations can be obtained. Its natural extraction leads to the undesirable formation of the toxic hydrocyanic acid, and for these reasons, the biotechnological production of benzaldehyde has been studied (Craig and Daugulis 2013; Kunjapur et al. 2014).

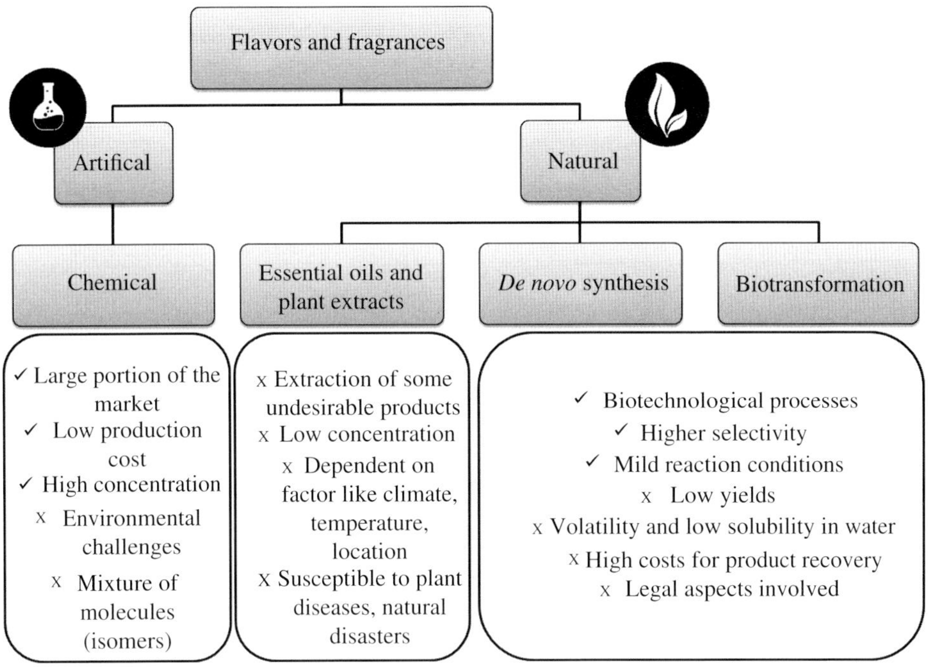


**Figure 5.2** Current routes for flavors production.

The methods for flavor production are summarized in Figure 5.2. Chemical processes are the most used nowadays, since they produce high yields of flavor compounds at low costs. These products cannot be labeled as natural, and their production generates high amounts of waste – for example, for the production of 1 kg of vanillin from lignin, it is necessary to remove 160 kg of waste. In addition, the reactions require high pressure and temperature and are not regio- and enantioselective, forming undesirable compounds (Akacha and Gargouri 2015; Gallage and Møller 2018). Natural flavors can be extracted directly from nature, but this process can result in complex compound mixtures and is restricted based on geographical availability, climatic changes, and seasonality. Moreover, these compounds usually occur at low concentrations, being necessarily a high quantity of biomass to obtain the necessary amount of aroma (Berger 2015; Felipe et al. 2017).

In this context, biotechnology rises as an attractive alternative for natural flavor production. This method provides high enantioselectivity, allowing aromas with high optical purity, pleasant sensory characteristics of the product, lower waste generation, and possibility of using agro-industrial residues as substrates. Production can supply the increasing demand of the consumer market for more natural additives (Felipe et al. 2017). The biotechnological processes can be divided into (i) de novo synthesis using microorganisms; and (ii) bioconversion or biotransformation of natural precursors with microorganisms or enzymes. The first strategy uses simple molecules as substrates, and complex substances are obtained through activation of several metabolic pathways. On the other hand, in the bioconversion/biotransformation, a single or few reactions are necessary to obtain a product structurally similar to the substrate, being economically advantageous (Bicas et al. 2016).

Many companies are already producing bioflavors through biotechnology. The company Amyris produces a sesquiterpene β-farnesene from sugarcane using a genetically modified yeast, and has also developed a patchouli oil substitute (named Clearwood) with better sensorial qualities than the natural oil in partnership with Firmenich. Evolva and Isobionics are producing Valencene, another sesquiterpene, and its derivative nootkatone. In these processes, Isobionics also uses a genetically modified strain of *Rhodobacter sphaeroides* that contains genes from the terpene biosynthesis pathway (Huembelin et al. 2014; Leffingwell and Leffingwell 2015; Schempp et al. 2018).

Regarding vanillin production, Rhovanil from Solvay was the first commercially available fermentation-derived product, obtained by bioconversion of ferulic acid by *Streptomyces setonii*. The same substrate is used for vanillin production by BASF and Shanghai Apple Flavor & Fragrance Group Co. Ltd., the first one using a genetically modified strain of *Pseudomonas* sp. and the later, a wild *Streptomyces* sp. V-1, isolated from orchard soil (Xu et al. 2009; Gallage and Møller 2015; Felipe et al. 2017). De Monchy Aromatics is responsible for the vanillin production from curcumin, and Evolva and International Flavors and Fragrances produce vanillin through de novo synthesis from glucose extracted of corn, using *Schizosaccharomyces pombe* (Hansen et al. 2009, 2014). In a different approach, Sense Capture produces vanillin by bioconversion of eugenol by a *Streptomyces* sp. strain, and this flavor is marketed by Mane (Gallage and Møller 2015; Leffingwell and Leffingwell 2015).

Although already used in industrial scale, the production of flavors by fermentation must still overcome some drawbacks, such as the high cost of fermentation process and downstream recovery of the final product, lack of a suitable catalyst that can resist to the substrate used, volatility and low solubility in water of many flavors, low yields and product concentration obtained, and legal aspects involving the use of microorganisms for food and feed applications (Bicas et al. 2010; Schempp et al. 2018). Therefore, several research groups are working to improve yields and selectivity of flavor production, using statistical methods to optimize process conditions, evaluate different low-cost raw materials, and develop new genetic modified strains to be more robust and with better performance (Bicas et al. 2008a,b; Schempp et al. 2018).

In this chapter, we provide an overview of the current studies for natural flavor production using biotechnology and microorganisms or enzymes as biocatalysts. We focus in the processes conditions, new technologies developed, improvements obtained, yields, and total flavor concentration obtained after fermentation.

## 5.2 Microbial Production of Bioflavors

### 5.2.1 Biotransformation of Terpenes

The application of terpenes as a substrate in the production of bioflavors has been described and reviewed in the last years and is considered an important approach in biotechnology (Sales et al. 2018a; Vespermann et al. 2017). The reasons for this are that terpenes, mainly monoterpenes, are abundantly found in industrial wastes – for example, D-limonene in citrus byproducts and pinenes in turpentine and can be bioconverted into value-added products, which are used as specialty biofuels and aroma compounds (Schempp et al. 2018; Mewalal et al. 2017; Vespermann et al. 2017). Therefore, these compounds can be considered low-cost substrates and interesting starting materials in biotransformation processes.

Limonene is a cyclic monoterpene commonly found in citrus fruits and its essential oils (Pérez-Mosqueda et al. 2015). Oxygenated derivatives of limonene such as carveol, carvone, menthol, perillyl alcohol, α-terpineol, and limone-1,2-diol are notable flavor compounds with wide use in cosmetics and in the food industry (Maróstica and Pastore 2007; Bicas et al. 2009). In addition, the production of these derivatives from limonene is possible using the biotransformation approach.

Recently, the biotransformation of limonene by an endophytic fungi strain identified as *Phomopsis* sp. was described using mineral medium and an orange residue-based media (Bier et al. 2017). Different biotransformation products were observed, depending on the medium applied. In this sense, $0.536\,g\,l^{-1}$ of carvone and $2.08\,g\,l^{-1}$ of limonene-1,2-diol were produced in synthetic medium after 192 hours, while $2.10\,g\,l^{-1}$ of limonene-1,2-diol was produced using an orange residue-based medium with a single fed-batch after 120 hours. Similar results were reported using endophytic strains of *Alternaria alternata* Eb03 and *Neofusicoccum* sp. Eb04 isolated from aerial parts of *Eupatorium buniifolium*, which were able to convert limonene into limonene-1,2-diol with 70% and 89% of yield, respectively (Cecati et al. 2018).

Other recent studies have shown the biotransformation of limonene isomers into limonene-1,2-diol by *Colletotrichum* sp. strains using mineral medium (Sales et al. 2018b). After 192 hours of processing, *Colletotrichum acutatum* produced up to $3.0\,g\,l^{-1}$ of limonene-1,2-diol, while *Colletotrichum nimphaeae* achieved approximately $4.0\,g\,l^{-1}$ in similar conditions. The difference of concentration of limonene-1,2-diol in these studies can be associated with the characteristics of the biotransformation procedure, including processes conditions, medium composition, and biocatalyst adopted.

Moreover, the use of limonene isomers as the sole carbon source and limitation of oxygen also have been described to influence different metabolic behavior in biotransformation of limonene by *Fusarium oxysporum* strain (Molina et al. 2015). The biotransformation of *S*-(−)-limonene to limonene-1,2-diol was not favored under anaerobic conditions, depending on the availability of oxygen to reach concentrations of $3.7\,g\,l^{-1}$ of product after 72 hours of process, while $4.0\,g\,l^{-1}$ of α-terpineol from *R*-(+)-limonene was achieved at 48 hours in anaerobic conditions.

Several strategies have been described for selective biotransformation of limonene in α-terpineol. In this context, Tai et al. (2016) performed the bioconversion of *R*-(+)-limonene carried with *Penicillium digitatum* DSM 62840 as biocatalyst reaching concentrations of α-terpineol ranging of $0.025–0.83\,g\,l^{-1}$ using 0.4% (v/v) different co-solvents. Previously, Rottava et al. (2010a) described the production of α-terpineol by wild strains selected in a screening study performed with more than 400 microorganisms. In this study, a *Penicillium* sp. strain (coded as 04.06.01) isolated from eucalyptus stem was able to produce around $3.45\,g\,l^{-1}$ of α-terpineol using *R*-(+)-limonene as substrate, while five yeast strains produced $0.52–1.70\,g\,l^{-1}$ of α-terpineol.

Rottava et al. (2011) described the optimization of α-terpineol production from *R*-(+)-limonene through a central composite rotatable design with three variables using two different yeasts isolated in screening studies. Evaluating the influence of substrate concentration (0.01–3.85%), ratio between substrate and ethanol (0–1 : 2.7 v/v), and inoculum biomass (0.32–3.68 g), it was observed that, for the first yeast (coded as 03.03.03), the best conditions were obtained in central points (1.75% of R-(+)-limonene with a 1 : 1 ratio between substrate and ethanol and 2 g of inoculum biomass) with concentration of α-terpineol ranging from 1.26 to 1.29 $g\,l^{-1}$. For the second yeast strain (coded as 05.01.35), the α-terpineol production achieved was about 1.7 $g\,l^{-1}$ also in central point conditions. Better results were described by Bicas et al. (2010), who reported the production of approximately 130 $g\,l^{-1}$ of α-terpineol by *Sphingobium* sp. in bioprocess carried with a biphasic system using sunflower oil as organic phase.

The production of others oxygenated compounds from limonene with potential in pharmaceutical and medical fields, such as perillic derivatives, also was described in previous studies (Bicas et al. 2009). For example, the biotransformation of limonene to perillic acid was reported by *Blastobotrys adeninivorans* (former *Arxula adeninivorans*) and *Yarrowia lipolytica,* which were able to produce 0.06 and 1.0 $g\,l^{-1}$, respectively (Van Rensburg et al. 1997). Ferrara et al. (2013) described the production of 0.56 $g\,l^{-1}$ of perillic acid from limonene after 48 hours of process and 0.855 $g\,l^{-1}$ with addition of substrate during the bioprocess.

More recently, Willrodt et al. (2017) reported the bioconversion of limonene in perillic acid through a catalytic biofilm formed by *Pseudomonas* sp. strains. In this study, 34 $g/L_{Tube}$ per day of perillic acid from limonene were produced by a wild-type *Pseudomonas putida* GS1 encoding the enzymes for limonene bioconversion when glycerol was supplied. Previously, the biotransformation of limonene into perillyl alcohol in different pH and temperature was described using a psychrotrophic fungi strain isolated from artic soil identified as *Mortierella minutissima,* with a maximal concentration of 0.125 $g\,l^{-1}$ at 15 °C and pH 6 (Trytek and Fiedurek 2005).

The most abundant bicyclic monoterpenes found in nature, α-pinene and β-pinene, are important precursors of value-added bioflavors widely used in cosmetic and food industries, such as α-terpineol, verbenol, and verbenone (Bicas et al. 2009; Vespermann et al. 2017). These compounds represent 75–90% of the total components of conifers essential oils and found as major components in a byproduct of the paper and cellulose industry called turpentine (Yoo and Day 2002; Vespermann et al. 2017). Thus, many studies describe the biotransformation of pinenes into flavor compounds using different strategies and biocatalysts (Vespermann et al. 2017).

The use of bacteria strains for biotransformation of pinenes was reported in the last years using some genera, including *Pseudomonas* sp., *Bacillus* sp., *Serratia* sp., and *Gluconobacter* sp. (Bicas et al. 2008a,b; Savithriry et al. 1998; Wright et al. 1986; Deepthi Priya et al. 2015). In addition, Schewe et al. (2011) studied the economic potential of bioconverting pinenes using bacterial monooxygenases as a green method for producing pinene derivatives.

In a study performed by Cecati et al. (2018) endophytic microorganisms isolated from aerial parts of *E. buniifolium* showed the potential for biotransformation of α-pinene isomers. The strain identified as *Fusarium solani* Eb01 was able to convert α-(−)-pinene and α-(+)-pinene into terpinen-4-ol, α-terpineol and *cis*-sabinene hydrate, while the strain of *Neofusicoccum* sp. Eb04 convert the pinene isomers in limonene-1,2-diol, borneol, *p*-menth-1-en-7-al, and *exo*-fenchol.

Other different derivatives have been reported for biotransformation of α-pinene; for example, the main products identified in the biotransformation process carried with strain *Pseudomonas* sp. PIN were limonene, *p*-cymene, α-terpinolene, camphor, terpinen-4-ol, α-terpineol, endo-borneol, and *p*-cymene-8-ol (Yoo et al. 2001). In this study, maximum concentrations of 130 mg $l^{-1}$ of *p*-cymene and 22 mg $l^{-1}$ of limonene were achieved after 24 hours in a process using mineral medium supplemented with 0.1% (v/v) of pinene as sole carbon source at 30 °C and 200 rpm.

Among the α-pinene derivatives, verbenone is a value-added oxygenated compound obtained after oxidation of verbenol, which acts as a multifunctional pheromone in bark beetles of *Dendroctonus* sp. (Burdock 2010; Hughes 1975; Sun et al. 2013). Moreover, this compound has been used in food and cosmetics due to flavor characteristics and in organic chemistry as a starting material for synthesis of bicyclic lactones (Vespermann et al. 2017; Sarmiento et al. 2011).

The production of verbenone by *Dendroctonus* species is related to oxidative metabolism of α-pinene from *Pinus* sp. oleoresin mediated by gut microbiota (Hughes 1975; Xu et al. 2016). The role of gut bacteria strains, including *Pseudomonas* sp., *Lactococcus* sp., *Enterococcus* sp., and *Rahnella* sp. in the bioproduction of verbenone, verbenol, and myrtenol has been studied (Xu et al. 2015). Hence, microorganisms belonging to these genera can be considered as potential biocatalysts for the development of biotechnological processes for bioconversion of α-pinene into aroma compounds. However, few studies have been developed, exploring the biotechnological potential of these microorganisms.

The use of fungal biocatalysts for biotransformation of pinenes has been cited as a promising approach (Vespermann et al. 2017). In this context, Sales et al. (2018b) reported the bioproduction of 0.04 g $l^{-1}$ of verbenone from α-pinene after 192 hours conduced in a mineral medium by a strain of *C. nimphaeae* at 30 °C and 150 rpm. Using cultures of *Stereum hirsutum,* the production of verbenone was observed as major compound (27.64%) and myrtenol (17.75%), camphor (8.49%), and isopinocarveol (3.10%) were noted as minor products after four days of cultivation. This profile was identified through qualitative analysis by gas chromatography-mass spectrometry (GC-MS) (Lee et al. 2015).

Previously, the production of verbenone and verbenol was described using strains of *Pleurotus sapidus, Chaetomium globosom, Stemphylium botryosum, Pleurotus eryngii,* and *Penicillium solitum* (Krings et al. 2009). In submerged cultures of these fungi, it was possible to obtain concentrations of verbenol of 0.5–19.9 mg $l^{-1}$ and verbenone 1.5–15.0 mg $l^{-1}$ in buffered medium at 24 °C and 150 rpm. Another study reported the biotechnological potential of a psychrotrophic fungi, identified as *Chrysosporium pannorum,* for production of 326 mg $l^{-1}$ verbenol, and 203 mg $l^{-1}$ verbenone using 1.5% (v/v) α-pinene as substrate, and 72-hours-old mycelium 1% (v/v) of substrate and 48-hours-old mycelium, respectively (Trytek et al. 2014).

In a screening study performed by Rottava et al. (2010b), 405 microorganisms were isolated from different industrial residues, soils of citric fruits, and vegetable tissues, and were tested for their ability to convert α-pinene. The results showed that 31 strains were able to produce verbenol from α-pinene through biotransformation. The maximum concentrations of verbenol produced were around 126 and 119 mg $l^{-1}$ by a yeast (coded as 05.01.05, isolated from industrial residue of orange juice) and a Gram-positive bacteria strain (coded as 04.05.05, isolated from orange tree stem), respectively. In addition, two other microorganisms were able to produce concentrations greater than 100 mg $l^{-1}$, a yeast strain (coded as 05.01.11) and a Gram-negative bacteria strain (coded as 05.01.19) also isolated from orange juice industrial residue.

Using the same approach, Rottava et al. (2010a) described the biotransformation of β-pinene into α-terpineol by five different microorganisms with concentrations of more than 140 mg $l^{-1}$. The best producer was an *Aspergillus* sp. strain (coded as 04.05.08) isolated from orange tree stem, which showed a production of 675.5 mg $l^{-1}$ of α-terpineol, while three yeasts strains (coded as 01.04.02, 01.04.03, and 01.04.07) isolated from the leaves of citric fruits were able to produce 175.1–281.3 mg $l^{-1}$ of α-terpineol. In optimized conditions, the *Aspergillus* sp. strain was able to produce concentrations higher than 735 mg $l^{-1}$ of α-terpineol, while the yeast strains coded as 01.04.02 and 01.04.03 showed concentration of its product more than 277 and 727 mg $l^{-1}$, respectively (Rottava et al. 2011).

Among the sesquiterpenes, valencene is the most studied substrate in biotransformation processes due to its low cost, since this compound is found as the main bicyclic sesquiterpene in orange peel oil (Misawa 2011; Elston et al. 2005). Valencene can be converted into nootkatone through biotransformation processes using whole cells of microorganisms and isolated enzymes (Fraatz et al. 2009a,b; Krügener et al. 2010).

A study performed by Omarini et al. (2014) evaluated the potential of two edible fungi, *Pleurotus florida* strain PFL-216 and *P. sapidus* strains PSA-69 and PSA-224, for biotransformation of valencene in (+)-nootkatone. These authors observed that after 16 hours of transformation, concentrations higher than 200 mg $l^{-1}$ of (+)-nootkatone were obtained by five new dikaryons strains derived from *P. sapidus* PSA-69.

Using different operating conditions, in a three-phase reactor model, it was possible to obtain concentrations around 600 mg $l^{-1}$ of (+)-nootkatone from valencene by whole cells of *Y. lipolytica* 2.2ab (Castilho-Araiza et al. 2017). This yeast strain was selected as the best biocatalyst in a previous screening study performed with six microorganisms (Palmerín-Carreño et al. 2015). In this study, *Y. lipolytica* 2.2ab was able to produce approximately 42 mg $l^{-1}$ of nootkatone, while strains of *Kluyveromyces marxianus*, *Botryodiplodia theobromae*, and *Phanerochaete chrysosporium* reached maximum concentrations of nootkatone ranging from 14.51 to 32.78 mg $l^{-1}$. Recently, Palmerín-Carreño et al. (2016) described the use of whole cell of *Y. lipolytica* 2.2ab for biotransformation of valencene into (+)-nootkatone using a stirred tank bioreactor with 100% orange essential oil as organic phase. The maximum (+)-nootkatone concentration reached was around 0.77 g $l^{-1}$ after 96 hours of process.

Despite advances in the biotransformation processes of terpenes in recent years, there is still a need for new studies to improve yield and focused in scale-up aiming the industrial application. In addition, it has become necessary to use statistical optimization as well as genetic tools to establishment of optimal conditions that allow the achievement of stable and economically viable processes.

### 5.2.2 De Novo Synthesis

The microbial production of bioflavors by de novo synthesis is an important approach in industrial biotechnology. Several aroma compounds can be produced using this strategy, such as patchoulol, farnesene, raspberry ketone, valencene, limonene, and santalol isomers (Sales et al. 2018a). Recent literature noted different examples of de novo synthesis of aroma compounds by wild and genetically engineered microorganisms (Etschmann et al. 2003; Celinska et al. 2018; Kim et al. 2014a,b).

The microbial generation of aroma compounds by de novo synthesis is related to the flavor characteristics of many food fermented products such as wines, beers, and cheeses (Martin et al. 2016; Denby et al. 2018; Afzal et al. 2017). Many terpenes, alcohols, phenylpropanoids, and amino-acid derived aroma compounds are considered interesting flavor compounds in the cosmetic, pharmaceutical, and food industry, which have been produced by biotechnological technologies using different microorganisms, including bacteria, filamentous fungi, and yeasts (Suzuki et al. 2014; Bennett and Inamdar 2015; Zhao et al. 2017a,b).

2-PE is an aromatic alcohol with a rose-like odor widely used in cosmetic, perfumery, and food products, and their microbial production has been reported in the last years, including by de novo synthesis (Hua and Xu 2011). This compound and its derivative 2-phenethyl acetate (2-PEA), which also is considered a value-added aroma compound, are synthetized from simple and low-cost sugars through the shikimate pathway (Martínez et al. 2018).

The production of 2-PE by genetically engineered *K. marxianus* using glucose as carbon source was reported by Kim et al. (2014a,b). In this study, the maximum concentration of 2-PE reached was of 1.3 g $l^{-1}$ from 20 g $l^{-1}$ glucose without adding phenylalanine in the medium after 72 hours cultivation.

Using a similar approach, Zhang et al. (2014) described the bioproduction of 2-PE by a wild strain identified as *Enterococcus* sp. CGMCC 5087 using glucose and $NH_4Cl$ as carbon and nitrogen source, respectively. After cloning and overexpressing of two genes of rate-limiting enzymes the engineered *Enterococcus* sp. strain was able to produce approximately 0.335 g $l^{-1}$ of PE in 12 hours of process, while the wild strain showed a production of 0.103 g $l^{-1}$ in same conditions. In addition, when phenylalanine was added to the production medium, production of 2-PE increased around 0.530 g $l^{-1}$. These results were better than those reported by Shen et al. (2016) for an engineered strain of *Saccharomyces cerevisiae* (BY4741), where the maximal concentration of 2-PE was 0.096 g $l^{-1}$ after 48 hours of fermentation using 20 g $l^{-1}$ of glucose as the sole carbon source.

Other studies showed the construction of an engineered *Escherichia coli* by the co-overexpression of the genes phenylpyruvate decarboxylase (*kdc*) from *Pichia pastoris* GS115 and alcohol dehydrogenase (*adh1*) *S. cerevisiae* S288c achieved around 0.130 g $l^{-1}$ of 2-PE using glucose as carbon source (Kang et al. 2014). In this case, the maximum concentration of 2-PE obtained was 0.285 g $l^{-1}$, when optimized the expression of four genes of key enzymes involved in the biosynthesis of this aroma compound. Recently, the production of 2-PE using two engineered *E. coli* strains was described by Guo et al. (2018a,b), which showed the production of around 0.58 and 1.01 g $l^{-1}$ of 2-PE from glucose in the fermentations carried with the engineered strains DG01 and DG02, respectively.

Nowadays, producing terpenes by de novo synthesis is closely related with genetically engineered microorganisms that exhibit interesting productivity characteristics (Carter et al. 2003; Binder et al. 2016; Du et al. 2014). Furthermore, the biosynthesis of mono- and sesquiterpenes is the most studied, mainly in regard to limonene, geranic acid, myrcene, perillyl alcohol, valencene, nootkatone, farnesene, patchoulol, α- and β-santalene. Some of these are produced at an industrial level (Carrol et al. 2016; Sales et al. 2018a).

As in biotransformation studies, limonene is the most studied monoterpene for the de novo synthesis approach. According to some authors, the principle for microbial production of limonene is dependent of expressing a plant limonene synthase in a host microorganism such as typical engineered strains of *E. coli* and *S. cerevisiae* reported in the last years (Carter et al. 2003; Jongedijk et al. 2016).

Willrodt et al. (2014) reported the production of *S*-limonene by seven different recombinant *E. coli* strains during batch cultivation. The final concentrations of limonene ranged from 15 to 25 $mg\,l^{-1}$ after 78 hours of process, while the maximum concentration of limonene reached was of 2.7 $g\,l^{-1}$ after 45 hours when glycerol was applied as sole carbon source using the *E. coli* BL21 (DE3) strain. Previously, Alonso-Gutierrez et al. (2013) described the production of 435 $mg\,l^{-1}$ of limonene by an engineered *E. coli* strain containing all mevalonate pathway genes in a single plasmid after 72 hours using 1% of glucose as carbon source.

In contrast to limonene and its oxygenated derivatives, scarce studies about the bioproduction of pinenes are found in literature. The most recent study was performed though the coexpression of a pinene synthase variant with mevalonate pathway enzymes in *E. coli* host strain, which reached 140 $mg\,l^{-1}$ of pinenes in flask culture experiments (Tashiro et al. 2016).

Regarding the production of sesquiterpenes by de novo synthesis, the focus of recent studies has been given to compounds such as farnesene, valencene, and nootkatone. These sesquiterpenes are considered interesting biotechnological targets for industrial production due to their chemical and flavor characteristics. Farnesene is an acyclic sesquiterpene found as a key volatile compound from apple peels and consists in a polymer of acetyl-CoA with potential applications in biofuel production, as well as in the food and cosmetics industries (Huelin and Murray 1966; Wang et al. 2011).

In this context, the efficiently of α-farnesene production in metabolically engineered *E. coli* by fusion of farnesyl diphosphate (FPP) synthase and α-farnesene synthase with the introduction of the mevalonate pathway was demonstrated previously (Wang et al. 2011). In this case, the engineered strain was able to produce around 380 $mg\,l^{-1}$ of α-farnesene using glycerol as carbon and energy source at 3.0% (v/v). The production of α-farnesene was also reported using engineered strains of *Y. lipolytica* (Yang et al. 2016). Using traditional yeast extract-peptone-dextrose (YPD) medium, the production of α-farnesene was around 57 $mg\,l^{-1}$ in shake flasks experiments, while about 260 $mg\,l^{-1}$ was reached using a complex medium composed by several minerals, vitamin B1, glucose, and fructose in bioreactor experiments.

Considering the influence of enzymatic co-factors and vitamins on regulation of gene expression and consequently the production efficiency of engineered microorganisms, Sandoval et al. (2014) demonstrated the role of pantothenate as a metabolic switch in the increase of farnesene bioproduction by an engineered *S. cerevisiae* strain. However, the improvement in terms of concentration was not exposed, but only described that the yield of farnesene after six days in bioreactor experiment increased 75% with the exclusion of pantothenate in the fermentation medium.

Advances in the improvement of farnesene production by metabolic engineered strains have also been described. Wang et al. (2011) described the bioproduction of 0.38 $g\,l^{-1}$ of α-farnesene by a *E. coli* strain using 30 $g\,l^{-1}$ of glycerol as substrate in the fermentation process, while Zhu et al. (2014) reported the production around 1.10 $g\,l^{-1}$ of farnesene by other engineered *E. coli* strain using 20 $g\,l^{-1}$ of glycerol as substrate.

More recently, You et al. (2017) evaluated the potential of byproduct of biodiesel production and glycerol as substrates for β-farnesene production using an engineered *E. coli* constructed by the heterologous expression of mevalonate pathway. In terms of yield, interesting results were obtained, depending of the scale and substrate utilized. In shaker flask experiments, $3.30\,\mathrm{g\,l^{-1}}$ of farnesene was achieved using $20\,\mathrm{g\,l^{-1}}$ of glycerol as substrate, while with the use of $23.5\,\mathrm{g\,l^{-1}}$ of biodiesel byproduct, around $2.90\,\mathrm{g\,l^{-1}}$ of farnesene was produced. Using the same concentration of substrates, the production of farnesene was approximately $2.80\,\mathrm{g\,l^{-1}}$ when biodiesel byproduct was utilized as substrate in a lab-scale bioreactor, while the use of glycerol produced around $8.70\,\mathrm{g\,l^{-1}}$ of sesquiterpene.

The biosynthesis of valencene and its oxygenated derivative, nootkatone, has been studied due to their unique odor characteristics, which has attracted attention of cosmetic, food, and pharmaceutical industries (Wriessnegger et al. 2014; Frohwitter et al. 2014). Moreover, these sesquiterpenes are found in only trace amounts in plant tissues, which makes their extraction from plant matrixes economically unviable (Zelena et al. 2012). Thus, the biotechnological production of valencene and nootkatone, especially by de novo synthesis, has been studied as an alternative (Frohwitter et al. 2014).

The expression of a valencene synthase (VS) from plants in different microorganisms is the most cited approach for valencene production in fermentation process (Beekwilder et al. 2014). In this context, Scholtmeijer et al. (2014) described the expression of VS in a wild-type and in a mutant strain of mushroom-forming fungi identified as *Schizophyllum commune* containing regulator of G-protein signaling (RGS) regulatory protein gene (*thn*) for valencene production. In terms of production efficiency, better results were found for the *thn* strain compared, which reached production levels around $0.016\,\mathrm{g\,l^{-1}}$.

Performing the expression of VS from the heartwood of Nootka cypress (*Callitropsis nootkatensis*) using *S. cerevisiae* and *R. sphaeroides* as host of gene expression, Beekwilder et al. (2014) reported a valencene production of 0.001 and $0.352\,\mathrm{g}^{-1}$, respectively. Frohwitter et al. (2014) also reported expression of VS from orange and Nootka cypress using *C. glutamicum* as a platform of expression. In this case, five strains were able to produce valencene in minimal medium-containing glucose as a carbon source, but only three exhibited concentrations of valencene superior to $0.15\,\mathrm{mg\,l^{-1}}$ and the maximum concentration achieved was around $2.40\,\mathrm{mg\,l^{-1}}$.

In a study carried with co-expression of *C. nootkatensis* valencene oxidase (VO) and VS in an engineered WAT11 yeast strain showed the production of valencene and its oxygenated derivatives trans-nootkatol and nootkatone (Cankar et al. 2014). Using in situ extraction with *n*-dodecane produced around $1.3\,\mathrm{mg\,l^{-1}}$ of valencene, $0.116\,\mathrm{mg\,l^{-1}}$ of *trans*-nootkatol and $3\,\mathrm{\mu g\,l^{-1}}$ of nootkatone, while applying an extraction procedure using ethyl acetate after fermentation resulted in a selective production of nootkatone, around $144\,\mathrm{\mu g\,l^{-1}}$.

Wriessnegger et al. (2014) described de novo production of valencene, *trans*-nootkatol, and nootkatone by four engineered strains of *P. pastoris*. In this study, the strain *Pp*ValS was able to produce about $50\,\mathrm{mg\,l^{-1}}$ of valencene, while in the strain *Pp*[HPO/CPR]ValS/ADH-C3/tHMG1 with overexpression of a *P. pastoris* alcohol dehydrogenase and truncated hydroxy-methylglutaryl-CoA reductase (tHMG1p) the maximum concentration of nootkatone achieved was around to $208\,\mathrm{mg\,l^{-1}}$. More recently, de novo heterologous production of nootkatone was reported with an engineered strain of *Y. lipolytica* using the co-expression of valencene synthase CnVS, codon-optimized nootkatone synthase opCYP706M1 and codon-optimized NADPH-cytochrome P450 reductase opAtCPR1 (Guo et al. 2018a,b). In this study, the concentration of nootkatone obtained using the engineered yeast strains ranged from approximately 0.045 to 0.980 $mg\,l^{-1}$, showing the potential of co-expression of different genes for improvement of nootkatone production.

Despite the challenges for establishing robust processes to produce aroma compounds by de novo synthesis, this approach is considered the most suitable for industrial purposes. The greatest drawbacks of these processes are related to yield and final concentrations of the aroma compound produced, which increase the costs in industrial scale. However, the construction of new engineered strains have shown interesting results and can provide promising ways to produce different aroma compounds.

## 5.3 Enzymatic Production of Bioflavors

Recently, some researchers have investigated the use of enzymes in flavor production. There are thousands of enzymes in the nature, and about 400 are available in the market, used for stereoselective organic synthesis and also for biotechnological production of flavors (Dubal et al. 2008). The world market of enzymes is more than US$1 billion and Novozymes is the largest player, followed by DSM and DuPont (Adrio and Demain 2014).

The increasing application of enzymes in industrial processes creates a growing demand for improved and new biocatalysts. The production of enzymes can be increased by using recombinant DNA techniques and genetic modification of microorganisms. Advances in genomics, metagenomics, proteomics, and development of expression systems have facilitated the discovery of new microbial enzymes and the creation of enzymes with improved catalytic properties by rational redesign of existing biocatalysts or combinatorial methods that search for the desired functionality in libraries (Adrio and Demain 2014; Patel et al. 2017).

Once isolated, enzymes offer the most direct access to a given flavor compound, provide high selectivity, high efficiency, high reaction speed, and catalytic activity in both reaction directions and when a natural substrate is used, the final aroma compound obtained can also be labeled as natural (Shaaban et al. 2016; Bel-Rhlid et al. 2018). The main enzymes classes used in biotechnological process are hydrolytic enzymes, transferases, oxidoreductases, and lyases (Menzel and Schreier 2007).

Hydrolitc enzymes present some advantages like low cost, wide availability, and substrate specificity, and they are active without the need of cofactors. They are widely used for ester synthesis, generally conducted in organic media (Shaaban et al. 2016). Among this class, lipases (EC 3.1.1.3) are serine hydrolases that perform the hydrolysis of lipids to fatty acids and glycerol. They can be used for esterification (acid and alcohol), transesterification (ester and alcohol), and interesterification (ester and acid) (Hasan et al. 2006; Geoffry and Achur 2018).

Lipases are mainly used in the production of esters and lactones. The use of lipases to produce esters of acids from acetic acid to hexanoic acid and alcohols from methanol to hexanol, geraniol, and citronellol has already been reported; however, the esterification of short-chain substrates are not interesting due to the low affinity of the enzymes with these substrates. Industrially, esters can be produced enzymatically using *Mucor miehei* and *Candida rugosa* lipases (Dubal et al. 2008; Akacha and Gargouri 2015).

Research on using commercial enzymes to produce aromatic esters enzymatically has increased as well. The main enzyme used is Novozym 435® from Novozymes, and the process condition, such as temperature and reaction time, varies according to the substrate used, amount of enzyme added, and molar ration between alcohol and acid (Chen et al. 2011; Hoang and Matsuda 2016; Santos et al. 2016; Sá et al. 2017).

Grosso et al. (2013) optimized the production of ethyl butyrate using an immobilized lipase from *Rhizopus oryzae*, obtaining a maximum concentration of 0.106 M, corresponding to a conversion of 47%. The authors employed a central composite rotatable design and evaluated the effects of temperature, initial butyric acid concentration, and initial molar ratio of ethanol/acid. In a similar study, response surface methodology was used to optimize ethyl butyrate production by *Candida antarctica* immobilized lipase. The optimal conditions identified were 0.04 M of substrate concentration, 7% of enzyme at 34 °C for 96 hours, achieving a reaction yield of 72.9% (Rodriguez-Nogales et al. 2005). Ethyl butyrate could also be produced by a lipase from *C. rugosa* immobilized in silica and coated with surfactant (Thakar and Madamwar 2005).

Short-chain flavor esters synthesis was also evaluated using whole-cell biocatalysts. In this study, a gene of a new lipase isolated from a metagenomic study was cloned into *E. coli* BL21 (DE3), and esters were produced by transesterification and esterification reactions in organic media (Brault et al. 2014).

Using the monoterpene α-terpineol and thioacetic acid as substrates, it is possible to obtain two α-terpineol thioacetate and α-terpineol thiols in a reaction catalyzed by nonimmobilized lipases. The compounds produced have characteristic aromas of exotic, sweet, roasted and green, fresh grapefruit, respectively, and can improve quality of cooked notes and tropical fruit aromas (Bel-Rhlid et al. 2015).

Lipases can also be used for separation of racemic mixtures. One example is the use of lipases of several microorganisms, such as *Penicillium* sp., *Rhizopus* sp., *Trichoderma* sp., and *Candida* sp. for purification of (−)-menthol. In this case, enzymes preferentially hydrolyze the (−)-menthyl esters enantiomer, whereas (+)-menthyl esters are not hydrolyzed (Akacha and Gargouri 2015; Toogood et al. 2015). Serra et al. (2005) reported the resolution of the commercially available racemic *trans*-jasmonate to (−)-*trans*-jasmonate by microbial lipase.

Another enzyme from hydrolases family is cutinase (EC 3.1.1.74). They act on a great variety of substrates, but are used in the flavor industry in the production of short-chain alkyl esters, while lipases have higher affinity for long-chain-length substrates (Shaaban et al. 2016). Short-chain alkyl esters are interesting for aroma industry due to their fruity flavor and are employed in beverages, candies, jelly, jam, yogurt, etc. A cutinase from *Burkholderia cepacia* NRRL B 2320 was used to catalyze the reaction of butanol with butyric or valeric acid in organic media (isooctane). The maximum conversion of 94.6% was obtained at 37 °C after 12 hours using butanol and butyric acid, while a conversion of 87.5% was obtained in the same conditions using valeric acid (Dutta and Dasu 2011). Similarly, Santos and Castro (2006) performed an optimization and achieved a conversion of 75% at 41 °C using an immobilized lipase from *C. rugosa* as biocatalyst and butanol and butyric acid as substrates.

Barros et al. (2009) studied the enzymatic production of ethyl esters in organic media, using a *F. solani pisi* cutinase. They observed yields of 84% for ethyl butyrate, 96% for ethyl valerate, and 97% for ethyl caproate after six hours of reaction and evaluated the best initial concentration of the substrates. Later, the same group evaluated cutinase activity according to initial acid: alcohol molar ratio in the production of hexyl octanoate in a miniemulsion environment. The highest conversion obtained was about 86% for an enzyme concentration of 5 mg ml$^{-1}$ and the highest activity was observed with an acid: alcohol molar ratio of 0.5 (Barros et al. 2011a). Using a central composite rotatable design, it was possible to optimize flavor esters production. The parameters, chain length of alcohols and acids, alcohol, and acid concentration, were evaluated; results show that cutinase is most effective for the synthesis of alkyl ester with the chain length of both substrates in the range of C4–C6, and for concentrations in the range of 80–180 mM, leading to yields of 95% and high initial activities (Barros et al. 2011b).

From the group of oxidoreductases, the most used enzymes are alcohol dehydrogenases, peroxidases, and lipoxygenases. These enzymes catalyze redox processes, which include electrons transfers, thus involving the need for cofactors like flavin, coenzymes, or transition metals (Menzel and Schreier 2007).

Alcohol dehydrogenases (ADH EC 1.1.1.1) catalyze the reversible conversion of aldehydes to the corresponding alcohol, depending on the presence of cofactors (Gargouri et al. 2008). In fruits, this enzyme is produced in a regulated manner, mainly during fruit ripening, and it could be purified from soybean seeds, barley and wheat seeds, cucumber fruit, grapes, pear, tomato, and melon (Akacha and Gargouri 2015; Tesnière and Verriès 2000; Fonseca et al. 2004; Pech et al. 2008). Schulz et al. (2015) were able to obtain (+)-nootkatone from (+)-valencene enzymatically, with a concentration of 360 ml l$^{-1}$ and productivity of 18 mg l$^{-1}$ h$^{-1}$ First, valencene was transformed to the secondary alcohol (nootkatol) by the action of a cytochrome P450 monooxygenase and then, an alcohol dehydrogenase oxidized this compound do nootkatone. This product is appreciated due to fruity sweet, citrusy, grapefruit peel oil-like aroma (Burdock 2010). The difficult of using this class of enzyme is the need of a strategy to regenerate cofactors during and after the reaction, by using co-substrates and the appropriate solvent; therefore, processes using cofactor-dependent enzymes still need to be more studied.

Lipoxygenases (EC 1.13.11.12) are responsible for the hydroperoxidation of polyunsaturated fatty acids (C18:2 and C18:3) containing a *cis, cis*-pentadiene structure. They can be found in animals and plants, in the later, being related to stress responses, wounding, pathogen defenses and also in the biosynthesis of fruits and vegetables volatiles compounds, such as hexanal, hexenal, and hexenol (Pech et al. 2008; Dubal et al. 2008). This enzyme is part of a multi-enzymatic system to produce C6- or C9 – aldehydes and alcohols with green leaf characteristic aroma, acting together with and hydroperoxide-lyase and an alcohol dehydrogenase, called lipoxygenase pathway (Akacha and Gargouri 2015). In some studies, these aldehydes are prepared by mixing polyunsaturated fatty acids with plant material, since plants produce the enzymes for lipoxygenase pathway, as reviewed by Shaaban et al. (2016); however, the large-scale production of these compounds is limited by the low stability of the enzymatic complex in the industrial processes conditions and the need of a cofactor for ADH enzyme, as already discussed.

Another strategy is to express plant enzymes from lipoxygenase pathway in yeasts and bacteria, where there is the possibility of regenerating cofactors and the enzymes can present more stability once they are inside living cells. In this approach, products like 3-hexenol, 2-hexenal can be obtained in a relatively fast process (Menzel and Schreier 2007).

Nacke et al. (2012) demonstrated for the first time a process for production and in situ separation of β-ionone from β-carotene. In their process, the authors employed a selective oxidative carotenoid cleavage dioxygenase (CCD), usually present in plants, and that requires only dissolved oxygen as co-substrate with no need of co-factors. The biotechnological process is innovative compared to the two current routes to obtain natural β-ionone, one using a nonspecific lipoxygenase for carotenoid cleavage and the second using a fungal peroxidase for direct cleavage (Wu and Robinson 1999; Zorn et al. 2003; Nacke et al. 2012). This same compound could be obtained using a carotenoid-cleavage enzyme from *Staphylococcus pasteuri* TS-82. The authors evaluated the activity of a crude enzyme in different carotenoids (β-carotene, lycopene, zeaxanthin, canthaxanthin, and astaxanthin), the optimal temperature, pH, and products obtained in each case (Zhu et al. 2016).

In plants, most of the aroma compounds are produced in the form of odorless glycosides, composed of a sugar moiety, generally glucose, and an aglycone-like monoterpenes, norisoprenoids, benzene derivatives, and long-chain-aliphatic alcohols (Schwab et al. 2015a,b). For this reason, juice and wine industries employ enzymes in their production processes in order to release and enhance flavor and aroma in their products. For example, glucosidases (EC 3.2.1.21) are generally added during wine fermentation to enhance aromas releasing, since the grape terpenols are produced linked to a glucose unit. They are also employed when it is necessary a controlled release of aroma, especially mono- and sesquiterpenes (Michlmayr et al. 2012; González-Pombo et al. 2014). There are few studies regarding the use of glucosidases industrially to produce different aroma molecules. Most research found relates to methods for improvement glucosidases activities and immobilization strategies to obtain better sensory characteristics in specific products (Schwab et al. 2015b).

Vanillin can also be produced enzymatically. Furuya et al. (2014) developed an enzymatic cascade system to produce vanillin from ferulic acid via 4-vinylguaiacol, without the need for coenzymes. In their first work, they identified a coenzyme-independent decarboxylase (Fdc) that converts ferulic acid to 4-vinylguaiacol and a subsequent step with a coenzyme-independent oxygenase (Cso2) that converts 4-vinylguaiacol to vanillin. The genes of these two enzymes were expressed in *E. coli* and the reaction was performed in one pot, obtaining $1.2\,g\,l^{-1}$ of vanillin, being that the second step (Cso2 reaction) limited the rate of vanillin production. In a second study, the same group optimized vanillin production in a two-pot process. First, *E. coli* cells expressing Fdc were used to convert ferulic acid to 4-vinilguaiacol in two hours at pH 9. Then the pH was adjusted to 10.5 and cells expressing Cso2 could efficiently produce vanillin from 4-vinilguaiacol, obtaining $7.8\,g\,l^{-1}$ in 24 hours (Furuya et al. 2015). More recently, they adapted vanillin production process using immobilized enzymes (Furuya et al. 2017).

Although the outlook is promising, there are still some difficulties of using enzymes as biocatalysts for flavor production, such as long and laborious steps for enzyme identification, isolation and purification, high cost of a tailor-made enzyme, and the difference of solubility between the reagents, when the reaction must occur in aqueous systems but the flavor precursors and products are not soluble in water. For these reasons, the enzymatic production of aroma compounds should be studied in more depth in order to find the best suitable catalyst with better performance and high specificity for each substrate (Shaaban et al. 2016; Bel-Rhlid et al. 2018).

## 5.4 Conclusion

It is remarkable how much consumers' preference has recently moved toward more natural additives in food, cosmetics, and pharmaceutical products. In this context, companies have been increasing their portfolio with naturally obtained compounds, most of them employing biotechnological processes in substitution to the formerly used chemical reactions. A great variety of compounds can be obtained in mild reaction using microorganisms or enzymes as biocatalysts, and to lower production costs, agro-industrial wastes can be employed as raw materials for these processes. In an attempt to increase yields, engineered strains have been developed and are preferable industrially in the production of natural flavors.

Although the production of natural aromas has already reached industrial scale, the drawbacks of low yields, high costs in downstream product recovery, regulatory and safety aspects, product labeling, and process robustness still need to be overcome. In this context, a joint effort between academia and industry in applied research for production of natural aromas is fundamental for the development of appropriate solutions to meet the increasing demand for natural food additives.

## References

Abe, F. and Horikoshi, K. (2005). Enhanced production of isoamyl alcohol and isoamyl acetate by ubiquination-deficient *Saccharomyces cerevisiae* mutants. *Cell. Mol. Biol. Lett.* 10: 383–388.

Adrio, J.L. and Demain, A.L. (2014). Microbial enzymes: tools for biotechnological processes. *Biomolecules* 4: 117–139.

Afzal, M.I., Ariceaga, C.C., Boulahya, K.A. et al. (2017). Biosynthesis and role of 3-methylbutanal in cheese by lactic acid bacteria: major metabolic pathways, enzymes involved, and strategies for control. *Crit. Rev. Food Sci. Nutr.* 22: 399–406.

Agrawal, R. and Joseph, R. (2000). Bioconversion of alpha pinene to verbenone by resting cells of *Aspergillus niger*. *Appl. Microbiol. Biotechnol.* 53: 335–337.

Agrawal, R., Deepika, N.U.A., and Joseph, R. (1999). Strain improvement of *Aspergillus* sp. and *Penicillium* sp. by induced mutation for biotransformation of α-pinene to verbenol. *Biotechnol. Bioeng.* 63: 249–252.

Aissou, M., Chemat-Djenni, Z., Yara-Varóns, E. et al. (2017). Limonene as an agro-chemical building block for the synthesis and extraction of bioactive compounds. *C.R. Chim.* 20: 346–358.

Akacha, B.N. and Gargouri, M. (2015). Microbial and enzymatic technologies used for the production of natural aroma compounds: synthesis, recovery modeling, and bioprocesses. *Food Bioprod. Process.* 94: 675–706.

Alonso-Gutierrez, J., Chan, R., Batth, T.S. et al. (2013). Metabolic engineering of *Escherichia coli* for limonene and perillyl alcohol production. *Metab. Eng.* 19: 33–41.

An, J.U. and Oh, D.K. (2013). Increased production of γ-lactones from hydroxy fatty acids by whole *Waltomyces lipofer* cells induced with oleic acid. *Appl. Microbiol. Biotechnol.* 97: 8265–8272.

Badgujar, K.C., Sasaki, T., and Bhanage, B.M. (2015). Synthesis of lipase nano-bioconjugatesas an efficient biocatalyst: characterization and activity – stability studies with potential biocatalytic applications. *RSC Adv.* 5 (68): 55238–55251.

Barros, D.P.C., Fonseca, L.P., Fernandes, P. et al. (2009). Biosynthesis of ethyl caproate and other short ethyl esters catalyzed by cutinase in organic solvent. *J. Mol. Catal. B: Enzym.* 60: 178–185.

Barros, D.P.C., Fernandes, P., Cabral, J.M.S., and Fonseca, L.P. (2011a). Synthetic application and activity of cutinase in an aqueous, miniemulsion model system: hexyl octanoate synthesis. *Catal. Today* 173: 95–102.

Barros, D.P.C., Azevedo, A.M., Cabral, J.M.S., and Fonseca, L.P. (2011b). Optimization off lavor esters synthesis by *Fusarium solani pisi* cutinase. *J. Food Biochem.* 1–10.

Beekwilder, J., van Houwelinger, A., Cancar, K. et al. (2014). Valencene synthase from the heartwood of Nootka cypress (*Callitropsis nootkatensis*) for biotechnological production of valencene. *Plant Biotechnol. J.* 12: 174–182.

Bel-Rhlid, R., Fleury, R.Y., Welti, D. et al. (2015). Chemo-enzymatic synthesis of α-terpineol thioacetate and thiol derivatives and their use as flavouring compounds. *Yeast* 32: 115–122.

Bel-Rhlid, R., Berger, R.G., and Blank, I. (2018). Bio-mediated generation of food flavors – towards sustainable flavor production inspired by nature. *Trends Food Sci. Technol.* 78: 134–143.

Bennett, J.W. and Inamdar, A.A. (2015). Are some fungal volatile organic compounds (VOCs) mycotoxins? *Toxins* 22: 3785–3804.

Berger, R.G. (2007). *Flavours and Fragrances: Chemistry, Bioprocessing and Sustainability.* Chemistry and Materials Science. Springer-Verlag Berlin Heidelberg.

Berger, R.G. (2009). Biotechnology of flavours—the next generation. *Biotechnol. Lett.* 31: 1651–1659.

Berger, R. (2015). Biotechnology as a source of natural volatile flavours. *Curr. Opin. Food Sci.* 1: 38–43.

Bhowmik, S.N. and Patil, R.T. (2018). Application of microbial biotechnology in food processing. In: *New and Future Developments in Microbial Biotechnology and Bioengineering. Crop Improvement Through Microbial Biotechnology* (eds. R. Prasad, S.S. Gill and N. Tuteja), 73–106. Elsevier.

Bicas, J.L., Barros, F.F.C., Wagner, R. et al. (2008a). Optimization of R-(+)-α-terpineol production by the biotransformation of R-(+)-limonene. *J. Ind. Microbiol. Biotech.* 35: 1061–1070.

Bicas, J.L., Fontanille, P., Pastore, G.M., and Larroche, C. (2008b). Characterization of monoterpene biotransformation in two pseudomonads. *J. Appl. Microbiol.* 105: 1991–2001.

Bicas, J.L., Dionísio, A.P., and Pastore, G.M. (2009). Bio-oxidation of terpenes: an approach for the flavor industry. *Chem. Rev.* 109: 4518–4531.

Bicas, J.L., Fontanille, P., Pastore, G.M., and Larroche, C. (2010). A bioprocess for the production of high concentrations of R-(+)-α-terpineol from R-(+)-limonene. *Process Biochem.* 45: 481–486.

Bicas, J.L., Maróstica, M.R. Jr., Barros, F.F.C. et al. (2016). Bioadditives produced by fermentation. In: *Fermentation Processes Engineering in the Food Industry* (eds. C.R. Soccol, A. Pandey and C. Larroche), 371–403. CRC Press.

Bier, M.C.J., Medeiros, A.B., and Soccol, C.R. (2017). Biotransformation of limonene by an endophytic fungus using synthetic and orange residue-based media. *Fungal Biol.* 121: 137–144.

Binder, D., Frohwitter, J., Mahr, R. et al. (2016). Light-controlled cell factories: employing photocaged isopropyl-β-D-thiogalactopyranoside for light-mediated optimization of *lac* promoter-based gene expression and (+)-valencene biosynthesis in *Corynebacterium glutamicum*. *Appl. Environ. Microbiol.* 82: 6141–6149.

Braga, A. and Belo, I. (2015). Production of γ-decalactone by *Yarrowia lipolytica*: insights into experimental conditions and operating mode optimization. *J. Chem. Technol. Biotechnol.* 90: 559–565.

Braga, A. and Belo, I. (2016). Biotechnological production of γ-decalactone, a peach like aroma, by *Yarrowia lipolytica*. *World J. Microbiol. Biotechnol.* 32: 169–175.

Brault, G., Shareck, F., Hurtubise, Y. et al. (2014). Short-chain flavor ester synthesis in organic media by an *E. coli* whole-cell biocatalyst expressing a newly characterized heterologous lipase. *PLoS One* 3: e91872.

Burdock, G.A. (2010). *Fenaroli's Handbook of Flavor Ingredients*, 6e. Boca Raton, FL: CRC Press.

Cankar, K., van Houwelingen, A., Goedbloed, M. et al. (2014). Valencene oxidase CYP706M1 from Alaska cedar (*Callitropsis nootkatensis*). *FEBS Lett.* 588: 1001–1007.

Carrol, A.L., Desai, S.H., and Atsume, S. (2016). Microbial production of scent and flavor compounds. *Curr. Opin. Biotechnol.* 37: 8–15.

Carter, O.A., Peters, R.J., and Croteau, R. (2003). Monoterpene biosynthesis pathway construction in *Escherichia coli*. *Phytochemistry* 64: 425–433.

de Carvalho, C.C.C.R. and de Fonseca, M.M.R. (2003). Towards the bio-production of *trans*- carveol and carvone from limonene: induction after cell growth on limonene and toluene. *Tetrahedron: Asymmetry* 14: 3925–3931.

Castilho-Araiza, C.O., Palmerín-Carreño, D.M., Prado-Barragán, A., and Huerta-Ochoa, S. (2017). On the conceptual design of a partitioning technology for the bioconversion of (+)-valencene to (+)-nootkatone on whole cells: experimentation and modelling. *Chem. Eng. Process.* 122: 493–507.

Cecati, F.M., Magallanes-Noguera, C., Tonn, C.E. et al. (2018). Ecofriendly chemical diversification of *Eupatorium buniifolium* essential oil by endophytic fungi. *Process Biochem.* 64: 93–102.

Celinska, E., Bonikowski, R., Biatas, W. et al. (2018). *Pichia cactophila* and *Kluyveromyces lactis* are highly efficient microbial cell factories of natural amino-acid derived aroma compounds. *Molecules* 23: 1–11.

Červeňanský, I., Mihaľ, M., and Markoš, J. (2017). Potential application of perfusion and pertraction for in situ product removal in biocatalytic 2-phenylethanol production. *Sep. Purif. Technol.* 183: 11–20.

Chantasuban, T., Santomauro, F., Gore-Lloyd, D. et al. (2018). Elevated production of the aromatic fragrance molecule, 2-phenylethanol, using Metschnikowia pulcherrima through both *de novo* and *ex novo* conversion in batch and continuous modes. *J. Chem. Technol. Biotechnol.* 93: 8.

Chen, H., Chen, J., Chang, C., and Shieh, C. (2011). Optimization of ultrasoundaccelerated synthesis of enzymatic caffeic acid phenethyl ester by response surface methodology. *Ultrason. Sonochem.* 18: 455–459.

Craig, T. and Daugulis, A.J. (2013). Polymer characterization and optimization of conditions for the enhanced bioproduction of benzaldehyde by *Pichia pastoris* in a two-phase partitioning bioreactor. *Biotechnol. Bioeng.* 110: 1098–1105.

Craig, T. and Daugulis, A.J. (2014). Strategies for improved bioproduction of benzaldehyde by *Pichia pastoris* and the use of hytrel as tubing material for integrated product removal by *in situ* pervaporation. *Biochem. Eng. J.* 82: 97–104.

Deepthi Priya, K., Petkar, M., and Chowdary, G.V. (2015). Bioproduction of aroma compounds from alpha pinene by novel strains. *Int. J. Biol. Sci. Appl.* 2: 15–19.

Denby, C.M., Li, R.A., Vu, V.T. et al. (2018). Industrial brewing yeast engineered for the production of primary flavor determinants in hopped beer. *Nat. Commun.* 9: 1–10.

Dionísio, A.P., Molina, G., Carvalho, D.S. et al. (2012). Natural flavorings from biotechnology for food and beverages. In: *Natural Food Additives, Ingredients and Flavourings* (eds. D. Baines and R. Seal), 231–259. Woodhead Publishing Series in Food Science, Technology and Nutrition.

Du, F.L., Yu, H.L., Xu, J.H., and Li, C.X. (2014). Enhanced limonene production by optimizing the expression of limonene biosynthesis and MEP pathway genes in *E. coli*. *Bioresour. Bioprocess.* 1: 10.

Dubal, S.A., Tilkari, Y.P., Momin, S.A., and Borkar, I.V. (2008). Biotechnological routes in flavor industries. *Adv. Biotechnol.* 8: 20–31.

Duetz, W.A., Fjällman, A.H.M., Ren, S. et al. (2001). Biotransformation of *D*-limonene to (+)- *trans*-carveol by toluene-grown *Rhodococcus opacus* PWD4 cells. *Appl. Environ. Microbiol.* 67: 2829–2832.

Dutta, K. and Dasu, V.V. (2011). Synthesis of short chain alkyl esters using cutinase from *Burkholderia cepacia* NRRL B2320. *J. Mol. Catal. B: Enzym.* 72: 150–156.

Elston, A., Lin, J., and Rouseff, R. (2005). Determination of the role of valencene in orange oil as a direct contributor to aroma quality. *Flavour Fragance J.* 20: 381–386.

Etschmann, M.M.W., Sell, D.J.S., and Schrader, J. (2003). Screening of yeasts for the production of the aroma compound 2-phenylethanol in a molasses-based medium. *Biotechnol. Lett.* 25: 531–536.

Fanaro, G.B., Fontanille, P., Maróstica, M.R., and Bicas, J.L. (2016). Biotechnological aroma compounds. In: *Biotechnological Production of Natural Ingredients for Food Industry* (eds. J.L. Bicas, M.R. Maróstica Jr. and G.M. Pastore), 255–287. Bentham Science Publishers.

Felipe, L.O., Oliveira, A.M., and Bicas, J.L. (2017). Bioaromas – perspectives for sustainable development. *Trends Food Sci. Technol.* 62: 141–153.

Ferrara, M.A., Almeida, D.S., Siani, A.C. et al. (2013). Bioconversion of R-(+)-limonene to perillic acid by the yeast *Yarrowia lipolytica*. *Braz. J. Microbiol.* 44: 1075–1080.

Fonseca, S., Hackler, L., Zvara, A. et al. (2004). Monitoring gene expression along pear fruit development, ripening and senescence using cDNA microarrays. *Plant Sci.* 167: 457–469.

Fraatz, M.A., Berger, R.G., and Zorn, H. (2009a). Nootkatone – a biotechnological challenge. *Appl. Microbiol. Biotechnol.* 83: 35–41.

Fraatz, M.A., Riemer, S.J.L., Stöber, R. et al. (2009b). A novel oxygenase from *Pleurotus sapidus* transforms valencene to nootkatone. *J. Mol. Catal. B: Enzym.* 61: 202–207.

Frohwitter, J., Heider, S.A.E., Peters-Wendisch, P. et al. (2014). Production of the sesquiterpene (+)-valencene by metabolically engineered *Corynebacterium glutamicum*. *J. Biotechnol.* 191: 205–213.

Furuya, T., Miura, M., and Kino, K. (2014). A coenzyme-independent decarboxylase/oxygenase cascade for the efficient synthesis of vanillin. *ChemBioChem* 13: 2248–2254.

Furuya, T., Miura, M., Kuroiwa, M., and Kino, K. (2015). High-yield production of vanillin from ferulic acid by a coenzyme-independent decarboxylase/oxygenase two-stage process. *New Biotechnol.* 32: 335–339.

Furuya, T., Kuroiwa, M., and Kino, K. (2017). Biotechnological production of vanillin using immobilized enzymes. *J. Biotechnol.* 243: 25–28.

Gallage, N.J. and Møller, B.L. (2015). Vanillin–bioconversion and bioengineering of the most popular plant flavor and its *de novo* biosynthesis in the vanilla orchid. *Mol. Plant* 8: 40–57.

Gallage, N.J. and Møller, B.L. (2018). Vanilla: the most popular flavour. In: *Biotechnology of Natural Products* (eds. W. Schwab, B. Lange and M. Wüst), 3–24. Cham: Springer.

Gargouri, M., Ben Akacha, N., Kotti, F., and Ben Rejeb, I. (2008). Lipoxygenase pathway: valorization of plant oils and aroma biosynthesis. *Biotechnol. Agron. Soc. Environ.* 12: 85–202.

Geoffry, K. and Achur, R.N. (2018). Screening and production of lipase from fungal organisms. *Biocatal. Agric. Biotechnol* 14: 241–253.

González-Pombo, P., Fariña, L., Carrau, F. et al. (2014). Aroma enhancement in wines using co-immobilized *Aspergillus niger* glycosidases. *Food Chem.* 143: 185–191.

Groeneveld, M., van Beek, H.L., Duetz, W.A., and Fraaije, M.W. (2016). Identification of a novel oxygenase capable of regiospecific hydroxylation of *D*-limonene into (+)-*trans*-carveol. *Tetrahedron* 72: 7263–7267.

Grosso, C., Ferreira-Dias, S., and Pires-Cabral, P. (2013). Modelling and optimization of ethyl butyrate production catalysed by *Rhizopus oryzae* lipase. *J. Food Eng.* 115: 475–480.

Guo, D., Zhang, L., Kong, S. et al. (2018a). Metabolic engineering of *Escherichia coli* for production of 2-phenylethanol and 2-phenylethyl acetate from glucose. *J. Agric. Food. Chem.* 66: 5886–5891.

Guo, X., Sun, J., Li, D., and Lu, W. (2018b). Heterologous biosynthesis of (+)-nootkatone in unconventional yeast *Yarrowia lipolytica*. *Biochem. Eng. J.* 137: 125–131.

Gupta, C., Prakash, D., and Gupta, S. (2015). A biotechnological approach to microbial based perfumes and flavours. *J Microbiol Exp.* 2: 34–42.

Han, S.Y., Pan, Z.Y., Huang, D.F. et al. (2009). Highly efficient synthesis of ethyl hexanoate catalyzed by CALB-displaying *Saccharomyces cerevisiae* whole-cells in non-aqueous phase. *J. Mol. Catal. B: Enzym.* 59: 168–172.

Hansen, E.H., Møller, B.L., Kock, G.R. et al. (2009). *De novo* biosynthesis of vanillin in fission yeast (*Schizosaccharomyces pombe*) and Baker's yeast (*Saccharomyces cerevisiae*). *Appl. Environ. Microbiol.* 75: 2765–2774.

Hansen J., Halkjaer E., Polur H., Sheridan J. M., Heal J. R., Hamilton W. D. O. 2014. Compositions and methods for the biosynthesis of vanillan or vanillin beta-dglucoside. US Patent 2014/0245496 A1.

Hasan, F., Shah, A.A., and Hameed, A. (2006). Industrial applications of microbial lipases. *Enzyme Microb. Technol.* 39: 235–251.

Hoang, H.N. and Matsuda, T. (2016). Expanding substrate scope of lipase-catalyzed transesterification by the utilization of liquid carbon dioxide. *Tetrahedron* 72: 7229–7234.

Hua, D. and Xu, P. (2011). Recent advances in biotechnological production of 2-phenylethanol. *Biotechnol. Adv.* 29: 654–660.

Huelin, F.E. and Murray, K.E. (1966). Alpha-farnesene in the natural coating of apples. *Nature* 210: 1260–1261.

Huembelin M., Beekwilder M.J., Gerardus J., Kierkels T. 2014. Rhodobacter for preparing terpenoids. WO2014014339A2.

Hughes, P.R. (1975). Pheromones of *Dendroctonus*: origin of α-pinene oxidation products present in emergent adults. *J. Insect Physiol.* 21: 687–691.

Jain, A.N., Khan, T.R., and Daugulis, A.J. (2010). Bioproduction of benzaldehyde in a solid-liquid tow-phase portioning bioreactor using *Pichia pastoris*. *Biotechnol. Lett* 32: 1649–1654.

Jongedijk, E., Cankar, K., Buchhaupt, M. et al. (2016). Biotechnological production of limonene in microorganisms. *Appl. Microbiol. Biotechnol.* 100: 2927–2938.

Kang, Z., Zhang, C., Du, G., and Chen, J. (2014). Metabolic engineering of *Escherichia coli* for production of 2-phenylethanol from renewable glucose. *Appl. Biochem. Biotechnol.* 172: 2012–2021.

Kim, T.-T., Lee, S.W., and Oh, M.-K. (2014a). Biosynthesis of 2-phenylethanol from glucose with genetically engineered *Kluyveromyces marxianus*. *Enzyme Microb. Technol.* 61-62: 44–47.

Kim, B., Cho, B.R., and Hahn, J.S. (2014b). Metabolic engineering of *Saccharomyces cerevisiae* for the production of 2-phenylethanol via Ehrlicj pathway. *Biotechnol. Bioeng.* 111: 115–124.

Koroch, A.R., Juliani, H.R., and Zygadlo, J.A. (2007). Bioactivity of essential oils and their components. In: *Flavors and Fragrances: Chemistry, Bioprocessing and Sustainability* (ed. R.G. Berger), 87–116. Berlin: Springer Verlag.

Krings, U., Lehnert, N., Fraatz, M.A. et al. (2009). Autoxidation versus biotransformation of α-pinene to flavors with *Pleurotus sapidus*: regioselective hydroperoxidation of α-pinene and stereoselective dehydrogenation of verbenol. *J. Agric. Food. Chem.* 57: 9944–9950.

Krishna, S.H., Manohar, B., Divakar, S. et al. (2000). Optimization of isoamyl acetate production by using immobilized lipase from *Mucor miehei* by response surface methodology. *Enzyme Microb. Technol.* 26: 131–136.

Krügener, S., Krings, U., Zorn, H., and Berger, R.G. (2010). A dioxygenase of *Pleurotus sapidus* transforms (+)-valencene regio-specifically to (+)-nootkatone via a stereo-specific allylic hydroperoxidation. *Bioresour. Technol.* 101: 457–462.

Kuis, A.J., Levisson, M., Mars, A.E. et al. (2017). Ethyl acetate production by the elusive alcohol acetyltransferase from yeast. *Metab. Eng.* 41: 91–101.

Kunjapur, A.M., Tarasova, Y., and Prather, K.L. (2014). Synthesis and accumulation of aromatic aldehydes in an engineered strain of *Escherichia coli*. *J. Am. Chem. Soc.* 136: 11644–11654.

Lee, S.Y., Kim, S.H., Hong, C.Y. et al. (2015). Biotransformation of (−)-α-Pinene by whole cells of white rot fungi, *Ceriporia* sp. ZLY-2010 and *Stereum hirsutum*. *Mycobiology* 43: 297–302.

Leffingwell & Associates: Flavor & Fragrance Industry Leaders. 2018. http://www.leffingwell.com/top_10.htm Accessed in 03/04/2018.

Leffingwell, J.C. and Leffingwell, D. (2015). Biotechnology – conquests and challenges in flavors & fragrances. *Leffingwell Reports* 7: 1–11.

Liu, J., Jiang, J., Bai, Y. et al. (2018). Mimicking a new 2-phenylethanol production pathway from *Proteus mirabilis* JN458 in *Escherichia coli*. *J. Agric. Food. Chem.* 66: 3498–3504.

Longo, M.A. and Sanromán, M.A. (2006). Production of food aroma compounds. *Food Technol. Biotechnol.* 44: 335–353.

Löser, C., Urit, T., Nehl, F., and Bley, T. (2011). Screening of *Kluyveromyces* strains for the production of ethyl acetate: design and evaluation of a cultivation system. *Eng. Life Sci.* 11: 369–381.

Löser, C., Urit, T., and Bley, T. (2014). Prespectives for the biotechnological production of ethyl acetate by yeasts. *Appl. Microbiol. Biotechnol.* 98: 5397–5415.

Maróstica, M.R. Jr. and Pastore, G.M. (2007). Biotransformation of limonene: a review of the main metabolic pathways. *Quim. Nova* 30: 382–387.

Martin, V., Giorello, F., Fariña, L. et al. (2016). *De novo* synthesis of benzenoid compounds by the yeast *Hanseniaspora vineae* increases the flavor diversity of wines. *J. Agric. Food. Chem.* 64: 4574–4583.

Martínez, O., Sánchez, A., Font, X., and Barrena, R. (2018). Bioproduction of 2-phenylethanol and 2-phenethyl acetate by *Kluyveromyces marxianus* through the solid-state fermentation of sugarcane bagasse. *Appl. Microbiol. Biotechnol.* 102: 4703–4716.

Medeiros, A.B.P., Pandey, A., Freitas, R.J.S. et al. (2000). Optimization of the production of aroma compounds by *Kluyveromyces marxianus* in solid-state fermentation using factorial design and response surface methodology. *Biochem. Eng. J.* 6: 33–39.

Medeiros, A.B.P., Pandey, A., Christen, P. et al. (2001). Aroma compounds produced by *Kluyveromyces marxianus* in solid state fermentation on a packed bed column bioreactor. *World J. Microbiol. Biotechnol.* 17: 767–771.

Menzel, M. and Schreier, P. (2007). Enzymes and flavour biotechnology. In: *Flavours and Fragrances* (ed. R.G. Berger). Berlin, Heidelberg: Springer.

Mewalal, R., Rai, D.K., Kainer, D. et al. (2017). Plant-derived terpenes: a feedstock for specialty biofuels. *Trends Biotechnol.* 35: 227–240.

Michlmayr, H., Nauer, S., Brandes, W. et al. (2012). Release of wine monoterpenes from natural precursors by glycosidases from *Oenococcus oeni*. *Food Chem.* 135: 80–87.

Misawa, N. (2011). Pathway engineering for functional isoprenoids. *Curr. Opin. Biotechnol.* 22: 627–633.

Molina, G. and Fanaro, G.B. (2016). Introductory overview of biotechnological additives. In: *Biotechnological Production of Natural Ingredients for Food Industry* (eds. J.L. Bicas, M.R. Maróstica Jr. and G.M. Pastore), 3–20. Bentham Science Publishers.

Molina, G., Pessôa, M.G., Pimentel, M.R. et al. (2014). Production of natural flavor compounds using monoterpenes as substrates. In: *New Developments in Terpenes Research* (ed. J. Hu), 1–24. New York: Nova Publishers.

Molina, G., Bution, M.L., Bicas, J.L. et al. (2015). Comparative study of the bioconversion process using R-(+)- and S-(−)-limonene as substrates for Fusarium *oxysporum* 152B. *Food Chem.* 174: 606–613.

Møller, K., Christensen, B., Förster, J. et al. (2002). Aerobic glucose metabolism *of Saccharomyces kluyveri*: growth, metabolite production, and quantification of metabolic fluxes. *Biotechnol. Bioeng.* 77: 186–193.

Moradi, H., Asadollahi, M.A., and Nahvi, I. (2013). Improved γ-decalactone production from castor oil by fed-batch cultivation of *Yarrowia lipolytica*. *Biocatal. Agric. Biotechnol.* 2: 64–68.

Nacke, C., Huettmann, S., Etschmann, M.M.W., and Schrader, J. (2012). Enzymatic production and in situ separation of natural ß-ionone from ß–carotene. *J. Ind. Microbiol. Biotechnol.* 39: 1771–1778.

Oda, S., Inada, Y., Kato, A. et al. (1995). Production of (S)-citronellic acid and (R)-citronellol with an interface bioreactor. *J. Ferment. Bioeng.* 80: 559–564.

Omarini, A.B., Plagemann, I., Schimanski, S. et al. (2014). Crosses between monokaryons of *Pleurotus sapidus* or *Pleurotus florida* show an improved biotransformation of (+)-valencene to (+)-nootkatone. *Bioresour. Technol.* 171: 113–119.

Palmerín-Carreño, D.M., Rutiaga-Quiñones, O.M., Verde Calvo, J.R. et al. (2015). Screening of microoganisms for bioconversion of (+)-valencene to (+)-nootkatone. *LWT Food Sci. Technol.* 64: 788–793.

Palmerín-Carreño, D.M., Castilho-Araiza, C.O.C., Rutiaga-Quiñones, O.M. et al. (2016). Whole cell bioconversion of (+)-valencene to (+)-nootkatone by *Yarrowia lipolytica* using a three phase partitioning bioreactor. *Chem. Technol. Biotechnol.* 91: 1164–1172.

Paludo, N., Alves, J.S., Altmann, C. et al. (2015). The combined use of ultrasound and molecular sieves improves the synthesis of ethyl butyrate catalyzed by immobilized *Thermomyces lanuginosus* lipase. *Ultrason. Sonochem.* 22: 89–94.

Patel, A.K., Simghania, R.R., and Pandey, A. (2017). Production, purification and application of microbial enzymes. In: *Biotechnology of Microbial Enzymes – Production, Biocatalysis and Industrial Applications* (ed. G. Brahmachari), 13–41. Academic Press.

Pech, J.C., Latché, A., and van der Rest, B. (2008). Genes involved in the biosynthesis of aroma volatiles and biotechnological applications. In: *Fruit and Vegetable Flavour – Recent Advances and Future Prospects* (eds. B. Brückenr and S.G. Wyllie), 254–271. Woodhead Publishing Series in Food Science, Technology and Nutrition.

Pérez-Mosqueda, L.M., Trujillo-Cayado, L.A., Carrillo, F. et al. (2015). Formulation and optimization by experimental design of eco-friendly emulsions based on D-limonene. *Colloids Surf., B* 128: 127–131.

Plata, C., Millán, C., Mauricio, J.C., and Ortega, J.M. (2003). Formation of ethyl acetate and isoamyl acetate by various species of wine yeasts. *Food Microbiol.* 20: 217–224.

Prasad, M.V., Narayana, A.V., Raju, C.A.I. et al. (2014). Optimum production of γ-decalactone by *Sporidiobolus salmonicolor* using response surface methodology. *Int. J. Chemtech Res.* 6: 3478–3486.

Ramos, A.S., Fiaux, S.B., and Leite, S.G.F. (2008). Production of 6-pentyl-α-pyrone by *Trichoderma harzianum* in solid-state fermentation. *Braz. J. Microbiol.* 39: 712–717.

Rodriguez-Nogales, J.M., Roura, E., and Contreras, E. (2005). Biosynthesis of ethyl butyrate using immobilized lipase: a statistical approach. *Process Biochem.* 40: 63–68.

Rojas, V., Gil, J.V., Pinaga, F., and Manzanares, P. (2001). Studies on acetate ester production by non-Saccharomyces wine yeasts. *Int. J. Food Microbiol.* 70: 283–289.

Rottava, I., Cortina, P.F., Zanella, C.A. et al. (2010a). Microbial oxidation of (−)-α-pinene to verbenol production by newly isolated strains. *Appl. Biochem. Biotechnol.* 162: 719–732.

Rottava, I., Cortina, P.F., Zanella, C.A. et al. (2010b). Microbial oxidation of (−)-α-pinene to verbenol production by newly isolated strains. *Appl. Biochem. Biotechnol.* 162: 2221–2231.

Rottava, I., Cortina, P.F., Martello, E. et al. (2011). Optimization of α-terpineol production by biotransformation of R-(+)-limonene and (−)-β-pinene. *Appl. Biochem. Biotechnol.* 164: 514–523.

Rozenbaum, H.F., Patitucci, M.L., Antunes, O.A.C., and Pereira, N. (2006). Production of aromas and fragrances through microbial oxidation of monoterpenes. *Braz. J. Chem. Eng.* 23: 273–279.
Sá, A.G.A., Meneses, A.C., Araújo, P.H.H., and Oliveira, D. (2017). A review on enzymatic synthesis of aromatic esters used as flavor ingredients for food, cosmetics and pharmaceuticals industries. *Trends Food Sci. Technol.* 69: 95–105.
Saerens, S.M.G., Delvaux, F., Verstrepen, K.J. et al. (2008). Parameters affecting ethyl ester production by *Saccharomyces cerevisiae* during fermentation. *Appl. Environ. Microbiol.* 74: 454–461.
Sales, A., Paulino, B.N., Pastore, G.M., and Bicas, J.L. (2018a). Biogeneration of aroma compounds. 19: 77–84.
Sales, A., Afonso, L.F., Americo, J.A. et al. (2018b). Monoterpene biotransformation by *Colletotrichum* species. *Biotechnol. Lett.* 40: 561–567.
Sandoval, C.M., Ayson, M., Moss, N. et al. (2014). Use of pantothenate as a metabolic switch increases the genetic stability of farnesene producing Saccharomyces cerevisiae. *Metab. Eng.* 25: 215–226.
Santos, J.C. and Castro, H.F. (2006). Optimization of lipase-catalysed synthesis of butyl butyrate using a factorial design. *World J. Microbiol. Biotechnol.* 22: 1007–1011.
Santos, P., Zabot, G.L., Meireles, M.A.A., and Mazutti, M.A. (2016). Synthesis of eugenyl acetate by enzymatic reactions in supercritical carbon dioxide. *Biochem. Eng. J.* 114: 1–9.
Sarmiento, G.P., Rouge, P.D., Fabian, L. et al. (2011). Efficient synthesis of chiral Δ2-1,3,4-thiadiazolines from α-pinene and verbenone. *Tetrahedron: Asymmetry* 22: 1924–1929.
Savithriry, N., Gage, D., Fu, W., and Oriel, P. (1998). Degradation of pinene by *Bacillus pallidus* BR425. *Biodegradation* 9: 337–341.
Schempp, F.M., Drummond, L., Buchhaupt, M., and Schrader, J. (2018). Microbial cell factories for the production of terpenoid flavor and fragrance compounds. *J. Agric. Food. Chem.* 14: 2247–2258.
Schewe, H., Mirata, M.A., Holtmann, D., and Schrader, J. (2011). Biooxidation of monoterpenes with bacterial monooxygenases. *Process Biochem.* 46: 1885–1899.
Scholtmeijer, K., Cankar, K., Beekwilder, J. et al. (2014). Production of (+)-valencene in the mushroom-forming fungus *S. commune*. *Appl. Microbiol. Biotechnol.* 98: 5059–5068.
Schulz, S., Girhard, M., Gassmeyer, S.K. et al. (2015). Selective enzymatic synthesis of the grapefruit flavor (+)-nootkatone. *Chem. Cat. Chem.* 7: 601–604.
Schwab, W., Fischer, T.C., Giri, A., and Wuest, M. (2015a). Potential applications of glucosyltransferases in terpene glucoside production: impacts on the use of aroma and fragrance. *Appl. Microbiol. Biotechnol.* 99: 165–174.
Schwab, W., Fischer, T., and Wuest, M. (2015b). Terpene glucoside production: improved biocatalytic processes using glycosyltransferases. *Eng. Life Sci.* 15: 376–386.
Serra, S., Fuganti, C., and Brenna, E. (2005). Biocatalytic preparation of natural flavours and fragrances. *Trends Biotechnol.* 23: 193–198.
Shaaban, H.A., Mahmoud, K.F., Amin, A.A., and Banna, H.E. (2016). Application of biotechnology to the production of natural flavor and fragrance chemicals. *Res. J. Pharm. Biol. Chem. Sci.* 7: 2670–2712.
Shen, L., Nishimura, Y., Matsuda, F. et al. (2016). Overexpressing enzymes of the Ehrlich pathway and deleting genes of the competing pathway in *Saccharomyces cerevisiae* for increasing 2-phenylethanol production from glucose. *J. Biosci. Bioeng.* 122: 34–39.

Soares, G.P.A., Souza, K.S.T., Vilela, L.F. et al. (2017). γ-Decalactone production by *Yarrowia lipolytica* and *Lindnera saturnus* in crude glycerol. *Prep. Biochem. Biotechnol.* 47: 633–637.

Stribny, J., Gamero, A., Pérez-Torrado, R., and Querol, A. (2015). *Saccharomyces kudriavzevii* and *Saccharomyces uvarum* differ from *Saccharomyces cerevisiae* during the production of aroma-active higher alcohols and acetate esters using their amino acidic precursors. *Int. J. Food Microbiol.* 205: 41–46.

Sun, J., Lu, M., Gillette, N.E., and Wingfield, M.J. (2013). Red turpentine beetle: innocuous native becomes invasive tree killer in China. *Annu. Rev. Entomol.* 58: 293–311.

Suzuki, S., Koeduka, T., Sugiyama, A. et al. (2014). Microbial production of plant specialized metabolites. *Plant Biotechnol.* 31: 465–482.

Tai, Y.N., Xu, M., Ren, J.N. et al. (2016). Optimization of α-terpineol production by limonene biotransformation using *Penicillium digitatum* DSM62840. *J. Sci. Food Agric.* 96: 954–961.

Tandon B. 2017. Flavors (Food & Beverages) Market-Global opportunity analysis and industry forecast 2017–2023. Allied Market Research. 133p.

Tashiro, M., Kiyota, H., Kawai-Noma, S. et al. (2016). Bacterial production of pinene by a laboratory-evolved pinene-synthase. *ACS Synth. Biol.* 5: 1011–1020.

Tesnière, C. and Verriès, C. (2000). Molecular cloning and expression of cDNAs encoding alcohol dehydrogenases from *Vitis vinifera* L. during berry development. *Plant Sci.* 157: 77–88.

Thakar, A. and Madamwar, D. (2005). Enhanced ethyl butyrate production by surfactant coated lipase immobilized on silica. *Process Biochem.* 40: 3263–3266.

Toogood, H.S., Cheallaigh, A.N., Tait, S. et al. (2015). Enzymatic menthol production: one-pot approach using engineered *Escherichia coli*. *ACS Synth. Biol.* 4: 1112–1123.

Trytek, M. and Fiedurek, J. (2005). A novel psychrotrophic fungus, *Mortierella minutissima*, for D-limonene biotransformation. *Biotechnol. Lett.* 27: 149–153.

Trytek, M., Jędrzejewski, K., and Fiedurek, J. (2014). Bioconversion of α-pinene by a novel cold-adapted fungus *Chrysosporium pannorum*. *J. Ind. Microbiol. Biotechnol.* 42: 181–188.

UBIC Consulting – The world biotech flavours market., 2014. http://ubicconsulting.com/wp-content/uploads/2015/08/Biotech-Flavor.pdf Accessed 03/04/20178.

Ullah, I., Khan, A.L., Ali, L. et al. (2015). Benzaldehyde as an insecticidal, antimicrobial, and antioxidant compound produced by *Photorhabdus temperata* M1021. *J. Microbiol.* 53: 127–133.

Van Rensburg, E., Moleleki, N., van der Walt, J.P. et al. (1997). Biotransformation of (+)limonene and (−)piperitone by yeast and yeast-like fungi. *Biotechnol. Lett.* 19: 779–782.

Vespermann, K.A., Paulino, B.N., Barcelos, M.C. et al. (2017). Biotransformation of α-and β-pinene into flavor compounds. *Appl. Microbiol. Biotechnol.* 101: 1805–1817.

Wang, C., Yoon, S.H., Jang, H.J. et al. (2011). Metabolic engineering of *Escherichia coli* for α-farnesene production. *Metab. Eng.* 13: 648–655.

Willrodt, C., David, C., Cornelissen, S. et al. (2014). Engineering the productivity of recombinant *Escherichia coli* for limonene formation from glycerol in minimal media. *Biotechnol. J.* 9: 1000–1012.

Willrodt, C., Halan, B., Karthaus, L. et al. (2017). Continuous multistep synthesis of perillic acid from limonene by catalytic biofilms under segmented flow. *Biotechnol. Bioeng.* 114: 281–290.

Wriessnegger, T., Augustin, P., Engleder, M. et al. (2014). Production of the sesquiterpenoid (+)-nootkatone by metabolic engineering of *Pichia pastoris*. *Metab. Eng.* 24: 18–29.

Wright, S.J., Caunt, P., Carter, D., and Baker, P.B. (1986). Microbial oxidation of alpha-pinene by *Serratia marcescens*. *Appl. Microbiol. Biotechnol.* 23: 224–227.

Wu, Z. and Robinson, D.S. (1999). Co-oxidation of beta-carotene catalyzed by soybean and recombinant pea lipoxygenase. *J. Agric. Food. Chem.* 47: 4899–4906.

Wu, J., Wang, J.L., Li, M.H. et al. (2010). Optimization of immobilization for selective oxidation of benzyl alcohol by *Gluconobacter oxydans* using response surface methodology. *Bioresour. Technol.* 101: 8936–8941.

Xu P., Hua D., Lin S., Song L., Ma C., Zhang, Z. 2009. *Streptomyces* strain and the method of converting ferulic acid to vanillin by using the same. US Patent 2009/0186399 A1.

Xu, L., Lou, Q., Cheng, C. et al. (2015). Gut-associated bacteria of *Dendroctonus valens* and their involvement in verbenone production. *Microb. Ecol.* 70: 1012–1023.

Xu, L., Shi, Z., Wang, B. et al. (2016). Pine defensive monoterpene α-pinene influences the feeding behavior of *Dendroctonus valens* and its gut bacterial community structure. *Int. J. Mol. Sci.* 17: 1734.

Yang, X., Nambou, K., Wei, L., and Hua, Q. (2016). Heterologous production of α-farnesene in metabolically engineered strains of *Yarrowia lipolytica*. *Bioresour. Technol.* 216: 1040–1048.

Yin, S., Zhou, H., Xiao, X. et al. (2015). Improving 2-Phenylethanol production via Ehrlich pathway using genetic engineered *Saccharomyces cerevisiae* strains. *Curr. Microbiol.* 70: 762–767.

Yoo, S.K. and Day, D.F. (2002). Bacterial metabolism of α- and β-pinene and related monoterpenes by *Pseudomonas* sp. strain PIN. *Process Biochem.* 37: 739–745.

Yoo, S.K., Day, D.F., and Cadwallader, K.R. (2001). Bioconversion of α- and β-pinene *Pseudomonas* sp. strain PIN. *Process Biochem.* 36: 925–932.

You, S., Yin, Q., Zhang, J. et al. (2017). Utilization of biodiesel by-product as substrate for high-production of β-farnesene via relatively balanced mevalonate pathway in *Escherichia coli*. *Bioresour. Technol.* 243: 228–236.

Yuan, J., Chen, X., Misha, P., and Ching, C.-B. (2017a). Metabolically engineered *Sacharomyces cerevisiae* for enhanced isoamyl alcohol production. *Appl. Microbiol. Biotechnol.* 101: 465–474.

Yuan, J., Misha, P., and Ching, C.-B. (2017b). Engineering the leucine biosymthetic pathway for isoamyl alcohol overproduction in *Saccharomyces cerevisiae*. *J. Ind. Microbiol. Biotechnol.* 44: 107–117.

Zelena, K., Krings, U., and Berger, R.G. (2012). Functional expression of a valencene dioxygenase from *Pleurotus sapidus* in *E.coli*. *Bioresour. Technol.* 108: 231–239.

Zhang, H., Cao, M., Jiang, X. et al. (2014). *De-novo* synthesis of 2-phenylethanol by *Enterobacter* sp. CGMCC 5087. *BMC Biotech.* 14: 30–37.

Zhao, G., Yin, G., Inamdar, A.A. et al. (2017a). Volatile organic compounds emitted by filamentous fungi isolated from flooded homes after Hurricane Sandy show toxicity in a Drosophila bioassay. *Indoor Air* 27: 518–528.

Zhao, C., Yan, X., Yang, S., and Chen, F. (2017b). Screening of *Bacillus* strains from Luzhou-flavor liquor making for high-yield ethyl hexanoate and low-yield propanol. *LWT Food Sci. Technol.* 77: 60–66.

Zheng, G.W., Yu, H.L., Zhang, J.D., and Xu, J.H. (2009). Enzymatic production of *l*-menthol by a high substrate concentration tolerable esterase from newly isolated *Bacillus subtilis* ECU0554. *Adv. Synth. Catal.* 351: 3.

Zheng, G.W., Pan, J., Yu, H.L. et al. (2010). An efficient bioprocess for enzymatic production of l-menthol with high ratio of substrate to catalyst using whole cells of recombinant *E. coli*. *J. Biotechnol.* 150: 108–114.

Zhu, F., Zhong, X., Hu, M. et al. (2014). In vitro reconstitution of mevalonate pathway and targeted engineering of farnesene overproduction in *Escherichia coli*. *Biotechnol. Bioeng.* 111: 1396–1405.

Zhu, M.M., Wang, S.L., Fan, M.T., and Li, J. (2016). In vitro study of the carotenoid cleavage enzyme from *Staphylococcus pasteuri* TS-82 revealed substrate specificities and generation of norisoprenoid flavors. *Food Sci. Biotechnol.* 25: 221–227.

Zorn, H., Langhoff, S., Schreibner, M. et al. (2003). A peroxidase from *Lepista irina* cleaves β-carotene to flavor compounds. *Biol. Chem.* 384: 1049–1056.

# 6

# Research and Production of Biosurfactants for the Food Industry

*Eduardo J. Gudiña and Lígia R. Rodrigues*

*CEB - Centre of Biological Engineering, Campus de Gualtar, University of Minho, 4710-057 Braga, Portugal*

## 6.1 Introduction

Biosurfactants are a structurally diverse group of surface-active compounds synthesized by different microorganisms, including bacteria, yeasts, and filamentous fungi. They are amphiphilic molecules containing both hydrophilic and hydrophobic moieties. Due to their structure, biosurfactants accumulate at the interface between fluid phases with different degrees of polarity, such as oil-water or air-water interfaces, reducing the surface or the interfacial tension, and forming and stabilizing emulsions (Abdel-Mawgoud et al. 2010; Gudiña et al. 2013). Biosurfactants are usually classified according to their chemical composition and microbial origin. Rosenberg and Ron (1999) suggested their division in low molecular weight compounds, which efficiently reduce the surface and the interfacial tension, and usually exhibit emulsifying activity, and high molecular weight polymers, which are more effective as emulsion stabilizing agents, but do not necessarily reduce the surface or the interfacial tension. The major classes of low-molecular-weight biosurfactants include glycolipids, lipopeptides, phospholipids, and neutral lipids, whereas the high-molecular-weight polymers comprise polymeric and particulate biosurfactants. Among the best-studied biosurfactants are rhamnolipids (glycolipids), produced by *Pseudomonas aeruginosa*, surfactin (lipopeptide) from *Bacillus subtilis*, emulsan (polymeric biosurfactant) from *Acinetobacter calcoaceticus* and sophorolipids (glycolipids) produced by yeast strains such as *Starmerella bombicola* (Bonmatin et al. 2003; Van Bogaert et al. 2007; Daverey and Pakshirajan 2009a; Abdel-Mawgoud et al. 2010; Geys et al. 2014).

The interest in biosurfactants has increased considerably in the last decades as environmentally friendly alternatives to the chemical surfactants (mainly obtained from petrochemical and oleo-chemical resources) for application in several fields, including agriculture, food, pharmaceutical, cosmetic and oil industries, among others, as it can be concluded from the increasing number of patents that have been issued claiming their application (reviewed by Shete et al. 2006). Biosurfactants exhibit similar or better performance when compared with their chemical counterparts, as well as lower toxicity, higher biodegradability, and stability at a wide range of temperatures, salinities, and pH values (Abdel-Mawgoud et al. 2010; Vaz et al. 2012; Pereira et al. 2013; Gudiña et al. 2015a). The market for these "green" alternatives to synthetic surfactants was

*Bioprocessing for Biomolecules Production,* First Edition.
Edited by Gustavo Molina, Vijai Kumar Gupta, Brahma N. Singh, and Nicholas Gathergood.

344 k tons in 2013, and it is expected to reach 462 k tons and 2308 million USD by 2020 (Grand View Research 2014).

Based on their interesting properties, biosurfactants find several applications in the food industry. They can be used to form and stabilize emulsions, as wetting agents, or to help in the mixing of food ingredients, due to their surface and emulsifying activities; because of their antimicrobial and anti-adhesive properties, they can be used as preservatives in food formulations, and as agents to inhibit biofilm formation and the adhesion of microorganisms to food contact surfaces; also, due to their nontoxic properties, they can be used as cleaning agents to remove and destroy pathogenic microorganisms from equipment and surfaces used in food processing (Nitschke and Costa 2007; Campos et al. 2013).

Nowadays, several companies commercialize biosurfactants, purified or incorporated in different formulations, for application in the food industry. Examples include SyntheZyme LLC., USA (www.synthezyme.com) and Rhamnolipid Companies Inc., USA (www.rhamnolipid.com), that commercialize sophorolipids and rhamnolipids, respectively, for food processing applications. This chapter will focus on the potential applications of biosurfactants in the food industry.

## 6.2 Biosurfactants as Food Additives

Surface-active compounds display a variety of functions in the food industry, especially as emulsifiers, foaming or wetting agents, as well as to promote the solubility of different food ingredients. Emulsification plays an important role in the consistency and texture of many food products. Emulsions are heterogeneous systems, consisting of at least two immiscible liquids (generally an aqueous and an oil phase) wherein one of them is dispersed in the other in the form of droplets. Emulsions are generally of two types: oil-in-water (O/W), when the aqueous phase is the continuous phase and the oil phase is the dispersed phase; and water-in-oil (W/O), when the aqueous phase is dispersed in the oil (continuous) phase (Garti 1999; Hasenhuettl and Hartel 2008; Patino et al. 2008; Kralova and Sjöblom 2009; McClements et al. 2017). Many natural and processed food products are emulsions. Examples of W/O emulsions include butter, margarine, or fat-based spreads, whereas O/W emulsions include milk, food dressings, mayonnaises, or ice cream, among other (Garti 1999; Patino et al. 2008; Kralova and Sjöblom 2009).

Emulsions are thermodynamically unstable systems that have a tendency to separate back into their components over time. To manufacture commercial products with long shelf life, the incorporation of agents to improve emulsion stability is required. Due to their ability to reduce the interfacial tension between the dispersed and the continuous phases, surface-active compounds reduce the surface energy and the amount of work required to disperse one phase into the other, preventing the coalescence of the particles and the subsequently phase separation (Garti 1999; Hasenhuettl and Hartel 2008; Patino et al. 2008; Kralova and Sjöblom 2009). One important parameter to determine the type of emulsion formed by a surface-active compound is its hydrophilic-lipophilic balance (HLB), which is defined as the ratio of the weight percentage of hydrophilic groups to the weight percentage of hydrophobic groups in the molecule. HLB values range between 1 and 20. Surface-active compounds with low HLB values (3–9) promote the formation of W/O emulsions, whereas those with values higher than 9 tend to stabilize O/W emulsions (Kralova and Sjöblom 2009). For instance, surfactin and rhamnolipids, according to their high HLB values, favor the formation of O/W emulsions (as reviewed by Gudiña et al. 2013).

Food emulsions are usually complex mixtures of numerous components, which difficult their stabilization (Garti 1999; Kralova and Sjöblom 2009). Furthermore, parameters such as pH or salt content, as well as the processing conditions, also have a significant effect on their stability. Several compounds have been used for many centuries as food emulsifiers, to facilitate the formation and stabilization of emulsions, and are essential ingredients in many food formulations to aid in their elaboration, to improve their sensory attributes (e.g. appearance, consistency, and texture), to enhance their shelf life, or to improve their long-term storage. Examples of naturally occurring emulsifiers used in the food industry include lecithin (from egg yolk or soybean) and its derivatives, plant gums (such as karaya, arabic, tragacanth, guar, locust bean, or fenugreek gum), xanthan gum (a polysaccharide secreted by the Gram-negative bacterium *Xanthomonas campestris*), phospholipids, and amphiphilic milk proteins. Among the synthetic and semi-synthetic emulsifiers currently used in food products are sorbitan esters, polyoxyethylene-sorbitan esters, cellulose derivatives (methyl cellulose, ethyl cellulose, carboxymethyl cellulose), mono- and di-glycerides of fatty acids, fatty alcohol ethoxylates, or sugar esters of fatty acids (Hasenhuettl and Hartel 2008; Freire et al. 2009; Kralova and Sjöblom 2009; Neta et al. 2012, 2015; McClements et al. 2017).

Regarding the synthetic emulsifiers used in the food industry, despite their excellent emulsifying properties and a wide spectrum of applications, they are gradually losing preference, due to an increasing demand among consumers for natural replacements to chemically synthesized additives. Furthermore, regulatory rules and restrictions imposed by food agencies and health authorities are constantly limiting their use (Nitschke and Costa 2007; Campos et al. 2013). Consequently, there is a growing tendency among food manufacturers to replace synthetic emulsifiers with natural alternatives. Natural food emulsifiers, such as lecithin or gum arabic, have been widely used and traditionally accepted by the food industry. However, they exhibit some functional limitations in modern processes involving irradiation or the use of microwaves (Campos et al. 2013). Plant gums are subjected to price fluctuations, mainly due to shortfalls in their supply caused by low crop yields, and the increasing energy and transportation costs. Furthermore, some of these natural emulsifiers are obtained from genetically modified crops (particularly soybean), which creates some limitations for food industries seeking to reduce their dependency on food additives derived from them, due to biological and biomedical concerns (Campos et al. 2013). These features, together with the possibility of using emulsifiers that provide simultaneously other favorable properties (e.g., antioxidant activity) resulted in an exhaustive search for alternative natural emulsifiers with similar or better performance than those traditionally used, suitable for application in new and advanced food formulations.

Emulsifiers are among the most frequently used food additives, as they are critical to the proper performance of any emulsion-based product, and their market showed a high increase in the last years. About 500 k tons of emulsifiers are produced each year for application in the food industry, and their sales in the European Union and the United States are estimated to be €300 million and US$275 million, respectively (Hasenhuettl and Hartel 2008). This market represents an opportunity for new emerging emulsifiers, including biosurfactants as the result of their excellent surface and emulsifying activities.

The emulsifying activity of biosurfactants has been studied with different hydrocarbons, due to their potential applications in bioremediation and in the oil industry (Vaz et al. 2012; Pereira et al. 2013), whereas studies to evaluate their emulsifying properties with oils or fats used in the food industry are less common.

For instance, lipopeptide biosurfactants synthesized by *B. subtilis* LB5a and rhamnolipids produced by *P. aeruginosa* 47T2 form stable emulsions with soybean oil, coconut fat, and linseed oil (Haba et al. 2002; Nitschke and Pastore 2006), which indicates their potential use as emulsifiers in the food industry. *Kluyveromyces marxianus* FRR 1586 produced a surface-active compound (glycoprotein) growing in a low-cost lactose-based medium, which form stable O/W emulsions with corn oil at different pH values (3–11) and salinities (2–50 g $NaCl\,l^{-1}$), suggesting its potential application as a natural emulsifier in processed foods such as mayonnaise, biscuits, crackers, cake, ice cream, and meat products such as sausages (Lukondeh et al. 2003). Also, the glycolipid biosurfactants produced by the yeasts *Trichosporon loubieri* CLV20, *Geotrichum* sp. CLOA40 and *Trichosporon montevideense* CLOA70 efficiently emulsify vegetable oils, remaining stable at high temperatures and salinities, as well as at a wide range of pH values (Monteiro et al. 2009, 2010).

The marine bacterium *Antarctobacter* sp. TG22 produces a high-molecular-weight biosurfactant (glycoprotein) with better emulsifying properties for different food oils (olive, sunflower, rapeseed, and groundnut oil) when compared with xanthan gum or gum arabic (Gutiérrez et al. 2007). Yasan and liposan, high-molecular-weight biosurfactants (glycoproteins) produced by the yeast *Yarowia lipolytica*, offer good emulsifying activities with different vegetable oils when compared with the commercial emulsifiers gum arabic, pectin, or hydroxypropyl-methyl cellulose (Cirigliano and Carman 1985; Trindade et al. 2008). Also, the biosurfactants produced by *Candida utilis, Hansenula anomala, A. calcoaceticus*, or *Klebsiella* sp. exhibit better emulsifying activities than the commonly used food emulsifiers gum arabic and carboxymethyl cellulose (Shepherd et al. 1995). Other examples of biosurfactants that efficiently emulsified different vegetable oils are those produced by *C. utilis* UFPEDA 1009 (Campos et al. 2014), *Candida lipolytica* UCP0988 (Santos et al. 2013), *Candida bombicola* (Daverey and Pakshirajan 2009a), *Klebsiella* sp. RJ-03 (Jain et al. 2013), or *Variovorax paradoxus* 7bCT5 (Franzetti et al. 2012).

The emulsifying activities exhibited by these biosurfactants, similar or better than those obtained with other commonly used food emulsifiers, as well as their high stability under conditions often used in the elaboration of food formulations, make them potential candidates to be used as emulsion-stabilizing agents in the food industry. This applicability could probably also spread to many other biosurfactants produced by different microorganisms, which exhibit an excellent emulsifying activity (Martínez-Checa et al. 2007; Zheng et al. 2011; Colin et al. 2013), although their application in the food industry has not been studied yet.

In addition to their obvious role as agents that promote the formation and stabilization of emulsions, surface-active compounds have several other applications in the food industry. For instance, they are used to control the agglomeration of fat globules in products such as ice cream or whipped toppings, to improve the texture and the shelf life of starch-containing products, or to improve the consistency and the texture of fat-based and dairy products. In the bakery industry, they are used to modify the rheological properties, improve the water retention capacity and handling characteristics of the dough, as well as to facilitate the mixture of the ingredients. Moreover, they are used as anti-spattering agents in margarine and during cooking of oil and fats, to inhibit the formation of crystals in salad oils, as anti-sticking agents in candies, to modify the viscosity of chocolate, to stabilize the freeze-thaw process in whipped toppings and coffee whiteners, and as gloss enhancers in confectionery coatings, canned, and moist pet foods (Garti 1999; Nitschke and Costa 2007; Hasenhuettl and Hartel 2008). Again, due to their properties, many biosurfactants could be used to perform these functions.

The addition of a lipopeptide biosurfactant produced by *B. subtilis* SPB1 to dough formulations resulted in a more cohesive dough, and significantly improved its textural properties and the gas retention capacity when compared with dough supplemented with the commercial emulsifier soy lecithin. As a consequence, the quality, texture, and volume of the bread were increased, thus suggesting the possible application of this biosurfactant as an additive in bakery products (Mnif et al. 2013). Furthermore, the biosurfactant also contributes to preserving the quality and the textural characteristics of the frozen dough, and allowed a decrease in the amount of yeast used and the time required for the fermentation (Mnif et al. 2013). The same biosurfactant was evaluated as an emulsifier in the elaboration of cookies using sesame peels flour. The addition of this biosurfactant to the dough results in a significant decrease in the dough hardness and an increase in the dough cohesion, resulting in softer cookies with better overall quality when compared with those containing the commercial emulsifier glycerol monosterate (Zouari et al. 2016).

Another lipopeptide biosurfactant produced by the marine actinobacteria *Nesterenkonia alba* MSA31 was evaluated as an alternative emulsifier in the elaboration of muffins. Muffins including this biosurfactant in their formulation exhibit decreased hardness, chewiness, and gumminess, as well as enhanced softness, springiness, and cohesiveness when compared with muffins prepared using conventional emulsifiers (Kiran et al. 2017). Similarly, the production of bread using wheat flour supplemented with sophorolipids improves gas retention during the fermentation and enlarges the bread volume, providing bread with better appearance and longer shelf life (Akira and Akira 1986). Van Haesendonck and Vanzeveren (2004) reported an improvement of the dough stability and enhancement of texture, volume, consistency, and conservation of baked products by the addition of rhamnolipids, and suggested their possible application to improve the properties of butter cream, croissants, and frozen confectionery products.

Considering the interesting properties demonstrated by biosurfactants, they have the potential to be used as alternatives to the synthetic surfactants in the food industry.

## 6.3 Biosurfactants as Powerful Antimicrobial and Anti-Adhesive Weapons for the Food Industry

As previously mentioned, based on their unique features, biosurfactants have a great potential for application in various sectors of the food industry not restricted to food processing. Among the most interesting properties of these compounds, their antimicrobial and anti-adhesive activities could represent powerful weapons for the assurance of food safety. Foodborne diseases, defined by the World Health Organization as illnesses caused by the ingestion of food or drink contaminated with microorganisms or chemicals, are a public health problem (Fernandes et al. 2014). Changes in eating habits, agriculture production, and food-processing practices have contributed to an increasing number of outbreaks of foodborne illnesses in many countries over the last decades (Heiman et al. 2015). Among the most common pathogenic microorganisms involved in foodborne diseases are *Escherichia coli*, *Staphylococcus aureus*, *Listeria monocytogenes*, *Salmonella* spp., *Shigella* spp., and *Yersinia* spp. (Fernandes et al. 2014; Heiman et al. 2015; Zhang et al. 2016; Silva et al. 2017).

Several biosurfactants have proven to act as antimicrobial agents against different bacteria, yeasts, fungi, algae, and viruses (Rodrigues et al. 2006a; Nitschke and Costa 2007; Rodrigues 2011). *B. subtilis* strains produce different lipopeptide biosurfactants that exhibit antimicrobial activity, including surfactin, fengycin, iturin, bacillomycins, and mycosubtilins (Ahimou et al. 2000; Bonmatin et al. 2003; Rahman et al. 2006; Fernandes et al. 2007; Das et al. 2008; Joshi et al. 2008; Roongsawang et al. 2010; Fan et al. 2017). Additionally, other *Bacillus* species have been reported to produce antimicrobial lipopeptide biosurfactants, such as *Bacillus licheniformis*, which produces lichenysin (Yakimov et al. 1995), *Bacillus pumilus*, which produces pumilacidin (Morikawa et al. 1992; Das et al. 2008), and *Bacillus amyloliquefaciens*, which produces iturin (Yu et al. 2002). Other examples of antimicrobial lipopeptides include gramicin S and polymyxins, produced by *Bacillus brevis* and *Bacillus polymyxa*, respectively (Gause and Brazhnikova 1944; Porter et al. 1949; Serrano et al. 2012). Most of the studies reported so far on the abovementioned lipopeptide biosurfactants have focused on biomedical applications mainly as the result of their biological activities. For instance, surfactin is known to exhibit antiviral and anticancer activity besides the antimicrobial activity (Gudiña et al. 2013; Duarte et al. 2014).

Rhamnolipids, produced mainly by *P. aeruginosa* strains, exhibit antimicrobial activity against several bacteria, yeasts, and filamentous fungi (Abalos et al. 2001; Rodrigues et al. 2017). Biosurfactants produced by yeasts with antimicrobial activity include sophorolipids produced by *S. bombicola* (Van Bogaert et al. 2007; Daverey and Pakshirajan 2009b; Valotteau et al. 2017), mannosylerythritol lipids (MEL-A and MEL-B) produced by *Candida antarctica* (Kitamoto et al. 1992, 1993), and the biosurfactants produced by *Candida sphaerica* UCP0995, *C. lipolytica* UCP0988 and *Wickerhamomyces anomalus* CCMA0358, which exhibit antimicrobial activity against several microbial pathogens (Luna et al. 2011; Rufino et al. 2011; Souza et al. 2017).

Biosurfactants comprise a very diverse group of molecules, which is explained by the great diversity of microbial producers that can be isolated from different environments. Indeed, there is an immense microbial potential that remains to be explored and consequently countless new molecules that are still to be discovered. Some new biosurfactants with interesting properties, including antimicrobial activity, have been reported in the last years. The biosurfactants produced by lactic acid bacteria (LAB) or bacteria present in dairy products, with structures that range from glycoproteins, glycopeptides, glycolipids, and lipoproteins, are included in such group of new molecules and have shown antimicrobial activity against pathogenic bacteria, yeasts and filamentous fungi. Among them stand out biosurfactants produced by *Streptococcus thermophilus* A, *Lactococcus lactis* 53, *Lactobacillus jensenii* $P_{6A}$, *Lactobacillus gasseri* $P_{65}$, *Lactobacillus agilis* CCUG31450 *Lactobacillus casei* 8/4, and *Lactobacillus paracasei* A20 (Rodrigues et al. 2004a, 2006b; Golek et al. 2009; Gudiña et al. 2010a,2015b; Morais et al. 2017) that show antimicrobial activity against a broad range of pathogenic microorganisms. Since LAB and bacteria isolated from dairy products are generally regarded as safe (GRAS), often considered probiotics, and most of them are already used in many food manufacturing and industrial processes, it is expected that their biosurfactants are also safe and non-toxic to mammals. Therefore, these compounds represent a powerful weapon to prevent food contamination directly, as food additives, or indirectly, as detergent formulations to clean surfaces and equipment that come in contact with the food. Consequently, food shelf life can be extended and food safety is assured.

Sophorolipids consist of a hydrophilic sophorose group attached to a fatty acid chain that can have a different number of carbons (16–18) and degrees of unsaturation. Furthermore, they can occur in free-acid form or in lactonic form. Those structural differences can result in different physicochemical properties. Zhang and co-workers (2017) studied the antimicrobial activity of different sophorolipid congeners against *E. coli* O157:H7, in the presence and absence of ethanol. The results obtained demonstrated that lactonic sophorolipids (with stearic or oleic acid as hydrophobic domain) at a concentration of $10\,g\,l^{-1}$ reduced *E. coli* O157:H7 populations to non-detectable levels after two hours of treatment in the presence of 20% of ethanol. On the contrary, the free-acid sophorolipids exhibited a weak antimicrobial activity (Zhang et al. 2017).

Rhamnolipids in combination with nisin were found to inhibit thermophilic spores in ultra-high-temperature (UHT) soymilk and also to extend its shelf life. Moreover, the use of formulations containing rhamnolipids and natamycin in the processing of salad dressing was found to inhibit mold growth, as well as to extend its shelf life. The same effects were observed when these formulations were used to prepare cottage cheese (Gandhi and Skebba 2007). Magalhães and Nitschke (2013) also demonstrated the rhamolipids antimicrobial activity against *L. monocytogenes* and its synergistic effect when combined with nisin.

Another perspective on the use of biosurfactants in food-related sectors is exemplified by the potential of using emulsions of rhamnolipids to treat the leaves of *Nicotiana glutinosa* infected with tobacco mosaic virus, as well as to control the potato virus-x disease (Desai and Banat 1997). Rhamnolipids have also been reported to cause zoospore lysis, zoospore inhibition, spore germination and hyphal growth inhibition in several phytopathogenic fungal species, including *Phytophthora capsici* and *Colletotrichum orbiculare* (Kim et al. 2000). Rodrigues and co-workers (2017) demonstrated that di-rhamnolipids (rhamnolipid congeners with two rhamnose moieties) inhibit the growth of *Aspergillus niger* and *Aspergillus carbonarius*, two fungal contaminants associated with the spoilage of agriculture products. Furthermore, these fungi produce mycotoxins, which have numerous and diverse toxic properties, including carcinogenic, mutagenic, teratogenic, neurotoxic, nephrotoxic, immunosuppressive, and estrogenic effects, even when ingested at low concentrations. The authors also demonstrated that the antifungal activity exhibited by di-rhamnolipids was considerably increased by NaCl (Rodrigues et al. 2017).

The company Jeneil Biosurfactant Corporation developed a rhamnolipid formulation to prevent crop contamination by pathogenic fungi. This product, which is considered safe, nonmutagenic, and of low toxicity to mammals, has been approved by the US Food and Drug Administration (FDA) to be directly used on vegetables, legumes, and fruits crops (Nitschke and Costa 2007). Rhamnolipids are also included in the formulation of the commercial biofungicide Zonix™ (NOP Supply LLC., USA), which can be used to prevent and control plant pathogenic fungi on agricultural crops (Rodrigues et al. 2017).

Given their antimicrobial activity and physicochemical properties, these biosurfactants can be viewed as powerful agents for food industry applications, increasing the food shelf life without concern to consumer health, excluding the need of adding synthetic compounds that often are not healthy.

Many microorganisms are able to colonize surfaces, establishing a sessile form of life known as a biofilm. Biofilms can be defined as microbial communities surrounded by a self-produced polymeric matrix, attached to biotic or abiotic surfaces. The main structural components of biofilm matrix are polysaccharides, proteins, lipids, and nucleic acids. Biofilms can exhibit different degrees of complexity, and may be formed by single species populations or, most commonly, by multiple species (Anderson 2001). Briefly, the biofilm formation starts with the conditioning step that consists in the adsorption at the surface of macromolecules from the bulk phase of a fluid containing microorganisms. Surface conditioning is dictated by the fluid bulk composition that will also greatly influence the subsequent bacterial adhesion. Bacterial adhesion generally occurs in a few minutes, while the real biofilm takes several hours or even days to be developed (Rodrigues 2011). Microorganisms organize themselves into biofilms as a survival strategy, i.e. to get protection from adverse environments, as well as facilitated access to nutrients (Maukonen et al. 2003). Indeed, it is well-known that the microorganisms on a biofilm are more resistant to the chemical and physical agents used for cleaning and disinfection than their planktonic counterparts (Chae and Schraft 2000; Jenkinson and Lappin-Scott 2001; Stepanovic et al. 2004; Rodrigues 2011).

Biofilms are an important concern among food industries, and can cause spoilage and contamination of food products by pathogenic microorganisms, resulting in serious food safety issues. Furthermore, their establishment can result in equipment damage (e.g. obstruction of pipelines, corrosion of equipment, or reduced efficiency of temperature transfer systems) (Lynch and Robertson 2008).

In the food industry, the lack of disinfection procedures may lead to the adhesion of residues on different surfaces. Furthermore, even with the application of cleaning and sanitization procedures consistent with good manufacturing practices, microorganisms can remain on equipment surfaces and survive for prolonged periods. Thus, under certain conditions, microbes can adhere to those surfaces and form a biofilm which is able to aggregate nutrients, residues, and other microorganisms (Bagge-Ravn et al. 2003). Since the food industry has zero tolerance levels to pathogens like *Salmonella* spp. and *L. monocytogenes*, a single cell may be as significant as a well-developed biofilm (Nitschke and Costa 2007). Food industry struggles to find ways to produce and supply safe, secure, and pleasant food; thus, an adequate microbial control is crucial to attaining such goal (Bagge-Ravn et al. 2003). Cleansing of all surfaces that make contact with food is a good start to prevent contamination (Somers and Wong 2004). Unfortunately, as previously mentioned, sanitization practices are less effective against microorganisms present in biofilms when compared with planktonic cells (Freire et al. 2009).

Strategies to prevent biofilm formation include physicochemical modification of the surface of the material to create anti-adhesive surfaces, incorporation of antimicrobial agents into the commonly used materials, mechanical design alternatives of the equipment, and release of antimicrobials. Indeed, to prevent the biofilm development it is crucial that the environmental factors at the food processing line are adequately controlled. Besides, the use of surface active agents, dispersants, enzymes, and new nontoxic or less-toxic biocide compounds comprises a direct approach toward that goal. Additionally, surfaces cleaning, staff training, good manufacturing practices and structure are essential factors to prevent sanitation issues in the food industry (Maukonen et al. 2003). The growing demand for sustainable technologies increased the interest in alternative bio-based or green solutions to control biofilms. Accordingly, the anti-adhesive activity of biosurfactants can represent an interesting tool for the food industry. Indeed, preconditioning different surfaces with biosurfactants have been shown to significantly reduce microbial contamination and prevent or reduce subsequent biofilm development (Rodrigues et al. 2004b; Meylheuc et al. 2006a). Such anti-adhesive activity indicates the potential application of biosurfactants either as coating agents for food related utensils and surfaces or to decrease antifouling rate or occurrence (Campos et al. 2013).

Preconditioning stainless steel surfaces with a biosurfactant from *Pseudomonas fluorescens* was shown to greatly inhibit the adhesion of *L. monocytogenes* L028 (Meylheuc et al. 2001). These authors also proved that the prior conditioning of the stainless steel with such biosurfactant favored the bactericidal effect of disinfectants (Meylheuc et al. 2006b). Moreover, this biosurfactant has been shown to inhibit the corrosion of stainless steel (Dagbert et al. 2006). Additionally, two lipopeptide biosurfactants produced by *Pseudomonas putida* PCL1445 were found to inhibit biofilm formation by several *Pseudomonas* species on PVC surfaces, and also to disrupt preformed biofilms (Kuiper et al. 2004).

Surfactin and rhamnolipids inhibited the adhesion (and biofilm formation) of *L. monocytogenes* and *P. fluorescens* on polystyrene and stainless steel surfaces, suggesting their potential as agents to control the adhesion of these microorganisms on food contact surfaces (Freire et al. 2009; Araujo et al. 2016). Pretreating stainless steel with surfactin was found to reduce the number of adhering cells of *Enterobacter sakazakii* ATCC 29004 (Nitschke and Costa 2007; Nitschke et al. 2009). Surfactin also proved to be efficient in removing *Salmonella enterica* from fruit surfaces with different roughness and hydrophobicity. Similar efficiencies of the sanitation treatments (between 92% and 94%) were obtained for *S. enterica* adhered to mango (rough and hydrophilic surface) and tomato (smooth and hydrophobic surface) using a solution containing 50 mg $l^{-1}$ of surfactin (Fernandes et al. 2014). Rhamnolipids removed up to 88% of *S. aureus* biofilms established on polystyrene surfaces. They proved to be efficient at low temperatures (4–37 °C), suggesting their potential application in dairy processing industries where low temperatures are applied (Silva et al. 2017). Also sophorolipids proved to be efficient in reducing the populations of *E. coli* O157:H7 on spinach leaves to not detectable levels after two hours of treatment, opening the possibility of their use as potential sanitizers (Zhang et al. 2016).

A dairy thermophilic *Streptococcus* sp. strain has been reported to produce a biosurfactant that can be used to control the fouling of heat exchanger plates in pasteurizers, as it delays the colonization by bacteria related to the fouling formation (Busscher et al. 1994). In our group, a great amount of work has been conducted on the effect of biosurfactants from probiotic LAB against multi-species biofilm formation. For instance, two biosurfactants produced by *L. lactis* and *S. thermophilus* were used to pre-condition the voices prostheses surfaces and evaluate their effect on the development of a mixed biofilm comprised of *Staphylococcus epidermidis, Streptococcus salivarius, S. aureus, Rothia dentocariosa, Candida albicans,* and *Candida tropicalis* (Rodrigues et al. 2004b). Both biosurfactants exhibited antimicrobial activity but depending on the microorganism, these compounds presented different effective concentrations. Furthermore, a biosurfactant produced by *L. paracasei* ssp. *paracasei* A20 was found to possess anti-adhesive activity against several microorganisms (Gudiña et al. 2010a,b). Indeed, the ability of biosurfactants produced by LAB to impede the adhesion of pathogens to metallic surfaces, silicone rubber, polystyrene, glass, surgical implants, voice prostheses, and epithelial cells has been widely demonstrated (Heinemann et al. 2000; Van Hoogmoed et al. 2000; Walencka et al. 2008; Golek et al. 2009; Fracchia et al. 2010; Spurbeck and Arvidson 2010; Brzozowski et al. 2011; Rodrigues 2011; Tahmourespour et al. 2011a,b). These results are encouraging and open the possibility of using such biosurfactants to inhibit biofilm formation on diverse surfaces in food industries.

## 6.4 Potential Role of Biosurfactants in New Nano-Solutions for the Food Industry

Currently, a great amount of knowledge on the development of drug delivery systems using common chemical surfactants is available; therefore, the surfactants replacement by similar molecules of microbial origin is expected to be simple and easy, as well as to provide novel opportunities given the existent wide variety of chemical structures (Singh et al. 2014). Among the different nano-sized delivery systems, biosurfactants could play a role in the development of the microemulsion-based colloidal ones. The key components to produce those systems include an aqueous phase, an oil phase, a surfactant, and often a co-surfactant or co-solvent. The (bio)surfactant self-aggregates to form varying structures that are able to encapsulate and/or solubilize hydrophobic or hydrophilic compounds in the presence of a dispersed phase (oil for O/W and water for W/O microemulsions) within its structural core, hence partitioning the dispersed phase from the continuous phase (Israelachvili 1994). Mannosylerythritol lipids have much higher emulsifying activity with soybean oil and tetradecane than polysorbate 80 (Kitamoto et al. 2009). Besides, these biosurfactants present a remarkable capacity to form stable W/O microemulsions without the addition of a co-surfactant or salt (Worakitkanchanakul et al. 2008). Moreover, other biosurfactants, namely rhamnolipids and sophorolipids can be mixed with lecithins to prepare biocompatible microemulsions (Nguyen et al. 2010). For instance, Liu and co-workers (2016) demonstrated the applicability of rhamnolipids to develop emulsion-based fish oil delivery systems containing high concentrations of ω-3 polyunsaturated fatty acids for incorporation into foods. Rhamnolipids offered better results when compared with other natural emulsifiers such as saponins regarding the protection of polyunsaturated fatty acids from oxidation (Liu et al. 2016).

In addition to their potential for the formulation of microemulsions, biosurfactants have been suggested as valuable molecules for the synthesis of nanoparticles and liposomes (Kiran et al. 2011). Recently, increased interest on biosurfactant-mediated processes has been reported, mainly due to their potential role on the synthesis of silver nanoparticles and NiO nanorods (Xie et al. 2006; Palanisamy 2008). Reddy and collaborators (2009) stabilized the synthesis of silver nanoparticles with surfactin. Rhamnolipids were also evaluated for their effect on the synthesis and stabilization of nanozirconia particles (Biswas and Raichur 2008).

Biosurfactants have been accepted as safe, versatile and useful for diverse applications including in the development of nano-solutions for the food industry.

Previous studies suggested the added value of biosurfactants both for the assembly of microemulsion-based drug formulations (Gudiña et al. 2013) or for the synthesis of nanoparticles (Kiran et al. 2011) for biomedical applications. Biosurfactants present high biocompatibility and low toxicity. Rhamnolipids, surfactin, iturin, and pumilacidin are some of the microbial surfactants generally used for preparing oral lipid-based formulations of therapeutic agents (Gudiña et al. 2013). Surfactin has been used to prepare a self-microemulsifying drug delivery system of vitamin E to enhance its biological performance. This system showed an extraordinary increase in the emulsification efficiency, dissociation rate, and consequently oral bioavailability (Singh et al. 2014). Moreover, cationic surfactin liposomes enhanced the delivery of siRNA in HeLa cells (Shim et al. 2009); mannosylerythritol lipid vesicles showed increased levels of gene transfection efficacy (Inoh et al. 2001; Imura et al. 2005; Igarashi et al. 2006); and rhamnolipids were used to prepare drug encapsulated lipid-polymer coated hybrid nanoparticles for biofilm control (Cheow and Hadinoto 2012). However, despite all the advantages of using biosurfactants for those purposes, this great potential remains unexplored, in particular for the food industry. Similar to the drug-delivery systems, we could envisage the development of such formulations to encapsulate or entrap food odors, aromas, colorants, prebiotics, vitamins, and other micronutrients for technical and functional improvement of foods, antimicrobials to prevent contamination, or even specific molecules that could signal a nonsafe or contaminated food.

## 6.5 Conclusions and Future Perspectives

Considering the interesting properties of biosurfactants, they have the potential to be used as multipurpose additives, due to their emulsifying, antimicrobial, and anti-adhesive activities, which make them suitable for many food applications. Indeed, they could be used either directly as food additives (through the use of adequate formulations) to enhance the appearance and the stability of foods, or indirectly as detergent formulations to clean surfaces that come in contact with food and prevent food contamination.

However, despite their promising properties, the application of biosurfactants as additives on a large scale in the food industry still remains a challenge. One of the reasons is the scant information available regarding their toxicity. Their applicability can only be ascertained after establishing them as food-grade additives, in order to certify their safety for utilization in food processes. Regulations regarding the approval of new food additives are required by governmental agencies, and these require long processes of research regarding their toxicological evaluation. Additionally, the pathogenic nature of some biosurfactant-producing microorganisms (such as the opportunistic pathogen *P. aeruginosa*) can restrict their use in the food industry. On the contrary, other biosurfactant-producing microorganisms are GRAS and have been traditionally used in many food-processing industries (such as the yeasts *Y. lipolytica* or *C. utilis*, as well as LAB), which allows the application of their metabolites in food-manufacturing processes, as they do not exhibit risks of toxicity or pathogenicity.

Furthermore, the high production costs of biosurfactants when compared with chemical surfactants and the difficulties to produce them in high amounts have hampered their wide application at an industrial level. Nowadays, the production costs for most biosurfactants do not compete with those of chemical surfactants. However, in the food industry, where high-value products are produced, this drawback could be less significant as compared with other industries. Also, different strategies can be applied to reduce their production costs, including the use of cheap sustainable resources (such as agro-industrial byproducts or wastes), the optimization of production and downstream processes, as well as the improvement of biosurfactant-producing microorganisms.

In summary, the current examples of biosurfactants applications in the food industry are limited, although – as discussed in this chapter – a plethora of opportunities for their exploration can be envisaged in the next years.

## Acknowledgments

The authors acknowledge the Portuguese Foundation for Science and Technology (FCT) for the financial support under the scope of the strategic funding of UID/BIO/04469/2013 unit and COMPETE 2020 (POCI-01-0145-FEDER-006684), and BioTecNorte operation (NORTE-01-0145-FEDER-000004) funded by the European Regional Development Fund under the scope of Norte2020 – Programa Operacional Regional do Norte. The authors also thank the FCT for the financial support under the scope of the Project MultiBiorefinery – multi-purpose strategies for broadband agro-forest and fisheries byproducts valorisation: a step forward for a truly integrated biorefinery (POCI-01-0145-FEDER-016403). E. J. Gudiña was supported by the Post-Doctoral grant CEB-BPD/01/2015/07 from the project UID/BIO/04469/2013, funded by FCT.

## References

Abalos, A., Pinazo, A., Infante, M.R. et al. (2001). Physicochemical and antimicrobial properties of new rhamnolipids produced by *Pseudomonas aeruginosa* AT10 from soybean oil refinery wastes. *Langmuir* 17: 1367–1371.

Abdel-Mawgoud, A.M., Lépine, F., and Déziel, E. (2010). Rhamnolipids: diversity of structures, microbial origins and roles. *Appl. Microbiol. Biotechnol.* 86: 1323–1336.

Ahimou, F., Jacques, P., and Deleu, M. (2000). Surfactin and iturin A effects on *Bacillus subtilis* surface hydrophobicity. *Enzyme Microb. Technol.* 27 (10): 749–754.

Akira S. and Y. Akira. 1986. Method of modifying quality of wheat flour product. Patent n° JP 61205449.

Anderson, J.M. (2001). Biological responses to materials. *Ann. Rev. Mater.* 31: 81–110.

Araujo, L.V., Guimarães, C.R., Marquita, R.L.S. et al. (2016). Rhamnolipid and surfactin: anti-adhesion/antibiofilm and antimicrobial effects. *Food Control* 63: 171–178.

Bagge-Ravn, D., Ng, Y., Hjelm, M. et al. (2003). The microbial ecology of processing equipment in different fish industries: analysis of the microflora during processing and following cleaning and disinfection. *Int. J. Food Microbiol.* 87: 239–250.

Biswas, M. and Raichur, A.M. (2008). Electrokinetic and rheological properties of nanozirconia in the presence of rhamnolipid biosurfactant. *J. Am. Ceram. Soc.* 91: 3197–3201.

Bonmatin, J.M., Laprevote, O., and Peypoux, F. (2003). Diversity among microbial cyclic lipopeptides: iturins and surfactins. Activity structure relationships to design new bioactive agents. *Comb. Chem. High Throughput Screening* 6: 541–556.

Brzozowski, B., Bednarski, W., and Golek, P. (2011). The adhesive capability of two *Lactobacillus* strains and physicochemical properties of their synthesized biosurfactants. *Food Technol. Biotechnol.* 49: 177–186.

Busscher, H.J., Neu, T.R., and Van der Mei, H.C. (1994). Biosurfactant production by thermophilic dairy streptococci. *Appl. Microbiol. Biotechnol.* 41: 4–7.

Campos, J.M., Stamford, T.L.M., Sarubbo, L.A. et al. (2013). Microbial biosurfactants as additives for food industries. *Biotechnol. Progr.* 29 (5): 1097–1108.

Campos, J.M., Stamford, T.L.M., and Sarubbo, L.A. (2014). Production of a bioemulsifier with potential application in the food industry. *Appl. Biochem. Biotechnol.* 172: 3234–3252.

Chae, M.S. and Schraft, H. (2000). Comparative evaluation of adhesion and biofilm formation of different *Listeria monocytogenes* strains. *Int. J. Food Microbiol.* 62: 103–111.

Cheow, W.S. and Hadinoto, K. (2012). Lipid–polymer hybrid nanoparticles with rhamnolipid-triggered release capabilities as anti-biofilm drug delivery vehicles. *Particuology* 10: 327–333.

Cirigliano, M.C. and Carman, G.M. (1985). Purification and characterization of liposan, a bioemulsifier from *Candida lipolytica*. *Appl. Environ. Microbiol.* 50: 846–850.

Colin, V.L., Pereira, C.E., Villegas, L.B. et al. (2013). Production and partial characterization of bioemulsifier from a chromium-resistant actinobacteria. *Chemosphere* 90: 1372–1378.

Dagbert, C., Meylheuc, T., and Bellon-Fontaine, M.N. (2006). Corrosion behavior of AISI 304 stainless steel in presence of a biosurfactant produced by *Pseudomonas fluorescens*. *Electrochim. Acta* 51: 5221–5227.

Das, P., Mukherjee, S., and Sen, R. (2008). Genetic regulations of the biosynthesis of microbial surfactants: an overview. *Biotechnol. Genet. Eng. Rev.* 25: 165–186.

Daverey, A. and Pakshirajan, K. (2009a). Production of sophorolipids by the yeast *Candida bombicola* using simple and low-cost fermentative media. *Food Res. Int.* 42: 499–504.

Daverey, A. and Pakshirajan, K. (2009b). Production, characterization, and properties of sophorolipids from the yeast *Candida bombicola* using a low-cost fermentative medium. *Appl. Biochem. Biotechnol.* 158 (3): 663–674.

Desai, J.D. and Banat, I.M. (1997). Microbial production of surfactants and their commercial potential. *Microbiol. Mol. Biol. Rev.* 61: 47–64.

Duarte, C., Gudiña, E.J., Lima, C.F., and Rodrigues, L.R. (2014). Effects of biosurfactants on the viability and proliferation of human breast cancer cells. *AMB Express* 4: 40.

Fan, H., Zhang, Z., Li, Y. et al. (2017). Biocontrol of bacterial fruit blotch by *Bacillus subtilis* 9407 via surfactin-mediated antibacterial activity and colonization. *Front. Microbiol.* 8: 1973.

Fernandes, P.A.V., Arruda, I.R., Santos, A.F.A.B. et al. (2007). Antimicrobial activity of surfactants produced by *Bacillus subtilis* R14 against multidrug-resistant bacteria. *Braz. J. Microbiol.* 38: 704–709.

Fernandes, P.E., São José, J.F.B., Zerdas, E.R.M.A. et al. (2014). Influence of the hydrophobicity and surface roughness of mangoes and tomatoes on the adhesion of *Salmonella enteric* serovar Typhimurium and evaluation of cleaning procedures using surfactin. *Food Control* 41: 21–26.

Fracchia, L., Cavallo, M., Allegrone, G., and Martinotti, M.G. (2010). A *Lactobacillus*-derived biosurfactant inhibits biofilm formation of human pathogenic *Candida albicans* biofilm producers. In: *Current Research, Technology and Education Topics in Applied Microbiology and Microbial Biotechnology* (ed. A. Méndez-Vilas), 827–837. Badajoz, Spain: Formatex.

Franzetti, A., Gandolfi, I., Raimondi, C. et al. (2012). Environmental fate, toxicity, characteristics and potential applications of novel bioemulsifiers produced by *Variovorax paradoxus* 7bCT5. *Bioresour. Technol.* 108: 245–251.

Freire, D.M., Araújo, L.V., Kronemberger, F.A., and Nitschke, M. (2009). Biosurfactants as emerging additives in food processing. In: *Innovation in Food Engineering: New Techniques and Products* (cap. 23) (eds. M.L. Passos and C.P. Ribeiro), 685–705. Boca Raton: CRC Press.

Gandhi N.R. and V.L.P. Skebba. 2007. Rhamnolipid compositions and related methods of use. W. O. International Application Patent (PCT) 2007/095258 A3, filed Feb. 12, 2007, and issued Aug. 23, 2007.

Garti, N. (1999). What can nature offer from an emulsifier point of view: trends and progress? *Colloids Surf. A Physicochem. Eng.* 152: 125–146.

Gause, G.F. and Brazhnikova, M.G. (1944). Gramicin S – origin and mode of action. *Lancet* 2: 715–716.

Geys, R., Soetaert, W., and Van Bogaert, I. (2014). Biotechnological opportunities in biosurfactant production. *Curr. Opin. Biotechnol.* 30: 66–72.

Golek, P., Bednarski, W., Brzozowski, B., and Dziuba, B. (2009). The obtaining and properties of biosurfactants synthesized by bacteria of the genus *Lactobacillus. Annals Microbiol.* 59: 119–126.

Grand View Research. 2014. Biosurfactants market by product (rhamnolipids, sophorolipids, MES, APG, sorbitan esters, sucrose esters) expected to reach USD 2308.8 Million by 2020. http://www.grandviewresearch.com/industry-analysis/biosurfactants-industry. (accessed 25.07.2014).

Gudiña, E.J., Rocha, V., Teixeira, J.A., and Rodrigues, L.R. (2010a). Antimicrobial and antiadhesive properties of a biosurfactant isolated from *Lactobacillus paracasei* ssp *paracasei* A20. *Lett. Appl. Microbiol.* 50: 419–424.

Gudiña, E.J., Teixeira, J.A., and Rodrigues, L.R. (2010b). Isolation and functional characterization of a biosurfactant produced by *Lactobacillus paracasei. Colloids Surf., B* 76: 298–304.

Gudiña, E.J., Rangarajan, V., Sen, R., and Rodrigues, L.R. (2013). Potential therapeutic applications of biosurfactants. *Trends Pharmacol. Sci.* 34 (12): 667–675.

Gudiña, E.J., Rodrigues, A.I., Alves, E. et al. (2015a). Bioconversion of agro-industrial by-products in rhamnolipids toward applications in enhanced oil recovery and bioremediation. *Bioresour. Technol.* 177: 87–93.

Gudiña, E.J., Fernandes, E.C., Teixiera, J.A., and Rodrigues, L.R. (2015b). Antimicrobial and anti-adhesive activities of cell-bound biosurfactant from *Lactobacillus agilis* CCUG31450. *RSC Adv.* 5: 90960–90968.

Gutiérrez, T., Mulloy, B., Bavington, C. et al. (2007). Partial purification and chemical characterization of a glycoprotein (putative hydrocolloid) emulsifier produced by a marine bacterium *Antarctobacter. Appl. Microbiol. Biotechnol.* 76: 1017–1026.

Haba, E., Pinazo, A., Jauregui, O. et al. (2002). Physicochemical characterization and antimicrobial properties of rhamnolipids produced by *Pseudomonas aeruginosa* 47T2 NCBIM 40044. *Biotechnol. Bioeng.* 81: 316–322.

Hasenhuettl, G.L. and Hartel, R.W. (2008). *Food Emulsifiers and Their Applications*. Springer Science & Business Media.

Heiman, K.E., Mody, R.K., Johnson, S.D. et al. (2015). *Escherichia coli* O157 outbreaks in the United States, 2003–2012. *Emerg. Infect. Dis.* 21 (8): 1293–1301.

Heinemann, C., Van Hylckama, V., Janssen, D.B. et al. (2000). Purification and characterization of a surface binding protein from *Lactobacillus fermentum* RC-14 that inhibits adhesion of *Enterococcus faecalis* 1131. *FEMS Microbiol. Lett.* 190: 177–180.

Igarashi, S., Hattori, Y., and Maitani, Y. (2006). Biosurfactant MEL-A enhances cellular association and gene transfection by cationic liposome. *J. Controlled Release* 112: 362–368.

Imura, T., Yanagishita, H., Ohira, J. et al. (2005). Thermodynamically stable vesicle formation from glycolipid biosurfactant sponge phase. *Colloids Surf., B* 43: 115–121.

Inoh, Y., Kitamoto, D., Hirashima, N., and Nakanishi, M. (2001). Biosurfactants of MEL-A increase gene transfection mediated by cationic liposomes. *Biochem. Biophys. Res. Commun.* 289: 57–61.

Israelachvili, J. (1994). The science and applications of emulsions – an overview. *Colloids Surf. A Physicochem. Eng.* 91: 1–8.

Jain, R.M., Mody, K., Joshi, N. et al. (2013). Production and structural characterization of biosurfactant produced by an alkaliphilic bacterium, *Klebsiella* sp.: evaluation of different carbon sources. *Colloids Surf., B* 108: 199–204.

Jenkinson, H.F. and Lappin-Scott, H.M. (2001). Biofilms adhere to stay. *Trends Microbiol.* 9: 9–10.

Joshi, S., Bharucha, C., and Desai, A.J. (2008). Production of biosurfactant and antifungal compound by fermented food isolate *Bacillus subtilis* 20B. *Bioresour. Technol.* 99: 4603–4608.

Kim, B.S., Lee, J.Y., and Hwang, B.K. (2000). *In vivo* control and in vitro antifungal activity of rhamnolipid B, a glycolipid antibiotic, against *Phytophthora capsici* and *Colletotrichum orbiculare*. *Pest Manag. Sci.* 56: 1029–1035.

Kiran, G.S., Selvin, J., Manilal, A., and Sujith, S. (2011). Biosurfactants as green stabilizers for the biological synthesis if nanoparticles. *Crit. Rev. Biotechnol.* 31: 354–364.

Kiran, G.S., Priyadharsini, S., Sajayan, A. et al. (2017). Production of lipopeptide biosurfactant by a marine *Nesterenkonia* sp. and its application in food industry. *Front. Microbiol.* 8: 1138.

Kitamoto, D., Fuzishiro, T., Yanagishita, H. et al. (1992). Production of mannosylerythritol lipids as biosurfactants by resting cells of *Candida antarctica*. *Biotechnol. Lett* 14 (4): 305–310.

Kitamoto, D., Yanagishita, H., Shinbo, T. et al. (1993). Surface active properties and antimicrobial activities of mannosylerythritol lipids as biosurfactants produced by *Candida antarctica*. *J. Biotechnol.* 29 (1–2): 91–96.

Kitamoto, D., Morita, T., Fukuoka, T. et al. (2009). Self-assembling properties of glycolipid biosurfactants and their potential applications. *Curr. Opin. Colloid Interface Sci.* 14: 315–328.

Kralova, I. and Sjöblom, J. (2009). Surfactants used in food industry: a review. *J. Dispersion Sci. Technol.* 30: 1363–1383.

Kuiper, I., Lagendijk, E.L., Pickford, R. et al. (2004). Characterization of two *Pseudomonas putida* lipopeptide biosurfactants, putisolvin I and II, which inhibit biofilm formation and break down existing biofilms. *Mol. Microbiol.* 51: 97–113.

Liu, F., Zhu, Z., Ma, C. et al. (2016). Fabrication of concentrated fish oil emulsions using dual-channel microfluidization: impact of droplet concentration on physical properties and lipid oxidation. *J. Agric. Food. Chem.* 64 (50): 9532–9541.

Lukondeh, T., Ashbolt, N.J., and Rogers, P.L. (2003). Evaluation of *Kluyveromyces marxianus* FII 510700 grown on a lactose-based medium as a source of a natural bioemulsifier. *J. Ind. Microbiol. Biotechnol.* 30: 715–720.

Luna, J.M., Rufino, R.D., Sarubbo, L.A. et al. (2011). Evaluation of antimicrobial and antiadhesive properties of biosurfactant lunasan produced by *Candida sphaerica* UCP 0995. *Curr. Microbiol.* 62 (5): 1527–1534.

Lynch, A.S. and Robertson, G.T. (2008). Bacterial and fungal biofilm infections. *Annu. Rev. Med.* 59: 415–428.

Magalhães, L. and Nitschke, M. (2013). Antimicrobial activity of rhamnolipids against *Listeria monocytogenes* and their synergistic interaction with nisin. *Food Control* 29: 138–142.

Martínez-Checa, F., Toledo, F.L., El Mabrouki, K. et al. (2007). Characteristics of bioemulsifier V2-7 synthesized in culture media added of hydrocarbons: chemical composition, emulsifying activity and rheological properties. *Bioresour. Technol.* 98: 3130–3135.

Maukonen, J., Matto, J., Wirtanen, G. et al. (2003). Methodologies for the characterization of microbes in industrial environments: a review. *J. Ind. Microbiol. Biotechnol.* 30: 327–356.

McClements, D.J., Bai, L., and Chung, C. (2017). Recent advances in the utilization of natural emulsifiers to form and stabilize emulsions. *Annu. Rev. Food Sci. Technol.* 8: 205–236.

Meylheuc, T., Van Oss, C.J., and Bellon-Fontaine, M.N. (2001). Adsorption of biosurfactant on solid surfaces and consequences regarding the bioadhesion of *Listeria monocytogenes* LO28. *J. Appl. Microbiol.* 91: 822–832.

Meylheuc, T., Methivier, C., Renault, M. et al. (2006a). Adsorption on stainless steel surfaces of biosurfactants produced by gram-negative and gram-positive bacteria: consequence on the bioadhesive behaviour of *Listeria monocytogenes*. *Colloids Surf., B* 52: 128–137.

Meylheuc, T., Renault, M., and Bellon-Fontaine, M.N. (2006b). Adsorption of a biosurfactant on surfaces to enhance the disinfection of surfaces contaminated with *Listeria monocytogenes*. *Int. J. Food Microbiol.* 109: 71–78.

Mnif, I., Besbes, S., Ellouze-Ghorbel, R. et al. (2013). Improvement of bread dough quality by *Bacillus subtilis* SPB1 biosurfactant addition: optimized extraction using response surface methodology. *J. Sci. Food Agric.* 93: 3055–3064.

Monteiro, A.S., Coutinho, J.O.P.A., Ary, C. et al. (2009). Characterization of new biosurfactant produced by *Trychosporon montevidense* CLOA72 isolated from dairy industry effluents. *J. Basic Microbiol.* 49: 1–11.

Monteiro, A.S., Bonfim, M.R.Q., Domingues, V.S. et al. (2010). Identification and characterization of bioemulsifier-producing yeasts isolated from effluents of a dairy industry. *Bioresour. Technol.* 101: 5186–5193.

Morais, I.M.C., Cordeiro, A.L., Teixeira, G.S. et al. (2017). Biological and physicochemical properties of biosurfactants produced by *Lactobacillus jensenii* $P_{6A}$ and *Lactobacillus gasseri* $P_{65}$. *Microb. Cell Factories* 16: 155.
Morikawa, M., Ito, M., and Imanaka, T. (1992). Isolation of a new surfactin producer *Bacillus pumilus* A-1, and cloning and nucleotide sequence of the regulator gene, psf-1. *J. Ferment. Bioeng.* 74 (5): 255–261.
Neta, N.S., Santos, J.C., Gonçalves, L.R. et al. (2012). Enzymatic synthesis of sugar esters and their potential as surface-active stabilizers of coconut milk emulsions. *Food Hydrocolloids* 27: 324–331.
Neta, N.S., Teixeira, J.A., and Rodrigues, L.R. (2015). Sugar ester surfactants: enzymatic synthesis and applications in food industry. *Crit. Rev. Food Sci. Nutr.* 55 (5): 595–610.
Nguyen, T.T.L., Edelen, A., Neighbors, B., and Sabatini, D.A. (2010). Biocompatible lecithin-based microemulsions with rhamnolipid and sophorolipid biosurfactants: formulation and potential applications. *J. Colloid Interface Sci.* 348: 498–504.
Nitschke, M. and Costa, S.G.V.A.O. (2007). Biosurfactants in food industry. *Trends Food Sci. Technol.* 18: 252–259.
Nitschke, M. and Pastore, G.M. (2006). Production and properties of a surfactant obtained from *Bacillus subtilis* grown on cassava wastewater. *Bioresour. Technol.* 97: 336–341.
Nitschke, M., Araújo, L.V., Costa, S.G. et al. (2009). Surfactin reduces the adhesion of foodborne pathogenic bacteria to solid surfaces. *Lett. Appl. Microbiol.* 49: 241–247.
Palanisamy, P. (2008). Biosurfactant mediated synthesis of NiO nanorods. *Mater. Lett.* 62: 743–746.
Patino, J.M.R., Sánchez, C.C., and Niño, N.M.R. (2008). Implications of interfacial characteristics of food foaming agents in foam formulations. *Adv. Colloid Interf. Sci.* 140: 95–113.
Pereira, J.F.B., Gudiña, E.J., Costa, R. et al. (2013). Optimization and characterization of biosurfactant production by *Bacillus subtilis* isolates towards microbial enhanced oil recovery applications. *Fuel* 111: 259–268.
Porter, J.N., Broschard, R., Krupka, G. et al. (1949). Isolation and production of polymyxin. *Ann. N.Y. Acad. Sci.* 51: 857–865.
Rahman, M.S., Ano, T., and Shoda, M. (2006). Second stage production of iturin A by induced germination of *Bacillus subtilis* RB14. *J. Biotechnol.* 125 (4): 513–515.
Reddy, S.A., Chen, C.Y., Baker, S.C. et al. (2009). Synthesis of silver nanoparticles using surfactin: a biosurfactant stabilizing agent. *Mater. Lett.* 63: 1227–1230.
Rodrigues, L.R. (2011). Inhibition of bacterial adhesion on medical devices. In: *Bacterial Adhesion: Biology, Chemistry, and Physics, Series: Advances in Experimental Medicine and Biology*, vol. 715 (eds. D. Linke and A. Goldman), 351–367. Berlin: Springer.
Rodrigues, L.R., Van der Mei, H.C., Teixeira, J.A., and Oliveira, R. (2004a). Biosurfactant from *Lactococcus lactis* 53 inhibit microbial adhesion on silicone rubber. *Appl. Microbiol. Biotechnol.* 66 (3): 306–311.
Rodrigues, L.R., Van der Mei, H.C., Teixeira, J.A. et al. (2004b). Influence of biosurfactants from probiotic bacteria on formation of biofilms on voice prostheses. *Appl. Environ. Microbiol.* 70 (7): 4408–4410.
Rodrigues, L.R., Banat, I.M., Teixeira, J.A., and Oliveira, R. (2006a). Biosurfactants: potential applications in medicine. *J. Antimicrob. Chemother.* 57 (4): 609–618.

Rodrigues, L.R., Van der Mei, H.C., Banat, I.M. et al. (2006b). Inhibition of microbial adhesion to silicone rubber treated with biosurfactant from *Streptococcus thermophilus* A. *FEMS Immunol. Med. Microbiol.* 46 (1): 107–112.

Rodrigues, A.I., Gudiña, E.J., Teixeira, J.A., and Rodrigues, L.R. (2017). Sodium chloride effect on the aggregation behaviour of rhamnolipids and their antifungal activity. *Sci. Rep.* 7: 12907.

Roongsawang, N., Washio, K., and Morikawa, M. (2010). Diversity of nonribosomal peptide synthetases involved in the biosynthesis of lipopeptide biosurfactants. *Int. J. Mol. Sci.* 12 (1): 141–172.

Rosenberg, E. and Ron, E.Z. (1999). High- and low-molecular-mass microbial surfactants. *Appl. Microbiol. Biotechnol.* 52: 154–162.

Rufino, R.D., Luna, J.M., Sarubbo, L.A. et al. (2011). Antimicrobial and antiadhesive potential of a biosurfactant Rufisan produced by *Candida lipolytica* UCP 0988. *Colloids Surf., B* 84 (1): 1–5.

Santos, D.K.F., Rufino, R.D., Luna, J.M. et al. (2013). Synthesis and evaluation of biosurfactant produced by *Candida lipolytica* using animal fat and corn steep liquor. *J. Pet. Sci. Eng.* 105: 43–50.

Serrano, N., Rodrigues, L.R., Hokka, C.O. et al. (2012). Optimal glucose and inoculum concentrations for the production of bioactive molecules by *Paenibacillus polymyxa* RNC-D. *Chem. Pap.* 66 (12): 1111–1117.

Shepherd, R., Rockey, J., Sutherland, I.W., and Roller, S. (1995). Novel bioemulsifiers from microorganisms for use in foods. *J. Biotechnol.* 40: 207–217.

Shete, A.M., Wadhawa, G., Banat, I.M., and Chopade, B.A. (2006). Mapping of patents on bioemulsifier and biosurfactant: a review. *J. Sci. Ind. Res.* 65: 91–115.

Shim, G.Y., Kim, S.H., Han, S.E. et al. (2009). Cationic surfactin liposomes for enhanced cellular delivery of siRNA. *Asian J. Pharm. Sci.* 4: 207–214.

Silva, S.S., Carvalho, J.W.P., Aires, C.P., and Nitschke, M. (2017). Disruption of *Staphylococcus aureus* biofilms using rhamnolipid biosurfactants. *J. Dairy Sci.* 100: 1–10.

Singh, B., Beg, S., Khurana, R.K. et al. (2014). Recent advances in self-emulsifying drug delivery systems (SEDDS). *Crit. Rev. Ther. Drug Carrier Syst.* 31: 89–153.

Somers, E.B. and Wong, A.C.L. (2004). Efficacy of two cleaning and sanitizing combinations of *Listeria monocytogenes* biofilms formed at low temperature on a variety of materials in the presence of ready-to-eat meat residue. *J. Food Prot.* 67: 2218–2229.

Souza, K.S.T., Gudiña, E.J., Azevedo, Z. et al. (2017). New glycolipid biosurfactants produced by the yeast strain *Wickerhamomyces anomalus* CCMA 0358. *Colloids Surf., B* 154: 373–382.

Spurbeck, R.R. and Arvidson, C.G. (2010). *Lactobacillus jensenii* surface-associated proteins inhibit *Neisseria gonorrhoeae* adherence to epithelial cells. *Infect. Immun.* 78: 3103–3111.

Stepanovic, S., Cirkovic, I., Ranin, L., and Svabic-Vlahovic, M. (2004). Biofilm formation by *Salmonella* spp. and *Listeria monocytogenes* on plastic surface. *Lett. Appl. Microbiol.* 38: 428–432.

Tahmourespour, A., Salehi, R., Kermanshahi, R.K., and Eslami, G. (2011a). The anti-biofouling effect of *Lactobacillus fermentum*-derived biosurfactant against *Streptococcus mutans*. *Biofouling* 27: 385–392.

Tahmourespour, A., Salehi, R., and Kermanshahi, R. (2011b). *Lactobacillus acidophilus*-derived biosurfactant effect on GTFB and GTFC expression level in *Streptococcus mutans* biofilm cells. *Braz. J. Microbiol.* 42: 330–339.

Trindade, J.R., Freire, M.G., Amaral, P.F.F. et al. (2008). Aging mechanisms of oil-in-water emulsions based on a bioemulsifiers produced by *Yarrowia lipolytica*. *Colloids Surf. A Physicochem. Eng.* 324: 149–154.

Valotteau, C., Banat, I.M., Mitchell, C.A. et al. (2017). Antibacterial properties of sophorolipid-modified gold surfaces against Gram positive and Gram negative pathogens. *Colloids Surf., B* 157: 325–334.

Van Bogaert, I.N.A., Saerens, K., De Muynck, C. et al. (2007). Microbial production and application of sophorolipids. *Appl. Microbiol. Biotechnol.* 76: 23–34.

Van Haesendonck I.P.H. and E.C.A. Vanzeveren. 2004. Rhamnolipids in bakery products. International Patent WO 2004/040984.

Van Hoogmoed, C.G., Van der Kuijl-Booij, M., Van der Mei, H.C., and Busscher, H.J. (2000). Inhibition of *Streptococcus mutans* NS adhesion to glass with and without salivary conditioning film by biosurfactant-releasing *Streptococcus mitis* strains. *Appl. Environ. Microbiol.* 66: 659–663.

Vaz, D.A., Gudiña, E.J., Alameda, E.J. et al. (2012). Performance of a biosurfactant produced by a *Bacillus subtilis* strain isolated from crude oil samples as compared to commercial chemical surfactants. *Colloids Surf., B* 89: 167–174.

Walencka, E., Rozalska, S., Sadowska, B., and Rozalska, B. (2008). The influence of *Lactobacillus acidophilus* derived surfactants on staphylococcal adhesion and biofilm formation. *Folia Microbiol.* 53: 61–66.

Worakitkanchanakul, W., Imura, T., Morita, T. et al. (2008). Formation of W/O microemulsion based on natural glycolipid biosurfactant. *J. Oleo Sci.* 57: 55–59.

Xie, Y., Ye, R., and Liu, H. (2006). Synthesis of silver nanoparticles in reverse micelles stabilized by natural biosurfactant. *Colloids Surf. A Physicochem. Eng.* 279: 175–178.

Yakimov, M.M., Timmis, K.N., Wray, V., and Fredrickson, H.L. (1995). Characterization of a new lipopeptide surfactant produced by thermotolerant and halotolerant subsurface *Bacillus licheniformis* BAS50. *Appl. Environ. Microbiol.* 61: 1706–1713.

Yu, G.Y., Sinclair, J.B., Hartman, G.L., and Bertagnolli, B.L. (2002). Production of iturin A by *Bacillus amyloliquefaciens* suppressing *Rhizoctonia solani*. *Soil Biol. Biochem.* 7: 955–963.

Zhang, X., Fan, X., Solaiman, D.K.Y. et al. (2016). Inactivation of *Escherichia coli* O157:H7 *in vitro* and on the surface of spinach leaves by isolated antimicrobial surfactants. *Food Control* 60: 158–165.

Zhang, X., Ashby, R.D., Solaiman, D.K.Y. et al. (2017). Antimicrobial activity and inactivation mechanism of lactonic and free acid sophorolipids against *Escherichia coli* O157:H7. *Biocatal. Agricult. Biotechnol.* 11: 176–182.

Zheng, C., He, J., Wang, Y. et al. (2011). Hydrocarbon degradation and bioemulsifier production by thermophilic *Geobacillus palidus* strains. *Bioresour. Technol.* 102: 9155–9161.

Zouari, R., Besbes, S., Ellouze-Chaabouni, S., and Ghribi-Aydi, D. (2016). Cookies from composite wheat-sesame peels flours: dough quality and effect of *Bacillus subtilis* SPB1 biosurfactant addition. *Food Chem.* 194: 758–769.

# 7

# Fermentative Production of Microbial Exopolysaccharides

*Jochen Schmid*[1,2] *and Volker Sieber*[1]

[1] *Chemistry of Biogenic Resources, Technical University of Munich, Campus Straubing for Biotechnology and Sustainability, Straubing, Germany*
[2] *Norwegian University of Science and Technology, Department of Biotechnology and Food Science, Sem Sælands vei 6-8, 7491 Trondheim, Norway*

## 7.1 Introduction

Microbial exopolysaccharides (EPSs) represent a highly diverse class of biopolymers which are produced by many different microorganisms from all genera such as fungi and bacteria (Schmid et al. 2016; Rühmann et al. 2015; Rehm 2010). Technically, they are normally produced in submerse culture and can then be harvested from the fermentation broth mainly by simple alcohol precipitation. They show a high diversity in their chemical structures and can be used in different applications, ranging from medical to food and feed applications as well construction chemistry or lubricants. The chemical structures can be the same in different microorganisms but can vary in chain length and molecular weight, depending on the production strain or the production conditions applied, such as carbon source, dissolved oxygen, or agitation.

The history of microbial exopolysaccharide production started in the early 1950s with the discovery of one of the most famous representatives, xanthan gum., which is produced by the bacteria *Xanthomonas campestris*. Xanthan is still the microbial exopolysaccharide with the highest production capacity worldwide and is a kind of benchmark for other and new polysaccharides. Xanthan gum shows highly interesting pseudoplastic behavior and is food approved by the FDA as well as in the European Union. The chemical structure is composed of glucose (Glc): mannose (Man): glucuronic acid (GlcA) in the ratio of 2 : 2 : 1 with additional decorations of acetate and pyruvate at the mannose monomers in the side chain. The backbone is made of glucose residues only, thus representing bacterial cellulose, which is made water-soluble by the attached sidechains. Table 7.1 summarizes the commercial polysaccharides that are applied in food applications, including their monosaccharide pattern, E-number, and charge of the biopolymer.

Other polysaccharides of commercial relevance include the different heteropolysaccharide sphingans, gellan, diutan, and welan or sanxan produced by different bacteria of the genus *Sphingomonas* (Schmid et al. 2014; Huang et al. 2016). The bacterial succinoglycan consists of mainly glucose with an additional galactose in its repeating unit and is produced by e.g. *Sinorhizobium meliloti* (Frey 2008). The bacterial alginates

*Bioprocessing for Biomolecules Production*, First Edition.
Edited by Gustavo Molina, Vijai Kumar Gupta, Brahma N. Singh, and Nicholas Gathergood.

**Table 7.1** Listing of microbial polysaccharides with commercial utilization in food, feed or medical and technical applications, including charge, monomer composition as well as E-numbers if available.

| Polysaccharide | Charge/structure | E-number | Monomers | References |
|---|---|---|---|---|
| Alginate | anionic/linear | E 400[a)] | GulA, ManA | Hay et al. (2013) |
| Cellulose | neutral/linear | E 460[b)] | Glc | Ullah et al. (2016) |
| Curdlan | neutral/linear | E 424 | Glc | Miwa et al. (1993) |
| Dextran | neutral/mostly linear | N.A.[c)] | Glc | Kothari et al. (2015) |
| Diutan | anionic/branched | N.A.[d)] | Glc, Rha, Man, GlcA | Schmid et al. (2014) |
| Gellan | anionic/linear | E 418 | Glc, Rha, GlcA | Omoto et al. (1999) |
| Levan | neutral/mostly linear | N.A. | Fru | Oner et al. (2016) |
| Scleroglucan | neutral/branched | N.A. | Glc | Schmid et al. (2011) |
| Succinoglycan | anionic/branched | N.A.[e)] | Glc, Gal | Frey (2008) |
| Schizophyllan | anionic/branched | N.A.[f)] | Glc | Zhang et al. (2013) |
| Pullulan | neutral/branched | E 1204 | Glc | Farris et al. (2014) |
| Xanthan | anionic/branched | E 415 | Glc, Man, GlcA | Sanderson (1981) |

Abbreviations: Fru – fructose, Gal, galactose, Glc – glucose, GlcA – glucuronic acid, GulA – guluronic acid, Man – mannose, ManA – mannuronic acid, Rha - rhamnose.

a) Up to now no bacterial alginic acid is available on the market for food applications, and the E-numbers for macroalgae derived products depend on the ions bound; therefore, alginate ranges from E 400–404.
b) Cellulose is defined as E 460, with further numbers on cellulose derivatives, but it is not clearly defined if there is a difference between bacterial and plant-based cellulose.
c) There is no E-number available, but the European Commission does not have any concerns on the utilization of dextran as a food additive.
d) There is no E-number available, but in the year 2016, diutan was approved by the US EPA for use as an inert ingredient in agricultural chemical applications, as save additive to agricultural applications.
e) Succinoglycan is allowed to be used in food only in Japan.
f) Schizophyllan is not allowed in food but has approval for medical and technical applications.

mainly produced by Pseudomonads or *Azotobacter* sp. (Hay et al. 2013), and the different glucans such as cellulose from *Gluconobacter* sp. (Ullah et al. 2016), curdlan from *Agrobacterium* sp. (Lee et al. 1999), and dextrans from e.g. *Leuconostoc mesenteroides* (Kothari et al. 2015). A representative of fructans is the bacterial levan, which is mainly linear but also exists as branched polymer with varying molecular weight and degree of branching as produced by many different bacteria (Oner et al. 2016). Of the fungal polysaccharides, pullulan as α-glucan has a long history in different applications. It is produced by the fungus *Aureobasidium pullulans* (Farris et al. 2014). For the branched β-glucans, the highly similar scleroglucan and schizophyllan produced by *Sclerotium rolfsii* and *Schizophyllum commune,* respectively, are of high relevance in different applications, especially in the Asian food industry.

Many further microbial polysaccharides exist in nature but have not been of commercial relevance up to now. The approval of novel microbial polysaccharides for food applications is a very cost-intensive procedure and limits the amount of polysaccharides used. Gellan was one of the last polysaccharides approved for food application in the United States and the European Union.

The principal production process of microbial polysaccharides comprises different steps, summarized in Figure 7.1. These steps include the fermentative production of

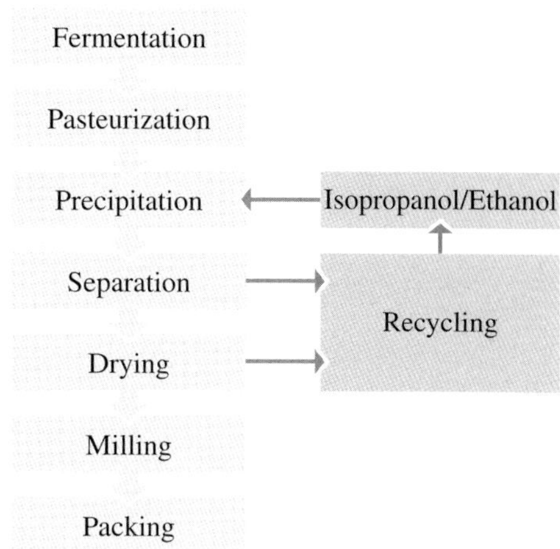


**Figure 7.1** Generalized workflow of microbial exopolysaccharide production. Starting from the fermentative production in a suitable bioreactor followed by a heat treatment at the end of the production process (normally around 72 hours). The heating step is realized by steam treatment to inactivate viable cells as well as hydrolytic enzymes. In the next step the polysaccharide is precipitated from the fermentation broth by use of water-miscible alcohols, to form a fibril precipitate. Following this, the dewatered polysaccharide is dried to a final water content of around 10% and the dry fibers are milled to get a defined particle size. The alcohol used for precipitation is recovered by distillation and rectification to realize process economics.

the specific polysaccharide, which usually is done in submers cultivation. Following the fermentation, the microbial cells are separated or inactivated by pasteurization and the polysaccharide is precipitated mainly by alcohol treatment. The precipitated polysaccharide is then dewatered until it can be milled and packed. In the case of xanthan gum, isopropanol is the alcohol of choice, due to its low price and good recyclability via distillation in rectification columns.

## 7.2 Cultivation Media and Renewable Resources

In most cases, polysaccharides are produced from pure sugar streams such as sucrose or glucose (depending on the polymer). But for efficient end economic production, mainly starch is used, which must be hydrolyzed via corresponding enzymes or in some cases can be directly used for the cultivation of the organisms. Starch therefore represents a low-cost substrate with high availability. But based on the permanent cost-pressure in the industry and the competition with oil-based additives such as polyacrylates etc. more economic substrates are being evaluated to lower the production cost of microbial polysaccharides.

Most important factor in polysaccharide production is the ratio of carbon: nitrogen (C:N ratio). High ratios favor the production of polysaccharides by different microorganisms (Sutherland 1998; Sutherland 2001). For some fungi, especially, the choice of the nitrogen source has revealed to be of most importance, such as nitrate for *S. rolfsii* (Castillo et al. 2015; Schmid et al. 2010; Fariña et al. 1998). In the case of Levan production, the utilization of sucrose is inevitable since the energy for the formation of the fructose polymer is gained from hydrolyzation of sucrose (Oner et al. 2016). For novel media, massive investigations are performed in the field of biomass and side products from renewable resources to lower the production cost from synthetic media. As for many microbial production processes, the cultivation media has switched from defined synthetic media toward cheaper alternatives such as starch, syrups, molasses, or wastewater from food production (Israilides et al. 1998; Kalogiannis et al. 2003; Survase et al. 2007; Bhatia et al. 2015). The main findings were recently summarized by Öner et al. (2013), and it is obvious that different sources of biomass, especially lignocellulose hydrolysates, are a current focus. In sum, the yields of the EPS produced from alternative resources such as biomass or side streams also reach yields comparable to fermentations based on starch or synthetic media and might represent a highly attractive alternative on an industrial level. But suitable media highly depend on the production strain used. Finally, the grade of wanted purity will clearly define the choice of the substrate.

## 7.3 Bioreactor Geometries and Design

Stirred tank reactors (STRs) still represent the most common form to produce microbial exopolysaccharides, and the industrial scale for xanthan production is up to 50–100 $m^3$ (Kuppuswami 2014; Hublik 2012). STRs provide the advantage of good mixing properties in contrast to high energy cost for agitation and oxygen supply (Hublik 2012). Large-scale fermenters enable culture conditions that can be carefully standardized, monitored, and controlled, and therefore can realize a consistent product quality. But, as for all processes, the up-scale required to guarantee stable physiological growth is one of the most challenging steps, which highly influences product yields as well as product quality of microbial exopolysaccharides (Garcia-Ochoa and Gomez 2009). One of the main challenges is the high intrinsic viscosity of most fermentation broths at the end of the process, thus limiting mass as well as oxygen transfer rates in the bioreactors (Herbst et al. 1992). Additionally, the conditions massively change within the process due to the low viscosity at the beginning of the process (Seviour et al. 2011; Fazenda et al. 2008; Garcia-Ochoa et al. 2000; Gibbs et al. 2000; McNeil and Harvey 1993). The high viscosity also is in need of high stirring speeds, which cause high shear rates in the bioreactor that can negatively influence cell viability and morphology (McNeil and Kristiansen 1987). The putative pseudoplastic behavior of the final fermentation broth can cause nonstirred areas for shear thinning polysaccharides, for example (Seviour et al. 2011). Since nearly all microbes used in industrial production of microbial exopolysaccharides are aerobes, the oxygen supply is the most crucial part of the entire process. Up to now, only a sparse number of specially designed bioreactors for exopolysaccharide production have been reported in the literature. Primarily, adaptions of general bioreactors have been used and are therefore the most represented compromises between the most suitable process parameters and process costs, as schematically depicted in Figure 7.2 (Garcia-Ochoa and Gomez 2009; Seviour et al. 2011). The increase in viscosity of the fermentation broth highly depends on the type of polysaccharide produced, but at the same time the molecular weight of the secreted polymer can vary by the agitation speed applied (Lee et al. 1999) or the oxygen level (Peña et al. 2000). Shifts in the culture morphology can influence the viscosity, the fermentation process itself, as well as the downstream processing, especially when filamentous fungi are used (Papagianni 2004). Therefore, the viscosity must be evaluated carefully over the production process. This is normally done offline, but new techniques such as measuring the heat transfer capacity have been evaluated in mid-scale STRs for xanthan production and might be a real alternative for industrial production (Wunderlich et al. 2016; Schelden et al. 2017).

Airlift bioreactors (ALRs) do not need any energy for stirring since they are agitated pneumatically, thus greatly reducing the energy input. Depending on the process conditions and final product, high amounts of energy can be saved. Additionally, the energy dissipation is uniform in ALRs compared to STRs, in which the energy dissipation rate

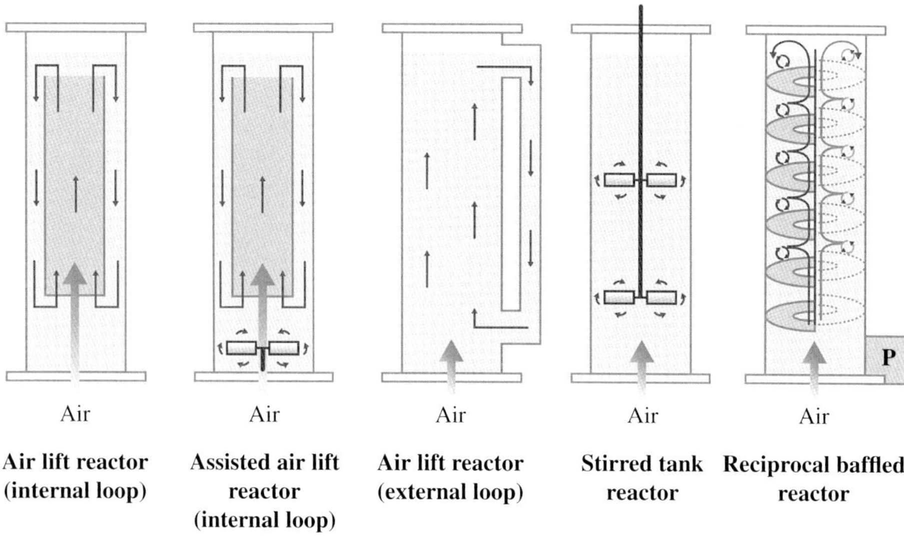


**Figure 7.2** Schematic drawings of the different reactor designs, including mixing profiles of the fermentation broth. From left to right; assisted air lift reactor with an additional impeller at the bottom of the reactor, air lift reactor with an internal loop, air lift reactor with an internal loop, classical stirred tank reactor with two impellers, scheme of an oscillating baffled reactor. The last reactor type can be realized by oscillating movement of the baffles, or by reciprocal movement (pumping) of the liquid. Abbreviation: P = pump.

greatly varies. This variation can be by a factor of about a hundred from the impeller edge to the reactor wall (Schugerl 1990). Thus, the ALRs are very well suited for shear sensitive microorganisms or cell cultures and filamentous fungi with a definite cell morphology.

In STRs, cell mass concentrations, volumetric productivities, and specific power inputs are higher than in ALRs, but the efficiency of oxygen-, heat-, and mass-transfer are in general higher in ALRs. Therefore, specific productivities and yield coefficients with regard to power input, substrate and oxygen consumption are generally higher in ALRs (Schugerl 1990). As a final advantage, startup and operational costs are lower than for STRs due to the absence of mechanical parts. The problem of lower oxygen transfer rates can be minimized by use of internal or external loops, which can enhance oxygen transfer.

ALRs are successfully used for EPS production by filamentous fungi, such as pullulan (Roukas and Mantzouridou 2001; Özcan et al. 2014) and scleroglucan (Kang et al. 2000). For scleroglucan production, a lower dissolved oxygen level is beneficial, and therefore ALR cultivation is well suited, with the additional effect that viscosity of the non-Newtonian fermentation broth is uniform due to more uniform energy dissipation (Kang et al. 2001). Additionally, mixed versions of STR and ALR by use of stirring elements in an ALR exist (assisted airlift reactor), as described for pullulan production (Gibbs and Seviour 1992).

The small number of specially designed bioreactors includes an oscillatory baffled bioreactor (OBR), as described for pullulan production (Gaidhani et al. 2003). This kind of bioreactor uses baffles that oscillate in the fermentation broth, combining the advantages of uniform mixing with a very low shear rate and energy consumption compared to STRs (Abbott et al. 2014). The principle is based on superimposing fluid oscillation in a cylindrical tube containing periodically spaced orifice baffles, resulting in enhanced residence time and more uniform bubble size, thus promoting better mass transfer (Hewgill et al. 1993). Additionally, the scale-up correlation of OBRs is described to be linear, which is nearly unique in microbial bioreactor systems (Ni and Gao 1996). Even though the advantages seem to be overwhelming, no industrial process on the use of OBRs is described for microbial exopolysaccharide production to our knowledge, which might depend on the mechanical oscillating movement of the baffles, as well as the increased energy input necessary for moving the baffles or the media in large scale. Additionally, for unmoved baffles, bioreactors of several hundred meters in length would be necessary to realize efficient bioprocesses. Just recently, these issues were discussed in detail by Abbott et al. (2013), offering some interesting strategies to solve the existing barriers and make useful of the existing advantages.

Other systems such as utilization of Ca-alginate immobilized levan producing *Zymomonas mobilis* cells in a packed-bed reactor show quite good productivities in batch as well as early continuous mode, but show strong limitations in reutilization of the cells, thus lowering the final yield massively (Silbir et al. 2014). Another immobilization of *X. campestris* cells was performed on porous celite particles suspended in the bioreactor with the intention to keep the polymer within the celite particle and therefore not increase the viscosity (Robinson and Wang 1987; Robinson and Wang 1988). They were able to obtain a final concentration of $50\,g\,l^{-1}$ after 140 hours within the particles and additionally $20\,g\,l^{-1}$ in the liquid phase. The main drawback of this approach is the choice of the particle size, since small particles cannot retain enough polymer strands to reduce the viscosity but enable an efficient oxygen supply. Larger particles, in contrast, can retain high amounts of polymer, but limit oxygen supply, thus causing reduced cell growth and xanthan yields.

Further examples for the utilization of immobilized cells are centrifugal fibrous bioreactors (Yang et al. 1996; Lo et al. 2001), where the cells are immobilized on a cylindrical fibrous matrix that is rotating in the bioreactor. By that rotation, the medium was radially forced through the matrix by introduction along the central axis and the produced xanthan polymer could easily be extracted as supported by the centrifugal force. Compared to free cells, the viability was reduced to around 40% and the specific productivity was $0.05–0.075\,g/gh^{-1}$ cell mass, which is far away from specific productivities reached in STRs ($0.3–0.45\,g/gh^{-1}$) (Ju 2007). Metabolic flux analysis of this centrifugal packed-bed reactor (CPBR), in contrast, identified the CPBR as the most favorable one for assembly of nucleotide sugars from glucose-6-phosphate as precursors for xanthan biosynthesis (Hsu and Lo 2003; Vorhölter et al. 2008). However, the low productivity in combination with the power consumption for rotating the bed and circulating the media in a highly complex system renders the process economically questionable, especially in large-scale production (Ju 2007) An overview of the different cultivation systems, including product titers and productivity, used for xanthan production is given in Table 7.2.

More promising are the approaches of testing different or novel stirrer designs, since it was shown very often that the stirrers massively influence the oxygen as well as mass transfer, especially in highly viscous, non-Newtonian fermentation broth (Garcia-Ochoa and Gomez 2009; Seviour et al. 2011; Lin and Thibault 2013). Relatively little information on industrial production is available, but several specially designed

**Table 7.2** Comparison of different xanthan gum yields and specific productivities obtained from different strains, reactor designs and process variants.

| Reactor/Technology | Titer (g l) | Volumetric productivity (g/lh$^{-1}$) | Specific productivity [g/g$_{glucose}$] | References |
|---|---|---|---|---|
| Stirred tank | 22 | 0.3–0.45 | 0.2–0.25 | Pons et al. (1990) |
| Bubble column | 25 | 0.17–0.35 | 0.002–0.11 | Pons et al. (1990) |
| Airlift reactor | 15 | < 0.15 | 0.02–0.05 | Suh et al. (1992) |
| Centrifugal fibrous bed reactor | 35 | 0.7–1.0 | 0.05–0.075 | Yang et al. (1996) |
| Batch fermentation (STR) | 30–50 | 0.4–0.7 | N.A. | Sieber et al. (2019) |
| Fed-batch fermentation (STR) | 62 | 0.72 | 0.82 | Amanullah et al. (1998) |
| Continuous fermentation (STR) | 7.27 | 0.36 | 0.67 | Roseiro et al. (1993) |
| W/O fermentation | 60–75 | 0.5 | 0.05 | Ju (2007), Ju and Zhao (1993) |

Abbreviations: STR = stirred tank reactor, W/O = water in oil emulsion.
Source: Adapted and amended from (Ju 2007).

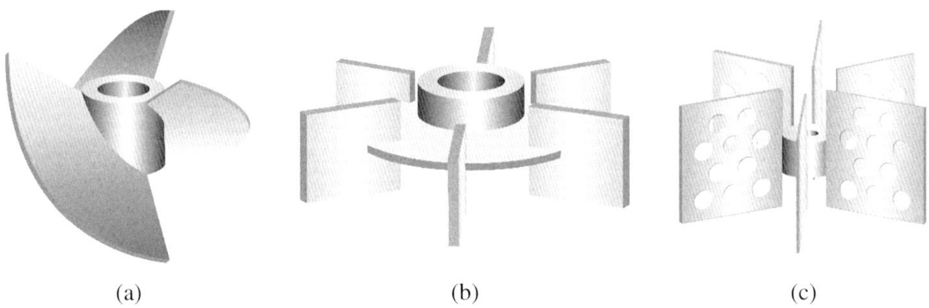


**Figure 7.3** (a) Schematic representation of a lower shear force provoking marine impeller with macromixing effect; (b) schematic drawing of a higher shear force provoking Rushton turbine with more local mixing regimes; and (c) special designed radial reciprocating plate impeller (RRPI) for exopolysaccharide production.

stirrer designs and process parameters exist. In many cases, especially for fungal producers, the low shear fans and marine propeller designs highly outperform the typical Rushton turbines as shown for schizophyllan (Gura and Rau 1993), the β-glucan produced by *Botrytis cinerea* (cinerean) (Crognale et al. 2007) and suggested for scleroglucan (Castillo et al. 2015). A typical Rushton turbine and a marine impeller, as well as a specially designed radial reciprocating plate impeller (RRPI), are schematically depicted in Figure 7.3. Impellers are described to be beneficial for enhanced mixing and oxygen supply for xanthan fermentation (Funahashi et al. 1987; Galindot and Nienow 1992). Additionally, the dimension of the impellers has to be carefully evaluated to adjust the optimal tip-speed, which is mainly responsible for the diameter of the micro- and macromixing regions (Funahashi et al. 1988). On the other hand, high shear rates were beneficial for curdlan yields, but at the same time minimized the product quality (lawford and Rousseau 1991).

Recently, a novel impeller type, the *rotational reciprocating perforated plate impeller*, was evaluated for pullulan production by *Aureobasidium pullulans* and showed final yields comparable to STRs equipped with Rushton turbines. However, it is in need of a novel reactor design (Lin and Thibault 2013). A relatively similar approach was performed with another type of close-clearance impellers, the so-called axial reciprocating plate impeller (ARPI). Promising results were obtained in *A. pullulans* fermentations using a laboratory-scale bioreactor equipped with an ARPI (Lounes et al. 1995; Audet et al. 1998). However, its adaptation to a large-scale industrial bioreactor could be difficult because of its characteristic oscillatory axial movement of a stack of perforated plates.

Next to reactor principles/design or stirrer geometries, atmospheric conditions such as overpressure can be applied on bioreactors to enhance the dissolved oxygen content, which is suitable for ALRs (Campani et al. 2015) as well as STRs (Wunderlich et al. 2016; Schelden et al. 2017; Meier et al. 2016). In principle, this is a very promising and highly efficient approach, but not on the industrial scale, since most of the industrial-scale fermenters are not compression proofed and the cost to realize overpressure fermentation would be too high. The same is true for utilization of pure oxygen to saturate the inlet air for aeration of the bioreactors. The use of pure oxygen is highly efficient to realize high levels of dissolved oxygen on a laboratory scale, but is too expensive for the industrial the production of cheap products such as food polysaccharides. Additionally, the handling of pure oxygen is risky and is thus not welcome in industrial-scale processes (Kong et al. 2009; Liu et al. 2016; Meyer et al. 2017).

## 7.4 Fermentation Strategies for Microbial Exopolysaccharide Production

The classical approach as for most biotechnical production processes is batch-fermentation, which is divided into the different phases of the fermentation process, namely lag-phase, log-phase, stationary phase, and starvation phase. This variant is the easiest to realize fermentation, but also includes the longest setup and changeover times. Other variants include fed-batch, repeated fed-batch, as well as continuous feeding strategies. Next to the classical batch fermentation, the fed-batch cultivation is applied by adding a single shot of carbon-source into the reactor, to elevate the C:N ratio. The repeated fed-batch cultivation will include several shots of carbon source, and eventually further supplements such as very low amounts of nitrogen or trace elements. The continuous cultivation is relatively rare in industrial processes, since the drawbacks of contamination or spontaneous mutation of the producing organisms preponderate the positive effects of reduced setup times. Additionally, the most batch-fermentations are characterized by very high substrate conversion rates of around 50–80% (Born et al. 2005). Current batch fermentations with *X. campestris* can reach up to $50\,g\,l^{-1}$ of xanthan gum (Hublik 2012). Different fed-batch strategies were compared, and it was shown that multiple pulses as well as a continuous glucose feeding could significantly enhance xanthan gum production up to $62\,g\,l^{-1}$ (Amanullah et al. 1998). The most important aspect was the addition of glucose at low nitrogen levels, and glucose levels around $30\,g\,l^{-1}$ to prevent substrate inhibition and osmotic stress. Thus, the fed-batch variant was superior to classical batch-fermentations and seemed not to decrease at the end of the different fed-batch approaches. Therefore, continuous cultivation can be a very promising approach to reach high product titers and substrate conversion rates. Different continuous cultivations were able to reach a constant product quality under constant cultivation conditions and high productivities. In the year 1979, Evans and co-workers reported a stable continuous process without any contamination for more than 2000 hours (Evans et al. 1979). In 1993, Roseiro et al. showed that continuous cultivation for xanthan production results in increased growth rates but similar lowered production rates compared to fed-batch cultivations (Roseiro et al. 1993).

For fungal EPS-producers also, batch as well as fed-batch or continuous processes exist but will be more influenced by the morphology changes over the time – thus, determining the choice of the process variant. In the case of *A. pullulans,* the morphology highly depends on the pH-value, and the yeast-like single cells produce higher product titers than the filamentous form (McNeil et al. 1989), Thus, a continuous cultivation was performed for more than 1000 hours without septic problems and increased formation rates of 0.35 g $(\mathrm{L\,h^{-1}})^{-1}$ compared to batch fermentation (0.16 g $(\mathrm{L\,h^{-1}})^{-1}$) at a dilution rate of 0.05 L $\mathrm{h^{-1}}$ (Schuster et al. 1993). But it was observed, that higher or lower dilution rates negatively influenced the product formation rate. Additionally, the biomass concentration was identified to be the limiting factor in continuous cultivation. Economic evaluation of the aseptic continuous production of scleroglucan by *S. rolfsii* was evaluated by Schilling in the year 2000 (Schilling 2000). This study revealed that continuous cultivation could result in volumetric productivities more than twice as high compared to batch cultivations, but at the same time lowers product yield and quality. The lowered product quality was linked to increased impeller tip-speed in continuous cultivations as well as the metabolic burden of increased volumetric productivity and specific growth rate on the cost of the molecular weight of the final product. The same effects were described for xanthan and schizophyllan (Schilling 2000). By that, it is obvious that the process variant has to be carefully evaluated for each exopolysaccharide concerning yield and quality and no general rule can be given.

## 7.5 Approaches to Reduce Fermentation Broth Viscosity

As described above, the viscosity of the fermentation broth massively changes over the entire production process of microbial exopolysaccharides and thus negatively influences the oxygen and mass transfer within the bioreactor. The viscosity of a 1% (w/v) xanthan solution is higher than 5 Pa s at a standard shear rate of one per second (Kelco 2007), and the broth is highly pseudoplastic with yield stress ranging from 0.2 to 1.5 Pa s for xanthan concentrations in the range of 10–50 g $\mathrm{l^{-1}}$ (Galindo et al. 1989). The fermentation broth of *X. campestris,* with a titer of around 50 g $\mathrm{l^{-1}}$ of xanthan can easily reach a dynamic viscosity of around 35 Pa s (Galindo et al. 1989), and thus limits the production capacities and efficiencies of the whole process. Additionally, the pseudoplastic behavior of xanthan and many other polysaccharides causes the so called "caverning" phenomenon which typically occurs in STRs and is defined by the well mixed area around the stirrer and solid non-moved areas in the surrounding (Seviour et al. 2011; Solomon et al. 1981). To circumvent this problem and raise the product titers above the maximally reached 50 g $\mathrm{l^{-1}}$ of xanthan at the end of the fermentation many efforts were preformed to cope with a further increase of the viscosity of the fermentation broth (Hublik 2012). Next to the above-described novel bioreactor designs, different approaches exist by use of water-in-oil cultivations (Ju 2007). The water-in-oil fermentation is based on complex multi-phase systems containing gas, oil, aqueous media, and cells of the production strain and can be simplified to a water in oil emulsion, where the cells grow in aqueous "minireactors" and produce locally very high polysaccharide concentrations, whereas the viscosity of the complete fermentation broth is still well manageable. Different organic compounds such as n-hexadecane, perfluorcarbon or vegetable oil were applied in the different approaches (Ju and Zhao 1993; Kuttuva et al. 2004) and all were able to remarkably decrease the viscosity of the whole fermentation broth with simultaneously high xanthan concentrations of up to 120 g $l^{-1}$ (Kreyenschulte et al. 2014). The main disadvantages of this highly promising approach are the difficulties of product recovery caused by the complex phase separation as well as the spontaneous phase conversions due to the addition of aqueous pH control agents or applied feeds. One rarely discussed advantage is the positive effect of the organic phase on foam avoidance, a not to be sneezed at issue in industrial scale. Thus, the water-in-oil cultivation represents a very promising technique in lab and pilot-scale but needs heavy adaption toward industrial scale concerning economics and practical application. The utilization of smart biosurfactants to stabilize the emulsions or initiate the targeted collapse of the reaction compartments is heavily investigated and shows promising results but will need further evaluation and research (Ju 2007).

## 7.6 Polymer Byproducts and Purity

It is obvious that a generalized strategy for microbial EPS production is not feasible, especially when talking about filamentous fungi, bacteria, or morphologies in-between them, such as the yeast-like fungus *A. pullulans*, which seems to produce pullulan mainly in the single cell morphology and not in filamentous state, as nicely reviewed by Seviour et al. (2011). Additionally, in filamentous state and depending on the nitrogen source as well as the specific production strain, *A. pullulans* can produce a β-glucan, and such a product that is different from the expected α-(1,6)-based pullulan (Witczak 1996). This fact impedes the clear comparison of many results of the past, since many processes were performed without comparative conditions, including choice of carbon and nitrogen source, agitation speed, and stirrer geometry (thus influencing the cell morphology), oxygen level, and finally, the whole reactor design and setup (Seviour et al. 2011). Within the last years the structure and properties of the β-glucan was described in more detail (Lotrakula et al. 2013) and just recently the NMR structure of the water-soluble β-glucan from *A. pullulans* was published, showing a six β-(1,3)-glucoses backbone with four β-(1,6) branching residues at the final glucose of the repeating unit, by cultivation on sucrose, rice bran, and ascorbic acid (Kono et al. 2017). In the meantime, many novel applications as well as patents have been identified for that "novel" β-(1,3)-(1,6)-glucan (Muramatsu et al. 2012). The same is true for many other EPS producers who might produce more than one chemically different exopolysaccharide under specific conditions, which will be discussed at the end of this chapter in detail, and the amounts may change with culture conditions and therefore can negatively influence the performance within the targeted industrial application (Rehm 2010; Rütering et al. 2016; Schmid et al. 2015). Another aspect is the simultaneous production of other biopolymers, such as polyhydroxybutyrate (PHB) by, e.g., the alginate-producing *Azotobacter vinelandii* (Yoneyama et al. 2015), or polyglutamate by, e.g., the levan-producing *Bacillus subtilis* ssp. *natto* (Shih et al. 2005). Additionally, depending on the organism, byproducts such as organic acids or pigments are described, which must be carefully removed depending on the targeted applications (Schmid et al. 2011; Sugumaran and Ponnusami 2017; Chaen 2009). In the case of *A. pullulans*, this was one of the main drawbacks, since melanin-like dark pigment is produced, which is very hard to remove from the final polymer and which is not accepted in transparent films or food applications. Many efforts were made to purify pullulan, but finally mutant strains, which are not capable of producing the pigment, have been used for pullulan production (Singh et al. 2008). Still, purification of the final polymers is one of the most important steps, especially for food or medical applications.

## 7.7 Downstream Processing of Microbial Exopolysaccharides

The downstream processing of microbial exopolysaccharides is highly dependent on numerous factors, such as cell morphology of the production strain used, the viscosity of the fermentation broth, amount of remaining substrate and salts from the fermentation broth, and byproducts as well as other polymers. The downstream processing of microbial polysaccharides from the viscous fermentation broth is a complex process that can account for up to 70% of the total production costs (Kreyenschulte et al. 2014; Torrestiana-Sanchez et al. 2007). In general, the purification process can be divided into three steps: (i) removal or inactivation of the cell mass; (ii) precipitation/concentration of the polysaccharide; and (iii) dewatering/drying and milling of the polysaccharide.

### 7.7.1 Removal of Cell Biomass

Removal of cell biomass is applied for most of the polysaccharides produced in laboratory scale as well as industrial scale, but there exist differences concerning the various applications as well as production strains. The classical process as used for technical grade xanthan avoids the removal of the biomass, since centrifugation or filtration of the highly viscous fermentation broth is time and energy consuming (Hublik 2012; Sieber et al. 2019). Reducing the viscosity by diluting the fermentation broth would mean that the volume of the bioreactors is multiplied (fourfold to fivefold) and this complete volume must be centrifuged or filtrated. This will result in high energy cost for the centrifugation step, and additional need of water. Thus, for technical applications, the fermentation broth is not centrifuged, and the small number of cells precipitated with the polysaccharide are accepted. For food-grade xanthan, lytic enzymes such as alcalase are added to the fermentation broth. By proteolysis of the cells a transparent xanthan can be obtained, and the number of cells is highly reduced (Pollock and Yamazaki 1993; da Silva et al. 2014). Other techniques for cell inactivation or removal can include thermal or chemical treatment, but these treatments might alter the chemical structure of the polysaccharide or must be further purified to remove the chemical compound. Thus, the typical treatment is pasteurization (20 minutes at 80 °C) or temperatures up to 90–95 °C, to minimize the number of viable cells in the final product and to inactivate enzymes, which might degrade the polymer (García-Ochoa et al. 2000; Freitas et al. 2011). Heating of the fermentation broth has the additional advantage of reducing viscosity in most cases and can improve product recovery as described for xanthan (Galindo and Albiter 1996). For lab-scale cell removal and detailed product characterization, dilution of the fermentation broth with a following centrifugation step still represent the mostly applied approach. The increase of the volume as caused by the dilution step can be regained by a cross-flow filtration by use of membranes with a suitable cutoff (Castillo et al. 2015; Haarstrick et al. 1991; Freitas et al. 2017). This method can be used to separate fungal biomass and supernatant, but is not successfully applied to bacterial polysaccharide producers, due to the different molecular masses of the EPS and the cell shape of the different microorganisms. Up to now, there has been no industrial application for tangential-flow filtration described in detail for purification/concentration of microbial exopolysaccharides. But the pilot-plant scale process for schizophyllan production, the fungal β-(1,3)-(1,6)-glucan as produced by the filamentous basidiomycete *S. commune*, which is used in enhanced oil recovery, is described to be filtrated and not precipitated. The concentrated polymer solution is transported in liquid state to the drilling hole thus indicating that a filtrated solution is used, since biomass would highly interfere with enhanced oil recovery process based on porous rock flooding (Leonhardt et al. 2014). Some special applications are described for cell removal in lab-scale, such as cell removal via adsorption on fibers, but they also lack economics or applicability on industrial scale (Yang et al. 1998). Removal of cell biomass via centrifugation in industrial scale can be performed by continuous centrifuges such as decanters but still requires enormous energy consumption.

### 7.7.2 Precipitation of the Polysaccharides

The most common strategy for many polysaccharides is precipitation by use of water-miscible alcohols, like isopropanol (Figure 7.4), ethanol, or acetone, which can make up to 75% (v/v) of the final volume (García-Ochoa et al. 2000; Freitas et al. 2011). The principle of the polysaccharide precipitation is the same as for DNA or RNA. The polar polysaccharide is soluble in water, but not in the less polar alcohols, and depending on the concentration of ions present in the solution, these can form stable ionic bonds which leads to precipitation of the polysaccharide. Therefore, the salts present in the fermentation broth, or added to the alcohol, can massively influence the precipitation efficiency (Smidsrød and Haug 1967). There are no general rules so the most efficient mixture of ions and alcohol must be evaluated for each polysaccharide itself (Flahive et al. 1994; García-Ochoa et al. 1993; Gonzales et al. 1990). Additionally, polyvalent cations like aluminum, calcium, or quaternary ammonium salts can be used to form complexes with polyanionic polysaccharides such as xanthan or alginate (Pace and Righelato 1980; Palaniraj and Jayaraman 2011). One drawback of these complexes is the reduced solubility after precipitation compared to alcohol precipitation. Additionally, the precipitation with higher amounts of alcohol can be beneficial to wash out impurities like certain salts, remaining monosaccharides or cell debris. Another procedure to precipitate some polysaccharides is the acidification of the fermentation broth to a pH of around 3.0, what can be done for, e.g., curdlan (Lee et al. 1999) or sanxan (Huang et al. 2016).

**Figure 7.4** Fibrous precipitate from isopropanol precipitation of xanthan from *Xanthomonas campestris* from lab scale-fermentation. Source: Photo by Moritz Gansbiller.

The addition of suitable electrolytes, such as potassium or sodium chloride (in the case of xanthan) to the fermentation broth, can minimize the amount of isopropanol from the ratio of 3 : 1 to 1.4 : 1 (v/v) (Galindo and Albiter 1996; García-Ochoa et al. 1993). Thus, cheap and efficient alcohols such as technical isopropanol are generally used in industrial-scale xanthan production (García-Ochoa et al. 1993; Becker et al. 1998). As mentioned before, the precipitation and purification can make up to 70% of the total cost, which mainly comes from the energy costs necessary for the recovery of the alcohol used for precipitation. Normally, a ratio of 2 : 1 or 3 : 1 of alcohol: fermentation broth is used, which renders recovery of the alcohol an economically reasonable step, which is mainly done by distillation of the supernatant or efflux after removal of the polysaccharide fibers. A further possibility is to reduce the volume of the fermentation broth before precipitation, which can be realized by cross-flow or ultra-filtration. For xanthan, it was calculated that an ultrafiltration step after heat treatment could reduce the isopropanol cost as well as energy cost for recovery up to 80% by minimizing the total volume, thus saving around 45% of the overall downstream costs (Lo et al. 1997). Using ultrafiltration to remove the fungal biomass from the EPS is already state of the art in lab-scale processes. This can be done by steel membranes or more loosely by membranes made from materials such as regenerated cellulose or polyethersulfone. Cross-flow filtration showed to be highly successful in the complete removal of inorganic content (around 32%) from the supernatant of the diluted and centrifuged fermentation broth of FucoPol production by using a membrane with a cutoff of 10 000 Da (Freitas et al. 2017). More promising for the large-scale concentration of microbial EPS might be materials such as ceramics, which can easily be cleaned, sterilized, and back-flushed by harsh conditions to avoid clogging and fouling (Schmid et al. 2011). One of the biggest drawbacks for applying ultra- or cross-flow filtration in industrial scale will be the additional time for the purification/concentration process, which will limit the profit when calculating for personal and operation unit costs.

Several novel filtration techniques were developed in lab-scale during the last years, such as pressure electro-filtration for EPS solutions (Hofmann and Posten 2003) and later transferred to pilot-scale for the purification and concentration of larger amounts of EPS solutions (Hofmann et al. 2006). This technique uses classical dead-end filtration, which is supported by an electrical field and results in guided migration of the charged polymers. This system was able to concentrate xanthan up to a final concentration of 220 $g\,l^{-1}$ from an initial 5 $g\,l^{-1}$ solution, by reducing the filtration time greater than 90% compared to classical dead-end filtration. Additional variants of cross-flow filtration, which combine precipitation and ultrafiltration in one-step by using alcohol and potassium chloride as separate additives, were developed and showed to reduce the flux resistance of the fouling layer by making it more porous (Torrestiana-Sanchez et al. 2007). All these methods and novel techniques have a high potential in the lab and pilot-scale purification of EPS, but suffer from time-consuming procedures (batchwise, low flow-rates) or huge operation units (large membrane areas) compared to classical precipitation method by use of alcohols. Thus, no real industrial scale process has been described up to now and will need massive economic optimization before becoming competitive with established techniques.

### 7.7.3 Dewatering/Drying of the Polysaccharides

After precipitation of the polysaccharide, the insoluble fibers are collected and concentrated via centrifugation or filtration (Hublik 2012). In lab-scale, this can easily be done by collecting the polysaccharide fibers with a mechanical stirrer, where most of the polysaccharides will be stuck to the stirrer. In the pilot and industrial scale, pressing through meshes or rotational filters will be applied (Hublik 2012; García-Ochoa et al. 2000; Palaniraj and Jayaraman 2011). Rotational and especially vacuum rotary filters allow a continuous filtration by adjusting the thickness of the filter cake to enable continuous filtration without losing efficiency over the process. In lab and pilot-scale, it must be considered that the suction still contains a high amount of alcohol, which might be dangerous if using a laboratory centrifuge without, e.g., a nitrogen supply. This could easily create an explosive atmosphere. Another possibility is using a spray-drying process by which the remaining liquid is evaporated, and dry powder is obtained. The last step includes milling the dry polysaccharide fibers to obtain particles of a defined size.

In lab-scale drying of the mainly manually collected fibers from the precipitation can be done in a lyophile or a vacuum oven to get rid of the remaining water/alcohol solution. Many different protocols exist for this procedure, but it must be mentioned that the drying process might have considerable influence on the resolubilization of the final polysaccharide and might highly depend on the polysaccharide itself (Castillo et al. 2015; García-Ochoa et al. 1993; Jeannin et al. 2001). In the author's experience, lyophilization of microbial polysaccharides often results in a paper-like structure, which limits the milling process and can hinder resolubilization, resulting in just partly solubilized polysaccharide solutions. Therefore, the drying in a vacuum oven results in a hardened dry mass of EPS, which can be milled easily toward a fine powder. These powders can be resolubilized very easily and give homogenous solutions without visible flocks or just swollen areas, as is often observed for lyophilizates of the same EPS.

## 7.8 Conclusions

Summarizing the various aspects, it is obvious that no general rule exists to optimally produce microbial exopolysaccharides. The most suitable process highly depends on the production strain and the physicochemical properties of the secreted polysaccharide. The optimal media composition must be evaluated for each strain, but general aspects such as C:N ratio might indicate the basic composition. Unfortunately, the genetic regulation of microbial exopolysaccharide biosynthesis has been only scarcely analyzed and differs from strain to strain. Some insights can be obtained by knowing the biosynthesis machinery with combined linkage of some of the information to the bioprocess, but only very little information is available on systems and synthetic biology approaches on microbial exopolysaccharide production (Ates 2015; Ruffing and Chen 2012). Further insights from different omics data might highly contribute to a better understanding of the regulation and targeted optimization of bioprocess engineering. Some highly attractive EPSs for technical applications are not suitable for industrial production due to the pathogenic character of the production strains. Synthetic biology might offer new possibilities here soon.

Within food industry, novel isolates from already accepted or approved biotechnological applications such as kefir or vinegar production can be used to isolate and identify novel EPS-producers, as recently shown for *Kozakia baliensis* (Brandt et al. 2018; Brandt et al. 2016; Schmid et al. 2014). Concerning bioreactor design and process parameters, there is enough space for optimization in lab and pilot-scale, but transfer toward industry will need a long time and convincing arguments, especially on the economic side. The downstream processing of microbial EPS is still the main drawback in economic production of microbial EPS and is one of the most promising as well as challenging tasks for improvement. As a conclusion, the problem of direct comparison of many results of EPS research must be mentioned. As seen by the challenges in byproduct formation or simultaneous production of a second biopolymer (such as levan or dextran), many published results must be carefully evaluated, since very often the complete analysis of the final product is missing. A comparison of different EPS titers from, e.g., different media or processes with different dissolved oxygen levels, without a complete product analysis (recovery, molecular weight etc.) might include higher protein or salt content, or even another polymer. Thus, one of the main drawbacks in exopolysaccharide research is the lack of defined analytical protocols of the final product(s) and a standardized comparison of specific production strains. For this, the authors recommend standardization of protocols for EPS production, purification, and analysis to enable comparable results.

## References

Abbott, M.S., Harvey, P., Perez, G.V. et al. (2013). *Biological processing in oscillatory baffled reactors: operation, advantages and potential. Interface Focus* 3 (1): 20120036.

Abbott, M.S.R., Valente Perez, G., Harvey, A.P. et al. (2014). *Reduced power consumption compared to a traditional stirred tank reactor (STR) for enzymatic saccharification of alpha-cellulose using oscillatory baffled reactor (OBR) technology. Chem. Eng. Res. Des.* 92 (10): 1969–1975.

Amanullah, A., Satti, S., and Nienow, A.W. (1998). *Enhancing xanthan fermentations by different modes of glucose feeding. Biotechnol. Prog.* 14 (2): 265–269.

Ates, O. (2015). *Systems biology of microbial exopolysaccharides production. Front. Bioeng. Biotechnol.* 3: 200.

Audet, J., Gagnon, H., Lounes, M. et al. (1998). *Polysaccharide production: experimental comparison of the performance of four mixing devices. Bioprocess Eng.* 19 (1): 45.

Becker, A., Katzen, F., Pühler, A. et al. (1998). *Xanthan gum biosynthesis and application: a biochemical/genetic perspective. Appl. Microbiol. Biotechnol.* 50 (2): 145–152.

Bhatia, S.K., Kumar, N., and Bhatia, R.K. (2015). *Stepwise bioprocess for exopolysaccharide production using potato starch as carbon source. 3 Biotech* 5 (5): 735–739.

Born, K., Langendorff, V., and Patrick, B. (2005). *Xanthan*. In: *Biotechnology of Biopolymers: From Synthesis to Patents* (eds. A. Steinbuchel and Y. Doi). Weinheim: Wiley-VCH.

Brandt, J.U., Jakob, F., Behr, J. et al. (2016). *Dissection of exopolysaccharide biosynthesis in Kozakia baliensis. Microb. Cell Fact.* 15: 170.

Brandt, J.U., Jakob, F., Wefers, D. et al. (2018). *Characterization of an acetan-like heteropolysaccharide produced by Kozakia baliensis NBRC 16680. Int. J. Biol. Macromol.* 106 (Supplement C): 248–257.

Campani, G., Ribeiro, M.P.A., Horta, A.C.L. et al. (2015). *Oxygen transfer in a pressurized airlift bioreactor. Bioprocess Biosyst. Eng.* 38 (8): 1559–1567.

Castillo, N.A., Valdez, A.L., and Farina, J.I. (2015). *Microbial production of scleroglucan and downstream processing. Front. Microbiol.* 6: 1106.

Chaen, H. (2009). *Pullulan*. In: *Food Stabilisers, Thickeners and Gelling Agents*, 266–274. Wiley-Blackwell.

Crognale, S., Bruno, M., Moresi, M. et al. (2007). *Enhanced production of β-glucan from Botryosphaeria rhodina using emulsified media or fan impellers. Enzyme Microb. Technol.* 41 (1): 111–120.

Evans, C.G.T., Yeo, R.G., and Ellwood, D.C. (1979). *Continuous culture studies on the production of extracellular polysaccharides by Xanthomonas juglandis*. In: *Microbial Polysaccharide and Polysaccharases* (eds. G.W. Gooday, B. RCW and D.C. Ellwood), 51–68. London: Academic Press.

Fariña, J.I., Siñeriz, F., Molina, O.E. et al. (1998). *High scleroglucan production by Sclerotium rolfsii: influence of medium composition. Biotechnol. Lett.* 20 (9): 825–831.

Farris, S., Unalan, I.U., Introzzi, L. et al. (2014). *Pullulan-based films and coatings for food packaging: present applications, emerging opportunities, and future challenges. J. Appl. Polym. Sci.* 131 (13): p. n/a-n/a.

Fazenda, M.L., Seviour, R., McNeil, B. et al. (2008). *Submerged culture fermentation of "higher fungi": the macrofungi. Adv. Appl. Microbiol.* 63: 33–103.

Flahive, J.J., Foufopoulos, A., and Etzel, M.R. (1994). *Alcohol precipitation of xanthan gum from pure solutions and fermentation broths. Sep. Sci. Technol.* 29 (13): 1673–1687.

Freitas, F., Alves, V.D., and Reis, M.A. (2011). *Advances in bacterial exopolysaccharides: from production to biotechnological applications. Trends Biotechnol.* 29 (8): 388–398.

Freitas, F., Torres, C.A.V., and Reis, M.A.M. (2017). *Engineering aspects of microbial exopolysaccharide production. Bioresour. Technol.* 245 (Pt B): 1674–1683.

Frey, H. (2008). *Handbook of Industrial Water Soluble Polymers*. ChemPhysChem, vol. 9 (ed. P. Williams), 1080–1080. WILEY-VCH Verlag.

Funahashi, H., Maehara, M., Taguchi, H. et al. (1987). *Effects of agitation by flat-bladed turbine impeller on microbial production of xanthan gum. J. Chem. Eng. Jpn.* 20 (1): 16–22.

Funahashi, H., Hirai, K.-I., Yoshida, T. et al. (1988). *Mixing state of xanthan gum solution in an aerated and agitated fermentor—effects of impeller size on volumes of mixing regions and circulation time distribution. J. Ferment. Technol.* 66 (1): 103–109.

Gaidhani, H.K., McNeil, B., and Ni, X.-W. (2003). *Production of pullulan using an oscillatory baffled bioreactor. J. Chem. Technol. Biotechnol.* 78 (2–3): 260–264.

Galindo, E. and Albiter, V. (1996). *High-yield recovery of xanthan by precipitation with isopropyl alcohol in a stirred tank. Biotechnol. Prog.* 12 (4): 540–547.

Galindo, E., Torrestiana, B., and Garcia-Rejón, A. (1989). *Rheological characterization of xanthan fermentation broths and their reconstituted solutions. Bioprocess Eng.* 4 (3): 113–117.

Galindot, E. and Nienow, A.W. (1992). *Mixing of highly viscous simulated xanthan fermentation broths with the Lightnin A-315 impeller. Biotechnol. Prog.* 8 (3): 233–239.

Garcia-Ochoa, F. and Gomez, E. (2009). *Bioreactor scale-up and oxygen transfer rate in microbial processes: an overview. Biotechnol. Adv.* 27 (2): 153–176.

García-Ochoa, F., Casas, J.A., and Mohedano, A.F. (1993). *Xanthan precipitation from solutions and fermentation broths. Sep. Sci. Technol.* 28 (6): 1303–1313.

Garcia-Ochoa, F., Castro, E.G., and Santos, V.E. (2000). *Oxygen transfer and uptake rates during xanthan gum production. Enzyme Microb. Technol.* 27 (9): 680–690.

García-Ochoa, F., Santos, V.E., Casas, J.A. et al. (2000). *Xanthan gum: production, recovery, and properties. Biotechnol. Adv.* 18 (7): 549–579.

Gibbs, P.A. and Seviour, R. (1992). *Influence of bioreactor design on exopolysaccharide production byAureobasidium pullulans. Biotechnol. Lett.* 14: 491–494.

Gibbs, P.A., Seviour, R.J., and Schmid, F. (2000). *Growth of filamentous fungi in submerged culture: problems and possible solutions. Crit. Rev. Biotechnol.* 20 (1): 17–48.

Gonzales, R., Johns, M.R., Greenfield, P.F. et al. (1990). *Phase equilibria for xanthan gum in ethanol-water solutions. Carbohydr. Polym.* 13 (3): 317–333.

Gura, E. and Rau, U. (1993). *Comparison of Agitators for the Production of Branched β-1,3-d-Glucans by Schizophyllum Commune*, vol. 27, 193–201.

Haarstrick, A., Rau, U., and Wagner, F. (1991). *Cross-flow filtration as a method of separating fungal cells and purifying the polysaccharide produced. Bioprocess Eng.* 6 (4): 179–186.

Hay, I.D., Rehman, Z.U., Moradali, M.F. et al. (2013). *Microbial alginate production, modification and its applications. Microb. Biotechnol.* 6 (6): 637–650.

Herbst, H., Schumpe, A., and Deckwer, W.-D. (1992). *Xanthan production in stirred tank fermenters: oxygen transfer and scale-up. Chem. Eng. Technol.* 15 (6): 425–434.

Hewgill, M.R., Mackley, M.R., Pandit, A.B. et al. (1993). *Enhancement of gas-liquid mass transfer using oscillatory flow in a baffled tube. Chem. Eng. Sci.* 48 (4): 799–809.

Hofmann, R. and Posten, C. (2003). *Improvement of dead-end filtration of biopolymers with pressure electrofiltration. Chem. Eng. Sci.* 58 (17): 3847–3858.

Hofmann, R., Käppler, T., and Posten, C. (2006). *Pilot-scale press electrofiltration of biopolymers. Sep. Purif. Technol.* 51 (3): 303–309.

Hsu, C.-H. and Lo, Y.M. (2003). *Characterization of xanthan gum biosynthesis in a centrifugal, packed-bed reactor using metabolic flux analysis. Process Biochem.* 38 (11): 1617–1625.

Huang, H., Wu, M., Yang, H. et al. (2016). *Structural and physical properties of sanxan polysaccharide from Sphingomonas sanxanigenens. Carbohydr. Polym.* 144: 410–418.

Hublik, G. (2012). *Xanthan*. In: *Polymer Science: A Comprehensive Reference* (eds. M. Moeller and K. Matyjaszewski), 221–229. Elsevier.

Israilides, C.J., Smith, A., Harthill, J.E. et al. (1998). *Pullulan content of the ethanol precipitate from fermented agro-industrial wastes. Appl. Microbiol. Biotechnol.* 49 (5): 613–617.

Jeannin, M., Rezzoug, S.A., Maache-rezzoug, Z. et al. (2001). *Solid-state 13C NMR study of Scleroglucan polysaccharide. Effect of the drying process and hydration on Scleroglucan structure and dynamics. Int. J. Polym. Anal. Charact.* 6 (1–2): 177–191.

Ju, L.-K. (2007). Water-in-oil cultivation technology for viscous xanthan gum fermentation. In: *Bioprocessing for Value-Added Products from Renewable Resources*, 397–419. Amsterdam: Elsevier.

Ju, L.-K. and Zhao, S. (1993). *Xanthan fermentations in water/oil dispersions. Biotechnol. Tech.* 7 (7): 463–468.

Kalogiannis, S., Iakovidou, G., Liakopoulou-Kyriakides, M. et al. (2003). *Optimization of xanthan gum production by Xanthomonas campestris grown in molasses. Process Biochem.* 39 (2): 249–256.

Kang, X., Wang, Y., Harvey, L.M. et al. (2000). *Effect of air flow rate on scleroglucan synthesis by Sclerotium glucanicum in an airlift bioreactor with an internal loop. Bioprocess Eng.* 23, 69–74.

Kang, X., Wang, H., Wang, Y. et al. (2001). *Hydrodynamic characteristics and mixing behaviour of Sclerotium glucanicum culture fluids in an airlift reactor with an internal loop used for scleroglucan production. J. Ind. Microbiol. Biotechnol.* 27 (4): 208–214.

Kelco (2007). *Xanthan Book*, 8e.

Kong, N., Mu, X., Han, H. et al. (2009). *Pilot-scale fermentation, purification, and characterization of recombinant human Oncostatin M in Pichia pastoris. Protein Expression Purif.* 63 (2): 134–139.

Kono, H., Kondo, N., Hirabayashi, K. et al. (2017). *NMR spectroscopic structural characterization of a water-soluble* β-*(1→3, 1→6)-glucan from Aureobasidium pullulans. Carbohydr. Polym.* 174 (Supplement C): 876–886.

Kothari, D., Das, D., Patel, S. et al. (2015). Dextran and food application. In: *Polysaccharides: Bioactivity and Biotechnology* (eds. K.G. Ramawat and J.-M. Mérillon), 1–16. Cham: Springer International Publishing.

Kreyenschulte, D., Krull, R., and Margaritis, A. (2014). *Recent advances in microbial biopolymer production and purification. Crit. Rev. Biotechnol.* 34 (1): 1–15.

Kuppuswami, G.M. (2014). *Fermentation (Industrial) | production of xanthan gum*. In: *Encyclopedia of Food Microbiology*, 2e (ed. M.L. Tortorello), 816–821. Oxford: Academic Press.

Kuttuva, S.G., Restrepo, A.S., and Ju, L.-K. (2004). *Evaluation of different organic phases for water-in-oil xanthan fermentation. Appl. Microbiol. Biotechnol.* 64 (3): 340–345.

Lee, J.H., Lee, I.Y., Kim, M.K. et al. (1999). *Optimal pH control of batch processes for production of curdlan by agrobacterium species. J. Ind. Microbiol. Biotechnol.* 23 (2): 143–148.

Lee, I.Y., Kim, M.K., Lee, J.H. et al. (1999). *Influence of agitation speed on production of curdlan by agrobacterium species. Bioprocess Eng.* 20 (4): 283–287.

Leonhardt, B., Ernst, B., Reimann, S., Steigerwald, A., and Lehr, F. (2014, April 12). Field Testing The Polysaccharide Schizophyllan: Results of The First Year. Society of Petroleum Engineers. doi:10.2118/169032-MS

Lin, Y. and Thibault, J. (2013). *Pullulan fermentation using a prototype rotational reciprocating plate impeller. Bioprocess Biosyst. Eng.* 36 (5): 603–611.

Liu, W.C., Gong, T., Wang, Q.H. et al. (2016). *Scaling-up fermentation of Pichia pastoris to demonstration-scale using new methanol-feeding strategy and increased air pressure instead of pure oxygen supplement. Sci. Rep.* 6: 18439.

Lo, Y.-M., Yang, S.-T., and Min, D.B. (1997). *Ultrafiltration of xanthan gum fermentation broth: process and economic analyses. J. Food Eng.* 31 (2): 219–236.

Lo, Y., Hsu, C., Yang, S. et al. (2001). *Oxygen transfer characteristics of a centrifugal, packed-bed reactor during viscous xanthan fermentation. Bioprocess Biosyst. Eng.* 24 (3): 187–193.

Lotrakula, P., Unhapattaratitikula, P., Seelananb, T. et al. (2013). *An aubasidan-like β-glucan produced by Aureobasidium pullulans in Thailand. Sci. Asia* 39 (4).

Lounes, M., Audet, J., Thibault, J. et al. (1995). *Description and evaluation of reciprocating plate bioreactors. Bioprocess Eng.* 13 (1): 1–11.

McNeil, B. and Harvey, L.M. (1993). *Viscous fermentation products. Crit. Rev. Biotechnol.* 13 (4): 275–304.

McNeil, B. and Kristiansen, B. (1987). *Influence of impeller speed upon the pullulan fermentation. Biotechnol. Lett.* 9 (2): 101–104.

McNeil, B., Kristiansen, B., and Seviour, R.J. (1989). *Polysaccharide production and morphology of Aureobasidium pullulans in continuous culture. Biotechnol. Bioeng.* 33 (9): 1210–1212.

Meier, K., Klöckner, W., Bonhage, B. et al. (2016). *Correlation for the maximum oxygen transfer capacity in shake flasks for a wide range of operating conditions and for different culture media. Biochem. Eng. J.* 109 (Supplement C): 228–235.

Meyer, H.-P., Minas, W., and Schmidhalter, D. (2017). *Industrial-scale fermentation.* In: *Industrial Biotechnology*, 1–53. Wiley-VCH Verlag GmbH & Co. KGaA.

Miwa, M., Nakao, Y., and Nara, K. (1993). Food applications of curdlan. In: *Food Hydrocolloids: Structures, Properties, and Functions* (eds. K. Nishinari and E. Doi), 119–124. Boston, MA: Springer US.

Muramatsu, D., Iwai, A., Aoki, S. et al. (2012). *β-Glucan derived from Aureobasidium pullulans is effective for the prevention of influenza in mice. PLoS One* 7: e41399.

Ni, X. and Gao, S. (1996). *Scale-up correlation for mass transfer coefficients in pulsed baffled reactors. Chem. Eng. J. Biochem. Eng. J.* 63 (3): 157–166.

Omoto, T., Uno, Y., and Asai, I. (1999). The latest technologies for the application of gellan gum. In: *Physical Chemistry and Industrial Application of Gellan Gum*, vol. 123–126 (ed. K. Nishinari). Berlin, Heidelberg: Springer Berlin Heidelberg.

Öner, E.T. (2013). *Microbial production of extracellular polysaccharides from biomass.* In: *Pretreatment Techniques for Biofuels and Biorefineries* (ed. Z. Fang), 35–56. Berlin, Heidelberg: Springer Berlin Heidelberg.

Öner, E.T., Hernandez, L., and Combie, J. (2016). *Review of Levan polysaccharide: from a century of past experiences to future prospects. Biotechnol. Adv.* 34 (5): 827–844.

Özcan, E., Sargın, S., and Göksungur, Y. (2014). *Comparison of pullulan production performances of air-lift and bubble column bioreactors and optimization of process parameters in air-lift bioreactor. Biochem. Eng. J.* 92 (Supplement C): 9–15.

Pace, G.W. and Righelato, R.C. (1980). Production of extracellular microbial polysaccharides. In: *Advances in Biochemical Engineering*, vol. 15, 41–70. Berlin, Heidelberg: Springer Berlin Heidelberg.
Palaniraj, A. and Jayaraman, V. (2011). *Production, recovery and applications of xanthan gum by Xanthomonas campestris. J. Food Eng.* 106 (1): 1–12.
Papagianni, M. (2004). *Fungal Morphology and Metabolite Production in Submerged Mycelial Process*, vol. 22, 189–259.
Peña, C., Trujillo-Roldán, M.A., and Galindo, E. (2000). *Influence of dissolved oxygen tension and agitation speed on alginate production and its molecular weight in cultures of Azotobacter vinelandii★. Enzyme Microb. Technol.* 27 (6): 390–398.
Pollock, T.J. and Yamazaki, M. (1993). *Clarification of microbial polysaccharides with enzymes secreted fromLysobacter species. J. Ind. Microbiol.* 11 (3): 187–192.
Pons, A., Dussap, C.G., and Gros, J.B. (1990). *Xanthan batch fermentations: compared performances of a bubble column and a stirred tank fermentor. Bioprocess Eng.* 5 (3): 107–114.
Rehm, B.H. (2010). *Bacterial polymers: biosynthesis, modifications and applications. Nat. Rev. Microbiol.* 8 (8): 578–592.
Robinson, D.K. and Wang, D.I.C. (1987). *A novel bioreactor system for biopolymer Productiona. Ann. N. Y. Acad. Sci.* 506 (1): 229–241.
Robinson, D.K. and Wang, D.I.C. (1988). *A transport controlled bioreactor for the simultaneous production and concentration of xanthan gum. Biotechnol. Prog.* 4 (4): 231–241.
Roseiro, J.C., Emery, A.N., Simões, P. et al. (1993). *Production of xanthan by in-flow cultures of Xanthomonas campestris. Appl. Microbiol. Biotechnol.* 38 (6): 709–713.
Roukas, T. and Mantzouridou, F. (2001). *Effect of the aeration rate on pullulan production and fermentation broth rheological properties in an airlift reactor. J. Chem. Technol. Biotechnol.* 76 (4): 371–376.
Ruffing, A.M. and Chen, R.R. (2012). *Transcriptome profiling of a curdlan-producing agrobacterium reveals conserved regulatory mechanisms of exopolysaccharide biosynthesis. Microb. Cell Fact.* 11: 17–17.
Rühmann, B., Schmid, J., and Sieber, V. (2015). *Methods to identify the unexplored diversity of microbial exopolysaccharides. Front. Microbiol.* 6: 565.
Rütering, M., Schmid, J., Rühmann, B. et al. (2016). *Controlled production of polysaccharides-exploiting nutrient supply for Levan and heteropolysaccharide formation in Paenibacillus sp. Carbohydr. Polym.* 148: 326–334.
Sanderson, G.R. (1981). *Applications of xanthan gum. Br. Polym. J.* 13 (2): 71–75.
Schelden, M., Lima, W., Doerr, E.W. et al. (2017). *Online measurement of viscosity for biological systems in stirred tank bioreactors. Biotechnol. Bioeng.* 114 (5): 990–997.
Schilling, B.M. (2000). *Sclerotium rolfsii ATCC 15205 in continuous culture: economical aspects of scleroglucan production. Bioprocess Eng.* 22 (1): 57–61.
Schmid, J., Muller-Hagen, D., Bekel, T. et al. (2010). *Transcriptome sequencing and comparative transcriptome analysis of the scleroglucan producer Sclerotium rolfsii. BMC Genomics* 11: 329.
Schmid, J., Meyer, V., and Sieber, V. (2011). *Scleroglucan: biosynthesis, production and application of a versatile hydrocolloid. Appl. Microbiol. Biotechnol.* 91 (4): 937–947.
Schmid, J., Sperl, N., and Sieber, V. (2014). *A comparison of genes involved in sphingan biosynthesis brought up to date. Appl. Microbiol. Biotechnol.* 98 (18): 7719–7733.

Schmid, J., Koenig, S., Pick, A. et al. (2014). *Draft genome sequence of Kozakia baliensis SR-745, the first sequenced Kozakia strain from the family Acetobacteraceae. Genome Announc.* 2 (3): e00594–e00514.

Schmid, J., Sieber, V., and Rehm, B. (2015). *Bacterial exopolysaccharides: biosynthesis pathways and engineering strategies. Front. Microbiol.* 6: 496.

Schmid, J., Farina, J., Rehm, B. et al. (2016). *Editorial: microbial exopolysaccharides: from genes to applications. Front. Microbiol.* 7: 308.

Schugerl, K. (1990). *Comparison of the performances of stirred tank and airlift tower loop reactors. J. Biotechnol.* 13 (4): 251–256.

Schuster, R., Wenzig, E., and Mersmann, A. (1993). *Production of the fungal exopolysaccharide pullulan by batch-wise and continuous fermentation. Appl. Microbiol. Biotechnol.* 39 (2): 155–158.

Seviour, R.J., McNeil, B., Fazenda, M.L. et al. (2011). *Operating bioreactors for microbial exopolysaccharide production. Crit. Rev. Biotechnol.* 31 (2): 170–185.

Shih, I.L., Yu, Y.T., Shieh, C.J. et al. (2005). *Selective production and characterization of Levan by Bacillus subtilis (Natto) Takahashi. J. Agric. Food Chem.* 53 (21): 8211–8215.

Sieber, V., Schmid, J., and Hublik, G. (2019). *Microbial polysaccharides.* In: *Industrial Microbiology* (eds. D.B. Wilson, H. Sahm and P. Stahmann). Heidelberg: Springer.

Silbir, S., Dagbagli, S., Yegin, S. et al. (2014). *Levan production by Zymomonas mobilis in batch and continuous fermentation systems. Carbohydr. Polym.* 2014, 99 (Supplement C): 454–461.

da Silva, C.R., Silva, M.L., Kamida, H.M. et al. (2014). *Lytic enzyme production optimization using low-cost substrates and its application in the clarification of xanthan gum culture broth. Food Sci. Nutr.* 2 (4): 299–307.

Singh, R.S., Saini, G.K., and Kennedy, J.F. (2008). *Pullulan: microbial sources, production and applications. Carbohydr. Polym.* 73 (4): 515–531.

Smidsrød, O. and Haug, A. (1967). *Precipitation of acidic polysaccharides by salts in ethanol–water mixtures. J. Polym. Sci. C Polym. Symposia* 16 (3): 1587–1598.

Solomon, J., Elson, T.P., Nienow, A.W. et al. (1981). *Cavern sizes in agitated fluids with a yield stress. Chem. Eng. Commun.* 11 (1–3): 143–164.

Sugumaran, K.R. and Ponnusami, V. (2017). *Review on production, downstream processing and characterization of microbial pullulan. Carbohydr. Polym.* 173: 573–591.

Suh, I.-S., Schumpe, A., and Deckwer, W.-D. (1992). *Xanthan production in bubble column and air-lift reactors. Biotechnol. Bioeng.* 39 (1): 85–94.

Survase, S.A., Saudagar, P.S., and Singhal, R.S. (2007). *Use of complex media for the production of scleroglucan by Sclerotium rolfsii MTCC 2156. Bioresour. Technol.* 98 (7): 1509–1512.

Sutherland, I.W. (1998). *Novel and established applications of microbial polysaccharides. Trends Biotechnol.* 16 (1): 41–46.

Sutherland, I.W. (2001). *Microbial polysaccharides from gram-negative bacteria. Int. Dairy J.* 11 (9): 663–674.

Torrestiana-Sanchez, B., Balderas-Luna, L., Brito-De la Fuente, E. et al. (2007). *The use of membrane-assisted precipitation for the concentration of xanthan gum. J. Membr. Sci.* 294 (1–2): 84–92.

Ullah, H., Santos, H.A., and Khan, T. (2016). *Applications of bacterial cellulose in food, cosmetics and drug delivery. Cellulose* 23 (4): 2291–2314.

Vorhölter, F.-J., Schneiker, S., Goesmann, A. et al. (2008). *The genome of Xanthomonas campestris pv. campestris B100 and its use for the reconstruction of metabolic pathways involved in xanthan biosynthesis. J. Biotechnol.* 134 (1): 33–45.
lawford, H.G. and Rousseau, J.D. (1991). *Bioreactor design considerations in the production of high-quality microbial exopolysaccharide. Appl. Biochem. Biotechnol.* 28 (1): 667–684.
Witczak, Z.J. (1996). *Polysaccharides in Medicinal Applications* (ed. S. Dumitriu). New York: Marcel Dekker, Inc. ISBN: 0-8247-9540-7, *794 pp. $195.00.* Journal of Carbohydrate Chemistry, 1997. 16(2): 245–247.
Wunderlich, M., Trampnau, P.P., Lopes, E.F. et al. (2016). *Online in situ viscosity determination in stirred tank reactors by measurement of the heat transfer capacity. Chem. Eng. Sci.* 152 (Supplement C): 116–126.
Yang, S.T., Lo, Y.M., and Min, D.B. (1996). *Xanthan gum fermentation by Xanthomonas campestris immobilized in a novel centrifugal fibrous-bed bioreactor. Biotechnol. Prog.* 12: 630.
Yang, S.-T., Lo, Y.-M., and Chattopadhyay, D. (1998). *Production of cell-free xanthan fermentation broth by cell adsorption on Fibers. Biotechnol. Prog.* 14 (2): 259–264.
Yoneyama, F., Yamamoto, M., Hashimoto, W. et al. (2015). *Production of polyhydroxybutyrate and alginate from glycerol by Azotobacter vinelandii under nitrogen-free conditions. Bioengineered* 6 (4): 209–217.
Zhang, Y., Kong, H., Fang, Y. et al. (2013). *Schizophyllan: a review on its structure, properties, bioactivities and recent developments. Bioact. Carbohydr. Diet. Fibre* 1 (1): 53–71.

# 8

# Research and Production of Microbial Polyunsaturated Fatty Acids

*Gwendoline Christophe, Pierre Fontanille, and Christian Larroche*

*Université Clermont Auvergne, CNRS, Sigma Clermont, Institut Pascal, F-63171, Aubière, France*

## 8.1 Introduction

The most important classes of lipids used for food supplement are polyunsaturated fatty acids or PUFAs. In this class of lipids, omega-6 and 3 are important in human health but have to be added in the diet because humans are unable to produce themselves some omega-6 and 3, such as eicosapentaenoic acid (EPA) or docosahexaenoic acid (DHA). Moreover, modern diets are unbalanced from its fat intake point of view (too important levels of fat, saturated fat and cholesterol) and a ratio of omega-6 to omega-3 up to 1 : 25 (Tocher 2015) whereas the right ratio is close to 1 : 1–4 (Ochsenreither et al. 2016; Tocher 2015) for health and well-being.

According to organizations such as the World Health Organization (WHO), the American Health Association (AHA), the Food Standards Agency in the United Kingdom, the US Food and Drug Administration (FDA), the French Agency for Food, Environmental and Occupational Health and Safety, and the European Food Safety Agency (EFSA), the recommended daily intakes of DHA and EPA, which are the most important PUFAs, are 200–500 $mg\ d^{-1}$ for healthy subjects (Bharathiraja et al. 2017).

With these recommendations, the potential demand of omega-3 fatty acid, to which DHA and EPA belong, is estimated to 1.274 million of tons, while fish production is only 0.84 million tons. This gap could be filled in by microbial oil as an alternate source to satisfy the increased needs with an expected annual growth rate of 12.8% between 2014 and 2019 (Aasen et al. 2016). Moreover, the amount of PUFAs in fatty fish is constantly decreasing and its use is more and more debatable because of the presence of environmental pollutants such as dioxins, PCBs, and heavy metals (Ratledge and Lippmeier 2017). Therefore, using oleaginous microorganisms represents a promising way to produce PUFAs. This competency creates a new prospect in the domain of food supplements, and the question of producing and administering PUFAs to a larger number of people is opened. In 2001, 2004, and 2015, the FDA evaluated levels of microbial oils used for human food (Bharathiraja et al. 2017; Ratledge and Lippmeier 2017). These microbial producers were established as GRAS (generally recognized as safe).

*Bioprocessing for Biomolecules Production*, First Edition.
Edited by Gustavo Molina, Vijai Kumar Gupta, Brahma N. Singh, and Nicholas Gathergood.

As demand increases for alternative sources of protein and nutrients, microorganisms offer several advantages compared to other oleaginous species cultures such as plants or fishes. Cultivations of these microorganisms are carried out without climatic or geographic constraints and do not require the use of arable land. They also present a better growth rate than oleaginous plants and can be cultivated of various types of substrates, even on industrial wastes. Yeasts, fungi, and microalgae are protein-rich microorganism cultures offering such potential advantages.

This chapter will discuss the use of PUFAs as a food supplement. Then it will focus on the biosynthesis of microbial lipids and will present the different types of microorganisms involved in the production of interesting lipids, their culture conditions, and the production strategies of these oils. The new ways of oil production will also be treated.

## 8.2 Lipids Used for Food Supplement

### 8.2.1 PUFAs: Omega-3 and Omega-6 Families

Fatty acids are composed of a hydrocarbon chain R giving the molecule its hydrophobic characteristics and carrying a terminal carboxylic acid function (-COOH). The fatty acids are distinguished by the length of the aliphatic chain, the degree of unsaturation, the location and conformation in cis or trans of their double bonds, as well as the presence and position of the modifications. Their nomenclature involves the designation of the number of carbons present in the aliphatic chain, the number of double bonds (Δ), their positions as well as their configurations. Some authors use the IUPAC (systematic) nomenclature and specify the position of the double bonds by taking as a reference the carboxylic acid function (Δ) of the molecule, while others prefer the so-called biochemical nomenclature and refer to the terminal methyl (indicated by n- or ω-).

Unsaturated fatty acids may have a double bond (monounsaturated) or several double bonds (polyunsaturated). Up to 30 carbons can be found in a chain in nature, but most FA consists in only 16–20 carbons. In general, the first double bond is inserted between C9 and C10 carbons (Δ9), C1 corresponding to the carbon of the carboxyl group. The multiple double bonds are generally not conjugated but separated by a methylene group. Most double bonds are of cis configuration, meaning that the carbon atoms adjacent to the double bond are located on the same side as the previous. Two classes of PUFAs are particularly important: n-3 (or ω-3) and n-6 (or ω-6) fatty acids.

These fatty acids are particularly interesting in nutrition and are described as essential. They cannot be synthesized by mammals because of the absence of Δ-12 and Δ-15 desaturases (Cf. Section 8.3.1) and must therefore be provided by food. It is now well accepted that the PUFAs present in the diet, by their nature and their abundance, influence human health and play a role in many pathologies (metabolic, neurodegenerative, cardiovascular, and inflammatory diseases, obesity).

Alpha-linolenic acid (ALA or all-*cis*-9,12,15-octadecatrienoic acid or C18:3*n*-3) is the precursor of the omega-3 families (also referred to as ω-3 or *n*-3) and is found mainly in rapeseed, nut and flax seeds, and oils. Once consumed, ALA leads to the specific synthesis of a longer chain through a process of successive elongation desaturations. The two major PUFAs thus formed are eicosapentaenoic acid (EPA or all-*cis*-5,8,11,14,17-eicosapentaenoic acid or C20:5*n*-3) and docosahexaenoic acid (DHA or all-*cis*-4,7,10,13,16,19-docosahexaenoic acid or C22:6*n*-3).

Linoleic acid (LA or all-*cis* 9,12-Octadecadienoic acid or C18:2*n*-6) is the precursor of the omega-6 family and is found mainly in sunflower and soybean seeds and oils. As for the n-3 series, LA is metabolized by successive desaturations and elongations into a longer chain fatty acid to form a γ-linolenic acid (GLA or all-*cis*-6,9,12-octadecatrienoic acid or C18:3*n*-6), and arachidonic acid (ARA or all-*cis*-5,8,11,14-eicosatetraenoic acid or C20:4*n*-6). LA is found in many foods such as meats, dairy, vegetables, vegetable oils, cereals, fruits, nuts, legumes, seeds, and bread (Whelan 2008), and ARA may also be brought directly by the diet through the consumption of animal products (egg, meat). GLA is produced in the body as a product of LA metabolism but can also be found in some plants oils such as evening primrose and borage oils.

### 8.2.2 Role of PUFAs in Health

Several studies have shown the importance of consuming omega-3/6 PUFAs. They have beneficial effects on insulin resistance, cardiovascular health, diabetes (type 2), autoimmune diseases, chronic inflammatory diseases, or for the prevention of degenerative diseases of the brain. They play a role in the health and well-being of women at various stages of their lives. They are particularly involved in menstrual pain, pregnancy, fetal development, or breastfeeding. PUFAs have a key role in the membrane structure because they influence the functional properties of the membrane particularly on the constituents of phospholipids. Indeed, they modify the physico-chemical properties and the functions of the proteins of the membrane (Tocher 2015).

ALA and LA are the direct precursors of cellular mediators such as eicosanoids, comprising prostaglandins, resolvins, leukotrienes, protectins, or thromboxanes regulating lipoprotein metabolism, blood rheology, leucocyte function, and platelet activation. ARA has anti-inflammatory activity (Kohli and Levy 2009; Sudheendran et al. 2010) by preventing platelet aggregations and by activating vasodilatation and inflammatory cells contain more omega-6 PUFAs than omega-3 PUFAs (Dunbar et al. 2014).

Some studies have highlighted the link between omega-3/6 consumption and cardiovascular disease and associated complications. Vascular lesions result from complex phenomena involving many parameters such as lipoproteins, platelets, and their oxidation products, endothelial cells and their metabolites. Low-density lipoprotein (LDL) fractions of lipoproteins drive the cholesterol from the liver to the other organs. When the concentration of cholesterol transported by the lipoproteins becomes too high, cells do not assimilate it anymore and it is deposited on the walls of the arteries. Step by step, real fat plaques, called atheroma, are formed. High-density lipoprotein (HDL fractions of lipoproteins are able to capture excess cholesterol back to the liver, where it is r)emoved. HDL/LDL balance is therefore very important for the regulation of lipid metabolism. The omega-3 fatty acids act on the lipid profile by decreasing hepatic synthesis of LDL and increasing that of HDL with, as a result, better purification of excess cholesterol (Chang and Singh 2009; Sudheendran et al. 2010). They also allow a significant reduction of circulating triglycerides. In addition, they not only have a beneficial effect on blood viscosity but also on endothelial structures.

DHA plays a major role in the brain (Birch et al. 2007). The brain is a very "fatty" organ, the first one after the adipose tissue. Its main constitutive lipids are sphingomyelin (essentially in the axon myelin), and glycerophospholipids in astrocyte and cellular parts of neuronal membranes, including phosphatidylethanolamines (PEs), phosphatidylcholines (PCs), and phosphatidylserines (PSs). The main acyl chains at position sn-2 correspond first to DHA and second to ARA. Because of its six cis/Z double bonds, DHA is responsible for the high fluidity of membranes. This is especially important in synapses to confer neuronal plasticity. It is recognized that DHA is required for neurogenesis in brain development and brain functions. Conversely, brain DHA deficits have been reported in neurodegenerative diseases, such as Alzheimer's disease (Ratledge and Lippmeier 2017).

Despite these requirements, the brain cannot make DHA from its essential precursor ALA, which is overall poorly transformed into DHA (1–2% in the whole body, except in pregnant woman where the conversion may attain 8%). The brain depends then on exogenous DHA induced after dietary intake. DHA and ARA are also beneficial in eyes diseases, especially in age-related macular degeneration and eye retina disease.

During pregnancy, essential fatty acids play a significant role for the health of the pregnant woman as well as for the development of the fetus, especially during the last months of pregnancy when the development of the fetal brain is most important and after childbirth. So, the consumption of omega-3/6 during pregnancy has a beneficial effect on the fetus and should be added to the diet of babies to ensure a normal development if the infant is not fed with mother's milk (source of DHA) (Lopez-Huertas 2010). DHA and ARA are absent from cow's milk used in place of mother's milk (Table 8.2).

EPA and DHA are often provided in dietary supplements because they have a similar benefit on the health. They have an important physiological role in the body, and some studies have shown their benefits in cancers treatments by inhibiting cancer cell proliferation (Nagao and Yanagita 2005; Sudheendran et al. 2010; Terano et al. 1999) or by protecting against some cancers (colon-rectal, breast, or prostate) (Gerber 2012).

GLA is incorporated in children's diet particularly to prevent or to treat atopic eczema, rheumatoid arthritis, multiple sclerosis, and premenstrual tension (Horrobin 1992).

The traditional sources of PUFAs are well known and are summarized in Table 8.2. The most important traditional sources of PUFAs are marine organisms.

## 8.3 Microbial Lipids

### 8.3.1 Biosynthesis in Oleaginous Microorganisms

The first step of the lipogenesis is a classical conversion of substrates into acetyl-CoA. This first part implies glycolysis for yeast and fungi in the cytoplasm and the Calvin cycle for the microalgae in the plastids. For microalgae, pyruvate is converted into acetyl-CoA by ACC (acetyl-CoA carboxylase) and into acyl-CoA by the action of plastidial FAS (fatty acid synthase). Photosystem I (PSI) brings NADPH necessary for this conversion. For yeast and fungi, pyruvate is transported into the mitochondrion via proton linked (Hildyard and Halestrap 2003). The pyruvate is converted into acetyl-CoA by a pyruvate decarboxylative dehydrogenase and acetyl-CoA is converted in the Krebs cycle into citrate. It has been shown for *Yarrowia lipolytica* that the NADPH used for lipids expression is brought by the oxidative pentose phosphate pathway (Wasylenko et al. 2015). Under lipids accumulation conditions, citrate is excreted in the cytoplasm by a citrate/malate translocase and converted into oxaloacetate (OAA) and acetyl-CoA by the ATP dependent citrate lyase (ACL) (Ratledge 2004; Ratledge and Lippmeier 2017). This enzyme is present in oleaginous yeasts but not in all yeast (Ratledge and Wynn 2002).

**Table 8.1** Traditional sources of PUFAs for 100 g of foodstuff portion and *for 100 g of oil, from Abedi and Sahari 2014; Rubio-Rodríguez et al. 2010.

| | Sources | PUFA | Omega-3 | | | Omega-6 | | |
|---|---|---|---|---|---|---|---|---|
| | | | ALA | EPA | DHA | LA | GLA | ARA |
| Marine organisms | Mackerel | | | 1.1 | 2.56 | | | |
| | Golden gray Mullet | | | 7.56 | 3.86 | | | |
| | Sardine | | | 0.62 | | | | |
| | Salmon | | | 0.5 | 1 | | | |
| | Tuna | | | 0.24 | 0.98 | | | |
| | Caviar | | | 2.74 | 3.8 | | | |
| Oils | Sardine oil | 55.1* | | | | | | |
| | Salmon oil | 63.1* | | | | | | |
| | Sunflower oil | 67.5* | | | | | | |
| | Corn oil | 55.6* | | | | | | |
| | Soya oil | 63.2* | 8.3* | | | 54.6* | | |
| | Cod liver oil | 56.8* | | | | | | |
| | Evening primrose | | | | | 65–85* | 8–14* | |
| Vegetables | Beet | | 29.44 | 0.54 | 0.65 | | 0.17 | 0.52 |
| | Curly dock | | 41.21 | 0.12 | | | | |
| | Hedge mustard | | 31.04 | 0.55 | 0.83 | | | 0.32 |
| Milk | Breast milk | 19.1 | 0.5 | | 0.19 | 16.29 | 0.12 | 0.51 |
| | Cow's milk | 3.6 | 2.9 | | 0 | 1.02 | – | 0 |
| Animal | Egg yolk | | 0.9 | | | 17.2 | | |
| | Animal liver | | 1.6 | 1.5 | 3.98 | 10.2 | | 10.5 |
| | Beef muscle | | 0.8 | 0.3 | – | 2.8 | | 0.5 |
| | Pork muscle | | 1.4 | 0.3 | 0.3 | 14.8 | | 1.1 |
| | Pork fat | | 1.5 | – | 0.2 | 14.8 | – | 0.2 |

It is essential for the accumulation of lipids, but its presence does not necessarily indicate an accumulation of lipids. Indeed, some non-oleaginous yeasts have ACL activity but do not accumulate lipids, which indicates that other enzymes are involved in the accumulation of lipids (Ratledge and Wynn 2002).

One of these enzymes is the malic enzyme. This enzyme generates NADPH (Wynn and Ratledge 1997) which is used for fatty acid synthesis by the fatty acid synthetase/elongase. Indeed, the incorporation of each acetyl-CoA requires 2 NADPH and therefore implies a supply of NADPH. If it is not present, the carbon flux from glucose to lipids is decreased and only the essential lipids are produced using another source of NADPH. According to several experiments by Wynn et al. (1999) the activity of the malic enzyme is the only one that can be correlated with the ability to accumulate lipids. In the same way, Wynn and Ratledge (1997) and Ratledge and Wynn (2002) established a direct link between the activity of the malic enzyme and the amount of lipid accumulation.

The second step of the lipogenesis is the conversion of acetyl-CoA into palmitic acid (Figure 8.1), which is the primer for longer fatty acids chains or unsaturated fatty acids. The fatty acid synthesis is the result of an iterative multienzyme complex named FAS (Schweizer and Hofmann 2004; Tehlivets et al. 2007; Ratledge and Lippmeier 2017). Acetyl-CoA is carboxylated into malonyl-CoA by an ACC. Then malonyl-CoA and Acetyl-CoA are linked thanks to an Acyl Carrier Protein (ACP) by a malonyl-CoA transacetylase and an acyl-CoA transacetylase, with the release of 2 CoA. A C2-unit is then added by a β-ketoacyl synthase to produce an acetoacetyl-ACP. A β-ketoacyl reductase leads to β-hydroxybutyryl-ACP regeneration of $NADP^+$ from NADPH $+H^+$. Dehydration of β-hydroxybutyryl-ACP by a β-ketoacyl dehydratase (DH), produces a crotonyl-ACP that is reduced by an enoyl-ACP reductase into butyryl-ACP, with regeneration of an $NAD^+$. Elongation of the fatty acid chain will continue by cycling addition to the acyl chain of a C2-unit from malonyl-ACP.

The overall equation is:

$$\text{acetylCoA} + 7\ \text{malonylCoA} + 14\ \text{NADPH} + 14\ \text{H}^+ \rightarrow \text{palmitic acid} + 7\ \text{CO}_2 \\ + 8\ \text{CoA} + 14\ \text{NADP}^+ + 6\text{H}_2 0$$

In most eukaryotic and prokaryotic cells, the last step of the PUFAs synthesis is to convert palmitic acid into unsaturated or PUFAs or into longer fatty acyl chains by specific enzymes: desaturase and/or elongase. Desaturase catalyzes the introduction of the double bond(s) into the fatty acid chain and elongase adds C2-unit to the molecule. In fungal producers of PUFAs, stearic acid is desaturated by Δ-9 desaturase to produce oleic acid, which is desaturated by Δ-12 desaturase to produce LA (Figure 8.2) (Certik and Shimizu 1999). LA is the first n-6 series PUFAs. LA can be either desaturated by the Δ-6 desaturase to produce the γ-linoleic acid or by the Δ-15 desaturase to produce α-linoleic acid (ALA). ALA is the first PUFA of the n-3 series. Then, a series of reactions of desaturation and elongation described in Figure 8.2 leads to the formation of others PUFAs. The n-3 or the n-6 fatty acid series implies the same elongases and desaturases, and it is possible with the presence of Δ-17 and/or Δ-19 desaturases to go from n-6 to n-3 series.

In microorganisms such as *Schizochytrium,* an alternative way of PUFAs synthesis is reported (Sun et al. 2014; Xie and Wang 2015; Aasen et al. 2016; Ratledge and Lippmeier 2017). DHA synthesis in *Schizochytrium* does not involve only desaturases and elongases enzymes such as those described before. It occurs via a polyketide synthase (PKS)-based pathway. PKSs are a family of multi-domain enzymes or enzyme complexes that produce polyketides, a large class of secondary metabolites such as antibiotics, in bacteria, fungi, plants, and a few animal lineages. The biosynthesis of polyketides shares similarities with fatty acid biosynthesis.

This PKS pathway consists of several modules with defined functions: a starting or loading module with acyltransferase (AT) and ACP domains, an elongation or extending modules with a keto-synthase (KS), a ketoreductase (KR), a DH, an enoylreductase (ER) and a methyltransferase (MT) domain, and a termination or releasing module with a thioesterase (TE) domain.

The starter group, usually acetyl-CoA or malonyl-CoA, is loaded on the ACP domain of the starter module catalyzed by the starter modules AT domain. Then, the elongation group, usually malonyl-CoA, is loaded on the current ACP domain.

The ACP-bound elongation group reacts with the KS-bound polyketide chain, leaving a free KS domain and an ACP-bound elongated polyketide chain. The reaction takes

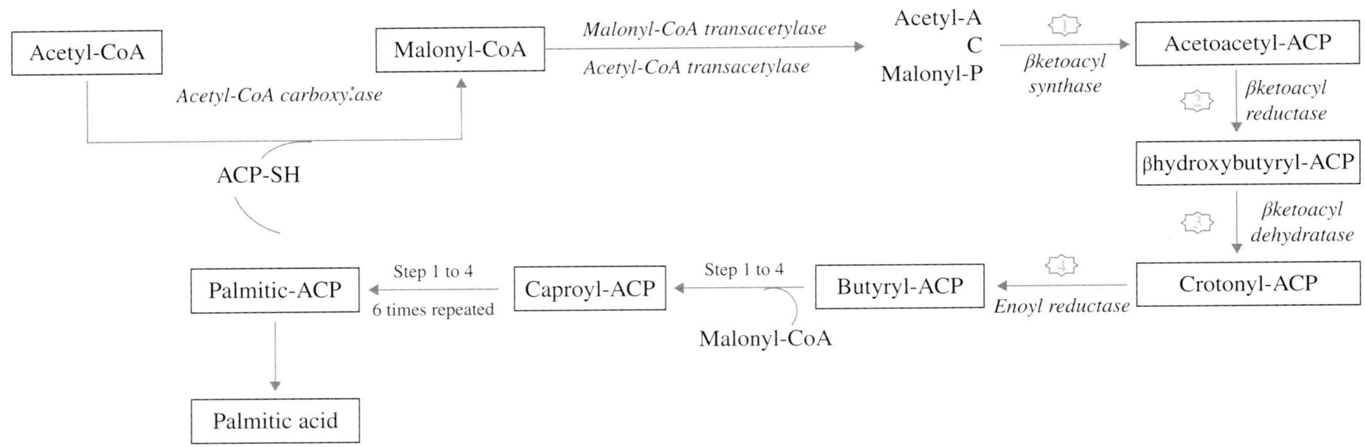


**Figure 8.1** Lipogenesis.

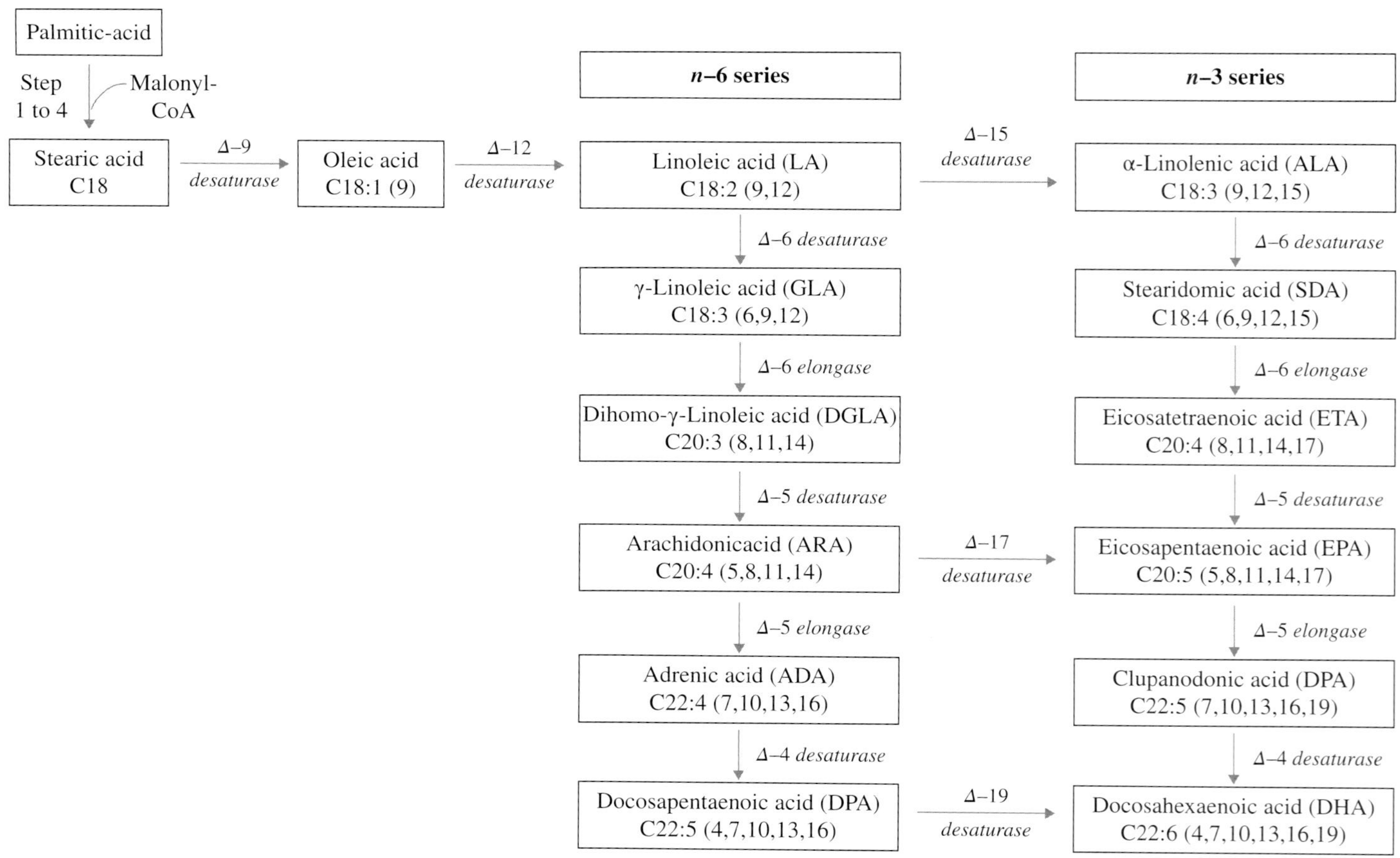


**Figure 8.2** PUFAs synthesis.

place at the $KS_n$-bound end of the chain, so that the chain moves out one position and the elongation group becomes the new bound group. This cycle is repeated for each elongation module. Finally, the TE domain hydrolyzes the completed polyketide chain from the ACP-domain (Qiu 2003) of the previous module.

### 8.3.2 Microorganisms Involved in PUFAs Production

Yeast, fungi, and microalgae have been studied for their potential to provide protein and other nutrients. In this section, we look at each of these more closely.

#### 8.3.2.1 Yeast

In yeast, the lipids synthesized do not stay free in the cytoplasm, but are aggregated in lipid bodies together with sterol esters, playing a key role in the regulation of energetic storage: storage of lipid and use of these de novo synthesized lipids (Murphy 2001). The lipids are embedded in a single-layer protein/phospholipid membrane. The proteins in these lipid bodies membrane may be enzymes implied in triacylglycerol building and protective against coalescence with neighboring lipid bodies (Murphy 2001; Mlickova et al. 2004). However, the range of fatty acids for yeasts is limited to oleic, linoleic, palmitic, and palmitoleic acids. These FAs are studied for biofuel production and will not be developed in this chapter.

In mammals, some desaturase (Δ-12 and Δ-15) are missing. Therefore, their diet must be supplemented by some PUFAs (ALA, GLA, DHA, or EPA), which are essentials for health and well-being. Only fungi and microalgae are able to produce PUFAs at levels of over 20% of the total fatty acids (TFAs) (Ratledge 2004). This is explained by the presence of specific enzymes: desaturase and/or elongase.

#### 8.3.2.2 Fungi

Fungi producers of PUFAs belongs to the order Mucorales. In this order, three genera produce a significant amount of PUFAs especially ARA and GLA on various substrates: *Mortierella, Cunninghamella,* and *Mucor* (Table 8.2).

The genus *Mortierella* is extensively studied for the production of ARA and GLA. Among this genus, the species *Mortierella isabellina, Mortierella alpina,* and *Mortierella elongata* are able to produce up to 45% of their DCW on glucose when the C/N ratio is high (around 245) (Dyal and Narinc 2005; Papanikolaou et al. 2007). They are the most important producers of ARA, especially *M. alpina* (Sakuradani and Shimizu 2009). Shimizu et al. (1988) tested several conditions of temperature to produce PUFAs by several *Mortierella* and reported an ARA production of 13.9% of TFAs for *M. elongata* and of 56.2% of TFA for *M. alpina* (Sakuradani and Shimizu 2009).

The production of PUFAs by *Mortierella* is done either by SSF (solid-state fermentation) (Jang and Yang 2008; Fakas et al. 2009) or in submerged culture (Shimizu et al. 1988; Sakuradani and Shimizu 2009; Xing et al. 2012; Demir et al. 2013). However, at industrial scale, companies like CABIO (Cargill Alking Co. Ltd) or DSM produce dietary supplement with ARA from *Mortierella alpine* using preferentially submerged cultures.

The genera *Cunninghamella* and *Mucor* are promising microorganisms for the production of GLA. Fakas et al. (2008) reported that *Cunninghamella echinulata* could produce GLA up to 20% TFA on various complex organic sources (Fakas et al. 2008). *Mucor (javanicus) circinelloides* was the first strain used to produce GLA and its production was

**Table 8.2** Fungi involved in PUFAs production.

| Strain | Substrate | Mode of culture | Experimental conditions | Lipid content %DCW w/w | PUFA content %TFA w/w | | | | | Ref. |
|---|---|---|---|---|---|---|---|---|---|---|
| | | | | | ALA | EPA | LA | GLA | ARA | |
| *Mucor circinelloides WJ11* | Glucose | Batch | Comparison of the two strains | 35.6 | | | 25.5 | 30.3 | | Tang et al. (2015) |
| *Mucor circinelloides CBS277* | | | | 14.7 | | | 19.6 | 38.3 | | |
| *Mucor circinelloides WJ11* | | Flask | | 40 | | | 27 | 29 | | Zhao et al. (2015) |
| *Mucor circinelloides CBS277* | | | | 22 | | | 20 | 30 | | |
| *Cunninghamella echinulata* | Corn gluten | Flask | Variation of the nitrogen concentrations | 11.2–13.3 | | | | 02-mars | | Fakas et al. (2008) |
| | Corn steep | | | 7.6–19.8 | | | | 12.6–14 | | |
| | Whey concentrate | | | 14.3–19.3 | | | | 13.8–14.4 | | |
| | Yeast extract | | | 12–16.6 | | | | 12.1–13.6 | | |
| | Tomato waste hydrolysate | | | 17.4–39.6 | | | | 11.5–18.3 | | |
| | Xylose | | C/N 285 | 23.8 | | | | 22 | | Fakas et al. (2009) |
| | | | C/N 78 | 53.6 | | | | 16.7 | | |
| *Mortierella alpina* | Rice bran | SSF | Nitrogen supplementation | | 2.7 | 16.9 | | | | Jang and Yang (2008) |
| | yeast extract | Fed-batch | Repeated fed-batch | 50 | | | | | 43 | Ji et al. (2014) |
| | | Flask | | | | | 14.4 | 10.9 | 69.7 | Shimizu et al. (1988) |

| | | | | | | | | | | |
|---|---|---|---|---|---|---|---|---|---|---|
| *Mortierella isabellina* | Xylose | Flask | C/N 285 | 64 | | | | 4.1 | | Fakas et al. (2009) |
| | Raw glycerol | | C/N 117 | 22.7 | | | | 11.6 | | |
| | | | C/N 157 | 21.7 | | | | 12.8 | | |
| | Xylose | Flask | | 71 | 10.9 | | | 2.5 | | Gao et al. (2013) |
| | | | 3 Substrates | 66.7 | 17.8 | | | | | Zeng et al. (2013) |
| | Glucose | | | 66.5 | 14.3 | | | | | |
| | Wheat straw hydrolysate | Flask | | 53 | – | | | | | |
| | Pear pomace | SFF dishes | Maturities of mycelia | 3.8 GLA | | | | | | Fakas et al. (2009) |
| | Glucose | Flask | 6 °C | | | 14.8 | 15 | 10 | 30.2 | Shimizu et al. (1988) |
| | | | 12 °C | | | 14.9 | 14.4 | 10.1 | 28.4 | |
| | | | 16 °C | | | 2.4 | 3.9 | 4.4 | 50.2 | |
| | | | 20 °C | | | 0 | 5.1 | 3.4 | 56.2 | |
| | | | 28 °C | | | 0 | 9.8 | 4.1 | 47.7 | |
| | Corn fiber hydrolysates | Flask | | 50.48 | | | 28.88 | 3.85 | | Xing et al. (2012) |
| *Mortierella elongata* | Soybean oil | Flask | | 24 | | 13 | | | 58 | Cheng et al. (1999) |
| | Glucose | | 6 °C | | | | 7.4 | 7.4 | 22.8 | Shimizu et al. (1988) |
| | | | 12 °C | | | 13.9 | 7.4 | 13.1 | 12.3 | |
| | | | 28 °C | | | 0 | 3.5 | 2.5 | 13.9 | |

more than twice the concentration extracted from evening primrose oil usually used as a dietary supplement (Ratledge and Lippmeier 2017). Evening primrose oil is still the main source of GLA but research continues.

#### 8.3.2.3 Thraustochytrids and Microalgae

Currently, most of the publications on PUFAs production concern PUFAs production by microalgae and Thraustochytrids. Two strains are commercially used to produce PUFAs and essentially DHA: *Crypthecodinium cohnii* and strains from the Traustochytrid marine protists (Table 8.3).

***Microalgae*** DHA from *C. cohnii* was the first marine product commercialized. This was possible thanks to the process developed by Martek Biosciences to include DHA in infant formula (Ratledge and Lippmeier 2017).

Liu et al. (2015) increased the cell proliferation and fatty acid production capacity of *C. cohnii* by the external addition of sesamol. De Swaaf et al. (1999 and 2003) tested *C. cohnii* for DHA production in batch culture on glucose and in high-density fed-batch. In these studies, they obtained 36.2% DHA of TFA on galactose (De Swaaf et al. 1999) and 38% DHA of TFA on glucose for the growth phase and acetic acid for the feeding medium (De Swaaf et al. 2003).

Jiang et al. (1999) compared four strains of *C. cohnii* and reported that the strain ATCC 30556 was the most efficient to produce DHA with a final concentration of 51.12% of TFA. Jiang and Chen (2000) used this strain for another study to define the effect of pH and glucose concentration on DHA production. Another strain (ATCC 30555) used by Safdar et al. (2017) shows its ability to produce DHA.

Other species are studied, such as *Isochrysis galbana,* which is a DHA-producing microalgae (Gorissen et al. 2012), *Porphyridium,* which is an EPA-producing microalgae, and *Chlamydomonas reinhardtii,* which is considered as the model to study lipids metabolism (Abedi and Sahari 2014). Lin et al. (2007) studied the influence of the growth phase on the production of PUFAs by *I. galbana* and concluded that the better phase of accumulation is the early stationary phase.

***Thraustochytrids*** *Schizochytrium* sp. and *Aurantiochytrium* sp. are the main Thraustochytrids organisms investigated for DHA production (Table 8.3). Most recent publications deal with the production of PUFAs by Thraustochytrids. Thraustochytrids are unicellular, eukaryote, heterotrophic, ubiquitous, and obligate marine microorganisms, and form a group of common marine microheterotrophs, originally classified as marine fungi and now taxonomically aligned with heterokont algae (Dick 2001). Thraustochytrid strains can produce important levels of the long-chain omega-3 (n-3) such as DHA (Aasen et al. 2016). High-density cultures ($100\,g\,l^{-1}$) done with these strains gave an accumulation of TFA of 50–70% of DCW with 30–70% DHA of TFA (Chang et al. 2013; Li et al. 2015). The US company OmegaTech (a part of DSM nutritional Product) commercializes DHA-rich oils from thraustochytrids (Barclay et al. 2010; Ratledge and Lippmeier 2017). *Aurantiochytrium* sp. can accumulate DHA up to 63% DCW under nitrogen limitation (Jakobsen et al. 2008). Abad and Turon (2015) tested this strain on glucose, pure glycerol et crude glycerol and obtained respectively 1.24, 1.23, and $1.33\,g\,l^{-1}$ of DHA. They confirmed that this strain is a suitable oleaginous microorganism to produce DHA from different carbon sources.

**Table 8.3** Microalgae and thraustochytrids involved in PUFAs production.

| Strain | Substrate | Mode of culture | Experimental conditions | Lipid content %DCW w/w | PUFA content %TFA w/w EPA | DHA | GLA | ARA | Ref. |
|---|---|---|---|---|---|---|---|---|---|
| *Schizochytrium* sp. SR21 | Glucose | Flask | | 60 | | 20 | | | Zhang et al. (2013) |
| | | | AEMR | 55 | | 15 | | | |
| | | Flask | | 70 | | 15 | | | Patil and Gogate (2015) |
| | Glucose + glycerol | | | 80.2 | | 18.3 | | | |
| | Glucose | Flask | Non limitation | | 0.97 | 42.45 | | | Sun et al. (2014) |
| | | | Glutamate limitation | | 0.93 | 41.43 | | | |
| | | | Phosphate limitation | | 1.27 | 44 | | | |
| | | | Glutamate and phosphate limitation | | 1.01 | 40.53 | | | |
| | | | Ammonium sulfate limitation | | 1.12 | 44.08 | | | |
| | Glucose + glycerol | Flask | | 62.4 | | 47.37 | | | Li et al. (2015) |
| | Glycerol | Fed batch | Modification of C/N | 70 | | 72.6 | | | Huang et al. (2012) |
| | Glucose | Fed batch | 1500 l $K_La$ fixed at 88.9 $h^{-1}$ | 49.35 | | 40.23 | | | Qu et al. (2013) |
| | | | 7000 l $K_La$ fixed at 88.9 $h^{-1}$ | 54.56 | | 40.11 | | | |
| | Sorghum sweet juice | Flask | 100% sorghum juice | 70.5 | 0.64 | 33.74 | | 1.07 | Liang et al. (2010) |
| | | Flask/Batch | 50% sorghum juice | 73.4 | 0.61 | 34.28 | | 1.01 | |
| | Glucose | Batch | + 3 g/L ascorbic acid | 55.98 | | 50.61 | | | Ren et al. (2017) |
| | | Unbaffled flask L | + 6 g $l^{-1}$ ascorbic acid | 60.32 | | 51.67 | | | |
| | | | + 9 g $l^{-1}$ ascorbic acid | 65.5 | | 52.81 | | | |
| | | | + 12 g $l^{-1}$ ascorbic acid | 47.08 | | 50.20 | | | |
| | | | C and N feeding strategy | 54.64 | | 18.5 | | | Ling et al. (2015) |

**Table 8.3** (Continued)

| Strain | Substrate | Mode of culture | Experimental conditions | Lipid content %DCW w/w | PUFA content %TFA w/w | | | | Ref. |
|---|---|---|---|---|---|---|---|---|---|
| | | | | | EPA | DHA | GLA | ARA | |
| *Schizochytrium* sp. LU310 | Glucose | Baffled flask | C and N feeding strategy | 64.25 | | 27.51 | | | Ling et al. (2015) |
| | Glycerol | Fed-batch | $K_{La} = 568$ | 52.9 | | 35.9 | | | |
| *Schizochytrium* sp. S31 | Glycerol | Fed-batch | Glycerol 75 g $l^{-1}$ | 52 | | 44 | | | Chang et al. (2013) |
| | | Flask | Glycerol 100 g $l^{-1}$ | 47.9 | | 35.3 | | | |
| | Soybean meal hydrolysate | | | | 0.46 | 46.3 | 0.27 | 0.32 | Song et al. (2015) |
| *Schizochytrium limanium* | Yeast extract | Flask | Oxygen variations | | 0.52 | 30–38.06 | 0.23 | 0.43 | Jakobsen et al. (2008) |
| *Aurantiochytrium limacinum* | Glucose | Flask | C/N 55 | 63.86 | | 23.93 | | | Rosa et al. (2010) |
| | | Batch | C/N 100 | 64.7 | | 22.8 | | | |
| | | | C/N 55 | 65.38 | | 32.65 | | | |
| | Glycerol | Fed-batch (10 L) | C/N 0.5 | 60.8 | | 55.2 | | | Huang et al. (2012) |
| | | Fed-batch (10 L) | C/N 1.25 | 72 | | 58.2 | | | |
| | | Flask | C/N 1.875 | 65.2 | | 66.6 | | | |
| | | | 5–150 g $l^{-1}$ | 43.95 | | 20.96 | | | |
| | | Flask | 5–150 g $l^{-1}$ | 28.3 | | 10.78 | | | |
| | | | Comparison of different strains | | | 55.2 | | | Lee Chang et al. (2015) |
| *Aurantiochytrium* sp. TC9 | Crude glycerol | Flask | Comparison of different strains | | | 33.9 | | | |
| | Pure glycerol | Fed batch | | | | 54.03 | | | |

| | | | | | | | | | |
|---|---|---|---|---|---|---|---|---|---|
| *Aurantiochytrium* sp. TC20 | Crude glycerol | | | | | 31 | | | |
| | Glucose | | | 15 | | 46 | | | |
| *Crypthecodinium cohnii* | Acetic acid | Fed batch | | 54 | | 38 | | | De Swaaf et al. (2003) |
| | Glucose | | | 13.4 | | 35.2 | | | De Swaaf et al. (1999) |
| | Acetic acid | Flask | | 11.4 | | 36.2 | | | |
| | Glucose | | 5–40 $g\,l^{-1}$ of glucose | | | 53.4–36.5 | | | Jiang and Chen (2000) |
| | | | pH 4–pH 10 | | | 44.8–56.8 | | | |
| | $CO_2$ and light | PBR | White light | | 28.2 | 1.83 | 2.07 | 1.56 | Yoshioka et al. (2012) |
| *Isochrysis galbana* | | | Red light | | 24.5 | 1.39 | 1.8 | 1.2 | |
| | | | Blue light | | 24.3 | 2.21 | 1.96 | 1 | |
| | Urea and light | Roux flask | Different growth phase | | | 14.4 | | | Lin et al. (2007) |
| | Glycerol, $CO_2$ and light | Flask | 2 periods of light | | 28.5 | | 0.8 | 4.4 | Shene et al. (2016) |
| *Nannochloropsis ocula* | | Batch | 100% of time light | | 8.2 | | | 2.6 | |
| | | | 50% of time light | | 7.9 | | | 4 | |

Ling et al. (2015) described how the production of biomass and DHA could be improved when a feeding strategy of carbon and nitrogen source is used. They increased the DHA production from 13 to 33% TFA in *Aurantiochytrium limacinum* by changing the C/N ratio and the inoculum age.

*Schizochytrium* sp. can accumulate DHA up to 40% of the total FA and is considered as a viable candidate for an alternative way to produce DHA (Ling et al. 2015). Huang et al. (2012) reported that this strain could produce 20.3 g $l^{-1}$ of DHA with an intermittent oxygen feeding. Sun et al. (2016) increase the production of DHA with continuous high oxygen supply and identify culture conditions, which involve DHA production.

Currently, DSM commercializes life'sDHA or life'sOMEGA, which are a nutraceutical containing DHA and/or EPA produced by *Schizochytrium* sp. and Aquafauna Bio-Marine (USA) commercializes Algamac® 2000 and Algamac3000, which are spray dried *Schizochytrium* sp. biomass to enrich aquaculture species in DHA (Fan and Chen 2007).

## 8.4 Production Strategies

### 8.4.1 Culture Conditions

Several parameters influence the production of lipids such as carbon source (Dyal and Narine 2005), nitrogen limitation (Papanikolaou 2004; Sun et al. 2014; Ling et al. 2015), C/N ratio, temperature (Dyal and Narine 2005; Chodchoey and Verduyn 2012), pH, the culture state (sporulation is associated with an increase of cellular lipids), and nutrition (Fan and Chen 2007).

#### 8.4.1.1 Nutritional Aspects

Some nutritional elements are known to influence lipid production. In mainly oleaginous microorganisms, lipid accumulation occurs when the carbon source is in excess and another component of the growth medium becomes limiting. The limitation of nitrogen in the culture medium has been used by many authors to induce lipid production by oleaginous microorganisms (Fakas et al. 2007; Sun et al. 2014; Ling et al. 2015). The effects of nitrogen limitation in lipid biosynthesis are well known. In oleaginous microorganisms, a decrease of AMP concentration occurs when lipid accumulation begins (Xie and Wang 2015). This decrease in AMP concentration occurs when nitrogen is exhausted and just before lipid accumulates (Ratledge and Wynn 2002). Nitrogen exhaustion induces cascades of biochemical events, leading to accumulation of lipids. The first of them is an increase of AMP desaminase activity to balance the lack of nitrogen feeding (Beopoulos et al. 2009). The same events were observed by Wynn et al. (2001) or by Zhao et al. (2015) in *Mucor circinelloides* cultures and by Safdar et al. (2017) in *C. cohnii* cultures under nitrogen limitation.

In fact, the parameter that significantly influences the lipid metabolism of the microorganism is the C/N ratio. When the C/N ratio is less than 10, growth of the microorganism will be favored, while a high C/N ratio will slow down the growth and switch the metabolism toward the synthesis of lipids (Rosa et al. 2010; Huang et al. 2012; Xie and Wang 2015). Several C/N ratios have been tested by authors to identify which ones are the most promising for lipid accumulation: 35 (Ryu et al. 2013), 49–50 (Chi et al.

2011; Fontanille et al. 2012), 70 (Ykema et al. 1988). Rosa et al. (2010) concluded that a C/N ratio of 55 resulted in maximum lipid content of 65% and DHA of 35% (TFA) in *A. limacinum* yeast grown on glucose. Ling et al. (2015) tested the influence of a carbon and nitrogen-feeding strategy on DHA production by *Schizochytrium* sp. This study describes how the production of DHA can be improved by a feeding strategy of carbon and nitrogen source. Nitrogen exhaustion is decreased using monosodium glutamate (MSG), and DHA production is increased with a moderate feeding and an intermittent feeding of glucose (Béligon et al. 2016, 2015).

Some authors have also investigated the influence of limitation by other nutrients. The effect of sulfate, glutamate, and phosphate limitations has been studied by Sun et al. (2014) on the lipid production by *Shizochytrium* sp. Other factors seem important for lipids accumulation, such as growth factors (inositol, biotin, vitamin $B_6$) or antioxidant (sesamol) (Dyal and Narine 2005; Tehlivets et al. 2007; Liu et al. 2015). Ren et al. (2017) tested the addition of ascorbic acid to improve cell growth and DHA production by *Shizochytrium* sp. In this study, DHA yields were improved with the increase of ascorbic acid concentration with a maximum of $9\,g\,l^{-1}$.

#### 8.4.1.2 Temperature

The culture's temperature plays a key role in the growth and the metabolism and especially in the production of lipids. According to the literature, oleaginous microorganisms can be classified according to the optimal temperature of lipid production.

Temperature also plays a greater key role in the fatty acid composition of lipids. Indeed, the decrease of the temperature leads to the increase in the unsaturation level of the lipids (De Swaaf et al. 1999; Dyal and Narine 2005). This phenomenon is observed in both oleaginous and nonoleaginous microorganisms and is associated with an adaptive response of the membrane to the temperature lowering to maintain membrane fluidity by increasing the amount of unsaturated fatty acids (Beltran et al. 2008). The modification of the composition concerns phospholipids, sphingolipids, and glycolipids, major constituents of the membrane (Kendrick and Ratledge 1992). Shimizu et al. (1988) showed that the temperature can be a good way to control the PUFAs composition of the genus *Mortierella*. They determined temperature of 28 °C for the production of ARA and a temperature of 20 °C to produce EPA (Shimizu et al. 1988). When the medium temperature of *Entomophtora exitalis* decreases from 30 to 20 °C, the content of PUFAs, and particularly ARA, increases from 18% to 27%. Under lipid accumulation conditions, a decrease of 15 °C of the temperature (35–20 °C) causes an increase of unsaturated fatty acids from 52% to 71.8% by *Schizochytrium limacinum* (Zhu et al. 2007). Similarly, Chodchoey and Verduyn (2012) showed an increase in DHA content from 29% to 42% when *A. mangegrovei* was cultivated at 30 and 10 °C.

#### 8.4.1.3 pH

pH also plays an important role in the growth and metabolism of microorganisms. Microorganisms are able to grow and produce lipids in large pH ranges. For example, *Cryptococcus albidus* oleaginous yeast can accumulate lipids on volatile fatty acids in a pH range between 5.5 and 7 (Fei et al. 2011) while *S. limacinum* (Zhu et al. 2007) or *C. cohnii* (Jiang and Chen 2000) can grow and produce lipids under pH ranging from 4 to 10.

The optimal pH for lipid production will depend mainly on the species and the carbon source used. For example, *Lipomyces starkeyi*, has an optimal pH at 5 on glucose (Angerbauer et al. 2008) and a maximum lipid content at pH 4 when grown on ethanol (De Swaaf et al. 2003). However, pH does not appear to influence the final composition of lipids.

The role of pH in lipid production can be translated in several ways. In the case of a culture of oleaginous yeasts on volatile fatty acids and in particular on acetate (De Swaaf et al. 2003; Béligon et al. 2016, 2015), the access of the acetate into the cell is mainly done by an acetate/proton pump. The pKa of the acetic acid/acetate is 4.75; then the acetate form will be predominant if the culture is carried at pH 7. This form can therefore be assimilated into the cell via the protomotive pump and thus be available for the synthesis of lipids.

#### 8.4.1.4 Oxygen

Most oleaginous microorganisms employed in DHA production are aerobic (Ren et al. 2010). Oxygen is an essential regulator of growth in aerobic organisms. A high concentration of dissolved oxygen ensures an optimal growth rate if essential nutrients for the growth are available. Oxygen can also influence the final lipid content in oleaginous organisms. Many examples are available in the literature (Huang et al. 2012; Jakobsen et al. 2008; Qu et al. 2013; Ren et al. 2010), but it seems to have no common rules on oxygen requirements between the species. Indeed, some studies reported that a high oxygen supply promotes cell growth and lipid accumulation (Jakobsen et al. 2008; Sun et al. 2016) during *Aurantiochytrid* sp. or *Shizochytrium* sp. cultures. In contrast, other authors show that an oxygen-rich environment causes a decrease in the final lipid content, especially the PUFAs, because they could be oxidized due to the accumulation of ROS (reactive oxygen species) (Guichardant et al. 2011; Johansson et al. 2016). Qu et al. (2013) developed a scale-up strategy based on the volumetric oxygen-transfer coefficient ($k_La$) to improve DHA production by *Shizochytrium* sp. and obtained better results in larger volumes when the $k_La$ is matched with $k_La$ obtained in the flask.

Oxygen can also mediate the biosynthesis of lipids during the desaturation. Δ9 desaturase requires an oxygen molecule to convert a molecule of stearic acid to oleic acid. Therefore, if oxygen is limiting in the medium during lipid accumulation phase, the degree of saturation of the lipids will be higher. For *Rhodotorula gracilis*, a decrease of stearic (C18:0, saturated) and oleic acids (C18:1, unsaturation) and an increase of PUFAs (C18: 2 and C18: 3) were observed when dissolved oxygen concentration increased (Choi et al. 1982). Davies et al. (1990) found an increase in saturated fatty acid content (mostly oleic and palmitoleic acids) when oxygen consumption rate decreased in *Apiotrichum curvatum* (Davies et al. 1990). The effect of oxygen on lipid production appears to be strain-dependent.

#### 8.4.1.5 Light

When phototrophic microorganisms are used, the light is a key parameter (Fernandes et al. 2015). Light has extra and multiple effects on growth and on lipid accumulation. Lipids productivities are affected by the quality (light photon flux) of the light and by the quantity (density) of the light (Fernandes et al. 2015). Yoshioka et al. (2012) studied the influence of intermittent light color on FA profile of *I. galbana*. They concluded that the cultivation under intermittent blue light promoted the productivity of biomass and lipids (Yoshioka et al. 2012). The time during which the microorganisms are lighted influences the accumulation of TFA and the ratio between the PUFAs produced on *Nannochloropsis oculate* (Shene et al. 2016).

## 8.5 Process Strategies

### 8.5.1 Modes of Culture

For lipid accumulation in oleaginous microorganisms, different modes of biomass production can be considered: batch, fed-batch, continuous culture modes (with light or not), and solid-substrate fermentation. The mode of culture is a crucial parameter for the improvement of lipid production and will therefore have a direct economic impact on process.

Most of the studies in the literature have been done using batch culture and nutrient limitation (usually nitrogen) (Lin et al. 2007). In the same way, most studies on phototrophic microorganisms occur in batch mode (Fernandes et al. 2015). In batch culture, the accumulation of lipids is generally divided into two main phases. During a first phase where biomass is produced (growth phase) nutrients are in excess in the medium (balance of nutrients). Then, during a second, stationary growth phase, some nutrients, frequently nitrogen, are limiting. (Beopoulos et al. 2009; Rosa et al. 2010; Wynn et al. 1999). During this phase, the synthesis of nucleic acids and proteins stopped, and the carbon source is metabolized and stored in lipids. Rosa et al. (2010), tested a two-stage strategy to improve DHA production with a variation of the C/N ratio. They show that a C/N ratio of 55 with an inoculum aged of 3.5 days were the best conditions to accumulated lipids (29.5 $g\,l^{-1}$) and DHA (7 $g\,l^{-1}$) using *A. limacinum*.

During the fed-batch modes, nutrients can be added in the medium at separate times. As in batch mode, the accumulation of lipids occurs in two phases: growth and accumulation. This mode allows better control of the substrate concentration in order, in most cases, to control the C/N ratio (Huang et al. 2012).

In continuous culture, the batch is extended by continuous feeding of the fresh nutrient medium. This mode of culture allows controlling a C/N ratio by varying the dilution rate (Ykema et al. 1986). Shen et al. (2013) show that lipid accumulation was increased at low dilution rates (Shen et al. 2013).

Only a few studies use fed-batch or continuous cultures (Muniraj et al. 2013; Qu et al. 2013; Sun et al. 2016), and most these works focused on lipids production for biodiesel (Béligon et al. 2016; Beopoulos et al. 2009; Meesters et al. 1996; Ratledge and Wynn 2002; Ykema et al. 1986).

SSF is carried out on a solid support in the absence of free water. The environment supplies the oxygen and the support provides the water. The support should be moist enough to provide nutrients for microorganism growth. The material as well as the substrates used in SSF are generally more simple and lower cost compared to those used for liquid cultures. Indeed, the substrates are in most cases agro-industrial residues such as pear pomace (Fakas et al. 2009) and cereal substrates (Čertík et al. 2013). The submerged modes of culture are well adapted from the culture of unicellular microorganisms such as yeast or microalgae, but culture on solid medium (SSF) is more adapted to the culture of mycelial organisms.

For microalgae cultivation, because the light is necessary the cultures are carried out in photobioreactors or open-air systems. Fernandes et al. (2015) provide an overview of continuous cultivation systems for photosynthetic microorganisms. The authors described how the system of culture could improve or guide the production.

Some studies tested the scale-up of the batch fermentation process (Song et al. 2010) on glucose soybean cake hydrolysate and fed-batch fermentation process on glucose to increase DHA production (Qu et al. 2013). These fermentation processes are well controlled at industrial scale, which is not the case for SSF for the production of PUFAs.

### 8.5.2 Substrates

Lots of studies focus on the production of PUFAs from glucose (Tables 8.1 and 8.3) and lots of studies focus on the production of biodiesel from glycerol. Some also show that glycerol can be used to produce PUFAs. Abad and Turon (2015) have reported that the productivities of DHA on glucose and pure glycerol are the same (55 mg $(\mathrm{l\,h^{-1}})^{-1}$) and are lower than the productivity on crude glycerol (60 mg $(\mathrm{l\,h^{-1}})^{-1}$) for the Thraustochytrids *A. limacinum*. Li et al. (2015) tested a mixed carbon source composed of glucose and glycerol and obtained DHA yield (around 30 g $\mathrm{l^{-1}}$) similar to those obtained on glycerol (Chang et al. 2013). Gupta et al. (2015) performed a culture of *Schizochytrium* sp. on sugar hydrolysate from saccharified hemp to produce PUFA and show production of DHA of 37.7% TFA. Other studies on complex substrates, such as crude glycerol from coconut water (Unagul et al. 2007), soap containing crude glycerol (Pyle et al. 2008), and sweet sorghum juice (Liang et al. 2010), show DHA production of, respectively, 33.6%, 38.2%, 24.5%, and 34.3% of TFA. In the same way, studies have reported the production of LA on whey pretreated with lacase (Demir et al. 2013) or on pear pomace (Fakas et al. 2009) by *M. isabellina*, the production of GLA on cereal substrates (Čertík et al. 2013) by *Mucor*, or the production of ARA by *M. elongata* and *Pythium irregulare* on industrial waste streams and crude soybean oil (Cheng et al. 1999).

All these studies confirm the possibility of producing low-cost PUFAs even if the productivities are slightly lower.

### 8.5.3 Metabolic Engineering

The microorganisms and their metabolism, which involve in PUFAs production, are today well-known, but production yields may be improved. These improvements may occur through metabolic engineering of PUFAs producers (Gong et al. 2014; Mao et al. 2017). Lots of publications deal with the production of oleaginous microorganism mutants. Several authors reported transcriptomic approach (Pei et al. 2017), phenotypic (Chutrakul et al. 2016), or metabolic fluxes analysis (Zhao et al. 2015), which increase the knowledge of lipid metabolism and enhance the metabolic engineering progress. The strategies employed have three main objectives (Yang et al. 2015): increase the productivities, eliminate or reduce the byproducts, and extend substrates range. Several metabolic engineering strategies have been used, such as the amplification of existing genes, the deletion of key genes of byproducts synthesis, the introduction of genes in nonoleaginous organisms, and the introduction of genes encoding for important enzymes to use complex substrates (Qi et al. 2017).

Qi et al. (2017) have isolated and cultivated a lignocellulosic hydrolysate tolerant mutant for *Aurantiochytrium* sp. to increase the production of DHA on sugarcane bagasse, and they obtained more lipids and DHA with this strain compared to the wild-type strain. In the same way, a mutant of *M. alpina* (Sakuradani and Shimizu 2009; Sakamoto et al. 2017) was constructed and used to increase the production of ARA.

## 8.6 Conclusions

In nutrition, PUFAs are known to be important for health and well-being. More recently, several publications show that the consumption of PUFAs can prevent some cancers and have positive effects on Alzheimer's disease (Ratledge and Lippmeier 2017). Currently, traditional sources of PUFAs (oily fishes: sardine, salmon, tuna; and vegetables: sunflower, soya, evening primrose) are the greatest contributors to PUFAs in dietary supplements. However, the use of these resources cannot be sufficient in term of quality (e.g., pollutants) and quantity (increase in consumption). This is especially true if we only consider the estimates of the World Health Organization in 2017 for 47.5 million people suffering from Alzheimer's disease, with a prospect of 135 million of people in 2050. So, it's necessary to find an alternative source of PUFAs to improve the quality and to fill the gap between consumption and production. Microorganisms are well known as lipid producers, and microbial oils have interesting potential for commercial applications in nutrition, pharmaceutical, or medical fields but also as biofuel. Commercial microbial oils appeared in the marketplace in 1985 and since that time, production has expanded.

To produce microbial PUFAs, the choice of the microorganism is essential. It is necessary to select the best-performing strains but also the best production strategy. In the last decades, most of the studies have dealt with non-genetically modified strains, in batch culture using glucose as substrate. However, the highest difficulty for microbial PUFAs to represent an alternative to the traditional PUFAs sources is the cost of overall production. Recently, some studies reported the use of microorganisms with the capacity to convert many cheap raw materials (Čertík et al. 2013; Demir et al. 2013; Fakas et al. 2009; Liang et al. 2010; Pyle et al. 2008; Unagul et al. 2007). More and more studies show that metabolic engineering allows the increase the production of such lipids (Qi et al. 2017; Sakamoto et al. 2017; Sakuradani and Shimizu 2009). This progress tends to decrease the production cost of microbial PUFAs. In the coming years, the main reluctance for the consumer to buy microbial PUFAs may be their nature and the use of genetically modified organisms.

## References

Aasen, I.M., Ertesvåg, H., Heggeset, T.M.B. et al. (2016). Thraustochytrids as production organisms for docosahexaenoic acid (DHA), squalene, and carotenoids. *Appl. Microbiol. Biotechnol.* 100: 4309–4321.

Abad, S. and Turon, X. (2015). Biotechnological production of docosahexaenoic acid using *Aurantiochytrium limacinum*: carbon sources comparison and growth characterization. *Mar. Drugs* 13: 7275–7284.

Abedi, E. and Sahari, M.A. (2014). Long-chain polyunsaturated fatty acid sources and evaluation of their nutritional and functional properties. *Food Sci. Nutr.* 2: 443–463.

Angerbauer, C., Siebenhofer, M., Mittelbach, M., and Guebitz, G.M. (2008). Conversion of sewage sludge into lipids by *Lipomyces starkeyi* for biodiesel production. *Bioresour. Technol.* 99: 3051–3056.

Barclay, W., Weaver, C., Metz, C., and Hansen, J. (2010). Development of a docosahexaenoic acid production technology using Schizochytrium: historical perspective and update. In: *Single Cell Oils: Microbial and Algal Oils*, 2e (eds. C. Ratledge and Z. Cohen), 75–96. Urbana, IL.

Béligon, V., Poughon, L., Christophe, G. et al. (2016). Validation of a predictive model for fed-batch and continuous lipids production processes from acetic acid using the oleaginous yeast Cryptococcus curvatus. *Biochem. Eng. J.* 111: 117–128.

Béligon, V., Poughon, L., Christophe, G. et al. (2015). Improvement and modeling of culture parameters to enhance biomass and lipid production by the oleaginous yeast Cryptococcus curvatus grown on acetate. *Bioresour. Technol.* 192: 582–591.

Beltran, G., Novo, M., Guillamón, J.M. et al. (2008). Effect of fermentation temperature and culture media on the yeast lipid composition and wine volatile compounds. *Int. J. Food Microbiol.* 121: 169–177.

Beopoulos, A., Chardot, T., and Nicaud, J.-M. (2009). *Yarrowia lipolytica*: a model and a tool to understand the mechanisms implicated in lipid accumulation. *Biochimie* 91: 692–696.

Bharathiraja, B., Sridharan, S., Sowmya, V. et al. (2017). Microbial oil – a plausible alternate resource for food and fuel application. *Bioresour. Technol.* 233: 423–432.

Birch, E.E., Garfield, S., Castañeda, Y. et al. (2007). Visual acuity and cognitive outcomes at 4 years of age in a double-blind, randomized trial of long-chain polyunsaturated fatty acid-supplemented infant formula. *Early Hum. Dev.* 83: 279–284.

Certik, M. and Shimizu, S. (1999). Biosynthesis and regulation of microbial polyunsaturated fatty acid production. *J. Biosci. Bioeng.* 87: 1–14.

Čertík, M., Adamechová, Z., and Guothová, L. (2013). Simultaneous enrichment of cereals with polyunsaturated fatty acids and pigments by fungal solid state fermentations. *J. Biotechnol.* 168: 130–134.

Chang, T.-S. and Singh, M.P. (2009). Mechanical model parameters for viscoelastic dampers. *J. Eng. Mech.* 135: 581–584.

Chang, G., Luo, Z., Gu, S. et al. (2013). Fatty acid shifts and metabolic activity changes of Schizochytrium sp. S31 cultured on glycerol. *Bioresour. Technol.* 142: 255–260.

Cheng, M.H., Walker, T.H., Hulbert, G.J., and Raman, D.R. (1999). Fungal production of eicosapentaenoic and arachidonic acids from industrial waste streams and crude soybean oil. *Bioresour. Technol.* 67: 101–110.

Chi, Z., Zheng, Y., Ma, J., and Chen, S. (2011). Oleaginous yeast Cryptococcus curvatus culture with dark fermentation hydrogen production effluent as feedstock for microbial lipid production. *Int. J. Hydrogen Energy* 36: 9542–9550.

Chodchoey, K. and Verduyn, C. (2012). Growth, fatty acid profile in major lipid classes and lipid fluidity of Aurantiochytrium mangrovei Sk-02 as a function of growth temperature. *Braz. J. Microbiol.* 43: 187–200.

Choi, S.Y., Ryu, D.D.Y., and Rhee, J.S. (1982). Production of microbial lipid: effects of growth rate and oxygen on lipid synthesis and fatty acid composition of *Rhodotorula gracilis*. *Biotechnol. Bioeng.* 24: 1165–1172.

Chutrakul, C., Jeennor, S., Panchanawaporn, S. et al. (2016). Metabolic engineering of long chain-polyunsaturated fatty acid biosynthetic pathway in oleaginous fungus for dihomo-gamma linolenic acid production. *J. Biotechnol.* 218: 85–93.

Davies, R.J., Holdsworth, J., and Reader, S. (1990). The effect of low oxygen uptake rate on the fatty acid profile of the oleaginous yeast *Apiotrichum curvatum*. *Appl. Microbiol. Biotechnol.* 33: 569–573.

De Swaaf, M.E., de Rijk, T.C., Eggink, G., and Sijtsma, L. (1999). Optimisation of docosahexaenoic acid production in batch cultivations by *Crypthecodinium cohnii*. *J. Biotechnol.* 70: 185–192.

De Swaaf, M.E., Sijtsma, L., and Pronk, J.T. (2003). High-cell-density fed-batch cultivation of the docosahexaenoic acid producing marine alga *Crypthecodinium cohnii*. *Biotechnol. Bioeng.* 81: 666–672.

Demir, M., Turhan, I., Kucukcetin, A., and Alpkent, Z. (2013). Oil production by *Mortierella isabellina* from whey treated with lactase. *Bioresour. Technol.* 128: 365–369.

Dick, M.W. (2001). Systematics. In: *Straminipolous Fungi: Systematics of the Peronosporomyletes* (ed. N.L. Dordrech), 267–432. Kluwer Academic Publishers.

Dunbar, B.S., Bosire, R.V., and Deckelbaum, R.J. (2014). Omega 3 and omega 6 fatty acids in human and animal health: an African perspective. *Mol. Cell. Endocrinol.* 398: 69–77.

Dyal, S.D. and Narine, S.S. (2005). Implications for the use of Mortierella fungi in the industrial production of essential fatty acids. *Food Res. Int.* 38: 445–467.

Fakas, S., Galiotou-Panayotou, M., Papanikolaou, S. et al. (2007). Compositional shifts in lipid fractions during lipid turnover in Cunninghamella echinulata. *Enzym. Microb. Technol.* 40: 1321–1327.

Fakas, S., Čertik, M., Papanikolaou, S. et al. (2008). γ-Linolenic acid production by Cunninghamella echinulata growing on complex organic nitrogen sources. *Bioresour. Technol.* 99: 5986–5990.

Fakas, S., Papanikolaou, S., Batsos, A. et al. (2009). Evaluating renewable carbon sources as substrates for single cell oil production by Cunninghamella echinulata and *Mortierella isabellina*. *Biomass Bioenergy* 33: 573–580.

Fan, K.-W. and Chen, F. (2007). Production of high-value products by marine microalgae Thraustochytrids. In: *Bioprocessing for Value-Added Products from Renewable Resources* (ed. S.-T. Yang), 293–323. Elsevier.

Fei, Q., Chang, H.N., Shang, L. et al. (2011). The effect of volatile fatty acids as a sole carbon source on lipid accumulation by *Cryptococcus albidus* for biodiesel production. *Bioresour. Technol.* 102: 2695–2701.

Fernandes, B.D., Mota, A., Teixeira, J.A., and Vicente, A.A. (2015). Continuous cultivation of photosynthetic microorganisms: approaches, applications and future trends. *Biotechnol. Adv.*.

Fontanille, P., Kumar, V., Christophe, G. et al. (2012). Bioconversion of volatile fatty acids into lipids by the oleaginous yeast *Yarrowia lipolytica*. *Bioresour. Technol.* 114: 443–449.

Gao, D., Zeng, J., Zheng, Y. et al. (2013). Microbial lipid production from xylose by *Mortierella isabellina*. *Bioresour. Technol.* 133: 315–321.

Gerber, M. (2012). Omega-3 fatty acids and cancers: a systematic update review of epidemiological studies. *Br. J. Nutr.* 107: S228–S239.

Gong, Y., Wan, X., Jiang, M. et al. (2014). Metabolic engineering of microorganisms to produce omega-3 very long-chain polyunsaturated fatty acids. *Prog. Lipid Res.* 56: 19–35.

Gorissen, L., De Vuyst, L., Raes, K. et al. (2012). Conjugated linoleic and linolenic acid production kinetics by bifidobacteria differ among strains. *Int. J. Food Microbiol.* 155: 234–240.

Guichardant, M., Chen, P., Liu, M. et al. (2011). Functional lipidomics of oxidized products from polyunsaturated fatty acids. *Chem. Phys. Lipids* 164: 544–548.

Gupta, A., Abraham, R.E., Barrow, C.J., and Puri, M. (2015). Omega-3 fatty acid production from enzyme saccharified hemp hydrolysate using a novel marine thraustochytrid strain. *Bioresour. Technol.* 184: 373–378.

Hildyard, J.C.W. and Halestrap, A.P. (2003). Identification of the mitochondrial pyruvate carrier in Saccharomyces cerevisiae. *Biochem. J.* 374: 607–611.

Horrobin, D.F. (1992). Nutritional and medical importance of gamma-linolenic acid. *Prog. Lipid Res.* 31: 163–194.

Huang, T.Y., Lu, W.C., and Chu, I.M. (2012). A fermentation strategy for producing docosahexaenoic acid in *Aurantiochytrium limacinum* SR21 and increasing C22:6 proportions in total fatty acid. *Bioresour. Technol.* 123: 8–14.

Jakobsen, A.N., Aasen, I.M., Josefsen, K.D., and Strøm, A.R. (2008). Accumulation of docosahexaenoic acid-rich lipid in thraustochytrid Aurantiochytrium sp. strain T66: effects of N and P starvation and $O_2$ limitation. *Appl. Microbiol. Biotechnol.* 80: 297–306.

Jang, H. and Yang, S. (2008). Polyunsaturated fatty acids production with a solid-state column reactor. *Bioresour. Technol.* 99: 6181–6189.

Ji, X.-J., Zhang, A.-H., Nie, Z.-K. et al. (2014). Efficient arachidonic acid-rich oil production by *Mortierella alpina* through a repeated fed-batch fermentation strategy. *Bioresour. Technol.* 170: 356–360.

Jiang, Y. and Chen, F. (2000). Effects of medium glucose concentration and pH on docosahexaenoic acid content of heterotrophic *Crypthecodinium cohnii*. *Process Biochem.* 35: 1205–1209.

Jiang, Y., Chen, F., and Liang, F.Z. (1999). Production potential of docosahexaenoic acid by the heterotrophic marine dinoflagellate Crypthecodinium cohnii. *Process Biochem.* 34: 633–637.

Johansson, M., Chen, X., Milanova, S. et al. (2016). PUFA-induced cell death is mediated by Yca1p-dependent and -independent pathways, and is reduced by vitamin C in yeast. *FEMS Yeast Res.* 16: fow007.

Kendrick, A. and Ratledge, C. (1992). Lipids of selected molds grown for production of n−3 and n−6 polyunsaturated fatty acids. *Lipids* 27: 15–20.

Kohli, P. and Levy, B.D. (2009). Resolvins and protectins: mediating solutions to inflammation: chemical mediators of inflammation resolution. *Br. J. Pharmacol.* 158: 960–971.

Lee Chang, K.J., Paul, H., Nichols, P.D. et al. (2015). Australian thraustochytrids: potential production of dietary long-chain omega-3 oils using crude glycerol. *J. Funct. Foods*.

Li, J., Liu, R., Chang, G. et al. (2015). A strategy for the highly efficient production of docosahexaenoic acid by *Aurantiochytrium limacinum* SR21 using glucose and glycerol as the mixed carbon sources. *Bioresour. Technol.* 177: 51–57.

Liang, Y., Sarkany, N., Cui, Y. et al. (2010). Use of sweet sorghum juice for lipid production by *Schizochytrium limacinum* SR21. *Bioresour. Technol.* 101: 3623–3627.

Lin, Y.-H., Chang, F.-L., Tsao, C.-Y., and Leu, J.-Y. (2007). Influence of growth phase and nutrient source on fatty acid composition of *Isochrysis galbana* CCMP 1324 in a batch photoreactor. *Biochem. Eng. J.* 37: 166–176.

Ling, X., Guo, J., Liu, X. et al. (2015). Impact of carbon and nitrogen feeding strategy on high production of biomass and docosahexaenoic acid (DHA) by Schizochytrium sp. LU310. *Bioresour. Technol.* 184: 139–147.

Liu, B., Liu, J., Sun, P. et al. (2015). Sesamol enhances cell growth and the biosynthesis and accumulation of docosahexaenoic acid in the microalga *Crypthecodinium cohnii*. *J. Agric. Food Chem.* 63: 5640–5645.

Lopez-Huertas, E. (2010). Health effects of oleic acid and long chain omega-3 fatty acids (EPA and DHA) enriched milks. A review of intervention studies. *Pharmacol. Res.* 61: 200–207.

Mao, X., Liu, Z., Sun, J., and Lee, S.Y. (2017). Metabolic engineering for the microbial production of marine bioactive compounds. *Biotechnol. Adv.* In press.

Meesters, P.A.E.P., Huijberts, G.N.M., and Eggink, G. (1996). High-cell-density cultivation of the lipid accumulating yeast Cryptococcus curvatus using glycerol as a carbon source. *Appl. Microbiol. Biotechnol.* 45: 575–579.

Mlickova, K., Roux, E., Athenstaedt, K. et al. (2004). Lipid accumulation, lipid body formation, and acyl coenzyme a oxidases of the yeast *Yarrowia lipolytica*. *Appl. Environ. Microbiol.* 70: 3918–3924.

Muniraj, I.K., Xiao, L., Hu, Z. et al. (2013). Microbial lipid production from potato processing wastewater using oleaginous filamentous fungi Aspergillus oryzae. *Water Res.* 47: 3477–3483.

Murphy, D.J. (2001). The biogenesis and functions of lipid bodies in animals, plants and microorganisms. *Prog. Lipid Res.* 40: 325–438.

Nagao, K. and Yanagita, T. (2005). Conjugated fatty acids in food and their health benefits. *J. Biosci. Bioeng.* 100: 152–157.

Ochsenreither, K., Glück, C., Stressler, T. et al. (2016). Production strategies and applications of microbial single cell oils. *Front. Microbiol.* 7: 1539.

Papanikolaou, S. (2004). Single cell oil (SCO) production by *Mortierella isabellina* grown on high-sugar content media. *Bioresour. Technol.* 95: 287–291.

Papanikolaou, S., Galiotou-Panayotou, M., Fakas, S. et al. (2007). Lipid production by oleaginous Mucorales cultivated on renewable carbon sources. *Eur. J. Lipid Sci. Technol.* 109: 1060–1070.

Patil, K.P. and Gogate, P.R. (2015). Improved synthesis of docosahexaenoic acid (DHA) using *Schizochytrium limacinum* SR21 and sustainable media. *Chem. Eng. J.* 268: 187–196.

Pei, G., Li, X., Liu, L. et al. (2017). De novo transcriptomic and metabolomic analysis of docosahexaenoic acid (DHA)-producing *Crypthecodinium cohnii* during fed-batch fermentation. *Algal Res.* 26: 380–391.

Pyle, D.J., Garcia, R.A., and Wen, Z. (2008). Producing docosahexaenoic acid (DHA)-rich algae from biodiesel-derived crude glycerol: effects of impurities on DHA production and algal biomass composition. *J. Agric. Food Chem.* 56: 3933–3939.

Qi, F., Zhang, M., Chen, Y. et al. (2017). A lignocellulosic hydrolysate-tolerant Aurantiochytrium sp. mutant strain for docosahexaenoic acid production. *Bioresour. Technol.* 227: 221–226.

Qiu, X. (2003). Biosynthesis of docosahexaenoic acid (DHA, 22:6-4, 7,10,13,16,19): two distinct pathways. *Prostaglandins Leukot. Essent. Fatty Acids* 68: 181–186.

Qu, L., Ren, L.-J., and Huang, H. (2013). Scale-up of docosahexaenoic acid production in fed-batch fermentation by Schizochytrium sp. based on volumetric oxygen-transfer coefficient. *Biochem. Eng. J.* 77: 82–87.

Ratledge, C. (2004). Fatty acid biosynthesis in microorganisms being used for single cell oil production. *Biochimie* 86: 807–815.

Ratledge, C. and Lippmeier, C. (2017). Microbial production of fatty acids. In: *Fatty Acids* (ed. M. Ahmad), 237–278. Elsevier.

Ratledge, C. and Wynn, J.P. (2002). The biochemistry and molecular biology of lipid accumulation in oleaginous microorganisms. *Adv. Appl. Microbiol.* 51: 1–51.

Ren, L.-J., Ji, X.-J., Huang, H. et al. (2010). Development of a stepwise aeration control strategy for efficient docosahexaenoic acid production by Schizochytrium sp. *Appl. Microbiol. Biotechnol.* 87: 1649–1656.

Ren, L.J., Sun, X.M., Ji, X.J. et al. (2017). Enhancement of docosahexaenoic acid synthesis by manipulation of antioxidant capacity and prevention of oxidative damage in Schizochytrium sp. *Bioresour. Technol.* 223: 141–148.

Rosa, S.M., Soria, M.A., Vélez, C.G., and Galvagno, M.A. (2010). Improvement of a two-stage fermentation process for docosahexaenoic acid production by *Aurantiochytrium limacinum* SR21 applying statistical experimental designs and data analysis. *Bioresour. Technol.* 101: 2367–2374.

Rubio-Rodríguez, N., Beltrán, S., Jaime, I. et al. (2010). Production of omega-3 polyunsaturated fatty acid concentrates: a review. *Innov. Food Sci. Emerg. Technol.* 11: 1–12.

Ryu, B.-G., Kim, J., Kim, K. et al. (2013). High-cell-density cultivation of oleaginous yeast Cryptococcus curvatus for biodiesel production using organic waste from the brewery industry. *Bioresour. Technol.* 135: 357–364.

Safdar, W., Shamoon, M., Zan, X. et al. (2017). Growth kinetics, fatty acid composition and metabolic activity changes of Crypthecodinium cohnii under different nitrogen source and concentration. *AMB Expr.* 7: 85–100.

Sakamoto, T., Sakuradani, E., Okuda, T. et al. (2017). Metabolic engineering of oleaginous fungus *Mortierella alpina* for high production of oleic and linoleic acids. *Bioresour. Technol.* 245: 1610–1615.

Sakuradani, E. and Shimizu, S. (2009). Single cell oil production by *Mortierella alpina*. *J. Biotechnol.* 144: 31–36.

Schweizer, E. and Hofmann, J. (2004). Microbial type I fatty acid synthases (FAS): major players in a network of cellular FAS systems. *Microbiol. Mol. Biol. Rev.* 68: 501–517.

Shen, H., Gong, Z., Yang, X. et al. (2013). Kinetics of continuous cultivation of the oleaginous yeast Rhodosporidium toruloides. *J. Biotechnol.* 168: 85–89.

Shene, C., Chisti, Y., Vergara, D. et al. (2016). Production of eicosapentaenoic acid by Nannochloropsis oculata: effects of carbon dioxide and glycerol. *J. Biotechnol.* 239: 47–56.

Shimizu, S., Shinmen, Y., Kawashima, H. et al. (1988). Fungal mycelia as novel source of eicosapentaenoic acid. *Biochem. Biophys. Res. Commun.* 150: 335–341.

Song, X., Zhang, X., Kuang, C. et al. (2010). Batch kinetics and modeling of DHA production by S. limacinum OUC88. *Food Bioprod. Process.* 88: 26–30.

Song, X., Zang, X., and Zhang, X. (2015). Production of high Docosahexaenoic acid by Schizochytrium sp. using low-cost raw materials from food industry. *J. Oleo Sci.* 64: 197–204.

Sudheendran, S., Chang, C.C., and Deckelbaum, R.J. (2010). N-3 vs. saturated fatty acids: effects on the arterial wall. *Prostaglandins Leukot. Essent. Fat. Acids (PLEFA)* 82: 205–209.

Sun, L., Ren, L., Zhuang, X. et al. (2014). Differential effects of nutrient limitations on biochemical constituents and docosahexaenoic acid production of Schizochytrium sp. *Bioresour. Technol.* 159: 199–206.

Sun, X.-M., Ren, L.-J., Ji, X.-J. et al. (2016). Adaptive evolution of Schizochytrium sp. by continuous high oxygen stimulations to enhance docosahexaenoic acid synthesis. *Bioresour. Technol.* 211: 374–381.

Tang, X., Chen, H., Chen, Y.Q. et al. (2015). Comparison of biochemical activities between high and low lipid-producing strains of Mucor circinelloides: an explanation for the high oleaginicity of strain WJ11. *PLoS One* 10: e0128396.

Tehlivets, O., Scheuringer, K., and Kohlwein, S.D. (2007). Fatty acid synthesis and elongation in yeast. *Biochim. Biophys. Acta BBA - Mol. Cell Biol. Lipids* 1771: 255–270.

Terano, T., Fujishiro, S., Ban, T. et al. (1999). Docosahexaenoic acid supplementation improves the moderately severe dementia from thrombotic cerebrovascular diseases. *Lipids* 34: S345–S346.

Tocher, D.R. (2015). Omega-3 long-chain polyunsaturated fatty acids and aquaculture in perspective. *Aquaculture* 449: 94–107.

Unagul, P., Assantachai, C., Phadungruengluij, S. et al. (2007). Coconut water as a medium additive for the production of docosahexaenoic acid (C22:6 n3) by Schizochytrium mangrovei Sk-02. *Bioresour. Technol.* 98: 281–287.

Wasylenko, T.M., Ahn, W.S., and Stephanopoulos, G. (2015). The oxidative pentose phosphate pathway is the primary source of NADPH for lipid overproduction from glucose in *Yarrowia lipolytica*. *Metab. Eng.* 30: 27–39.

Whelan, J. (2008). The health implications of changing linoleic acid intakes. *Prostaglandins Leukot. Essent. Fatty Acids* 79: 165–167.

Wynn, J.P. and Ratledge, C. (1997). Malic enzyme is a major source of NADPH for lipid accumulation by Aspergillus Nidulans. *Microbiology* 143: 253–257.

Wynn, J.P., Hamid, A.b.A., and Ratledge, C. (1999). The role of malic enzyme in the regulation of lipid accumulation in filamentous fungi. *Microbiology* 145: 1911–1917.

Wynn, J.P., Hamid, A.A., and Ratledge, C. (2001). Biochemical events leading to the diversion of carbon into storage lipids in the oleaginous fungi Mucor circinelloides and Mortierella alpina. *Microbiol* 147: 2857–2864.

Xie, Y. and Wang, G. (2015). Mechanisms of fatty acid synthesis in marine fungus-like protists. *Appl. Microbiol. Biotechnol.* 99: 8363–8375.

Xing, D., Wang, H., Pan, A. et al. (2012). Assimilation of corn fiber hydrolysates and lipid accumulation by *Mortierella isabellina*. *Biomass Bioenergy* 39: 494–501.

Yang, X., Jin, G., Gong, Z. et al. (2015). Recycling microbial lipid production wastes to cultivate oleaginous yeasts. *Bioresour. Technol.* 175: 91–96.

Ykema, A., Verbree, E.C., van Verseveld, H.W., and Smit, H. (1986). Mathematical modelling of lipid production by oleaginous yeasts in continuous cultures. *Antonie Van Leeuwenhoek* 52: 491–506.

Ykema, A., Verbree, E.C., Kater, M.M., and Smit, H. (1988). Optimization of lipid production in the oleaginous yeast *Apiotrichum curvatum* in whey permeate. *Appl. Microbiol. Biotechnol.* 29: 211–218.

Yoshioka, M., Yago, T., Yoshie-Stark, Y. et al. (2012). Effect of high frequency of intermittent light on the growth and fatty acid profile of *Isochrysis galbana*. *Aquaculture* 338–341: 111–117.

Zeng, J., Zheng, Y., Yu, X. et al. (2013). Lignocellulosic biomass as a carbohydrate source for lipid production by *Mortierella isabellina*. *Bioresour. Technol.* 128: 385–391.

Zhang, L., Zhao, H., Lai, Y. et al. (2013). Improving docosahexaenoic acid productivity of Schizochytrium sp. by a two-stage AEMR/shake mixed culture mode. *Bioresour. Technol.* 142: 719–722.

Zhao, L., Zhang, H., Wang, L. et al. (2015). 13C-metabolic flux analysis of lipid accumulation in the oleaginous fungus Mucor circinelloides. *Bioresour. Technol.* 197: 23–29.

Zhu, L., Zhang, X., Ji, L. et al. (2007). Changes of lipid content and fatty acid composition of *Schizochytrium limacinum* in response to different temperatures and salinities. *Process Biochem.* 42: 210–214.

# 9

# Research and Production of Organic Acids and Industrial Potential

*Sandeep Kumar Panda[1], Lopamudra Sahu[2], Sunil Kumar Behera[3], and Ramesh Chandra Ray[4]*

[1] *School of Biotechnology, KIIT University, Bhubaneswar, 751024, India*
[2] *Department of Botany, Utkal University, Bhubaneswar, 751004, India*
[3] *Department of Bioscience & Bioinformatics, Khallikote University, Berhampur, Odisha, 760001, India*
[4] *ICAR-Regional Center of Central Tuber Crops Research Institute, Bhubaneswar, 751019, India*

## 9.1 Introduction: History and Current Trends

Organic acids have long been used in their natural forms. These are low-molecular-weight compounds that contain one or more carboxyl groups and are observed in almost every microorganism. They have numerous applications in industries related to biocommodities such as food, cosmetics, surfactants, and textile industries (Panda et al. 2016). Production of organic acids for commercial exploitation is carried out either by chemical synthesis or fermentation, with preference for fermentation because of its high yield in the microbiological processes. The initial production practices of acetic acid date back to 1823 and citric acid to 1913. But commercial fermentation of citric acid production via microbe-assisted processing started around 1920. Over the years myriad organic acids of biosynthetic origin have been produced, most of which are natural bioproducts of microbes, or may be expected as intermediary in important metabolic pathways (Alonso et al. 2015). Because of their functional groups such as keto and hydroxyl groups, organic acids are enormously valuable as initial substances for chemical industries. Besides their traditional usage in food, feed, and pharmaceuticals of late, production of organic acids as monomers with bifunctional characteristic has been triggered due to the resurgence of bioplastics. Indulgence of recombinant DNA technology and metabolic engineering techniques have efficiently engineered organic acid-producing microorganisms to enhance and speed up process developments for preferred bioproducts at high concentration, quality, and productivity. Even novel microbial isolates demonstrating potential production properties of interest are effectively engineered for the overproduction of organic acid on a commercial scale (Li and Borodina 2015; Cho et al. 2015).

This chapter provides an overview of the latest developments in organic acid production, its market, and its future.

*Bioprocessing for Biomolecules Production*, First Edition.
Edited by Gustavo Molina, Vijai Kumar Gupta, Brahma N. Singh, and Nicholas Gathergood.

## 9.2 Current and Future Markets for Organic Acids

Production of organic acids through microbial processing has long been traditionally practiced. With the increasing demand for more sustainable biobased processes for fuel, biochemical, and material production, market expansion for microbe-assisted organic acid production has been triggered owing to the fact that they constitute to key building block in chemicals produced from renewable resources. In 2015, the market share of organic acids was at US$6.55 billion with expectations that it could reach as high as US$9.29 billion by 2021, at 6.0% AAGR (average annual growth rate) (during the period (Markets and Markets 2015).

In 2015, the largest market for organic acids was the Asia-Pacific region. China is the largest manufacturer of organic acids owing to the maintenance of its large livestock population. It is the leading market for acetic acid production, which plays a vital role in the production of vinyl, polyesters, and engineering plastics. With the improvisation of small- and medium-scale industries in India, its market is also predicted to rise at the maximum CAGR (compound annual growth rate) (Vandenberghe et al. 2018).

Though both chemical and microbial processes are used to produce all the organic acids, microbial fermentation processes have been preferred due to their higher yields by using highly productive strains. The organic acid with highest annual production is citric acid at 1.6 million tons by AAGR of 3.5–4.0%, followed by acetic and lactic acid with 0.19 and 0.15 million tons, respectively (Vandenberghe et al. 2018). The manufacture of succinic acid is anticipated to increase to more than 7.0 million tons by the year 2020 due to the fact that it is a starting chemical for the production of various valuable chemical compounds (Choi et al. 2015).

The demand for green plastic has driven researchers toward bio-based production of organic acids for, e.g., adipic acid (AA), which is among the prime building blocks in plastic industries with a market of approximately 3 million tons per annum (Yu et al. 2014; Polen et al. 2013). Of late, itaconic acid has gained importance as a chemical building block for its diversified utilities in surfactants, coatings, and rubber industry (Huang et al. 2014). It has a yearly market of around 80 000 tons, mostly produced by fermentation using fungus, i.e., *Aspergillus* sp. producers found in nature (Choi et al. 2015; Werpy and Petersen 2004). But it is estimated that market sales will go far beyond 410 000 tons per year by 2020 (Choi et al. 2015). Predominantly, the translation of itaconic acid to methyl methacrylate, also called Plexiglass, has a vast market prospective, reaching 3.2 million tons (Choi et al. 2015).

## 9.3 Types of Organic Acids

Organic acids are segregated into various clusters. Common organic acids constitutes the carboxylic acids, containing one or more carboxyl groups (–COOH), for example acetic acid ($CH_3COOH$), possess one carboxyl group. Malic acid, or cis-butenedioic acid ($C_4H_4O_4$), has two carboxyl groups, and citric acid ($C_6H_8O_7$) bears three carboxyl groups. As most of the organic acids are weak acids, in aqueous solutions they work as a buffer, and this property is used in the production of food and feed (Panda and Ray 2015).

Organic acids and their derived products have also been used as antioxidative agents or synergists. Phenolics, such as gallic acid (Pokorny et al. 2001; Aruoma et al. 1993; Yen et al. 2002) or ferulic acid (Pokorny et al. 2001) and derivatives are the most important antioxidants. Ascorbic acid and citric acid along with their derivatives are regarded as potential antioxidants (Pokorny et al. 2001; Yen et al. 2002). The role of organic acids such as ascorbic acid (Arrigoni and De Tullio 2002) and lactobionic acid (Alonso et al. 2013; Gutiérrez et al. 2012) as health-promoting ingredients has been quite promising for the food industry.

Organic acids can be used for both food preservatives and also as leavening agents in bakery products (Leavening agent 2013). The popularly used organic acids such as acetic, propionic, lactic, and citric acids are generally regarded as safe (GRAS) and are used for miscellaneous purposes. Acetic acid, the key constituent of vinegar, is mostly applicable for its flavor enhancing attributes (Mani-López et al. 2012) and is also employed as a feedstock in production of vinyl acetate monomer (Rasrendra et al. 2011). Propionic acid is an intermediate organic compound with applications in the manufacture of cellulose fibers, herbicides etc. (Zhang and Yang 2009). The sodium, potassium, and calcium of propionic acid inhibit mold growth and find its usage in food and feed preservatives (Silva and Lidon 2016). Lactic acid is used as a flavoring agent and color stabilizer, while citric acid has applications in food and beverage manufacturing units as a potential acidulant. A few others, like pyruvic acid, are extensively used as food additives, weight managing supplements, in cosmetics, and as preparatory elements for the biosynthesis of pharmaceuticals such as L-tyrosine, N-acetyl-D-neuraminic acid, and (R)-phenylacetylcarbinol (Song et al. 2016); succinic acid is applied in innumerable ways, such as surfactant, foaming agents, and an ion chelator. Also, it has applications in the food and health-promoting industrial sectors (Sauer et al. 2008). Most organic acids are naturally obtained in the form of intermediates of important metabolic pathways of microorganisms. Various fungal as well as bacterial species are utilized for the commercial production of organic acids. These include the fungus *Rhizopus* sp., *Aspergillus* sp., *Penicillium* sp., *Yarrowia lipolytica* and related yeast species, and the bacteria *Arthrobacter paraffinensis, Bacillus* sp., *Lactobacillus* sp., *Streptococcus thermophillus,* have been used for the production of organic acids (Shaikh and Qureshi 2013). Sucrose, glucose, fructose, maltose, xylose, whey lactose, and glycerol are commonly used carbon sources for organic acid production (Liu et al. 2012). The fermentation of different organic acids is discussed below.

### 9.3.1 Citric Acid

Citric acid (CA), an intermediary compound of tricarboxylic acid or TCA cycle (Swain et al. 2011), has wide applications, ranging from food and beverages to polymer synthesis and cosmetics (Soccol et al. 2006). Though initially CA bioproduction was carried out by surface fermentation, later on the commercial production was conducted by submerged fermentation of starch or sucrose-based medium by *Aspergillus niger*. Recently, Panda et al. 2016 reviewed the production of citric acids from fruit and vegetable wastes where in *A. niger* has shown higher yields than other strains, as most of them do not satisfy the commercial conditions. CA being an intermediary biochemical in TCA cycle, it is amassed in higher levels in drastic conditions (Soccol et al. 2006). *A. niger,* a GRAS organism, remains the best choice for commercial production because (i) it is easy to handle and harvest; (ii) it has the potential to exploit a diversity of cheap substrates and agro-wastes; and (iii) it has high rates of production (Dhillon et al. 2011).

Currently, the major quantity of CA, i.e., around 80% of the total demand, is met by *A. niger* in submerged fermentation. The accumulation of CA is ascertained under ideal conditions of high concentrations of sugar, acidity, dissolved oxygen, and aeration. Also, the appropriate concentration of trace metals, nitrogen, and phosphate are factors affecting the accumulation of CA.

### 9.3.2 Acetic Acid

Acetic acid used in vinegar worldwide is fermented both by aerobic and anaerobic processes. The concentration of acetic acid in table vinegar varies from 4% to 8% depending on the regulation of the producing country. *Clostridium* and *Acetobacterium* are known for the conversion of glucose to acetic acid in a single-step anoxic microbial processing. However, food-grade acetic acid production is mostly produced through oxidative fermentation using *Acetobacter*. The final quality of acetic acid depends on the cumulative effect of the bacterial isolate, wood contact, and aging process.

Silva et al. (2007) illustrated the manufacture of acetic acid using cashew apple wine as substrate in a stirred batch reactor. Optimization was done in a $2^2$ factorial experimental design. The highest productivity obtained was 0.55 $g\,lh^{-1}$ and the yield was more than 75%. Similarly, in another study, Awad et al. (2012) carried out submerged fermentation using *Acetobacter aceti* in media of different compositions. The standardized broth formulation was composed of glucose, 100 $g\,l^{-1}$; yeast extract, 12 $g\,l^{-1}$ and peptone 5 $g\,l^{-1}$ and could produce 53 $g\,l^{-1}$ acetic acid after 144 hours of fermentation. Further, fermentation was conducted in the optimized formulation in a pilot scale-16L stirred tank bioreactor. Under fully aerobic and controlled pH conditions, the production of acetic acid was 76 $g\,l^{-1}$. Effect of Pyrroquinoline quinone-dependent alcohol dehydrogenase (PQQ-ADH), which is an enzyme of the ethanol oxidase respiratory chain of acetic acid bacteria, was investigated for the production of acetic acid by *Acetobacter pasteurianus*. Subunits I (adhA) and II (adhB) of PQQ-ADH were overexpressed and the engineered strain showed enhanced acetic acid production with lower residual ethanol.

Along with the high production and demand of acetic acid, it has become quite essential to reform the current production technologies to eco-friendly mode as the longstanding conventional practices involve chemical routes and also huge energy-consuming operations such as distillation, evaporation, absorption, and crystallization (Pal and Nayak 2017).

### 9.3.3 Propionic Acid (PA)

Propionic acid is a key transitional chemical of metabolic cycles, and it applies in the manufacture of cellulose fibers, herbicides, and perfumes (Hsu and Yang 1991). It is conventionally produced via petrochemical routes. Nevertheless, fermentation from renewable resources has become a substitute technology for the generation of PA. With the shift from relying on fossil fuels to seeking more sustainable production, researchers have turned their attention to microbial fermentation. Species of the *Propionibacteria* used in the microbial fermentation of PA include *Propionibacterium acidipropionici, P. freudenreichii* subsp. *freudenreichii,* and *P. freudenreichii* subs.,

*P. shermanii*. Propionibacteria can use various chemicals, such as glucose, fructose, xylose, lactate, whey lactose, hemicellulose, and glycerol as its carbon sources for PA production. Though various PA fermentation methods are used, in the past few decades batch and fed-batch culture methods have significantly improved PA production (Liu et al. 2012). Rasrendra et al. (2011) reported the highest PA production of 136 $g\,l^{-1}$, from constant, fed-batch fermentation in a plant fibrous bed bioreactor (FBB) with immobilized *Propionibacterium freudenreichii* (Rasrendra et al. 2011) and it is found to be an effective bioprocesses for generation of organic acids (Wallenius et al. 2015). Jiang et al. (2015) also reported that by using a developed FBB, a continuous production of PA can be achieved with immobilized *P. acidipropionici* in the reactor using renewable carbon substrates such as sugarcane bagasse hydrolysate or whey lactose.

### 9.3.4 Succinic Acid

Succinic acid (1,4-butanedioic acid) is a four-carbon dicarboxylic acid and is a component of almost all living tissues and microorganisms. Succinic acid (Latin *succinum*, amber) was first isolated in 1546 by Agricola through dry distillation of amber. It has uses in the production of polymers, clothing fibers, plasticizers, solvents, and paints, for example, and in various other commercial products.

Though chemical methods were used for succinic acid production in the past, microbial fermentation processes have shown comparatively higher yields using fungi, yeast, and Gram-positive bacteria from various carbon sources (Song and Lee 2006). *Anaerobiospirillum succiniciproducens* and *Actinobacillus succinogenes* are the most preferred organisms (Lee et al. 2001). Various strains such as *Escherichia coli, Lactobacillus plantarum* have been metabolically engineered to produce succinic acid (Jantama et al. 2008; Tsuji et al. 2013). Industrial systems biology tools have been used to convert *Saccharomyces cerevisiae* into a novel SA cell factory (Otero et al. 2013). Glucose, glycerol, sucrose, maltose, lactose, and fructose are used as carbon resources for SA production. Jiang et al. (2010) reported the highest SA production through dual-phase fermentation by metabolically improved *E. coli* using a conjugation of aerobic growth and anaerobic production to the extent of 101.2 $g\,l^{-1}$. *Y. lipolytica,* which is one of the best-studied strains for overproduction of organic acids, has been utilized for SA by redirecting metabolic pathways. Yuzbashev et al. (2010) constructed a temperature-susceptible mutant with a damaged SDH1 (succinate dehydrogenase 1) followed by the deletion of SDH2 gene to enhance the cell viability. Chemical mutagenesis was applied, leading to a maximum SA concentration of 45.5 $g\,l^{-1}$ with pH control and 17.4 $g\,l^{-1}$ without pH control. Jost et al. (2015) reported that the alteration of the native promoter of the SHD2 gene by inducible promoters resulted in accumulation of more SA from glycerol by the recombinant strains. Genetic engineering was applied to strains of *E. coli* and *Corynebacterium glutamicum* for obtaining improved production (up to 127 $g\,l^{-1}$ in *E. coli* and 146 $g\,l^{-1}$ in *C. glutamicum*) (Becker and Wittmann 2015).

Shortage in resources and environment concerns has prompted researchers to develop new technologies and strategies that are being adopted for enhancing the yield of SA. Uses of cheap raw materials such as agriculture and industrial residues over costly carbon sources have tremendously lowered the fermentation costs. Moreover, SA as a fermentation product is an eco-friendly technology involving consumption of $CO_2$ during the process.

### 9.3.5 Lactic Acid

Lactic acid (LA) is a three-carbon (2-hydroxypropionic acid) carboxylic acid present in plants and microorganisms. It is a widely available organic acid and is generated by lactic acid bacteria, which have a long history in the food industry and microbial processes over the last century (Magnuson and Lasure 2004).Worldwide production of LA is mostly contributed by bacterial fermentation, which is more eco-friendly with superior yields, comparatively fast, and produces one of the two stereoisomers of LA as well as their racemic mixture (Axelsson 2004).

LA bacteria require complex nutrients for growth and other metabolic activities, and they ferment sugars through various pathways. As a consequence, different patterns of fermentations are observed, such as homo-, hetero-, or mixed-acid fermentation (Hofvendahl and Hahn-Hagerdal 2000); homo-fermentation leading to nearly 90% pure LA (Vijayakumar et al. 2008).

LA is safe for human consumption and as a food additive. It has been utilized for the manufacture of eco-friendly biocompatible polylactate polymers such as polylactic acid (Li et al. 2015) that has boosted the commercial exploration of this acid.

The filamentous fungus *Rhizopus oryzae* is also a natural producer (Magnuson and Lasure 2004, Koutinas et al. 2007,) and can grow on mineral medium and carbon sources such as starch or xylose. Large-scale production of LA has been carried out using bacterial as well as fungal strains. Agro-industrial waste residues in bioprocessing are mostly preferred to reduce the cost of raw materials. Fundamental to selecting the appropriate microbes rendering high productivity, the optimal fermentation conditions, such as, temperature, pH, aeration, and agitation are also vital for high yields.

Traditional lactic acid production using lactic acid bacteria is not cost-effective and feasible for commercial-scale manufacturing due to the nutritionally rich media and maintaining adequate pH conditions (Sauer et al. 2008). Immobilized *R. oryzae* have been manipulated for LA manufacture in fed-batch culture to a high titer of $231\,g\,l^{-1}$ of LA with yield of 0.92 g/g glucose and $1.83\,g\,(l\,h^{-1})^{-1}$ (Yamane and Tanaka 2013). *R. oryzae* has not been commercially viable due to aeration requirement and maintenance of pH above 4.5, thereby increasing the costs of operation, for efficient production. Lee et al. (2015) explored various expressions of genes and metabolic pathways of *S. cerevisiae* for LA production that resulted in a strain capable of generating $117\,g\,l^{-1}$ of L-LA in a fed-batch mode with pH maintained at 3.5.

Recently, Ozaki et al. (2017) designed a metabolically engineered *Schizosaccharomyces pombe* strain using a CRISPR-Cas9 system for production of D-LA from glucose and cellobiose. The engineered strain could significantly produce D-LA at $25.2\,g\,l^{-1}$ from $35.5\,g\,l^{-1}$ of glucose utilized and a yield rate of 0.71 g D-LA/g glucose was recorded. Though lactic acid bacteria produced LA with higher yield (95% >) and titer ($100\,g\,l^{-1}$>) (Sauer et al. 2017), however, *S. pombe* exhibited high resistance to low extracellular pH, and it was found to be advantageous over the former for the production of organic acid (Suyama et al. 2017).

### 9.3.6 Other Organic Acids

Various other organic acids have also been produced, such as $\alpha$-ketoglutarate ($\alpha$-KG), which is an intermediary in the TCA cycle that can be generated through deamination of glutamate (Liu et al. 2013) and finds wide applications in food and agriculture industry and is also used in medicines. Some important constituents of medicines, such as apressinum and nepresol, are known to be prepared from pyridasines (manufactured from $\alpha$-KG derivatives). (Barrett and Yousaf 2008). Though large-scale manufacturing of $\alpha$-KG is mainly achieved by the chemical route, the drawbacks are low yield and the difficulties in handling environmentally hazardous residues. Microbial processing for $\alpha$-KG production are well studied, and the most commonly used production strains are *Y. lipolytica* and *Candida glabrata* (Zhang et al. 2009). However, the natural strains of these two organisms were not preferred due to the high cost of production. A glutamate-overproducing mutant of *C. glutamicum* was used to engineer a 2-ketoglutarate-producer, which showed a concomitant increase in 2-ketoglutarate in a fermentation broth containing glucose, molasses, glutamate, and soybean hydrolysate (Jo et al. 2012).

Another such organic acid is fumaric acid (2-butenedioic acid *trans*; 1,2-ethylenedicarboxylic acid), which is an important TCA cycle intermediate. Though earlier fumaric acid was produced by the fungal sp. of *Rhizopus,* later on it was substituted by petrochemical means of production, being economically more attractive. Chemically, fumaric acid is produced by processing maleic anhydride (or maleic acid), a petroleum derivative. However, this is not cost effective, and there has been a renovated interest in microbial fermentation. Of the various strains used, *Rhizopus* species are the most worthy and economically feasible strains to yield fumaric acid in aerobic as well as anaerobic conditions.

Itaconic acid is a dicarboxylic acid and is one of the top 12 chemicals by the US Department of Energy. It is an exceptional polymer precursor and has wide industrial applications viz. production of plastics, resins, coatings, etc. (Willke and Vorlop 2001) and can be synthesized to methacrylic acid by chemical means (Choi et al. 2015). Itaconic acid is commercially produced by filamentous fungi *Aspergillus terreus* and *Aspergillus itaconicus* from carbon sources like sucrose, xylose, and glucose (Okabe et al. 2009; Willke and Vorlop 2001). Under optimal conditions, fermentation by the fungal strain reported productivity of 0.48 g $(l\,h^{-1})^{-1}$ and a yield of 0.81 mol/(mol glucose) (Hevekerl et al. 2014). Ever since the cis-aconitic acid decarboxylase (cadA) gene has been identified and characterized by Kanamasa et al. (2008), it has been expressed in many microbes such as *E. coli* (Okamoto et al. 2014), *S. cerivisae* (Blazeck et al. 2014), *C. glutamicum* (Otten et al. 2015), and *Y. lipolytica* (Blazeck et al. 2015). The engineered strains could produce itaconic acid titers ranging from 0.17 g $l^{-1}$ for *S. cerivisiae* (Blazeck et al. 2014), 4.3 g $l^{-1}$ for *E. coli* (Okamoto et al. 2014) to 7.8 g$l^{-1}$ for *C. glutamicum* (Otten et al. 2015). Harder et al. (2016) incorporated a plasmid (pCadCS) carrying genes representing itaconic acid production in *E. coli* and proposed a model-based approach to engineer a overproducing strain with maximum yield of 0.45 g $(l\,h^{-1})^{-1}$.

## 9.4 Metabolic/Genetic Engineering: Trends in Organic Acid Technology

The last decade has witnessed an increase in the productivity of organic acids with the advancement in protein and metabolic engineering. With a better understanding of the relations of important cellular molecules, quorum sensing, and the mechanism of dynamic and complex behavior of biological systems, technological acumen in the engineering of microbial cell factories has grown, leveraging the biochemical mode of action of proteins (Olson and Tabor 2014). Such modern engineering techniques have resulted in cheaper and sustainable manufacturing of organic acids. This section discusses a few organic acids (Table 9.1).

As mentioned, *Propionibacteria* species are major producers of PA production. However, the high-priced complex nitrogen sources used for immobilization during fermentation has limited the economic viability of manufacturing of propionic acid at the industrial/commercial level (Feng et al. 2011). Therefore, finding a metabolic engineering solution for producing *Propionibacteria* in conjugation with the most cost-effective biocatalytic system is key to attaining higher PA productivity (Es et al. 2017). By engineering the metabolic pathways, propionic acid production has been enhanced. It has been reported that a recombinant *Propionibacterium* grew faster and the PA titer was significantly enhanced by expressing Phosphoenolpyruvate carboxylase (PPC), pyruvate carboxylase (PYC), methylmalonyl-CoA decarboxylase (MMD), and methylmalonyl-CoA carboxyltransferase (MMC) to increase carbon flux toward oxaloacetate (OAA) in the dicarboxylic acid pathway (Ammar et al. 2014;). Further, genome shuffling and metabolic engineering of *Propionibacteria* was directed to develop acid tolerance and PA manufacture. In this context, Guan and colleagues have identified two vital components of acid resistance, viz. arginine deaminase and glutamate decarboxylase resulting a tenfold increase in its production w.r.t. its natural counterpart (Guan et al. 2016).

Proteins purified from *E. coli* can be used directly for substrate catalysis. Further, immobilized enzymes or whole cells can act as instruments to create economic and ecological catalysts. Keeping in view the intermediates of TCA cycle, *E.coli* can be metabolically engineered to overproduce several organic acids (Yu et al. 2011). Liu et al. (2017a,2017b) reviewed that the deletion or knocking out of certain genes of the metabolic pathways or their overexpression in the yeasts and bacteria has enhanced the performance of the production of SA. 3-Hydroxypropionic acid (3-HP) is a likely key chemical, which is harnessed for manufacturing many high-valued chemicals such as acrylic acid, acrylamide, and malonic acid (Chen and Liu 2016;) Various types of genetic manipulations were incorporated in the engineered *C. glutamicum* strain, which can produce 62.6 $g\,l^{-1}$ 3-HP at a yield of 0.51 g/g glucose in fed-batch fermentation.

## 9.5 Research Gaps and Techno-Economic Feasibility

Although several studies in the field of fermentation for the manufacture of organic acids are being conducted, most of the technologies have been unable to land in the market. The major research gap of the latest findings lies in the "lack to fit the cost economics." Several types of research have been conducted to establish the techno-economic feasibility but they are not adequate to the requirement.

In a study conducted by Christodoulou and Velasquez-Orta (2016), evaluation of investments and production cost of acetic acid were done through microbiological (microbial electrosynthesis and anaerobic fermentation) and chemical processes (methanol carbonylation and ethane direct oxidation). Microbial electrosynthesis and anaerobic fermentation showed the high price for production and investment costs

**Table 9.1** Most relevant engineered microorganisms reporting the highest acid production titers.

| Organic acid | Metabolically engineered microorganisms | Substrate used | References | Uses |
|---|---|---|---|---|
| Acetic acid | *Clostridium thermoaceticum* | Glucose | Parekh and Cheryan (1994) | Production of vinegar |
| Citric acid | *Aspergillus niger*<br>*Yarrowia lipolytica* | Glucose<br>Inulin | Soccol et al. (2006)<br>Liu et al. (2013) | Additive, detergents, cosmetic, pharmaceutical, food and beverages |
| Lactic acid | *Saccharomyces cerevisiae*<br>*Corynebacteriumglutamicum* | Glucose | Tsuge et al. (2015) | Food, cosmetic, pharmaceutical, and leather industries |
| Succinic acid | *Candida krusei*, and *S. cerevisiae*<br>*E. coli* and *C. glutamicum* | Glucose<br>Glycerol | Jansen and van Gulik (2014), Yan et al. (2014)<br>Becker and Wittmann (2015) | Production of fine chemicals, such as perfume esters or as a neutralizing agent used in the food industry |
| Itaconic acid | *E.coli*<br>*Yarrowialipolytica*<br>*Aspergillusterreus* | Glucose,<br>Glycerol | Harder et al. (2016)<br>Blazeck et al. (2015)<br>Huang et al. (2014)<br>Hevekerl et al. (2014) | Detergents, coatings, and rubber |
| Propionic acid | *Propionibacterium* sp. | Glycerol | Ammar et al. (2014), Wang et al. (2015). | Food, perfume, paint, and pharmaceutical industries |
| Glycolic acid | *C. glutamicum*<br>*S. cerevisiae and Kluyveromyces lactis*<br>*E. coli* | Carbohydrates, glycerol | Zahoor et al. (2014)<br>Koivistoinen et al. (2013)<br>Pilot scale | Packaging material |
| Fumaric acid | *Rhizopus nigricans, Rhizopus arrhizus, Rhizopus oryzae, Rhizopus formosa* | Glucose | Engel et al. (2008) | Lubricating oil, inks, lacquers,carboxylating agent for styreneb- utadiene rubber, food and beverage additive |
| 3-Hydroxypropionic acid | *E. coli* and *Klebsiella pneumoniae* | Glucose<br>Glycerol | Kim et al. (2014) | Platform for the production of acrylamide, acrylonitrile, ethyl 3-HP, 1,3-propanediol, and malonic acid |
| Malic acid | *Ustilago trichophora* TZ1<br>*S. cerevisiae* | Glucose<br>Glycerol | Zambaninia et al. (2017) | Used to prepare maleate salts for pharmaceuticals |

i.e. 1.44£/kg, 1770£/t and 4.14£/kg, 1598£/t, respectively, as compared to methanol carbonylation (0.26£/kg, 261£/t) and ethane direct oxidation (0.11£/kg, 258£/t). However, it was observed that integration of anaerobic fermentation and microbial electrosynthesis could decrease the release of carbon dioxide, double production rates (200 t/y), and reduce investment cost by 9% (1366£/t). Research on economic feasibility of succinic acid by taking bakery waste as substrates was conducted. The study was conducted in a pilot plant in Hong Kong, with a capacity of converting 1 tonne of bakery waste per day. The total capital investment for the plant was US$1 118 243 $yr^{-1}$ and the total production cost was US$230 750 $yr^{-1}$. The return on investment was 12.8% whereas the payback period and the internal rate of return of the project were 7.2 years and 15.3%, respectively.

## 9.6 Conclusion

In the modern era, interdisciplinary research has become necessary for addressing the various issues of organic acid production. For instance, systems metabolic engineering, a developing science, utilizes the application of omics data for synthetic biology and evolutionary engineering for strain breeding and process improvisation. High-throughput technologies and computational methods have already established the enormous possibilities in providing genomic data and cues for synthetic biology and evolutionary engineering. Recently, systems biology, synthetic biology, and evolutionary biology have been efficaciously applied for metabolic engineering of industrial strains, and it is important that steps should be taken toward transformation of laboratory-scale research into industrial scale. Further such advances in information and high throughput technologies have the potential to prompt the development of innovative, environmentally friendly, and techno-economically viable fermentation processes for organic acid production.

## References

Alonso, S., Rendueles, M., and Díaz, M. (2013). Bio-production of lactobionic acid: current status, applications and future prospects. *Biotechnol. Adv.* (8): 1275–1291.

Alonso, S., Rendueles, M., and Diaz, M. (2015). Microbial production of specialty organic acids from renewable and waste materials. *Crit. Rev.Biotechnol.* 35 (4): 497–513.

Ammar, E.M., Jin, Y., Wang, Z., and Yang, S.-T. (2014). Metabolic engineering of *Propionibacterium freudenreichii*: effect of expressing phosphoenolpyruvate carboxylase on propionic acid production. *Appl. Microbiol. Biotechnol.* 98: 7761–7772.

Arrigoni, O. and De Tullio, M.C. (2002). Ascorbic acid: much more than just an antioxidant. *Biochim. Biophys. Acta* 1569: 1–9.

Aruoma, O.I., Murcia, A., Butler, J., and Halliwell, B. (1993). Evaluation of the antioxidant and prooxidant actions of gallic acid and its derivatives. *J. Agric. Food Chem.* 41: 1880–1885.

Awad, H.M., Diaz, R., Malek, R.A. et al. (2012). Efficient production process for food grade acetic acid by *Acetobacter aceti* in shake flask and in bioreactor cultures. *J. Chem.* 9 (4): 2275–2286.

Axelsson, L. (2004). Lactic acid bacteria: classification and physiology. In: *Microbiological and Functional Aspects* (ed. S. Salminen, A. Von Wright and A. Ouwehand), 1–67. New York: Marcel Dekker, Inc.

Barrett, D.G. and Yousaf, M.N. (2008). Poly (triol a-ketoglutarate) as biodegradable, chemoselective, and mechani- cally tunable elastomers. *Macromolecules* 41: 6347–6352.

Becker, J. and Wittmann, C. (2015). Advanced biotechnology: metabolically engineered cells for the bio-based production of chemicals and fuels, materials, and health-care products. *Angew. Chem. Int. Ed. Engl.* 54: 3328–3350.

Blazeck, J., Miller, J., Pan, A. et al. (2014). Metabolic engineering of *Saccharomyces cerevisiae* for itaconic acid production. *Appl. Microbiol. Biotechnol.* 98: 8155–8164.

Blazeck, J., Hill, A., Jamoussi, M. et al. (2015). Metabolic engineering of *Yarrowia lipolytica* for itaconic acid production. *Metab. Eng.* 32: 66–73.

Chen, Z. and Liu, D. (2016). Toward glycerol biorefinery: metabolic engineering for the production of biofuels and chemicals from glycerol. *Biotechnol. Biofuels* 9: 205.

Cho, C., Choi, S.Y., Luo, Z.W., and Lee, S.Y. (2015). Recent advances in microbial production of fuels and chemicals using tools and strategies of systems metabolic engineering. *Biotechnol. Adv.* 33 (7): 1455–1466.

Choi, S., Song, C.W., Shin, J.H., and Lee, S.Y. (2015). Biorefineries for the production of top building block chemicals and their derivatives. *Metab. Eng.* 28: 223–239.

Christodoulou, X. and Velasquez-Orta, S.B. (2016). Microbial electrosynthesis and anaerobic fermentation: an economic evaluation for acetic acid production from $CO_2$ and CO. *Environ. Sci. Technol.* 50 (20): 11234–11242.

Dhillon, G.S., Brar, S.K., Verma, M., and Tyagi, R.D. (2011). Recent advances in citric acid bio-production and recovery. *Food Bioprocess Technol.* 4: 505–529.

Engel, C.A.R., Straath, A.J., Zijlmans, T.W. et al. (2008). Fumaric acid production by fermentation. *Appl. Microbiol. Biotechnol.* 78: 379–389.

Es, I., Khaneghah, A.M., Hashemi, S.M.B., and Koubaa, M. (2017). Current advances in biological production of propionic acid. *Biotechnol. Lett.* 1–11.

Feng, X., Chen, F., Xu, H. et al. (2011). Green and economical production of propionic acid by *Propionibacterium freudenreichii* CCTCC M207015 in plant fibrous-bed bioreactor. *Bioresour. Technol.* 102: 6141–6146.

Guan, N., Li, J., Shin, H.D. et al. (2016). Metabolic engineering of acid resistance elements to improve acid resistance and propionic acid production of *Propionibacterium jensenii. Biotechnol. Bioeng.* 113: 1294–1304.

Gutiérrez, L.F., Hamoudi, S., and Belkacemi, K. (2012). Lactobionic acid: a high value-added lactose derivative for food and pharmaceutical applications. *Int. Dairy J.* 26: 103–111.

Harder, B.-J., Bettenbrock, K., and Klamt, S. (2016). Model-based metabolic engineering enables high yield itaconic acid production by *Escherichia coli*. *Metab. Eng.* 38: 29–37.

Hevekerl, A., Kuenz, A., and Vorlop, K.D. (2014). Influence of the pH on the itaconic acid production with *Aspergillus terreus*. *Appl. Microbiol. Biotechnol.* 98: 10005–10012.

Hofvendahl, K. and Hahn-Hagerdal, B. (2000). Factors affecting the fermentative lactic acid production from renewable resources. *Enzyme Microb. Technol.* 26: 87–107.

Hsu, S. and Yang, S.T. (1991). Propionic acid fermentation of lactose by *Propionibacterium acidipropionici*:effects of pH. *Biotechnol. Bioeng.* 38: 571–5788.

Huang, X., Lu, X., Li, Y. et al. (2014). Improving itaconic acid production through genetic engineering of an industrial *Aspergillus terreus* strain. *Microb. Cell Fact.* 13: 119.

Jansen, M.L.A. and van Gulik, W.M. (2014). Towards large scale fermentative production of succinic acid. *Curr. Opin. Biotechnol.* 30: 190–197.

Jantama, K., Haupt, M., Svoronos, S.A. et al. (2008). Combining metabolic engineering and metabolic evolution to develop nonrecombinant strains of *Escherichia coli* C that produce succinate and malate. *Biotechnol. Bioeng.* 99: 1140–1153.

Jiang, M., Liu, S., Ma, J. et al. (2010). Effect of growth phase feeding strategies on succinate production by metabolically engineered *Escherichia coli*. *Appl. Environ. Microbiol.* 1298–1300.

Jiang, L., Cui, H., Zhu, L. et al. (2015). Enhanced propionic acid production from whey lactose with immobilized *Propionibacterium acidipropionici* and the role of trehalose synthesis in acid tolerance. *Green Chem.* 17: 250–259.

Jo, J.H., Seol, H.Y., Lee, Y.B. et al. (2012). Disruption of genes for enhanced biosynthesis of $\alpha$-ketoglutarate in *Corynebacterium glutamicum*. *Can. J. Microbiol.* 58: 278–286.

Jost, B., Holz, M., Aurich, A. et al. (2015). The influence of oxygen limitation for the production of succinic acid with recombinant strains of *Yarrowia lipolytica*. *Appl. Microbiol. Biotechnol.* 99 (4): 1675–1686.

Kanamasa, S., Dwiarti, L., Okabe, M., and Park, E.Y. (2008). Cloning and functional characterization of the cis-aconitic acid decarboxylase (CAD) gene from *Aspergillus terreus*. *Appl. Microbiol. Biotechnol.* 80: 223–229.

Kim, K., Kim, S.K., Park, Y.C., and Seo, J.H. (2014). Enhanced production of 3-hydroxypropionic acid from glycerol by modulation of glycerol metabolism in recombinant *Escherichia coli*. *Bioresour. Technol.* 156: 170–175.

Koivistoinen, O.M., Kuivanen, J., Barth, D. et al. (2013). Glycolic acid production in the engineered yeasts Saccharomyces cerevisiae and Kluyveromyces lactis. *Microb. Cell Fact.* 12 (1): 82. https://doi.org/10.1186/1475-2859-12-82.

Koutinas, A.A., Malbranque, F., Wang, R. et al. (2007). Development of an oat-based biorefinery for the production of L(+)-lactic acid by *Rhizopus oryzae* and various value-added coproducts. *J. Agric. Food Chem.* 55: 1755–1761.

Leavening agent 2013. Encyclopædiabritannica online. Accessed 24 Jun 2013 http://www.britannica.com/EBchecked/topic/334116/leavening-agent

Lee, P.C., Lee, W.G., Lee, S.Y., and Chang, H.N. (2001). Succinic acid production with reduced by-product formation in the fermentation of *Anaerobiospirillum succiniciproducens* using glycerol as a carbon source. *Biotechnol. Bioeng.* 72: 41–48.

Lee, J.Y., Kang, C.D., Lee, S.H. et al. (2015). Engineering cellular redox balance in *Saccharomyces cerevisiae* for improved production of L-lactic acid. *Biotechnol. Bioeng.* 112: 751–758.

Li, M. and Borodina, I. (2015). Application of synthetic biology for production of chemicals in yeast *Saccharomyces cerevisiae*. *FEMS Yeast Res.* 15 (1): 1–12.

Liu, L., Zhu, Y., Li, J. et al. (2012). Microbial production of propionic acid from propionibacteria: current state, challenges and perspectives. *Crit. Rev. Biotechnol.* 32: 374–381.

Liu, X.Y., Chi, Z., Liu, G.L. et al. (2013). Both decrease in ACL1 gene expression and increase in ICL1 gene expression in marine-derived yeast *Yarrowia lipolytica* expressing INU1 gene enhance citric acid production from inulin. *Mar. Biotechnol.* 15: 26–36.

Liu, J., Li, J., Shin, H.-D. et al. (2017a). Protein and metabolic engineering for the production of organic acids. *Bioresource Technol.* 239: 412–421.

Liu, J., Li, J., Shin, H.-d. et al. (2017b). Protein and metabolic engineering for the production of organic acids. *Bioresour. Technol.* 239: 412–421.

Magnuson, J.K. and Lasure, L.L. (2004). Organic acid production by filamentous fungi. In: *Advances in Fungal Biotechnology for Industry, Agriculture, and Medicine* (ed. J.S. Tkacz and L. Lange), 307–340. Kluwer Academic/Plenum Publishers.

Mani-López, E., García, H., and López-Malo, A. (2012). Organic acids as antimicrobials to control *Salmonella* in meat and poultry products. *Food Res. Int.* 45: 713–721.

Markets and Markets, 2015. http://www.marketsandmarkets.com/PressReleases/organic-acid.asp.

Okabe, M., Lies, D., Kanamasa, S., and Park, E.Y. (2009). Biotechnological production of itaconic acid and its biosynthesis in *Aspergillusterreus*. *Appl. Microbiol. Biotechnol.* 84 (4): 597–606.

Okamoto, S., Chin, T., Hiratsuka, K. et al. (2014). Production of itaconic acid using metabolically engineered *Escherichia coli*. *J. Gen. Appl. Microbiol.* 60: 191–197.

Olson, E.J. and Tabor, J.J. (2014). Optogenetic characterization methods overcome key challenges in synthetic and systems biology. *Nat. Chem. Biology* 10 (7): 502–511.

Otero, J.M., Cimini, D., Patil, K.R. et al. (2013). Industrial systems biology of *Saccharomyces cerevisiae* enables novel succinic acid cell factory. *PLoS One* 8: e54144.

Otten, A., Brocker, M., and Bott, M. (2015). Metabolic engineering of *Corynebacterium glutamicum* for the production of itaconate. *Metab. Eng.* 30: 156–165.

Ozaki, A., Konishi, R., Otomo, C. et al. (2017). Metabolic engineering of *Schizosaccharomyces pombe* via CRISPR-Cas9 genome editing for lactic acid production from glucose and cellobiose. *Metabol. Eng. Commun.* 5: 60–67.

Pal, P. and Nayak, J. (2017). Acetic acid production and purification: critical review towards process intensification. *Sep. Purif. Rev.* 46: 44–61.

Panda, S.K. and Ray, R.C. (2015). Microbial processing for valorization of horticultural wastes. In: *Environmental Microbial Biotechnology* (ed. L.B. Sukla, N. Pradhan, S. Panda and B.K. Mishra), 203–221. Heidelberg, Germany: Springer.

Panda, S.K., Mishra, S.S., Kayitesi, E., and Ray, R.C. (2016). Microbial-processing of fruit and vegetable wastes for production of vital enzymes and organic acids: biotechnology and scopes. *Environ. Res.* 146: 161–172.

Parekh, S.R. and Cheryan, M. (1994). High concentrations of acetate with a mutant strain of *C. thermoaceticum*. *Biotechnol. Lett.* 16: 139–142.

Pokorny, J., Yanishlieva, N., and Gordon, M. (2001). *Antioxidants in Food*. Boca Raton, FL: CRC Press.

Polen, T., Spelberg, M., and Bott, M. (2013). Toward biotechnological production of adipic acid and precursors from biorenewables. *J. Biotechnol.* 167: 75–84.

Rasrendra, C., Girisuta, B., Van de Bovenkamp, H. et al. (2011). Recovery of acetic acid from an aqueous pyrolysis oil phase by reactive extraction using tri-*n*-octylamine. *Chem. Eng. J.* 176: 244–252.

Sauer, M., Porro, D., Mattanovich, D., and Branduardi, P. (2008). Microbial production of organic acids: expanding the markets. *Trends Biotechnol.* 26: 100–108.

Sauer, M., Russmayer, H., Grabherr, R. et al. (2017). The efficient clade: lactic acid bacteria for industrial chemical production. *Trends Biotechnol.* 35 (8): 756–769.

Shaikh, Z. and Qureshi, P. (2013). Screening and isolation of organic acid producers from samples of diverse habitats. *Int. J. Curr. Microbiol. Appl.* 2 (9): 39–44.

Silva, M.M. and Lidon, F. (2016). Food preservatives–An overview on applications and side effects. *Emir J Food Agric.* 26 (6): 366–373.

Silva, M.E., Torres Neto, A.B., Silva, W.B. et al. (2007). Cashew wine vinegar production: alcoholic and acetic fermentation. *Braz. J. Chem. Eng.* 24 (2): 163–169.

Soccol, C.R., Vandenberghe, L.P.S., Rodrigues, C., and Pandey, A. (2006). New perspectives for citric acid production and application. *Food Technol. Biotechnol.* 44: 141–149.

Song, H. and Lee, S.Y. (2006). Production of succinic acid by bacterial fermentation. *Enzyme Microb. Technol.* 39: 352–361.

Song, Y., Li, J., Shin, H.-D. et al. (2016). Biotechnological production of alpha-keto acids: current status and perspectives. *Bioresour. Technol.* 219: 716–724.

Suyama, A., Higuchi, Y., Urushihara, M. et al. (2017). Production of 3-hydroxypropionic acid via the malonyl-CoA pathway using recombinant fission yeast strains. *J. Biosci. Bioeng.* 124 (4): 392–399.

Swain, M.R., Ray, R.C., and Patra, J.K. (2011). Citric acid: microbial production and applications in food and pharmaceutical industries. In: *Citric Acid* (ed. D.A. Vargas and J.V. Medina), 97–118. New York: Nova Science Publisher.

Tsuge, Y., Yamamoto, S., Kato, N. et al. (2015). Overexpression of the phosphofructokinase encoding gene is crucial for achieving high production of D-lactate in *Corynebacterium glutamicum* under oxygen deprivation. *Appl. Microbiol. Biotechnol.* 99: 4679–4689.

Tsuji, A., Okada, S., Hols, P., and Satoh, E. (2013). Metabolic engineering of *Lactobacillus plantarum* for succinic acid production through activation of the reductive branch of the tricarboxylic acid cycle. *Enzyme Microbiol. Technol.* 53: 97–103.

Vandenberghe, L.P.S., Karp, S.G., de Oliveira, P.Z. et al. (2018). Solid-state fermentation for the production of organic acids. *Curr. Dev. Biotechnol. Bioengineer.* 415–434. https://doi.org/10.1016/B978-0-444-63990-5.00018-9.

Vijayakumar, J., Aravindan, R., and Viruthagiri, T. (2008). Recent trends in the production, purification and application of lactic acid. *Chem. Biochem. Eng.* 22 (2): 245–264.

Wallenius, J., Pahimanolis, N., Zoppe, J. et al. (2015). Continuous propionic acid production with *Propionibacterium acidipropionici* immobilized in a novel xylan hydrogel matrix. *Bioresour. Technol.* 197: 1–6.

Wang, Z., Lin, M., Wang, L. et al. (2015). Metabolic engineering of *Propionibacteriumfreudenreichii* subsp. *shermanii* for enhanced propionic acid fermentation: effects of overexpressing three biotin-dependent carboxylases. *Process Biochem.* 50: 194–204.

Werpy T., and G. Petersen 2004.Top value-added chemicals from biomass, vol 1: results of screening for potential candidates from sugars and synthesis gas. 2004, US Department of Energy http://www.osti.gov/bridge

Willke, T. and Vorlop, K.D. (2001). Biotechnological production of itaconic acid. *Appl. Microbiol. Biotechnol.* 56: 289–295.

Yamane, T. and Tanaka, R. (2013). Highly accumulative production of L(+)- lactate from glucose by crystallization fermentation with immobilized *Rhizopusoryzae*. *J. Biosci. Bioeng.* 115: 90–95.

Yan, D., Wang, C., Zhou, J. et al. (2014). Construction of reductive pathway in *Saccharomyces cerevisiae* for effective succinic acid fermentation at low pH value. *Bioresour. Technol.* 156: 232–239.

Yen, G.C., Duh, P.D., and Tsai, H.L. (2002). Antioxidant and pro-oxidant properties of ascorbic acid and gallic acid. *Food Chem.* 79: 307–313.

Yu, C., Cao, Y., Zou, H., and Xian, M. (2011). Metabolic engineering of *Escherichia coli* for biotechnological production of high-value organic acids and alcohols. *Appl. Microbiol. Biotechnol.* 89: 573–583.

Yu, J.-L., Xia, X.-X., Zhong, J.-J., and Qian, Z.-G. (2014). Direct biosynthesis of adipic acid from a synthetic pathway in recombinant *Escherichia coli*. *Biotechnol. Bioeng.* 111: 2580–2586.

Yuzbashev, T.V., Yuzbasheva, E.Y., Sobolevskaya, T.I. et al. (2010). Production of succinic acid at low pH by a recombinant strain of the aerobic yeast *Yarrowia lipolytica*. *Biotechnol.Bioeng.* 107: 673–682.

Zahoor, A., Otten, A., and Wendisch, V.F. (2014). Metabolic engineering of *Corynebacterium glutamicum* for glycolate production. *J. Biotechnol.* 366–375.

Zambaninia, T., Tehrania, H.H., Geisera, E. et al. (2017). Metabolic engineering of *Ustilago trichophora* TZ1 for improved malic acid production. *Metabol. Eng. Commun.* 4: 12–21.

Zhang, A. and Yang, S. (2009). Propionic acid production from glycerol by metabolically engineered *Propionibacterium acidipropionici*. *Process Biochem.* 44: 1346–1351.

Zhang, C., Yang, H., Yang, F., and Ma, Y. (2009). Current progress on butyric acid production by fermentation. *Curr. Microbiol.* 59: 656–663.

# 10

# Research and Production of Microbial Polymers for Food Industry

*Sinem Selvi Selvi*[1], *Edina Eminagic*[1], *Muhammed Yusuf Kandur*[1], *Emrah Ozcan*[1,2], *Ceyda Kasavi*[1], *and Ebru Toksoy Oner*[1]

[1] *Department of Bioengineering, Industrial Biotechnology and Systems Biology (IBSB) Research Group, Marmara University, Istanbul, Turkey*
[2] *Department of Bioengineering, Gebze Technical University, Kocaeli, Turkey*

## 10.1 Introduction

Natural polymers can be classified as microbial-, plant-, and animal-derived based on their sources. High cost of downstream processes of plant and algal gums drives the polymer industry toward microbial derived polymers. Furthermore, microbial polysaccharides generally have higher molecular weights than plants, which affect their properties (Oner et al. 2016). Current discoveries in microbial polymer biosynthesis have initiated new areas for medical and industrial applications. Novel molecular mechanisms and production processes have been discovered. These molecular mechanisms have formed important tools for process engineering and applications, which are getting popular in pharmaceutical, agriculture, and particularly, in food industry. Microbial polymers are long-chained, natural, biodegradable, biocompatible, nontoxic materials and are easy to handle compared to synthetic polymers. Xanthan gum, dextran, alginate, bacterial cellulose, gellan, curdlan, levan, pullulan, glycogen are important microbial polysaccharides that can be of bacterial or fungal origin (Vijayendra and Shamala 2014). They are produced through sugar fermentation from various microorganisms such as *Pseudomonas elodea, Alcaligenes faecalis, Xanthomonas campestris, Halomonas smyrnensis, Zymomonas mobilis*, and *Acetobacter xylinum*.

Generally, water-soluble polymers are used as suspending, gelling, and thickening agents in food industry. Polymers can also add characteristics such as sweetening, cryoprotection, antioxidant, anticaking, flavoring, antifoaming, chelating, stabilizing, preservative, and coating (Rosalam and England 2006). Polymer additives modify the rheological properties and increase the stability of emulsion. These properties generally depend on the structure and the interactions that polymers exhibit with other substances such as ionic effect and hydrogen bonding.

This chapter focuses on the microbial polymers that are commonly used in food industry – namely, polysaccharides like levan, pullulan, alginate, curdlan, gellan, scleroglucan, xanthan, and dextran as well as the polyhydroxyalkanoates (PHAs). After a general view

*Bioprocessing for Biomolecules Production*, First Edition.
Edited by Gustavo Molina, Vijai Kumar Gupta, Brahma N. Singh, and Nicholas Gathergood.

on the biosynthesis mechanisms of microbial polymers, structural features and general properties of each polymer are introduced, followed by microbial production processes and food applications.

### 10.1.1 Biosynthesis of Microbial Polymers

Microbial polymers can be divided into four major groups: polysaccharides, polyesters, polyamides, and inorganic polyanhydrides (Rehm 2010; Ates 2015; Schmid et al. 2015). Biosynthesis of these microbial polymers involves complex pathways and secretion mechanisms.

Although several types of polysaccharides are produced intracellularly such as storage polysaccharides (e.g. glycogen) and capsular polysaccharides (structural polymers that act as surface antigens and responsible for virulence factors) (Whitfield 2006; Rehm 2010) only extracellular polysaccharides (EPSs) are considered industrially relevant.

EPSs are produced via four known mechanisms: (i) Wzx/Wzy-dependent pathway, (ii) the ATP binding cassette (ABC) transporter-dependent pathway, (iii) the synthase-dependent pathway, and (iv) the extracellular synthesis (Rehm 2010; Ates 2015; Schmid et al. 2015). While EPSs synthesized via the first three mechanisms are generated intracellularly and exported to the extracellular environment, EPSs that are synthesized via mechanism (iv) are secreted directly into the extracellular environment using enzymes.

The precursor molecules formed intracellularly from intermediates of the central carbon metabolism (such as fructose-6-P) are transformed to activated sugars or sugar acids in the first three mechanisms (Rehm 2010).

In the Wzx/Wzy-dependent mechanism, activated sugars are assembled at the cytoplasmic region of the inner membrane by several glycosyltransferases and transferred across the cytoplasmic membrane, mediated by a polysaccharide-specific transport protein, namely Wzx. Then the polymerization occurs at the periplasm by the Wzy protein before repeating units are exported to the cell surface. In the ABC transporter-dependent pathway, polymerization occurs at the cytoplasmic side of the inner membrane and transferred by ABC transporter proteins (Rehm 2010; Ates 2015). In the synthase-dependent pathway, complete polymer strands are secreted across the membranes and the cell wall (Rehm 2010; Ates 2015; Schmid et al. 2015). In extracellular synthesis, as the name suggests, polymerization takes place at the extracellular environment by use of a single sucrase enzyme that catalyzes the formation of disaccharides from monosaccharides, which leads up to an assembling polysaccharide chain (Rehm 2010; Ates 2015). Dextran, which is one of the oldest commercial polymers, is produced by the extracellular synthesis mechanism. *Leuconostoc mesenteroides* produces dextran by binding glucose molecules to each other in order to form a sucrose molecule, which will lead up to be a dextran chain, mediated by dextransucrase in extracellular environment (Dols et al. 1997).

The most popular biopolyester class is PHAs, which is also a bacterial bioplastic produced through PHA synthase, which has a broad substrate specificity. Any organic molecule containing a carboxyl group such as acetyl-coA and a hydroxyl group can be converted into PHA (Rehm 2010). Poly-$\gamma$-glutamate (PGA) is a type of polyamide polymer. Biosynthesis of PGA takes place when glutamate units connect to each other through amide linkages between $\alpha$-amino and $\gamma$-carboxylic acid groups (Luo et al. 2016).

Precursor metabolites and related biosynthesis mechanisms of related polymers are depicted in Table 10.1 (Rehm 2010; Schmid et al. 2015).

## 10.2 Levan

### 10.2.1 General Properties of Levan

Hypocaloric functional oligosaccharides find various application areas in food industry due to their prebiotic, nondigestible properties. Among oligosaccharides; fructooligosaccharides (FOSs) display adequate results to be used as potential prebiotics, and these fructans actually constitute the most commonly used prebiotics in the functional food market (Roberfroid 1998). Levan is an unusual fructan homopolysaccharide composed of β-(2,6)-fructofuranosyl residues with a terminal D-glucopyranosyl unit. Figure 10.1 shows the chemical structure of levan.

Levan is an EPS produced by several microorganisms and few plants (Han and Clarke 1990). There are multifunctional bacterial enzymes fructosyltransferases that are capable of converting sucrose substrate into different polymerized fructans. The final products of inulosucrase enzymes (EC 2.4.1.9) and levansucrase enzymes (EC 2.4.1.10) are inulin and levan, respectively (Oner et al. 2016). This mechanism takes place in three steps: sucrose hydrolization, transfructosylation of the hexosyl groups, and levan polymerization. Levansucrase enzyme utilizes the free energy from cleavage of sucrose substrate to transfer fructosyl part to newly developing *β*-(2–6)-levan polymer chain instantly (Inthanavong et al. 2013; González-Garcinuño et al. 2017). Polymer diameter, molecular weight, viscosity, stability, and branching degree vary with the microorganism, synthesized levansucrase and production conditions (Jakob et al. 2013). Levan is a nonionic polymer and can be produced by *Acetobacter, Aerobacter, Azotobacter, Bacillus, Erwinia, Gluconobacter, Pseudomonas, Streptococcus, Zymomonas*, and *Halomonas smyrnensis* (Kazak Sarilmiser et al. 2015).

Levan has high solubility in water and oil and other distinctive properties that include biocompatibility, film forming capability, anti-tumor, cholesterol lowering, high adhesive strength and viscosity lowering compared to other polysaccharides with similar molecular weights (Yamamoto et al. 1999; Kazak et al. 2010; Oner et al. 2016). These unique properties with high application possibilities made levan a focus of scientific and industrial interest.

### 10.2.2 Production Processes for Levan

Levan is synthesized from various bacteria such as *Acetobacter, Aerobacter, Azotobacter, Bacillus, Erwinia, Gluconobacter, Pseudomonas, Streptococcus, Zymomonas* and *Halomonas sp.* (Kazak Sarilmiser et al. 2015). Extremophilic and gram- negative *Halomonas sp.* was reported as first levan producer in 2009 by Poli et al. This system is very promising compared to mesophilic producers because it enables unsterile, low-cost production (Oner et al. 2016). *Halomonas* levan and its derivatives can be used as bioflocculating agent (Sam et al. 2011), adhesive nanostructured multilayer films (Costa et al. 2013), heparin-mimetic bioactive material (Erginer et al. 2016), and temperature sensitive hydrogels for drug-releasing systems with pNIPA (Osman et al. 2017) among many others.

**Table 10.1** General properties of polymers that used in food industry.

| Polymer | Microorganisms | Structure | Precursor metabolites/ biosynthesis mechanisms | Food Industry | References |
|---|---|---|---|---|---|
| Alginate | *Azotobacter sp.*<br>*Pseudomonas sp.* | (1–4)β-D-mannuronate and α-L- guluronate | GDP–mannuronic acid / Synthase dependent | Gelling agent, stabilizer, moisture retention | (Hay et al. 2013) (Rehm 2010, Schmid et al. 2015) |
| Curdlan | *Alcaligenes faecalis*<br>*Agrobacterium* sp. | β (1–3) D-glucose | UDP–glucose / Synthase dependent | Stabilizer, viscoelasticity and moisture improvement | (Kai et al. 1993) (Rehm 2010, Schmid et al. 2015) |
| Dextran | Lactic acid bacteria | α (1–6) glucan | Sucrose / Extracellular synthesis | Stabilizer, packaging, thickener | (Naessens et al. 2005, Rehm 2010, Schmid et al. 2015) |
| Gellan | *Pseudomonas elodea*<br>*Sphingomonas paucimobilis* | Heteropolysaccharide | UDP–glucose, dTDP–rhamnose and UDP–glucuronate / Wzx/Wzy dependent | Edible coating, antibrowning agent, microencapsulation agent | (Banik and Santhiagu 2006)<br>(Rehm 2010, Schmid et al. 2015) |
| Levan | *Bacillus* sp.<br>*Halomonas smyrnensis* AAD6<br>*Leuconostoc mesenteroides*<br>*Zymomonas mobilis* | β (2–6) D-fructose | Sucrose / Extracellular synthesis | Prebiotic, taste improvement, coating agent, packaging | (Poli et al. 2009)<br>(de Oliveira et al. 2007)<br>(Han 1989)<br>(Patel et al. 2012) (Rehm 2010, Schmid et al. 2015) |
| PHAs | *Bacillus* spp.<br>*Pseudomonas* sp. | 3-hydroxybutyrate | (*R*)-3-hydroxyacyl<br>CoA | Vapor barrier, food packaging, film forming | (Shamala et al. 2003)<br>(Singh and Mallick 2009) (Rehm 2010, Schmid et al. 2015) |
| Pullulan | *Aureobasidium pullulans* | α (1–6) glucan | UDP-glucose | Stabilizer, filler, intensifier | (Göksungur et al. 2011) (Rehm 2010, Schmid et al. 2015) |
| Scleroglucan | *Sclerotinum rolfsii* | β (1–3) glucan | UMP<br>UDP-glucose | Thickener, gelling agent, stabilizer | (Castillo et al. 2015) (Rehm 2010, Schmid et al. 2015) |
| Xanthan Gum | *Xanthomonas campestris* | β (1–4) glucan | UDP–glucose,<br>GDP–mannose and UDP–glucuronate / Wzx/Wzy dependent | Thickener, emulsifier, stabilizer, gelling agent | (Velu et al. 2016) (Rehm 2010, Schmid et al. 2015) |

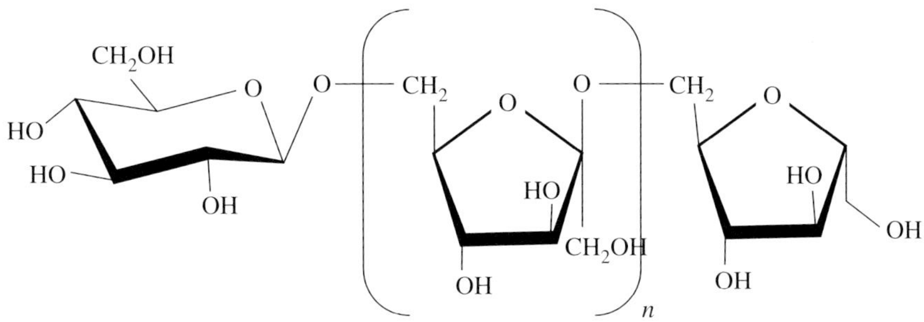


**Figure 10.1** Chemical structure of linear levan polysaccharide.

Production and downstream processes for bacterial levan begin with fermentation of microorganisms in a sucrose-based medium. Biomass separation from medium follows the fermentation process and polymer precipitation occurs generally by ethanol or acetone. Purification process is performed subsequently and after drying step levan is obtained (Oner et al. 2016). Levan production is affected from several conditions such as pH, temperature, oxygen level, culture medium, and bioreactor type. During production, glucose is released to the medium, which, in turn, is utilized for microbial growth. The fermentation system may be run in batch, fed-batch, and continuous modes for levan production. Also, syrups and molasses are low-cost sucrose sources that can be utilized for microbial fermentation for levan production (Özcan and Öner 2015). For instance, levan produced by *Halomonas smyrnensis* AAD6T cultures on starch molasses and sugar beets (Küçükaşik et al. 2011), by *Microbacterium laevaniformans* PTCC 1406 cultures on date syrup (Moosavi-Nasab et al. 2010), by *Zymomonas mobilis* ATCC 31821 cultures on sugarcane molasses and syrup (de Oliveira et al. 2007) have been published.

For each microorganism, optimum parameters and different production conditions that required for higher levan production yields are investigated. Different parameters such as sucrose, nitrogen, and inoculant concentrations were studied using *Acetobacter xylinum* NCIM2526 for levan synthesis. Optimum parameters for concentrations of nitrogen, sucrose, and inoculant were concluded as 10, 50–60, and 1.49 $g\,l^{-1}$, respectively. When pH of the production medium was set to 6.8, yield increased after 24 hours and maximum levan yield was measured at the 122nd hour. When sucrose concentration was increased to 40–50 $g\,L^{-1}$, levan yield increased; on the other hand, it decreased at sucrose levels of 70–80 $g\,l^{-1}$ and no important change was recorded at succrose concentrations of 20–40 $g\,l^{-1}$ (Srikanth et al. 2015).

Another important study on levan production was performed with *Halomonas smyrnensis* AAD6T bacteria. Batch bioreactor system with different feeding scenarios was applied in shake flask and maximum levan yield was achieved at pH 7 with sucrose concentration at 50 $g\,l^{-1}$. In further experiments, nitrogen and phosphorus limitations were performed leading to a decrease in biomass values, while increasing the yield of levan in terms of biomass ($Y_{p/x}$). The most effective result of the study was performed under the boric acid conditions where levan yield increased up to 8.84 $g\,L^{-1}$ in bioreactor. Yield increase was attributed to quorum-sensing mechanisms where boron atoms are used (Kazak Sarilmiser et al. 2015; Abbamondi et al. 2016).

### 10.2.3 Food Applications of Levan

Levan polysaccharide has extensive properties for food industry applications. In 1989, Mays and Dally worked up a stabilized colloid system with levan polymer as foam, aerosol, and emulsion, which can be used in food applications. Levan cannot be detected by taste sensors due to its high molecular weight and very low volatility, which makes levan an important fat substitute. In two different studies, levan phosphate and fructans were proposed as a fat alternative to increase mouthfeel taste (Booten et al. 1998; Roberts and Garegg 1998). Levan was used as a yogurt additive for functional food recently (Xiao et al. 2014). Levan has prebiotic effects, making it an important food additive. Levan is hydrolyzed by gastric acid, and low-molecular-weight levan oligosaccharides can be used by lumen bacteria (Gupta et al. 2015). Recent studies have shown the role of levan-type fructans in microbiota in the intestinal system (Sonnenburg et al. 2010; Adamberg et al. 2015; Visnapuu et al. 2015).

## 10.3 Pullulan

### 10.3.1 General Properties of Pullulan

Pullulan is a natural, water soluble, linear homopolysaccharide composed out of maltotriose units. Maltose molecules are linked by α (1→4) glycosidic bonds, while consecutive maltotriose units are linked by α (1→6) glycosidic bonds. Pullulan was discovered in the late 1950s and isolated from a polymorphic fungus called *Aureobasidium pullulans*. It has been commercially produced since 1976. This homopolysaccharide has been used in many studies and applications involving cosmetics, pharmaceuticals, and food industries.

Structural flexibility, high solubility in water, variable molecular weight (between $3.6 \times 10^5$ and $6 \times 10^5$ Da depending on the production conditions), mechanical strength, adhesive properties, fiber-forming ability, and biodegradability are some of the properties of pullulan that made it attractive in the food industry. Pullulan is also tasteless, odorless, nonmutagenic, noncarcinogenic, impermeable to oxygen (which is ideal for film applications), and insoluble in organic solvents, with the exceptions of formamide and dimethyl sulfoxide. Additional advantages are that this biopolymer is edible but not affected by digestive enzymes, which in turn is the property that makes pullulan a good prebiotic since it also promotes the growth of beneficial bifidobacteria in human intestines and inhibits the growth of fungi. Low viscosity, which does not vary with pH or heat in aqueous solutions, allows pullulan to be used as filler in beverages and sauces. Some studies have blended pullulan with other biopolymers such as chitosan, cellulose, and alginate to improve thermal and mechanical properties (Cheng et al. 2011; Oğuzhan and Yangilar 2013; Prajapati et al. 2013; Giavasis 2014; Ferreira et al. 2016) (Figure 10.2).

### 10.3.2 Production Processes of Pullulan

Pullulan is a biopolymer that is synthesized within the cell and then excreted on the outer layer after production. Like many biopolymers, the main disadvantage is a high production cost. Therefore, the research has been shifted to the use of

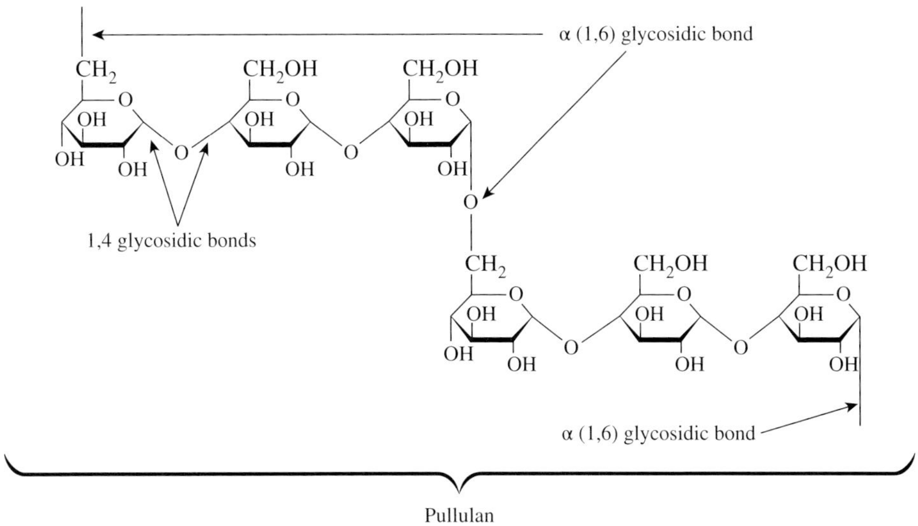


**Figure 10.2** Pullulan monomer composed out of three consecutive Maltose units.

different substrates, conversion technologies, and changing fermentation conditions for the production process (Prajapati et al. 2013; Wang et al. 2015; Wu et al. 2016). Characteristics of pullulan are highly dependent on fermentation parameters, fungal strain, and its morphology. Even though many studies have been carried out to find a relationship between the morphology of the fungus and the characteristics of pullulan, no definitive evidence has been found yet. The content of the fermentation medium is crucial for the optimal polymer yield. Commercial fermentation media are composed of peptone, phosphate, and basal salts. For industrial production, pH of the environment is initially set to 6.5, but pH decreases to 3.5 within 24 hours. Maximal culture growth and polymer production are obtained within 75 and 100 hours, respectively (Simon et al. 1993; Oğuzhan and Yangilar 2013; Prajapati et al. 2013). Pullulan can be produced using various carbon sources such as beet molasses, coconut byproducts, and agro-industrial wastes, which would decrease the commercial production cost. Even though the production mechanism of pullulan is not yet well understood, there are some ideas about the mechanism and how it is affected by the environment. For example, it is known that initial pH levels around 2.5 will inhibit the pullulan synthesis; however, the optimal pH level for biomass growth is 4.5 or lower. Ideal pH levels for pullulan production have been suggested between 5.5 and 7.5 by various studies. Pullulan is stable up to 250 °C, but an ideal temperature for culture growth is 25–30 °C. Carbon and nitrogen play crucial roles in pullulan synthesis yield and various combinations of components have been tested. It is now known that high sucrose concentrations can decrease the production yield more than 50%, Nitrogen limitation can increase pullulan yield and 10 : 1 ratio of C : N is most favorable for pullulan production (Cheng et al. 2011; Prajapati et al. 2013; Wang et al. 2015; Wu et al. 2016). *A. pullulans* can express an undesired black color due to a pigment called melanin, but this color can be removed by using charcoal, solvent-salt, or solvent-solvent blends (Cheng et al. 2011).

### 10.3.3 Food Applications of Pullulan

Hayashibara Company located in Okayama, Japan, has been the leader of commercial pullulan production since 1976. Shandong Jinmei Biotechnology Co. Ltd., which is located in Zucheng, China, is also a major pullulan producer; however, its focus is on pullulan production in powder and capsule forms to make oral-dissolving membranes such as candy coating (Prajapati et al. 2013). Pullulan is currently used as edible and protective food-packaging material, low-viscosity fillers in beverages, prebiotic dietary fiber, binders, intensifiers, protective glazes, cholesterol, and fatty acid substitutes. These protective films are biodegradable, heat sealable, do not let oxygen through, and can inhibit fungus growth in food. Although it has been suggested that using pullulan in baked goods could improve the quality of food, the use of this biopolymer outside Japan is quite limited (Cheng et al. 2011; Oğuzhan and Yangilar 2013; Prajapati et al. 2013; Ana et al. 2016).

## 10.4 Alginate

### 10.4.1 General Properties of Alginate

Alginate is a polysaccharide composed of β-D-mannuronate and α-L-guluronate linked by 1,4 glycosidic bonds. Figure 10.3 shows the chemical structure of alginate. Alginate was initially collected from brown seaweeds and has been commercially available since the beginning of the twentieth century. Alginate is produced by several different species of brown seaweed and two different species of bacteria; *Pseudomonas* and *Azotobacter*. Mannuronate:guluronate ratio in the polymer chain depends on the alginate substrate and the temperature of the environment. Recent studies reveal that alginate is a biocompatible and biodegradable polymer. Alginate has significant physical properties, allowing it to be used as a stabilizer, condenser, and gelling agent in food, paper, and pharmaceutical industry. Molecular weight of the alginate varies from 32 000 to 400 000 $g\,mol^{-1}$.

### 10.4.2 Production Processes for Alginate

Microbial production has benefits over algal production such as low cost, ability of production in small scales and applied in different fields. As mentioned previously, bacterial alginate can be obtained using *Pseudomonas* and *Azotobacter*; for commercial alginate production, human pathogen *Pseudomonas aeruginosa* and soil bacteria *Azotobacter vinelandii* are most widely preferred (Sabra and Zeng 2009; Hay et al. 2013; Ahmad et al. 2015). Molecular production mechanisms of these two bacteria are almost identical but the alginate produced show different properties. Several *P. aeruginosa* strains synthesize alginate to form thicker biofilms (from 2 to 40 μm) (Nivens et al. 2001), while *Azotobacter* produces much thinner (rod shape, 1 μm) alginate due to high guluronate content, and this firmer film provides drying resistance to cell (Drury et al. 1993; O'Toole et al. 2015).

Microbial production of alginate can be obtained through batch, fed-batch, and continuous fermentation. Epimerases lyases and acetyase enzymes are the important alginate modifying enzymes that were reported previously (Høidal et al. 2000). As the pH of the process medium decreases, alginate viscosity increases and reach a maximum value between pH 3 and 3.5. During alginate production, different types of peptone in

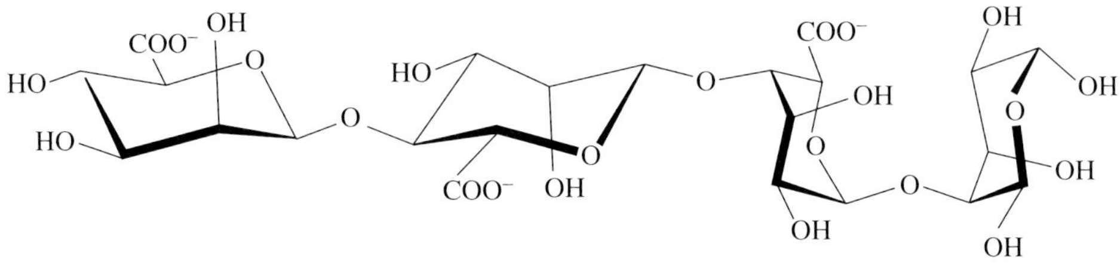


**Figure 10.3** Chemical structure of alginate with mannuronate and guluronate residues, respectively.

culture medium can change the alginate yield by 30% (Brivonese and Sutherland 1989). Also, medium with ammonium inhibits the alginate production for *A. vinelandii* (Sabra 1999).

Extraction of alginate starts with the acidification step generally by mineral acid, followed by the neutralization of alginic acid with alkali solutions by sodium hydroxide to form sodium alginate. Sodium alginate is water soluble and for further purification phases, precipitation, centrifugation, and filtration are performed. Sodium alginate is precipitated for isolation from byproducts formed using alcohol or calcium chloride. In the final step, alginate is dried and granulated (Sabra and Zeng 2009).

### 10.4.3 Food Applications of Alginate

Alginate is considered safe as a food additive and has an E number assigned (Food Standards Agency 2002). Alginates have been used for many years in food industry for many purposes; including thickening, gelling, stabilizing, and as colloidal agents. Thickening property of alginate helps moisture retention, acts as a gelling agent, and makes the biopolymer useful for jams, sauces, syrups, and ice creams. The most prevalent application of alginates is the reformation of onion rings and pimento to use in olives. Alginates provide the size and shape of the food and give consistency. Alginate is being used in bread and other pastry products, meat, vegetables and seafood to increase the shelf life and retention of moisture.

Propylene glycol alginates (PGAs) are esters of alginates and different esterification amounts and viscosity can be achieved. PGAs are also widely used in the food industry for the same purposes as pure alginates and they are numbered as E number E405 (Smith and Hong-Shum 2011). PGAs are applied in soft drinks, milks, sorbet, noodles, and ice cream because they can be used as a suspension, foaming, and acid-stabilizing agent (Aliste et al. 2000; Brownlee et al. 2005, 2009).

## 10.5 Curdlan

### 10.5.1 General Properties of Curdlan

Curdlan is a linear bacterial exopolysaccharide and classified as (1, 3) β-glucans. Curdlan is a special polysaccharide due to its rheological properties, solubility, and biomedical characteristics. Curdlan is named after its "curdle" competence when heated. In 1966, it was discovered that curdlan was produced by *Alcaligenes faecalis* var. *myxogenes* 10C3 strain (Harada et al. 1966). Currently, curdlan is produced from mutant *Agrobacterium* strains, but it can also be obtained by *A. radiobacter*, *A. rhizogenes* strains, which have high curdlan yield and can be beneficial for industrial production.

Although curdlan is insoluble in water, it is soluble in alkaline solutions and organic solvents such as dimethylsulfoxide (DMSO). When curdlan is heated, two different gels can be observed; around 55 °C a reversible gel is formed and around 80 °C a nonreversible gel is formed. Gel formation depends on the helical structure of curdlan at these temperatures. Curdlan has both single and triple helices structure at room temperature and condensed triple helices at higher temperatures.

Unusual thermal dependent gelation properties make curdlan an important applicable polysaccharide in food industry. Curdlan is used as thickening and fat-like agents in food. Also, in recent studies curdlan potential in biomedical applications was investigated because of this biopolymers immunostimulatory effect. Similar to other polysaccharides, curdlan can be hydrolyzed in acidic solutions to obtain curdlan oligomers that can be used in the biomedical or food industry (Zhang and Edgar 2014). Modifications of curdlan, such as sulfation, enhance the wound healing and anti-tumor effects (Stone and Clarke 1992; Bohn and BeMiller 1995).

### 10.5.2 Production Processes for Curdlan

Parameters such as pH, nitrogen, carbon, oxygen, and phosphate levels affect the production yields of curdlan. Curdlan is an extracellular polymer and biosynthesis occurs in three different steps. Substrate utilization, followed by intracellular metabolism of utilized substrate and finally excretion of polymer out of the cell membrane (Sutherland 1977, Zhang and Edgar 2014). Investigations with C-labeled glucose revealed that curdlan is generally polymerized directly from glucose instead of cleaved triose after glycolysis (Kai et al. 1993).

For commercial polymers, the response surface methodology (RSM) and artificial neural network are considered as the two important cost-effective optimization methods for bacterial growth and production. Nowadays these methods are used to develop and enhance the curdlan production (Shih et al. 2009; Salah et al. 2010; Rafigh et al. 2014; Yang et al. 2016).

Optimum conditions of curdlan production have been determined by RSM; at pH 7.5, urea concentration $0.52\,g\,l^{-1}$, and sucrose concentration $142.9\,g\,l^{-1}$ for *Agrobacterium* sp. Under these optimum conditions, the production of curdlan increased from 2.4 to $5.02\,l^{-1}$. Highest production rate was found to be 0.84 g/l/h at pH 7.48 with a sucrose concentration of $150\,l^{-1}$ (Jang et al. 2001). Fermentations were carried out at 30 °C with 250 rpm shaking for five days and final curdlan polymer had a molecular weight of 160 000 Da.

Curdlan production has high cost due to glucose and sucrose substrates, consequently, alternative carbon substrates are being investigated. Byproducts of palm industry are significant aspirant for an alternative carbon source due to their high sugar concentrations. Moreover, *Rhizobium radiobacter* strain was used to produce curdlan using byproducts of palm juice and $22.83\,g\,l^{-1}$ curdlan was produced from $120\,g\,l^{-1}$ glucose rich date juice via 51 hours fermentation carried out at pH 7. Final curdlan polymer had a molecular weight of 180 000 Da (Salah et al. 2011).

Another study reveals the effect of changing the nitrogen sources (e.g., ammonium chloride, urea, yeast extract, sodium nitrate, ammonium nitrate, potassium nitrate and ammonium carbonate) using *Alcaligenes faecalis* ATCC 31749 and fermentation was carried out with 0.25 $g\,l^{-1}$ nitrogen at pH 7 and 30 °C for six days. Highest curdlan yield (24.32 $g\,l^{-1}$) was reached via fermentation with urea for four days. Curdlan production rate was found to be 0.25 g/l/h. Ammonium chloride is a common source of nitrogen in curdlan production process but results revealed that urea is better for production (Jiang 2013).

In a recent study, optimum conditions for *Pseudomonas* sp. were determined. Bacteria were isolated from the soil sample and identified. Then optimizations were performed by RSM and conventional single-factor techniques. Shake flask fermentation was carried out for 72 hours at 30 °C with 200 rpm shaking. A sucrose concentration of 30.11 $g\,l^{-1}$ and yeast extract concentration of 5.94 $g\,l^{-1}$ gave the maximum yield (5.92 $g\,l^{-1}$) at pH 8.03. Obtained yield was the highest value for *Pseudomonas* sp. Final curdlan polymer had a molecular weight of 618,000 Da (Yang et al. 2016).

### 10.5.3 Food Applications of Curdlan

Curdlan is suitable for food industry due to its rheological and thermal properties. It is used as thickening and fat-like material. This effect is due to the gelation properties of curdlan and is used as an oil and moisture barrier with heating (Funami et al. 1999). Curdlan is also used in noodles to improve flexibility, toughness, and chewiness properties (Ji et al. 2010). Curdlan is tasteless and colorless polymer and due to these properties is used commonly in meat, dairy products and baking industry as a protective film layer (Zhang and Edgar 2014).

## 10.6 Gellan Gum

### 10.6.1 General Properties of Gellan Gum

Gellan is a bacterial polysaccharide produced by *Sphingomonas elodea*. It belongs to a group of polysaccharides called sphingans, named by the bacteria from which it is produced. This biopolymer is an anionic, linear polysaccharide with high molecular weight composed out of D-glucose, Lrhamnose, and D-glucuronic acid in molar ratios of 2 : 1 : 1 (Tako 2015). The primary structure of the gellan is made of tetrasaccharide repeating units (Figure 10.4). Native gellan gum can form weak gels under appropriate conditions. Gellan can form highly viscous solutions in water. Physical properties such as, number and distribution along the repeating tetrasaccharide backbone of acyl groups are crucial for gellans gel-forming abilities. For example, acetylated gellan is soluble in hot water, while deacetylated gellan or gellan with less acyl groups, are soluble in both hot and cold water. Gellan has a wide range of applications and can be used in pharmaceutical, cosmetics, and food industries (Matricardi et al. 2009; Iurciuc et al. 2016).

### 10.6.2 Production Processes for Gellan Gum

Production of gellan begins with the isolation of the bacterium from the surface of a plant belonging to *Elodea* genus. Gellan production is accomplished via fermentation with immersion method. The medium used for incubation contains nitrogen, carbon sources, and some crucial trace minerals. Ideal temperature for producing gellan is 30 °C, as higher temperatures can reduce production rate of the biopolymer. Suitable

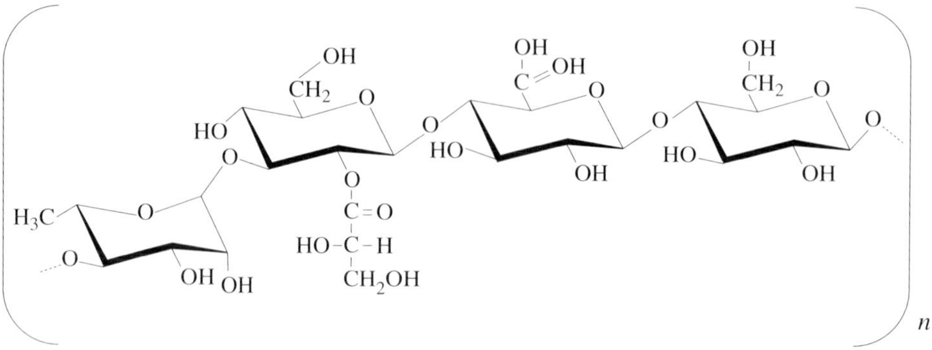

Figure 10.4 Structure of the gellan gum.

pH range for production is between 6.5 and 7 (Meng et al. 2013). Alkaline or acidic conditions can decrease bacterial growth and subsequently gellan production. Composition of medium is of high importance in gellan production. Carbohydrates, such as glucose, fructose, maltose, sucrose, and mannitol can be used as carbon sources. The amount of carbon source should be in the range of 2–4% of the fermentation medium mass. Increased amounts of gellan can be produced with excess carbon and limited nitrogen sources (Hasheminya and Dehghannya 2013; Meng et al. 2013; Iurciuc et al. 2016).

### 10.6.3 Food Applications of Gellan Gum

Gellan has a broad range of applications in various industries, such as pharmacy, food, and biomedicine. It can be used as thickening, gelling, emulsifying, stabilizing, and food film formatting agents in food industry. Also gellan can be used to replace undesired ingredients, such as fats, obtaining low-calorie foods, or satiety increasing products (Hasheminya and Dehghannya 2013). One of the important roles is an ability to give support to encapsulation of various bioactive compounds, allowing them to be protected, used in controlled release systems and functional foods to elicit numerous benefits such as prevention and treatment of chronic diseases. Gellan provides better structure, texture, and flavor in many foods compared to gelatine. This bioploymer also can form gels in a short period of time and offers an optimal structure and texture to the products. For example, gellan is used in baking industry by replacing effectively jellies made of starch. In addition, gellan gels prevent loss of moisture in sweet foods, and offers greater clearness of the gelatine desserts. The melting temperature of a gellan gel can be altered, which helps to keep foods soft and juicy. The water-binding characteristic of gellan increases food stability because of the modified starch. The heating–cooling process includes the removal of the starch effect on food flavor by giving shape and structure. This can be applied to many foods such as meat, fruit, and sweets (Hasheminya and Dehghannya 2013; Iurciuc et al. 2016).

Production of liquid gels with the aid of biopolymers is a modern method that gives the opportunity of creation of a unique texture and properties in food products. Liquid gels are viscous solutions that act like strong gels under low shear speeds, and as liquids under high shear speeds. The possibility of creating tailored textures with unique characteristics have led to the production of liquid gels from gellan, which are utilized in the food industry. Liquid gels that are obtained from gellan are fluid or paste-like materials, which show soft texture with a yield stress. Liquid gels with these characteristics give a suitable surrounding for a good texture formation and inhibition of sedimentation or suspension of particles. This is used in production of cacao, fibers, fruit juices with pulps, and nonsoluble minerals. Particles that are suspended in the production process are trapped in liquid gel structures. Liquid gel can be produced through three different methodologies: gellan heating at 70–95 °C followed by gradual cooling and stirring, solution heating and sudden cooling and gellan dissolution in cold water and gel formation by adding ions and stirring (Hasheminya and Dehghannya 2013).

Mixed gels are another application of gellan gels used in food industry. They are obtained from numerous hydrocolloids. Studies have shown that food containing mixed gels show better characteristics and a mixture of such gels is much more efficient than using a single type of gel. These gels are used to produce functional foods, which have economic and health benefits. One study performed on the nutritional gels with carrot pulp proved that they can be prepared with gellan (Hasheminya and Dehghannya 2013).

Together with the gellan gels already described, food films can be produced from the gellan as well. Edible films are defined as a thin layer of biopolymers coated on the external surface of food products. The reasons why they are useful in food industry are their biodegradability, ability to control respiration and moisture loss, retarding inappropriate textural changes and forming a suitable barrier to inhibit oil uptake in different foods (Meng et al. 2013).

Gellan is found to be an economic biopolymer since it is suitable for producing gelatine desserts, jams, and other sweets. Additional advantages are easy gelling properties and an adjustable moisture content. Desserts prepared with gellan polymer tend to have a great transparency. They can be kept at room temperature because of the high melting point of gellan, which is a great advantage over other polymers. For preparation of such products the generally used gellan concentration is 0.3%. Moreover, pectin can be replaced with gellan for the production of jam. Gellan offers products with less calories than pectin and much lower syneresis with desired sensory properties (Hasheminya and Dehghannya 2013). Besides applications in food industry, gellan is a suitable biopolymer for coating and protecting of probiotics. One such role is in microencapsulation due to its biodegradability, heat stability, high resistance to pH, gel strength control, and resistance to many enzymes in the gastrointestinal tract. Scientists have used gellan in combination with xanthan for coating bifidobacterium species and found that the microencapsulated bacteria were stable in acidic conditions (Iurciuc et al. 2016). It was found that *B. infantis* microencapsulated in gellan gum and xanthan survived five weeks in pasteurized yogurt at the pH range 1.5–2.5 (Sun and Griffiths 2000).

## 10.7 Polyhydroxyalkanoates (PHAs)

### 10.7.1 General Properties of PHAs

PHA is a polymer produced in nature through bacterial fermentation of sugars or lipids. PHAs are linear polyesters that can be modified to make many different useful products. Their biodegradability and naturality are beneficial in application of single-use packaging and agriculture. Poly(3-hydroxybutyrate) (PHB) is the most widespread and comprehensively characterized member of the PHA family. It is a homo-polymer

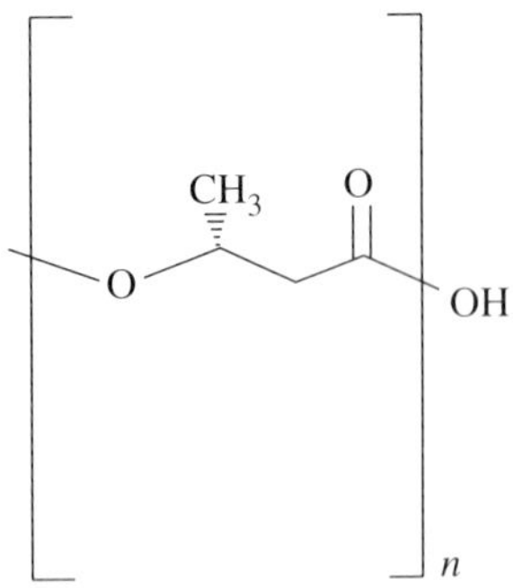


**Figure 10.5** Structure of PHB.

of 3-hydroxybutyrate. PHAs applications include packaging of materials, including films, boxes, coatings, fibers, and foam materials, biofuels, medical implants, and drug delivery carriers. In order to use this polymer for large-scale applications two main limiting factors that have to be overcome are the cost and properties of PHAs (Bugnicourt et al. 2014)

PHB was the first discovered polymer (and hence the most extensively studied and characterized) member of the PHA family. Inside a membrane-enclosed inclusion, this polymer can accumulate up to 80% of the dry cell weight in many bacteria. Properties of this polymer are very similar to conventional plastics. PHB is a linear polyester of D-(−)-3-hydroxybutyric acid, which was first discovered in bacteria (Figure 10.5) (Pötter et al. 2004). The aforementioned accumulation of PHB in intracellular granules, is observed in a wide variety of Gram-positive and Gram-negative bacteria and takes place when nutrient limitation other than the carbon source takes place. Depending on the organism, conditions of growth and method of extraction, molecular weight of PHB varies, but usually PHB is known as high molecular weight homopolymer. Some of the most important properties of this polymer are thermoplasticity and biodegradability. Due to these properties, it has attracted considerable commercial interest, and with the immense research studies completed PHB can be extruded, molded, spun into fibers, made into films and used to make heteropolymer. One of the PHB copolymers, poly(hydroxybutyrate-co-hydroxyvalerate) (PHBV), (Figure 6) (Boufarguine et al. 2012), is less brittle than PHB, making PHBV potentially more usable in a wider range of applications. Research has shown that PHB has the valuable property of being degraded into D-3-hydroxybutyrate (HB), which is a natural constituent of human blood (Pötter et al. 2004). As a result, this homopolymer is biocompatible and is applicable in biomedical applications, such as tissue engineering scaffolds and drug carriers (Bugnicourt et al. 2014; Khosravi-Darani and Bucci 2015).

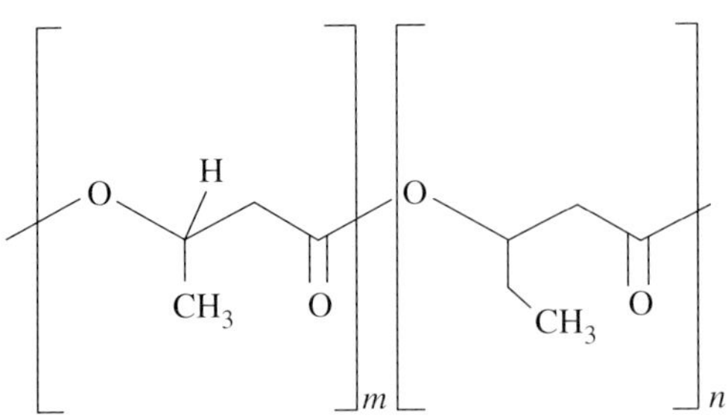


**Figure 10.6** Structure of PHBV.

### 10.7.2 Food Applications of PHAs

PHAs are known to have various applications in packaging, medical, and disposal usage. Copolymer, PHBV can be used for packaging of films, blow-molded bottles, and as a coating on paper. Furthermore, PHBV medical applications in reconstructive surgery are extremely important. It can be used as a scaffold due to slow hydrolytic degradation and biocompatibility. Some PHAs are applicable in coatings and medical temporary implants (e.g., scaffolding for the regeneration of nerve axons and arteries). On the other hand, amphiphilic PHA copolymers have significant applications in drug delivery, tissue engineering and cardiovascular area (e.g., heart valves, artery augments, vascular grafts, cardiologic stents, implants, pericardial patches, microparticulate carriers, and dressing tablets) (Bugnicourt et al. 2014).

According to the literature PHA-based films showed good characteristics in food packaging applications due to their renewability, biodegradability, and water vapor barrier properties. PHBs are a good light barrier in both visible and ultraviolet light regions. Comparison of PHBs with synthetic thermoplastics identifies many similarities, but widespread usage of PHB has been limited because of (i) high cost; (ii) narrow melt processing window due to brittleness; and (iii) low thermal stability in the molten state (Pötter et al. 2004). Low thermal stability of PHB during process decreases the viscosity and molar mass. PHB production from various cheap carbon sources has been reported; however, the cost of carbon source is still approximately 40% of the total operating cost (Bugnicourt et al. 2014; Khosravi-Darani and Bucci 2015).

## 10.8 Scleroglucan

### 10.8.1 General Properties of Scleroglucan

Scleroglucan is a polysaccharide produced by *Sclerotium* fungi. The polymer backbone is built of (1, 3)-linked β-D-glucopyranosyl (glcp) units with single glcp side chains linked with β-(1, 6) to every third residue in the main chain (Figure 10.7). It is linked to the C-2 atoms of the glucose units on the backbone in the strands that are parallel and the side chains are directed away from the helix. It is linked to the C-2 atoms of the glucose units on the backbone (O-2(A), O-2(B), and O-2(C) in the strands that are aligned parallel

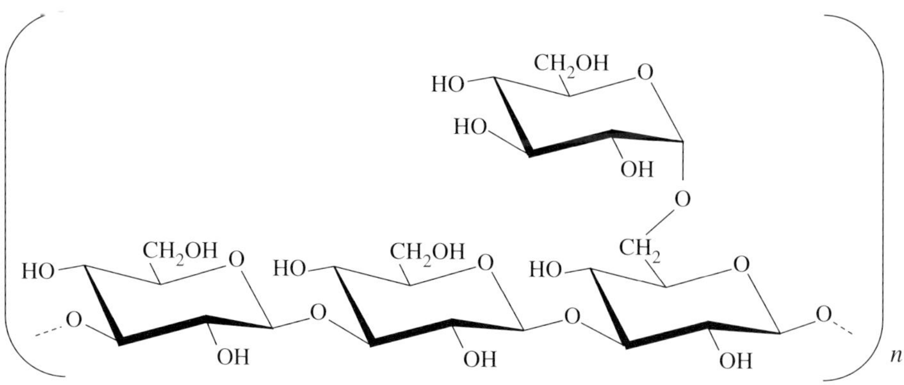


**Figure 10.7** Scleroglucan structure.

and the side chains are directed away from the helix core. Scleroglucan has a triple helix conformation (triplex) both in aqueous solution and in the solid state (Castillo et al. 2015). It is used in various applications such as food, secondary oil recovery, cosmetics, ceramic glazes, paints, etc. (Palleschi et al. 2005).

### 10.8.2 Production Processes for Scleroglucan

Production of scleroglucan is usually carried out in stirred-tank reactors using a sterile medium under aseptic management of the culture under submerged aerobic conditions with the selected producing strain. A successful culture broth develops a gel-like consistency, and scleroglucan synthesis proceeds along with mycelial growth (Coviello et al. 2005). A sharp drop in pH, which is usually around 2–2.5, is typically observed during the first 12–24 h of cultivation, mostly due to the accumulation of oxalic acid (Castillo et al. 2015). In order to obtain high yields of a consistent polymer, it is critical to standardize a large-scale production process with a given strain under controlled conditions. Making changes in culture medium composition, process parameters or even the downstream processing can affect the scleroglucan recovery rate and quality, with eventual variations in biopolymer chemical, physical, or biological properties (Coviello et al. 2005). Scleroglucan solutions can be stable up to 100–120 °C, and across a wide pH range (e.g., pH 1–13). These polymers are triplex and show the tendency to form thermos-reversible gels at temperatures around 7 °C, due to a weakly interacting triple-helix cross-linking mechanism (Castillo et al. 2015).

### 10.8.3 Food Applications of Scleroglucans

Scleroglucans have various industrial applications. The production value of scleroglucan over xanthan is greater efficiency and stability, which makes it more suitable for different application areas. Initially, scleroglucan has been used in the oil industry for thickening, enhanced oil recovery, and discharge of drilling muds.

According to literature, scleroglucan could be used for stabilization of dressings and ice creams in food industry. However, scleroglucan has not been approved yet by food safety legislation agencies in Europe, and in United States, its usage is limited. Research that was conducted in Japan showing that numerous Japanese patents describe quality improvements of frozen or heat-treated edibles, such as Japanese cakes, steamed foods, rice crackers, and bakery products in which scleroglucan had been used (Viñarta et al. 2006). Furthermore, scleroglucan shows important value in cosmetic industry where it can be applied in hair control compositions and in various skin care preparations, creams, and protective lotions (Schmid et al. 2011).

## 10.9 Xanthan Gum

### 10.9.1 General Properties of Xanthan Gum

Xanthan is a complex exopolysaccharide synthesized by plant-pathogenic bacterium *Xanthomonas campestris*. Exopolysaccharides produced by these pathogenic bacteria have a characteristic feature of protection against adverse environmental conditions

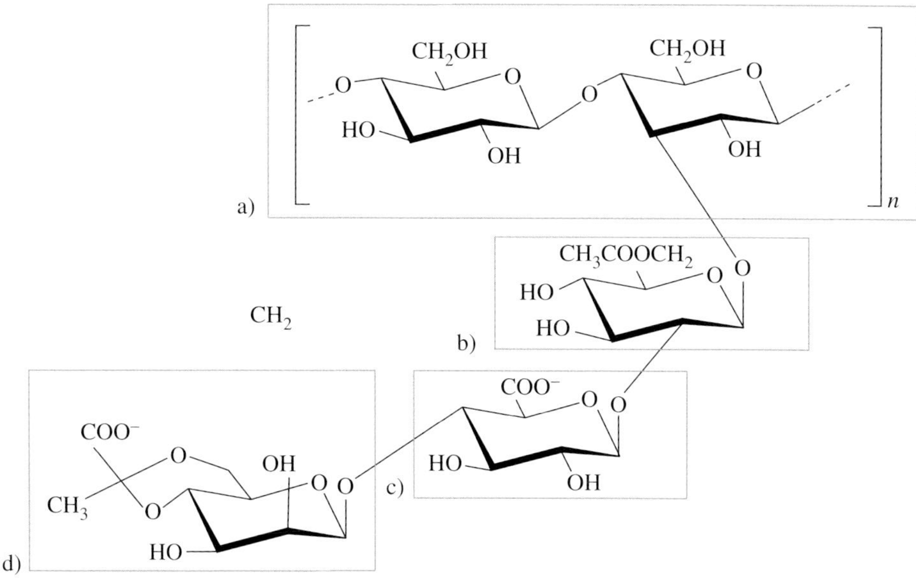


**Figure 10.8** Structure of xanthan polymer.; a) glucose backbone, b) mannose c) glucuronic acid d) terminal mannose.

such as drying, temperature oscillations, radiation, and adhesion (Luvielmo et al. 2016). Xanthan gum is commonly applied as a thickening and stabilizing agent in different types of food and industrial products. Primary structure of this polymer is composed of repeating pentasaccharide units made out of two glucose, two mannose, and one glucuronic acid, in the molar ratio 2.8 : 2.0 : 2.0 (Figure 10.8) (Garcia-Ochoa et al. 2000; Moffat et al. 2016).

### 10.9.2 Production Processes of Xanthan Gum

The process of production of xanthan includes several steps. First, the chosen microbial is grown on solid media or in liquid media and used to inoculate the culture in large bioreactors. The mode of operation, medium composition, type of bioreactor, temperature, pH, and dissolved oxygen concentration influence both the microorganism growth and xanthan production. At the end of the fermentation, cells are usually removed via filtration or centrifugation operations from the culture broth that contains xanthan, bacterial cells, and numerous other chemicals. Next step is purification, where precipitation can also be included by using water-miscible nonsolvents, followed by the addition of certain salts and pH adjustments. The product is then mechanically dewatered and dried. The dried product is milled and packed into containers with a low permeability to water.

### 10.9.3 Food Applications of Xanthan Gum

Xanthan gum has a broad spectrum of applications ranging from the food industry to oil drilling. Typical food industries where xanthan gum is used are; salad dressings, sauces, gravies, dairy products, desserts, low-calorie foods, and convenience foods. Xanthan gum is also used in cosmetics, coatings, polishes, and in agriculture. The importance of the use of xanthan gum in food industry comes from the fact that when it is dispersed either in hot or cold water, the resulting aqueous dispersions are thixotropic. The weak gel-like structure results in an unusually low shear rate viscosity at low polymer concentrations, which is widely used to solidify aqueous samples and helps in stabilization of emulsions, froths, and some types of suspensions (Garcia-Ochoa et al. 2000). Reversible shear thinning behavior of the sample allows manipulation of processes such as spreading, pumping, pouring, and spraying (Katzbauer 1998).

Furthermore, this polysaccharide is very important in the bakery industry. Increasing water binding during baking and storage accelerates the process as well. It also extends the time in which baked and frozen dough products can be used. Xanthan gum can replace eggs in soft-baked products. Here egg content can be minimized without affecting taste and appearance of the product. Also, xanthan prevents the filling of baked products from being absorbed by the pastry. Xanthan gum is also good at controlling cake shape during baking. It can be found in many gluten-free products and in some mixes where this biopolymer promotes thickening of the batter or dough. Due to xanthan's high alcohol tolerance, it can be used to enhance the uniformity of cream liqueurs (Garcιa-Ochoa et al. 2000; Sharma et al. 2006)

Crystal formation in ice creams and frozen foods can be prevented by adding xanthan in the mixture. Suspension of fruit pulps and other particles in liquids is also one of the characteristics of this polymer. In products that contain fruit fillings, xanthan is also used for prevention of migration of filling. Xanthan gum mixed with guaran or locust bean gum can improve the flavor and texture of reduced fat cream cheese. Furthermore, it is added as an element that maintains the smoothness of yoghurts, desserts, and salad dressings. Xanthan can prolong the stability of a suspension up to a year. It can also be used in combination with propylene glycol alginate or pectin for different mixtures (Garcιa-Ochoa et al. 2000; Sharma et al. 2006).

As already noted, xanthan gum can be used as an emulsifier. Blending oil and water into a more unified mix prevents separation over shelf life. In processed meats, xanthan also helps to bind water, emulsify, and improve tenderness. In pet food, xanthan gum enhances homogeneity of soft pet treats and works well with other gums to produce gravy. Besides these, buttered syrups and chocolate toppings also contain xanthan. Xanthan is an ideal choice for products where thickness is needed since it has high acid stability (Sharma et al. 2006).

## 10.10 Dextran

### 10.10.1 General Properties of Dextran

Dextrans are a group of homopolysaccharides composed of a linear chain of α-(1, 6) glycosidic linkages that may form branches on the main chain. It was first observed by Louis Pasteur, but this bioploymer's potential in food industry was not investigated until the 1950s. Dextran is one of the oldest bacterial polysaccharides with a multitude of functions. Structural properties of dextrans, such as chain length, viscosity, and molecular mass, are highly dependent on the strain-producing dextran, dextransucrase

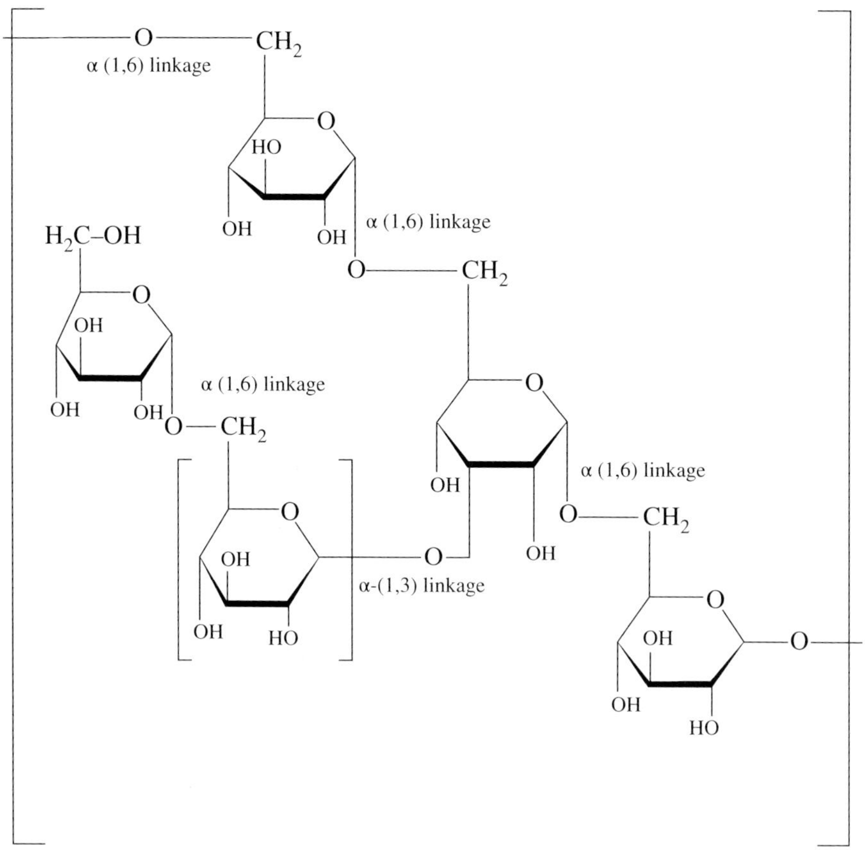


Figure 10.9 Structure of Dextran polymer.

enzyme conformation and the process parameters. Some dextrans are composed of almost solely 1,6 linkages, whereas others can contain 1,2, 1,3 and/or 1,4 linkages given the polymer-specific properties (Ioan et al. 2000; Galle and Arendt 2014; Wolter et al. 2014; Baruah et al. 2017). Dextran synthesis can be inhibited when salt in the environment is removed. Even though numerous studies have been conducted in order to determine and evaluate dextran solution properties, scientists still find it a challenge to prepare dextran with precisely controlled properties (Zannini et al. 2016) (Figure 10.9)

### 10.10.2 Production Processes of Dextran

Dextran is an exopolysaccharide synthesized by *Streptococcus, Lactobacillus,* and some *Weisella* species and is very sensitive to environmental conditions like substrate concentration, pH, temperature, and salinity. Because different strains of bacteria belonging to the same species can produce dextran with varying structures, it is, in theory, possible to produce dextran according to specific needs. For example, keeping the substrate levels low tends to give dextran a higher molecular weight (Das and Goyal 2014; Zannini et al. 2016; Baruah et al. 2017). Ideal environmental conditions vary according to the species used to produce dextran, but optimal temperature and pH values tend to vary between 30–45 °C and 3.6–6.0, respectively. Amounts of nitrogen, phosphorus, ash in the media, viscosity of the environment, and solubility of components are other physicochemical properties that affect dextran quality. For food, pharmaceutical, and medical applications, dextran with lower molecular weight is usually preferred, whereas dextran polymers with high molecular weight tend to be used in industrial applications. Dextran-producing lactic acid bacteria can be isolated from fermented foods and food products such as yogurt, wine, cheese, sauerkraut, and tomato juice (Das and Goyal 2014; Du et al. 2017; Zarour et al. 2017). Industrial dextran is generally produced through fermentation of sucrose-rich substrates. After fermentation, the biopolymer is isolated through centrifugation, drying, dissolving, dialyzing, and lyophilization (Das and Goyal 2014; Du et al. 2017).

### 10.10.3 Food Applications of Dextran

The backbone of dextran is dominantly composed of α-1,6 linkages resulting in a high water absorption property, which can increase the polymer volume up to 10%. Dextran can also be used as a texturing agent, emulsifier, preserver, thickener, and stabilizer for frozen foods. In some cases, iron can be bound to dextran, which can be used for treatment of anemia (Baruah et al. 2017; Zarour et al. 2017). As an ingredient in bakery products, dextran can soften foods by improving moisture retention, prevent crystallization, and increase rheological properties. It can also be used to preserve protein-based food (cheese, vegetable, fish, meat, etc.) surfaces to obtain a tight seal. Sourdough is a predough product obtained through fermentation from yeasts and bacteria, and using dextran in a sourdough mixture can improve shelf life, volume, stability, crumb texture, aroma, nutritional value, and texture of the final product (Galle and Arendt 2014; Wolter et al. 2014; Baruah et al. 2017). Dextrans with high molecular weight and linear chain structures have been found to be more effective in bakery applications. Moreover, dextran can be metabolized by gut bacteria (making it easier for digestion), where it is broken down into acetate, propionate, and butyrate, which can reduce cholesterol and triglyceride levels, increase insulin sensitivity, and improve gut fauna (Galle and Arendt 2014; Baruah et al. 2017). Dextran has similar nutritional values to starch and is widely used in bakery, confectionery, frozen, and dried foods products to prevent oxidation, crystallization, and chemical changes, and improve viscosity, stability, volume, and flavor (Galle and Arendt 2014; Zannini et al. 2016; Baruah et al. 2017).

## 10.11 Conclusions

The use of biopolymers in food industry is significant, and microbial production processes are favored over plants due to the lower costs and easily controlled fermentation conditions of production, as well as the fact that microbial systems are devoid of any risks due to seasonal changes. Moreover, productions of biopolymers are eco-friendly compared to synthetic polymers, since microbial processes can be driven by natural enzymatic and metabolic pathways or by the use of genetically engineered systems. These developed biotechnological techniques enable skipping intermediate reactions and decrease the production time. These are very important advantages in lowering the production cost and increasing applicability. As a food ingredient, the main problem of polysaccharides is high manufacturing costs, and many polysaccharides still cannot be produced on an industrial scale. In order to overcome this problem, new methods such as enzymatic productions or genetic engineering methods should be developed. Also, alternative carbon sources could be utilized from waste products in order to decrease cost. There are various biopolymers and several characteristics, so each polymer should be investigated individually.

Levan is an unusual fructan-type homopolysaccharide and is an important fat substitute due to its tasteless characteristic. In recent years, numerous functional food studies with levan have focused on its unique properties, like biocompatibility, biodegradability, and colorlessness. High production costs have limited the industrial applications, but among other microbial levan producers, extremophilic *Halomonas* production systems provide advantages such as enabling unsterile and low-cost levan production. This alternative system could enable the production of levan at an industrial scale and increase its applications particularly in food industry.

The flexible structure of pullulan makes this polymer water-soluble, which allows it to be used in various applications, including food. Being odorless, nonmutagenic, non-carcinogenic, tasteless, and edible has certainly made pullulan widely applicable in the food industry. However, production costs are a serious problem in industrial scale and have so far prevented the polymer from reaching its full potential. Improved production processes and expanded competition could open up new pathways for pullulan.

Alginate is a safe food additive and has an E number. Like many other polymers, alginate has thickening, gelling, stabilizing, and colloidal properties that make it an excellent food ingredient. For many years, researchers have focused on alginate production to extend the shelf life of bread, meat, and vegetables. It is certain that, as a cheap and nontoxic ingredient, alginate will be in food industry for a long time.

Curdlan is used as thickening and fat-like agents in food industry, although it is insoluble in water. This decreases its application, but its gelling property makes it an important candidate as a food ingredient. As a tasteless and flexible polymer, applications will increase and the disadvantages due to its insolubility will be overcome with further studies.

Gellan has been a common additive in the food, beverage, and personal care since its discovery more than three decades ago. Gellan has been found to be an economic biopolymer, since it is suitable for producing gelatine desserts, jams, and other different sweets in food industry.

PHAs are polymers that can be modified to make many types of useful products. Their biodegradability is beneficial in application of single-use packaging and agriculture. PHAs application includes packaging of materials, such as films, boxes, coatings, fibers, and foam materials. The properties of PHAs must be improved and production costs must be reduced in order to expand this polymer to large-scale applications.

The great advantage of scleroglucan is its stability, which makes it suitable for different application areas. Scleroglucan is used in the oil industry for thickening. It has the potential to be used to stabilize dressings and ice creams in food industry but has not been approved yet by food safety legislation, and studies have focused on this in recent years.

Xanthan has a characteristic feature of protection against adverse environmental conditions such as drying, temperature oscillations, radiation, and adhesion. This polysaccharide is very important in the bakery industry, where it increases water binding during baking and storage. Furthermore, it can be found in much gluten-free products, which gives it great market value.

Being one of the oldest biopolymer discovered, it is not surprising that dextran has been utilized in numerous industries with different purposes. It can not only be used to improve shelf life of protein-based, frozen, and pastry food products, but it can also enhance the gut microbiota of the consumer. Considering how widely dextran is used today, there is no doubt that this biopolymer will continue to shape various processes and industries in the future.

Consequently, applications of microbial polymers in food industry are rising in global market. Different approaches are required to decrease the production costs with increasing the yields in order to discard all hazardous synthetic polymers from food industry.

One very exciting area for microbial production is the use of extremophiles. This largely overlooked area has come under increased attention of both researchers and industry. Extremophiles like thermophiles could be advantageous due to their fast metabolism at high production temperatures as well as lower risk of contamination by mesophiles. Likewise, halophiles that require high concentrations of salt are also gaining increased popularity due to their production conditions, enabling open nonsterile production systems.

## References

Abbamondi, G.R., Suner, S., Cutignano, A. et al. (2016). Erratum to: Identification of N-Hexadecanoyl-L-homoserine lactone (C16-AHL) as signal molecule in halophilic bacterium *Halomonas smyrnensis* AAD6. *Ann. Microbiol.* 66: 1335–1335.

Adamberg, K., Tomson, K., Talve, T. et al. (2015). Levan enhances associated growth of Bacteroides, Escherichia, Streptococcus and Faecalibacterium in fecal microbiota. *PLoS One* 10, e0144042.

Ahmad, N.H., Mustafa, S., and Che Man, Y.B. (2015). Microbial polysaccharides and their modification approaches: a review. *Int. J. Food Prop.* 18: 332–347.

Aliste, A.J., Vieira, F.F., and Del Mastro, N.L. (2000). Radiation effects on agar, alginates and carrageenan to be used as food additives. *Radiat. Phys. Chem.* 57: 305–308.

Ana, R.V., Ferreira, V.D.A., and Coelhoso, I.S.A.B.E.L.M. (2016). Polysaccharide based membranes in food packaging applications. *Membranes* 6: 1–17.

Ates, O. (2015). Systems biology of microbial exopolysaccharides production. *Front. Bioeng. Biotechnol.* 3.

Banik, R. and Santhiagu, A. (2006). Improvement in production and quality of gellan gum by Sphingomonas paucimobilis under high dissolved oxygen tension levels. *Biotechnol. Lett* 28: 1347.

Baruah, R., Maina, N.H., Katina, K. et al. (2017). Functional food applications of dextran from Weissella cibaria RBA12 from pummelo (*Citrus maxima*). *Int. J. Food Microbiol.* 242: 124–131.

Bohn, J.A. and Bemiller, J.N. (1995). (1→ 3)-β-D-Glucans as biological response modifiers: a review of structure-functional activity relationships. *Carbohydr. Polym.* 28: 3–14.

Booten, K., De Soete, J., and Frippiat, A. (1998). Fructan-and/or polydextrose-containing dairy powders, process for preparing same and their use. *Eur. Patent Appl. EP0821885A1.*

Boufarguine, M., Guinault, A., Miquelard-Garnier, G., and Sollogoub, C. (2012). PLA/PHBV Films with Improved Mechanical and Gas Barrier Properties. *Macromolecular Materials and Engineering* 298: 1065–1073.

Brivonese, A.C. and Sutherland, I.W. (1989). Polymer production by a mucoid stain of Azotobacter vinelandii in batch culture. *Appl. Microbiol. Biotechnol.* 30: 97–102.

Brownlee, I., Allen, A., Pearson, J. et al. (2005). Alginate as a source of dietary fiber. *Crit. Rev. Food Sci. Nutr.* 45: 497–510.

Brownlee, I.A., Seal, C.J., Wilcox, M. et al. (2009). *Applications of Alginates in Food.* Alginates: *Biology and applications.* Springer.

Bugnicourt, E., Cinelli, P., Lazzeri, A. er al. (2014). Polyhydroxyalkanoate (PHA): Review of synthesis, characteristics, processing and potential applications in packaging.

Castillo, N.A., Valdez, A.L., and Fariña, J.I. (2015). Microbial production of scleroglucan and downstream processing. *Front. Microbiol.* 6.

Cheng, K.-C., Demirci, A., and Catchmark, J.M. (2011). Pullulan: biosynthesis, production, and applications. *Appl. Microbiol. Biotechnol.* 92: 29.

Costa, R.R., Neto, A.I., Calgeris, I. et al. (2013). Adhesive nanostructured multilayer films using a bacterial exopolysaccharide for biomedical applications. *J. Mater. Chem. B* 1: 2367–2374.

Coviello, T., Palleschi, A., Grassi, M. et al. (2005). Scleroglucan: a versatile polysaccharide for modified drug delivery. *Molecules* 10: 6–33.

Das, D. and Goyal, A. (2014). Characterization and biocompatibility of glucan: a safe food additive from probiotic Lactobacillus plantarum DM5. *J. Sci. Food Agric.* 94: 683–690.

De Oliveira, M.R., Da Silva, R.S.S.F., Buzato, J.B., and Celligoi, M.A.P.C. (2007). Study of levan production by Zymomonas mobilis using regional low-cost carbohydrate sources. *Biochem. Eng. J.* 37: 177–183.

Dols, M., Chraibi, W., Remaud-Simeon, M. et al. (1997). Growth and energetics of Leuconostoc mesenteroides NRRL B-1299 during metabolism of various sugars and their consequences for dextransucrase production. *Appl. Environ. Microbiol.* 63: 2159–2165.

Drury, W.J., Characklis, W.G., and Stewart, P.S. (1993). Interactions of 1 μm latex particles with pseudomonas aeruginosa biofilms. *Water Res.* 27: 1119–1126.

Du, R., Xing, H., Yang, Y. et al. (2017). Optimization, purification and structural characterization of a dextran produced by L. mesenteroides isolated from Chinese sauerkraut. *Carbohydr. Polym.* 174: 409–416.

Erginer, M., Akcay, A., Coskunkan, B. et al. (2016). Sulfated levan from Halomonas smyrnensis as a bioactive, heparin-mimetic glycan for cardiac tissue engineering applications. *Carbohydr. Polym.* 149: 289–296.

Ferreira, A.R., Alves, V.D., and Coelhoso, I.M. (2016). Polysaccharide-based membranes in food packaging applications. *Membranes* 6: 22.

Food Standards Agency (2002). *Food additives legistlation: guidance notes.* London: Food Standards Agency.

Funami, T., Funami, M., Tawada, T., and Nakao, Y. (1999). Decreasing oil uptake of doughnuts during deep-fat frying using curdlan. *J. Food Sci.* 64: 883–888.

Galle, S. and Arendt, E.K. (2014). Exopolysaccharides from sourdough lactic acid bacteria. *Crit. Rev. Food Sci. Nutr.* 54: 891–901.

Garcia-Ochoa, F., Santos, V., Casas, J., and Gomez, E. (2000). Xanthan gum: production, recovery, and properties. *Biotechnol. Adv.* 18: 549–579.

Giavasis, I. (2014). Bioactive fungal polysaccharides as potential functional ingredients in food and nutraceuticals. *Curr. Opin. Biotechnol.* 26: 162–173.

Göksungur, Y., UZUNOĞULLARI, P., and DAĞBAĞLI, S. (2011). Optimization of pullulan production from hydrolysed potato starch waste by response surface methodology. *Carbohydr. Polym.* 83: 1330–1337.

González-Garcinuño, Á., Tabernero, A., Domínguez, Á. et al. (2017). Levan and levansucrases: Polymer, enzyme, micro-organisms and biomedical applications. *Biocatal. Biotransform.* 1–12.

Gupta, S.K., Pal, A.K., Sahu, N.P. et al. (2015). Effects of dietary microbial levan on growth performance, RNA/DNA ratio and some physio-biochemical responses of Labeo rohita (Hamilton) juveniles. *Aquacult. Nutr.* 21: 892–903.

Han, Y.W. (1989). Levan production byBacillus polymyxa. *J. Ind. Microbiol. Biotechnol.* 4: 447–451.

Han, Y.W. and Clarke, M.A. (1990). Production and characterization of microbial levan. *J. Agric. Food. Chem.* 38: 393–396.

Harada, T., Masada, M., Fujimori, K., and Maeda, I. (1966). Production of a firm, resilient gel-forming polysaccharide by a mutant of Alcaligenes faecalis var. myxogenes 10 C3. *Agri. Biol. Chem.* 30: 196–198.

Hasheminya, S.-M. and Dehghannya, J. (2013). An overview on production and applications of gellan biopolymer. *Int. J. Agric. Crop Sci.* 5: 3016.

Hay, I.D., Rehman, Z.U., Moradali, M.F. et al. (2013). Microbial alginate production, modification and its applications. *Microb. Biotechnol.* 6: 637–650.

Høidal, H.K., Svanem, B.I.G., Gimmestad, M., and Valla, S. (2000). Mannuronan C-5 epimerases and cellular differentiation of Azotobacter vinelandii. *Environ. Microbiol.* 2: 27–38.

Inthanavong, L., Tian, F., Khodadadi, M., and Karboune, S. (2013). Properties of Geobacillus stearothermophilus levansucrase as potential biocatalyst for the synthesis of levan and fructooligosaccharides. *Biotechnol. Progr.* 29: 1405–1415.

Ioan, C.E., Aberle, T., and Burchard, W. (2000). Structure properties of dextran. 2. Dilute solution. *Macromolecules* 33: 5730–5739.

Iurciuc, C., Savin, A., Lungu, C. et al. (2016). Gellan food applications. *Cellul. Chem. Technol.* 13.

Jakob, F., Pfaff, A., Novoa-Carballal, R. et al. (2013). Structural analysis of fructans produced by acetic acid bacteria reveals a relation to hydrocolloid function. *Carbohydr. Polym.* 92: 1234–1242.

Jang, K.-H., Song, K.-B., Kim, C.H. et al. (2001). Comparison of characteristics of levan produced by different preparations of levansucrase from Zymomonas mobilis. *Biotechnol. Lett* 23: 339–344.

Ji, W.-K., Lan, W., Dong, X. et al. (2010). Application in noodle of the curdlan. *China Food Addit.* 1: 38–43.

Jiang, L. (2013). Effect of nitrogen source on curdlan production by Alcaligenes faecalis ATCC 31749. *Int. J. Biol. Macromol.* 52: 218–220.

Kai, A., Ishino, T., Arashida, T. et al. (1993). Biosynthesis of curdlan from culture media containing 13C-labeled glucose as the carbon source. *Carbohydr. Res.* 240: 153–159.

Katzbauer, B. (1998). Properties and applications of xanthan gum. *Polym. Degrad. Stab.* 59: 81–84.

Kazak Sarilmiser, H., Ates, O., Ozdemir, G. et al. (2015). Effective stimulating factors for microbial levan production by Halomonas smyrnensis AAD6T. *J. Biosci. Bioeng.* 119: 455–463.

Kazak, H., Oner, E., and Dekker, R.F. (2010). Extremophiles as sources of exopolysaccharides. In: *Handbook of Carbohydrate Polymers: Development, Properties and Applications* (ed. R. Ito and Y. Matsuo), 605–619.

Khosravi-Darani, K. and Bucci, D. (2015). Application of poly (hydroxyalkanoate) in food packaging: Improvements by nanotechnology. *Chem. Biochem. Eng. Q.* 29: 275–285.

Küçükaşik, F., Kazak, H., Güney, D. et al. (2011). Molasses as fermentation substrate for levan production by Halomonas sp. *Appl. Microbiol. Biotechnol.* 89: 1729–1740.

Luo, Z., Guo, Y., Liu, J. et al. (2016). Microbial synthesis of poly-$\gamma$-glutamic acid: current progress, challenges, and future perspectives. *Biotechnol. Biofuels* 9: 134.

Luvielmo, M.D.M., Borges, C.D., Toyama, D.D.O. et al. (2016). Structure of xanthan gum and cell ultrastructure at different times of alkali stress. *Brazilian J. Microbiol.* 47: 102–109.

Matricardi, P., Cencetti, C., Ria, R. et al. (2009). Preparation and characterization of novel gellan gum hydrogels suitable for modified drug release. *Molecules* 14: 3376–3391.

Meng, Y.C., Hong, L.B., and Jin, J.Q. (2013). A study on the gelation properties and rheological behavior of gellan gum. *App. Mechan. Mater.*. Trans Tech Publ, 20-24.

Moffat, J., Morris, V.J., Al-Assaf, S., and Gunning, A.P. (2016). Visualisation of xanthan conformation by atomic force microscopy. *Carbohydr. Polym.* 148: 380–389.

Moosavi-Nasab, M., Layegh, B., Aminlari, L., and Hashemi, M.B. (2010). Microbial production of levan using date syrup and investigation of its properties. *World Acad. Sci. Eng. Technol.* 44: 1248–1254.

Naessens, M., Cerdobbel, A., Soetaert, W., and Vandamme, E.J. (2005). Leuconostoc dextransucrase and dextran: production, properties and applications. *J. Chem. Technol. Biotechnol.* 80: 845–860.

Nivens, D.E., Ohman, D.E., Williams, J., and Franklin, M.J. (2001). Role of alginate and its O acetylation in formation of *Pseudomonas aeruginosa* microcolonies and biofilms. *J. Bacteriol.* 183: 1047–1057.

Oğuzhan, P. & Yangilar, F. (2013). Pullulan: Production and usage in food industry.

Oner, E.T., Hernandez, L., and Combie, J. (2016). Review of Levan polysaccharide: From a century of past experiences to future prospects. *Biotechnol. Adv.* 34: 827–844.

Osman, A., Oner, E.T., and Eroglu, M.S. (2017). Novel levan and pNIPA temperature sensitive hydrogels for 5-ASA controlled release. *Carbohydr. Polym.* 165: 61–70.

O'toole, A., Ricker, E.B., and Nuxoll, E. (2015). Thermal mitigation of *Pseudomonas aeruginosa* biofilms. *Biofouling* 31: 665–675.

Özcan, E. and Öner, E.T. (2015). Microbial production of extracellular polysaccharides from biomass sources. *Polysacc. Bioact. Biotechnol.* 161–184.

Palleschi, A., Bocchinfuso, G., Coviello, T., and Alhaique, F. (2005). Molecular dynamics investigations of the polysaccharide scleroglucan: first study on the triple helix structure. *Carbohydr. Res.* 340: 2154–2162.

Patel, S., Majumder, A., and Goyal, A. (2012). Potentials of exopolysaccharides from lactic acid bacteria. *Ind. J. Microbiol.* 52: 3–12.

Poli, A., Kazak, H., Gürleyendağ, B. et al. (2009). High level synthesis of levan by a novel Halomonas species growing on defined media. *Carbohydr. Polym.* 78: 651–657.

Pötter, M., Müller, H., Reinecke, F. et al. (2004). The complex structure of polyhydroxybutyrate (PHB) granules: four orthologous and paralogous phasins occur in Ralstonia eutropha. *Microbiology* 150: 2301–2311.

Prajapati, V.D., Jani, G.K., and Khanda, S.M. (2013). Pullulan: an exopolysaccharide and its various applications. *Carbohydr. Polym.* 95: 540–549.

Rafigh, S.M., Yazdi, A.V., Vossoughi, M. et al. (2014). Optimization of culture medium and modeling of curdlan production from Paenibacillus polymyxa by RSM and ANN. *Int. J. Biol. Macromol.* 70: 463–473.

Rehm, B.H. (2010). Bacterial polymers: biosynthesis, modifications and applications. *Nat. Rev. Microbiol.* 8: 578–592.

Roberfroid, M. (1998). Prebiotics and synbiotics: concepts and nutritional properties. *Br. J. Nutr.* 80: S197–S202.

Roberts, E. & Garegg, P. (1998). Levan derivatives, their preparation, composition and applications including medical and food applications. *World Intellectual Property Organization Patent Application No. WO,* 98, 03184.

Rosalam, S. and England, R. (2006). Review of xanthan gum production from unmodified starches by Xanthomonas comprestris sp. *Enzyme and Microbial Technology* 39: 197–207.

Sabra, W. (1999). Microaerophilic production of alginate by Azotobacter vinelandii. Ph. D. dissertation. Technische Universität Braunschweig, Braunschweig, Germany.

Sabra, W. and Zeng, A.P. (2009). Microbial production of alginates: physiology and process aspects. In: *Alginates: Biology and Applications* (ed. B.H.A. Rehm), 153–173. Springer.

Salah, R.B., Chaari, K., Besbes, S. et al. (2010). Optimisation of xanthan gum production by palm date (Phoenix dactylifera L.) juice by-products using response surface methodology. *Food Chem.* 121: 627–633.

Salah, R.B., Jaouadi, B., Bouaziz, A. et al. (2011). Fermentation of date palm juice by curdlan gum production from Rhizobium radiobacter ATCC 6466™: Purification, rheological and physico-chemical characterization. *LWT-Food Sci. Technol.* 44: 1026–1034.

Sam, S., Kucukasik, F., Yenigun, O. et al. (2011). Flocculating performances of exopolysaccharides produced by a halophilic bacterial strain cultivated on agro-industrial waste. *Bioresour. Technol.* 102: 1788–1794.

Schmid, J., Meyer, V., and Sieber, V. (2011). Scleroglucan: biosynthesis, production and application of a versatile hydrocolloid. *Appl. Microbiol. Biotechnol.* 91: 937–947.

Schmid, J., Sieber, V., and Rehm, B. (2015). Bacterial exopolysaccharides: biosynthesis pathways and engineering strategies. *Front. Microbiol.* 6: 496.

Shamala, T., Chandrashekar, A., Vijayendra, S., and Kshama, L. (2003). Identification of polyhydroxyalkanoate (PHA)-producing Bacillus spp. using the polymerase chain reaction (PCR). *J. Appl. Microbiol.* 94: 369–374.

Sharma, B., Naresh, L., Dhuldhoya, N. et al. (2006). Xanthan gum-A boon to food industry. *Food Promotion Chron.* 1: 27–30.

Shih, L., Yu, J.-Y., Hsieh, C., and Wu, J.-Y. (2009). Production and characterization of curdlan by Agrobacterium sp. *Biochem. Eng. J.* 43: 33–40.

Simon, L., Caye-Vaugien, C., and Bouchonneau, M. (1993). Relation between pullulan production, morphological state and growth conditions in *Aureobasidium pullulans*: new observations. *Microbiology* 139: 979–985.

Singh, A.K. and Mallick, N. (2009). Exploitation of inexpensive substrates for production of a novel SCL–LCL-PHA co-polymer by *Pseudomonas aeruginosa* MTCC 7925. *J. Ind. Microbiol. Biotechnol.* 36: 347–354.

Smith, J. and Hong-Shum, L. (2011). *Food additives data book.* Wiley.

Sonnenburg, E.D., Zheng, H., Joglekar, P. et al. (2010). Specificity of polysaccharide use in intestinal bacteroides species determines diet-induced microbiota alterations. *Cell* 141: 1241–1252.

Srikanth, R., Siddartha, G., Reddy, C.H.S. et al. (2015). Antioxidant and anti-inflammatory levan produced from Acetobacter xylinum NCIM2526 and its statistical optimization. *Carbohydr. Polym.* 123: 8–16.

Stone, B.A. and Clarke, A.E. (1992). *Chemistry and Biology of 1, 3-β-Glucans.* Intl Specialized Book Service Inc.

Sun, W. and Griffiths, M. (2000). Survival of *bifidobacteria* in yogurt and simulated gastric juice following immobilization in gellan–xanthan beads. *Int. J. Food Microbiol.* 61: 17–25.

Sutherland, I. (1977). *Microbial exopolysaccharide synthesis.* ACS Publications.

Tako, M. (2015). The principle of polysaccharide gels. *Adv. Biosci. Biotechnol.* 6: 22–36.

Velu, S., Velayutham, V. and Manickkam, S. (2016). Optimization of fermentation media for xanthan gum production from Xanthomonas campestris using Response Surface Methodology and Artificial Neural Network techniques.

Vijayendra, S.V.N. and Shamala, T.R. (2014). Film forming microbial biopolymers for commercial applications—A review. *Crit. Rev. Biotechnol.* 34: 338–357.

Viñarta, S., Molina, O., Figueroa, L., and Fariña, J. (2006). A further insight into the practical applications of exopolysaccharides from *Sclerotium rolfsii*. *Food Hydrocolloids* 20: 619–629.

Visnapuu, T., Mardo, K., and Alamaee, T. (2015). Levansucrases of a *Pseudomonas syringae* pathovar as catalysts for the synthesis of potentially prebiotic oligo-and polysaccharides. *New Biotechnol.* 32: 597–605.

Wang, D., Chen, F., Wei, G. et al. (2015). The mechanism of improved pullulan production by nitrogen limitation in batch culture of Aureobasidium pullulans. *Carbohydr. Polym.* 127: 325–331.

Whitfield, C. (2006). Biosynthesis and assembly of capsular polysaccharides in *Escherichia coli*. *Annu. Rev. Biochem.* 75: 39–68.

Wolter, A., Hager, A.-S., Zannini, E. et al. (2014). Influence of dextran-producing Weissella cibaria on baking properties and sensory profile of gluten-free and wheat breads. *Int. J. Food Microbiol.* 172: 83–91.

Wu, S., Lu, M., Chen, J. et al. (2016). Production of pullulan from raw potato starch hydrolysates by a new strain of *Auerobasidium pullulans*. *Int. J. Biol. Macromol.* 82: 740–743.

Xiao, M., Feng, F. and Lu, L. (2014). Preparation method of levan-contained yoghourt. Google Patents.

Yamamoto, Y., Takahashi, Y., Kawano, M. et al. (1999). In vitro digestibility and fermentability of levan and its hypocholesterolemic effects in rats. *J. Nutr. Biochem.* 10: 13–18.

Yang, M., Zhu, Y., Li, Y. et al. (2016). Production and optimization of curdlan produced by Pseudomonas sp. QL212. *Int. J. Biol. Macromol.* 89: 25–34.

Zannini, E., Waters, D.M., Coffey, A., and Arendt, E.K. (2016). Production, properties, and industrial food application of lactic acid bacteria-derived exopolysaccharides. *Appl. Microbiol. Biotechnol.* 100: 1121–1135.

Zarour, K., Llamas, M.G., Prieto, A. et al. (2017). Rheology and bioactivity of high molecular weight dextrans synthesised by lactic acid bacteria. *Carbohydr. Polym.* 174: 646–657.

Zhang, R. and Edgar, K.J. (2014). Properties, chemistry, and applications of the bioactive polysaccharide curdlan. *Biomacromolecules* 15: 1079–1096.

# 11

# Research and Production of Microbial Functional Sugars and Their Potential for Industry

*Helen Treichel, Simone Maria Golunski, Aline Frumi Camargo, Thamarys Scapini, Tatiani Andressa Modkovski, Bruno Venturin, Eduarda Roberta Bordin, Vanusa Rossetto, and Altemir José Mossi*

*Laboratory of Microbiology and Bioprocesses, Federal University of Fronteira Sul, Erechim, Rio Grande do Sul, Brazil*

## 11.1 Introduction

Several agencies such as the United Nations (UN) and the Food and Agriculture Organization of the United Nations (FAO) have highlighted concerns over food security, mainly regarding the prevention of foodborne diseases that represent millions of deaths worldwide (Cook et al. 2016).

Consumers' interest in healthy food has progressively evolved over the last decade. The recognition that dietary behavior directly affects individuals' quality of life has resulted in a tendency in the choice of functional foods that have a great market potential in the food industry (Martins and Ferreira 2017).

Due to the advances related to food science and technology, new concepts were introduced in this area, one being functional foods. Essentially, functional foods are consumed daily, are part of a normal diet, and have high nutritional value. In addition, they are natural compounds that exert beneficial effects, such as reducing the risk of some diseases and increasing quality of life (El Sohaimy 2012; Ghosh et al. 2014; Reis et al. 2017).

Granato et al. (2017) share the idea that functional foods have a positive effect on health, and some plant varieties can be considered functional foods because they have the potential to prevent diseases that can cause health complications.

The use of enzymes in the food industry is directly related to the functionality of food products, such as improving the quality and shelf life of the product. These enzymes can be produced by a variety of microorganisms from fermentative processes to meet specific applications of operation such as pH and extreme temperatures (Kirk et al. 2002; Van Oort 2010; Simpson et al. 2012; Ermis 2017).

Biomolecules have several metabolic activities. For example, proteins and sugars show properties with the potential to induce certain hormonal activities. Especially, the bioactive ones have the capacity to act in functional food, favorable to health (Agyei et al. 2016).

*Bioprocessing for Biomolecules Production*, First Edition.
Edited by Gustavo Molina, Vijai Kumar Gupta, Brahma N. Singh, and Nicholas Gathergood.

Some sugars considered functional are not easily found in nature, and their focus is on biological and functional use. These sugars can be produced from enzymatic processes and have great potential in the food and pharmaceutical industries (Zhang et al. 2017).

The growing interest in the discovery of new products has led to expansion of the market for microbial polysaccharides, which are used to improve the characteristics of various products in the food industry (Barcelos et al. 2017).

In addition, bioactive compounds, including probiotics and prebiotics, have been shown to be directly applicable in the food industry, due to their functional characteristics, which are the strain, growth, storage time, and temperature. For these reasons, several experimental investigations are being undertaken in the area to evaluate the probiotic and prebiotic viability of these compounds (Castro et al. 2014; Mizock 2015; Coghetto et al. 2016; Reid 2016; Dias et al. 2017).

The study of the production of bioactive molecules is the basis for the commercialization of functional foods, such as sugars produced by microorganisms. It is necessary to analyze the composition of these sugars and the potential enzymes to perform protein breakdown in smaller molecules (bioactive) for large-scale production to be possible.

Accordingly, this chapter is focused on the current research pertaining to production of functional foods in order to broaden studies in the evaluation of the functional properties of microbial sugars. We also discuss and propose the potential of these foods for industry.

## 11.2 Bioactive Compounds

Currently, in addition to the concern with the basic role of nutrition that is to provide nutrients necessary for the growth and development of the body, there is growing interest in some additional aspects such as health maintenance and disease control. In this context, several studies report the benefits to human health of the use of probiotics and prebiotics in nutrition (Markowi and Katarzyna 2017).

### 11.2.1 Probiotics

Probiotic foods are processed products that contain live microorganisms, which provide benefits to human health when administered in adequate amounts (Guarner and Schaafsma 1998). The most commonly used probiotics in the food industry are *Lactobacillus*, *Bifidobacterium*, and *Saccharomyces* (Castro et al. 2014; Coghetto et al. 2016; Mizock 2015; Reid 2016). These foods have several health benefits, such as prevention of intestinal diseases, inhibition of pathogens, reduction of lactose intolerance and cholesterol levels, and anticancer activities (Daliri and Byong 2015; Miremadi et al. 2016).

Probiotics act by different mechanisms, such as providing barriers, enhancing immunity, and eliminating harmful bacteria from the gastrointestinal tract (Elmer 2001). Some probiotics have mechanisms capable of preventing the colonization of pathogenic microorganisms by inhibiting the adhesion of microorganisms to the surfaces of host cells, being this mechanism developed by specific proteins present on the surface of probiotic agents (Ragione et al. 2004). Figure 11.1 illustrates the effects of probiotics on body metabolism, microflora of body, and immunomodulatory properties.

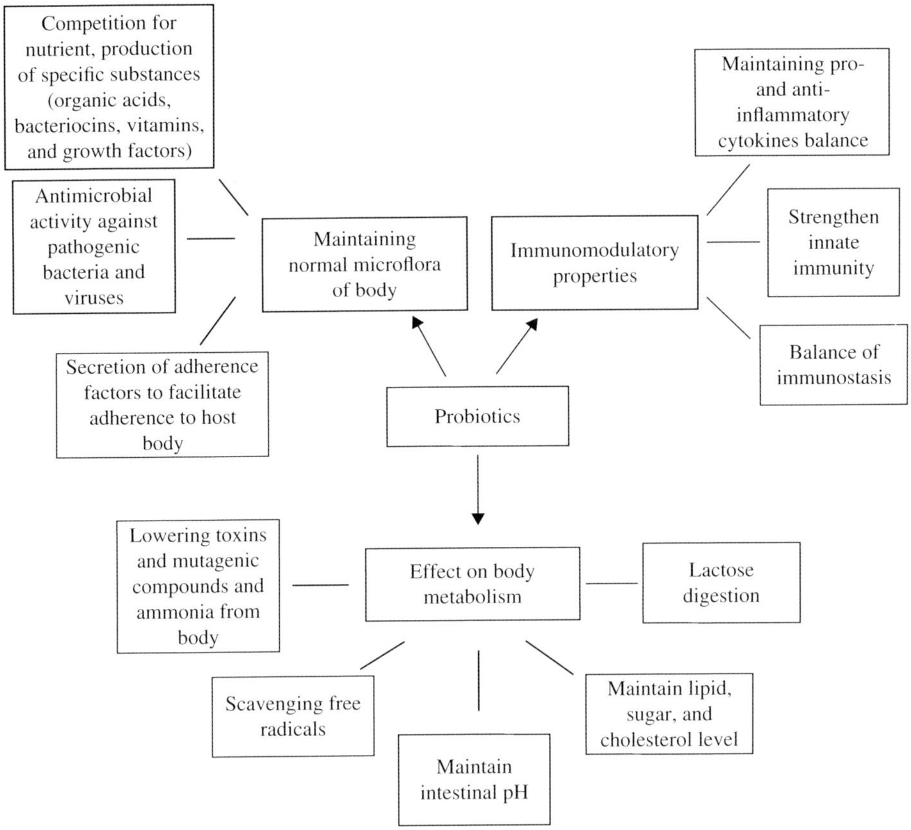


**Figure 11.1** Health benefits of consuming probiotics. Source: Gupta et al. 2018.

Probiotics are available in capsules, sachets, pills, and in some foods like dairy products, including yogurts, cheeses, and ice creams (Prisco and Gianluigi 2016; Reid 2015; Kandylis 2016). In addition, these are found in fruit juices, cereal bars, breads, sausages, chocolates, and pate (Champagne et al. 2015; Farnworth and Champagne 2016; Kandylis 2016; Konar 2016; Shori 2015).

### 11.2.2 Prebiotics

Gibson and Roberfroid (1995) defined prebiotic as "a nondigestible food ingredient that beneficially affects the host by selectively stimulating the growth and/or activity of one or a limited number of bacteria in the colon, and thus improves host health," as illustrated in Figure 11.2. Nondigestible carbohydrates (oligo- and polysaccharides), certain peptides and proteins, and certain lipids (both ethers and esters) are among the food ingredients considered prebiotic, since due to their chemical structure, they are not absorbed in the upper gastrointestinal tract or hydrolyzed by human digestive enzymes. Among natural, nondigestible oligosaccharides that meet the criteria of a prebiotic food, fructooligosaccharides are the products with the greatest field of research (Lopes et al. 2017).

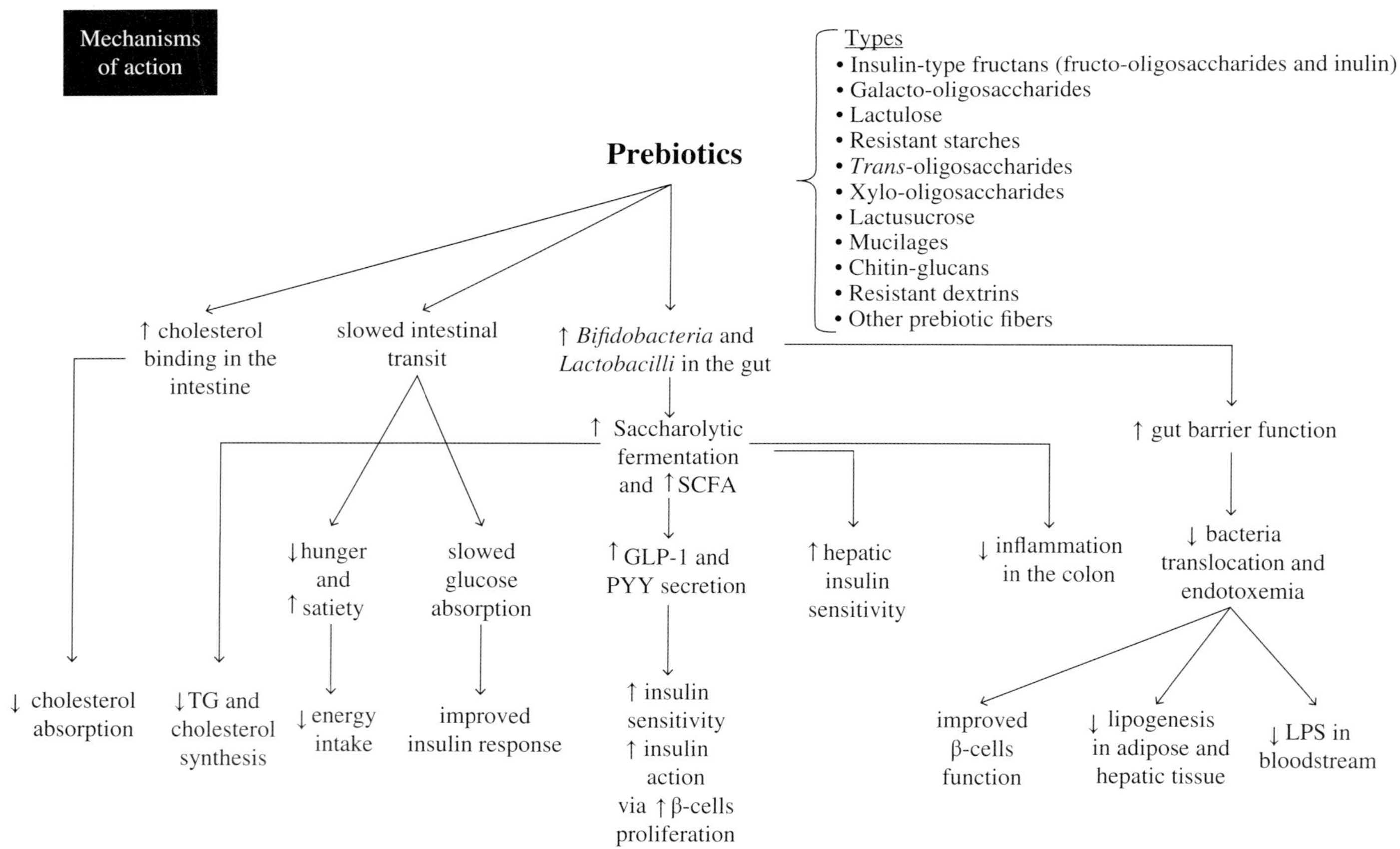


**Figure 11.2** Proposed mechanisms of action of prebiotics. Abbreviation: SCFA (short-chain fatty-acids), GLP-1 (Glucagon-like peptide-1), PYY (Peptide YY), TG (Triglycerides), LPS (Lipopolysaccharides). Source: Adapted O' Connor et al. 2017.

The prebiotics usually employed and emergent in new searches are fructooligosaccharides (FOS), galacto-oligosaccharides (GOS), galactooligosaccharides/transglucosylases oligosaccharides (GOS/TOS), inulin, isomalto-oligosaccharides, lactulose, pyrodextrins, soy-oligosaccharides (SOS) (Anadón et al. 2016).

Inulin and fructooligosaccharides, also called fructans, are the most widely used prebiotics in the food industry. The fructans provide a combination of nutritional properties and technological applications in the food industry and is widely used as a fat substitute and sugar in low-calorie foods (Lopes et al. 2016). Some recent studies found in the literature about FOS properties were developed by Lopes et al. (2016) and Park et al. (2016). Lopes et al. (2016) isolated and characterized fructo-oligosaccharides (FOSs) from *S. rebaudiana* roots and *in vitro* adventitious root cultures and evaluated the potential prebiotic effect of these molecules. The *in vitro* adventitious root cultures were obtained using a roller bottle system. Chemical analyses (gas chromatography–mass spectrometry, 1H nuclear magnetic resonance, and off-line electrospray ionization-mass spectrometry) revealed similar chemical properties of FOSs that were obtained from different sources.

Park et al. (2016) showed the potential prebiotic effects of FOSs (that were isolated from *S. rebaudiana* roots) enhanced the growth of both bifidobacteria and lactobacilli, with strains specificity in their fermentation ability. The baking quality of frozen doughs containing different levels of fructo-oligosaccharides (FO) or isomalto-oligosaccharides (IMO) (3–9%, w/w flour), and stored for 0–8 weeks at −18 °C, was examined. The addition of FO or IMO increased the proof volume of the dough and the loaf volume of bread prepared from frozen dough. A 6% addition of FO or IMO was optimum, giving the highest proof volume and bread loaf volume, but a higher concentration than 6% induced low baking quality, including lower proof volume and bread loaf volume. The bread crumb was moister and softer after the addition of FO or IMO before, and even after, frozen storage. Darker crumb color was observed in the bread after the addition of FO or IMO. The oligosaccharides added to the frozen dough were effective in improving the quality of bread made from frozen dough, except for resulting in a darker bread crumb.

Prebiotics can also be used as a supplement, providing and increasing the consumption of prebiotic fibers. Prebiotics are generally soluble and imperceptible in water, so they can be easily incorporated into food and beverages. These food items often contain fructooligosaccharides, inulins, or starch resistant for their fiber content and prebiotic advantages (Anadón et al. 2016).

## 11.3 Production Technology for Probiotic Strains

The large-scale preparation of bacterial strains is difficult, time-consuming, and expensive. In addition, most probiotic strains show poor growth rates in milk-based media. Therefore, it can be difficult to achieve high numbers of viable cells after fermentation (Ross et al. 2005). The media composition and type of substrate used for fermentation can have a great impact on probiotic strain viability during production and downstream processing (Lacroix and Yildirim 2007).

The foremost disadvantage of probiotic strains is the rapid growth and acidification rate of typical starter cultures. Therefore, they are added to dairy products after fermentation (Champagne et al. 2005). The cultivation conditions are directly proportional to growth, culture stability, and activity as well as drying and subsequent storage. The stability and function of probiotics can be enhanced by changing the culture conditions. Usually probiotic cells are produced by large-scale fermentation. It is important to design growth media and change the fermentation technology to improve biomass yield and enhance cell stability. MRS broth (De Man et al. 1960) is the most widely used medium for cultivation of LAB or bifidobacteria. Higher cell counts of bifidobacterial strains were produced in ultra-filtered skim milk with various protein concentrations or inclusion of milk with nitrogenous substrates such as whey and casein fractions from human or cow milk when compared to growth in skim milk only (Ventling and Mistry 1993). Soy milk or animal-product-free vegetable medium (based on soy peptone, glucose, and yeast extract) resulted in lower yields or lower viability of bifidobacteria during storage because of the low buffer capacity of the vegetable medium (Heenan et al. 2002). Redox-reducing compounds (e.g., cysteine) were used in bifidobacteria to enhance growth (Doleyres and Lacroix 2005). In some cases, the disulfide bonds were reduced and hence these supplements may lose their growth-enhancing properties (Ibrahim and Bezkorovainy 1994).

## 11.4 Stabilization Technology for Probiotic Strains

### 11.4.1 Microencapsulation

Encapsulation maintains the bioavailability of the bioactive compounds, avoiding the formation of undesirable or harmful compounds due to chemical changes over time while also masking undesirable sensorial characteristics (Dias et al. 2017).

Microencapsulation is a fascinating field of biopharmacy that originated and then developed rapidly in the past decade. Several microorganisms have been immobilized within semi-permeable and biocompatible materials that modulate the delivery of cells using this technique (Vidhyalakshmi et al. 2009). Microencapsulation is a technology for packaging solids, liquids, or gaseous materials in tiny, sealed capsules that can release their contents at a controlled rates under specific conditions (Shahidi and Han 1993). Microencapsulation of probiotics in hydrocolloid beads has been found to enhance their viability and activity in food products and the intestinal tract by entrapping the cells inside a bead matrix, thus separating them from harsh environmental conditions, as well as protecting them against bacteriophages (Krasaekoopt et al. 2003). A microcapsule consists of a semi-permeable, spherical, thin, strong membrane surrounding a solid or liquid core, with a diameter varying from a few microns to 1 mm (Anal and Singh 2007). Hence, the bacterial cells are retained within the microcapsule (Jankowski et al. 1997). The most widely used materials in microencapsulation of probiotic bacteria include polysaccharides originated from seaweed (alginate, $\kappa$-carrageenan), other plants (starch and its derivatives, gum arabic), bacteria (gellan, xanthan), and animal proteins (milk, gelatin). The chief techniques for microencapsulation of probiotics are extrusion techniques, spray drying, and spray coating (Chavarri et al. 2012).

The microencapsulation techniques applied to probiotic cells for use in fermented milk products or biomass production can be classified into two groups, based on the method used to form the beads: extrusion (droplet method) and emulsion or two-phase system. Both extrusion and emulsion techniques show high survival of probiotic bacteria by up to 80–95% (Kebary et al. 1998). Lack of oxygen in the interior of the capsule

**Table 11.1** Different types of microencapsulation matrices, its polymer system, and which are the highest values obtained for some compounds analyzed.

| Matrix | Polymer systems | Potentialized compounds | References |
|---|---|---|---|
| Acerola pulp<br>Residue extracts | Gum arabic<br>Maltodextrin | Contained higher concentrations of these compounds: phenolic compounds, total anthocyanins, total flavonoids and antioxidant activities | Rezende et al. (2018) |
| Tamarillo juice | Maltodextrin<br>Octenyl succinic anhydride modified starch<br>Low viscosity gum Arabic alternative Resistant maltodextrin<br>Gum arabic | Higher concentrations of the carotenoids.<br>Total phenolic content, total flavonoid content and antioxidant activity exhibited decreases during the storage | Ramakrishnan et al. (2018) |
| Nutmeg oleoresin | Gum arabic<br>Octenyl succinic anhydride<br>Succinylated sorghum starches | The proportions of 25% octenyl succinic anhydride sorghum starch with 75% gum arabic and 75% native sorghum starch along with 25% gum arabic resulted in high oil retention and encapsulation efficiency | Arshad et al. (2018) |
| Brewers's spent grain | Capsul® | Higher concentration of the polyphenols, flavonoids, and with a better antioxidant activity than the control sample | Spinelli et al. (2016) |

Source: The authors.

may result in cell death, but conversely is advantageous to anaerobic bacteria such as bifidobacteria (Kailasapathy 2002).

Entrapment of probiotic microorganisms in a biodegradable polymer matrix has various advantages. Once entrapped in matrix beads, the cells are easier to handle than a suspension or slurry. The number of cells in each bead can be quantified, allowing dosages to be readily controlled. Cryo- and osmoprotective components can be incorporated in the polymer matrix, increasing survival of cells during processing and storage. Ultimately, once the matrix beads have been dried, a surface coating can be applied, as can be observed in Table 11.1, which refers to different types of microencapsulation matrices, its polymer system, and which are the highest values obtained for some compounds analyzed in each study. This outer layer can be used to alter the esthetic and sensory properties of the product and may also be functional, providing an extra level of protection for the cells. In addition, the coating layer can have preferable dissolution properties that permit overdue release of the cells or release upon, for example, change in pH. Cells produced during continuous immobilized cell cultures show changes in cell membranes that may induce formation of cell aggregates and tolerance to bile salts and aminoglycosidic antibiotics. Cell immobilization, in combination with continuous culture, can be used to produce probiotic microorganisms with improved stress tolerance (Anandharaj et al. 2017).

### 11.4.2 Spray Drying

In general, dairy starter and probiotic cultures are preserved and distributed in frozen or dried form. High costs of storage, shipping, and energy are the primary reason to preserve probiotic cultures in dried forms (Johnson and Etzel 1995). Bacterial cells require water activity (aw) of about 0.98 in the resultant matrix for their growth and survival. Hence, the physical removal of water to convert the bacterial cells into a dried form is a risky process. It is necessary to maintain either high aw values for metabolic activity or low aw values to preserve the bacteria in live state, so that they can survive in a dormant state in powders (Paul et al. 1993).

Freeze drying or spray drying is the predominant method used to dry the bacterial suspension. The common method used to formulate starter and probiotic cultures is freeze drying. However, the major detriment is that this is an expensive process with low yields. By contrast, in the case of inexpensive spray drying, higher production rates result (Zamora et al. 2006). Spray drying is a well-organized tool in food industries for the production of milk powders and instant coffee. Although there are several restrictive process conditions for microorganisms (inlet reaching ≥180 °C), the rapidity of drying combined with the ability to dry large amounts of bacterial cultures has caught the attention of research and industry (Meng et al. 2008; Zamora et al. 2006).

### 11.4.3 Freeze Drying

Freeze drying is a commonly used method that provides higher survival rates compared with spray drying. Generally, freeze drying is achieved by three important steps: freezing, primary, and secondary drying. To increase the survival ratio of bacterial cultures, they are typically frozen at −196 °C in liquid nitrogen. After freezing, the solid samples are sublimated under high vacuum conditions to complete the primary freezing. In this step, high temperatures under pressure cause the phase transition from solid to gas. After the primary drying step, almost 95% of the water content in the sample is removed. However, secondary drying is also important to remove the remaining hydrogen bond water molecules to achieve a final water content below 4%, thus improving survival rates and long-term storage efficiency and preventing spoilage (Santivarangkna et al. 2007).

Freeze drying is an expensive method with a low yield. When bacterial cells are exposed to extremely low temperatures (i.e., freezing), osmotic pressure across the membrane is increased due to formation of extracellular ice and the cells dehydrate until an eutectic point is reached (Fowler and Toner 2005). During the freeze-drying process, the bacteria face various stresses that damage the cell; generally, longer rod-shaped lactobacilli are more susceptible to damage than small, round enterococci, because of their larger surface area (Fonseca et al. 2000). The cell membrane lipids are more sensitive and easily damaged during freezing, and destabilization of nucleic acids also limits several important growth functions, including replication of DNA, transcription, and translation (Van de Guchte et al. 2002).

To reduce these problems during freeze drying, several approaches have been developed, including the addition of protectants such as trehalose, betaine, adonitol, sucrose, skim milk powder, lactose, commercial cryoprotectants (e.g., Unipectine, Satialgine), and several commercially available products (Burns et al. 2008).

### 11.4.4 Fluidized Bed and Vacuum Drying

Fluidized bed dryers use an upward-moving flow of heated air and mechanical shaking to create a fluidized effect in a solid product. Particles are freely suspended in air and are dried by rapid heat exchange (Santivarangkna et al. 2007). This method is more economical than others, including spray drying. In this method, the length of bacterial exposure time to heat is easily controlled, which reduces the risk of thermal inactivation. Using the fluidized bed drying method, several yeast strains have been successfully dried, and this method is also employed for lactic acid bacteria (LAB) (Bayrock and Ingledew 1997). It is suitable for granular particles, so bacterial solutions are encapsulated with alginate, potato starch, skim milk, or casein before being introduced into the drier. The desired moisture content is achieved by drying the granulated particles.

The vacuum drying method is used to dry heat-sensitive compounds; here, the water molecules in the samples are removed at low temperature under vacuum conditions. Due to the high vacuum conditions, oxidation reactions are reduced in the vacuum drying method, which is more suitable for oxygen-sensitive bacteria, but this method is not extensively used for drying LAB strains. It has several limitations, including long drying time (10–100 hours) compared with the spray drying or fluidized bed method (Santivarangkna et al. 2007). However, these problems can be overcome by modifications such as using continuous vacuum drying. Continuous vacuum drying is used in large-scale industries for drying enzymes, food additives, and other pharmaceutical products (Hayashi et al. 1983).

### 11.4.5 Other Technologies

The action of depolymerization in the ultrasonic system is generally attributed to the cavitation process generated by the system. This process consists of two possible mechanisms. The first is the formation of microbubbles, which, on collapse, result in polymer degradation, while the second mechanism is chemical degradation. The latter is a result of the chemical reaction between the polymer and high-energy molecules, such as hydroxyl radicals produced during cavitation (Gomes et al. 2017).

Gomes et al. (2017) demonstrated feasible results for the application of ultrasonic system and high-pressure processes for the treatment of cranberry prebiotic juice, showing an alternative to the heat treatment process, maintaining the good preservation of the fructooligosaccharides. This group also observed that there was no significant change in pH, soluble solids content, organic acids, bioactive compounds and antioxidant capacity. In addition, the cranberry prebiotic juice maintained its liveliness and characteristic color after processing. Thus, the juice processing using ultrasound and high pressure were demonstrated as adequate to preserve the quality and functionality of the cranberry prebiotic juice containing fructooligosaccharides.

Nonthermal technologies have the objective of producing foods with the minimum of nutritional changes as in the case of functional foods, where the compounds responsible for their functionalities must be preserved (Almeida et al. 2017).

Almeida et al. (2017) studied the advantages and limitations of nonthermal processes applied to prebiotic fructooligosaccharides (FOS). The authors showed that when treated by exposure to direct cold atmospheric plasma (ACP), the orange juice sample did not have its functionality compromised, as it presented the lowest changes in the degree of FOS polymerization throughout the treatment time. Moreover, when high-pressure processing (HPP) and indirect ACP were used, degrees of depolymerization were found for the orange juice sample, and in the use of ACP of indirect exposure the depolymerization was much lower when compared to HPP. However, after the statistical analysis the authors observed that the difference between direct and indirect ACP was not significant for the orange juice samples.

## 11.5 Study of Scale-Up Process: Advances, Difficulties, and Limitations Achieved

A number of prevailing factors must be considered and optimized when attempting to take a technological process into full commercial production. A major ramp-up in design, development, demonstration, and deployment is needed to engineer predictable, controllable, and cost-effective processing systems. Engineering principles have been and will continue to be critical for fulfillment of ever-increasing needs for environmentally efficient production of food or feed (Sanford et al. 2016). To meet these objectives, a discovery to delivery approach that integrates a synthetic bio-operating system, chemical engineering practices, process safety, sustainability, and socioeconomic considerations must be implemented (Walker 2015). Scale-up challenges are usually dependent on whether the product is a specialty or commodity chemical, where the difference in volumes of production can be of several orders of magnitude.

When used in dairy and bakery foods, inulin and fructooligosaccharides become good substitutes for sugar (Coussement 1999). In addition, they are considered low-intensity sweeteners and are industrially used in combination with high intensity sweeteners to replace sugars, making them a well-balanced sweetener and masking the bitter taste of aspartame or acesulfame K (Wiedmann and Jager 1997). They are also used in ice cream to replace all the sugar and reduce the fat content (Sangeetha et al. 2005).

In 1984, Meiji Seika Co., Japan, added fructooligosaccharides on the market as food products, and to this day they are widely used in the confectionery and dairy industries. Among the current intrusive applications of fructooligosaccharides are the use in probiotic yogurts and dairy drinks; baking, including biscuits, breads, and pastries; confectionery and marmalades; and milk formulations (Crittenden and Playne 1996).

Using inulin or fructooligosaccharides in fruit soft drinks and jellies is not possible because in acidic foods with a long shelf life, both inulin and fructooligosaccharides are unstable and efficiently hydrolyzed to fructose (Coussement 1999).

Regarding the manufacturing process of sugary kefir beverages, the same is not well established industrially, and the production equipment must still be designed and tested. In this sense, the difficulties associated with fermentation relate primarily to transport and contamination by the microbiota of the natural starter. Furthermore, it is necessary to stabilize the microbial growth and ensure that in all batches the same groups of microorganisms are present (Fiorda et al. 2017).

## 11.6 Potential Development of the Area and Future Prospects

For the future of modern industrial processes, it is essential to establish the correct monitoring of all the steps involved, resulting in a well-controlled development (Fiorda et al. 2017).

Jaffé (2015) reports that there is a tendency to recognize noncentrifugal cane sugar as a promising functional food, which has seen an increasing research effort to expand its use. A comprehensive characterization is necessary in order to analyze the presence and content of vitamins and antioxidants, given their potential health effects. An additional factor is that the contents of the noncentrifugal sugarcane components show wide variations within the same product in a country or region and between products from different countries.

Consumers today are looking for low-fat foods that offer additional health benefits. Fructooligosaccharides are in focus because they fulfill these conditions with their well-established functional properties and broad applications in the food industry (Singh and Singh 2010). In this way, the demand for fructooligosaccharides is increasing day by day, boosting their production. There is a large market that evaluates using sophisticated analytical techniques to isolate new strains producing these compounds (Sangeetha et al. 2005).

Most studies related to the consumption of fructooligosaccharides are short-term, with a dearth of data for the consequences of their long-term consumption. These should be investigated. Moreover, the structure–function effects of the prebiotic fructooligosaccharides and the mechanisms behind their specific metabolism are not known. If these mechanisms are elucidated, a large field of study opens up to produce tailor-made prebiotics in the future (Beine et al. 2008) because these compounds are ingredients of the future meeting the needs of the food industry today and are among the emerging trends for functional foods.

King et al. (2017) bring some advances in technologies that can be used for functional food research, such as genetic sequencing metagenomics that is a study in which more complex microbial communities are understood, diverse, and dynamic in food. In addition, a microbiology predictive also stands out because it involves responses in relation to microbial growth to environmental factors that work with curricula, with equations or mathematical models, facilitating the study of processes involving functional microbial sugars.

Some varieties of sugars, such as oligosaccharides, have potential in the food industry because they act as prebiotics. For instance, Zhang et al. (2017) study a branch of research related to the specific physiological functions of oligosaccharides, since they are a source of rare functional sugars.

## 11.7 Conclusion

In discussing and presenting the industrial potential of the functional microbial sugars, it is possible to verify that the fructooligosaccharides and inulin (besides being used by the pharmaceutical industry and as supplements for athletes and in formulations for infant feeding) are also used in the food industry as ingredients in many foods because they have physicochemical properties of interest.

The market for oligosaccharides, in particular the fructooligosaccharides, has been highlighted due to the numerous benefits and the increase of consumer demand for these foods. The search for products that can improve quality of life has increasing appeal in the food industry due to the prevalence of high-fat and low-fiber diets, especially among children and adolescents. The use of microbial sugars with functional properties in food should certainly be considered an alternative to meet these needs.

However, it is still necessary to advance the technology used in the processes of obtaining different oligomers for the development of even more efficient functional food products. Awareness and appropriate eating habits are also very important so that all the benefits of these compounds are gained.

## References


Agyei, D., Ongkudon, C.M., Wei, C.Y. et al. (2016). Bioprocess challenges to the isolation and purification of bioactive peptides. *Food Bioprod. Process.* 98: 244–256.

Almeida, F.D.L., Gomes, W.F., Cavalcante, R.S. et al. (2017). Fructooligosaccharides integrity after atmospheric cold plasma and high-pressure processing of a functional orange juice. *Food Res. Int.* 110: 289–298.

Anadón, A., Martínez-Larrañaga, M.R., Arés, I., and Martínez, M.A. (2016). *Prebiotics and Probiotics: An Assessment of Their Safety and Health Benefits: Bioactive Foods Health Promotion*, vol. 1, 3–23. San Diego, CA: Academic Press, Elsevier.

Anal, A.K. and Singh, H. (2007). Recent advances in microencapsulation of probiotics for industrial applications and targeted delivery. *Trends Food Sci. Technol.* 18: 240–251.

Anandharaj, M., Rani, R.P., and Swain, M.R. (2017). Production of high-quality probiotics by fermentation. *Microb. Funct. Foods Nutr.* 235.

Arshad, H., Ali, T.M., and Hasnain, A. (2018). Native and modified Sorghum starches as wall materials in microencapsulation of nutmeg oleoresin. *Int. J. Biol. Macromol.* https://doi.org/10.1016/j.ijbiomac.2018.03.082.

Barcelos M., K.A.C. Vespermann, B. Marise, F.M Pelissari, and G. Molina. (2017). Screening and selection of polysaccharide-producer microorganisms from Brazilian cerrado fruits. In: XXI Simpósio Nacional de Bioprocessos e XII Simpósio de Hidrólise Enzimática.

Bayrock, D. and Ingledew, W.M. (1997). Mechanism of viability loss during fluidized bed drying of baker's yeast. *Food Res. Int.* 30: 417–425.

Beine, R., Moraru, R., Nimtz, M. et al. (2008). Synthesis of novel fructooligosaccharides by substrate and enzyme engineering. *J. Biotechnol.* 138: 33–41.

Burns, P., Vinderola, G., Molinari, F., and Reinheimer, J. (2008). Suitability of whey and buttermilk for the growth and frozen storage of probiotic lactobacilli. *Int. J. Dairy Technol.* 61: 156–164.

Castro, J.M., Tornadijo, M.E., Fresno, J.M., and Sandoval, H. (2014). Biocheese: a food probiotic carrier. *Biomed. Res. Int.* .

Champagne, C.P., Gardner, N.J., and Roy, D. (2005). Challenges in the addition of probiotic cultures to foods. *Crit. Rev. Food Sci. Nutr.* 45: 61–84.

Champagne, C.P., Raymond, N., Guertin, N., and Bélanger, G. (2015). Effects of storage conditions, microencapsulation and inclusion in chocolate particles on the stability of probiotic bacteria in ice cream. *Int. Dairy J.* 47: 109–117.

Chavarri, M., Maranon, I., and Villara, M.C. (2012). Encapsulation technology to protect probiotic bacteria. In: *Probiotics* (ed. E.C. Rigobelo), 501–540. InTech.

Coghetto, C.C., Flores, S.H., Brinques, G.B., and Ayub, M.A.Z. (2016). Viability and alternative uses of a dried powder, microencapsulated *Lactobacillus plantarum* without the use of cold chain or dairy products. *Food Sci. Technol.* 71: 54–59.

Cook, N., D'Agostinho, M., and Thompson, K.C. (2016). *Molecular Microbial Diagnostic Methods: Pathways to Implementation for the Food and Water Industries*. London: Elsevier Inc.

Coussement, P.A.A. (1999). Inulin and oligofructose: Safe intakes and legal status. *J. Nutr.* 129: 1412–1712.

Crittenden, R.G. and Playne, M.J. (1996). Production, properties and applications of food-grade oligosaccharides. *Trends Food Sci. Technol.* 7: 353–361.

Daliri, E.B. and Byong, L.E.E. (2015). New perspectives on probiotics in health and disease. *Food Sci. Hum. Wel.* 4: 56–65.

De Man, J.D., Rogosa, M., and Sharpe, M.E. (1960). A medium for the cultivation of Lactobacilli. *J. Appl. Bacteriol.* 23: 130–135.

Dias, D.R., Botrel, D.A., Fernandes, R.V.B., and Borges, S.V. (2017). Encapsulation as a tool for bioprocess ingoffunctional foods. *Cur. Opin. Food Sci.* 13: 31–37.

Doleyres, Y. and Lacroix, C. (2005). Technologies with free and immobilized cells for probiotic bifidobacteria production and protection. *Int. Dairy J.* 15: 973–988.

El Sohaimy, S.A. (2012). Functional foods and nutraceuticals - modern approach to food science. *World Appl. Sci. J.* 20: 691–708.

Elmer, G.W. (2001). Probiotics: living drugs. *Am. J. Health Syst. Pharm.* 58: 1101–1109.

Ermis, E. (2017). Halal status of enzymes used in food industry. *Trends Food Sci. Technol.* 64: 69–73.

Farnworth, E.R. and Champagne, C.P. (2016). Production of probiotic cultures and their incorporation into foods. *Probio, Prebiot, and Synbiot.* 303–318.

Fiorda, F.A., Pereira, G.V.M., Thomaz-Soccol, V. et al. (2017). Microbiological, biochemical, and functional aspects of sugary kefir fermentation - A review. *Food Microbiol.* 66: 86–95.

Fonseca, F., Beal, C., and Corrieu, G. (2000). Method of quantifying the loss of acidification activity of lactic acid starters during freezing and frozen storage. *J. Dairy Res.* 67: 83–90.

Fowler, A. and Toner, M. (2005). Cryo-injury and bio-preservation. *Ann. N.Y. Acad. Sci.* 1066: 119–135.

Ghosh, D., Bagchi, D., and Konishi, T. (2014). *Clinical Aspects of Functional Foods and Nutraceuticals*. Boca Raton, FL: CRC Press.

Gibson, G.R. and Roberfroid, M. (1995). Dietary modulation of the human colonic microbiota: introducing the concept of prebiotics. *J. Nut.* 125: 1401–1412.

Gomes, W.F., Tiwari, B.K., Barragán, O.R. et al. (2017). Effect of ultrasound followed by high pressure processing on prebiotic cranberry juice. *Food Chem.* 218: 261–268.

Granato, D., Nunes, D.S., and Barba, F.J. (2017). An integrated strategy between food chemistry, biology, nutrition, pharmacology, and statistics in the development of functional foods: A proposal. *Trends Food Sci. Technol.* 62: 13–22.

Guarner, F. and Schaafsma, G.J. (1998). Probiotics. *Int. J. Food Microbiol.* 39: 237–238.

Gupta, V.K., Treichel, H., Shapaval, V. et al. (2018). *Microbial Functional Foods and Nutraceuticals*, 1ste. Wiley.

Hayashi, H., Kumazawa, E., Saeki, Y., and Lahicka, Y. (1983). Continuous vacuum dryer for energy saving. *Drying Technol.* 1: 275–284.

Heenan, C.N., Adams, M.C., Hosken, R.W., and Fleet, G.H. (2002). Growth medium for culturing probiotic bacteria for applications in vegetarian food products. *Food Sci. Technol.* 35: 171–176.

Ibrahim, S.A. and Bezkorovainy, A. (1994). Growth promoting factors for Bifidobacterium longum. *J. Food Sci.* 59: 189–191.

Jaffé, W.R. (2015). Nutritional and functional components of non centrifugal cane sugar: A compilation of the data from the analytical literature. *J. Food Compos. Anal.* 43: 194–202.

Jankowski, T., Zielinska, M., and Wysakowska, A. (1997). Encapsulation of lactic acid bacteria with alginate/starch capsules. *Biotechnol. Tech.* 11: 31–34.

Johnson, J.A.C. and Etzel, M.R. (1995). Properties of *Lactobacillus helveticus* CNRZ-32 attenuated by spray drying, freeze drying, or freezing. *J. Dairy Sci.* 78: 761–768.

Kailasapathy, K. (2002). Microencapsulation of probiotic bacteria: technology and potential applications. *Curr. Iss. Intest. Microbiol.* 3: 39–48.

Kandylis, P. (2016). Dairy and non-dairy probiotic beverages. *Curr. Opin. Food Sci.* 7: 58–63.

Kebary, K.M.K., Hussein, S.A., and Badawi, R.M. (1998). Improving viability of Bifidobacteria and their effect on frozen ice milk. *Egypt. J. Dairy Sci.* 26: 319–337.

King, T., Cole, M., Farber, J.M. et al. (2017). Food safety for food security: Relationship between global megatrends and developments in food safety. *Trends Food Sci.Technol.* 68: 160–175.

Kirk, O., Borchert, T.V., and Fuglsang, C.C. (2002). Industrial enzyme applications. *Curr. Opin. Biotechnol.* 13: 345–351.

Konar, N. (2016). Improving functionality of chocolate: A review on probiotic, prebiotic, and/or synbiotic characteristics. *Trends Food Sci.Technol.* 49: 35–44.

Krasaekoopt, W., Bhandari, B., and Deeth, H.F. (2003). Evaluation of encapsulation techniques of probiotics for yoghurt. *Int. Dairy J.* 13: 3–13.

Lacroix, C. and Yildirim, S. (2007). Fermentation technologies for the production of probiotics with high viability and functionality. *Curr. Opin. Biotechnol.* 18: 176–183.

Lopes, S.M.S., Francisco, M.G., Higashi, B. et al. (2016). Chemical characterization and prebiotic activity of fructo-oligosaccharides from *Stevia rebaudiana* (Bertoni) roots and in vitro adventitious root cultures. *Carbohy. Polymer.* 152: 718–725.

Lopes, S.M.S., Krausová, G., Carneiro, J.W.P. et al. (2017). A new natural source for obtainment of inulin and fructo-oligosaccharides from industrial waste of Steviare baudiana Bertoni. *Food Chem.* 225: 154–161.

Markowi, P. and Katarzyna, S. (2017). Effects of Probiotics, Prebiotics, and Synbiotics on Human Health. *Nutrients* 9: 1021.

Martins, N. and Ferreira, I.C.F.R. (2017). Mountain food products: A broad spectrum of market potential to be exploited. *Trends Food Sci. Technol.* 67: 12–18.

Meng, X.C., Stanton, C., Fitzgerald, G.F. et al. (2008). Anhydrobiotics: the challenges of drying probiotic cultures. *Food Chem.* 106: 1406–1416.

Miremadi, F., Frank, S., and Stojanovska, L. (2016). Hypocholesterolaemic effect and anti-hypertensive properties of probiotics and prebiotics: A review. *J. Funct. Foods* 25: 497–510.

Mizock, B.A. (2015). Probiotics. *Disease-a-Month* 61: 259–290.

O' Connor, S., Chouinard-Castonguay, S., Gagnon, C., and Rudkowska, I. (2017). Prebiotics in the management of components of the metabolic syndrome. *Maturitas* https://doi .org/10.1016/j.maturitas.2017.07.005.

Park, E.Y., Jang, S.B., and Lima, S.T. (2016). Effect of fructo-oligosaccharide and isomalto-oligosaccharide addition on baking quality of frozen dough. *Food Chem.* 213: 157–162.

Paul, E., Fages, J., Blanc, P. et al. (1993). Survival of alginate-entrapped cells of *Azospirillum lipoferum* during dehydration and storage in relation to water properties. *Appl. Microbiol. Biotechnol.* 40: 34–39.

Prisco, A. and Gianluigi, M. (2016). Probiotication of foods: a focus on microencapsulation tool. *Trends Food Sci. Technol.* 48: 27–39.

Ragione, R.M.L.A., Narbad, A., Gasson, M.J., and Woodward, M.J. (2004). In vivo characterization of *Lactobacillus johnsonii* FI9785 for use as a defined competitive exclusion agent against bacterial pathogens in poultry. *Let. In Appl. Microbiol.* 38: 197–205.

Ramakrishnan, Y., Adzahan, N.M., Yusof, Y.A., and Muhammad, K. (2018). Effect of wall materials on the spray drying efficiency, powder properties and stability of bioactive compounds in tamarillo juice microencapsulation. *Powder Technol.* 328: 406–414.

Reid, G. (2015). The growth potential for dairy probiotics. *Int. Dairy J.* 49: 16–22.

Reid, G. (2016). Probiotics: definition, scope and mechanisms of action. *Best Pract. Res. Clin. Gastroenterol.* 30: 17–25.

Reis, F.S., Martins, A., Vasconcelos, M.H. et al. (2017). Functional foods based on extracts or compounds derived from mushrooms. *Trends Food Sci. Technol.* 66: 48–62.

Rezende, Y.R.R.S., Nogueira, J.P., and Narain, N. (2018). Microencapsulation of extracts of bioactive compounds obtained from acerola (*Malpighia emarginata* DC) pulp and residue by spray and freez drying: Chemical, morphological and chemometric characterization. *Food Chem.* 254: 281–291.

Ross, R.P., Desmond, C., Fitzgerald, G.F., and Stanton, C. (2005). Overcoming the technological hurdles in the development of probiotic foods. *J. Appl. Microbiol.* 98: 1410–1417.

Sanford, S., Chotani, G., Danielson, N., and Zahn, J.A. (2016). Scaling up of renewable chemicals. *Curr. Opin. Biotechnol.* 38: 112–122.

Sangeetha, P.T., Ramesh, M.N., and Prapulla, S.G. (2005). Recent trends in microbial production, analysis, and applications of fructooligosaccharides. *Trends Food Sci. Technol.* 16: 442–457.

Santivarangkna, C., Kulozik, U., and Foerst, P. (2007). Alternative drying processes for the industrial preservation of lactic acid starter cultures. *Biotechnol. Progr.* 23: 302–315.

Shahidi, F. and Han, X.Q. (1993). Encapsulation of food ingredients. *Crit. Rev. Food Sci. Nutr.* 33: 501–547.

Shori, A. (2015). The potential applications of probiotics on dairy and non-dairy foods focusing on viability during storage. *Biocatal. Agric. Biotechnol.* 4: 423–431.

Simpson, B.K., Rui, X., and Klomklao, S. (2012). Enzymes in food processing. *Food Biochem. Food Process* 8: 181–206.

Singh, R.S. and Singh, R.P. (2010). Production of fructooligosaccharides from inulin by endo inulinases and their prebiotic potential. *Food Technol. Biotechnol.* 48: 435–450.

Spinelli, S., Conte, A., and Del Nobile, M.A. (2016). Microencapsulation of extracted bioactive compounds from brewer's spent grain to enrich fish-burgers. *Food Bioprod. Process.* 100: 440–456.

Van de Guchte, M., Serror, P., Chervaux, C. et al. (2002). Stress responses in lactic acid bacteria. *Antonie Van Leeuwenhoek* 82: 187–216.

Van Oort, M. (2010). *Enzymes in Food Technology: Introduction*. Chichester, West Sussex: Wiley-Blackwell.

Ventling, B.L. and Mistry, V.V. (1993). Growth characteristics of bifidobacteria in ultrafiltered milk. *J. Dairy Sci.* 76: 962–971.
Vidhyalakshmi, R., Bhakyaraj, R., and Subhasree, R.S. (2009). Encapsulation "The future of probiotics": a review. *Adv. Biol. Res.* 39: 96–103.
Walker, L. (2015). Engineering principles provide the foundation for industrial biotechnology development. *Ind. Biotechnol.* 11: 181–182.
Wiedmann, M. and Jager, M. (1997). Synergistic sweeteners. *Food Integr. Anal.* 11: 51–56.
Zamora, L.M., Carretero, C., and Pares, D. (2006). Comparative survival rates of lactic acid bacteria isolated from blood, following spray-drying and freeze-drying. *Food Sci. Technol. Int.* 12: 77–84.
Zhang, W., Zhang, T., Jiang, B., and Mu, W. (2017). Enzymatic approaches to rare sugar production. *Biotechnol. Adv.* 35: 267–274.

# 12

# Research and Production of Ingredients Using Unconventional Raw Materials as Alternative Substrates

*Susana Rodríguez-Couto*

*IKERBASQUE, Basque Foundation for Science, Maria Diaz de Haro 3, 48013, Bilbao, Spain*

## 12.1 Introduction

Huge amounts of vegetable wastes are generated worldwide by the agricultural, forestry, and food-processing industries, causing serious environmental pollution, including a significant amount of greenhouse gas emission (Oelofse and Nahman 2013). This is particularly problematic in countries whose economy is mainly based on agriculture and farming. Most of these wastes are used as animal feed or burned (Mussatto et al. 2012). However, they are a good source of bioactive compounds (Ayala-Zavala and González-Aguilar 2011; Gornas and Rudzinska 2016; Ravindran and Jaiswal 2016; Banarjee et al. 2017), which can have potential as food additives and/or nutraceuticals. In addition, the demand of food additives from natural sources is growing considerably due to the increasing trend in consuming foods and supplements with health benefit properties. Table 12.1 lists the annual worldwide generation of several biological wastes.

The European 2020 growth strategy, launched in 2010, has set the goal of shifting the models of production and consumption from linear to circular (Imbert 2017). The European Union (EU) generates 1000 million tons of agricultural waste, 500 million tons of garden and forestry waste, and 250 million tons of food waste from the food-processing industry per year (Stabnikova et al. 2010). In this context, the management of industrial biological wastes presents a great challenge. Current management of biological wastes is costly and has a negative impact on the environment. Therefore, valorization of such wastes in an environmentally friendly and cost-effective way is needed. In this sense, the recovery of added-value compounds from biological wastes by solid-state fermentation (SSF) using fungal strains is considered as a promising approach (Farinas 2015). In addition, the bioproducts obtained can be labeled as natural, which would make them more attractive for the consumers.

The potential of biological wastes as raw materials to produce added-value compounds has been reviewed by several authors. Mirabella et al. (2014) published an extensive review on the potential use of wastes from the food industry. The authors concluded that the sustainability of the whole recovery process must be assessed before considering its implementation. Also, Gowe (2015) analyzed the potential of fruit and vegetable wastes to obtain bioactive compounds of commercial interest for the food

*Bioprocessing for Biomolecules Production,* First Edition.
Edited by Gustavo Molina, Vijai Kumar Gupta, Brahma N. Singh, and Nicholas Gathergood.

**Table 12.1** Major biological wastes produced worldwide.

| Waste | Annual production | Reference |
|---|---|---|
| Rice (*Oryza sativa*) straw | 731 million tons | Karimi et al. (2006) |
| Wheat (*Triticum aestivum*) straw | 529 million tons | Govumoni et al. (2013) |
| Sugarcane (*Saccharum officinarum*) bagasse | 200 million tons | Lachos-Pérez et al. (2016) |
| Wheat (*Triticum aestivum*) bran | 150 million tons | Prückler et al. (2014) |
| Corn (*Zea mays*) cobs | 144 million tons | da Silva et al. (2015) |
| Banana (*Musa paradisiaca*) peel | 57.6 million tons | Ahmad and Danish (2018) |
| Citrus wastes | 15 million tons | Marin et al. (2007) |
| Potato (*Solanum tuberosum*) waste | 12 million tons | El-Boushy and Van der Poel (1994) |
| Apple (*Malus domestica*) pomace | 8 million tons | FAO (2011) |
| Grape (*Vitis vinifera*) pomace | 6 million tons | Wadhwa et al. (2013) |
| Tomato (*Solanum lycopersicum*) waste | 320 000 tons | Encinar et al. (2008) |

industry. Similarly, Helkar et al. (2016) assessed the potential of the food industry byproducts as raw materials for the development of functional food ingredients. They indicated that the efficient utilization of these byproducts could make the food industry sustainable. More recently, Kowalska et al. (2017) reviewed the potential of agro-wastes as sources of new natural ingredients for the food, pharmaceutical, and cosmetic industry. They pointed out that despite the need to reduce agro-wastes and exploit their potential effectively, at present few agro-wastes are utilized properly in the food industry as a source of new natural ingredients. Therefore, there is a gap for research and innovation with commercial potential in this area.

## 12.2 Solid-State Fermentation (SSF)

SSF is a fermentation process in which the microorganisms grow on a solid support in an environment with none or very little free-flowing water (Pandey 2003). SSF is known from ancient times and has been mainly used for food processing (e.g. cheese, koji, soya) and is still used to produce important biomolecules and products for many industries, including food, pharmaceutical, textile, biochemical, and bioenergy (Pandey 2003; Soccol and Vandenberghe 2003).

SSF has received more attention by researchers in the last decades due to the increasing interest in producing added-value products from organic wastes to replace non-renewable materials and make the industrial chemical processes cleaner. The interest in SSF comes from its simplicity and closeness to the natural living habitats of many microorganisms (Wang and Yang 2007). In addition, several studies have shown that SSF processes can lead to higher yields, higher productivities, or better product characteristics than those attained by submerged fermentation (SmF)

(a) (b)

**Figure 12.1** The white-rot fungus *Trametes pubescens* grown on solidstate fermentation (SSF) conditions using sunflower seed shells as a support-substrate (a) and on submerged fermentation (SmF) forming pellets (b).

methods. Moreover, the low water content of the SSF reduces the fermenter size, downstream processing, stirring, and sterilization, thus, making the process more economical (Pandey 2003; Raghavarao et al. 2003; Hölker and Lenz 2005; Nigam 2009). In Figure 12.1 photographs of a white-rot fungus grown under SSF and SmF conditions are shown.

However, although the scale-up methods for SmF are well developed, this is not the case for SSF. The main difficulties to the scale-up of SSF processes arise from the intense heat generation, the water loss, and the heterogeneity of the system (Mitchell et al. 2000; di Luccio et al. 2004). Hence several solid-state bioreactor designs have been developed to overcome these problems, but few have been used at large-scale (Durand 2003).

The successful development of SSF processes depends on several aspects such as the type of microorganism, the type of substrate, and the process parameters. In Figure 12.2, the general processes involved in the valorization of organic waste by SSF are summarized (Yazid et al. 2017). Soccol et al. (2017) reviewed recent developments and innovations in SSF and asserted that the practicability and advantages of SSF must be evaluated for each process, and the economic viability will depend on a thorough comparison between SSF and SmF processes. Table 12.2 presents the advantages and disadvantages of SSF over SmF (Pérez-Guerra et al. 2003).

## 12.3 Production of Food Ingredients from Unconventional Raw Materials by SSF

In Table 12.3 the production of different food ingredients from unconventional raw materials by SSF processes in the last five years is presented.

### 12.3.1 Organic Acids

Organic acids are one of the most versatile additives in the food and beverage industry due to their solubility, hygroscopy, buffering, and chelation abilities. Thus, they are

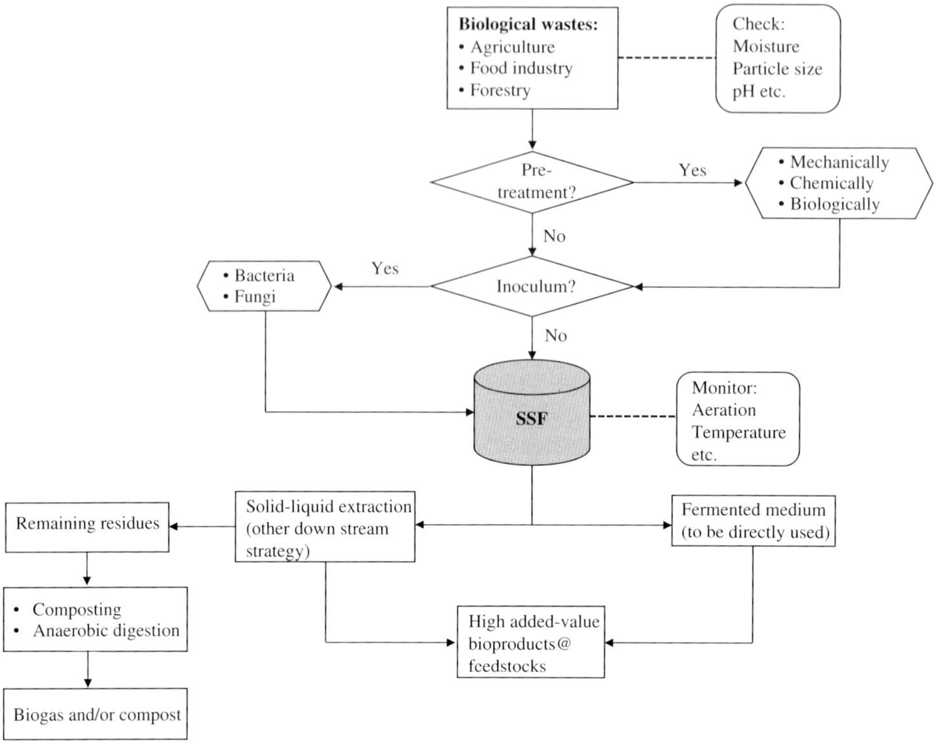


**Figure 12.2** Flowchart of valorization of biological wastes to produce added-value products by solidstate fermentation (SSF) (after Yazid et al. 2017).

widely used in food and beverage industries as preservatives and acidulants. Usually, organic acids are produced commercially by either chemical synthesis or SmF. However, SSF is a very promising technique where high concentrations of the product can be attained with the use of low-cost substrates, such as those from the agricultural, food, and forestry industries, resulting in processes with economic and environmental advantages (Vandenberghe et al. 2018). In fact, SSF has been successfully used for many years to produce citric and lactic acid on a large scale (Mussatto et al. 2012).

Certik et al. (2013) tested four *Mucor* strains and different cereal substrates (wheat bran, rye bran, oat flakes, barley groats, and spent malt grain) to produce gamma-linoleic acid under SSF conditions. *Mucor circinelloides* was the best producer of gamma-linoleic acid among the tested strains when grown on rye bran/spent mal grains at a ratio of 3 : 1. Furthermore, they observed that the addition of sunflower oil at 30% led to the highest amount of gamma-linoleic acid in the fermented substrate (24.2 g/kg).

Dhillon et al. (2013) produced citric acid by cultivation of *Aspergillus niger* under SSF conditions in a 12-L rotating drum bioreactor using fresh apple pomace, provided by an apple juice company, as a substrate. They obtained a production of 220.6 ± 13.9 g citric acid/kg dry solids operating under optimized conditions (i.e. 3% (v/v) methanol, intermittent agitation of 1 hour every 12 hours at 2 rpm, 1 vvm of aeration rate and 120 hours of incubation time). Also, Yadegari et al. (2013) studied the production of citric acid by

**Table 12.2** Advantages and disadvantages of solid-state fermentation (SSF) over submerged fermentation (SmF).

| Advantages | Disadvantages |
|---|---|
| Similar or higher yields are obtained than those obtained in the corresponding submerged cultures. | Only microorganisms that can grow at low moisture levels can be used. |
| The low availability of water reduces the possibilities of contamination by bacteria and yeast. This allows working in aseptic conditions in some cases. | Usually the substrates require pretreatment (size reduction by grinding, rasping or chopping, homogenization, physical, chemical or enzymatic hydrolysis, cooking or vapor treatment). |
| Similar environment conditions are present to those of the natural habitats for fungi that constitute the main group of microorganisms used in SSF. | Biomass determination is very difficult. |
| Higher levels of aeration, especially adequate in those processes demanding an intensive oxidative metabolism. | The solid nature of the substrate causes problems in the monitoring of the process parameters (pH, moisture content, and substrate, oxygen, and biomass concentration). |
| The inoculation with spores (in those processes that involve fungi) facilitates their uniform dispersion through the medium | Agitation may be very difficult. For this reason, static conditions are preferred |
| Culture media are often quite simple. The substrate usually provides all the nutrients necessary for growth | Frequent need of high inoculum volumes |
| Simple design reactors with few spatial requirements can be used due to the concentrated nature of the substrates | Many important basic scientific and engineering aspects are yet poor characterized. Information about the design and operation of reactors on a large scale is scarce |
| Low energetic requirements (in some cases autoclaving or vapor treatment, mechanical agitation and aeration are not necessary) | Possibility of contamination by undesirable fungi |
| Small volumes of polluting effluents. Fewer requirements of dissolvents are necessary for product extraction due to its high concentration | The removal of metabolic heat generated during growth may be very difficult |
| The low moisture availability may favor the production of specific compounds that may not be produced or may be poorly produced in SmF | Extracts containing products obtained by leaching of fermented solids are often of viscous nature |
| In some cases, the products obtained have slightly different properties (e.g. more thermotolerance) when produced in SSF in comparison to SmF | Mass transfer limited to diffusion |
| Due to the concentrated nature of the substrate, smaller reactors in SSF with respect to SmF can be used to hold the same amounts of substrate | In some SSF, aeration can be difficult due to the high solid concentration |
| | Spores have longer lag times due to the need for germination |
| | Cultivation times are longer than in SmF |

Source: Extracted from Pérez-Guerra et al. (2003).

**Table 12.3** Some food ingredients produced by solid-state fermentation (SSF) of biological wastes in the last five years.

| Waste | Microorganism | Product | Reference |
|---|---|---|---|
| *Organic acids* | | | |
| Wheat bran, rye bran, oat flakes, barley groats, spent malt grain | *Mucor petrinsularis*, *Mucor dimorphosporus*, *Mucor circinelloides*, *Mucor hiemalis* | Gamma-linolenic acid | Certik et al. (2013) |
| Apple pomace | *Aspergillus niger* | Citric acid | Dhillon et al. (2013) |
| Sugarcane bagasse | *A. niger* | Citric acid | Yadegari et al. (2013) |
| Oil palm empty fruit bunch | *Trichoderma reesei* | Humic acids | Motta and Santana (2014) |
| Nut shells, rice bran, rice husk, nut oil cake, sugar cane bagasse, fresh orange fruit wastes | *Ustilago maydis* | Itaconic acid | Rafi et al. (2014) |
| Apple pomace | *Rhizopus oryzae* | Fumaric acid | Das et al. (2015) |
| Rice straw, faba bean straw | *Aspergillus oryzae*, *Azotobacter chroococcum* | Organic acids | Saber et al. (2015) |
| Apple pomace and peanut shell | Consortium of *Aspergillus ornatus* and *Alternaria alternata* | Citric acid | Ali et al. (2016) |
| Pulp and solid paper waste | *R. oryzae* | Fumaric acid | Das et al. (2016) |
| Corn cob powder | *A. niger* | Oxalic acid | Mai et al. (2016) |
| Citric pulp | *Fusarium moniliforme*, *Gibberella fukikuroi* | Gibberellic acid | De Oliveira et al. (2017) |
| Wheat bran and sugarcane bagasse | Endophytic fungi | Organic acids | Dezam et al. (2017) |
| *Phenolic compounds* | | | |
| Coffee wastes | *A. niger*, *Aspergillus ustus*, *Mucor* sp., *Penicillium purpurogenum*, *Neurospora crassa* | Phenolics | Machado et al. (2013) |

| | | | |
|---|---|---|---|
| Wheat bran | *R. oryzae* | Phenolics | Dey and Kuhad (2014) |
| Citrus waste | *Paecilomyces variotii* | Phenolics | Madeira et al. (2014) |
| Rice bran | *R. oryzae* | Phenolics | Schmidt et al. (2014) |
| Berry pomaces | *A. niger* | Polyphenols, lipids | Dulf et al. (2015) |
| Tangerine residues | *Lentinus polychrous* | Antioxidants | Nitayapat et al. (2015) |
| Rice bran | *Rhizopus oligosporus* and *Monascus purpureus* | Antioxidants | Razak et al. (2015) |
| Soybean mean | *Bacillus amyloliquefaciens, Lactobacillus* spp., *Saccharomyces cerevisiae* | Phenolics | Chi and Cho (2016) |
| Plum by-products | *A. niger*<br>*R. oligosporus* | Polyphenols, lipids | Dulf et al. (2016) |
| Cauliflower outer leaves | *A. niger, A. oryzae, A. sojae, R. oryzae, R. azygosporus. Phanerochaete chrysoporium* | Kaemferol metabolites | Huynh et al. (2016) |
| Fig by-products | *R. oryzae,*<br>*Trichoderma* sp., *A. niger* | Phenolics | Buenrostro-Figueroa et al. (2017) |
| Apricot pomace | *A. niger*<br>*R. oligosporus* | Phenolics | Dulf et al. (2017) |
| Buckwheat | *Agaricus* strains | Phenolics | Kang et al. (2017) |
| *Psidium guajava* leaves | *Monascus anka* and *Bacillus* sp. | Polyphenols | Wang et al. (2017) |
| Chokeberry pomace | *A. niger*<br>*R. oligosporus* | Phenolics | Dulf et al. (2018) |
| Grape, apple and pitahaya residues | *Rhizomucor miehei* | Phenolics | Zambrano et al. (2018) |
| *Aroma compounds* | | | |
| Food waste mixtures | *S. cerevisiae, Kluyveromyces marxianus,* kefir | $\varepsilon$-pinene | Aggelopoulos et al. (2014) |
| Sugarcane bagasse | *Trichoderma viride* | Coconut aroma | Fadel et al. (2015) |
| Orange peels | *S .cerevisiae* | Aroma esters | Mantzouridou et al. (2015) |

*(Continued)*

**Table 12.3** (Continued)

| Waste | Microorganism | Product | Reference |
|---|---|---|---|
| Apple pomace | *S. cerevisiae, Hanseniaspora valbyensism, Hanseniaspora uvarum* | Aroma compounds | Rodriguez-Madrera et al. (2015) |
| Sugracane bagasse, sugar beet molasses | *K. marxianus* | Aroma compounds | Martinez et al. (2017) |
| *Pigments* | | | |
| *Wheat bran, rye bran, oat flakes, barley groats, spent malt grain* | *M. petrinsularis, M. dimorphosporus, M. circinelloides, M. hiemalis* | *β-Carothene* | *Certik et al.* (2013) |
| Whole grain, dehulled grain, bran substrates | *M. purpureus* | Pigments | Srianta and Harijono (2015) |
| Wheat wastes | *Yamadazyma guilliermondii, Yarrowia lipolytica, Xanthophylomyces dendrorhous, Sporidiobulus salmonicolor* | Astaxanthin | Dursun and Dalgic (2016) |
| Olive pomace | *X. dendrorhous, S. salmonicolor* | Astaxanthin | Erilymaz et al. (2016) |
| Rice, corn, whole sorghum grain, dehulled sorghum grain, sorghum bran | *M. purpureus* | Pigments | Srianta et al. (2016) |
| Rice, corn, whole sorghum grain, dehulled sorghum grain, sorghum bran | *M. purpureus* | Pigments | Srianta et al. (2017) |

*A. niger* under SSF conditions using sugarcane bagasse, provided by a sugar factory, as a substrate. They found that pretreating the sugarcane bagasse with sodium hydroxide increased the citric acid production from 75.34 g $kg^{-1}$ substrate to 97.81 g $kg^{-1}$ substrate.

Motta and Santana (2014) reported the production of humic acid by SSF of *Trichoderma reesei* in fixed-bed columns using empty palm bunch, an underutilized by-product of the palm oil industry, as a substrate. They found a humic acid productivity of 0.73 mg/100 g of substrate after 72 hours of incubation.

Rafi et al. (2014) tested different agro-food wastes such as ground nut shells, rice bran, rice husk, orange pulp, ground nut oil cake, and sugarcane bagasse as substrates for itaconic acid production by the fungus *Ustilago maydis* under SSF conditions. All the substrates led to promising yields of itaconic acid, rice bran, orange pulp and sugarcane bagasse producing the highest ones (around 26–27 g $kg^{-1}$ substrate) operating at optimal conditions (i.e. pH 3, moisture 60%, temperature 32 °C and a cultivation time of five days).

Das et al. (2015) reported a production of $52 \pm 2.1$ g of fumaric acid per kg of dry apple pomace (with 50% moisture content) by the fungus *Rhizopus oryzae* after 14 cultivation days in tray bioreactors. Later (Das et al. 2016), they tested pulp and paper solid waste from a paper industry as a substrate. They found that the particle size highly influenced the fumaric acid production, a particle size range of $850\ \mu m < x \leq 300\ \mu m$ leading to the highest production (41.65 g $kg^{-1}$ dry substrate).

Saber et al. (2015) investigated the synergistic effect of the filamentous fungus *Aspergillus oryzae* and the non-symbiotic nitrogen fixing bacterium *Azotobacter chroococcum* as an innovative technique to convert rice straw, faba bean straw, and rock phosphate into organic acids under SSF conditions. They found that citric (15.3 mg $g^{-1}$ fermented biomass) and succinic acid (13.7 mg $g^{-1}$ fermented biomass) were the main organic acids contained in the fermented biomass, corresponding to the 98.87% of the total organic acids detected.

Ali et al. (2016) used different agro-wastes (apple pomace, peanut shell and a mixture of both apple pomace and peanut shell at 50 : 50 ratio) as substrates for SSF to enhance the citric acid production from single and co-culture consortium of *Aspergillus ornatus* and *Alternaria alternata*. Partial optimisation of the co-culture (arginine addition, 30 °C, 25 g apple pomace at 50% moisture, pH 5 and a cultivation time of 48 hours) showed a maximum citric acid yield of $2.644 \pm 0.99$ mg $ml^{-1}$.

Mai et al. (2016) studied the production of oxalic acid by a methanol-resistant *A. niger* strain under semi SSF conditions using corncob as a substrate. A maximum productivity of 123.0 g $kg^{-1}$ dry weight of corncob was attained operating at optimal conditions (i.e., $10^6$ spores $g^{-1}$ dry weight and 5% (w/v) corncob in 0.1 N NaOH solution).

De Oliveira et al. (2017) investigated the production of the important phytohormone gibberellic acid by the fungi *Fusarium moniliforme* and *Gibberella fjikuroi* using different cultivation techniques (SSF, semi SSF, and SmF) at flask and bioreactor scale operating with citric pulp, a subproduct obtained from the extraction of orange juice, as a substrate. SSF led to the highest production of gibberellic acid at both flask (7.60 g $kg^{-1}$ dry substrate) and bioreactor scale (7.34 g $kg^{-1}$ dry substrate).

Dezam et al. (2017) investigated the potential of endophytic fungi, isolated from mangrove areas in Sao Paulo (Brazil), to produce organic acids by SSF using wheat bran and sugarcane bagasse as substrates. The highest yield of organic acids (135.5 mg $g^{-1}$ dry substrate), corresponding mainly to citric acid (106.8 mg $g^{-1}$ dry substrate), was produced by the isolate *Aspergillus awamori* when grown on a mixture of wheat bran and sugarcane bagasse at a ratio of 1 : 3. These results pointed out the potential of endophytic fungi for organic acid production using agro-industrial wastes as feedstock under SSF conditions, thus, providing a sustainable alternative to the organic acid manufacture and contributing to the circular economy.

### 12.3.2 Phenolic Compounds

Phenolic compounds have been extensively studied for their application in the food industry as preservatives to improve the shelf life of perishable products. The current concern about the impact of food on health influences the consumer choice of food based on its formulation. Consumers prefer phenolic compounds from natural sources.

Antioxidants have become increasing popular during the last decades due to their potential properties to prevent chronic diseases such as cardiovascular diseases, cancer, osteoporosis, diabetes mellitus, and neurodegenerative diseases (Pandey and Rizvi 2009). Consequently, various sources of different antioxidant phenolic compounds, such as fruits, vegetables, wine, coffee, tea, and cereals, have been investigated to replace health hazard antioxidant compounds like butylated hydroxyanisole, butylated hydroxytoluene, and tertiary butyl hydroquinone in different food products.

Phenolic compounds are commonly extracted from plant materials by organic solvents such as methanol, ethyl acetate, acetone, and *n*-hexane using various conventional extraction methods (Ignat et al. 2011). However, these methods do not allow the complete release of bound phenolics from plant materials and, in addition, they are not environmentally friendly. Thus, microbial fermentation processes are considered very promising to produce antioxidant phenolic compounds due to their cost-effectiveness and environmental advantages. Table 12.3 shows the production of phenolic compounds from different biological wastes by SSF processes.

Machado et al. (2013) found that the strains *Penicillium purpurogenum* GH2, *Neurospora crassa* ATCC10337 and *Mucor* sp. 3P presented great ability to grow and release phenolic compounds from coffee silverskin and spent coffee grounds, abundant coffee wastes from the coffee industry, under SSF conditions. Thus, these findings offer new possibilities for the use of underutilized wastes.

Dey and Kuhad (2014) showed that the extraction of phenolic compounds from wheat improved by SSF with *R. oryzae*. Therefore, fermented wheat may serve as a powerful source of natural antioxidants. Madeira et al. (2014) developed a process for the biotransformation of phenolic compounds from citrus wastes, provided by a local industrial company, by SSF with *Paecilomyces variotii*. This process increased by 73% the antioxidant capacity of the waste. In addition, the commercially interesting bioactive compounds hesperetin, naringenin, and ellagic acid were also produced.

Schmidt et al. (2014) reported the increased in free phenolics by more than 100% of rice bran after SSF with *R. oryzae*. In addition, the profile of the phenolics changed, with gallic and ferulic acid showing the highest increase (170 and 765 mg/g, respectively).

Dulf et al. (2015) showed the increase in phenolic content and antioxidant activity of berry pomaces by SSF of *A. niger*. Later, they observed the same for SSF of plum, apricot, and chokeberry pomaces with *A. niger* and *Rhizopus oligosporus* (Dulf et al. 2016, 2017, 2018). Also, Nitayapat et al. (2015) increased the phenolic content and the antioxidant activity of tangerine wastes by SSF with *Lentinus polychrous*.

Razak et al. (2015) proved that the phenolic content and antioxidant activity of rice bran could be increased by SSF with fungi such as *R. oligosporus* and *Monascus purpureus*. Likewise, Chi and Cho (2016) showed that the SSF of soybean meal with *Bacillus amyloliquefaciens* significantly improved its total phenolic content and antioxidant activity.

Huynh et al. (2016) tested cauliflower outer leaves as a substrate to produce phenolic compounds by different filamentous fungi, namely *A. niger*, *A. oryzae*, *Aspergillus sojae*, *R. oryzae*, *Rizhopus. azygosporus* and *Phanerochaete chrysoporium* under SSF conditions. *A. sojae* led to the highest level of total phenolic compounds (321 mg rutin equivalents/100 g fresh weight) after 1 fermentation day, this value being three-fold higher than that found in the unfermented substrate. Also, Bei et al. (2017) reported that the SSF of oats with *Monascus anka* considerably increased their content in phenolic compounds and, in addition, they showed higher antioxidant activities.

Buenrostro-Figueroa et al. (2017) investigated the valorization of fig byproducts from a jam and wine-making company, by SSF with the following fungal strains: *R. oryzae*, *Trichoderma* sp., *A. niger* HT4 and *A. niger* GH1. They found that *A. niger* HT4 led to the highest release of total polyphenols. Further, the optimized process (36 h, 40 °C, pH 5.0, moisture content 60%, $NaNO_3$ 0.57 g $l^{-1}$, $KH_2PO_4$ 3.04 g $l^{-1}$, $MgSO_4 \cdot 7H_2O$ 1.52 g $l^{-1}$ and KCl 5.37 g $l^{-1}$) released 10.37 mg of gallic acid equivalents/g of dry matter.

Kang et al. (2017) assessed the effect of SSF of buckwheat with three *Agaricus* strains (*Agaricus blazei* Murrill SH26, *Agaricus bisporus* AS2796, and *Agaricus bisporus* G1) on its total phenolic content and antioxidant properties. *A. blazei* supported the highest phenolic content in buckwheat (18.07 mg $g^{-1}$) after 21 fermentation days. Also, the antioxidant properties of buckwheat increased, save for the *A. bisporus* G1 cultures.

Wang et al. (2017) showed that the co-fermentation of guava leaves with *Monascus anka* and *Bacillus* sp. under SSF conditions promoted the release of insoluble-bound polyphenol components. In addition, these polyphenols presented higher antioxidant activities than those extracted from unfermented guava leaves. Moreover, they observed that the antioxidant capacities of the soluble polyphenols were considerably increased by this microbial co-fermentation. Recently, Zambrano et al. (2018) also reported that the SSF of different agro-wastes (black grape pomace, and apple and yellow pitahaya peel, core, peduncle, and seed mixtures) with *Rhizomucor miehei* increased their extractable phenolic content and improved their phenolic antioxidant properties.

### 12.3.3 Flavor and Aroma Compounds

Flavors are largely used in the food industry to improve food organoleptic properties. Most flavoring compounds are produced *via* chemical synthesis or extraction from natural materials. However, since consumers prefer food free of chemical substances, the production of aroma compounds by SSF would likely be well received (Table 12.3).

Aggelopoulos et al. (2014) reported the production of high amounts of the aroma volatile compound $\varepsilon$-pinene by kefir grown on a mixture of food wastes (orange pulp, molasses, potato pulp and whey) under SSF conditions at a yield of 4 kg per ton of treated substrate. Also, Fadel et al. (2015) assessed the ability of *Trichoderma viride* to produce coconut aroma in SSF using sugarcane bagasse as a substrate. The analysis revealed that 6-pentyl-$\alpha$-pyrone was the main volatile compound produced contributing to the coconut aroma, the highest production of 3.62 mg $g^{-1}$ of dry matter after five cultivation days being attained.

Mantzouridou et al. (2015) reported for the first time the feasibility of SSF of orange peel waste to produce yeast volatile esters with fruity-like character at a high yield (250 mg $kg^{-1}$ of fermented orange peel waste). Likewise, Rodriguez-Madrera et al. (2015) showed that the SSF of apple pomace with autochthonous yeast strains generated different volatile compounds with application in the food industry. More recently, Martinez et al. (2017) proved the efficiency of the generally recognized as safe (GRAS) strain *Kluyveromyces marxianus* to produce aroma compounds by SSF using a mixture of sugarcane bagasse and sugar beet molasses as the only substrate.

### 12.3.4 Pigments

Synthetic colorants are considered hazardous to human health and, thus, only few classes are acceptable to be used in the food industry. Therefore, the production of non-hazardous colorants suitable for use in the food industry is needed. In this context, natural pigments secreted by certain fungal species such as *Aspergillus, Fusarium, Penicillium, Paecilomyces,* and *Trichoderma* appear as a promising approach (Akilandeswari and Pradeep 2016). Furthermore, several natural pigments have antioxidant properties which is very interesting since there is an increasing trend toward the development of nutraceutical food. In addition, these natural pigments can be produced in SSF using agro-industrial wastes as substrates for the microorganism, thus, making the process more cost-efficient and ecological. In Table 12.3 different natural pigments produced under SSF are presented.

Certik et al. (2013) tested four *Mucor* strains and different cereal substrates (wheat bran, rye bran, oat flakes, barley groats and spent malt grain) to produce the pigment $\beta$-carotene under SSF conditions. *Mucor circinelloides* was the best producer of $\beta$-carotene (9.5 mg $kg^{-1}$) among the tested strains when grown on rye bran and spent malt grains at a ratio of 3 : 1 with glucose addition.

Srianta and Harijono (2015) showed that *M. purpureus* was able to produce yellow, orange, and red pigments when grown on whole sorghum grain, dehulled sorghum grain, and sorghum bran substrates with soaking under SSF conditions. The authors also observed that the whole sorghum grain cultures led to the highest ethanol soluble pigments, while the fermented bran with soaking cultures contained the highest water-soluble pigments. They (Srianta et al. 2016) also studied the production and composition of the pigments produced by *M. purpureus* under SSF grown on different cereal substrates (i.e. rice, corn, whole sorghum grain, dehulled sorghum grain, and sorghum bran). The highest pigment production was achieved on rice, followed by dehulled sorghum grain, whole sorghum grain, corn, and finally sorghum bran. Twelve pigments were detected on the *Monascus*-fermented products at different levels, the red pigment rubropunctamine being the major one (57–87%) except for sorghum bran cultures, which produced Yellow II as the major one. Later (Srianta et al. 2017), they found that the pigments produced by *M. purpureus* when grown on rice, corn, and sorghum under SSF conditions presented considerable antioxidant activities. However, the specific pigments responsible for these activities were not identified.

Dursun and Dalgic (2016) proved the successful use of wheat wastes, one of the main agro-industrial wastes worldwide, to produce the carotenoid pigment astaxanthin by

*Yamadazyma guilliermondii*, *Yarrowia lipolytica*, *Xanthophyllomyces dendrorhous*, and *Sporidiobolus salmonicolor* under SSF conditions. The maximal astaxanthin yield (109.23 mg $g^{-1}$ substrate) was produced by *X. dendrorhous* when cultured at optimized conditions (i.e. 20 °C, pH 5.5 and 90% moisture). Also, Eryılmaz et al. (2016) studied the production of astaxanthin by the yeasts *X. dendrorhous* and *S. salmonicolor* under SSF conditions using olive pomace as a substrate. The former led to a maximum astaxanthin yield of 220.24 ± 17.47 μg $g^{-1}$ dry substrate.

## 12.4 Outlook

Many industries generate huge quantities of biological wastes, causing a serious environmental concern. Thus, there is an urgent need to change the global perception toward industrial biological wastes, as they can be effectively utilized in an eco-friendly manner to produce added-value products (e.g., food additives). In this sense, many industrial and academic laboratories are focusing on the valorization of industrial biowastes. However, further research and optimization studies on valorization technologies focused on full rather than laboratory studies are needed. Although SSF is a very promising green technology for biowaste valorization, research on improving the scale-up facilities in SSF, by the development of robust designs and configurations of bioreactors, process automation, and online monitoring of parameters, is required.

## References

Aggelopoulos, T., Katsieris, K., Bekatorou, A. et al. (2014). Solid state fermentation of food waste mixtures for single cell protein, aroma volatiles and fat production. *Food Chem.* 145: 710–716.

Ahmad, T. and Danish, M. (2018). Prospects of banana waste utilization in wastewater treatment: a review. *J. Environ. Manage.* 206: 330–348.

Akilandeswari, P. and Pradeep, B.V. (2016). Exploration of industrially important pigments from soil fungi. *Appl. Microbiol. Biotechnol.* 100: 1631–1643.

Ali, S.R., Anwar, Z., Irshad, M. et al. (2016). Bio-synthesis of citric acid from single and co-culture-based fermentation technology using agro-wastes. *J. Radiat. Res. Appl. Sci.* 9: 57–62.

Ayala-Zavala, J.F. and González-Aguilar, G.A. (2011). Use of additives to preserve the quality of fresh-cut fruits and vegetables. In: *Advances in Fresh-Cut Fruits and Vegetables Processing* (ed. O. Martín-Belloso and R. Soliva-Fortuny), 231–254. Boca Raton, FL, USA: CRC Press.

Banarjee, J., Singh, R., Vijayaraghavan, R. et al. (2017). Bioactives from fruit processing wastes: green approaches to valuable chemicals. *Food Chem.* 225: 10–22.

Bei, Q., Liu, Y., Wang, L. et al. (2017). Improving free, conjugated, and bound phenolic fractions in fermented oats (Avena sativa L.) with Monascus anka and their antioxidant activity. *J. Funct. Foods* 32: 185–194.

Buenrostro-Figueroa, J.J., Velázquez, M., Flores-Ortega, O. et al. (2017). Solid state fermentation of fig (*Ficus carica* L.) by-products using fungi to obtain phenolic compounds with antioxidant activity and qualitative evaluation of phenolics obtained. *Process Biochem.* 62: 16–23.

Certík, M., Adamechová, Z., and Guothová, L. (2013). Simultaneous enrichment of cereals with polyunsaturated fatty acids and pigments by fungal solid state fermentations. *J. Biotechnol.* 168: 130–134.

Chi, C.H. and Cho, S.J. (2016). Improvement of bioactivity of soybean meal by solid-state fermentation with *Bacillus amyloliquefaciens versus Lactobacillus* spp. and *Saccharomyces cerevisiae. LWT-Food Sci. Technol.* 68: 619–625.

Das, R.K., Brar, S.K., and Verma, M. (2015). A fermentative approach towards optimizing directed biosynthesis of fumaric acid by *Rhizopus oryzae* 1526 utilizing apple industry waste biomass. *Fungal Biol.* 119: 1279–1290.

Das, R.K., Brar, S.K., and Verma, M. (2016). Potential use of pulp and paper solid waste for the bio-production of fumaric acid through submerged and solid state fermentation. *J. Cleaner Prod.* 112: 4435–4444.

Dey, T.B. and Kuhad, R.C. (2014). Enhanced production and extraction of phenolic compounds from wheat by solid-state fermentation with *Rhizopus oryzae* RCK2012. *Biotechnol. Rep.* 4: 120–127.

Dezam, A.P.G., Vasconcellosa, V.M., Lacavab, P.T., and Farinasa, C.S. (2017). Microbial production of organic acids by endophytic fungi. *Biocatal. Agric. Biotechnol.* 11: 282–287.

Dhillon, G.S., Kaur, S., Sarma, S.J., and Brar, S.K. (2013). Integrated process or fungal citric acid fermentation using apple processing wastes and sequential extraction of chitosan from waste stream. *Ind. Crops Prod.* 50: 346–351.

Di Luccio, M., Capra, F., Ribeiro, N.P. et al. (2004). Effect of temperature, moisture, and carbon supplementation on lipase production by solid state fermentation of soy cake by *Penicillium simplicissimum. Appl. Biochem. Biotechnol.* 113: 173–180.

Dulf, F.V., Vodnar, D.C., Dulf, E.H., and Toşa, M.I. (2015). Total phenolic contents, antioxidant activities, and lipid fractions from berry pomaces obtained by solid-state fermentation of two *Sambucus* species with *Aspergillus niger. J. Agric. Food. Chem.* 63: 3489–3500.

Dulf, F.V., Vodnar, D.C., and Socaciu, C. (2016). Effects of solid-state fermentation with two filamentous fungi on the total phenolic contents, flavonoids, antioxidant activities and lipid fractions of plum fruit (*Prunus domestica* L.) by-products. *Food Chem.* 209: 27–36.

Dulf, F.V., Vodnar, D.C., Dulf, E.H., and Pintea, A. (2017). Phenolic compounds, flavonoids, lipids and antioxidant potential of apricot (*Prunus armeniaca* L.) pomace fermented by two filamentous fungal strains in solid state system. *Chem. Cent. J.* 11: 92. https://doi .org/10.1186/s13065-017-0323-z.

Dulf, F.V., Vodnar, D.C., Dulf, E.H. et al. (2018). Liberation and recovery of phenolic antioxidants and lipids in chokeberry (*Aronia melanocarpa*) pomace by solid-state bioprocessing using *Aspergillus niger* and *Rhizopus oligosporus* strains. *LWT-Food Sci. Technol.* 87: 241–249.

Durand, A. (2003). Bioreactor designs for solid state fermentation. *Biochem. Eng. J.* 13: 113–125.

Dursun, D. and Dalgic, A.C. (2016). Optimization of astaxanthin pigment bioprocessing by four different yeast species using wheat wastes. *Biocatal. Agric. Biotechnol.* 7: 1–6.

El-Boushy, A.R.Y. and Van der Poel, A.F.B. (1994). *Poultry Feed from Waste Processing and Use.* London, UK: Chapman and Hall.

Encinar, J.M., González, J.F., and Martínez, G. (2008). Energetic use of the tomato plant waste. *Fuel Process. Technol.* 89: 1193–1200.

Eryılmaz, E.B., Dursun, D., and Dalgıc, A.C. (2016). Multiple optimization and statistical evaluation of astaxanthin production utilizing olive pomace. *Biocatal. Agric. Biotechnol.* 7: 224–227.

Fadel, H.H.M., Mahmoud, M.G., Asker, M.M.S., and Lotfy, S.N. (2015). Characterization and evaluation of coconut aroma produced by *Trichoderma viride* EMCC-107 in solid state fermentation on sugarcane bagasse. *Electron. J. Biotechnol.* 18: 5–9.

FAO Statistical Database. http://faostat.fao.org/site/368/DesktopDefault.aspx?PageID368#ancor; 2011.

Farinas, C.S. (2015). Developments in solid-state fermentation for the production of biomass-degrading enzymes for the bioenergy sector. *Renewable Sustainable Energy Rev.* 52: 179–188.

Gornas, P. and Rudzinska, M. (2016). Seeds recovered from industry by-products of nine fruit species witha high potential utility as a source of unconventional oil for biodiesel and cosmetic and pharmaceutical sectors. *Ind. Crops Prod.* 83: 329–338.

Govumoni, S.P., Koti, S., Kothagouni, S.Y. et al. (2013). Evaluation of pretreatment methods for enzymatic saccharification of wheat straw for bioethanol production. *Carbohydr. Polym.* 91: 646–650.

Gowe, A. (2015). Review on potential use of fruit and vegetables by-products as a valuable source of natural food additives. *Food Sci. Quality Manage.* 45: 47–61.

Helkar, P.B., Sahoo, A.K., and Patil, N.J. (2016). Review: food industry by-products used as a functional food ingredients. *Int. J. Waste Resour.* 6. https://doi.org/10.4172/2252-5211.1000248.

Hölker, U. and Lenz, J. (2005). Solid-state fermentation: are there any biotechnological advantages? *Curr. Opin. Microbiol.* 8: 301–306.

Huynh, N.T., Smagghe, G., Gonzales, G.B. et al. (2016). Extraction and bioconversion of kaempferol metabolites from cauliflower outer leaves through fungal fermentation Biochem. *Eng. J.* 116: 27–31.

Ignat, I., Volf, I., and Popa, V.I. (2011). A critical review of methods for characterisation of polyphenolic compounds in fruits and vegetables. *Food Chem.* 26: 1821–1835.

Imbert, E. (2017). Food waste valorization options: opportunities from the bioeconomy. *Open Agric.* 2: 195–204.

Kang, M., Zhai, F.H., Li, X.X. et al. (2017). Total phenolic contents and antioxidant properties of buckwheat fermented by three strains of Agaricus. *J. Cereal Sci.* 73: 138–142.

Karimi, K., Emtiazi, G., and Taherzadeh, M.J. (2006). Ethanol production from dilute-acid pretreated rice straw by simultaneous saccharification and fermentation with *Mucor indicus, Rhizopus oryzae,* and *Saccharomyces cerevisiae. Enzyme Microb. Technol.* 40: 138–144.

Kowalska, H., Czajkowska, K., Cichowska, J., and Lenart, A. (2017). What's new in biopotential of fruit and vegetable by-products applied in the food processing industry. *Trends Food Sci. Technol.* 67: 150–159.

Lachos-Perez, D., Martinez-Jimenez, F., Rezende, C.A. et al. (2016). Subcritical water hydrolysis of sugarcane bagasse: an approach on solid residues characterization. *J. Supercrit. Fluids* 108: 69–78.

Machado, E.M.S., Rodriguez-Jasso, R.M., Teixeira, J.A., and Mussatto, S.I. (2013). Growth of fungal strains on coffee industry residues with removal of polyphenolic compounds. *Biochem. Eng. J.* 60: 87–90.

Madeira, J.V., Nakajima, V.M., Macedo, J.A., and Macedo, G. (2014). Rich bioactive phenolic extract production bymicrobial biotransformation of Brazilian citrus residues. *Chem. Eng. Res. Des.* 92: 1802–1810.

Mai, H.T.N., Lee, K.M., and Choi, S.S. (2016). Enhanced oxalic acid production from corncob by amethanol-resistant strain of *Aspergillus niger* using semi solid-satefermentation. *Process Biochem.* 51: 9–15.

Mantzouridou, F.T., Paraskevopoulou, A., and Lalou, S. (2015). Yeast flavour production by solid state fermentation of orange peel waste. *Biochem. Eng. J.* 101: 1–8.

Marin, F.R., Soler-Rivas, C., Benavente-Garcia, O. et al. (2007). By-products from different citrus processes as a source of customized functional fibres. *Food Chem.* 100: 736–741.

Martinez, O., Sánchez, A., Font, X., and Barrena, R. (2017). Valorization of sugarcane bagasse and sugar beet molasses using *Kluyveromyces marxianus* for producing value-added aroma compounds via solid-state fermentation. *J. Cleaner Prod.* 158: 8–17.

Mirabella, N., Castellani, V., and Sala, S. (2014). Current options for the valorization of food manufacturing waste: a review. *J. Cleaner Prod.* 65: 28–41.

Mitchell, D.A., Krieger, N., Stuart, D.M., and Pandey, A. (2000). New developments in solid-state fermentation: II. Rational approaches to the design, operation and scale-up of bioreactors. *Process Biochem.* 35: 1211–1225.

Motta, F.L. and Santana, M.H.A. (2014). Solid-state fermentation for humic acids production by a *Trichoderma reesei* strain using an oil palm empty fruit bunch as the substrate. *Appl. Biochem. Biotechnol.* 172: 2205–2217.

Mussatto, S.I., Ballesteros, L.F., Martins, S., and Teixeira, J.A. (2012). Use of agro-industrial wastes in solid-state fermentation processes. In: *Industrial Waste* (ed. K.Y. Show and X. Guo), 121–140. Rijeka, Croatia: InTech.

Nigam, P.S. (2009). Production of bioactive secondary metabolites. In: *Biotechnology for Agro-Industrial Residues Utilization* (ed. P.S. Nigam and A. Pandey), 129–145. Amsterdam, The Netherlands: Springer.

Nitayapat, N., Prakarnsombut, N., Lee, S.J., and Boonsupthip, W. (2015). Bioconversion of tangerine residues by solid-state fermentation with *Lentinus polychrous* and drying the final products. *LWT-Food Sci. Technol.* 63: 773–779.

Oelofse, S.H.H. and Nahman, A. (2013). Estimating the magnitude of food waste generated in South Africa. *Waste Manage. Res.* 31: 80–86.

de Oliveira, J., Rodrigues, C., Vandenberghe, L.P.S. et al. (2017). Gibberellic acid production by different fermentation systems using citric pulp as substrate/support. *BioMed Res. Int.* https://doi.org/10.1155/2017/5191046.

Pandey, A. (2003). Solid-state fermentation. *Biochem. Eng. J.* 13: 81–84.

Pandey, K.B. and Rizvi, S.I. (2009). Plant polyphenols as dietary antioxidants in human health and disease. *Oxid. Med. Cell Longv.* 2: 270–278.

Pérez-Guerra, N., Torrado-Agrasar, A., López-Macias, C., and Pastrana, L. (2003). Main characteristics and applications of solid substrate fermentation. *Electron. J. Environ. Agric. Food Chem.* 2: 343–350.

Prückler, M., Siebenhandl-Ehn, S., Apprich, S. et al. (2014). Wheat bran-based biorefinery 1: composition of wheat bran and strategies of functionalization. *LWT-Food Sci. Technol.* 56: 211–221.

Rafi, M.M., Hanumanthu, M.G., Rao, D.M., and Venkateswarlu, K. (2014). Production of itaconic acid by *Ustilago maydis* from agro wastes in solid state fermentation. *J. Biosci. Biotech.* 3: 163–168.

Raghavarao, K.S.M.S., Ranganathan, T.V., and Karanth, N.G. (2003). Some engineering aspects of solid state. Fermentation. *Biochem. Eng. J.* 13: 127–135.

Ravindran, R. and Jaiswal, A.K. (2016). Exploitation of food industry waste for high-value products. *Trends Biotechnol.* 34: 58–69.

Razak, D.L.A., Rashid, N.Y.A., Jamaluddin, A. et al. (2015). Enhancement of phenolic acid content and antioxidant activity of rice bran fermented with *Rhizopus oligosporus* and *Monascus purpureus*. *Biocatal. Agric. Biotechnol.* 4: 33–38.

Rodríguez, M.R., Pando Bedriñana, R., and Suárez Valles, B. (2015). Production and characterization of aroma compounds from apple pomace by solid-state fermentation with selected yeasts. *LWT-Food Sci. Technol.* 64: 1342–1353.

Saber, W.I.A., El-Naggar, N.E., El-Hersh, M.S., and El-Khateeb, A.Y. (2015). An innovative synergism between *Aspergillus oryzae* and *Azotobacter chroococcum* for bioconversion of cellulosae biomass into organic acids under restricted nutritional conditions using multi-response surface optimization. *Biotechnology* 14: 47–57.

Schmidt, G.C., Gonçalves, L.M., Prietto, L. et al. (2014). Antioxidant activity and enzyme inhibition of phenolic acids from fermented rice bran with fungus *Rizhopus oryzae*. *Food Chem.* 146: 371–377.

da Silva, J.C., de Oliveira, R.C., da Silva Neto, A. et al. (2015). Extraction, addition and characterization of hemicelluloses from corn cobs to development of paper properties. *Procedia Mater. Sci.* 8: 793–801.

Soccol, C.R. and Vandenberghe, L.P.S. (2003). Overview of applied solid-state fermentation in Brazil. *Biochem. Eng. J.* 13: 205–218.

Soccol, C.R., da Costa, E.S.F., Letti, L.A.J. et al. (2017). Developments and innovations in solid state fermentation. *Biotechnol. Res. Innov.* https://doi.org/10.1016/j.biori.2017.01.002.

Srianta, I. and Harijono (2015). *Monascus*-fermented sorghum: pigments and monacolin K produced by *Monascus purpureus* on whole grain, dehulled grain and bran substrates. *Int. Food Res. J.* 22: 377–382.

Srianta, I., Zubaidah, E., Estiasih, T. et al. (2016). Comparison of *Monascus purpureus* growth, pigment production and composition on different cereal substrates with solid-state fermentation. *Biocatal. Agric. Biotechnol.* 7: 181–186.

Srianta, I., Zubaidah, E., Estiasih, T. et al. (2017). Antioxidant activity of pigments derived from *Monascus purpureus*-fermented rice, corn, and sorghum. *Int. Food Res. J.* 24: 1186–1191.

Stabnikova, O., Wang, J.Y., and Ivanov, V. (2010). Value-added biotechnological products from organic wastes. In: *Environmental Biotechnology*, Handbook of Environmental Engineering, vol. 10 (ed. L. Wang, V. Ivanov and J.H. Tay), 343–394. Totowa, NJ: Humana Press.

Vandenberghe, L.P.S., de Carvalho, J.C., Libardi, N. et al. (2018). Microbial enzyme factories: current trends in production processes and commercial aspects. In: *Agro-Industrial Wastes as Feedstock for Enzyme Production* (ed. G.S. Dhillon and S. Kaur), 1–22. The Netherlands: Elsevier.

Wadhwa, M., Bakshi, M.P.S., and Makkar, H.P.S. (2013). *Utilization of Fruit and Vegetable Wastes as Livestock Feed and as Substrates for Generation of Other Value-Added Products*, 15. FAO publications. ISBN: 978-92-5- 107631-6.

Wang, L. and Yang, S.T. (2007). Solid state fermentation and its applications. In: *Bioprocessing for Value-Added Products from Renewable Resources: New Technologies and Applications* (ed. S.T. Yang), 465–489. Amsterdam, The Netherlands: Elsevier B.V.

Wang, L., Bei, Q., Wu, Y. et al. (2017). Characterization of soluble and insoluble-bound polyphenols from *Psidium guajava* L. leaves co-fermented with *Monascus anka* and *Bacillus* sp. and their bio-activities. *J. Funct. Foods* 32: 149–159.

Yadegary, M., Hamidi, A., Alavi, S.A. et al. (2013). Citric acid production from sugarcane dagasse through solid state fermentation method using *Aspergillus niger* mold and optimization of citric acid production by Taguchi method. *Jundishapur J. Microbiol.* 6: https://doi.org/10.5812/jjm.7625.

Yazid, N.A., Barrena, R., Komilis, D., and Sánchez, A. (2017). Solid-state fermentation as a novel paradigm for organic waste valorization: a review. *Sustainability* 9: 224. https://doi.org/10.3390/su9020224.

Zambrano, C., Kotogán, A., Bencsik, O. et al. (2018). Mobilization of phenolic antioxidants from grape, apple and pitahaya residues via solid state fungal fermentation and carbohydrase treatment. *LWT-Food Sci. Technol.* 89: 457–465.

# Part III

# Biotechnological Research and Production of Biomolecules

# 13

# Genetic Engineering as a Driver for Biotechnological Developments and Cloning Tools to Improve Industrial Microorganisms

*Cíntia Lacerda Ramos*[1], *Leonardo de Figueiredo Vilela*[2], *and Rosane Freitas Schwan*[2]

[1] *Department of Basic Science, Federal University of Vales do Jequitinhonha e Mucuri, CEP 39100-000, Diamantina, Brazil*
[2] *Biology Department, Federal University of Lavras, CEP 37200-000, Lavras, Brazil*

## 13.1 Introduction

Bacteria, yeasts, and fungi are commonly used in the food industry as well as in the production of biofuels and pharmaceutical products. These microorganisms are rich repositories of genetic material encoding many activities of potential interest. Efforts have been made to search for novel metabolic treasures of culturable and uncultured microorganisms with industrial and economic interest. This chapter describes the recent findings that have provided a great advance for researches in the industrial fields by producing several products, such as enzymes, bioactive compounds, biofuels, and foods by microbial activity. It has been demonstrated that culturable microorganisms are limited (less than 1% of the diversity from nature) and screening studies for industrial characteristics are laborious and time-consuming, making it difficult to find suitable industrial strains. An important advance found by researchers is the generation of genetically modified microorganisms (GMOs), which may substantially increase the range of strains with improved industrial characteristics. Nowadays, an important requirement for microbiologists is the ability to generate and screen many GMOs as well as to identify and isolate the rare cells with superior or interesting characteristics from a large and heterogeneous population in a short time and low cost. Multiplex automated genome engineering (MAGE) and CRISPR/Cas systems are recent advanced tools which significantly increased the throughput of genome editing as more genes and cells can be engineered simultaneously. Another high-throughput technique is global transcription machinery engineering (gTME), which may induce genetic variation, providing a new source of microbial genetic diversity. This chapter presents the microbial potential as a source of genetic material for industrial applications and advances in the techniques used to study and improve industrial GMOs.

## 13.2 Microorganisms and Metabolites of Industrial Interest

Microorganisms are essential for life because of several reasons including the production of value industrial products in a variety of fields such as pharmaceutical, foods,

*Bioprocessing for Biomolecules Production*, First Edition.
Edited by Gustavo Molina, Vijai Kumar Gupta, Brahma N. Singh, and Nicholas Gathergood.

and biofuels. In general, compounds of interest may be obtained from plant and animal source as well as synthetized by chemical ways. However, microorganisms might be used as important substitutes for these traditional sources. Microorganisms can carry out a wide variety of reactions and have a high ratio surface area and volume, which facilitates the rapid uptake of nutrients required to support high rates of metabolism and biosynthesis. Furthermore, microbial strains are able to adapt to a large variety of environments, allowing a culture to be transplanted from the environment to the laboratory flask and to the industrial bioreactor. Microorganisms may grow on inexpensive carbon and nitrogen sources producing valuable compounds. An important advantage in using microorganisms is their ability to increase the production up to a 1000-fold and modify structures and activities by genetic manipulation. The microbial metabolism releases products, which may be of great interest for the industry and are grouped into primary (e.g., amino acids, vitamins, organic acids, and ethanol) and secondary metabolites (e.g., antibiotic, pharmaceutical compounds, toxins, pesticides, animal, and plant growth factor).

### 13.2.1 Primary Metabolites

Primary metabolites are molecules of all living cells that are directly involved in normal growth. They are intermediate or end products of the pathways of intermediary metabolism, or are building blocks for essential macromolecules, or are converted into coenzymes. Some primary metabolites such as amino acids, vitamins, nucleotides, organic acids, and ethanol are of great interest of industrial applications and may be overproduced by GMOs. The amino acid glutamate is an important example produced by fermentation using species of the genera *Corynebacterium* and *Brevibacterium* (Demain and Adrio 2008). Glutamic acid overproduction would not naturally occur because of feedback regulation. However, modification in the cell membrane allows glutamate to be pumped out of the cell and, consequently, its biosynthesis may constantly proceed. Another important industrial amino acid produced via GMOs is lysine. Its biosynthetic pathway is controlled very tightly in an organism like *Escherichia coli,* which contains three aspartate kinases, each of which is regulated by a different end product (lysine, threonine, and methionine). In addition, after each branch point, their respective end products inhibit the initial enzymes. However, in GMOs for lysine fermentation (e.g., mutants of *Corynebacterium glutamicum*), there is only a single aspartate kinase, which is regulated via concerted feedback inhibition by threonine plus lysine. By genetic removal of homoserine dehydrogenase, a glutamate-producing wild-type *C. glutamicum* is converted into a lysine-overproducing mutant that cannot grow unless methionine and threonine are added to the medium (Demain and Adrio 2008).

Other primary metabolites obtained by the microbial source are vitamins, organic acids, and ethanol. Vitamins such as riboflavin (vitamin $B_2$) and vitamin $B_{12}$ have been produced industrially by yeast-like molds *Eremothecium ashbyii* and *Ashbya gossypii* (riboflavin), and bacteria *Propionibacterium shermanii* or *Pseudomonas denitrificans* (vitamin $B_{12}$). Organic acids have been commercially produced by filamentous fungi. *Aspergillus niger* is industrially employed to produce citric acid using media deficient in iron and manganese. Alternative processes have been developed to produce citric acid by *Candida* yeasts, especially from hydrocarbons since these yeasts are able to convert n-paraffins to citric and isocitric acids in high yields (150–170% on a weight basis). Other acids produced by microbes at high titers include lactic, acetic, pyruvic, gluconic, succinic, shikimic, and oxalic acid (Erickson et al. 2012).

The yeast *Saccharomyces cerevisiae* is the main microorganism employed for different purposes including production of alcoholic beverages, pharmaceutical products, and biofuel. However, some differences should be mentioned. For example, for alcoholic beverages production, besides ethanol, other compounds such as esters, acids, and higher alcohols are also important for specific flavor formation. Although the same yeast species (*S. cerevisiae*) are employed in different alcoholic processes, the sensorial properties are strain-specific (Duarte et al. 2010). The distinctive aromas of wine, beer, sake, spirits, and all other fermented foodstuffs are highly affected by many yeast-associated compounds, including esters, higher alcohols, ketones, phenolic compounds, sulfuric compounds, and terpenes. Strains with improved flavor production cannot be easily selected from large pools of variants, and therefore, many improvement strategies targeting this phenotype, such as genome shuffling or directed evolution, require testing of individual clones. Consequently, studies have focused on the use of genetic modification to produce strains with improved ester and higher alcohol profiles (Rossouw et al. 2008; Zhang et al. 2013). It is important to emphasize that the use of GMOs in food-fermentation processes is still controversial and heavily debated. Nonetheless, numerous research groups and companies are using genetic modification to alter industrial yeast properties.

Ethanol for pharmaceutical and biofuel purposes are generally produced via fermentation of sugars (or a polysaccharide that can be depolymerized to a fermentable sugar) mainly using *S. cerevisiae* in the case of hexoses, and *Kluyveromyces fragilis* or *Pichia* species with lactose or a pentose, respectively. The yeast *S. cerevisiae* is a eukaryotic model microorganism, and it is commonly utilized in bioethanol and brewery industries as well as bakeries worldwide. It presents a high level of ethanol tolerance (i.e., product inhibition is absent or minimal) and ability to grow under different aeration conditions, including those strictly anaerobic (which makes the process more easily controlled). In addition, this yeast has little nutrition requirements and shows high tolerance to toxic compounds as well as low pH tolerance, which also contributes to preventing bacterial contamination (Ramos et al. 2013). Nowadays, studies have focused on the search for either wild or genetically modified yeasts able to produce ethanol from sugars that are present in lignocellulosic hydrolysates (Alper and Stephanopoulos 2009). These biomasses are byproducts generated from agricultural activities that may be sources of carbon for fermenting. Biomass use for fuel ethanol production would contribute to the greenhouse effect and environmental sustainability by replacing petroleum use as well as by using residues to produce value-added products.

### 13.2.2 Secondary Metabolites

Secondary metabolites from microbial resources are of great importance for humans. This group includes antibiotics, toxins, pesticides, and both animal and plant growth factors. These metabolites are usually produced after growth has slowed down and had no function in cell growth. Antibiotics are the most known secondary metabolites used for pharmaceutical purposes. They are defined as low-molecular-weight organic natural products made by microorganisms that are active at low concentrations against other microorganisms. Actinomycetes, especially the genus *Streptomyces*, and filamentous fungi are the main producers of natural antibiotics. The search for new antibiotics is continuous due to the emergence of evolving pathogens, naturally resistant bacteria and fungi, and previously susceptible microbes that have developed resistance. To increase the antibiotic action spectrum, their derivatives may be obtained by performing minor modifications to the chemical structure of the drug without altering structures critical to drug action, such as for β-lactam peptide antibiotics and others.

Another important group of functional and healthy metabolites, produced by fungal strains, are the cholesterol-lowering agents, which include lovastatin, pravastatin, and others that act as inhibitors of 3-hydroxy-3-methylglutaryl-coenzyme A reductase, the regulatory and rate-limiting enzyme of cholesterol biosynthesis in the liver. Lovastatin was approved by the FDA in 1987. Doxorubicin, daunorubicin, mitomycin, bleomycin, and taxol are also secondary metabolites and act as natural anticancer compounds. The first four mentioned metabolites are produced by actinomycetes, while Taxol, which was originally discovered in plants, is also produced by the fungus *Taxomyces andreanae* (Stierle et al. 1993). Taxol compound is approved for breast and ovarian cancer and is the only commercial antitumor drug known to act by blocking depolymerization of microtubules. Taxol also promotes tubulin polymerization and inhibits rapidly dividing mammalian cancer cells. Another studied antitumor agent is epothilone, which is produced by myxobacteria. This compound acts similarly to Taxol and is a very important agent against Taxol-resistant tumors.

Secondary microbial metabolites may also be used as bioinsecticides, as demonstrated by the proteins produced by the spore-forming bacterium *Bacillus thuringiensis* (BT). The most known are crystal proteins (Cry), which are produced during the sporulation phase. Several proteins have been identified acting on different orders of insects. There is a diversity of protein families organized by numerical codes such as Cry1 family, which acts on Lepidoptera; Cry3, which acts on Coleopteran; and Cry4, which acts on Dipterans. Crops such as corn, cotton, and soybean have been genetically modified by insertion of genes coding for these proteins in order to better control the insect pests.

### 13.2.3 Microbial Enzymes

Enzymes are known to play a crucial role as metabolic catalysts, leading to their use in various industries and applications. Microbes are a useful source of enzymes because they can be economically cultured in small spaces (fermenters or bioreactors) and in inexpensive media. Bacterial glucose isomerase (xylose isomerase) in conjunction with fungal α-amylase and glucoamylase are examples of current microbial enzymes used to convert starch into mixtures of glucose and fructose known as "high fructose corn syrup." Furthermore, yeast enzymes have been successfully employed in the food industry such as invertase for candy and jam manufacture, β-galactosidase (lactase) for hydrolysis of lactose from milk or whey, and α-galactosidase for crystallization of beet sugar. The major application for proteases is in the dairy industry for the manufacture of cheese. Calf rennin has been preferred in cheese making due to its high specificity, but microbial proteases produced by generally regarded as safe (GRAS) microorganisms like *Mucor miehei, Bacillus subtilis,* and *Endothia parasitica* have been replacing it gradually.

The use of enzymes as detergent additives still represents the largest application of industrial enzymes. Proteases, lipases, amylases, oxidases, peroxidases, and cellulases are added to the detergents to catalyze the breakdown of chemical bonds with the addition of water. To be suitable, these enzymes must be active under thermophilic (60 °C) and alkalophilic (pH 9–11) conditions and in the presence of the various components of washing powders.

In the paper and textile industries, the use of enzymes has increased to develop cleaner processes and reduce the use of raw materials and production of waste. An alternative enzymatic process in the manufacturing of cotton was developed based on a pectate lyase (Tzanov et al. 2001). This process is performed at very low temperatures and uses less water than the classical method. Enzymes are also used in a wide range of agro-biotechnological processes, and the major application is the production of feed supplements to improve feed efficiency. An important advance in feed enzymes involves the application of phytases to improve plant phosphorus uptake by monogastric animals. Phytases produced by fungi with higher specific activities or improved thermo-stability have been identified (Kirk et al. 2002).

As observed, environmental samples provide a vast amount of microbial enzyme resources. The ability to tap into such immense biodiversity depends on the tools available to expand the search for new enzymes by metagenome screening (Ferrer et al. 2009; Uchiyama and Miyazaki 2009; Gilbert and Dupont 2011), and genome mining in a great number of sequenced microbial genomes (Ahmed 2009; Kaul and Asano 2012), thus exploring the diversity of extremophiles (Schiraldini and de Rosa 2002; Kumar et al. 2011; Yin et al. 2015).

## 13.3 The Culture-Independent Method for Biotechnological Developments

Microorganisms are the most abundant and diverse group of organisms in the world. Microorganisms have been found to colonize even the harshest natural environments with extreme temperature, pH, salinity, and others. The reason for this wide colonization is due to their ability to utilize a wide range of organic and inorganic molecules as sources of energy and nutrients. As already mentioned, microorganisms produce a great variety of enzymes and metabolic pathways leading to the production of a range of primary and secondary metabolites that have been employed in a diversity of industrial processes. In general, the microbial resources used in industrial processes have been traditionally studied and obtained from laboratory cultures. Microorganisms may be isolated from a variety of environments, cultured in the laboratory and screened for target activities. For this purpose, a great number of isolates should be evaluated in order to find strains with desired properties, which, many times, is laborious and time-consuming. Furthermore, it has been estimated that cultured microorganisms represent less than 1% of the microbial diversity thriving in nature. Many efforts are being made to cultivate different kinds of microorganisms from complex habitats and only a few types of research have obtained success in finding novel and promising strains (Ling et al. 2015).

An alternative to searching for the novel promising microbial pathway is to use culture independent techniques such as metagenomics. Metagenomics involves directly isolating and analyzing microbial DNA from environmental samples and has been successfully used to discover many new enzyme activities (Coughlan et al. 2015). Analyses can be based on the functional screening of libraries constructed from environmental DNA or the similarity of metagenomic DNA to sequences with known structures or functions. Suitable bioinformatics analyses may help to elucidate functional traits of microorganisms, which colonize environments. The main advantage of this technique is that it does not require prior knowledge of the gene encoding activities of interest; thus, it may lead to the discovery of activities encoded by genes lacking homologies to known sequences. Many novel enzymatic activities such as dioxygenases, esterases, proteases, and nitrilases, as well as lignocellulosic biomass depolymerizing activities and the production of bioactive compounds have been discovered by functional screening (Culligan et al. 2014; Armstrong et al. 2015; Coughlan et al. 2015; Montella et al. 2015). These discoveries are of great industrial importance as they can be used to produce compounds of interest in GMOs. However, although novel microbial functional pathways have been discovered and studied, issues such as improving screening methods for activities of interest and gene expression in a host must be carefully considered for a successful industrial application (Terrón-González et al. 2014).

## 13.4 Tools and Methodologies Applied to GMOs Generation

Microorganisms are rich repositories of genetic material that encode many activities of potential interest. However, despite the immense diversity of natural microorganisms already mentioned, such as safety conditions and industrial specifications, there is a current need for a combination of phenotypic characteristics that cannot be found in nature. Most industrial applications require the maximization of processes and features that may not be beneficial in natural environments. Thus, various techniques have been developed to artificially increase the genetic diversity of the microorganisms in order to generate variants that might have better characteristics than strains that are selected in environments (Wang et al. 2012; Steensels et al. 2014).

Recent studies have provided great advances in research by producing several products, such as enzymes, bioactive compounds, biofuels, and foods by using GMOs. Nowadays, an important feature is the ability to generate and screen many GMOs as well as to identify and isolate the unique cells with interesting characteristics in a short time and at low cost. Multiplex automated genome engineering (MAGE) and CRISPR/Cas systems are recent tools that significantly increased the throughput of genome editing, so more genes and cells can be engineered simultaneously. Another high-throughput technique is gTME, which may induce genetic variation, providing a new source of microbial genetic diversity (Vervoort et al. 2017).

MAGE established a new approach to genetic engineering – genomic engineering. Thus, this tool made it possible to modify whole genomes by manipulating several genes in parallel. Before that, these analyses were restricted only to altering isolated genes. This new technology also allows for almost endless applications that extend far beyond medicine or industry by offering quickness and low-cost procedures (Wang et al. 2009). MAGE makes it possible to generate independent or combined genome modifications in a living cell population. The resulting cells carry different mutations of a single site and/or multisite. The technique consists in efficient recombination with single-stranded

DNA (ssDNA) oligonucleotides, based on the recognition of ssDNA oligos in the chain during DNA replication and thus introducing mutation points or small insertions and deletions into the genome with up to 20% efficiency (Dalia et al. 2014). The distinguishing feature of MAGE from alternative methods is the ability to generate combinatorial mutations rapidly and specifically or degenerate mutations at each target site. In addition, its multiplexing ability is higher than other genome-editing technologies based on targeted nucleases. The MAGE technology was developed in the bacteria *E. coli* and has been widely and successfully used in "accelerated evolution" of this species, which has been exploited for metabolic and phenotypic engineering applications (Dalia et al. 2014; Gallagher et al. 2014).

The diversity generated by MAGE is adjustable and with high specificity. Oligos with specific sequences produce specific modification, whereas oligos with degenerate sequences produce high-diversity modifications to further explore the sequence (Wang et al. 2009). Although some mutations would normally disappear in the population via natural selection or because of the other factors, MAGE can also represent a platform that permits the introduction and maintenance of mutations in the population and accelerates the rate of their accumulation. Thus, this technique increases the likelihood of finding sets of mutations to produce a beneficial phenotype and thus provide accelerated evolution (Cropp and Schultz 2004; Gibson et al. 2008; Wang et al. 2009).

The MAGE projects involve six stages. First, to describe the desired phenotype or genotype. Second, to choose target loci. In the third stage, to design ssDNAs to modify target sites and predict the required cycle time for the fourth. More cycles increase the prevalence of allelic replacements and population diversity. MAGE cycling is conducted in the fifth stage. This cycling involves all stages of the process namely inducing competence, electroporation to transform cells with mutagenic ssDNA and to permit recombination of ssDNAs. Finally, growth and clone identification with desirable genotype or phenotype comprise the sixth and last stage (Dalia et al. 2014; Gallagher et al. 2014). MAGE protocol is shown in Figure 13.1. In general, *E. coli* cells are grown to the mid log phase at 42 °C, the temperature at which the beta protein expression is induced.

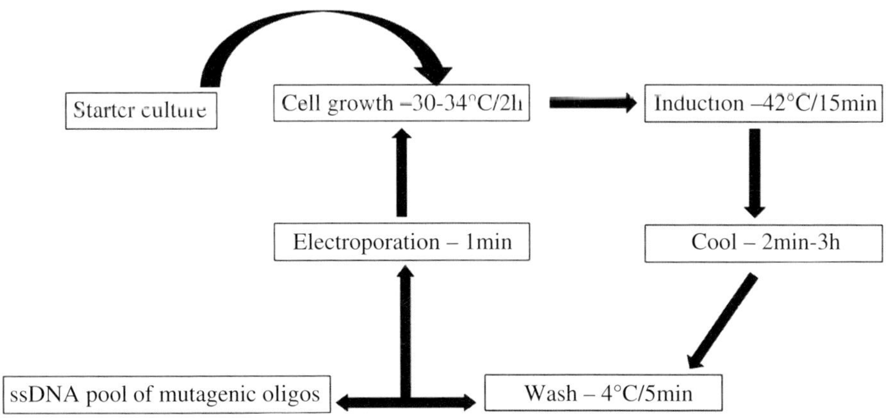


**Figure 13.1** Flowchart of MAGE protocol. A grown culture is submitted to a heat shock to induce recombination of proteins. After that, the cells are cooled and washed to remove salts. Then, the cells are mixed with ssDNA, electroporated for transformation, and submitted to a selection of the desirable phenotype characteristic.

Cells are chilled and the solution containing the oligonucleotides is added, followed by electroporation. Cells are chosen based on positive markers. Church et al. (2012) have optimized the *E. coli* strain EcNR2 to work with MAGE. EcNR2 contains a plasmid with the λ phage that when expressed, phage genes help keep the oligonucleotide annealed to the lagging strand of the DNA during replication. Using an improved technique called co-selection MAGE (CoS-MAGE), the researchers created EcHW47, the successor to EcNR2 (Wang et al. 2009, 2012).

Some important characteristics of oligos, including their length and structure, should be taken into account. The optimal mutagenic oligo length is 90 nucleotides (nt), because short ssDNAs do not bind well to the β-protein and possess reduced homology to the chromosome. On the other hand, oligos with more than 90-nt show decreased efficiency because of errors in synthetic DNA or an increased likelihood of secondary structure formation that prevents oligo binding at its genomic target. Regarding the oligo structure, it is recommended ssDNAs with higher predicted ΔG, which recombine at higher frequencies. Unstructured ssDNAs can easily hybridize with target genomic. The ssDNAs with significant homology to multiple genetic loci will be recombined less frequently, presumably because off-target sites compete with target sites to regain ssDNA. Mismatch position, mismatch identities, and base modifications of oligos should also be considered for the success of this technique (Wang et al. 2009, 2012).

The main advantages of MAGE are quickness and efficiency. The systems allow the generation of billions of mutants by repeated insertion, deletion, or mutation of DNA at multiple chromosomal targets in just a few hours. For example, each round of MAGE requires approximately 2.5 hours and may be highly scaled up, thus achieving greater functional diversity. MAGE also provides great control of the mutagenesis extent, and consequently genetic diversity giving higher possibilities to find more efficient microorganisms. Further, this method is targeted, multiplexed, and cheap. However, there are some disadvantages of the technique that confer a limited industrial application: (i) MAGE has been described only in *E. coli*, except for a recent paper that applied the technique to *C. glutamicum*; (ii) low sampling compared to the data obtained, since the majority of the diversity generated by MAGE has never been sampled, thus new high-throughput ways of analyzing are required; (iii) the need of detailed prior knowledge since target selection for MAGE mutagenesis is dependent on a priori knowledge for design MAGE oligos; (iv) low frequency of multisite mutants because, for mutagenesis targeting many sites, the replacement frequency for any single site can be low (Wang et al. 2009, 2012; Dalia et al. 2014; Gallagher et al. 2014). Despite this, the MAGE technology has the ability to modify a genome that confers an immense power and unlimited possibilities that certainly change the genomic science (Vervoort et al. 2017).

Another technique that has been used in high-throughput microbial genome engineering is the CRISPR-Car system. This system consists of the clustered and regularly interspaced short palindromic repeats (CRISPR), together with CRISPR-associated genes (*cas*) (Barrangou 2012). This technique enables genome edition at specific sites using RNA-guided nuclease activity to target DNA and allows faster, cheaper, and more efficient genome editing than the traditional techniques (Kim 2016; Wright et al. 2016). Researchers have successfully engineered microorganisms using CRISPR/Cas system, as demonstrated by the increase in compound production of *B. subtilis* and its resistance to spore formation while reducing its foam production (Zhang et al. 2016). Other researchers edited *S. cerevisiae* genome to increase mevalonate (Jakočiūnas et al. 2015) or (R-R)-2,3-butanediol production and xylose utilization (Shi et al. 2016; Vervoort et al. 2017).

The CRISPR/Cas system is an adaptive immune system that helps protect bacteria and archaea from phages or conjugative plasmids, thus protecting from potential invaders (Barrangou and Marraffini 2014). This system has been classified into three major types: type I, type II, and type III, and 12 subtypes, given their genetic content, and structural and functional differences. The type II CRISPR/Cas9 system is composed of the Cas9 endonuclease and two RNA molecules, namely crRNA and tracrRNA (Sander and Joung 2014), as shown in Figure 13.2. The crRNA and tracrRNA require RNAseIII enzyme for processing and maturation of the crRNA molecule, which guides the Cas9 endonuclease to its DNA target, in contrast to other applied nuclease classes that use target DNA through protein/DNA interactions (Gasiunas et al. 2012; Jinek et al. 2012;

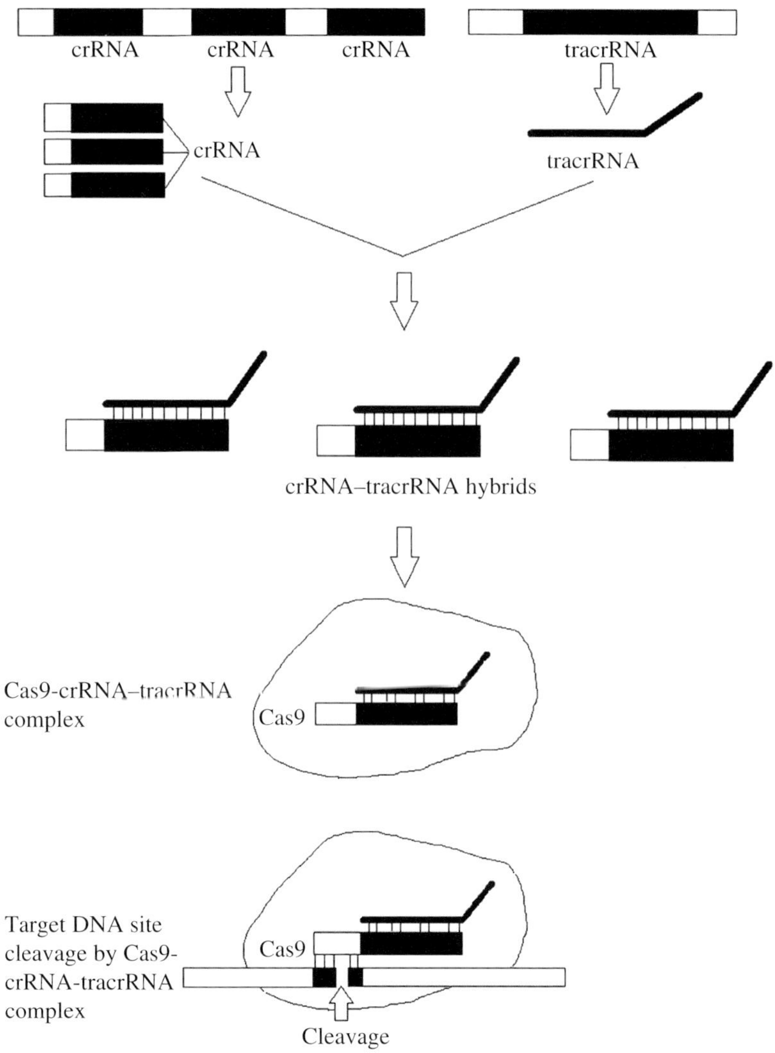


**Figure 13.2** The CRISPR-Cas9 systems incorporate DNA sequences into CRISPR arrays, which then produce crRNAs, which are complementary to the target DNA site. The crRNAs hybridize to tracrRNAs (also encoded by the CRISPR system) and this pair of RNAs associate with the Cas9 nuclease. This complex recognizes and cleaves target DNAs.

Mahfouz et al. 2014). This system can recognize and cleave the target locus with high specificity. Repair of the break can be directed to create a variety of targeted DNA sequence modifications such as DNA deletions/ insertions. However, besides being used for creating knockouts, it is also applied in the ectopic regulation of gene expression by using the catalytically inactive version of Cas9 (Piatek et al. 2014).

The CRISPR/Cas has been engineered to be a tool for genetic engineering. In contrast to other nuclease classes that use target DNA through protein/DNA interactions, CRISPR/Cas is extremely simple and requires the cloning of a 20-nt sequence, complementary to a target DNA sequence (Kamthan et al. 2016). The limitations of this system are the possible off-target cleavage (Curtin et al. 2013; Qi et al. 2013), and the nuclease size, because the Cas9 is one of the largest sequence-specific nucleases making it difficult to be delivered by the vectors (Kamthan et al. 2016).

Other genome-editing tools have been developed, including zinc finger nucleases (ZFNs) and transcription activator-like effector nucleases (TALENs). The ZFNs and TALENs have been utilized for specific deletion or inactivation of some genes (Cradick et al. 2010; Chen et al. 2013). These techniques are programmable protein-based genome-editing tools, consisting of laborious and time-consuming techniques, since the targeting specific DNA sequences should be customized for the DNA-binding domains. On the contrary, CRISPR/Cas9 system can be conveniently redirected to target any desired DNA sequence by simply redesigning the sequence of guide RNA (gRNA), which is complementary to the target DNA sequence. This system provides a powerful and flexible strategy to specifically cleave DNA sequences (Cong et al. 2013; Mali 2013; Yang and Chen 2017).

Despite the immense diversity of natural microorganisms and their importance for industrial development already discussed in this chapter, the extremely selective and specific conditions of industry sometimes lead to phenotypic traits that might not be commonly encountered in nature. Therefore, several techniques have been developed to artificially increase the microorganism diversity and generate variants that may perform better in industrial settings. Alper et al. (2006) suggested that the huge phenotypic variation could be realized by (random) transcription reprogramming followed by a selection of those variants that show improved properties. The technique gTME is based on the generation of mutants of the SPT15 gene encoding TATA-binding protein (TBP) by error-prone polymerase chain reaction (PCR), followed by a selection of mutants under the desired conditions. The TBP is a transcription factor that binds specifically to a DNA sequence called TATA BOX. This DNA sequence is found in 25–30 base pairs prior to the transcription initiation site in some promoters of eukaryotic genes. TBP-associated factors that bind to class I, II, and III promoters are called TAFIs, TAFIIs, and TAFIIIs, respectively. Among the eight factors that interact with TBP, four (TAFI, TAFIIs, TAFIIIs, and PTF/SNAPc) work in promoter selection. The other four (SAGA, Mot1, NC2, and Nots) work in conjunction with the TAFIIs regulation and expression of the protein by gene encoding (Lee and Young 1998). TBP together with a variety of factors associated with TBP makes transcription factor II D (TFIID) a general transcription factor which, in turn, forms part of the preinitiation complex of RNA polymerase II (Kornberg 2007). Since this technique involves the modification of only one gene that affects the expression of most microorganism genes, it is therefore an ideal approach to be applied to increase the expression of desired industrial characteristic as demonstrated for the industrial yeast *S. cerevisiae*. The technique was successfully applied for obtaining and selecting strains with increased ethanol tolerance, a desired characteristic for the biofuel industry (Alper et al. 2006).

## 13.5 Conclusion

Microorganisms colonize a great diversity of environments, including those with extreme conditions of temperature, pH, salinity, and others. It makes them versatile organisms and a very important source of genetic material, encoding many activities of potential interest for industries from different fields. Traditionally, microorganisms are isolated from nature and studied in the laboratory prior to employment on an industrial scale. For this purpose, a great number of isolates should be evaluated, which is frequently laborious and time-consuming. Moreover, it has been estimated that cultured microorganisms represent less than 1% of the microbial diversity from nature. An alternative to searching for a novel promising microbial pathway in a deep and fast way is to use culture independent techniques such as metagenomics, as well as to obtain novel strains by genetic engineering. MAGE, CRISPR/Cas, ZFNs and TALENs, gTME are genetic tools with different principles, however, with the same objective: to obtain GMOs with improved and desired characteristics. These techniques have been described and used by researchers obtaining microbial strains with desired industrial characteristics. It is important to consider that the strain was successfully obtained and selected on the laboratory scale. However, to dominate and persist during the harsh conditions imposed by the industrial process is another challenge for these GMOs and should thus be evaluated.

## References

Ahmed, N. (2009). A flood of microbial genomes—do we need more? *PLoS One* 4: 1–5.

Alper, H. and Stephanopoulos, G. (2009). Engineering for biofuels: exploiting innate microbial capacity or importing biosynthetic potential. *Nat. Rev. Microbiol.* 7: 715–723.

Alper, H., Moxley, J., Nevoigt, E. et al. (2006). Engineering yeast transcription machinery for improved ethanol tolerance and production. *Science* 314: 1565–1568.

Armstrong, Z., Mewis, K., Strachan, C., and Hallam, S.J. (2015). Biocatalysts for biomass deconstruction from environmental genomics. *Curr. Opin. Chem. Biol.* 29: 18–25.

Barrangou, R. (2012). RNA-mediated programmable DNA cleavage. *Nat. Biotechnol.* 30: 836–838.

Barrangou, R. and Marraffini, L.A. (2014). CRISPR-Cas systems: prokaryotes upgrade to adaptive immunity. *Mol. Cell* 54: 234–244.

Chen, J., Zhang, W., Lin, J. et al. (2013). An efficient antiviral strategy for targeting hepatitis B virus genome using transcription activator-like effector nucleases. *Mol. Ther.* 22: 303–311.

Church, G.M., Gao, Y., and Kosuri, S. (2012). Next-generation digital information storage in DNA. *Science* 337: 1628.

Cong, L., Ran, F.A., Cox, D. et al. (2013). Multiplex genome engineering using CRISPR/Cas systems. *Science* 339: 819–823.

Coughlan, L.M., Cotter, P.D., Hill, C., and Alvarez-Ordóñez, A. (2015). Biotechnological applications of functional metagenomics in the food and pharmaceutical industries. *Front. Microbiol.* 6: 672.

Cradick, T.J., Keck, K., Bradshaw, S. et al. (2010). Zinc-finger nucleases as a novel therapeutic strategy for targeting hepatitis B virus DNAs. *Mol. Ther.* 18: 947–954.

Cropp, T.A. and Schultz, P.G. (2004). An expanding genetic code. *Trends Genet.* 20: 625–630.

Culligan, E.P., Sleator, R.D., Marchesi, J.R., and Hill, C. (2014). Metagenomics and novel gene discovery: promise and potential for novel therapeutics. *Virulence* 5: 399–412.

Curtin, S.J., Anderson, J.E., Starker, C.G. et al. (2013). Targeted mutagenesis for functional analysis of gene duplication in legumes. *Methods Mol. Biol.* 1069: 25–42.

Dalia, A.B., McDonough, E., and Camillia, A. (2014). Multiplex genome editing by natural transformation. *Proc. Natl. Acad. Sci. U.S.A.* 111: 8937–8942.

Demain, A.L. and Adrio, J.L. (2008). Contributions of microorganisms to industrial biology. *Mol. Biotechnol.* 38: 41–55.

Duarte, W.F., Dragone, G., Dias, D.R. et al. (2010). Fermentative behavior of *Saccharomyces* strains during microvinification of raspberry fruit (*Rubus idaeus* L.). *Int. J. Food Microbiol.* 143: 173–182.

Erickson, B., Nelson, J.E., and Winters, P. (2012). Perspective on opportunities in industrial biotechnology in renewable chemicals. *Biotechnol. J.* 7: 176–185.

Ferrer, M., Beloqui, A., Timmis, K.M., and Golyshin, P.N. (2009). Metagenomics for mining new genetic resources of microbial communities. *J. Mol. Microbiol. Biotechnol.* 16: 109–123.

Gallagher, R.R., Li, Z., Lewis, A.O., and Isaacs, F.J. (2014). Rapid editing and evolution of bacterial genomes using libraries of synthetic DNA. *Nat. Protoc.* 9: 2301–2316.

Gasiunas, G., Barrangou, R., Horvath, P., and Siksnys, V. (2012). Cas9-crRNA ribonucleoprotein complex mediates specific DNA cleavage for adaptive immunity in bacteria. *Proc. Natl. Acad. Sci. U.S.A.* 109: 2579–2586.

Gibson, D.G., Benders, G.A., Andrews-Pfannkoch, C. et al. (2008). Complete chemical synthesis, assembly, and cloning of a *Mycoplasma genitalium* genome. *Science* 319: 1215–1220.

Gilbert, J.A. and Dupont, C.L. (2011). Microbial metagenomics: beyond the genome. *Annu. Rev. Mar. Sci.* 3: 347–371.

Jakočiūnas, T., Bonde, I., Herrgård, M. et al. (2015). Multiplex metabolic pathway engineering using CRISPR/Cas9 in *Saccharomyces cerevisiae*. *Metab. Eng.* 28: 213–222.

Jinek, M., Chylinski, K., Fonfara, I. et al. (2012). A programmable dual-RNA-guided DNA endonuclease in adaptive bacterial immunity. *Science* 337: 816–821.

Kamthan, A., Chaudhuri, A., Kamthan, M., and Datta, A. (2016). Genetically modified (GM) crops: milestones and new advances in crop improvement. *Theor. Appl. Genet.* 129: 1639–1655.

Kaul, P. and Asano, Y. (2012). Strategies for discovery and improvement of enzyme function: state of the art and opportunities. *Microbiol. Biotechnol.* 5: 18–33.

Kim, J.-S. (2016). Genome editing comes of age. *Nat. Protoc.* 11: 1573–1578.

Kirk, O., Borchert, T.V., and Fulgsang, C.C. (2002). Industrial enzyme applications. *Curr. Opin. Microbiol.* 13: 345–351.

Kornberg, R.D. (2007). The molecular basis of eukaryotic transcription. *Proc. Natl. Acad. Sci. U.S.A.* 104: 12955–11261.

Kumar, L., Awasthi, G., and Singh, B. (2011). Extremophiles: a novel source of industrially important enzymes. *Biotechnology* 10: 1–15.

Lee, T.I. and Young, R.A. (1998). Regulation of gene expression by TBP associated proteins. *Genes Dev.* 12: 1398–1408.

Ling, L.L., Schneider, T., Peoples, A.J. et al. (2015). A new antibiotic kills pathogens without detectable resistance. *Nature* 517: 455–459.

Mahfouz, M.M., Piatek, A., and Stewart, C.N. Jr. (2014). Genome engineering via TALENs and CRISPR/Cas9 systems: challenges and perspectives. *Plant Biotechnol. J.* 12: 1006–1014.

Mali, P. (2013). RNA-guided human genome engineering via Cas9. *Science* 339: 823–826.

Montella, S., Amore, A., and Faraco, V. (2015). Metagenomics for the development of new biocatalysts to advance lignocellulose saccharification for bioeconomic development. *Crit. Rev. Biotechnol.* 36: 998–1009.

Piatek, A., Ali, Z., Baazim, H. et al. (2014). RNA-guided transcriptional regulation in planta via synthetic dCas9-based transcription factors. *Plant Biotechnol. J.* 13: 578–589.

Qi, Y., Li, X., Zhang, Y. et al. (2013). Targeted deletion and inversion of tandemly arrayed genes in *Arabidopsis thaliana* using zinc finger nucleases. *G3 Genes Genom. Genet.* 3: 1707–1715.

Ramos, C.L., Duarte, W.F., Freire, A.L. et al. (2013). Evaluation of stress tolerance and fermentative behavior of indigenous *Saccharamyces cerevisiae*. *Braz. J. Microbiol.* 44: 935–944.

Rossouw, D., Naes, T., and Bauer, F.F. (2008). Linking gene regulation and the exo-metabolome: a comparative transcriptomics approach to identify genes that impact on the production of volatile aroma compounds in yeast. *BMC Genomics* 9: 530.

Sander, J.D. and Joung, J.K. (2014). CRISPR-Cas systems for genome editing, regulation and targeting. *Nat. Biotechnol.* 32: 347–355.

Schiraldini, C. and de Rosa, M. (2002). The production of biocatalysts and biomolecules from extremophiles. *Trends Biotechnol.* 20: 515–521.

Shi, S., Liang, Y., Zhang, M.M. et al. (2016). A highly efficient single-step, markerless strategy for multi-copy chromosomal integration of large biochemical pathways in *Saccharomyces cerevisiae*. *Metab. Eng.* 33: 19–27.

Steensels, J., Snoek, T., Meersman, E. et al. (2014). Improving industrial yeast strains: exploiting natural and artificial diversity. *FEMS Microbiol. Rev.* 38: 947–995.

Stierle, A., Strobel, G., and Stierle, D. (1993). Taxol and taxane production by *Taxomyces andreanae*, an endophytic fungus of Pacific yew. *Science* 260: 214–216.

Terrón-González, L., Genilloud, O., and Santero, E. (2014). Potential and limitations of metagenomic functional analyses. In: *Metagenomics, Methods, Applications and Perspectives* (ed. C. Benedetti), 1–43. New York: Nova Publishers.

Tzanov, T., Calafell, M., Guebitz, G.M., and Cavaco-Paulo, A. (2001). Bio-preparation of cotton fabrics. *Enzyme Microb. Technol.* 29: 357–362.

Uchiyama, T. and Miyazaki, K. (2009). Functional metagenomics for enzyme discovery: challenges to efficient screening. *Curr. Opin. Biotechnol.* 20: 616–622.

Vervoort, Y., Linares, A.G., Roncoroni, M. et al. (2017). High-throughput system-wide engineering and screening for microbial biotechnology. *Curr. Opin. Biotechnol.* 46: 120–125.

Wang, H.H., Isaacs, F.J., Carr, P.A. et al. (2009). Programming cells by multiplex genome engineering and accelerated evolution. *Nature* 460: 894–898.

Wang, H.H., Kim, H., Cong, L. et al. (2012). Genome-scale promoter engineering by coselection MAGE. *Nat. Methods* 9: 591–593.

Wright, A.V., Nunez, J.K., and Doudna, J.A. (2016). Biology and applications of CRISPR systems: harnessing nature's toolbox for genome engineering. *Cell* 164: 29–44.

Yang, H.-C. and Chen, P.-J. (2017). The potential and challenges of CRISPR-Cas in eradication of hepatitis B virus covalently closed circular DNA. *Virus Res.* 244: 304–310.

Yin, J., Chen, J.-C., Wu, Q., and Chen, G.-Q. (2015). Halophiles, coming stars for industrial biotechnology. *Biotechnol. Adv.* 33: 1433–1442.

Zhang, C.Y., Liu, Y.L., Qi, Y.N. et al. (2013). Increased esters and decreased higher alcohols production by engineered brewer's yeast strains. *Eur. Food Res. Technol.* 236: 1009–1014.

Zhang, K., Duan, X., and Wu, J. (2016). Multigene disruption in undomesticated *Bacillus subtilis* ATCC 6051 using the CRISPR/Cas9 system. *Sci. Rep.* 6: 27943.

# 14

# Advances in Biofuel Production by Strain Development in Yeast from Lignocellulosic Biomass

*Aravind Madhavan[2], Raveendran Sindhu[1], K.B. Arun[2], Ashok Pandey[3], Parameswaran Binod[1], and Edgard Gnansounou[4]*

[1] *Microbial Processes and Technology Division, CSIR-National Institute for Interdisciplinary Science and Technology (CSIR-NIIST), Thiruvananthapuram 695 019, India*
[2] *Rajiv Gandhi Center for Biotechnology, Thiruvananthapuram 695 014, India*
[3] *CSIR-Indian Institute of Toxicology Research (CSIR-IITR), Lucknow 226 001, India*
[4] *Ecole Polytechnique Federale de Lausanne, ENAC GR-GN, CH-1015, Lausanne, Switzerland*

## 14.1 Introduction

Microbial conversion of sugars from lignocellulose biomass to fuel is a source of ethanol and has produced using *Saccharomyces cerevisiae* worldwide. There is great interest in the use of lignocellulose as a feedstock for future production of ethanol, since this source is much more abundant than crop biomass and, most important, it does not compete with the available food supply (Caspeta et al. 2013; Tsai et al. 2015). The major bottleneck associated with biomass conversion to ethanol is that the whole production process should be economical, which is dependent on the performance of the yeast cells, since the strain used should be compatible with the high concentrations of toxic compounds, chemicals, and harmful process conditions. The manipulation of other conditioning parameters for yeast tolerance is uneconomical (Caspeta et al. 2014). Thus, yeast cells are exposed to inhibitory concentrations of toxic chemicals and low pH, resulting from ther mochemical pretreatment of lignocellulose. Furthermore, saccharification and fermentation of sugar polymers exposed *S. cerevisiae* to high temperatures, elevated osmolarity, and high concentrations of ethanol (Garay-Arroyo et al. 2004; Caspeta et al. 2014).

*Saccharomyces cerevisiae* has been widely used in the field of food and biotechnology industry and is generally regarded as safe (GRAS). Molecular genetics and cell biology of *S. cerevisiae* has been studied in detail for decades as a model eukaryotic system. Therefore, yeast is a good platform microorganism for producing biofuels and a good candidate for all types of gene manipulation, including metabolic engineering to synthetic biology. The main advantages of *S. cerevisiae* are: (i) unmatched homologous recombination ability allowing stable integration of foreign DNA fragments of long-length *in vivo*; and (ii) the availability of many genetic tools for heterologous overexpression and knockout of desired genes that have developed. Unlike *Escherichia coli, S. cerevisiae* has multiple organelles providing different environments to perform compartmentalized biosynthesis. Finally, *S. cerevisiae* exhibits high tolerance against products and toxic

*Bioprocessing for Biomolecules Production*, First Edition.
Edited by Gustavo Molina, Vijai Kumar Gupta, Brahma N. Singh, and Nicholas Gathergood.

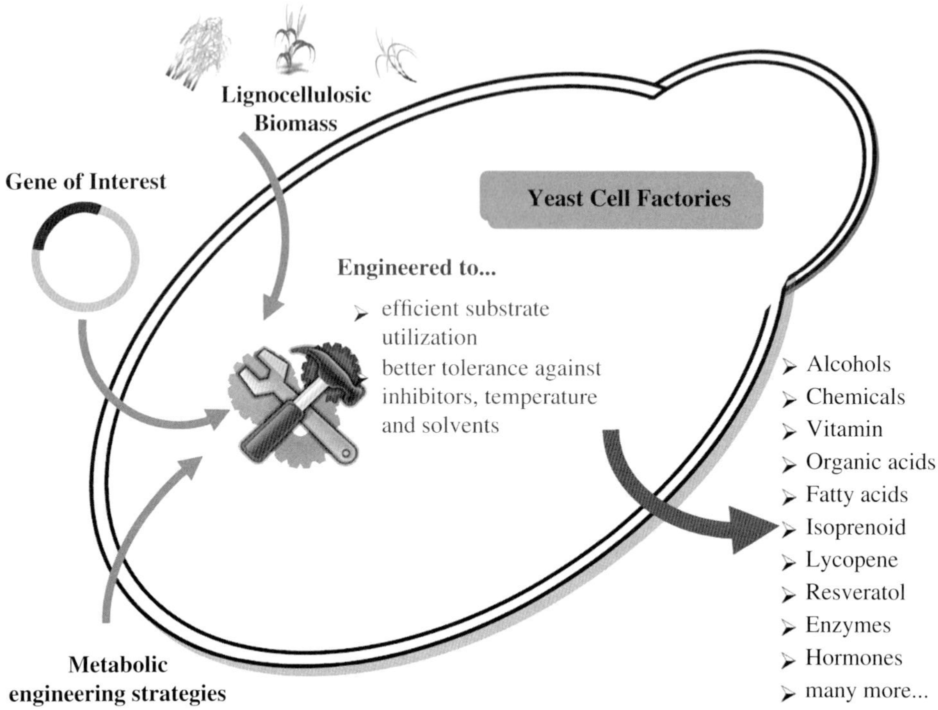


**Figure 14.1** Schematic outline of yeast cell engineering for enhanced production.

inhibitors present in most cellulosic hydrolysates (Madhavan et al. 2017; Blount et al. 2012; Bao et al. 2015).

Apart from this, *S. cerevisiae* has been engineered extensively to produce biofuels and important chemicals of great economic value (Figure 14.1). Tsai et al. (2015) reported that the manufacture of biofuels and chemicals using engineered *S. cerevisiae* face challenges mainly in four forms: (i) to engineer ethanol tolerance; (ii) to widen the substrate range; (iii) to generate efficient biofuels by developing innovative synthetic pathways; and (iv) to reduce product inhibition by toxic products. In addition to this, advancements in yield, process performance, cellular properties, and widening the product range has been quoted as challenges by Ostergaard et al. (2000a,b). Herein, we review the current status on design-based engineering of tolerant and efficient *S. cerevisiae* strains for ethanol production from lignocellulose.

## 14.2 Improvement of Ethanol Tolerance in *Saccharomyces cerevisiae*

The fermentation stage is one of the major bottlenecks that lead to decreased ethanol production and stuck fermentation (Gibson et al. 2007; Thatipamala et al. 1992; Zhao and Bai 2009). The presence of a high concentration of ethanol (>10%) in fermentation broth may denature cellular protein, change plasma membrane permeability, mitochondrial dysfunction and subsequent inhibition of yeast growth, and ethanol fermentation and sugar to alcohol conversion (Ansanay-Galeote et al. 2001; Moon et al. 2012). Several attempts were initiated to enhance the ethanol tolerance of *S. cerevisiae* by developing engineered strains. Gene expression and protein expression profiling of tolerant and nontolerant strains revealed the existence of target genes involved in ethanol tolerance. Some genes involved in this tolerance are the global TF MSN2, some genes of the cAMP-PKA signaling pathway, genes related to the cellular wall integrity, and some genes encoding enzymes of lipids and carbohydrates metabolism (Lewis et al. 2010). Researchers reported that manipulation of ion transport system can improve ethanol tolerance. For example, changing potassium ion and proton electrochemical forces improved the yeast tolerance to ethanol (Lam et al. 2014). Overexpression of the TRK1 gene, a member of the potassium transport system, and the H(+)-ATPase gene, PMA1, in laboratory strains increased ethanol production by around 30% compared to the laboratory strain S288C and by 10% compared to industrial strains (Lam et al. 2014). The introduction of TPS1 (6-phosphate-trehalose synthase) from *Saccharomycopsis fibuligera* in *S. cerevisiae* resulted in increased tolerance in 18% (v/v) ethanol (Cao et al. 2014). In another attempt, reduction of ATH1 (acid trehalase) expression increased yeast survival in 8% (v/v) ethanol (Jung and Park 2005). Another study reported ~18% more ethanol accumulation than the wild type through UV-C mutagenesis (Thammasittirong et al. 2013).

The genetic basis of the ethanol tolerance was solved using transposon mutagenesis and single-gene knockout (SGKO) (van Voorst et al. 2006; Teixeira et al. 2009) and subsequent mapping of regions responsible for ethanol tolerance using different genomic technologies. This led to the identification of potential genetic loci. MKT1 gene was initially identified to encode a positive regulator of HO expression (Tadauchi et al. 2004) and was subsequently identified as a gene responsible for ethanol tolerance by QTL mapping (Swinnen et al. 2012). Recently, evolutionary engineering has been applied to *S. cerevisiae* for improving ethanol tolerance. Evolutionary engineering is usually a phenomenon of long-term adaptation of cells under particular selective pressure in laboratory conditions, where the cell with desired trait exponentially grows over the initially dominating cells (Berg 1995). Researchers showed that the simple *in vivo* evolutionary engineering technique is a very efficient method for the development of highly ethanol tolerant *S. cerevisiae*. This technique involves a gradual increase of ethanol in the culture medium, followed by selection of the best tolerant cells. Error-prone polymerase chain reaction (PCR) was employed to engineer the subunit Rpb7 of RNAP II to improve yeast ethanol tolerance and production. The ethanol titers of the strain were ~122 g $l^{-1}$ under laboratory very high gravity (VHG) fermentation, and tolerance of 8–10% ethanol. DNA microarray assay showed that 369 genes had differential expression in M1 after 12 hours VHG fermentation, which are involved in glycolysis, alcoholic fermentation, oxidative stress response, etc. (Turanl*ı*-Y*ı*ld*ı*z et al. 2017).

## 14.3 Engineering of Substrate Utilization in *Saccharomyces cerevisiae*

In order to overcome the scarcity of fossil energy source and associated climate change due to the extensive usage of fossil energy, microbial production utilizing renewable and economical carbon sources has been strongly promoted worldwide. Lignocellulosic biomass has recently become attractive as an efficient carbon since it can be converted to various forms of monosaccharide and acetate, which are well exploited by microorganisms (Bhatia et al. 2012; Tashiro et al. 2015). Cellulose, hemicellulose, and lignin are the components of lignocellulosic biomass. Glucose molecules (60–70%) contribute to cellulose, while xylose and arabinose (30–40%) form the hemicelluloses part (Moyses et al. 2016). The maximum exploitation of the biomass substrate is possible only when the microorganism can metabolize glucose as well as pentose sugars. *S. cerevisiae*, the commonly used organism for bioethanol production, lacks the ability to metabolize L-arabinose and D-xylose (Kuhad et al. 2011). During the scale-up process of desired products from lignocellulosic biomass, microbial production faces several challenges. One such challenge is the usage of mixed sugars by the organisms. Due to the carbon catabolite repression phenomena, microbes are either forced for sequential or selective usage of these mixed sugars, which, in turn, diminishes the carbon conversion efficiency during microbial production. Various metabolic engineering techniques have been employed to overcome this issue.

*Saccharomyces cerevisiae* in native form can utilize only glucose (not xylose) as a carbon source. Xylose is an important pentose sugar with immense economic value, as it is the prevalent form present in lignocellulosic biomass (Sonderegger and Sauer 2003). Various metabolic strategies have been adapted in engineering in *S. cerevisiae* to make it possible to utilize both glucose and xylose simultaneously. One strategy is the overexpression of xylose utilization pathway, which is achieved by the overexpression of primary enzymes involved in pathway – mainly xylose reductase, xylitol dehydrogenase, and xylulokinase (Gonçalves et al. 2014). The overexpression of these specified genes allows the *S. cerevisiae* to co-utilize glucose and xylose efficiently thereby enhancing ethanol production by this strain. Moreover, the overexpression of hexose transporters – HXT1 and HXT7 results in maximum utilization of xylose and glucose, which significantly enhances ethanol production (Wu et al. 2016). D-ribulose-5-phosphate 3-epimerase deletion is yet another way to enhance co-utilization of glucose and xylose (Shen et al. 2015). The deletion reduces the carbon flow from glucose to pentose phosphate pathway.

Oh et al. (2013) had showed the effective co-utilization of cellobiose and xylose for bioethanol production by introducing cellobiose metabolic pathway and xylose reductase in *S. cerevisiae*. Wisselink et al. (2009) subjected a *S. cerevisiae* strain to repeated batch cultivation with cycles of consecutive growth in three different media with different compositions, and this novel method evolved a new strain with more efficient ethanol production. Bettiga et al. (2009) developed a *S. cerevisiae* strain by introducing, L-arabitol dehydrogenase from *Trichoderma reesei*, NADH-dependent L-xylulose reductase from *Ambrosiozymamonospora*, and xylitol dehydrogenase from *Pichia stipitis* in aldose reductase mutant strain, which efficiently ferment pentose sugars – Arabinose and xylose. *S. cerevisiae* engineered with cellodextrin transporter and β-glucosidase effectively co-fermented glucose, cellobiose, and galactose in red algae, which is another renewable and cost-effective source for microbial production (Ha et al. 2011). The metabolic engineering strategies adopted for widening the substrate range of *S. cerevisiae* are summarized in Table 14.1. The data clearly show that optimization of gene expression in heterologous pathway (including the rate-limiting enzyme) enhances the ability of *S. cerevisiae* to utilize monosaccharides other than glucose, which plays significant role in optimization of bioethanol production.

**Table 14.1** Metabolic engineering strategies adopted for enhancing substrate range.

| | Strategy | Substrate | Reference |
|---|---|---|---|
| 1. | Overexpression of xylose isomerase, XKS1, RPE1, RKI1, TAL1, and TKL1; Deletion of GRE3 and COX4 genes; Adaptive evolution, endogenous phosphoglucomutase | Glucose and xylose | Shen et al. 2012; Bro et al. 2005 |
| 2. | Protein engineering for altering cofactor preference of *Candida tenuis* XR and *Scheffersomyces stipitis* XR | Xylose | Petschacher and Nidetzky 2008; Watanabe et al. 2007 |
| 3. | Overexpressed *Clostridium phytofermentans* XI, *Piromyces* XI, *S. stipitis* XK, and pentose phosphate pathway | Xylose | Brat et al. 2009; Zhou et al. 2012 |
| 4. | Overexpressed heterologous fungal l-arabinose assimilating pathway from *Trichoderma reesei, S. stipitis, Bacillus subtilis, E. coli, Lactobacillus plantarum* | Arabinose | Richard et al. 2003; Becker and Boles 2003; Wisselink et al. 2007 |
| 5. | Deletion of D-ribulose-5-phosphate 3-epimerase and negative regulators of GAL genes | Glucose, xylose, galactose | Shen et al. 2015; Ostergaard et al. 2000a,b |
| 6. | Integration of the fermentation pathways of cellobiose and xylose and an acetic acid reduction pathway | Cellobiose, xylose, and acetic acid | Wei et al. 2015 |
| 7. | Introduction of heterologous l-arabinose transporters from *Arabidopsis thaliana* and *S. stipitis* | Arabinose | Subtil and Boles 2011 |
| 8. | Introduction of cellodextrin transporter and intracellular β-glucosidase from *Neurospora crassa* | Xylose | Haa et al. 2011 |
| 9. | Evolutionary engineering strategy based on repeated batch cultivation with repeated cycles of consecutive growth; and via continuous culture using xylose and arabinose as limiting carbon sources | Glucose, xylose, and arabinose | Wisselink et al. 2009; Sanchez et al. 2010 |
| 10. | Introduction of β-glucanase gene of *Trichoderma reesei* lactose permease and β-galactosidase of *Kluyveromyces marxianus* | β glucans<br>Lactose | Penttila et al. 1987; Dominques et al. 1999 |

## 14.4 Engineering Tolerance Against Inhibitors, Temperature, and Solvents

The generation of inhibitors from lignocellulosic biomass is one of the problems that required a solution so that biofuel production can become a more economically feasible process. The severe pretreatment steps in bioethanol production results in the production of inhibitors, which significantly hinder subsequent fermentation

(Klinke et al. 2004). *S. cerevisiae* usually prefers glucose to other sugars and hence fails to co-utilize hexoses and pentoses simultaneously (Oreb et al. 2012). The inhibition caused by glucose was overcome by the co-utilization strategies that were discussed earlier in this chapter. Toxicity induced by the products of fermentation by engineered yeast negatively regulates the fermentation process and reduces the titer value of desired products (Tsai et al. 2015). The biofuels possess a hydrophobic nature that affects the membrane integrity of the cells (Dunlop 2011).

Spruce, one of the resources used for the bioethanol production, pretreated with sulfur dioxide/sulfuric acid results in formation of various degraded products affecting further fermentation sequences (Stenberg et al. 1998). Inhibitors are usually categorized under the headings – carboxylic acids, furan aldehydes, and phenolic compounds (Klinke et al. 2004; Taherzadeh et al. 1997; Almeida et al. 2007). The source and pretreatment techniques followed usually determine the amount of these inhibitors formed during fermentation process (Westman et al. 2014). Inhibitors reduce fermentation rate mainly by (i) affecting the performance of catalytic enzymes and biologically important proteins; (ii) increased lipid fluidity (Hong et al. 2010); (iii) generating free radicals; (iv) reducing pH and ATP production (Jeffries and Jin 2000); and (v) reducing the membrane integrity of cells (Almeida et al. 2007; Dunlop 2011).

The *in situ* detoxification strategy involves maintaining inhibitors at a lower level relative to the number of metabolically active cells (Ylitervo et al. 2013). Another technique is the fed-batch process by which inhibitor toxicity is reduced, ensuring that inhibitor concentration is less inside the bioreactor (Taherzadeh et al. 1999). The tolerance of yeast to toxic inhibitors can also be enhanced by evolutionary engineering, which mainly leads to mutation of specific genes that enable the cells to withstand the toxic effect of inhibitors present in the fermentation medium (Koppram et al. 2012). Overexpression of specific gene of interest is another method to enhance the resistance to inhibitor toxicity. Ask et al. (2013) has shown that enhancing the *S. cerevisiae* cells to perform well in saccharification and fermentation of pretreated spruce after the overexpression of genes involved in glutathione synthetic pathway.

Temperature plays a significant role in metabolic functioning of yeast cells, mainly enzymes like hexokinase, glyceraldehyde-3-phosphate dehydrogenase; antioxidant enzymes like thioredoxin reductase, and porin; and heat shock proteins like HSP104, HSP82, HSP60, HSP42, HSP30 (Caspeta et al. 2015; Kültz 2003). When temperature increases from 25 to 40 °C, a reduction in protein synthesis followed by trehalose accumulation occurs inside cells. These are signals for thermotolerance, and due to these factors heat shock proteins are transcribed consecutively (Hottiger et al. 1987); mutation in CRY2 genes increases the thermal tolerance as it continuously produces trehalose (Hottiger et al. 1989). Thermotolerance changes the lipid composition of membrane, further signaling the overexpression of antioxidants and transcription factors responsible for stress response usually involves – Msn2 and Msn4 (Imazu and Sakurai 2005). Researchers like Yona et al. (2012) and Caspeta et al. (2014) were successful in selecting thermotolerant *S. cerevisiae* strains that work at 39 °C. Wallace-Salinas and Gorwa-Grauslund used a long-term adaptation strategy with a stable *S. cerevisiae* isolate, which ferments spruce hydrolysate at 39 °C with enhanced yield of ethanol (2013). These studies confirmed that an evolutionary engineering approach enables us to isolate *S. cerevisiae* strains that are thermostable and have an improved yield.

Even after the optimization of a metabolic pathway in yeast engineering further hurdles raise after the primary conversion. The solvent toxicity is one such hurdle which limits the production. Yeast has been extensively used for the production of solvent-like chemicals, including hydrocarbons. Organic solvents have been used for the extraction of desired products after fermentation. These solvents are usually not harmful to the yeast. However, this occasion demands studies on tolerance toward solvents, as yeast survival is important. Researchers have classified the solvents based on the toxicity toward yeast cell. In general carboxylic acids are the most toxic to yeast cells, followed by alcohols, esters, amines, ketones, ethers, and finally hydrocarbons (Offeman et al. 2008; Roddy 1981; Munson and King 1984; Cabral 1991). Synchronization of the fermentation and extraction process using organic solvent, followed by distillation to produce ethanol, is one of the strategies attempted to reduce solvent toxicity (Sirkar 2011). Solvent toxicity usually alters membrane integrity following membrane lipid modification badly affecting ATP generation, protein misfolding, and upregulation of chaperones and ubiquitin complex for degradation. There are not many reports on the solvent toxicity on yeast cells, as solvent bubbling through the broth causes emulsification, which drastically affects the revival of desired products (Roy et al. 2013).

General strategies based on various literature sources for enhancing the tolerance of *S. cerevisiae* toward inhibitors, temperature, and solvent are depicted in Table 14.2.

## 14.5 Future Perspectives and Conclusions

As the price of nonrenewable fuels increases and the global level of $CO_2$ increases, the production of lignocellulose based fuels have attracted the attention of researchers as an alternative source of fuel. Biofuels can be used as fuel for combustion engines and should be compatible. Lignocellulosic biomass is the major source of biofuel since it is economical and easily available. The yeast cell factory *S. cerevisiae* is a well-known producer of ethanol and fatty acids (Tsai et al. 2015).

The greatest challenge in the uptake of valuable fuels and chemicals produced by industrial biotechnology is the inability to cross the gap between the research laboratory and the commercial market. This failure is primarily because engineered strains do not meet the required standards for commercialization. The major challenge regarding conversion of lignocellulose to bioethanol is to develop robust *S. cerevisiae* strains capable of withstanding the inhibitory conditions during the fermentation process while maintaining satisfactory enzymatic functions for biomass conversion to ethanol. Major inhibitory conditions in the fermentation operations required for the biomass conversion technology include: (i) the generation of toxic compounds using pretreatment of biomass and sugar fermentation; (ii) the high temperature during the simultaneous saccharification step; (iii) the very high osmotic conditions; and (iv) high solids loadings at the beginning of the fermentation. Addressing problems associated with these unit operations is essential for decreasing the production costs and energy requirements.

In conclusion, the advent of different technologies for the generation of targeted mutations, high-throughput screening of gene expression, cells adaptation and selection strategies, evolutionary engineering of enzymes, and the new tools for editing gene expression have revolutionized the field of yeast strain engineering. These developments have accelerated the development of improved yeast strains with desirable fermentation

**Table 14.2** Strategies adopted for enhancing tolerance against inhibitors, temperature, and solvents.

| Strategy | Reference |
|---|---|
| **Inhibitors** | |
| Muscle bovine lactate dehydrogenase gene (*LDH-A*) is expressed in *S. cerevisiae* for lactic acid tolerance. | Dominques et al. 1999 |
| Mutated TATA-binding protein (SPT15) leads to high ethanol concentration. | Alper et al. 2006 |
| Overexpression of proteasome genes increases butanol tolerance. | Gonzalez-Ramos et al. 2013 |
| Overexpression of INO1, DOG1, HAL1 increase tolerance to iso-butanol and ethanol. | Hong et al. 2010 |
| Increase in oleic acid content enhances alcohol tolerance. | Yazawa et al. 2011 |
| Alkane tolerance enhanced by overexpression of<br>• Alkane-induced genes SNQ2 and PDR5<br>• heterologous pumps ABC2 and ABC3 from *Yarrowia lipolytica* | Ling et al. 2013; Chen et al. 2013 |
| Fed-batch strategy assists in xylose metabolism by limiting the glucose concentration. | Olofsson et al. 2008 |
| Mutagenesis and evolutionary engineering enhances xylose utilization and tolerance to ethanol. | Smith et al. 2014 |
| Increasing proline and myo-inositol enhances tolerance under furfural, acetic acid and phenol. | Wang et al. 2015 |
| Transcriptional activator Msn2 overexpression shows tolerance to oxidative stress, furfural resistance, and enhanced ethanol fermentation. | Sasano et al. 2012 |
| **Temperature** | |
| Thermostable β-galactosidase encoded by *lacA* from *Aspergillus niger* was expressed from an *ADH1* promoter in *S. cerevisiae*. | Kumar et al. 1992. |
| *CYR1-2* mutation-produced trehalose are more temperature tolerant. | Hottiger et al. 1989 |
| Long-term adaptation strategy: a stable *S. cerevisiae* isolate ferments spruce hydrolysate at 39 °C. | Wallace-Salinas and Gorwa-Grauslund 2013 |
| **Solvents** | |
| Ethyl oleate and ethyl benzene are potential solvents for in situ extraction of ethanol; Decanol and ethyl caprylate offers toxicity toward yeast. | Roy et al. 2013 |

characteristics – especially higher performance in the inhibitory conditions present in the lignocellulose ethanol fermentation processes.

## Acknowledgments

Aravind Madhavan acknowledges DBT for Research Associateship. Raveendran Sindhu acknowledges DST for sanctioning a project under DST-WOS-B scheme

(SR/WOS-B/740/2016). Parameswaran Binod and Raveendran Sindhu acknowledge EPFL, Lausanne for visiting fellowship. K B Arun acknowledges DST-SERB for National Post-Doctoral Fellowship.

## References

Almeida, J.R.M., Modig, T., Petersson, A. et al. (2007). Increased tolerance and conversion of inhibitors in lignocellulosic hydrolysates by *Saccharomyces cerevisiae*. *J. Chem. Technol. Biotechnol.* 82: 340–349.

Alper, H., Moxley, J., Nevoigt, E. et al. (2006). Engineering yeast transcription machinery for improved ethanol tolerance and production. *Science* 314: 1565–1568.

Ansanay-Galeote, V., Blondin, B., Dequin, S., and Sablayrolles, J.M. (2001). Stress effect of ethanol on fermentation kinetics by stationary-phase cells of *Saccharomyces cerevisiae*. *Biotechnol. Lett.* 23: 677–681.

Ask, M., Mapelli, V., Hock, H. et al. (2013). Engineering glutathione biosynthesis of *Saccharomyces cerevisiae* increases robustness to inhibitors in pretreated lignocellulosic materials. *Microb. Cell Fact.* 12: 87.

Bao, Z., Xiao, H., Liang, J. et al. (2015). Homology-integrated CRISPR-Cas (HI-CRISPR) system for one-step multigene disruption in *Saccharomyces cerevisiae*. *ACS Synth. Biol.* 4: 585–594.

Becker, J. and Boles, E. (2003). A modified *Saccharomyces cerevisiae* strain that consumes L-Arabinose and produces ethanol. *Appl. Environ. Microbiol.* 69: 4144–4150.

Berg, O.G. (1995). Periodic selection and hitchhiking in a bacterial population. *J. Theor. Biol.* 173: 307–320.

Bettiga, M., Bengtsson, O., Hahn-Hagerdal, B., and Gorwa-Graushund, M.E. (2009). Arabinose and xylose fermentation by recombinant *Saccharomyces cerevisiae* expressing a fungal pentose utilization pathway. *Microb. Cell Fact.* 8: 40.

Bhatia, L., Johri, S., and Ahmad, R. (2012). An economic and ecological perspective of ethanol production from renewable agro waste. *AMB Express.* 2 (1): 65.

Blount, B.A., Weenink, T., and Ellis, T. (2012). Construction of synthetic regulatory networks in yeast. *FEBS Lett.* 586: 2112–2121.

Brat, D., Boles, E., and Wiedemann, B. (2009). Functional expression of a bacterial xylose isomerase in *Saccharomyces cerevisiae*. *Appl. Environ. Microbiol.* 75: 2304–2311.

Bro, C., Knudsen, S., Regenberg, B. et al. (2005). Improvement of galactose uptake in *Saccharomyces cerevisiae* through overexpression of phosphoglucomutase: example of transcript analysis as a tool in inverse metabolic engineering. *Appl. Environ. Microbiol.* 71: 6465–6472.

Cabral, J.M.S. (1991). *Extractive Bioconversions* (ed. B. Mattiasson and O. Holst), 207–235. New York: Marcel Dekker.

Cao, T.S., Chi, Z., Liu, G.L., and Chi, Z.M. (2014). Expression of TPS1 gene from *Saccharomycopsis fibuligera* A11 in *Saccharomyces* sp. W0 enhances trehalose accumulation, ethanol tolerance, and ethanol production. *Mol. Biotechnol.* 56: 72–78.

Caspeta, L., Buijs, N.A.A., and Nielsen, J. (2013). The role of biofuels in the future energy supply. *Energy Environ. Sci.* 6: 1077–1082.

Caspeta, L., Chen, Y., Ghiaci, P. et al. (2014). Altered sterol composition renders yeast thermotolerant. *Science* 346: 75–78.

Caspeta, L., Castillo, T., and Nielsen, J. (2015). Modifying yeast tolerance to inhibitory conditions of ethanol production processes. *Front. Bioeng. Biotechnol.* 11: 184.

Chen, B., Ling, H., and Chang, M.W. (2013). Transporter engineering for improved tolerance against alkane biofuels in *Saccharomyces cerevisiae. Biotechnol. Biofuels* 6: 21.

Dominques, L., Dantas, M.M., Lima, N., and Teixeira, J.A. (1999). Continuous ethanol fermentation of lactose by a recombinant flocculating *Saccharomyces cerevisiae* strain. *Biotechnol. Bioeng.* 64: 692–697.

Dunlop, M.J. (2011). Engineering microbes for tolerance to nextgeneration biofuels. *Biotechnol. Biofuels* 4: 32.

Garay-Arroyo, A., Covarrubias, A.A., Clark, I. et al. (2004). Response to different environmental stress conditions of industrial and laboratory Saccharomyces cerevisiae strains. *Appl. Microbiol. Biotechnol.* 63 (6): 734–741.

Gibson, B.R., Lawrence, S.J., Leclaire, J.P. et al. (2007). Yeast responses to stresses associated with industrial brewery handling. *FEMS Microbiol. Rev.* 31: 535–569.

Gonçalves, D.L., Matsushika, A., de Sales, B.B. et al. (2014). Xylose and xylose/glucose co-fermentation by recombinant *Saccharomyces cerevisiae* strains expressing individual hexose transporters. *Enzyme Microb. Technol.* 63: 13–20.

Gonzalez-Ramos, D., van den Broek, M., van Maris, A.J.A. et al. (2013). Genome-scale analyses of butanol tolerance in *Saccharomyces cerevisiae* reveal an essential role of protein degradation. *Biotechnol. Biofuels* 6: 48.

Ha, S.J., Wei, Q.S., Kim, S.R. et al. (2011). Cofermentation of cellobiose and galactose by an engineered *Saccharomyces cerevisiae* strain. *Appl. Environ. Microbiol.* 77: 5822–5825.

Haa, S.J., Galazkac, J.M., Kima, S.R. et al. (2011). Engineered *Saccharomyces cerevisiae* capable of simultaneous cellobiose and xylose fermentation. *PNAS* 108 (2): 504–509.

Hong, M.E., Lee, K.S., Yu, B.J. et al. (2010). Identification of gene targets eliciting improved alcohol tolerance in *Saccharomyces cerevisiae* through inverse metabolic engineering. *J. Biotechnol.* 149: 52–59.

Hottiger, T., Schmutz, P., and Wiemken, A. (1987). Heat-induced accumulation and futile cycling of trehalose in *Saccharomyces cerevisiae. J. Bacteriol.* 169: 5518–5522.

Hottiger, T., Boller, T., and Wiemken, A. (1989). Correlation of trehalose content and heat resistance in yeast mutants altered in the RAS/adenylate cyclase pathway: is trehalose a thermoprotectant? *FEBS Lett.* 255: 431–434.

Imazu, H. and Sakurai, H. (2005). *Saccharomyces cerevisiae* heat shock transcription factor regulates cell wall remodeling in response to heat shock. *Eukaryot. Cell* 4: 1050–1056.

Jeffries, T.W. and Jin, Y.S. (2000). Ethanol and thermotolerance in the bioconversion of xylose by yeasts. *Adv. Appl. Microbiol.* 47: 221–268.

Jung, Y.J. and Park, H.D. (2005). Antisense-mediated inhibition of acid trehalase (ATH1) gene expression promotes ethanol fermentation and tolerance in *Saccharomyces cerevisiae. Biotechnol. Lett.* 27: 1855–1859.

Klinke, H.B., Thomsen, A.B., and Ahring, B.K. (2004). Inhibition of ethanol-producing yeast and bacteria by degradation products produced during pre-treatment of biomass. *Appl. Microbiol. Biotechnol.* 66 (1): 10–26.

Koppram, R., Albers, E., and Olsson, L. (2012). Evolutionary engineering strategies to enhance tolerance of xylose utilizing recombinant yeast to inhibitors derived from spruce biomass. *Biotechnol. Biofuels* 5: 32.

Kuhad, R.C., Gupta, R., Khasa, Y.P. et al. (2011). Bioethanol production from pentose sugars: current status and future prospects. *Renewable Sustainable Energy Rev.* 15: 4950–4962.

Kumar, V., Ramakrishnan, S., Teeri, T.T. et al. (1992). *Saccharomyces cerevisiae* cells secreting an *Aspergillus niger* β-galactosidase grow on whey permeate. *Biotechnology* 10: 82–85.

Kültz, D. (2003). Evolution of the cellular stress proteome: from monophyletic origin to ubiquitous function. *J. Exp. Biol.* 206: 3119–3124.

Lam, F.H., Ghaderi, A., Fink, G.R., and Stephanopoulos, G. (2014). Engineering alcohol tolerance in yeast. *Science* 346: 71–75.

Lewis, J.A., Elkon, I.M., Mcgee, M.A. et al. (2010). Exploiting natural variation in *Saccharomyces cerevisiae* to identify genes for increased ethanol resistance. *Genetics* 186: 1197–1205.

Ling, H., Chen, B., Kang, A. et al. (2013). Transcriptome response to alkane biofuels in *Saccharomyces cerevisiae*: identification of efflux pumps involved in alkane tolerance. *Biotechnol. Biofuels* 6: 95.

Madhavan, A., Jose, A.A., Binod, P. et al. (2017). Synthetic biology and metabolic engineering approaches and its impact on non-conventional yeast and biofuel production. *Front. Energy Res.* 5: 8.

Moon, M.H., Ryu, J., Choeng, Y.H. et al. (2012). Enhancement of stress tolerance and ethanol production in *Saccharomyces cerevisiae* by heterologous expression of a trehalose biosynthetic gene from *Streptomyces albus*. *Biotechnol. Bioprocess Eng.* 17: 986–996.

Moyses, D.N., Reis, V.C., de Almeida, J.R. et al. (2016). Xylose fermentation by *Saccharomyces cerevisiae*: challenges and prospects. *Int. J. Mol. Sci.* 17 (3): 207.

Munson, C.L. and King, C.J. (1984). Factors influencing solvent selection for extraction of ethanol from aqueous solutions. *Ind. Eng. Chem. Process Des. Dev.* 23: 109–115.

Offeman, R.D., Stephenson, S.K., Franqui, D. et al. (2008). Extraction of ethanol with higher alcohol solvents and their toxicity to yeasts. *Sep. Purif. Technol.* 63: 444–451.

Oh, E.J., Ha, S.J., Kim, S.R. et al. (2013). Enhanced xylitol production through simultaneous co-utilization of cellobiose and xylose by engineered *Saccharomyces cerevisiae*. *Metab. Eng.* 15: 226–234.

Olofsson, K., Rudolf, A., and Lidén, G. (2008). Designing simultaneous saccharification and fermentation for improved xylose conversion by a recombinant strain of *Saccharomyces cerevisiae*. *J. Biotechnol.* 134: 112–120.

Oreb, M., Dietz, H., Farwick, A., and Boles, E. (2012). Novel strategies to improve co-fermentation of pentoses with D-glucose by recombinant yeast strains in lignocellulosic hydrolysates. *Bioengineered* 3 (6): 347–351.

Ostergaard, S., Olsson, L., and Nielsen, J. (2000a). Metabolic engineering of *Saccharomyces cerevisiae*. *Microbiol. Mol. Biol. Rev.* 64 (1): 4–50.

Ostergaard, S., Olsson, L., Johnston, M., and Nielsen, J. (2000b). Increasing galactose consumption by *Saccharomyces cerevisiae* through metabolic engineering of the GAL gene regulatory network. *Nat. Biotechnol.* 18: 1283–1286.

Penttila, M.E., Suihko, U., Lehtinen, M. et al. (1987). Construction of brewer's yeasts secreting fungal endo-β-glucanase. *Curr. Genet.* 12: 413–420.

Petschacher, B. and Nidetzky, B. (2008). Altering the coenzyme preference of xylose reductase to favor utilization of NADH enhances ethanol yield from xylose in a metabolically engineered strain of *Saccharomyces cerevisiae*. *Microb. Cell Fact.* 7: 9.

Richard, P., Verho, R., Putkonen, M. et al. (2003). Production of ethanol from L-arabinose by *Saccharomyces cerevisiae* containing a fungal L-arabinose pathway. *FEMS Yeast Res.* 3: 185–189.

Roddy, J.W. (1981). Distribution of ethanol–water mixtures to organic liquids. *Ind. Eng. Chem. Process Des. Dev.* 20: 104–108.

Roy, S.R., Bhattacharya, P., and Sirkar, A. (2013). Studies on toxicity effect of solvents on growth of *Saccharomyces cerevisiae* (NCIM 3186). *Indian Chem. Eng.* 55 (4): 247–257.

Sanchez, R.G., Karhumaa, K., Fonseca, C. et al. (2010). Research improved xylose and arabinose utilization by an industrial recombinant *Saccharomyces cerevisiae* strain using evolutionary engineering. *Biotechnol. Biofuels* 3: 13.

Sasano, Y., Watanabe, D., Ukibe, K. et al. (2012). Overexpression of the yeast transcription activator Msn2 confers furfural resistance and increases the initial fermentation rate in ethanol production. *J. Biosci. Bioeng.* 113 (4): 451–455.

Shen, Y., Chen, X., Peng, B.Y. et al. (2012). An efficient xylose-fermenting recombinant *Saccharomyces cerevisiae* strain obtained through adaptive evolution and its global transcription profile. *Appl. Microbiol. Biotechnol.* 96: 1079–1091.

Shen, M.H., Song, H., Li, B.Z., and Yuan, Y.J. (2015). Deletion of D-ribulose-5-phosphate 3-epimerase (RPE1) induces simultaneous utilization of xylose and glucose in xylose-utilizing *Saccharomyces cerevisiae*. *Biotechnol. Lett.* 37: 1031–1036.

Sirkar, A. (2011). *Upgraded Chemical Process Technology and Efficient Equipment*. New Delhi: New Age, Section I, Chapter 1.

Smith, J., van Rensburg, E., and Görgens, J.F. (2014). Simultaneously improving xylose fermentation and tolerance to lignocellulosic inhibitors through evolutionary engineering of recombinant *Saccharomyces cerevisiae* harbouring xylose isomerase. *BMC Biotechnol.* 14: 41.

Sonderegger, M. and Sauer, U. (2003). Evolutionary engineering of *Saccharomyces cerevisiae* for anaerobic growth on xylose. *Appl. Environ. Microbiol.* 69: 1990–1998.

Stenberg, K., Tengborg, C., Galbe, M., and Zacchi, G. (1998). Optimisation of steam pretreatment of $SO_2$-impregnated mixed softwoods for ethanol production. *J. Chem. Technol. Biotechnol.* 71: 299–308.

Subtil, T. and Boles, E. (2011). Improving L-arabinose utilization of pentose fermenting *Saccharomyces cerevisiae* cells by heterologous expression of L-arabinose transporting sugar transporters. *Biotechnol. Biofuels* 4: 38.

Swinnen, S., Schaerlaekens, K., Pais, T. et al. (2012). Identification of novel causative genes determining the complex trait of high ethanol tolerance in yeast using pooled-segregant whole genome sequence analysis. *Genome Res.* 22: 975–984.

Tadauchi, T., Inada, T., Matsumoto, K., and Irie, K. (2004). Posttranscriptional regulation of HO expression by the Mkt1-Pbp1 complex. *Mol. Cell. Biol.* 24: 3670–3681.

Taherzadeh, M.J., Eklund, R., Gustafsson, L. et al. (1997). Characterization and fermentation of dilute-acid hydrolyzates from wood. *Ind. Eng. Chem. Res.* 36: 4659–4665.

Taherzadeh, M.J., Niklasson, C., and Liden, G. (1999). Conversion of dilute-acid hydrolyzates of spruce and birch to ethanol by fed-batch fermentation. *Bioresour. Technol.* 69: 59–66.

Tashiro, Y., Desai, S.H., and Atsumi, S. (2015). Two-dimensional isobutyl acetate production pathways to improve carbon yield. *Nat. Commun.* 6: 7488.

Teixeira, M.C., Raposo, L.R., Mira, N.P. et al. (2009). Genome-wide identification of *Saccharomyces cerevisiae* genes required for maximal tolerance to ethanol. *Appl. Environ. Microbiol.* 75: 5761–5772.

Thammasittirong, S.N.R., Thirasaktana, T., Thammasittirong, A., and Srisodsuk, M. (2013). Improvement of ethanol production by ethanol-tolerant *Saccharomyces cerevisiae* UVNR56. *SpringerPlus* 2: 583.

Thatipamala, R., Rohani, S., and Hill, G.A. (1992). Effects of high product and substrate inhibitions on the kinetics and biomass and product yields during ethanol batch fermentation. *Biotechnol. Bioeng.* 40: 289–297.

Tsai, C.S., Kwak, S., Turner, T.L., and Jin, Y.S. (2015). Yeast synthetic biology toolbox and applications for biofuel production. *FEMS Yeast Res.* 15 (1): 1–15.

Turanlı-Yıldız, B., Benbadis, L., Alkım, C. et al. (2017). In vivo evolutionary engineering for ethanol-tolerance of *Saccharomyces cerevisiae* haploid cells triggers diploidization. *J. Biosci. Bioeng.* 124: 309–318.

van Voorst, F., Houghton-Larsen, J., Jønson, L. et al. (2006). Genome-wide identification of genes required for growth of *Saccharomyces cerevisiae* under ethanol stress. *Yeast* 23: 351–359.

Wallace-Salinas, V. and Gorwa-Grauslund, F.M. (2013). Adaptive evolution of an industrial strain of Saccharomyces cerevisiae for combined tolerance to inhibitors and temperature. *Biotechnol. Biofuels* 6: 151.

Wang, X., Bai, X., Chen, D.F. et al. (2015). Increasing proline and myo-inositol improves tolerance of *Saccharomyces cerevisiae* to the mixture of multiple lignocellulose-derived inhibitors. *Biotechnol. Biofuels* 8: 142.

Watanabe, S., Abu Saleh, A., Pack, S.P. et al. (2007). Ethanol production from xylose by recombinant *Saccharomyces cerevisiae* expressing protein-engineered NADH preferring xylose reductase from *Pichia stipitis*. *Microbiology* 153: 3044–3054.

Wei, N., Oh, E.J., Million, G. et al. (2015). Simultaneous utilization of cellobiose, xylose, and acetic acid from lignocellulosic biomass for biofuel production by an engineered yeast platform. *ACS Synth. Biol.* 4: 707–713.

Westman, J.O., Mapelli, V., Taherzadeh, M.J., and Franzén, C.J. (2014). Flocculation causes inhibitor tolerance in *Saccharomyces cerevisiae* for second generation bioethanol production. *Appl. Environ. Microbiol.* 80: 6908–6918.

Wisselink, H.W., Toirkens, M.J., del Rosario Franco Berriel, M. et al. (2007). Engineering of *Saccharomyces cerevisiae* for efficient anaerobic alcoholic fermentation of L-arabinose. *Appl. Environ. Microbiol.* 73: 4881–4891.

Wisselink, H.W., Toirkens, M.J., Wu, Q. et al. (2009). Novel evolutionary engineering approach for accelerated utilization of glucose, xylose, and arabinose mixtures by engineered *Saccharomyces cerevisiae* strains. *Appl. Environ. Microbiol.* 75: 907–914.

Wu, Y., Shen, X., Yuan, Q., and Yan, Y. (2016). Metabolic engineering strategies for co-utilization of carbon sources in microbes. *Bioengineering* 3: 10.

Yazawa, H., Kamisaka, Y., Kimura, K. et al. (2011). Efficient accumulation of oleic acid in *Saccharomyces cerevisiae* caused by expression of rat elongase 2 gene (rELO2) and its contribution to tolerance to alcohols. *Appl. Microbiol. Biotechnol.* 91: 1593–1600.

Ylitervo, P., Franzén, C., and Taherzadeh, M. (2013). Impact of furfural on rapid ethanol production using a membrane bioreactor. *Energies* 6: 1604–1617.

Yona, A.H., Manor, Y.S., Herbst, R.H. et al. (2012). Chromosomal duplication is a transient evolutionary solution to stress. *PNAS* 109: 21010–21015.

Zhao, X.Q. and Bai, F.W. (2009). Mechanisms of yeast stress tolerance and its manipulation for efficient fuel ethanol production. *J. Biotechnol.* 144: 23–30.

Zhou, H., Cheng, J.S., Wang, B.L. et al. (2012). Xylose isomerase overexpression along with engineering of the pentose phosphate pathway and evolutionary engineering enable rapid xylose utilization and ethanol production by *Saccharomyces cerevisiae*. *Metab. Eng.* 14: 611–622.

# 15

# Fermentative Production of Beta-Glucan: Properties and Potential Applications

*Rafael Rodrigues Philippini, Sabrina Evelin Martiniano, Júlio César dos Santos, Silvio Silvério da Silva, and Anuj Kumar Chandel*

*Department of Biotechnology, Engineering School of Lorena (EEL), University of São Paulo, Lorena 12602-810, Brazil*

## 15.1 Introduction

Microbial exopolysaccharides (EPS) are favorable substitutes for the natural plant polysaccharides due to their unique biochemical properties (Zhu et al. 2016; Singh et al. 2008). EPS include a wide variety of polysaccharides such as lasiodiplodan, pullunan, and xanthan from *Lasiodiplodia theobromae, Aureobasidium pullulans,* and *Xanthomonas campestris* (Madhuri and Prabhakar 2014). Naturally, beta-glucan are a class of carbohydrate polymer abundant in cereals, oat bran, legumes, and fruits. Besides this, microorganisms also produce EPS, which potentially can fulfill a large-scale requirement.

The properties of beta-glucan depend upon the source from which it has been obtained. Beta-glucans consist of glucose units linked with beta (1→3) glycoside linkage and differ based on molecular mass, viscosity, and solubility, thus having different physiological effects in animals. EPS are pro-bioactive molecules that provide functional effects to foods and thus offer important health benefits (Kodali et al. 2009), including antitumor, immuno-stimulatory, antioxidant, antiulcer, and reducing the risk of cardiovascular disease and cholesterol levels. Beside these health benefits, EPS provide texture, good mouthfeel, and stabilize dairy products (Welman and Maddox 2003). Microbial derived polysaccharides are also useful in infectious disease control and modulate the immune systems, stimulating macrophages (Toklu et al. 2006). In cosmetic and nutraceutical products, beta-glucan provides texture-supplementing soluble fiber requirement (Kodali et al. 2009).

Microorganisms are considered an asset in beta-glucan commercial production. They can grow on natural agro-residues and produce beta-glucan in a high amount in order to fulfill the surplus demand (Philippini et al. 2018). Rather than summarizing all of the existing literature on beta-glucan production, this chapter presents the important parameters for beta-glucan production. This chapter summarizes the properties of beta-glucan, the various microorganisms producing beta-glucan, fermentation parameters, recovery, and potential applications.

*Bioprocessing for Biomolecules Production*, First Edition.
Edited by Gustavo Molina, Vijai Kumar Gupta, Brahma N. Singh, and Nicholas Gathergood.

## 15.2 Beta-Glucan Structure and Properties

Glucans are composed of monomeric sugars linked with α- or β-glycosidic bonds in a polymeric form. Beta-glucans are present in plants such as oats and can be produced by many organisms such as bacteria, fungi, and lichens (fungi and algae/cyanobacteria symbiosis) (Shih et al. 2009; Synytsya and Novak 2013; De Souza et al. 2015). These polymers may be located internally (as carbohydrate storage), within the cell wall (promoting structural properties), or as EPS (in the form of capsule, slime, or biofilm), depending on the microorganism.

The cell wall beta-glucans contain β (1 → 3) and β (1 → 6) glucan structural features. The chemical structure, physico-chemical properties and biosynthesis of beta-glucan has been extensively studied since 1950 (Grün et al. 2005). It is determined to be a six-fold helix structure, stabilized by hydrogen bonds and has a degree of polymerization (DP) of approximately 1500 (Jelsma and Kreger 1975). The β (1 → 4) linkage is similar to cellulose, which is usually not presented in fungal structure, although the linear combination of β (1 → 3) is observed in lichenized fungi (Carbonero et al. 2005; Synytsya and Novak 2013). They are associated to proteins, lipids, and other carbohydrates (e.g. mannan), forming a robust barrier, which protects cells and prevents mannoproteins to be secreted into the extracellular medium (Corradi da Silva et al. 2005; Vasconcelos et al. 2008; Cabib and Arroyo 2013).

Microbial beta-glucans can be divided into linear glucans and branched glucans. Linear beta-glucans may be composed of single β (1 → 3) or β (1 → 6) glucose units or mixed linkages. Same linkage of beta-glucans can be found in laminaran β (1 → 3), pachyman β (1 → 3), and pustulan β (1 → 6) (Hoffmann et al. 1971; Alquini et al. 2004; Huang and Zhang 2011). Branched beta-glucans containing β (1 → 3) and β (1 → 6) linkages are present in the cell wall and have been widely studied.

The proposal developed by Cabib and Arroyo (2013) for cellular wall synthesis of *Saccharomyces cerevisiae* is shown in Figure 15.1. Initially, it is synthesized by the backbone of β (1 → 3)-glucan from the UDP-glucose by the enzyme β-1,3-glucan synthase (GS), added of β (1 → 6)-glucan. Next mannoprotein (MP) together with the glycosylphosphatidylinositol remnant is transferred to the β (1 → 6)-glucan. Finally, the chitin is synthetized from UDP-N-acetylglucosamine (UDP-GlcNAc) by chitin synthase III (CSIII) and also transferred to the β (1 → 6)-glucan. The role of mannan is not completely understood in cell walls but it is considered that the later presents a structural role in cell wall, aiding in its maintenance and hardness, while chitin is presented as an essential material for the development of the periplasmic space, where several enzymes acts (Cabib and Arroyo 2013).

Other branched beta-glucans are presented as β (1 → 3) or β (1 → 6) backbone. Degree of polymerization (DP) and branching distribution may vary as well, as complexity of other cell wall polysaccharides, primarily with chitin. It is available in concentration from 5% to 20%, connecting the β (1 → 3) and chitin to the proteins of the cell wall. The DP may also occur in O-4 and O-2 position according to the microorganism species, isolation method, and polysaccharide purification (Synytsya and Novak 2013). Botryosphaeran, lentilan, and scleroglucan are some examples of branched beta-glucans produced by filamentous fungi (Barbosa et al. 2004; Coviello et al. 2005; Bhunia et al. 2011). The major structure of major beta glucan types is shown in Figure 15.2.

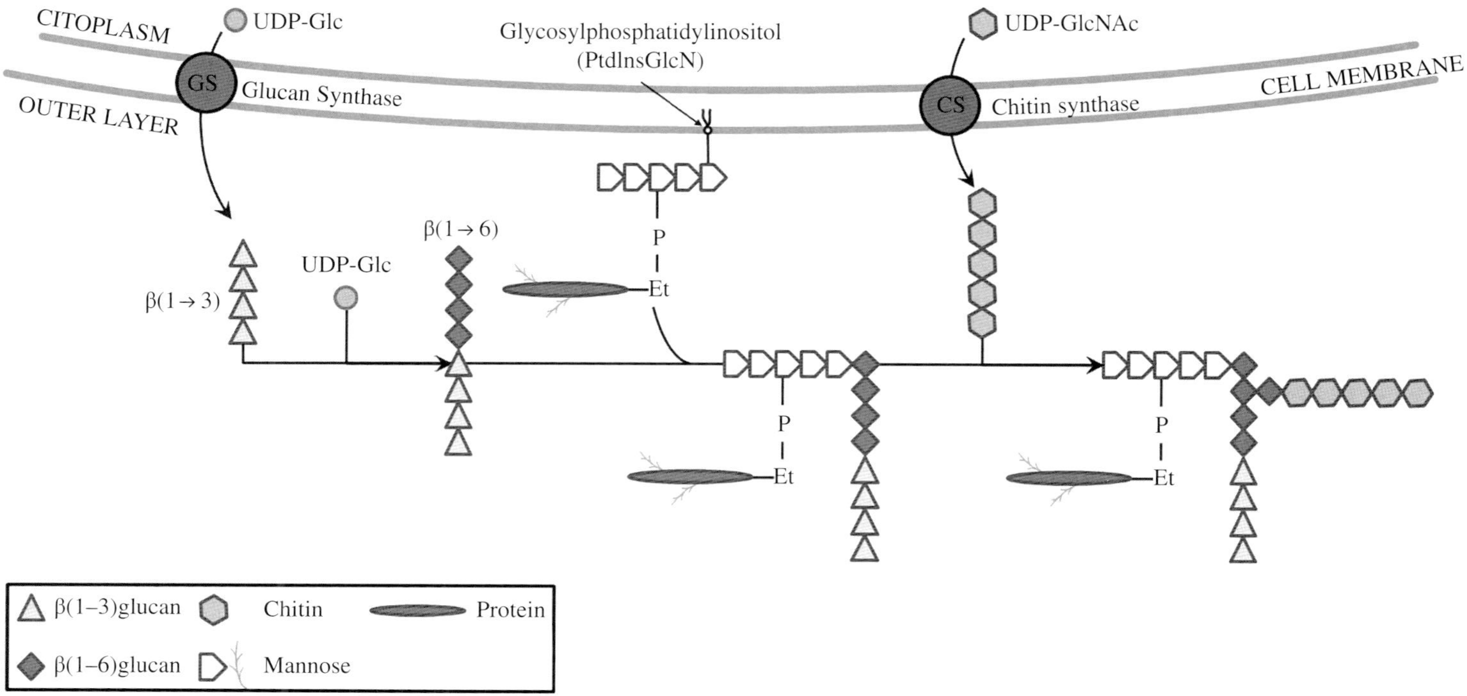


**Figure 15.1** Depiction of yeast cell wall assembly. Source: Modified from Cabib and Arroyo 2013.

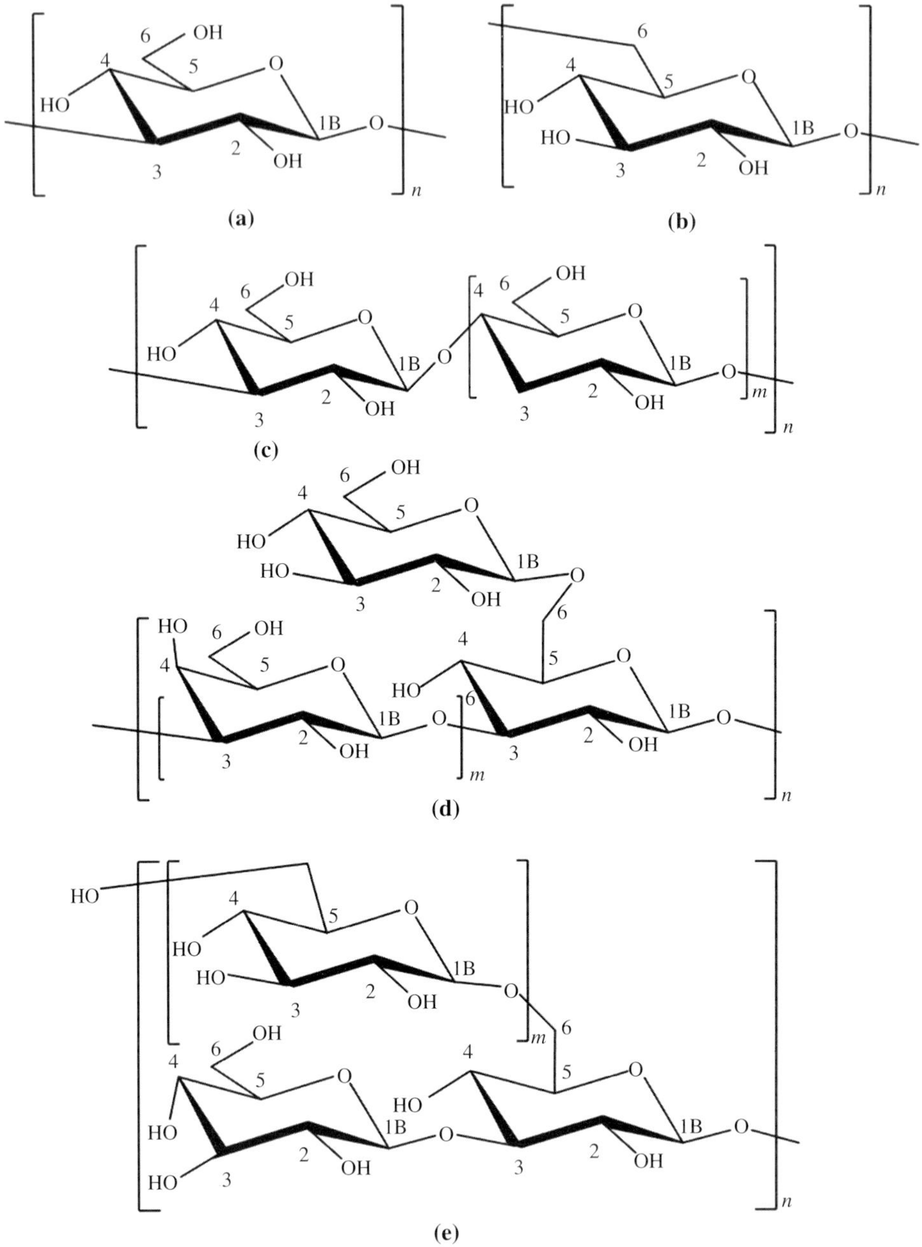


**Figure 15.2** Structure of some fungal beta-glucans. (a): β (1 → 3)-glucan; (b): β (1 → 6)-glucan; (c): β (1 → 3),(1 → 4)-glucan; (d): β (1 → 3),(1 → 6)-glucan; (e): β (1 → 6)-(1 → 3)-glucan.

Beta-glucans also present in alpha-glucans, constituting side-chains, as shown in the fungus *Piptoporus betulinus* (Olennikov et al. 2012). A mixed backbone with both alpha and beta-glucans, is observed in *Pleurotus florida* (Rout et al. 2008). Mixed side-chained alpha and beta-glucans are found in *Pleurotus sajor-caju* (Pramanik et al. 2007)

EPS are natural polymers excreted by some bacteria and fungi species. In some cases, the excretion could be in the form of capsule formation (intrinsic with cell wall structure); while in other cases it could be slime formation (with diffusion capacity for the liquid during the fermentation process). In natural environments, these biomolecules promote the adhesion to the host (providing better fixation), cellular protection against external agents (dryness, bacteriophage, and protozoa), and as carbohydrate storage.

The physiological properties of beta-glucans may vary according to its chemical composition, glycosidic bonds, degree of polymerization, and rheological properties, which are directly related to microorganisms that are synthesizing the biomolecule (Cunha et al. 2012; Mahapatra and Banerjee 2013). Sometimes the EPS production is related to the microorganism virulence to the host tissue of plants and animals (Barbosa et al. 2004; Synytsya and Novak 2014). EPS have wide applications in food/feed, cosmetic, and pharmaceutical products due to its gelling and viscous properties (Kagimura et al. 2015).

Beta-glucans can be also chemically modified for increasing properties, bioactivities, and applications, in processes like acetylation, carboxymethylation, phosphorylation, and sulfonation. These processes of chemical derivatization can be evaluated by techniques such as glucan methylation, FTIR, and NMR as well as the degree of substitution (DS) of chemical groups of the molecule (Kagimura et al. 2015). Acetylation of beta-glucans may increase the antioxidant capacity, decreasing hardness, and adhesiveness of polymers (Chen et al. 2014; De Souza et al. 2015). The carboxymethylated polysaccharides present anti-oxidant properties, increasing the ferric ion reducing capacity (Kagimura et al. 2015). The phosphorylated derivatives also presented anti-inflammatory, anti-proliferative and antiviral properties (Ye et al. 2013). Sulfonated polysaccharides present anti-inflammatory, anticoagulant, and antithrombotic properties, and it is reported as being similarly structured as heparin after sulfonating (Vasconcelos et al. 2013).

## 15.3 Microorganisms: Assets in Beta-Glucan Production

Beta-glucan can be synthesized by many species of bacteria, fungi, and yeasts; however, most of them remain unexplored (Mahapatra and Banerjee 2013). In bacteria, EPS are secreted into the culture medium, but in fungi and yeasts beta-glucan is also present in the structure of cell wall. For example, beta-glucan is secreted in Basidiomycetes mushrooms fruit bodies, mycelium, and in some species, beta-glucan can also be secreted into the medium (Nie et al. 2018).

Beta-glucan is the major component of fungal cell wall, corresponding to 30–60% of total dry mass (Kwiatkowski and Kwiatkowski 2012). Various fungi, including basidiomycetes, filamentous fungi, and yeasts, are known for their capacity to produce beta-glucan and is highly dependent on medium composition, physical conditions, and fungal strain (Mahapatra and Banerjee 2013).

Many fungal beta-glucans have been studied for their antitumor effect. Fungal beta glucans also demonstrate other properties in food/feed, cosmetic, and pharmaceutical products (Cunha et al. 2012; Vasconcelos et al. 2013; Kagimura et al. 2015). Some species of bacteria are able to produce beta-glucan as capsules that can have linear, chained, or cyclic structures, but in most species, it is associated with pathogenicity or related to symbiotic interactions with plants (McIntosh et al. 2005). The cyclic beta-glucans have the capacity to form complexes and have been studied for encapsulation of drugs and natural compounds, such as dexamethasone, reserpine, curcumin, etc. (Kambhampati et al. 2018). Table 15.1 summarizes the beta-glucan production profile from various microorganism, glycosidic bonding composition and their applications.

## 15.4 Strain Improvement Methods for Beta-Glucan Production

Although many microorganisms are able to produce β-glucan, only a few microorganisms produce it in a short time and with high yields. Besides, some microbial species show pathogenicity to human, animals, and plants.

Studies have been developed over the years to find effective β-glucan microorganism producers and to improve production and downstream processes. One of these is the improvement of β-glucan production from natural wild strains by mutation.

Exposing microorganisms to UV light causes mutations in DNA and, depending on the conditions, can lead to cell death. Nevertheless, these changes can enhance β-glucan production of yeast. For instance, when fungus *A. pullulans* colonies were exposed under a 20 W UV lamp at 60 cm of distance for 10 minutes, it resulted in a β-glucan with β-($1 \rightarrow 3, 1 \rightarrow 6$)-glycosidic bonds. Production increased threefold higher than observed using the original strain. In addition, the product had a desirable light color due to low melanin content. This was significant, as high melanin content cause poor glucan recovery (Moriya et al. 2013).

Chemical mutagenesis can also improve β-glucan production. When *Agrobacterium* species were exposed to *N*-methyl-*N*-nitro-nitrosoguanidine (MNNG) solution, an improvement in the β-glucan production was observed (Kim et al. 2003). In "cauliflower" mushroom *Sparassis crispa*, the treatment of basidiospores with an alkaline agent and subsequent mating resulted in mutants that produce about 2.5 folds more β-glucan than the native wild strain (Kim et al. 2013).

## 15.5 Fermentation: Methods and New Formulations

Determining an optimal set of fermentation conditions are necessary for increasing production of beta-glucans from microorganisms. Only a few metabolic pathways have been studied; there is a dearth of information about the synthesis of target biomolecules by different species (Mahapatra and Banerjee 2013). Thus, beta-glucan production research is related to optimization of each studied species, affecting directly the productivity. There are three major factors which have been studied widely: (i) medium composition: optimization of carbon, nitrogen, micronutrients (supplementary salts),

**Table 15.1** Some beta-glucan producing microorganism and its applications.

| Microorganism | Type | Glycosidic bond | Application | Reference |
|---|---|---|---|---|
| *Agrobacterium* sp. | L | β (1 → 3) | Food and feed ingredients | Shih et al. 2009 |
| *Sphingomonas* sp. | L | β (1 → 3)-GAcβ (1 → 3)-β (1 → 3)-L-Rhaα(1 → 3) | Food and feed ingredients | Zhang et al. 2015 |
| *Xanthomonas* spp. | C | β (1 → 4),β (1 → 3)-GAc β (1 → 3)-β (1 → 3) | Food and feed ingredients, oil industry, cosmetics | Habibi and Khosravi-Darani 2017 |
| *Agaricus brasiliensis* | L | β (1 → 6) | Immunostimulatory | Gern et al. 2015 |
| *Agrobacterium* sp. | L | β (1 → 3) | Gelling and thickening agent | McIntosh et al. 2005 |
| *Botryosphaeria rhodina* | C | β (1 → 3),(1 → 6) | Hypoglycemic | Giese et al. 2011 |
| *Botrytis cinerea* | C | β (1 → 3),(1 → 6) | Antitumor | Barbosa et al. 2004 |
| *Bradyrhizobium* sp. | C, Cy | β (1 → 3),(1 → 6) | Drug encapsulation | Kambhampati et al. 2018 |
| *Ganoderma lucidum* | C | β (1 → 3),(1 → 6) | Antitumor | Amaral et al. 2008 |
| *Lasiodiplodia theobromae* | L | β (1 → 6) | Anticoagulant | Cunha et al. 2012 |
| *Phanerochaete chrysosporium* | C | β (1 → 3),(1 → 6) | Enzymatic immobilization | Buchala 1987 |
| *Phellinus* sp. | C | β (1 → 4)-(1 → 6), β (1 → 3) | Antitumor | Cao et al. 2013 |
| *Phlebia radiata* | L | β (1 → 3) | Enzymatic immobilization | Krcmar et al. 1999 |
| *Phoma herbarum* | C | β (1 → 3),(1 → 6) | Free radical scavenging | Yang et al. 2005 |
| *Pleurotus ostreatus* | L | β (1 → 3) | Antitumor | Rosado et al. 2003 |
| *Saccharomyces cerevisiae* | C | β (1 → 3),(1 → 6) | Food and feed ingredients | Kwiatkowski and Kwiatkowski 2012 |
| *Schizophyllum commune* | C | β (1 → 3),(1 → 6) | Antitumor | Hao et al. 2010 |
| *Sclerotium* sp. | C | β (1 → 3),(1 → 6) | Drug release | Coviello et al. 2005 |

L: linear; C: chained; Cy: cyclic.

additives, and vitamins; (ii) physical parameters: pH, temperature, and fermentation time; and (iii) Fermentation methods: agitated and static flasks fermentation and bioreactors applications.

### 15.5.1 Carbon Sources

Different carbon sources have been studied for beta-glucan production by different microorganisms. This includes monomeric sugars (such as arabinose, galactose, and glucose) and disaccharides (celobiose, lactose, maltose, and sucrose). Some polyalcohols like inositol, mannitol, sorbitol, and xylitol were also tested (Steluti et al. 2004; Rusinova-Videva et al. 2011; Lee et al. 2015; Gientka et al. 2016). Glucose, maltose, and sucrose are the most utilized substrates due to their low costs and are widely availability. Vegetal carbon sources such as corn starch, cassava and potato flour, and sugarcane molasses were also used as carbon source by different fungi (Selbmann et al. 2002). Although in most of the cases, the utilized carbohydrate for beta-glucan production is different for each study, the production intensity depends on carbohydrate concentration. The presented concentrations varied from 1.5% to 12% with preferences for concentrations between 3% and 6% (Mahapatra and Banerjee 2013). Higher carbohydrate concentrations may cause osmotic stress and inhibit microorganisms' functions.

### 15.5.2 Nitrogen Sources

A nitrogen source is another major media component that affects directly the beta-glucan production. Nitrogen source can be organic: yeast extract, L-asparagine, glutamate, peptone, and urea; or inorganic: ammonium, potassium and sodium nitrates, ammonium phosphate, and ammonium sulfate (Barbosa et al. 2004; Cunha et al. 2012; Mahapatra and Banerjee 2013; Oliveira et al. 2015). Among, the various nitrogen sources, ammonium salts-based nitrogen sources, showed improved beta-glucan production when compared to other salts (Gibbs and Seviour 1998). Lower nitrogen concentrations stimulates the beta-glucan concentration,s while higher concentrations inhibits beta-glucan production (Barbosa et al. 2004). Concentrations of nitrogen sources from 0.1% to 1% have been found effective in beta-glucan production. Several observations suggest that filamentous fungi produce lower glucan concentrations when utilizing inorganic nitrogen sources compared to organic sources (Mahapatra and Banerjee 2013). Although this mechanism regulation depends on the utilized nitrogen source for media supplementation for the microorganism.

### 15.5.3 Micronutrients, Additives, and Vitamins

Micronutrients are microelements utilized in small quantities by microorganisms, actively influencing metabolites production (Barbosa et al. 2004). Among the major salts utilized for microbial supplementation, potassium phosphates, magnesium sulfate, calcium chloride, and sodium chloride have been studied for beta-glucan production. Other additives like amino acids (glutamate) vegetal oils (cotton, olive, sesame, and soybean), surfactants (Tween 80), and vitamins (A and D) have been added to medium for increased beta-glucan production (Yang and He 2008; Xiao et al. 2010).

### 15.5.4 pH, Temperature, and Fermentation Time

The medium pH is an important factor for microorganism beta-glucan production. Usually, the glucan production is favored by lowering pH of fermentation medium ranging to between 3 and 6.5. Although some fungi gave a maximum production from neutral to alkaline pH ranges (Mahapatra and Banerjee 2013). The literature also reports that lower pH values may affect the molecular weight of glucans. Glucans at lower pH values are produced in lower concentrations with higher molecular weight, while higher pH values showed better production of beta-glucan of lower molecular weight (Shu and Lung 2004). Abdel-Aziz et al. (2012) found that abrupt pH reduction increased the beta-glucan production by *Mucor rouxii* for the protection of fungal cells.

The temperature is a limiting factor for the survival of living beings and has a great significance on microorganisms, affecting its vegetative growth, pathogenicity, reproduction, and physiological aspects. In general, fungi are able to grow in a wide temperature range, which lead them to be disseminated in different environmental conditions around the globe. Optimal temperatures for beta-glucan production are generally 22–30 °C, with a few reported cases of favored production below 20 °C (Kim and Yun 2005).

Fermentation time is another fundamental factor for beta-glucan production. Studies demonstrate that beta-glucan production is maximized both in last exponential growth stage and stationary phase, once the glucans production is associated with secondary microorganism metabolism (Zhu et al. 2012; Gientka et al. 2016). In contrast to bacterial beta-glucan, fungi require higher incubation time for maximization production, about 4–15 days of fermentation (Kim and Yun 2005; Pavlova et al. 2011). Figure 15.3 provides an overview of microbial beta glucan production.

### 15.5.5 Fermentation Methods

Beta-glucan producing microorganisms are majorly aerobic or facultative aerobic, proving that the oxygen limitation is not affordable for its production (Sandford 1979). Fungal beta-glucan production is primarily studied in shake flasks than bioreactors (Gibbs and Seviour 1998; Kim and Yun 2005; Cunha et al. 2012). The aeration conditions are easily achieved in flasks, improving the beta-glucan production over static flasks and thus reducing the necessary time for its maximum production (Leal and Rupérez 1978). Aeration rates for beta-glucan production in stirred-tank bioreactor (STB) and/or airlift might present significant variation, presenting values from 0.05 to 3.5 vvm for different microorganisms (Buchala 1987; Xu and Yun 2004). High aeration rates in bioreactors for beta-glucan production might be justified due to the intense mycelial growth and

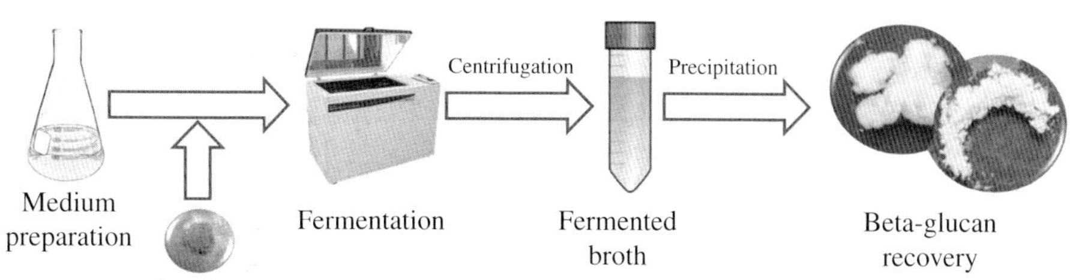


**Figure 15.3** General depiction of fermentative production of beta-glucan and recovery.

the increased viscosity due to the beta-glucan production inside the vessel, which can reduce the dissolved oxygen transfer in the medium (Cunha et al. 2012).

## 15.6 Beta-Glucan Recovery Methods

Polysaccharides from fungi are usually extracted by aqueous solvents and vary according to temperature, pressure, extraction time, and ratio between water and raw material. However, the efficiency of beta-glucan extraction process depends on the type of microorganism and the applied method to recover the biopolymer. In general, hot water is the most common method utilized for cell wall polysaccharides recovery due to its low cost, and the extract obtained is separated and collected by centrifugation (Nie et al. 2018).

In mushrooms, beta-glucan can be extracted from fruiting body and in filamentous fungi from mycelium (Zhu et al. 2016) or precipitated directly from culture broth with absolute ethanol at low temperatures (Cunha et al. 2012). However, beta-glucan extraction from filamentous fungi still has difficulties, since these microorganisms produce other kinds of polysaccharides in concomitance and water extraction yield is generally low (Nie et al. 2018).

Cell lysis is necessary for the extraction of beta-glucan from yeasts, as it is present in the cell wall, connected to the other components by covalent bonds (Kwiatkowski and Kwiatkowski 2012). After cell rupture, beta-glucan can be recovered and separated from other cell wall polysaccharides by alkali extraction, followed by neutralization and ethanol precipitation (Kim and Yun 2005). Even though beta-glucan extraction from yeasts requires simple methods, it is difficult and expensive to obtain isolated product with more than 65% purity (Kwiatkowski and Kwiatkowski 2012).

Some bacteria are able to produce beta-glucan. Curdlan, a beta-glucan produced from species of the genus *Agrobacterium* sp., was recovered from commercial-scale cultures by alkali extraction by a weak base solution and separated from cells by filtration (McIntosh et al. 2005). Figure 15.3 shows a general depiction of beta-glucan production and recovery.

## 15.7 Potential Applications of Beta-Glucan

Beta-glucans are products of great importance, demonstrating vast versatility for new product development for animal and human health (Kagimura et al. 2015). Beta-glucan has profound applications in feed/food, pharmaceutical, and chemical industries. The extensive use of beta-glucan for product generation in industries such as cosmetics for skin care products, pharmaceuticals for immunity enhancer, and in food and beverages industry, has been a critical factor in beta-glucan's high-market growth. Figure 15.4 summarizes the main applications of beta-glucan.

### 15.7.1 Food Applications

Customer awareness of functional foods consumption has created an increasing global trend in recent years, allowing the development of new market niches. Beta-glucan consumption is one of the leading supplement markets and is primarily consumed by the

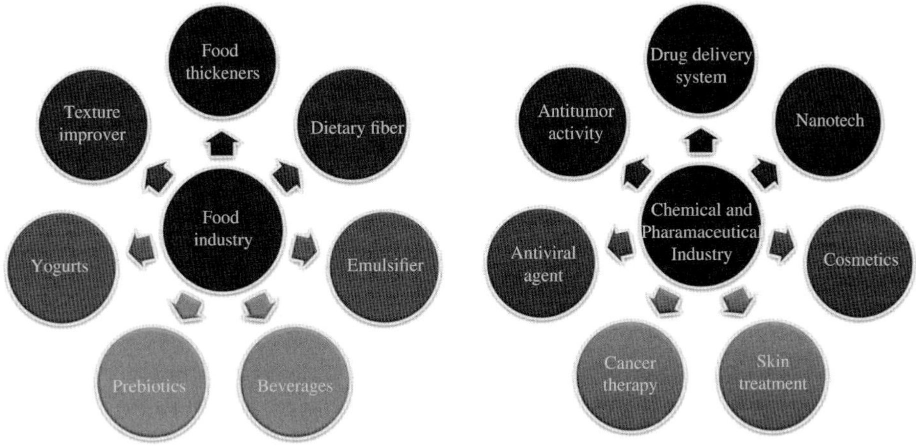


**Figure 15.4** Main applications for beta-glucan in chemical, food, and pharmaceutical industries. Source: Modified from Kagimura et al. 2015.

elderly population due to its cholesterol-reducing properties, weight control, and immunity improvement (Kagimura et al. 2015).

Biopolymers such as curdulan, gellan, and xanthan gum are the most commercially relevant EPS produced by bacteria in the food industry (Shih et al. 2009; Zhang et al. 2015; Habibi and Khosravi-Darani 2017). *S. cerevisiae* derived beta-glucan has probiotic properties and can control both cholesterol and blood sugar levels. Beta-glucans may also be applied as edible coating and edible film in substitution of starch, preventing moisture loss and promoting gas transmission and surface sterility to increase the product's shelf life (Paviath et al. 2009). Chitosan is obtained from the deacetylation of chitin and extracted from fungal cell walls and crustacean exoskeleton, it is used as film and coating in prevention of microbial pathogens growth (Romanazzi et al. 2002). Xanthan gum, produced by *X. campestris*, is a widely utilized beta-glucan showing both thickening and stabilizing properties and employed in a variety of food and beverages (Coviello et al. 2005; Habibi and Khosravi-Darani 2017). Beta-glucans can be applied for enzymes immobilization, acting as a potential support for enzymes and protecting against denaturation (Sutherland 2001).

For beta-glucans to become more commercially viable, factors such as production costs, mechanical properties (elasto-plasticity), and biochemical stability must be addressed by commercial manufactures. Glucans must be nontoxic and safely carry antioxidants, odors, pigments, and other antimicrobial and/or nutritional additives (Diab et al. 2001).

### 15.7.2 Chemical Applications

Beta-glucan may be utilized as biosensors to detect metal ions, such as lead (PbIII), cadmium (CdII), copper (CuII), and manganese (MnII) and shows ceramic properties due to its elastoplasticity (Cunha et al. 2017). Because of elastoplastic properties, beta-glucan could be used in agriculture. It can be formulated with fungicides in prevention of *Bipolaris sorokiniana*, for barley cultivars protection (Antoniazzi and Deschamps 2006).

Another application of beta-glucans is in the cosmetic sector. Anti-aging lotions, facial creams, gels, hair lotions, and makeup are some of the products providing several benefits for skin treatment (Coviello et al. 2005; Mahapatra and Banerjee 2013; Gientka et al. 2016). The utilization of glucans in cosmetics varies according to its action for the formulation of functional polysaccharides (emulsifier, gelling, thickener, and viscosity adjusting properties) and second, active polysaccharides (where the biomolecule has the role of an ingredient presenting beneficial effects on the skin) (Kanlayavattanakul and Lourith 2014). Other important applications of beta-glucan are in wound healing, relief in insect bite, eczema treatment, psoriasis, and protection against ultraviolet radiation (Du et al. 2014).

### 15.7.3 Pharmaceutical Applications

Beta-glucan therapeutic activity is associated with immune system and production of defense cells which fight and destroy pathogens (Kwiatkowski and Kwiatkowski 2012). Beta-glucan can also act as encapsulating agents and have been studied for encapsulating of low water-soluble compounds, such as betulinic acid, dexamethasone, reserpine, curcumin, etc. (Kambhampati et al. 2018). Scleroglucan, a fungal beta-glucan produced by *Sclerotium* species can also act as drug delivery due to its swelling properties when it comes in contact with an aqueous medium which enables the drug release through diffusion (Coviello et al. 2005). Some beta-glucan also presents free radical scavenging along with antioxidant properties, increasing health benefits and preventive effects against diseases associated with the damage caused by free radicals (Sun et al. 2004).

Studies have been reported of anti-inflammatory and anticarcinogenic activities of beta-glucans, as well as the hypocholesterolemic and hypoglycemic properties. Botryosphaeran and lasiodiplodan are two EPS produced by *Botryosphaeria rhodina*, a $\beta$ $(1 \rightarrow 3)$, $\beta$ $(1 \rightarrow 6)$ glucan and *L. theobromae*, a $\beta$ $(1 \rightarrow 6)$ glucan where antimutagenic, anticoagulant antithrombotic, and antiproliferative effects were evaluated (Cunha et al. 2012; Kagimura et al. 2015).

### 15.7.4 Utilization of Agroindustrial Byproducts as Carbon and Nitrogen Sources

There are few reports in the literature where agroindustrial biomass was utilized in growth media as a carbon and nitrogen source substituting synthetic media for beta-glucan production (Kim et al. 2007; Bzducha-Wróbel et al. 2018). The production of the EPS lasiodiplodan by *L. theobromae* was achieved by using the corn-bran hydrolysate (CBH) as an inexpensive media for both carbon and nitrogen sources (Philippini et al. 2018). The utilization of agricultural residues in substitution of synthetic media might provide proper supplementation, development of low-cost and eco-friendly media formulations for fungal exopolymers synthesis. The nitrogen source supplementation substitution for a more feasible source is necessary, as it represent high value costs in bioprocesses, thus increasing production costs. Biomass hydrolysates must fulfill major fungal requirements for EPS production for example:

generating synthetic beta-glucans without toxic compounds production eventually decreasing the final production costs of the biomolecule.

### 15.7.5 Future Commercial Prospects

According to Grand View Research (GVR) report of 2017, which evaluates future market estimations and trend analysis, beta-glucan may achieve more than US$700 million in demand by 2025. This increasing commercial demand is because of several commercial applications of beta glucans in industrial sectors led by formulation of beverages and cosmetics. North America and Asia Pacific represents the major beta-glucan market share and it is expected to have a positive impact on industry growth. Europe is the largest market share holder of the global beta-glucan industry in 2016, having accounted for 31.9% of the overall volume demand, led by France (AIT Ingredients, India (Triveni Interchem), DSM NV (Netherlands), Norway (Biotec Pharmacon, Immunomedic, Natural Artic, Nutramunity), and the United Kingdom (Tate & Lyle Oat Ingredients). Some other industries around the globe are also into the business of beta-glucan, such as in Brazil (Zilor Inc. – Biorigin), China (Shanghai Greaf Biotech), Japan (Umeken), and the United States (Biothera Pharmaceuticals, Immunechoice, Super Beta Glucan Inc.), among others.

GVR also reported that nutraceutical and health benefits properties provided by beta-glucan attracts the attention of both customers and industries. Moreover, the pharmaceutical utilization of beta-glucans in the development of new drugs for cancer treatment, for in-vitro applications and its medicinal properties leading to health benefits calls for a high market share, as the consumption of beta glucan is increasing due to the awareness of the population concerning the health benefits of this biomolecule.

## 15.8 Conclusions

Beta-glucan is one of the most important industrial carbohydrate polymers, and is gaining attention in the world due to its unique properties. This biopolymer has shown several health benefits. To cater to the growing demand, microbial production of beta glucan offers a promising alternative. Using this method, optimization of fermentation parameters and suitable fermentation media play a pivotal role. Exploring economic and available surplus carbon and nitrogen sources is also essential for beta glucan production on a commercial scale. Continuous improvements in fermentation strategies, biomass processing to recover high amount of sugars, strain improvement, and downstream processing will eventually pave the way for economic beta-glucan production. Efforts are underway to explore new and potential applications of beta-glucan in pharmaceutical and health care area.

## Acknowledgment

Authors would like to thank CAPES, CNPq, and FAPESP for their financial support.

## References


Abdel-Aziz, S.M., Hamed, H.A., Mouafi, F.E., and Gad, A.S. (2012). Acidic pH-shock induces the production of an exopolysaccharide by the fungus *Mucor rouxii*: utilization of beet-molasses. *N. Y. Sci. J.* 5: 52–61.

Alquini, G. et al. (2004). Polysaccharides from the fruit bodies of the basidiomycete Laetiporus sulphureus (Bull.: Fr.) Murr. *FEMS Microbiology Letters* 230 (1): 47–52.

Amaral, A.E., Carbonero, E.R., Simão, R.C.G., and Kadowaki, M.K. (2008). An unusual water-soluble β-glucan from the basidiocarp of the fungus *Ganoderma resinaceum*. *Carbohydr. Polym.* 72: 473–478.

Antoniazzi, N. and Deschamps, C. (2006). Growth analysis of two barley cultivars after elicitors and fungicides treatment. *Cienc. Rural* 36: 1065–1071.

Barbosa, A.M., Cunha, P.D.T., Pigatto, M.M., and Silva, M.L.C. (2004). Produção e aplicações de exopolissacarídeos fúngicos. *Semina Ciênc. Exatas Tecnol.* 25: 29–41.

Bhunia, S.K., Dey, B., Maity, K.K. et al. (2011). Isolation and characterization of an immunoenhancing glucan from alkaline extract of an edible mushroom, *Lentinus squarrosulus* (Mont.) singer. *Carbohydr. Res.* 346: 2039–2044.

Buchala, A. (1987). Structure of the β-D-glucan secreted by *Phanerochaete chrysosporium* in continuous culture. *Carbohydr. Res.* 165: 146–149.

Bzducha-Wróbel, A., Błażejak, S., Kieliszek, M. et al. (2018). Modification of the cell wall structure of *Saccharomyces cerevisiae* strains during cultivation on waste potato juice water and glycerol towards biosynthesis of functional polysaccharides. *J. Biotechnol.* 281: 1–10.

Cabib, E. and Arroyo, J. (2013). How carbohydrates sculpt cells: chemical control of morphogenesis in the yeast cell wall. *Nat. Rev. Microbiol.* 11 (9): 648–655.

Cao, C., Peng, F., and Cui, B.-K. (2013). Chemical characterization and structure of exopolysaccharides from submerged culture of new medicinal mushroom from China, Phellinus mori (higher Basidiomycetes). *International journal of medicinal mushrooms* 15 (1): 57–69.

Carbonero, E.R., Montai, A.V., Mellinger, C.G. et al. (2005). Glucans of lichenized fungi: significance for taxonomy of the genera *Parmotrema* and *Rimelia*. *Phytochemistry* 66 (8): 929–934.

Chen, Y., Zhang, H., Wang, Y. et al. (2014). Acetylation and carboxymethylation of the polysaccharide from *Ganoderma atrum* and their antioxidant and immunomodulating activities. *Food Chem.* 156: 279–288.

Corradi da Silva, M.L., Izeli, N.L., Martinez, P.F. et al. (2005). Purification and structural characterisation of (1→3;1→6)-β- D-glucans (botryosphaerans) from *Botryosphaeria rhodina* grown on sucrose and fructose as carbon sources: a comparative study. *Carbohydr. Polym.* 61: 10–17.

Coviello, T., Palleschi, A., Grassi, M. et al. (2005). Scleroglucan: a versatile polysaccharide for modified drug delivery. *Molecules* 10 (1): 6–33.

Cunha, M.A.A., Turmina, J.A., Ivanov, R.C. et al. (2012). Lasiodiplodan, an exocellular (1→6)-β-d-glucan from *Lasiodiplodia theobromae* MMPI: production on glucose, fermentation kinetics, rheology and anti-proliferative activity. *J. Ind. Microbiol. Biotechnol.* 39: 1179–1188.

Cunha, M.A.A., Albornoz, S.L., Santos, V.A.Q. et al. (2017). Structure and biological functions of D-Glucans and their applications. In: *Studies in Natural Products Chemistry* (ed. A.U. Rahman), 309–336. Elsevier.
De Souza, N.L., Bartz, J., Zavareze, E.R. et al. (2015). Functional, thermal and rheological properties of oat β-glucan modified by acetylation. *Food Chem.* 178: 243–250.
Diab, T., Biliaderis, C.G., Gerasopoulos, D., and Sfakiotakis, E. (2001). Physicochemical properties and application of pullulan edible films and coatings in fruit preservation. *J. Sci. Food Agric.* 81: 988–1000.
Du, B., Bian, Z., and Xu, B. (2014). Skin health promotion effects of natural beta-glucan derived from cereals and microorganisms: a review. *Phytother. Res.* 28: 159–166.
Gern, J.C., Santos, L.F., Habu, S. et al. (2015). Exopolysaccharide from *Agaricus brasiliensis* LPB and its scale up studies in a stirred tank fermenter. *Rom. Biotechnol. Lett.* 20: 10945–10950.
Gientka, I., Bzducha-Wróbel, A., and Stasiak-Różańska, L. (2016). The exopolysaccharides biosynthesis by *Candida* yeast depends on carbon sources. *Electron. J. Biotechnol.* 22: 31–37.
Giese, E.C., Dekker, R.F.H., Scarminio, I.S. et al. (2011). Comparison of β-1,3-glucanase production by *Botryosphaeria rhodina* MAMB-05 and *Trichoderma harzianum* Rifai and its optimization using a statistical mixture-design. *Biochem. Eng. J.* 53: 239–243.
Grand View Research (2017). *Market Estimates & Trend Analysis*. Beta-glucan Market. USA: Grand View Research Inc.
Gibbs, P.A. and Seviour, R.J. (1998). The production of exopolysaccharides by Aureobasidium pullulans in fermenters with low-shear configurations. *Applied Microbiology and Biotechnology* 49 (2): 168–174.
Grün, C.H., Hochstenbach, F., Hummel, B.M. et al. (2005). The structure of cell wall α-glucan from fission yeast. *Glycobiology* 15: 245–257.
Habibi, H. and Khosravi-Darani, K. (2017). Effective variables on production and structure of xanthan gum and its food applications: a review. *Biocatal. Agric. Biotechnol.* 10: 130–140.
Hao, L.M., Xing, X.H., Li, Z. et al. (2010 Jan). Optimization of effect factors for mycelial growth and exopolysaccharide production by Schizophyllum commune. *Appl Biochem Biotechnol.* 160 (2): 621–631.
Hoffmann, G.C., Simson, B.W., and Timell, T.E. (1971). Structure and molecular size of pachyman. *Carbohydr. Res.* 20: 185–188.
Huang, Q. and Zhang, L. (2011). Preparation, chain conformation and anti-tumor activities of water-soluble phosphated (1→3)-β-d-glucan from *Poria cocos* mycelia. *Carbohydr. Polym.* 83: 1363–1369.
Jelsma, J. and Kreger, D.R. (1975). Ultrastructural observations on (1→3)-β-glucan from fungal cell-walls. *Carbohydr. Res.* 43: 200–203.
Kagimura, F.Y., da Cunha, M.A.A., Theis, T.V. et al. (2015). Carboxymethylation of (1→6)-β-glucan (lasiodiplodan): preparation, characterization and antioxidant evaluation. *Carbohydr. Polym.* 127: 390–399.
Kanlayavattanakul, M. and Lourith, N. (2014). Biopolysaccharides for skin hydrating cosmetics. *Polysaccharides* 1–23.
Kambhampati, N.S.V., Kar, S., Pinnepalli, S.S.K. et al. (2018). Microbial cyclic β-(1→3),(1→6)-Glucans as potential drug carriers: interaction studies between cyclic

β-glucans isolated from *Bradyrhizobium japonicum* and betulinic acid. *Spectrochim. Acta* 203: 494–500.

Kim, H.O. and Yun, J.W. (2005). A comparative study on the production of exopolysaccharides between two entomopathogenic fungi *Cordyceps militaris* and *Cordyceps sinensis* in submerged mycelial cultures. *J. Appl. Microbiol.* 99: 728–738.

Kim, M.K., Ryu, K.E., Choi, W.A. et al. (2003). Enhanced production of (1→3)-β-D-glucan by a mutant strain of *Agrobacterium* species. *Biochem. Eng. J.* 16: 163–168.

Kim, Y.H., Kang, S.W., Lee, J.H. et al. (2007). High cell density fermentation of *Saccharomyces cerevisiae* JUL3 in fed-batch culture for the production of β-Glucan. *J. Ind. Eng. Chem.* 13: 153–158.

Kim, S.R., Kang, H.W., and Ro, H.S. (2013). Generation and evaluation of high β-glucan producing mutant strains of *Sparassis crispa*. *Mycobiology* 41: 159–163.

Kodali, V.P., Das, S., and Sen, R. (2009). An EPS from a probiotic: biosynthesis dynamics, composition and emulsifying activity. *Food Res. Int.* 42: 695–699.

Krcmar, P., Novotny, C., Marais, M.-F., and Joseleau, J.-P. (1999). Structure of extracellular polysaccharide produced by lignin-degrading fungus *Phlebia radiata* in liquid culture. *Int. J. Biol. Macromol.* 24: 61–64.

Kwiatkowski, S. and Kwiatkowski, S.E. (2012). Yeast (*Saccharomyces cerevisiae*) glucan polysaccharides – occurrence, separation and application in food, feed and health industries. In: *The Complex World of Polysaccharides* (ed. D.N. Karunaratne), 47–70. InTech.

Leal, J.A. and Rupérez, P. (1978). Extracellular polysaccharide production by *Aspergillus nidulans*. *Trans. Br. Mycol. Soc.* 70: 115–120.

Lee, M.J., Liu, H., Barker, B.M. et al. (2015). The fungal exopolysaccharide galactosaminogalactan mediates virulence by enhancing resistance to neutrophil extracellular traps. *PLoS Pathog.* 11: 1–22.

Madhuri, K.V. and Prabhakar, K.V. (2014). Microbial exopolysaccharides: biosynthesis and potential applications. *Orient. J. Chem.* 30: 1401–1410.

Mahapatra, S. and Banerjee, D. (2013). Fungal exopolysaccharide: production, composition and applications. *Microbiol. Insights* 6: 1–16.

McIntosh, M., Stone, B.A., and Stanisich, V.A. (2005). Curdlan and other bacterial (1→3)-β-D-glucans. *Appl. Microbiol. Biotechnol.* 68: 163–173.

Moriya, N., Moriya, Y., Nomura, H. et al. (2013). Improved β-glucan yield using an *Aureobasidium pullulans* M-2 mutant strain in a 200-L pilot scale fermentor targeting industrial mass production. *Biotechnol. Bioprocess Eng.* 18: 1083–1089.

Nie, S., Cui, S.W., and Xie, M. (2018). Beta-glucans and their derivatives. In: *Bioactive Polysaccharides* (ed. S. Nie, S.W. Cui and M. Xie), 99–141. London: Academic Press Elsevier.

Olennikov, D.N., Agafonova, S.V., Rokhin, A.V. et al. (2012). Branched glucan from the fruiting bodies of *Piptoporus betulinus* (Bull.:Fr) karst. *Appl. Biochem. Microbiol.* 48: 65–70.

Oliveira, K.S.M., Bastiani, M.D., Cordeiro, L.M.C. et al. (2015). (1→6)- and (1→3)(1→6)-β-glucans from *Lasiodiplodia theobromae* MMBJ: structural characterization and pro-inflammatory activity. *Carbohydr. Polym.* 133: 539–546.

Paviath, A.E., Orts, W., and Pavlath, A. (2009). Edible films and coatings: why, what, and how? In: *Edible Films and Coatings for Food Applications* (ed. K.C. Huber and M.E. Embuscado), 1–23. Springer.

Pavlova, K., Rusinova-Videva, S., Kuncheva, M. et al. (2011). Synthesis and characterization of an exopolysaccharide by antarctic yeast strain *Cryptococcus laurentii* AL100. *Appl. Biochem. Biotechnol.* 163: 1038–1052.

Philippini, R.R., Martiniano, S.E., Marcelino, P.R.F. et al. (2018). *Corn-Bran Hydrolysate as Low-Cost Media for Lasiodiplodan Biopolymer by Filamentous Fungi Lasiodiplodia theobromae* (ed. A. Mendez-Vilas). Universal-Publishers.

Pramanik, M., Chakraborty, I., Mondal, S., and Islam, S.S. (2007). Structural analysis of a water-soluble glucan (Fr.I) of an edible mushroom, *Pleurotus sajor-caju*. *Carbohydr. Res.* 342: 2670–2675.

Romanazzi, G., Nigro, F., Ippolito, A. et al. (2002). Effects of pre- and postharvest chitosan treatments to control storage grey mold of table grapes. *J. Food Sci.* 67: 1862–1866.

Rosado, F.R., Germano, S., Carbonero, E.R. et al. (2003). Biomass and exopolysaccharide production in submerged cultures of *Pleurotus ostreatoroseus* Sing. and *Pleurotus ostreatus* "florida" (Jack.: Fr.) Kummer. *J. Basic Microbiol.* 43: 230–237.

Rout, D., Mondal, S., Chakraborty, M., and Islam, S.S. (2008). Chemical analysis of a new (1→3)-,(1→6)-branched glucan from an edible mushroom, *Pleurotus florida*. *Carbohydr. Res.* 340: 2533–2539.

Rusinova-Videva, S., Pavlova, K., and Georgieva, K. (2011). Effect of different carbon sources on biosynthesis of exopolysaccharide from antarctic strain *Cryptococcus laurentii* AL 62. *Biotechnol. Biotechnol. Equip.* 25: 80–84.

Sandford, P.A. (1979). Exocellular, microbial polysaccharides. *Adv. Carbohydr. Chem. Biochem.* 36: 265–313.

Selbmann, L., Crognale, S., and Petruccioli, M. (2002). Exopolysaccharide production from *Sclerotium glucanicum* NRRL 3006 and *Botryosphaeria rhodina* DABAC-P82 on raw and hydrolysed starchy materials. *Lett. Appl. Microbiol.* 34: 51–55.

Shih, I.-L., Yu, J.-Y., Hsieh, C., and Wu, J.-Y. (2009). Production and characterization of curdlan by *Agrobacterium sp*. *Biochem. Eng. J.* 43: 33–40.

Shu, C.H. and Lung, M.Y. (2004). Effect of pH on the production and molecular weight distribution of exopolysaccharide by *Antrodia camphorata* in batch cultures. *Process Biochem.* 39: 931–937.

Singh, R.S., Saini, G.K., and Kennedy, J.F. (2008). Pullulan: microbial sources, production and applications. *Carbohydr. Polym.* 73: 515–531.

Steluti, R.M., Giese, E.C., Piggato, M.M. et al. (2004). Comparison of botryosphaeran production by the ascomyceteous fungus *Botryosphaeria* sp., grown on different carbohydrate carbon sources, and their partial structural features. *J. Basic Microbiol.* 44: 480–486.

Sun, C., Wang, J.-W., Fang, L. et al. (2004). Free radical scavenging and antioxidant activities of EPS2, an exopolysaccharide produced by a marine filamentous fungus *Keissleriella* sp. YS 4108. *Life Sci.* 75: 1063–1073.

Sutherland, O.W. (2001). Biofilm exopolysaccharides: a strong and sticky framework. *Microbiology* 147: 3–9.

Synytsya, A. and Novak, M. (2013). Structural diversity of fungal glucans. *Carbohydr. Polym.* 92: 792–809.

Synytsya, A. and Novak, M. (2014). Structural analysis of glucans. *Ann. Transl. Med.* 2: 17.

Toklu, H.S., Seirli, O., Velioglu-Ogunc, A. et al. (2006). Acetaminophen-induced toxicity is prevented by β-D-Glucan treatment in mice. *Eur. J. Pharmacol.* 543: 133–140.

Vasconcelos, A.F.D., Monteiro, N.K., Dekker, R.F.H. et al. (2008). Three exopolysaccharides of the β-(1→6)-D-glucan type and a β-(1→3;1→6)-D-glucan produced by strains of *Botryosphaeria rhodina* isolated from rotting tropical fruit. *Carbohydr. Res.* 343 (14): 2481–2485.

Vasconcelos, A.F.D., Dekker, R.F.H., Barbosa, A.M. et al. (2013). Sulfonation and anticoagulant activity of fungal exocellular β-(1→6)-D-glucan (lasiodiplodan). *Carbohydr. Polym.* 92 (2): 1908–1914.

Welman, A.D. and Maddox, I.S. (2003). Exopolysaccharides from lactic acid bacteria: perspectives and challenges. *Trends Biotechnol.* 21: 269–274.

Xiao, J.-H., Xiao, D.-M., Xiong, Q. et al. (2010). Nutritional requirements for the hyperproduction of bioactive exopolysaccharides by submerged fermentation of the edible medicinal fungus *Cordyceps taii*. *Biochem. Eng. J.* 49: 241–249.

Yang, H. and He, G. (2008). Influence of nutritional conditions on exopolysaccharide production by submerged cultivation of the medicinal fungus *Shiraia bambusicola*. *World J. Microbiol. Biotechnol.* 24: 2903–2907.

Yang, X.B., Gao, X.D., Han, F. et al. (2005). Purification, characterization and enzymatic degradation of YCP, a polysaccharide from marine filamentous fungus *Phoma herbarum* YS4108. *Biochimie* 87: 747–754.

Ye, M., Yuan, R., He, Y. et al. (2013). Phosphorylation and anti-tumor activity of exopolysaccharide from *Lachnum* YM120. *Carbohydr. Polym.* 97: 690–694.

Xu, C.P. and Yun, J.W. (2004). Influence of aeration on the production and the quality of the exopolysaccharides from Paecilomyces tenuipes C240 in a stirred-tank fermenter. *Enzyme and Microbial Technology* 35 (1): 33–39.

Zhang, J., Dong, Y., Fan, L. et al. (2015). Optimization of culture medium compositions for gellan gum production by a halobacterium *Sphingomonas paucimobilis*. *Carbohydr. Polym.* 115: 694–700.

Zhu, H., Tian, B., Liu, W. et al. (2012). A three-stage culture process for improved exopolysaccharide production by *Tremella fuciformis*. *Bioresour. Technol.* 116: 526–528.

Zhu, F., Du, B., and Xu, B. (2016). A critical review on production and industrial applications of betaglucans. *Food Hydrocolloids* 52: 275–288.

# 16

# Extremophiles for Hydrolytic Enzymes Productions: Biodiversity and Potential Biotechnological Applications

*Divjot Kour[1], Kusam Lata Rana[1], Tanvir Kaur[1], Bhanumati Singh[2], Vinay Singh Chauhan[2], Ashok Kumar[3], Ali A. Rastegari[4], Neelam Yadav[5], Ajar Nath Yadav[1], and Vijai Kumar Gupta[6]*

[1] *Department of Biotechnology, Akal College of Agriculture, Eternal University, Baru Sahib, Sirmour, Himachal Pradesh 173101, India*
[2] *Department of Biotechnology, Bundelkhand University, Jhansi, India*
[3] *Department of Biotechnology and Bioinformatics, Jaypee University of Information Technology, Waknaghat, Solan, India*
[4] *Department of Molecular and Cell Biochemistry, Falavarjan Branch, Islamic Azad University, Isfahan, Iran*
[5] *Gopi Nath P.G. College, Veer Bahadur Singh Purvanchal University, Jaunpur, Uttar Pradesh, India*
[6] *ERA Chair of Green Chemistry, Department of Chemistry and Biotechnology, School of Science, Tallinn University of Technology, Tallinn 12618, Estonia*

## 16.1 Introduction

Microbial communities are found worldwide in diverse conditions, including extremes of temperature, salinity, water deficiency and pH. These organisms have developed adaptive features to function under such extreme conditions. These microorganisms, referred to as extremophiles, grow optimally under one or more environmental extreme, while polyextremophiles grow optimally under multiple extremes. The extremophiles can grow optimally in some of the Earth's most hostile environments of pH (<4 Acidophiles and >9; Alkaliphiles), salinity (2–5 M NaCl; Halophiles), temperature (−2 to 20 °C; Psychrophiles; and 60–115 °C; Thermophiles) (Yadav et al. 2015b).

Extremophiles are members of the archaea, although extremophilic members of the bacterial and eukarya domain are also known. Extremophiles are a source of enzymes which are known as extremozymes. An extremozymes is an enzyme, often created by archaea and other extremophilic microbes, that can function under extreme environments (highly acidic/basic conditions, high/low temperatures, high salinity, or other factors). Due to increased stability and effective activities of extremozymes at extreme conditions, these enzymes are of interest to a variety of potential biotechnological applications in the agricultural, energy, environmental, food, health, pharmaceutical, and textile industries.

Microbial communities in extreme habitats have undergone the physiological adaptations to extreme stress of low/high temperature, salinity, and chemical stress. Recently, possible applications of microbial communities from extreme habitats have focused on diverse sectors such as agriculture, medicine, and industry (Saxena et al. 2016; Yadav 2015b). Extremophiles are classified as living organisms able to survive and proliferate in environments with extreme conditions of pH (Acidophile, Alkaliphile), pressure

*Bioprocessing for Biomolecules Production*, First Edition.
Edited by Gustavo Molina, Vijai Kumar Gupta, Brahma N. Singh, and Nicholas Gathergood.

(piezophile), radiation (radioresistant), redox potential (xerophile), salinity (Halophile) and temperature (psychrotolerant, psychrophile, thermophile, hyperthermophiles) (Kumar et al. 2018a; Kumar et al. 2014; Sahay et al. 2017; Saxena et al. 2016; Shukla et al. 2016; Suman et al. 2015; Verma et al. 2015a,b, 2016a,b; Yadav et al. 2015a,b, 2017a–d). The microbes grown, which have the capability to survive and proliferate at two or more extreme environmental conditions, are known as polyextremophilic microbes. Extracellular hydrolytic enzymes such as β-glucosidase, β-galactosidase, xylanase, protease, pectinase, lipase, laccase, chitinase, cellulase, and amylase have diverse potential applications in areas such as agriculture, bioconversion of hemicellulose, biodegradation, bioethanol production, biorefinery, chemical industry, composting, dairy industry, detergent industry, feed supplement, food industry, feed industry, glucose feedstock from cellulose, improving rumen digestion, leather industry, molecular biology, paper and pulp industry, peptide synthesis, pharmaceutical industry, and therapeutic agent synthesis (Coker and Brenchley 2006; Hagihara et al. 2001; Lorentzen et al. 2006; Sahay et al. 2017; Saxena et al. 2016; Suman et al. 2015; Zeng et al. 2003; Kumar et al. 2018b).

Biotechnology in the food and feed industry makes use of extremophilic microbes for the preservation of food and for the production of a range of value-added products such as microbial cultures, enzymes, flavor compounds, food ingredients, and vitamins. Microbial biotechnology has opened up new possibilities concerning the applications of extremophilic microbes for different agricultural, industrial, and allied sectors. An understanding of extremophilic microbial diversity from a wide range of environmental habitats underpins potential applications. This chapter describes the biodiversity of extremophilic microbes, abundance of microbes from diverse habitats, production of different extremozymes, and their applications in agriculture, medicine, pharmaceuticals, and different industry.

## 16.2 Enumeration and Characterization of Extremophiles

The extremophilic microbes from diverse natural extreme habitats can be isolated and enumerated using different culture media. For instance, ammonium mineral salt for methylotrophic microbes and Congo red yeast mannitol for *Rhizobium*. DSMZ-97, DSMZ-823, DSMZ-1184, Halophilic medium, chemically defined medium, complex media, and OS media with NaCl concentrations ranging from 10% to 25% (w/v) is used for halophilic archaea and bacteria. There are different selective media used for isolation of different groups of microbes, e.g., Jensen's agar for $N_2$-fixing bacteria; King'B agar for Pseudomonads; nutrient agar for heterotrophic microbes; nutrient agar with crystal violet for Gram −ve bacteria; nutrient agar with methylene blue for Gram +ve bacteria; rose bengal and potato dextrose agar for fungi; soil extract agar for soil-specific microbes; trypticase soya agar for *Arthrobacter* (Sahay et al. 2017; Suman et al. 2015; Verma et al. 2017; Yadav 2009, 2015a; Yadav et al. 2015a, 2018b). Along with these media some specific media for extremophilic may be used such as R2A agar, trypticase soya agar for psychrophilic; thermus agar, thermos peptone meat extract yeast extract medium (TPMY), yeast extract tryptone, Thermus *thermophilus* medium for thermophiles; halophilic and Horikoshi media for Halophilic microbes. To isolate different groups of extremophilic microbes, all medium and conditions can be used such as for psychrophilic (incubation at >5 °C temperature); thermophilic (incubation at >45 °C temperature); halophilic (with 5–20% NaCl concentration); drought tolerant (7–10% polyethylene glycol) and acidophilic (pH 3–5); alkaliphilic (pH 8–11) (Verma et al. 2017; Yadav et al. 2018a; Yadav et al. 2018c; Yadav et al. 2018d) (Figure 16.1).

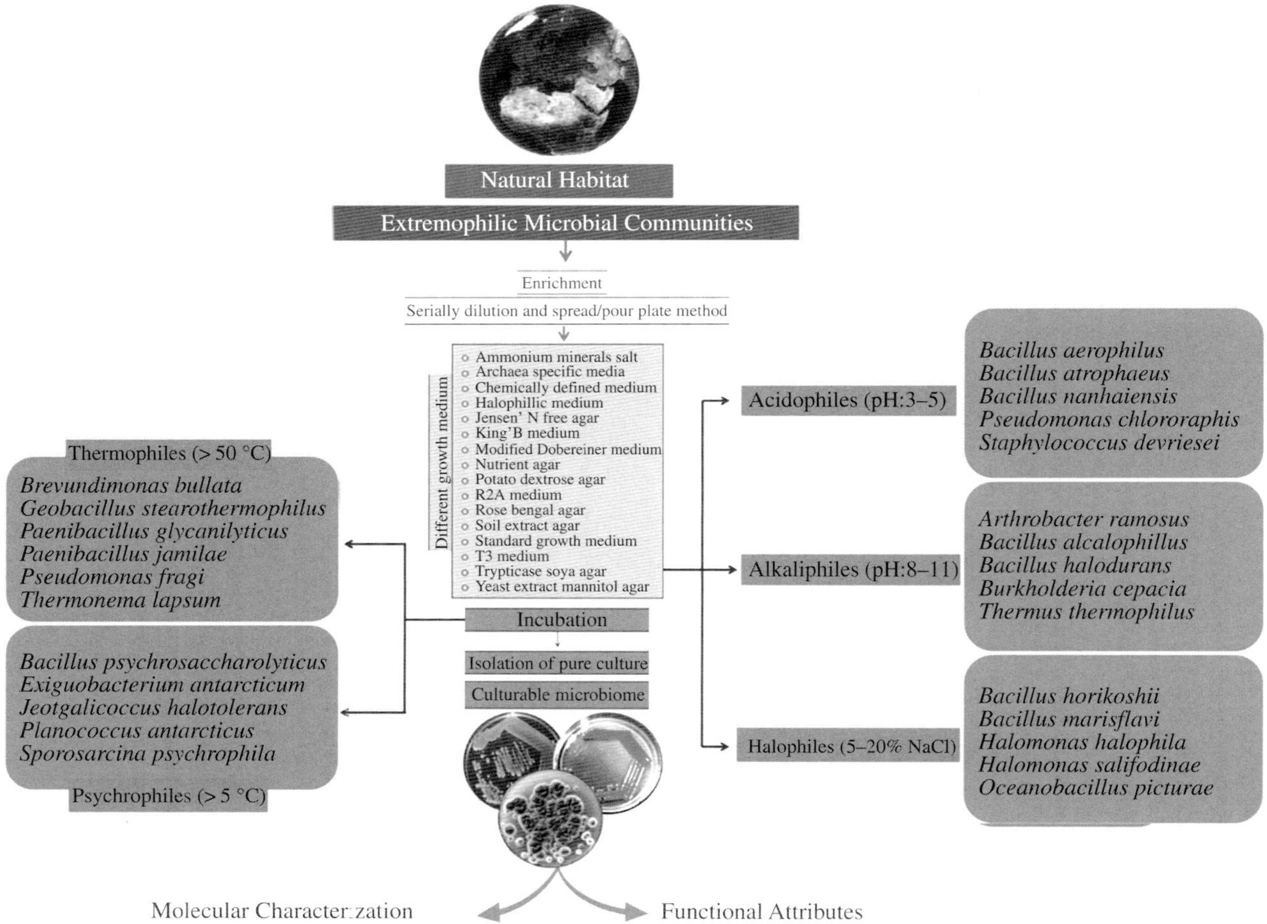


**Figure 16.1** A schematic representation of the isolation, characterization, identification, and Biotechnological potential application of extremophilic microbes.

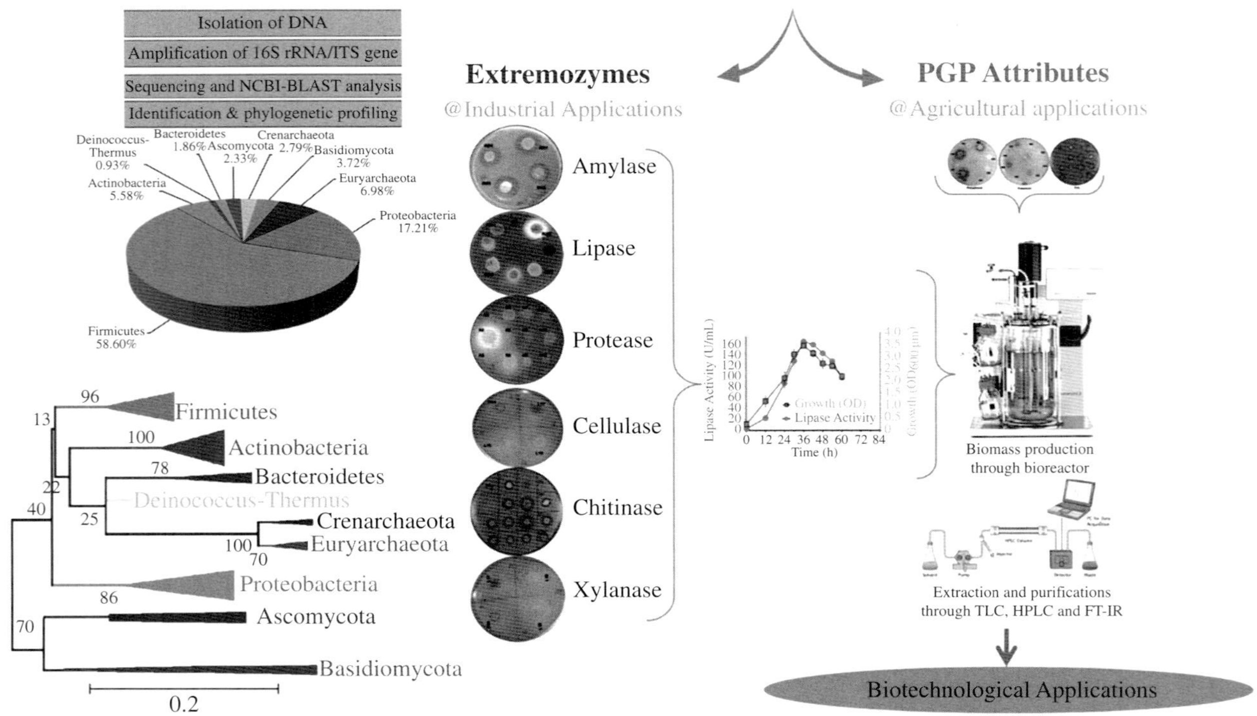


Figure 16.1 (*Continued*)

The culturable extremophilic microbes can be isolated through enrichment using the standard serial dilution plating technique. The culturable extremophilic *Bacillus* and *Bacillus* derived genera (BBDG) can be isolated using heat enrichment technique along with a selective enrichment technique (0.25 and 0.75 M sodium acetate buffer with LB broth and $T_3$ agar) (Yadav et al. 2015b).

For identification of microbes, genomic DNA can be isolated using Zymo Research Fungal/Bacterial DNA MicroPrep™ following the standard protocol prescribed by the manufacturer. After the gDNA isolation and quantification using agarose gel electrophoresis and spectrophotometry techniques, the amplification of 16S rRNA gene for archaea and bacteria while 18S rRNA gene for fungi can be done using different universal primers. The polymerase chain reaction (PCR) product may be quantified using agarose gel electrophoresis and PCR amplified 16S/18S rRNA genes can be purified and sequenced. The partial 16S or 18S rRNA gene sequences may be compared with sequences available in the NCBI database. The phylogenetic tree can be constructed on aligned data sets using the neighbor joining (NJ) method and the program MEGA 4.0.2.

The extremozyme production from extremophilic microbe can be completed using the standard method and media. Microbes for enyzme production under normal as well as extreme conditions can be screened using diffusion agar plate of basal medium (1 g yeast extract, 1 g $KH_2PO_4$, 0.1 g $MgSO_4$.7H2O, 0.05 g $CaCl_2$.H2O, 5 g NaCl, 1 g $NaCO_3$ and 18 g agar per liter) supplemented with different substrates such as starch (0.25% g), xylan (0.1%), carboxy methyl cellulose (0.5%), skimmed milk (10% w/v), and pectin (0.2%) for the production of amylase, xylanase, cellulase, protease, and pectinase, respectively. Plates may be observed for the formation of a clear zone for protease, pectinase, and chitinase. The clear zone around the colony may be observed in plates flooded with 0.1% Congo red solution for 15 minutes at room temperature followed by further treatment with 1 M NaCl for cellulase (Zhou et al. 2004) and xylanase (Wejse et al. 2003) activity. Lipase activity can be detected using basal medium containing 1% tributyrin (v/v) and filter sterilized fluorescent dye Rhodamine B (0.005%). Orange-colored fluorescent halos around colonies may be observed in plates exposed to UV light at 350 nm (Kouker and Jaeger 1987). β-galactosidase activity can be examined on basal medium plates containing 0.1% (w/v) lactose, 0.002% (w/v) X-gal and 0.1 mM IPTG (isopropyl-thio-β-D-galactopyranoside). The appearance of blue-colored colonies indicates β-galactosidase activity (Karasova et al. 2002). The β-glucosidase activity was examined on basal medium plates amended with 1% of esculin. The appearance of a black to reddish color around the bacterial cultures in plates is indication of β-glucosidase activity (Lebbink et al. 2000). Laccase activity was examined on basal medium supplemented with 0.2 mM 2, 2′-Azino-bis (3-Ethylbenzthiazoline-6-Sulfonic Acid) and 25 mg $l^{-1}$ $CuSO_4$. The green halos around bacterial colony indicate laccase activity (Soden et al. 2002).

## 16.3 Biodiversity and Abundance of Extremophiles

The diversity of microorganisms inhabiting extreme environments has been comprehensively investigated in recent decades, with a focus on culture-dependent and culture-independent techniques. Extremophiles from a wide range of natural extreme environments have been characterized, and many biotechnological applications have been realized, stemming from efficient microbial strain and its products. There are many reports on microbial diversity of extremophilic microbes, including acidophilies (Verma et al. 2013), alkaliphiles (Saxena et al. 2016), halophilies (Gaba et al. 2017; Sahay et al. 2017; Verma et al. 2015b; Yadav and Saxena 2017; Yadav et al. 2015a), psychrophiles (Singh et al. 2016; Yadav 2015a; Yadav et al. 2017c,d), thermophiles (Kumar et al. 2014; Suman et al. 2015; Verma et al. 2016a,b), and xerophiles (Yadav et al. 2015b). The microbes from extreme habitat have potential applications in agriculture, medicine, pharmaceutical, and allied sectors (Saxena et al. 2015; Yadav et al. 2018a). Extremophilic microbes have hence been reported from all three domain archaea, bacteria, and eukarya, which included different phylum – mainly Actinobacteria, Deinococcus-Thermus, Bacteroidetes, Ascomycota, Crenarchaeota, Basidiomycota, Euryarchaeota, Proteobacteria, and Firmicutes of diverse genera including *Alkalibacillus, Arthrobacter, Bacillus, Burkholderia, Desemzia, Exiguobacterium, Flavobacterium, Geobacillus, Halobacillus, Haloferax, Halomonas, Jeotgalicoccus, Lysinibacillus, Nitrincola, Oceanobacillus, Paenibacillus, Penicillium, Pontibacillus, Pseudomonas, Psychrobacter, Rhodococcus, Sediminibacillus, Sporosarcina, Staphylococcus, Streptomyces, Thalassobacillus, Thermobacillus,* and *Virgibacillus* (Figure 16.2 a–d). Overall, the distribution

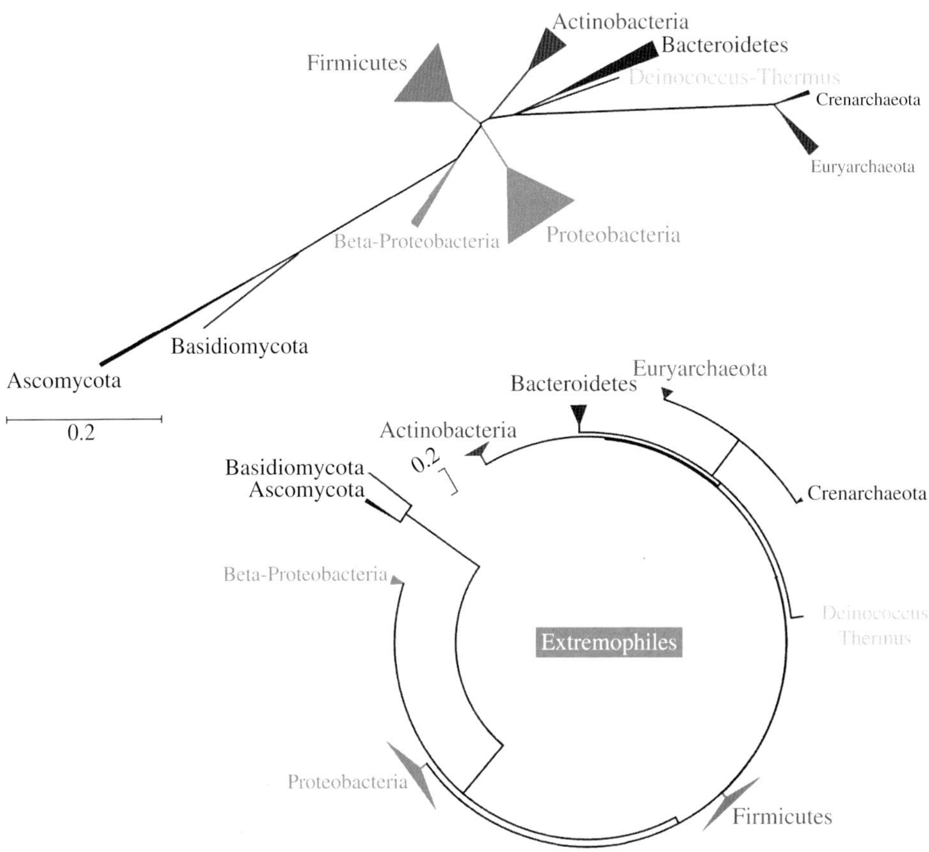


**Figure 16.2** Phylogenetic tree shows the relationship among different groups of extremophilic microbes isolated from diverse extreme habitats worldwide.

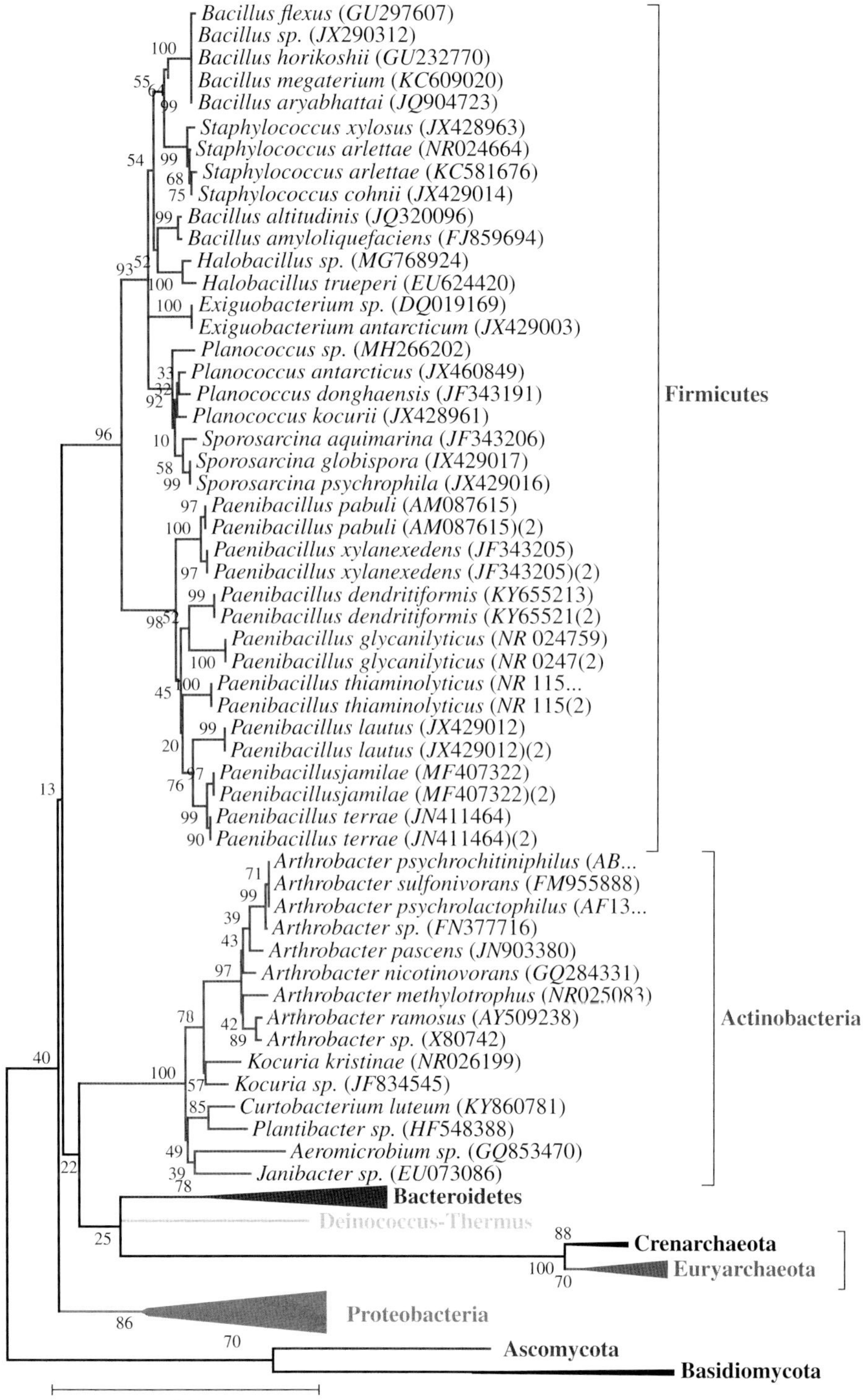


**Figure 16.2** *(Continued)*

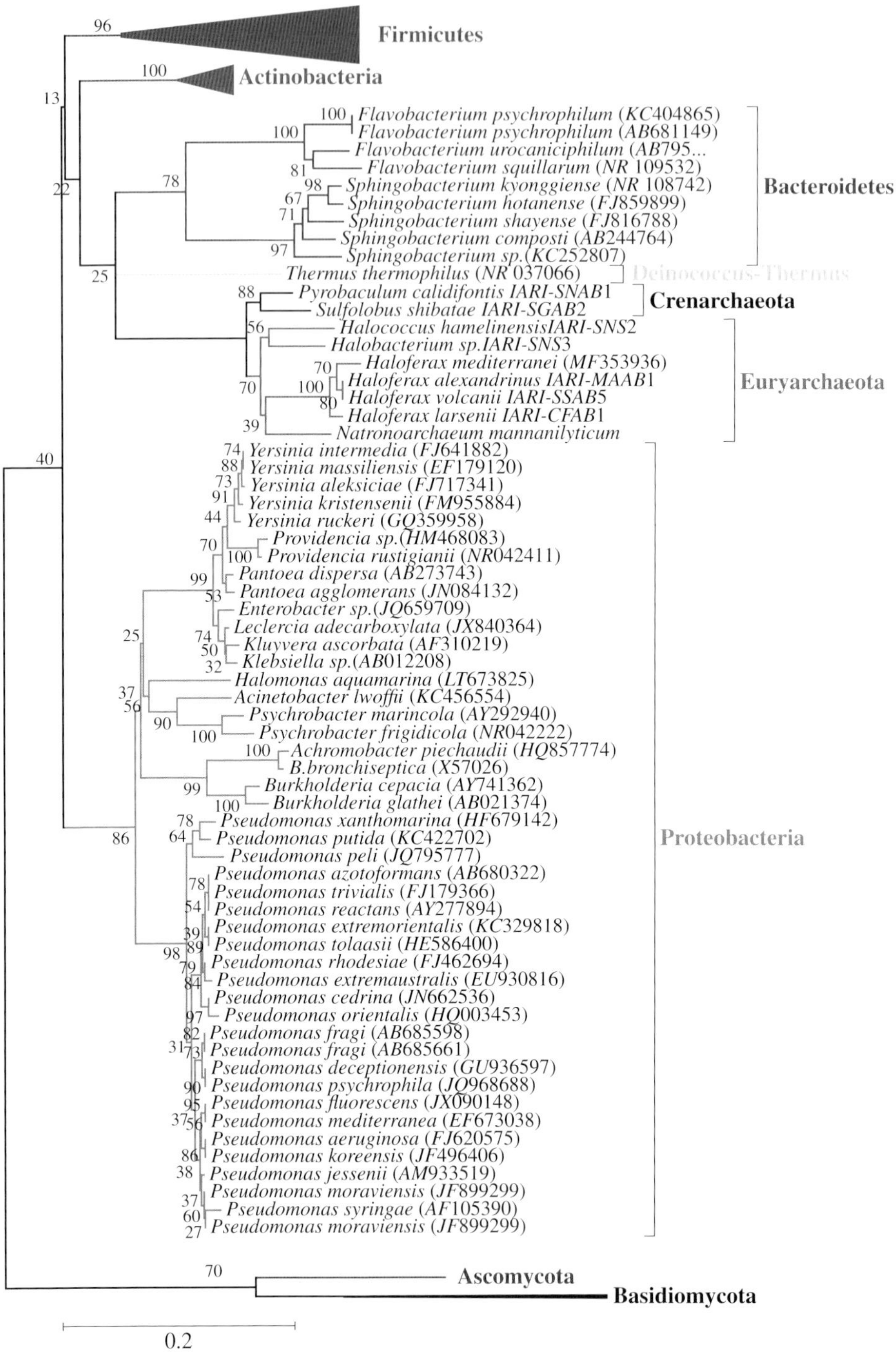


**Figure 16.2** (*Continued*)

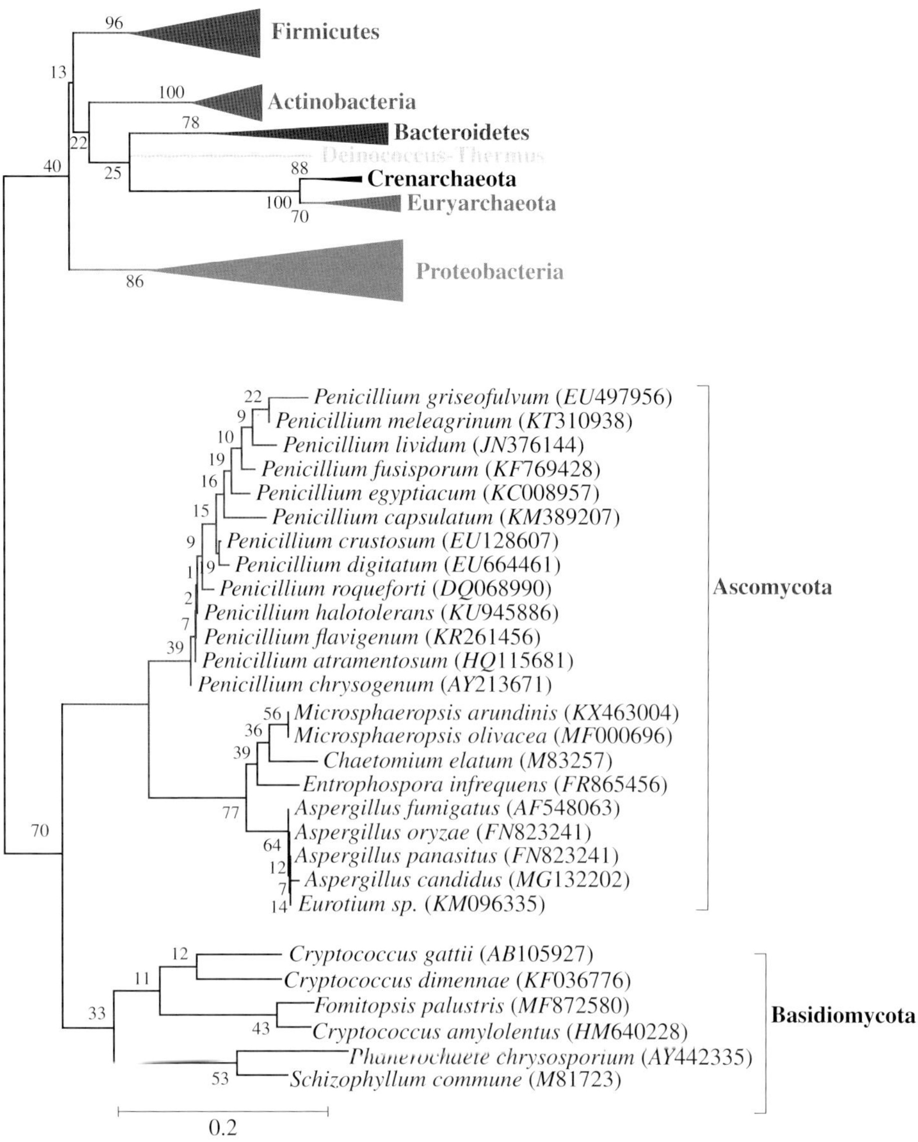


**Figure 16.2** (*Continued*)

of extremophilic microbes varied in all microbial phyla. Firmicutes were the dominant phylum, followed by proteobacteria and actinobacteria. Least number of microbes has been reported from phylum Deinococcus-Thermus, followed by Bacteroidetes (Figure 16.3).

By reviewing the literature, it can be concluded that dominant genera from all extremophilic environments is BBDG followed by *Halomonas*, *Pseudomonas*, and *Staphylococcus* (Figure 16.4). The genus *Bacillus* consists of a heterogenic group of Gram positive rods. These are able to form endospores that allow microbes to

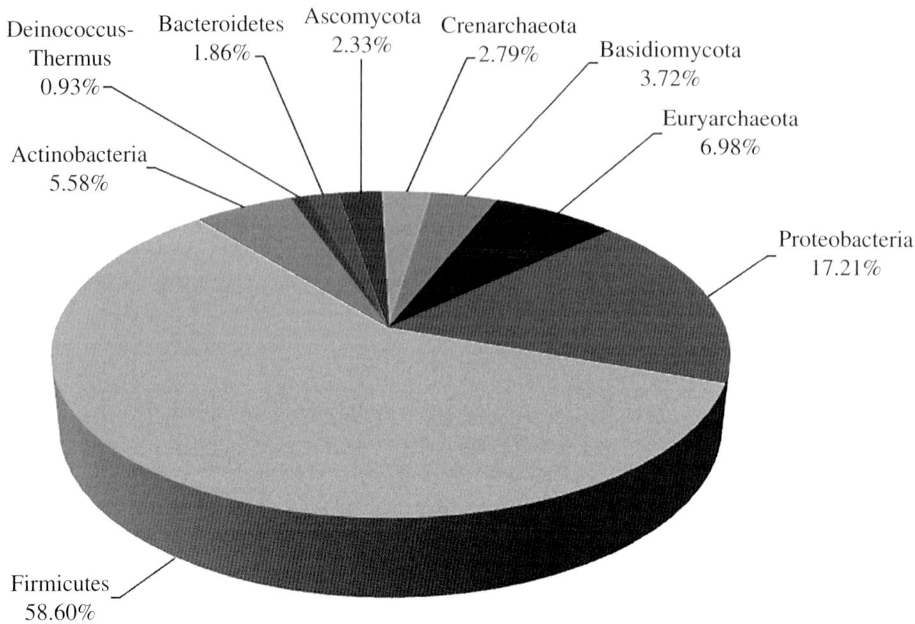


**Figure 16.3** Extremophilic microbes of diverse phylum isolated from a wide range of extreme habitats worldwide.

survive for extended periods under adverse environmental conditions, better than nonsporulating bacterial enteropathogens. Endospore formation is the dominant feature in the characterization of *Bacillus*. There is a boundary that separates this genus from other genera in which endospores are produced. The BBDG belong to the phylum Firmicutes, which is further distributed into several families including Bacillaceae, Bacillales Incertae Sedis, Clostridia, Panenibacillaceae, Planococcaceae, and Staphylococcaceae, including genera *Alicyclobacillus, Amphibacillus, Bacillus, Brevibacillus, Clostridium, Desemzia, Exiguobacterium, Geobacillus, Halobacillus, Halothermothrix, Jeotgalicoccus, Lactobacillus, Lysinibacillus, Oceanobacillus, Paenibacillus, Planococcus, Pontibacillus, Rummeliibacillus, Salimicrobium, Sediminibacillus, Sinobaca, Sporosarcina, Staphylococcus, Thalassobacillus, Thermoactinomyces, Thermobacillus, Thermococcus,* and *Virgibacillus*. The Phylum Actinobacteria is divided into six classes, namely, Actinobacteria, Acidimicrobiia, Coriobacteriia, Nitriliruptoria, Rubrobacteria, and Thermoleophilia, which contain 29 orders and 67 families, 391 genera with 3900 distinct species (Yadav et al. 2018c). Studies on extremophilic microbial diversity and application in different sectors concluded that the member belonging to phylum actinobacteria included diverse genera, in particular *Arthrobacter, Curtobacterium, Micrococcus, Nesterenkonia, Nocardiopsis, Rhodococcus,* and *Streptomyces*. The members of proteobacteria are one of most important bacteria utilized in agriculture, and industry have been reported from diverse extreme environment and belong to different genera including *Acinetobacter, Alcaligenes, Brevundimonas, Burkholderia, Enterobacter, Halomonas, Marinobacter, Nitrincola, Pseudalteromonas, Pseudomonas, Psychrobacter, Salicola, Shewanella,* and *Vibrio* (Figure 16.5).

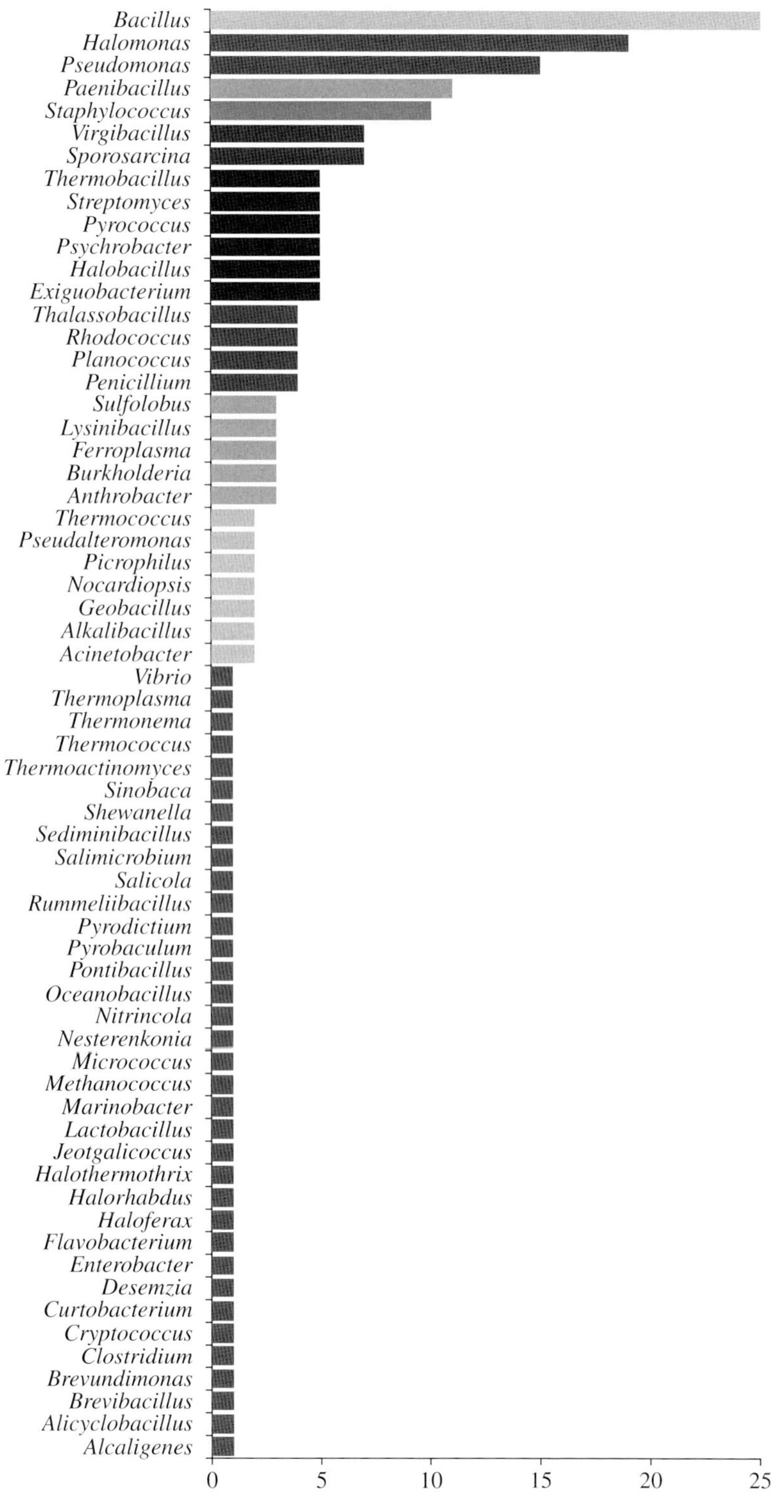


**Figure 16.4** Abundance of extremophilic microbes isolated from a wide range of extreme habitats worldwide.

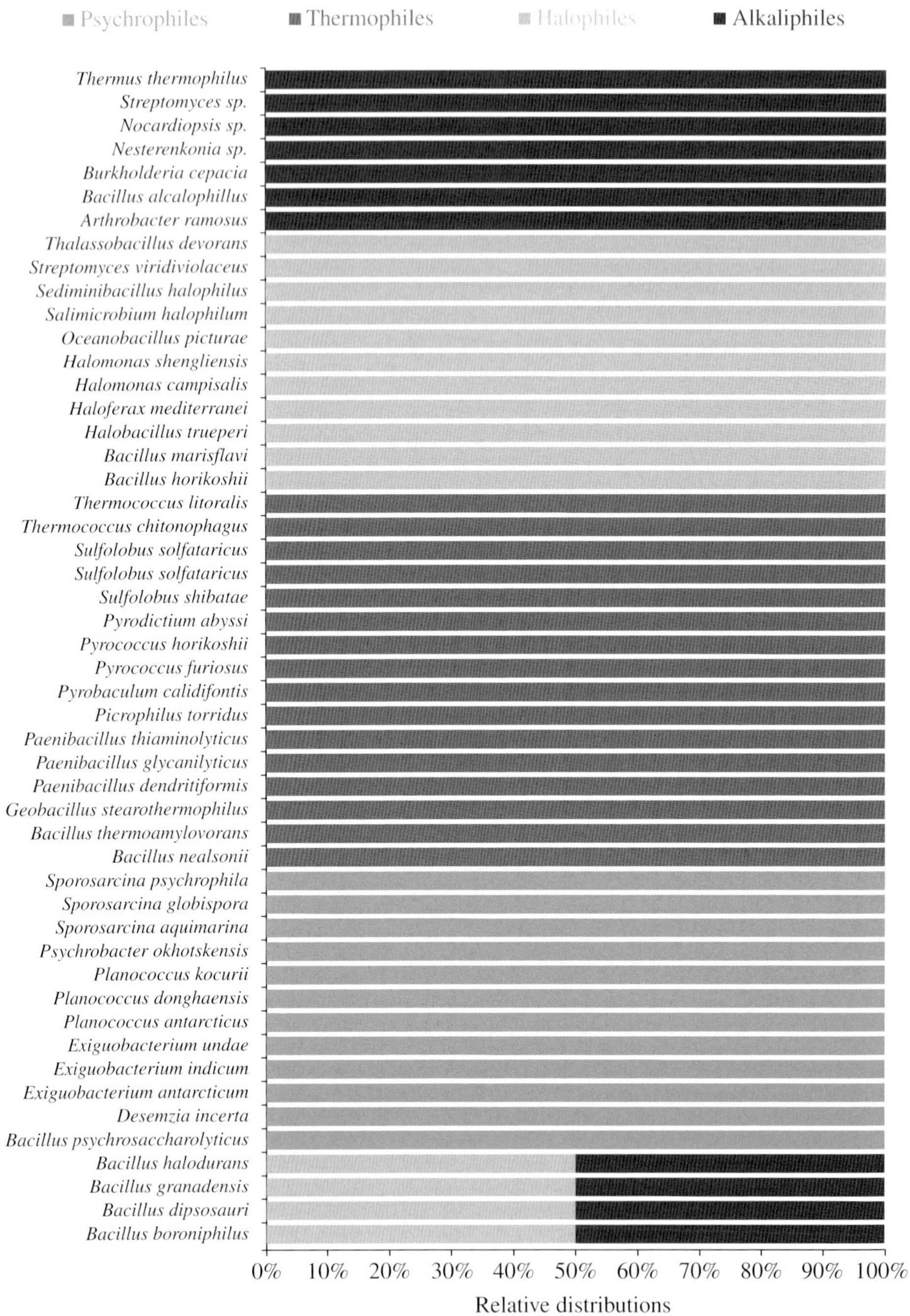

Psychrophiles
Thermophiles
Halophiles
Alkaliphiles
Thermus thermophilus
Streptomyces sp.
Nocardiopsis sp.
Nesterenkonia sp.
Burkholderia cepacia
Bacillus alcalophillus
Arthrobacter ramosus
Thalassobacillus devorans
Streptomyces viridiviolaceus
Sediminibacillus halophilus
Salimicrobium halophilum
Oceanobacillus picturae
Halomonas shengliensis
Halomonas campisalis
Haloferax mediterranei
Halobacillus trueperi
Bacillus marisflavi
Bacillus horikoshii
Thermococcus litoralis
Thermococcus chitonophagus
Sulfolobus solfataricus
Sulfolobus solfataricus
Sulfolobus shibatae
Pyrodictium abyssi
Pyrococcus horikoshii
Pyrococcus furiosus
Pyrobaculum calidifontis
Picrophilus torridus
Paenibacillus thiaminolyticus
Paenibacillus glycanilyticus
Paenibacillus dendritiformis
Geobacillus stearothermophilus
Bacillus thermoamylovorans
Bacillus nealsonii
Sporosarcina psychrophila
Sporosarcina globispora
Sporosarcina aquimarina
Psychrobacter okhotskensis
Planococcus kocurii
Planococcus donghaensis
Planococcus antarcticus
Exiguobacterium undae
Exiguobacterium indicum
Exiguobacterium antarcticum
Desemzia incerta
Bacillus psychrosaccharolyticus
Bacillus halodurans
Bacillus granadensis
Bacillus dipsosauri
Bacillus boroniphilus
0%
10%
20%
30%
40%
50%
60%
70%
80%
90%
100%
Relative distributions

**Figure 16.5** Diversity and distribution of microbiomes from diverse extreme habitats.
Source: Psychrophiles: (Coker and Brenchley 2006; Collins et al. 2002; Ferrer et al. 2005a,b; Hu et al. 2007; Lee et al. 2010; Luo et al. 2006; Okuda et al. 2004; Sakamoto et al. 2003; Secades et al. 2003; Singh et al. 2016; Yadav 2015a, 2016, 2017c,d; Yazdi et al. 2008; Yoshimune et al. 2005; Zeng et al. 2003); Thermophiles: (Andrade et al. 2001; Andronopoulou and Vorgias 2003; Cannio et al. 2004; Fang et al. 2004; Fukushima et al. 2005; Gao et al. 2003; Gomes et al. 2003; Hatada et al. 2005; Hotta et al. 2002; Hutcheon et al. 2005; Kang et al. 2007; Kim et al. 2001; Lorentzen et al. 2006; Palmieri et al. 2006; Sahay et al. 2017; Saxena et al. 2016; Silva et al. 2005; Suman et al. 2015; Sutrisno et al. 2004; Tatur et al. 2006; Taylor et al. 2004); Halophiles: (Amoozegar et al. 2003; Deutch 2002; Jeon et al. 2009; Mijts and Patel 2002; Pérez-Pomares et al. 2003; Sahay et al. 2011, 2012; Sanchez-Porro et al. 2003; Wainø and Ingvorsen 2003; Wang et al. 2010; Yadav and Saxena 2018); Alkaliphilic (Chang et al. 2004; Haddar et al. 2009; Hagihara et al. 2001; Hashim et al. 2004; Kanekar et al. 2002; Mitsuiki et al. 2002; Miyazaki 2005; Pretz et al. 2005; Shi et al. 2006; Solingen et al. 2001; Zeng et al. 2006).

## 16.4 Diversity of Extremozymes and Their Biotechnological Applications

The current challenge besetting the study of biological molecules produced by extremophiles is that their potential applications may not be well known. This, however, offers an immense potential for future development. To this end, evidence from past research suggests biotechnologically significant roles ranging from microbial strains and their secondary metabolites. Among the most explored applications of extremophiles are their productions of enzymes, fatty acids and proteins, antibiotics, or biomedicines (Kaur et al. 2017; Kumar et al. 2016; 2017; Suman et al. 2015; Verma et al. 2015b). Among microbial groups the most predominant microbes in extreme environments are arguably the richest source of small molecule, with widespread global and environmental dispersal. As the survival strategies of extremophiles are often novel and unique, the necessity for microbial cell components to adapt to extreme environments (natural or synthetic) implies that a broad range of cellular products (genes and metabolites) are available for biotechnological applications. The enzymes produced by extremophilic microbes are extremozymes, including keratinases, lipases, pectinases, phytases, proteases, xylanases, esterases, cellulases, catalases, amylases, β-galactosidase, β-glucosidase, laccase, and chitinase, which have great potential for applications in various biotechnological processes (Figure 16.6). Extremozymes have a great economic potential in many important processes, including agricultural, pharmaceutical, and chemical applications. The discovery of new extremophilic microbes will offer novel opportunities for industrially important enzymes for diverse applications in agriculture, medical, pharmaceutical, energy, and environments (Figure 16.7).

### 16.4.1 Amylase

An enzyme that catalyzes the hydrolysis of starch into sugars is referred to as an amylase. Diastase, an amylase enzyme, was first reported by a French chemist, Anselme Payen (Hill and Needham 1970; Ramasubbu et al. 1996; Silverman 2002). The term *amylase* was used initially to entitle enzymes accomplished of hydrolyzing α-1,4- glucosidic bonds of amylose, amylopectin, glycogen, and their degradation products (Aiyer 2005). Currently, the α-amylase is of boundless importance in food, fermentation, textile to pulp and paper industries, medicinal and analytical chemistry, and in automatic dishwashing detergents.

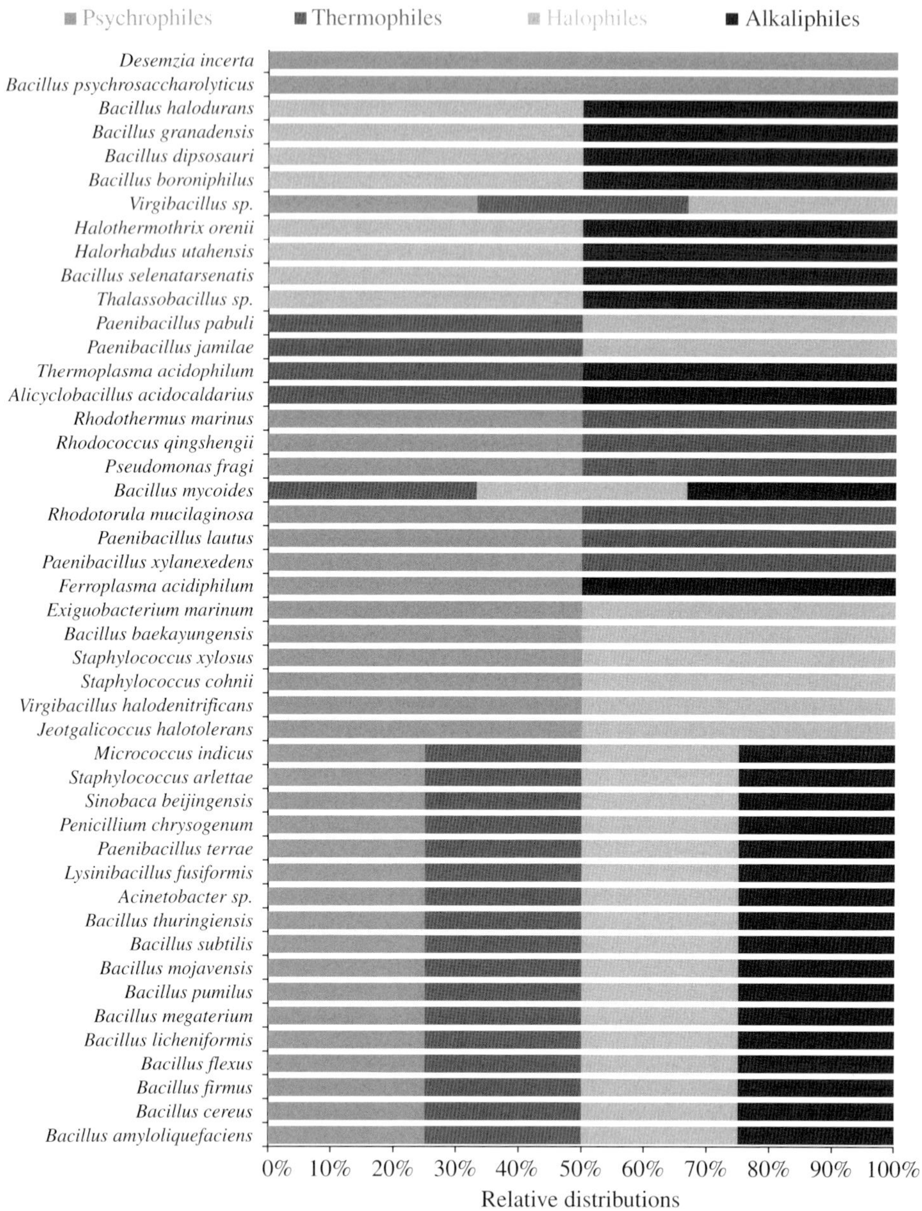


**Figure 16.5** (*Continued*)

The enzyme amylases can be divided into two groups: endoamylases and exoamylases. In the center of the starch molecule endoamylases catalyze hydrolysis in a random mode. From the nonreducing end, exoamylases hydrolyze gradually developing in short end products. The enzyme is further divided into α-amylases, β-amylase, and γ-Amylase. α-amylase catalyzes the hydrolysis of α-1,4 glycosidic linkages in starch to produce glucose, dextrins, and limit dextrins. The process of scattering indissoluble

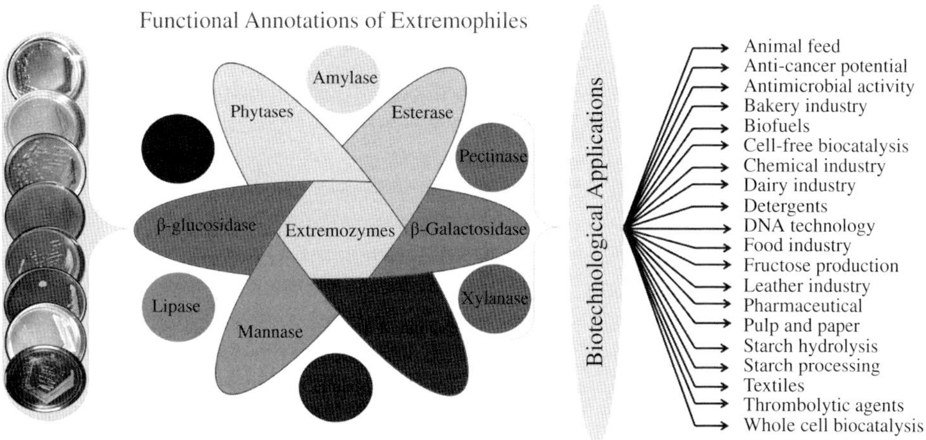


**Figure 16.6** Extremozymes producing microbes and its biotechnological application in food and allied sectors.

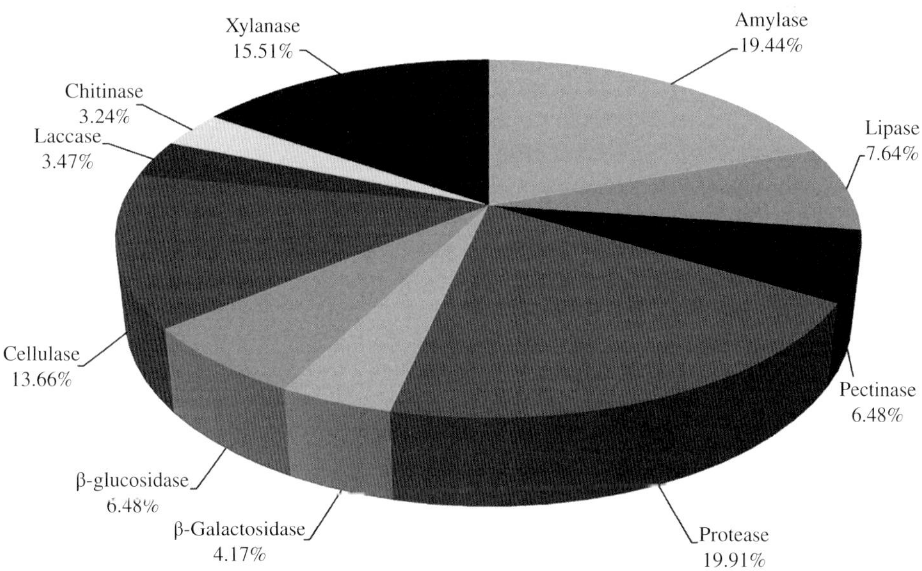


**Figure 16.7** Distribution of extremozymes for diverse biotechnological applications.

starch particles in aqueous solution monitored by partial hydrolysis by means of thermostable amylases is referred to as liquefaction. Thermostable α-amylase is used as a thinning agent, which carries lessening in viscosity and fractional hydrolysis of starch. In starch processing technology, two thermostable α-amylases are economically accessible and used thoroughly. On a large scale, the amylase of *Bacillus amyloliquefaciens* was the first liquefying α-amylase. A heat-stable enzyme from *Bacillus licheniformis* is now also used economically (Madsen et al. 1973).

A variety of biochemical aspects are well-documented as influencing the making of α-amylase by submerged fermentation (SmF) and solid-state fermentation (SSF), which comprises of composition of the growth medium, pH of the medium, phosphate concentration, inoculum age, temperature, aeration, carbon source, and nitrogen source (Lonsane and Ramesh 1990). Different microbes reported to support the production of α-amylase using different sources, such as yeast extract, ammonium sulphate in the case of *Bacillus subtilis* (Dercova et al. 1992), beef extract, peptone, and corn steep liquor favored greatest α-amylase production by bacterial strains soybean meal and casamino acids by *Aspergillus oryzae* (Ueno et al. 1987) various inorganic salts such as ammonium sulphate, ammonium nitrate, and Vogel salts have been reported to support better α-amylase production in fungi. Phosphate, one of the macronutrient, plays a vital part in the development of the organism and production of α-amylase (Cheng et al. 1989).

Acidophilic microorganisms grow ideally at pH values within 1–4. They have adjusted to the acidic situation by lasting the pH of their cytoplasm near to neutrality. The α-amylase from the Gram-positive *Alicyclobacillus acidocaldarius* ATCC27009 produces the enzyme, which is thermoacidophilic, was an illustrative of an acidophilic protein, with optimum of temperature 75 °C and pH of 3, respectively (Matzke et al. 1997). Asoodeh et al. (2010), stated a rare acidophilic α-amylase produced from newly isolated *Bacillus* sp. Ferdowsicous. The molecular weight of the enzyme comprises of 53 kDa and ranges between 3.5 and 7 pH with an optimum around 4.5. The thermostable, acidophilic α-amylases were generally utilized in the food processing industry. In the starch industry, the natural pH of starch slurry is generally around 4.5 (Sivaramakrishnan et al. 2006). Thermo and alkaline stable α-amylases is employed as cumulative in detergents for washing processes, since these processes are generally achieved on the alkaline pH in both hot and cold water. Alkaline and thermotolerant α-amylases have been cleansed from *Bacillus* species, *B. licheniformis,* and *Bacillus halodurans* (Hayashi et al. 1988; Hmidet et al. 2008; Kim et al. 1995; Medda and Chandra 1980).

Psychrophiles or cryophiles are extremophilic organisms capable of growth and reproduction in low temperatures, ranging from −20 to +10 °C. They are detected in places that are lastingly cold, such as the Polar regions and the deep sea (Singh et al. 2016; Verma et al. 2015a; Yadav 2015a). In SSF thermophilic *Bacillus coagulans* reported to produce extracellular, thermostable α-amylase (Babu and Satyanarayana 1995). The *B. subtilis* JS-2004 strain described to produce high levels of thermostable α-amylase with peculiarities appropriate for utilization in processing of starch and food industries. The enzyme was prompted by $Ca^{2+}$and was completely consistent for one hour at 60 and 70 °C (Asgher et al. 2007). Zhang and Zeng (2008), reported an actinomycetes isolated from the deep sea sediment of Prydz Bay, Antarctic, which is identified as *Nocardiopsis* sp. and produces α-amylase. There are many reports on production of cold-adapted amylase (Yadav 2015a; Yadav et al. 2016, 2018d).

The commercial application of amylases predecessor with first use in 1984, as a pharmaceutical benefit for the cure of digestive diseases. For all starch-based industries, amylases are among the most important hydrolytic enzymes. Chemical hydrolysis of starch processing has been entirely substituted by microbial amylases. In the pharmaceutical and fine chemical industries enzymes can also be of huge potential. For centuries, enzymes in particular malt and microbial α-amylases have been broadly used in the baking industry (Hamer 1995). In bread and rolls to provide higher volume to these products, these enzymes were used to provide improved color. The α-amylases used in baking have been cereal enzymes from barley malt and microbial enzymes from fungi and bacteria (Hebeda et al. 1990, 1991). The use of α-amylase for the production of low viscosity, high molecular weight starch for covering of paper is reported. Since 1975, α-amylases have been used in powder laundry detergents. Currently, 90% of all liquid detergents contain α-amylase (Kottwitz et al. 1994). With the introduction of new frontiers in biotechnology, the scope of amylase uses has extended into several other fields, such as clinical, medicinal, and analytical chemistry.

### 16.4.2 Proteases

An enzyme that performs the proteolysis: catabolism of protein by hydrolysis of peptide bonds is referred to as proteases. Proteases occur in all organisms, from prokaryotes to eukaryotes to viruses. Several microbial strains including fungi (*Aspergillus flavus, Aspergillus melleu, Aspergillus niger, Chrysosporium keratinophilum, Fusarium graminarum, Penicillium griseofulvin, Scedosporium apiosermum*) and bacteria (*B. licheniformis, Bacillus firmus, Bacillus alcalophilus, B. amyloliquefaciens, Bacillus proteolyticus, B. subtilis, Bacillus thuringiensis*) were reported to produce proteases (Panjiar et al. 2017; Prakasham et al. 2006; Yadav et al. 2016, 2017b, 2018c).

Increasing consideration has been given in recent times to enzymes from psychrophilic microorganisms, which are categorized by a particular activity in the range of 0–20 °C. Extracellular degradative enzymes comprising proteases, released by psychrophilic marine bacteria, characteristically direct activity targets at temperatures well above the upper growth boundary of the producing strain. Yet, a few particularly psychrophilic bacterial isolates have been shown to produce proteases with activity optimized at 20 °C (Huston et al. 2000). The cold-active anaerobic bacteria producing extracellular proteases can potentially be employed for biodegradation of organic wastes rich in protein, such as waste from humans. Alam et al. (2005), reported the psychrotrophic proteolytic bacterium *Clostridium* sp. SPA3 isolated from the lake sediment of Antarctica forming maximum cell mass of 5–10 °C and produced extracellular protease. Growth was observed in the pH range of 6.5–8.5 with optimum at pH 8.

An aerobic Gram-positive, thermophilic in nature *Bacillus* species (P-OOlA) has been isolated from an alkaline hot spring at Wondo Genet, Ethiopia, by Atalo and Gashe (1993). The protease activity against casein was 65 U $ml^{-1}$. Activity of enzyme was identified between 30 °C and 70 °C and pH of 4.5–11.5. Starch was the best substrate, followed by trisodium citrate, citric acid, and sucrose as carbon sources for the production of protease utilized by *Bacillus* sp. strain SMIA-2. Ammonium nitrate was found to be the best nitrogen source. *Bacillus* sp. strain SMIA-2 thermophilic bacteria were reported as Protease producer. The enzyme was stable for two hours at 30 °C, while at 40 °C and 80 °C, 14% and 84% of the original activities were lost (Nascimento and Martins 2004).

### 16.4.3 Pectinase

Pectinases are a heterogeneous group of linked enzymes that hydrolyze the pectic substances, generally existing in plants. The enzymes are widely circulated in higher plants and microorganisms. They are of primary significance for plants as they support cell wall extension (Ward et al. 1989) and softening of some plant tissues (Aguilar and Huitrón 1990). They also support in keeping ecological balance by causing decomposition and recycling of waste plant materials. Plant pathogenicity and spoilage of fruits and vegetables by rotting are some other foremost indicators of pectinolytic enzymes (Fraissinet-Tachet and Fevre 1996; Palomäki and Saarilahti 1997; Perombelon and Kelman 1980). Pectin substance is the generic name used for the compounds that are acted upon by the pectinolytic enzymes. They are present as the major components of middle lamella between the cells in the form of calcium pectate and magnesium pectate (Rastogi 1998). The enzymes hydrolyzing these pectic substances are broadly known as pectinases, and include polygalacturonases, pectin esterases, pectin lyases, and pectatelyases, depending on their mode of action (Alkorta et al. 1998).

*Aspergillus niger* is the most commonly used fungal species for industrial production of pectinolytic enzymes (Barnby et al. 1990; Naidu and Panda 1998). Alkaline pectinases are produced predominantly from the genus *Bacillus* (Cao et al. 1992; Kapoor and Kuhad 2002), and *Pseudomonas* sp. (Hayashi et al. 1997), although there have also been several reports on alkaline pectinase production from actinomycetes (Beg et al. 2000), and fungi (Baracat-Pereira et al. 1993). The acidophilic pectinases have extensive applications in the extraction and clarification of fruit juices and wine (Alkorta et al. 1998; Pretel et al. 1997). Alkaline pectinases are being used for the pretreatment of wastewater from vegetable food processing industries containing pectinacious material (Tanabe et al. 1987), and processing and degumming of plant fibers such as ramie (*Boehmeria nivea*), sunn hemp (*Crotalaria juncea*), buel (*Grewia optiva*), flax (*Linum usitatissimum*), and jute (*Chorchorus capsularis*) (Brühlmann et al. 2000; Cao et al. 1992; Henriksson et al. 1999; Kashyap et al. 2001b), as well as depolymerizing and debarking (Viikari et al. 2001). Psychrophiles are microbes exhibiting optimal growth at 15 °C or lower temperatures (Morita 1975), and have attracted attention as sources of enzymes with potential for low-temperature catalysis. The yeast strains were isolated from soil from forest in Abashiri (Hokkaido, Japan). The isolated strains were able to grow on pectin at below 5 °C, and showed the activities of several cold-active pectinolytic enzymes. The sequences of 28S rDNA D1/D2 of strains pectinolytic–psychrophilic yeast (PPY) strains PPY-3, 4, 5 and 6, PPY-3 and 4 indicated a taxonomic affiliation to *Cryptococcus cylindricus* and *Mrakia frigida*, respectively, strains PPY-5 and 6 belonged to *Cystofilobasidium capitatum*. It is possible that the cold-active pectinolytic enzymes from the isolated strains can be applied to the food industry, e.g., the clarification of fruit juice below 5 °C (Nakagawa et al. 2004). Gummadi et al. (2007), reported *Debaryomyces nepalensis* NCYC 3413 isolated from rotten apple was studied for its halotolerance. The specific growth rate of *D. nepalensis* was not affected by KCl, even up to a concentration of 1 M, whereas NaCl and LiCl affected the growth of *D. nepalensis*. Pectinase production by *D. nepalensis* was noted at all high salt concentrations, namely, 2 M NaCl, 2 M KCl, and 0.5 M LiCl, and the maximum specific activity was observed when the strain was grown in 2 M NaCl.

Over the years, pectinases have been used in several conventional industrial processes, such as textile, plant fiber processing, tea, coffee, oil extraction, treatment of industrial wastewater, containing pectinacious material, etc. They have also been reported to work on purification of viruses (Salazar and Jayasinghe 1999) and in making of paper (Ricard and Reid 2004). The largest industrial application of pectinases is in fruit juice extraction and clarification. Treatment of fruit pulps with pectinases also showed an increase in fruit juice volume from banana, grapes, and apples (Kaur et al. 2004). Pectinases have been used in conjunction with amylases, lipases, cellulases, and hemicellulases to remove sizing agents from cotton in a safe and ecofriendly manner, replacing toxic caustic soda used for the purpose earlier (Hoondal et al. 2002). Biotechnological degumming using pectinases in combination with xylanases presents an ecofriendly and economic alternative (Kapoor et al. 2001).

### 16.4.4 Cellulase

Cellulases are enzymes that hydrolyze the α-1,4-glucosidic linkages of celluloses. Cellulases hydrolyze the α-1,4-D-glucan linkages in cellulose and yield primary products such as glucose, cellobiose, and cello-oligosaccharides. Endo-glucanases generate nicks in the cellulose polymer revealing reducing and non-reducing ends. Cellobiohydrolases acts upon these reducing and non-reducing ends to liberate cello-oligosaccharides and cellobiose units (Sukumaran et al. 2005). Cellulases are currently the third largest industrial enzyme worldwide, because of their wide applications in cotton processing; paper recycling, in juice extraction, as detergent enzymes and animal feed additives. However, cellulases may develop into the largest volume industrial enzyme, if ethanol from lignocellulosic biomass through enzymatic route becomes a major transportation fuel. Ethanol from lignocellulosic biomass seems hopeful, as the raw material is ubiquitous, abundant, and could play a major role in reduction of greenhouse gases. There are several companies involved in production of cellulase for textile, detergent, paper industries, and other industries. In the production of cellulase microbes has been used including aerobic and anaerobic bacteria (Kumar et al. 2004; Thirumale et al. 2001; Wang et al. 2009); white rot and soft rot fungi (Lo et al. 2010; Shrestha et al. 2009; Tanaka et al. 2009) and anaerobic fungi (Dashtban et al. 2009; Ljungdahl 2008). Most of the cellulases employed for industrial uses are from filamentous fungi such as *Trichoderma, Penicillium, Fusarium, Humicola,* and *Phanerochaete* (Bak et al. 2009; de Siqueira et al. 2010; Fang et al. 2010; Sukumaran et al. 2009).

Aerobic, thermophilic, and cellulolytic microorganisms include several species of fungi (Loginova et al. 1983) and a few species of filamentous bacteria belonging to the family Actinomycetaceae (Zeikus 1979). Among the fungi, *Myceliophthora thermophila* is of significant interest; it develops well on cellulose in submerged culture at 50 °C and pH 4.5 and synthesizes cellulolytic enzymes (Loginova et al. 1983). Thus, the organism is acidophilic, as well as cellulolytic and thermophilic. Among the thermophilic, aerobic bacteria, only a few actinomycetes are actively cellulolytic, mostly *Thermomonospora curvata* (Stutzenberger 1979) and *Thermoactinomyces cellulosae* (Bellamy 1977; Su and Paulavicius 1975). Mohagheghi et al. (1986), reported the isolation of 12 isolates of thermophilic, acidophilic, cellulolytic bacteria from three different primary enrichment cultures from acidic hotsprings at Yellowstone National Park, Wyoming. The three isolates showed the highest cellulolytic activity. Cellulase producing psychrotrophic microorganisms have been isolated from various environments. Cold-adapted microbial strains have been isolated generally from the Antarctica and Polar Regions, which characterize permanently cold environments (Yadav et al. 2016, 2017a,d). Other probable sources of cold-active cellulases are the mud and deep-sea sediment microorganisms. Cellulose-producing psychrotrophic bacteria belonging to the genera *Paenibacillus* and *Pseudomonas* have been isolated from the cold environments in the Western Himalaya. The cellulase of *Acremonium alcalophilum* showed maximal activity at 40 °C and pH 7.0, the enzyme was cold active, holding more than 20% activity even at 0 °C (Hayashi et al. 1996).

Microbial enzymes are often more useful than enzymes from plants and animals because of the great variety of catalytic activities available, higher yields, possible ease of genetic manipulation, and regular supply. Microbial enzymes are also more stable than plant and animal enzymes, and their production is comparatively more suitable and harmless (Kasana and Gulati 2011). Since the early 1990s, cellulases have been integrated into detergents. During textile washing, cellulases remove cellulose microfibrils formed during manufacturing and washing of the cotton-based cloth (Ito 1997).

### 16.4.5 Xylanases

Xylanases are glycosidases that comprise endo-1,4-b-xylanase and β-xylosidase and catalyzing the endohydrolysis of 1,4-b- D-xylosidic linkages in xylan. These enzymes basically cause the hydrolysis of the xylan present in hemicelluloses of plants, converting them into the monomeric sugars. The function is performed with the assistance of certain other hydrolytic enzymes such as acetyl xylan esterase, α-L-arabinofuranosidase, α-glucuronidase, and phenolic acid including ferulic and p-coumaric acid esterase (Collins et al. 2005; Thomas et al. 2017). They were originally called pentosanases, and in 1961 were recognized by the International Union of Biochemistry and Molecular Biology (IUBMB) and assigned code EC 3.2.1.8. They have been officially named as endo-1,4-β-xylanase, but various other terms are also used, including β –xylanase, β –1,4-xylanase, 1,4- β -D-xylan-xylanohydrolase, endoxylanase, endo-1,4- β b-D-xylanase, and xylanase (Collins et al. 2005).

On the basis of the similarities of amino acid sequence, as well as hydrophobic cluster analysis, xylanases is mainly categorized into two glycosyl hydrolase families, including family 10 (>30 kDa and low pI values) and family 11(<30 kDa and high pI values), although there are other families such as 5, 7, 8, and 43, which also consist of some xylanolytic enzymes (Thomas et al. 2017). The major substrate of xylanases is xylan, which is the major structural polysaccharide of plant cells, and also the second-most abundant polysaccharide in nature, it accounts for approximately one-third of all renewable organic carbon on earth (Collins et al. 2005; Prade 1996). Xylanases have a lot of applications in animal feed, baking industries, biofuels, food and feed, and de-inking as well as the paper and pulp industries (Polizeli et al. 2005; Thomas et al. 2017). Microbes are important source of the xylanases for industries because they offer a lot of advantages including controlled production in closed systems, fast multiplication, cost-effective production, they are easy to handle, can utilize cheap sources of carbon, require less space, are capable of biodegradation (Thomas et al. 2017). The microbial sources of xylanases include *Actinomadura, Aspergillus, Aureobasidium, Bacillus, Caldicellulosiruptor, Cellulomonas, Cerrena, Chaetomium, Clostridium, Coprinus, Cryptococcus, Cryptococcus, Dictyoglomus, Flavobacterium, Humicola, Melanocarpus, Nonomuraea, Paenibacillus, Pseudoalteromonas, Pyrodictium, Rhizopus, Rhodothermus, Streptomyces, Sulfolobus, Talaromyces, Thermoascus, Thermococcus, Thermomonospora, Thermotoga* and *Trichosporon* (Tables 16.1 and 16.2).

Xylanases have also been obtained from microbes of extreme environments such as thermophiles including *Bacillus stearothermophilus, Caldicellulosiruptor* sp., *Chaetomium thermophilum, Clostridium thermocellum, Dictyoglomus thermophilum, Geobacillus thermoleovorans, Lichtheimia ramose, Malbranchea cinnamomea, Nonomuraea flexuosa, Paecilomyces varioti, Rhodothermus marinus, Thermoascus*

**Table 16.1** Extremophilic microbes with multifarious extremozymes production activities.

| Extremophiles | Amylase | Lipase | Pectinase | Protease | β-Galactosidase | β-glucosidase | Cellulase | Laccase | Chitinase | Xylanase |
|---|---|---|---|---|---|---|---|---|---|---|
| Psychrophiles | | | | | | | | | | |
| *Bacillus amyloliquefaciens*, IARI-R-25 | − | + | + | + | − | + | + | − | − | + |
| *Bacillus baekryungensis*, IARI-ABL-36 | + | − | − | + | − | + | + | − | − | + |
| *Bacillus cereus*, IARI-L-73 | + | − | + | + | − | + | − | − | − | − |
| *Bacillus firmus*, IARI-L-21 | + | − | + | + | + | + | − | − | + | − |
| *Bacillus flexus*, IARI-ABL-37 | + | − | + | + | − | + | + | − | − | + |
| *Bacillus licheniformis*, IARI-ABL-38 | + | − | + | + | − | + | − | − | − | − |
| *Bacillus marisflavi*, IARI-ABL-39 | + | − | + | + | + | + | + | − | − | + |
| *Bacillus mojavensis*, IARI-ABL-40 | + | − | + | + | − | + | − | − | − | − |
| *Bacillus muralis*, IARI-L-74 | − | − | + | + | − | + | − | + | − | − |
| *B. psychrosaccharolyticus*, IARI-R-2 | + | − | + | + | − | − | − | − | + | − |
| *Bacillus pumilus*, IARI-L-54 | − | − | − | − | − | + | + | + | − | − |
| *Bacillus simplex*, IARI-L-118 | − | − | − | + | − | − | − | − | − | − |
| *Bacillus subtilis*, IARI-L-69 | + | − | + | + | − | + | − | − | − | + |
| *Desemzia incerta*, IARI-L-46 | + | − | + | − | − | + | + | − | − | − |
| *Exiguobacterium antarcticum*, IARI-L-70 | + | − | + | + | − | + | − | − | − | + |
| *Exiguobacterium indicum*, IARI-R-137 | + | + | + | + | + | + | + | + | − | + |
| *Exiguobacterium marinum*, IARI-R-40 | + | − | + | + | − | − | + | − | − | − |
| *Exiguobacterium undae*, IARI-L-116 | + | − | + | + | − | + | − | − | − | − |
| *Jeotgalicoccus halotolerans*, IARI-ABR-5 | + | − | − | + | − | + | − | − | − | − |
| *Lysinibacillus fusiformis*, IARI-R-8 | − | − | − | + | − | − | − | − | − | − |
| *Paenibacillus lautus*, IARI-R-27 | + | − | + | + | − | − | − | − | − | − |
| *Paenibacillus terrae*, IARI-L-127 | + | − | + | + | + | + | + | − | + | + |
| *Paenibacillus xylanexedens*, IARI-L-76 | + | − | + | + | − | + | − | − | − | − |
| *Planococcus antarcticus*, IARI-ABL-9 | + | − | − | + | + | − | + | − | − | + |
| *Planococcus donghaensis*, IARI-L-39 | − | + | + | + | − | + | − | + | − | − |
| *Planococcus kocurii*, IARI-ABL-3 | + | − | + | + | − | + | − | − | − | − |
| *Pontibacillus* sp., IARI-ABL-41 | + | − | + | + | + | + | + | − | + | + |
| *Sinobaca beijingensis*, IARI-ABL-18 | + | − | + | − | − | + | − | − | − | − |
| *Sporosarcina aquimarina*, IARI-L-77 | + | − | + | + | − | + | − | − | − | − |
| *Sporosarcina globispora*, IARI-R-111 | + | − | + | + | + | + | + | + | − | − |
| *Sporosarcina psychrophila*, IARI-R-110 | + | − | + | + | − | + | + | − | − | − |
| *Staphylococcus arlettae*, IARI-ABL-33 | + | − | + | + | − | + | − | − | − | − |
| *Staphylococcus cohnii*, IARI-R-29 | + | − | + | + | − | − | − | − | − | − |
| *Staphylococcus xylosus*, IARI-ABR-1 | + | − | + | + | + | + | − | − | − | − |
| *Virgibacillus halodenitrificans*, IARI-ABR-18 | + | − | − | − | − | + | + | − | − | + |

(*continued*)

**Table 16.1** (Continued)

| Extremophiles | Amylase | Lipase | Pectinase | Protease | β-Galactosidase | β-glucosidase | Cellulase | Laccase | Chitinase | Xylanase |
|---|---|---|---|---|---|---|---|---|---|---|
| **Thermophilic Microbes** | | | | | | | | | | |
| *Lysinibacillus* sp. M-2 | − | − | − | − | − | − | + | − | − | + |
| *Enterobacter cloacae* M-3 | − | − | − | − | − | − | + | − | − | + |
| *Rhodococcus qingshengii* M-9 | − | − | − | − | − | − | + | − | − | + |
| *Paenibacillus pabuli* M-10 | − | − | − | − | − | − | + | − | − | + |
| *Bacillus pumilus* M-14 | − | − | − | − | − | − | + | − | − | + |
| *Micrococcus indicus* M-17 | − | − | − | − | − | − | + | − | − | + |
| *Pseudomonas fragi* M-20 | − | − | − | − | − | − | + | − | − | + |
| *Bacillus amyloliquefaciens,* NBM82 | + | − | − | + | − | − | − | − | − | − |
| *Bacillus licheniformis,* NBM60 | + | − | − | + | − | − | − | − | − | − |
| *Bacillus mycoides,* NBM19 | + | − | − | + | − | − | + | − | − | − |
| *Bacillus nealsonii,* NBM23 | + | − | − | + | − | − | − | − | − | − |
| *Bacillus pumilus,* NBM31 | + | − | − | + | − | − | + | − | − | + |
| *Bacillus* sp., NBM43 | + | − | − | + | − | − | + | − | − | |
| *Bacillus subtilis,* NBM45 | + | − | − | + | − | − | − | − | − | − |
| *Bacillus subtilis,* NBM48 | − | − | − | + | − | − | + | − | − | + |
| *Bacillus thermoamylovorans,* NBM38 | + | − | − | + | − | − | − | − | − | − |
| *Brevundimonas bullata,* NBM35 | − | − | − | − | − | − | + | − | − | + |
| *Geobacillus* sp., NBM49 | − | − | − | + | − | − | + | − | − | + |
| *Paenibacillus dendritiformis,* NBM41 | + | − | − | + | − | − | − | − | − | − |
| *Paenibacillus glycanilyticus,* NBM30 | + | − | − | + | − | − | − | − | − | − |
| *Paenibacillus glycanilyticus,* NBM75 | + | − | − | | − | − | + | − | − | + |
| *Paenibacillus jamilae,* NBM25 | + | − | − | + | − | − | − | − | − | − |
| *Paenibacillus thiaminolyticus,* NBM40 | − | − | − | − | − | − | + | − | − | + |
| *Paenibacillus thiaminolyticus,* NBM71 | + | − | − | + | − | − | + | − | − | + |
| *Planococcus* sp., NBM37 | + | − | − | − | − | − | − | − | − | + |
| *Thermonema lapsum,* NBM28 | + | − | − | + | − | − | − | − | − | − |
| *Bacillus amyloliquefaciens,* NBY37 | + | − | − | + | − | − | − | − | − | + |
| *Bacillus cereus,* NBY23 | + | − | − | − | − | − | − | − | − | + |
| *Bacillus licheniformis,* NBY5 | | − | − | + | − | − | + | − | − | − |
| *Bacillus megaterium,* NBY38 | + | − | − | + | − | − | + | − | − | + |
| *Bacillus pumilus,* NBY4 | + | − | − | + | − | − | + | − | − | + |
| *Bacillus simplex,* NBY1 | − | − | − | + | − | − | + | − | − | − |
| *Bacillus* sp., NBY16 | − | − | − | + | − | − | + | − | − | − |
| *Bacillus subtilis,* NBY44 | + | − | − | + | − | − | + | − | − | + |
| *Bacillus thermoamylovorans,* NBY26 | + | − | − | − | − | − | + | − | − | − |
| *Brevibacillus* sp., NBY28 | + | − | − | − | − | − | + | − | − | − |
| *Burkholderia glathei,* NBY12 | − | − | − | + | − | − | + | − | − | − |

(*continued*)

**Table 16.1** (Continued)

| Extremophiles | Amylase | Lipase | Pectinase | Protease | β-Galactosidase | β-glucosidase | Cellulase | Laccase | Chitinase | Xylanase |
|---|---|---|---|---|---|---|---|---|---|---|
| *Paenibacillus thiaminolyticus*, NBY3 | + | – | – | + | – | – | – | – | – | + |
| *Thermoactinomyces* sp., NBY60 | + | – | – | + | – | – | – | – | – | |
| *Thermobacillus* sp., NBY36 | – | – | – | + | – | – | – | – | – | + |
| **Halophiles** | | | | | | | | | | |
| *Halobacillus* sp., SL3 | – | – | – | + | – | – | – | – | – | |
| *Bacillus marisflavi*, SL7 | – | – | – | + | – | – | – | – | – | – |
| *Halobacillus trueperi*, SL10 | + | – | – | + | – | – | + | – | – | + |
| *Bacillus baekryungensis*, SL13 | – | – | – | – | – | – | – | – | – | + |
| *Thalassobacillus devorans*, SL14 | – | – | – | – | – | – | – | – | – | + |
| *Virgibacillus* sp., SL15 | – | – | – | – | – | – | – | – | – | + |
| *Bacillus aquimaris*, SL17 | – | – | – | + | – | – | + | – | – | + |
| *Bacillus licheniformis*, SL20 | – | – | – | + | – | – | + | – | – | + |
| *Bacillus flexus*, SL21 | + | – | – | – | – | – | + | – | – | |
| *Bacillus* sp. SL23 | + | – | – | + | – | – | – | – | – | + |
| *Exiguobacterium* sp. SL34 | + | – | – | + | – | – | – | – | – | |
| *Sediminibacillus halophilus*, SL35 | + | – | – | + | – | – | + | – | – | + |
| *Thalassobacillus* sp. SL37 | + | – | – | – | – | – | + | – | – | – |
| *Bacillus selenatarsenatis*, SL39 | + | – | – | + | – | – | + | – | – | + |
| *Oceanobacillus picturae*, SL42 | + | – | – | – | – | – | – | – | – | + |
| *Alkalibacterium* sp., SL109 | + | – | – | – | – | – | + | – | – | + |
| *Amphibacillus* sp., SL117 | + | – | – | – | – | – | – | – | – | |
| *Halomonas* sp., SL26 | + | – | – | + | – | – | + | – | – | + |
| *Halomonas salifodinae*, SL40 | – | – | – | – | – | – | – | – | – | + |
| *Marinobacter*, SL101 | + | – | – | – | – | – | – | – | – | + |
| *Nitrincola* sp., SL111 | – | – | – | – | – | – | – | – | – | + |
| *Halomonas variabilis*, SL121 | + | – | – | – | – | – | – | – | – | – |
| *Halomonas campisalis*, SL125 | – | – | – | – | – | – | – | – | – | – |
| *Halomonas salina*, SL128 | + | – | – | + | – | – | – | – | – | – |
| *Salicola marasensis*, SL131 | + | – | – | + | – | – | + | – | – | + |
| *Halomonas caseinilytica*, SL136 | + | – | – | – | – | – | + | – | – | + |
| *Halomonas aquamarina*, SL139 | – | – | – | – | – | – | – | – | – | + |
| *Halomonas muralis*, SL141 | + | – | – | – | – | – | + | – | – | + |
| *Halomonas meridian*, SL145 | – | – | – | – | – | – | – | – | – | + |
| *Alcaligenes* sp., SL107 | + | – | – | + | – | – | – | – | – | + |
| *Streptomyces viridiviolaceus*, SL2 | – | – | – | – | – | – | + | – | – | + |
| *Streptomyces radiopugnans*, SL19 | – | – | – | – | – | – | – | – | – | + |

(*continued*)

**Table 16.1** (Continued)

| Extremophiles | Amylase | Lipase | Pectinase | Protease | β-Galactosidase | β-glucosidase | Cellulase | Laccase | Chitinase | Xylanase |
|---|---|---|---|---|---|---|---|---|---|---|
| *Bacillus flexus*, PL 1 | – | – | – | + | – | – | – | – | – | – |
| *Bacillus megaterium*, PL 5 | – | – | – | + | – | – | – | – | – | + |
| *Bacillus flexus*, PL 4 | + | – | – | – | – | – | – | – | – | – |
| *Bacillus thuringiensis*, PL 10 | + | – | – | + | – | – | – | – | – | – |
| *Bacillus aquimaris*, PL 29 | + | – | – | – | – | – | – | – | – | – |
| *Halomonas shengliensis*, PL 207 | + | – | – | – | – | – | + | – | – | – |
| *Bacillus altitudinis*, PL 14 | – | – | – | + | – | – | – | – | – | + |
| *Halomonas salina*, PL 33 | – | – | – | – | – | – | + | – | – | – |
| *Halomonas salifodinae*, PL 43 | + | – | – | – | – | – | – | – | – | + |
| *Halomonas pacifica*, PL 51 | – | – | – | + | – | – | + | – | – | – |
| *Halomonas salina*, PL 205 | – | – | – | + | – | – | – | – | – | – |
| *Bacillus granadensis*, PL 12 | – | – | – | + | – | – | + | – | – | – |
| *Bacillus thuringiensis*, PL A | – | – | – | – | – | – | + | – | – | + |
| *Halomonas shengliensis*, PL 16 | + | – | – | – | – | – | – | – | – | – |
| *Virgibacillus* sp., PL 31 | + | – | – | + | – | – | – | – | – | + |
| *Bacillus horikoshii*, PL 23 | + | – | – | + | – | – | – | – | – | + |
| *Halomonas aquamarina*, PL 34 | – | – | – | – | – | – | – | – | – | – |
| *Bacillus horikoshii*, PL 42 | – | – | – | – | – | – | – | – | – | + |
| *Halomonas shengliensis*, PL 15 | – | – | – | – | – | – | – | – | – | – |
| *Salimicrobium halophilum*, PL F | + | – | – | – | – | – | – | – | – | – |
| *Bacillus boroniphilus*, PL 68 | – | – | – | + | – | – | – | – | – | – |
| *Bacillus boroniphilus*, PL 69 | + | – | – | + | – | – | – | – | – | – |
| *Alkalibacillus* sp. PL 203 | – | – | – | + | – | – | – | – | – | + |
| *Virgibacillus halonitrificans*, PL 18 | + | – | – | + | – | – | – | – | – | + |
| *Halomonas pacifica*, PL 36 | – | – | – | + | – | – | – | – | – | – |
| *Rummeliibacillus stabekisii*, PL 71 | + | – | – | – | – | – | – | – | – | – |
| *Halomonas halophila*, PL D | + | – | – | + | – | – | – | – | – | – |
| *Halobacillus truperi*, PL 202 | – | – | – | + | – | – | + | – | – | + |
| *Halobacillus* sp., PL 209 | – | – | – | – | – | – | + | – | – | + |

Sources: **Psychrophiles**: (Coker and Brenchley 2006; Collins et al. 2002; Ferrer et al. 2005a,b; Hu et al. 2007; Lee et al. 2010; Luo et al. 2006; Okuda et al. 2004; Sakamoto et al. 2003; Secades et al. 2003; Singh et al. 2016; Yadav 2015a, 2016, 2017c,d; Yazdi et al. 2008; Yoshimune et al. 2005; Zeng et al. 2003); **Thermophiles**: (Andrade et al. 2001; Andronopoulou and Vorgias 2003; Cannio et al. 2004; Fang et al. 2004; Fukushima et al. 2005; Gao et al. 2003; Gomes et al. 2003; Hatada et al. 2005; Hotta et al. 2002; Hutcheon et al. 2005; Kang et al. 2007; Kim et al. 2001; Lorentzen et al. 2006; Palmieri et al. 2006; Sahay et al. 2017; Saxena et al. 2016; Silva et al. 2005; Suman et al. 2015; Sutrisno et al. 2004; Tatur et al. 2006; Taylor et al. 2004); **Halophiles**: (Amoozegar et al. 2003; Deutch 2002; Jeon et al. 2009; Mijts and Patel 2002; Pérez-Pomares et al. 2003; Sahay et al. 2011, 2012; Sanchez-Porro et al. 2003; Wainø and Ingvorsen 2003; Wang et al. 2010; Yadav and Saxena 2018); **Acidophiles/Alkaliphiles**: Chang et al. 2004; Haddar et al. 2009; Hagihara et al. 2001; Hashim et al. 2004; Kanekar et al. 2002; Mitsuiki et al. 2002; Miyazaki 2005; Pretz et al. 2005; Shi et al. 2006; Solingen et al. 2001; Zeng et al. 2006).

**Table 16.2** Production of extremozymes by extremophilic microbes.

| Extremophiles | Extremozymes | Temp | pH | References |
|---|---|---|---|---|
| **Alkaliphiles** | | | | |
| *Bacillus halodurans* | Amylase | 55–65 | 10.0 | Hashim et al. (2004) |
| *Burkholderia cepacia* ST-200 | Cholesterol | 45 | 12.0 | Shi et al. (2006) |
| *Streptomyces* sp. | Endocellulase | 50 | 8.0 | Solingen et al. (2001) |
| *Bacillus* sp. JAMB-750 | Mannase | 55 | 10.0 | Zeng et al. (2006) |
| *Bacillus alcalophillus* | Pectate lyase | 45 | 9–10 | Zhai et al. (2003) |
| *Nesterenkonia* sp. AL-20 | Protease | 74 | 10.0 | Bakhtiar et al. (2003) |
| *Bacillus pumilus* | Protease | 50–60 | 10 | Kumar (2002) |
| *Arthrobacter ramosus* | Protease | 65 | 10–11 | Kanekar et al. (2002) |
| *Nocardiopsis* sp. | Protease | 70–75 | 10–11 | Mitsuiki et al. (2002) |
| *Bacillus mojavensis* A21 | Protease | 60 | 10 | Haddar et al. (2009) |
| *Bacillus subtilis*DM-04 | Protease | 45 | 10.5 | Rai et al. (2009) |
| *Bacillus halodurans* | Pullulanase | 55–65 | 10.0 | Hashim et al. (2004) |
| *Thermus thermophilus* | Sulfite oxidase | 60 | 11.0 | Pretz et al. (2005) |
| *Bacillus firmus* | Xylanases | 70 | 5.0–9.5 | Chang et al. (2004) |
| *Bacillus* KSM-K38 | α-Amylase | 55–60 | 8.0–9.0 | Hagihara et al. (2001) |
| *Nocardiopsis* sp. F96 | β-Glucanase | 70 | 9.0 | Miyazaki (2005) |
| *Bacillus subtilis* RM-01 | β-Keratinase | 50 | 8.0 | Rai et al. (2009) |
| **Psychrophiles** | | | | |
| *Ferroplasma acidiphilum* | 2α- Glucosidases | 50 | 1.5 | Golyshina et al. (2006) |
| *Pseudomonas strain* DY-A | Alkaline protease | 60 | 2.5–4.0 | Golyshina et al. (2006) |
| *Vibrio salmonicida* | Catalase | 92 | 3.5–9.5 | Hu et al. (2007) |
| *Bacillus subtilis* A-53 | Cellulase | 35 | 6.8 | Lee et al. (2010) |
| *Pseudoalteromonas* sp. DY3 | Cellulase | 25 | 5.3–8.0 | Collins et al. (2002) |
| *Shewanella* sp. Ac10 | DnaK protein | 10–40 | 7.5 | Yoshimune et al. (2005) |
| *Penicillium chrysogenum* | Endo-arabinanase | 40–45 | 10–11.0 | Okuda et al. (2004) |
| *Rhodotorula mucilaginosa* | Esterase | 20 | | Luo et al. (2006) |
| *Psychrobacter* sp. Ant300 | Esterase | 30–40 | 6.0–7.0 | Sakamoto et al. (2003) |
| *Ferroplasma acidiphilum* | Esterase | 40 | 7.5 | Yazdi et al. (2008) |
| *Acinetobacter* sp. | Esterase | 50 | 7.8 | Suzuki et al. (2001) |
| *Acinetobacter* sp. | Lipase | 20 | 7.0 | Suzuki et al. (2001) |
| *Psychrobacter okhotskensis* | Lipase | 40 | 10.0 | Zeng et al. (2003) |
| *Curtobacterium luteum* | Metalloprotease | 24 | 6.5 | Secades et al. (2003) |
| *F. psychrophilum* | Metalloprotease | 50 | 6.0 | Secades et al. (2003) |
| *Cryptococcus cylindricus* | Pectinase | 40–60 | 3.0–11.0 | Ferrer et al. (2005b) |
| *Bacillus* sp. | Subtilisin | 60 | 2.5–3.0 | Ferrer et al. (2005a) |
| *Ferroplasma acidiphilum* | α- Glucosidases | 0–30 | 7.5 | Zimmer et al. (2006) |
| *Arthrobacter* sp. SB | β-Galactosidase | 0–10 | 4.0 | Nakagawa et al. (2006) |
| *Arthrobacter* sp. C2–2 | β-Galactosidase | 45–50 | 5.0–5.0 | Gomes et al. (2003) |

(*continued*)

**Table 16.2** (Continued)

| Extremophiles | Extremozymes | Temp | pH | References |
|---|---|---|---|---|
| **Thermophiles** | | | | |
| *Rhodothermus marinus* | Amylase | 85 | 6.5 | Gomes et al. (2000) |
| *Pyrobaculum calidifontis* | Carboxyl esterase | 90 | 7 | Gao et al. (2003) |
| *Pyrococcus horikoshii* | Cellulase | 100 | 5.6 | Fukushima et al. (2005) |
| *Pyrococcus furiosus* | Chitinase | 95 | 6 | Hutcheon et al. (2005) |
| *Thermoplasma acidophilum* | Chitinase | 70 | 5 | Taylor et al. (2004) |
| *Picrophilus torridus* | Chitinase | 70 | 7 | Sutrisno et al. (2004) |
| *A. acidocaldarius* | Endoglucanase | 80 | 4 | Hatada et al. (2005) |
| *Pyrococcus horikoshii* | Endoglucanase | 95 | 7 | Andrade et al. (2001) |
| *Sulfolobus solfataricus* | Esterase | 95 | 6 | Silva et al. (2005) |
| *Pyrococcus furiosus* | Ferritin | 120 | 7 | Gomes et al. (2003) |
| *Thermococcus chitonophagus* | L-Aminoacylase | 85 | 8 | Palmieri et al. (2006) |
| *Pyrococcus horikoshii* | L-Threonine | 100 | 10 | Tatur et al. (2006) |
| *Pyrodictium abyssi* | Xylanase | 105 | 6 | Lorentzen et al. (2006) |
| *Sulfolobus solfataricus* | Xylanase | 100 | 7 | Kang et al. (2007) |
| *Geobacillus* sp. | α-amylase | 80 | 5 | Kim et al. (2001) |
| *Methanococcus jannaschii* | α-amylase | 120 | 5–8 | Hotta et al. (2002) |
| *Sulfolobus shibatae* | α-Glucosidase | 120 | 4.5 | Cannio et al. (2004) |
| *Thermococcus litoralis* | α-Glucosidase | 98 | 5.5 | Fang et al. (2004) |
| **Halophiles** | | | | |
| *Lactobacillus reuteri* | Glutaminase | 40 | 7.5 | Jeon et al. (2009) |
| *Pseudalteromonas* sp. CP76 | Protease CPI | 55 | 8.5 | Sanchez-Porro et al. (2003) |
| *Bacillus* sp. NTU-06 | Xylanase | 40 | 8.0 | Wang et al. (2010) |
| *Halothermothrix orenii* | α-Amylase | 65 | 7.5 | Mijts and Patel (2002) |
| *Bacillus dipsosauri* | α-Amylase | 60 | 6.5 | Deutch (2002) |
| *Halobacillus* sp. MA-2 | α-Amylase | 50 | 7.5–8.5 | Amoozegar et al. (2003) |
| *Haloferax mediterranei* | α-Amylase | 50–60 | 7–8 | Pérez-Pomares et al. (2003) |
| *Halorhabdus utahensis* | β-Xylanase | 55.70 | — | Wainø and Ingvorsen (2003) |

*aurantiacus, Thermobifida alba, Thermomyces lanuginosus, Thermotoga maritime, Thermotoga petrophila, Thermotoga* sp., *Thielavia terrestris* (Table 16.1); psychrophiles such as *Alternaria alternate, Clostridium* sp., *Coprinus psychromorbidus, Cryptococcus adeliae, Flavobacterium frigidarium, Penicillium* sp., *Phoma* sp., *Pseudoalteromonas haloplanktis* (Table 16.2).

Xylanases can be produced either by solid-state fermentation or SmF. SSF has certain advantages over SmF in terms of nature of the substrates, requirement of aeration, energy and water, physical space, etc. Moreover, SSF provides an environment that is close to the natural habitat of fungi, and thus they can easily be stimulated to produce more and more xylanolytic enzymes. More than 75% of the industrial enzymes are produced by SmF due to a lack of bioreactors that are required for the production under SSF (Thomas et al. 2017). The production of xylanases requires a consistent distribution of nutrients and regular supply of oxygen in the medium. Furthermore, the carbon sources play an important role in production of the enzymes. Various cheap sources of carbon are available and have been utilized for the production of the enzymes including apple, banana waste, citrus fruits, corncob, corn stover, potato, rice straw, soy oil cake, sugarcane bagasse, tea, wheat bran (Archana and Satyanarayana 1997; Thomas et al. 2017). Different inorganic and organic sources of nitrogen used for the production of xylanases include ammonium chloride, ammonium sulfate, sodium nitrate, corn steep solids, peptone, soybean meal, tryptone, yeast extract (Thomas et al. 2017). Finally, purification is done using different chromatographic techniques including adsorption chromatography, electroelution, expanded-bed, hydrophobic interaction column chromatography, ion exchange chromatography and among all ion exchange chromatography is usually preferred (Dos Santos et al. 2002; Thomas et al. 2017). The presence of xylanase is qualitatively tested by Congo red plate assay and Remazol brilliant blue xylan plates assay (Thomas et al. 2017). Yang et al. (2006), reported the production of extracellular xylanase from thermophilic fungus, *Paecilomyces themophila* J18, by SSF on wheat straw by optimizing certain conditions, including growth temperature, initial moisture level, initial pH of the culture, particle size of wheat straw, nitrogen source and yield up to 18 580 U $g^{-1}$ of carbon source of xylanase has been achieved. Pirota et al. (2013) produced xylanase from *A. oryzae* (P6B2) isolated from the Amazon Rain Forest by SSF using wheat bran as solid substrate and by maintaining proper operational conditions yield up to 2830.7 IU $g^{-1}$ has been achieved. Izidoro and Knob (2014), obtained xylanases from *A. niger* using various agro-industrial wastes including apple peel, brewer's spent grain, corn straw, orange peel, passion fruit peel, pineapple peel, soybean peel, soybean waste, sugarcane bagasse, sugarcane straw, wheat bran. Among all these waste used, brewer's spent grain has been demonstrated to be the best inducer for xylanase production by $3.63 \pm 0.03$ U $mL^{-1}$ and 41.34 U $mg^{-1}$ protein. Marques et al. (2018), used 14 strains of endophytic fungi including *Cladosporium cladosporioides Phomopsis stipata, Trichoderma viridae, Botryosphaeria* sp., *Saccharicola* sp., *Paecilomyces* sp., *Ustilaginoidea* sp. for the production of the cellulases and xylanases by SSF. Among all these *Phomopsis stipata* and *Botryosphaeria* sp. have been demonstrated to be the best xylanase and β-xilosidase producers, yielding up to 694.33 and 4.87 U $g^{-1}$, respectively.

Xylanases have a lot of industrial applications. These are used in baking industry to improve the strength of the gluten and ultimately the quality of the bread as it has potential to absorb water and interact with gluten (Butt et al. 2008; Nuyens et al. 2001). Furthermore, they improve the tolerance of dough to different flour quality and to variations in parameters used during processing. They also make dough softer, reducing the sheeting work requirements and considerably increases the volume of the baked bread (Butt et al. 2008; Harbak and Thygesen 2002). Xylanases are also used in clarifying juices and wines, for extraction of coffee, plant oils and starches, for improvement of the nutritional properties of agricultural silage and grain feed (Butt et al. 2008; Malathi and Devegowda 2001). Sugars like xylose, xylobiose, and xylo oligomers can be prepared by the enzymatic hydrolysis of xylan. The endoxylanase obtained from *A. niger* IBT-90 has been used for improving wheat-rye and whole meal bread. The addition of this xylanase in the range of 500–1000 U $kg^{-1}$ flour improved kneading and increased the loaf volume in all kinds of bread (Butt et al. 2008; Romanowska et al. 2003). Addition of xylaase, XYL22 from *Aspergillus* sp. FP 470, to bread dough resulted in increased loaf volume and also improvement of crumb structure of the baked product (Butt et al. 2008; Hilhorst et al. 1999; Rouau et al. 2006). Cunha et al. (2018), utilized recombinant xylanase produced by *Pichia pastoris* in the bread-making process, the decrease in firmness, stiffness, and consistency, and enhancement of specific volume and reducing sugar content has been recorded. Xylanase is utilized in biscuit-making, to make cream crackers lighter and improve the texture, palatability, and uniformity of the wafers (Harris and Ramalingam 2010; Polizeli et al. 2005). Xylanases, in combination with other enzymes including cellulases, amylases, and pectinases, improves the yield of juices by means of liquefaction of fruit and vegetables; stabilization of the fruit pulp; enhances the recovery of aromas, edible dyes, essential oils, hydrolysis of substances, mineral salts, pigments vitamins, reduces the viscosity, that hamper the physical or chemical clearing of the juice, or which otherwise may lead to the cloudiness in the concentrate (Polizeli et al. 2005). Xylanases are also used to cause the hydrolysis of arabino xylans to lower oligosaccharides, which diminishes the viscosity of beer thereby eliminating its muddy aspect (Harris and Ramalingam 2010). Further, these are also used in feed and food industries. Canola seeds are used as cattle feed after oil extraction. The endoxylanase from *Neocallimastix patriciarum* expressed as a fusion protein with the membrane protein oleosin of the canola seed has been added to feed so as to increase its digestibility in the rumen of cattle thus reducing the feed conversion rate (Juturu and Wu 2012).

### 16.4.6 Lipases

Lipases are triacylglycerol acyl hydrolases involved in the hydrolysis of fats and oils yielding glycerol and free fatty acids. They break down ester bonds by means of catalytic triad composed of serine, histidine, and aspartate (Gaur et al. 2017). These belong to serine hydrolases and do not require any cofactor and are involved in diverse conversion reactions including acidolysis, alcoholysis, aminolysis, esterification, interesterification, and transesterification (Gopinath et al. 2013). Lipases possess the sole feature of acting at the interface between an aqueous and a nonaqueous phase (Aravindan et al. 2007). These are secreted by bacteria, yeasts, protozoans, and molds, and even viruses are known to encode genes for lipases (Gaur et al. 2017). Extracellular secretion of lipases has been well demonstrated in fungi, chiefly in zygomycetes, hyphomycetes, (Gopinath et al. 2013). The production of lipase has also been known in certain ascomycetes, coelomycetes (Gopinath et al. 2013). Lipolytic activity has been demonstrated in *Aspergillus* sp., *Acremonium strictum, Candida rugosa, Cunninghamella verticillata, Humicola lanuginose, Lipomyces starkeyi, Mucor* sp., *Pencillium* sp., *Rhizopus* sp., yeasts including *Candida antarctica, C. rugosa, Pichia burtonii, Rhodotorula glutinis Saccharomyces lipolytica, Yarrowia lipolytica* and bacterial sources of lipases include *Acinetobacter* sp. *B. coagulans, Bacillus pumilus, B. stearothermophilus, B. subtilis, Bacillus thermocatenulatus, Burkholderia glumae, Chromobacterium viscosum, Lactobacillus* sp., *Pseudomonas aeruginosa, Pseudomonas cepacia* further they have also been obtained from actinomycetes including *Streptomyces fradiae, Streptomyces* sp. Microbial lipases are of commercial importance due to broader availability, greater stability and low production cost as compared to plant and animal lipases (Tables 16.1 and 16.2).

The lipases have also been obtained from microbes of extreme environment such as from thermophilic *Bacillus thermoleovorans* ID-1 (Lee et al. 1999), from a thermophilic *Rhizopus oryzae* (Hiol et al. 2000), *Bacillus* sp. J33 (Nawani and Kaur 2000), thermostable lipases from thermophilic bacterium *B. thermoleovorans* ID-1 (Lee et al. 2001) thermostable alkaline lipase from thermophilic *Bacillus* sp. RSJ-1 (Sharma et al. 2002), from *Geobacillus* sp. TW1 (Li and Zhang 2005), thermophilic and alkaliphilic *B. coagulans* BTS-3 (Kumar et al. 2005), thermo-alkaline lipase thermophile *Geobacillus thermodenitrificans* (AV-5) (Christopher et al. 2015) and thermophilic alkaline lipases from *Thermosyntropha lipolytica*, (Moh'd and Wiegel 2007).

The microbes from cold habitats exhibited the lipase activity and reported from different psychrophiles, including *Acinetobacter calcoaceticus* LP009 (Pratuangdejkul and Dharmsthiti 2000), facultative psychrophile *Psychrobacter okhotskensis* (Yumoto et al. 2003), alkaline lipase from psychrotrophic *Acinetobacter* sp. (Kasana et al. 2008), cold-adapted organic solvent tolerant alkalophilic lipase from an Antarctic *Pseudomonas* sp. (Ganasen et al. 2016), and organic solvent-tolerant and cold-adapted lipase from *Psychrobacter* sp. ZY124 (Zhang et al. 2018). Further, psychrophilic lipase has been obtained from *Pseudomonas fluorescens* (Luo et al. 2006), cold-adapted lipase from *Photobacterium lipolyticum* (Ryu et al. 2006), and psychrophilic alkaline lipase from *Pseudomonas* sp. (MSI057) (Kiran et al. 2008).

The lipase production has been also reported from microbes with two or more abiotic stress conditions such as thermophilic lipase from *B. licheniformis* Ht7v (Ugras 2017), halotolerant and thermotolerant lipase from a *Brevibacterium halotolerans* PS4 (Sharma et al. 2017), an alkaline and surfactant-tolerant lipase from *Trichoderma lentiforme* (ACCC30425) (Wang et al. 2018), cold-active lipases from *Cryptococcus* sp. Y-32 and *Rhodococcus erythropolis* N149 (Maharana and Singh 2018).

The production of the microbial lipases is mostly carried out by submerged culture, but SSF is also used. Most lipases are extracellular enzymes, and production is directly correlated with microbial growth, and the maximum production is achieved during the late log phase; additionally, medium composition as well as the fermentation conditions plays a vital role (Gaur et al. 2017). Lipid-rich sources including detergents, fatty acids, glycerol hydrolyzable esters, oil, and triacylglycerols are known to stimulate lipase production in the bacterial culture broth (Gaur et al. 2017) and nitrogen sources including peptone and yeast extract (Gaur et al. 2017), whereas inorganic sources known to induce the production of lipases include ammonium chloride and diammonium hydrogen phosphate (Gaur et al. 2017). Further, divalent cations also affect the lipase production. Calcium and magnesium ions have both been reported to stimulate lipase production in *Burkholderia* sp. (Gaur et al. 2017), and calcium alone in *Bacillus* sp. (Gaur et al. 2017). Other important parameters, including pH, temperature, proper aeration, and agitation, are also important to maintain for production of lipase enzymes. Lipases are purified by hydrophobic-interaction chromatography, ion-exchange chromatography, aqueous two-phase systems, reversed micellar extraction, immune purification, and affinity chromatography (Gaur et al. 2017).

Lipases are used in the food industry. Fats and oils form one of the most important constituents of foods (Ray 2012). Lipase modified butter fat has wide application in various food processes (Aravindan et al. 2007). Chocolates with cocoa butter substitutes, bread, structured lipids like human milkfat replacers, low-calorie health oils, and nutraceuticals are some of the lipase mediated food products (Aravindan et al. 2007).

The commercially prepared lipases are used for the development of flavor in dairy products, in processing of foods including baked products, beer, fruits, meat, milk products, and vegetables (Aravindan et al. 2007). Phospholipases have been used in treatment of egg yolk that is useful for processing of custard, mayonnaise, baby foods, dressings, and in dough preparation, sauces, like Hollandaise, Béarnaise, and cafe de Paris (Aravindan et al. 2007). Lipases are also utilized in hydrolysis of milkfat, acceleration of cheese ripening, and the lipolysis of butter, fat, and cream (Aravindan et al. 2007; Sharma et al. 2001). Further, lipase is also used in enhancing the flavor content of bakery products (Ray 2012). Texture and softness can be improved by lipase catalyzation (Ray 2012). Lipases have also been used for the production of fruit juices and in vegetable fermentation (Ray 2012). Adding lipases to noodles result in appreciably softer textural characteristics (Undurraga et al. 2001).

### 16.4.7 L-Glutaminase

Glutaminase is an amidase enzyme that converts glutamine to glutamate with the release of ammonia. There are two isozymes of glutaminase found in humans and kidney type (GLS1) (Vijayan et al. 2017). A number of studies have demonstrated that GLS2 is expressed not only in the liver but also in other tissues, including brain. However, the ratio of GLS2 to GLS1 is very small in the brain, which indicates that most of the brain glutaminase is GLS1 (Huang et al. 2011).

Glutaminase is also found in axonal terminals of neurons (Vijayan et al. 2017). These can be obtained from microbes, plants, and animal tissues. A number of bacterial as well as fungal strains have been reported to be the source of this enzyme, including *Actinomucorelegans* sp., *Aerobacter aerogenes, A. oryzae, Aspergillus sojae, Candida scottii, Candida utilis, Cryptococcus albidus, Cryptococcus laurentii, Enterobacter cloacae, Erwinia aroideae, Erwinia carotovora, Escherichia coli, Hansenula sp., Klebsiella aerogenes, Proteus morganii, Proteus vulgaris, Providencia* sp., *P. aeruginosa, Pseudomonas aurantiaca, Pseudomonas aureofaciens, P. fluorescens, Rhodotorula* sp., *Serratia marcescens, Streptomyces enissocaesilis, Streptomyces griseus, Streptomyces* sp., *Torulopsis candida, Vibrio* sp., *Xanthomonas juglandis* (Tables 16.1 and 16.2). Furthermore, a salt-tolerant and thermostable glutaminase has been obtained from *Cryptococcus nodaensis* (Sato et al. 1999). Kumar et al. (2012) obtained halotolerant, thermostable alkaline L-glutaminase from a *Bacillus* sp. isolated from Gangotri region of Uttarakhand Himalaya. Ye et al. (2013) obtained salt-tolerant and acidity-tolerant glutaminase from *B. amyloliquefaciens*.

Glutaminases from microbial sources have more stability compared to those from plants and animals (Vijayan et al. 2017), due to which microbial glutaminases are utilized in various therapeutic fields. *E. coli* is among the most studied in this context, though various other microbes have also been utilized for the production of glutaminases including *Pseudomonas* sp., *Acinetobacter* sp., *Bacillus* sp., *Hansenula* sp., *C. utilis, C. scottii, Rhodotorula* sp. (Sarada 2013). It has been produced from fungal strains like *Beauveria bassiana* under SSF. Solid-state fermentation offers certain advantages including high yield, lower capital, and lower wastewater output and has been employed for the production of the enzyme in *P. fluorescens, Vibrio costicola, Vibrio cholerae, Zygosaccharomyces rouxii,* utilizing wheat bran (Sabu et al. 2002; Vijayan et al. 2017). Commercially, glutaminases is produced by SmF. The cultural conditions play an essential role for the production of the glutaminase from microbes. Optimization of various parameters such as aeration rate, concentration of carbon and nitrogen sources, fermentation period, inoculum size, inoculum age, as well as pH, temperature is very essential (Binod et al. 2017). Many microbes utilize L-glutamine as both carbon and nitrogen sources for the production of glutaminase, e.g., *Beauveria* sp., *Pseudomonas* sp., *Providencia* sp., *Streptomyces rimosus, Trichoderma koningii* (Binod et al. 2017; Sabu et al. 2000). It has been found that the addition of glucose enhanced the production of glutaminase (Iyer and Singhal 2008; Binod et al. 2017). Various inorganic as well as organic sources of nitrogen including yeast extract, ammonium sulphate, and urea have been reported to enhance the production.

Furthermore, glutaminase is one of the chief enzymes controlling the taste of fermented foods, including soy sauce, especially in Asian countries (Nandakumar et al. 2003). It increases the flavor and aroma of fermented foods by increasing the glutamic acid content (Löliger 2000; Vijayan et al. 2017). Glutamic acid and aspartic acid, well known as important amino acids, contributes to fine taste, umami and sharp taste, sour simultaneously making contribution to nutritional effects to food (Nandakumar et al. 2003; O'Mahony and Ishii 1986). Salt-tolerant glutaminase obtained from *Stenotrophomonas maltophilia* is extensively used in Japanese soy sauce fermentation (Vijayan et al. 2017; Wakayama et al. 2005).

### 16.4.8 β-Galactosidase

β-D-galactosidase (also referred to as lactase) is an enzyme that catalyze the hydrolysis of β-1,4-D-galactosidic linkages. This enzyme hydrolyzes lactose into glucose and galactose which are sweeter, more soluble, and more digestible carbohydrates than lactose (Anisha 2017 β-D-galactosidase has many applications in the industrial sector including dairy, fermentation, and food industries (Anisha 2017). Though the enzyme can be obtained from plants, animals, and microbes, the higher yields are obtained from microbial sources. It has been produced from bacterial sources, including *Streptococcus thermophilus, Saccharomyces cerevisiae, Propionibacterium acidipropionici, Lactobacillus reuteri, Lactobacillus plantarum, Lactobacillus fermentum, Lactobacillus delbrueckii* subsp. *bulgaricus, Lactobacillus crispatus, Kluyveromyces marxianus, Kluyveromyces lactis, E. cloacae, Bifidobacterium longum, Bifidobacterium infantis, Beijerinckia indica, Bacillus megaterium, Bacillus circulans* and fungal including, *Paecilomyces aerugineus, Penicillium chrysogenum, Penicillium expansum, Aspergillus aculeatus* (Tables 16.1 and 16.2). *B. stearothermophilus, Pyrococcus woesei, Thermus* sp., have been reported to produce thermostable β-galactosidase (Anisha 2017; Chen et al. 2008). Cold-active and cold-adapted β-galactosidase have been reported from psychrophile including *Arthrobacter psychrolactophilus, P. haloplanktis* (Van De Voorde et al. 2014), thermostable β-galactosidase from *A. niger* (Chen et al. 2008), a novel acidophilic intracellular β-galactosidase from *Teratosphaeria acidotherma* (Chiba et al. 2015).

The enzyme is mostly produced by SmF, as it is found in most cases to be produced either intracellular or cell bound, although extracellular β-galactosidase production is also reported (Anisha 2017). Associative growth of various microbial cultures is another technique, which is used in the production of β-galactosidase. (Tari et al. 2009) assessed the growth *S. thermophilus* and *L. delbrueckii* subsp. *bulgaricus* with respect to lactic acid, biomass, and β-galactosidase enzyme production using response surface methodology. The study reported that more β-galactosidase activity could be obtained with associative growth of both strains rather than using them singly.

The enzyme is used chiefly in industrial-scale production of milk and dairy products. Milk and dairy products are treated with lactase to reduce their lactose content. This is an appropriate method for enhancing possible uses and also to deal with the problems of lactose insolubility and lack of sweetness (Panesar et al. 2010). In addition, this treatment could make milk, a suitable food, available to a large number of lactose intolerant adults and children. Further, the hydrolysis of whey converts lactose into beneficial products like sweet syrup, which can be easily utilized in a range of processes such in the baking, confectionary, dairy, and soft-drink industries (Panesar et al. 2010; Pivarnik et al. 1995). The enzyme β-galactosidase could also be used in transglycosylation of lactose for the synthesis of galacto-oligosaccharides (GOSs), which have been known to be nondigestible oligosaccharides, neither hydrolyzed nor absorbed in the upper intestinal tract, and they pass onto the colon where beneficial bacteria in the intestines ferments them (Grosová et al. 2009; Maischberger et al. 2008; Panesar et al. 2010).

### 16.4.9 Tannases

Tannases consist of two classes of enzymes, the tannin acyl hydrolases (or classical tannases), and the ellagitannin acyl hydrolases (also called ellagitannases). Tannin acyl hydrolases are utilized in beverage, food industries, leather, and pharmaceutical (González et al. 2017). Tannase can easily be obtained from vegetable and animal tissues but on an industrial scale, microbial sources are preferred. Tannases have been obtained from fungi including *Aspergillus P. variotii, Penicillium* (González et al. 2017).

SmF has been mainly for its production of fungi including *Aspergillus* sp. and *Penicillium* sp. though solid-state fermentation is also gaining attention (González et al. 2017). In SmF, 0.1–10% tannic acid is required for tannase production, whereas it has been estimated that for solid-state fermentation, a concentration of up to 10% is required (González et al. 2017). The most common inducer substrate of tannase is tannic acid. In addition, glucose has also been shown to enhance tannase activity (González et al. 2017). After production downstream processing for concentration and purification of product is done. Purification of tannase is achieved by protein concentration followed by ion-exchange or gel-filtration chromatography (Bhardwaj et al. 2003; González et al. 2017).

### 16.4.10 Aminopeptidases

Aminopeptidase or α-amino-acyl-peptide hydrolase, are enzymes that hydrolyze peptide bonds at the N-terminus of proteins and polypeptides. These are believed to be one of the earliest proteases discovered (Nandan and Nampoothiri 2017). They are distributed throughout the plant and animal kingdoms and even the microbial aminopeptidases have been characterized. They play a wide range of roles in protein degradation, nutrient utilization, downstream processing of cytosolic proteins, recycling of amino acids and regulation during chemical-induced stress (Méndez et al. 2014). Microbial aminopeptidases have important roles in the utilization of exogenous proteins as a source of essential amino acids that can be further used for synthesis of proteins, metabolic energy generation and the recycling of reduced cofactors (Nandan and Nampoothiri 2017).

Thermostable aminopeptidisaes has been known from hyperthermophilic archaeon *Pyrococcus horikoshii* strain OT3 (Ando et al. 1999), solvent resistant leucine amino peptidase from hyperthermophile *Aquifex aeolicus* (Khan et al. 2000), a thermostable leucine aminopeptidase from *Bacillus kaustophilus* CCRC 11223 (Lin et al. 2004), a solvent stable aminopeptidase from *P. aeruginosa* (Gaur et al. 2010), thermostable lysine aminopeptidase from *P. aeruginosa* (Wu et al. 2014), aminopeptidase from thermophilic archaea *Sulfolobustokodaii* (Yu et al. 2015), halotolerant aminopeptidase from *Mesorhizobium* SEMIA 3007 (Sierra et al. 2017) and salt-tolerant aminopeptidase from *B. licheniformis* SWJS33 (Lei et al. 2017).

Furthermore, bacterial aminopeptidases have been categorized as metalloaminopeptidases, serine aminopeptidases, and cysteine aminopeptidases (Nandan and Nampoothiri 2017; Andersson et al. 1992; Arora and Lee 1992; Nandan and Nampoothiri 2017; Yoshimoto et al. 1994). Most of the microbial enzymes of industrial importance are produced by aerobic SmF, as it is advantageous in terms of purity of the product, greater control of growth conditions and capability to scale up the process (Nandan and Nampoothiri 2017; Wiseman 1985). The SSF is also used with wheat bran reported to be one of the best substrates for the production of many of the antibiotics, bioactives, microbial enzymes (Nandan and Nampoothiri 2017; Pandey et al. 2000).

For the production of proline aminopeptidase from *Streptomyces lavendulae* by SSF, wheat bran has been reported to be good agro-residue supporting production (Nandan and Nampoothiri 2013, 2017), further soybean meal has been reported to be a good carbon source for the production of leucine aminopeptidase from *Streptomyces gedenensis* (Rahulan et al. 2009) glucose and yeast extract as carbon and nitrogen sources respectively increased the production of aminopeptidases from *Lactococcus* sp. (Pérez-Guzmán et al. 2004).

The major application of aminopeptidases in the food industry include the manufacturing of protein hydrolysates and protein-rich fermented products derived from soy, meat, milk, cereals. These food protein hydrolysates are further used in, e.g., fortification of foods and beverages, generation of various bioactive peptides, and even manufacture of health-care products. These aminopeptidases not only improve the nutritional value but also enhance the flavor of the final product by degrading hydrophobic peptides with undesirable tastes (Nandan and Nampoothiri 2017). Prolyl aminopeptidase activity of *Penicillium camemberti* is significant for the ripening of Camembert cheese as it decreases the amount of bitter-tasting peptides (Fuke and Matsuoka 1993; Nandan and Nampoothiri 2017).

There are many commercially available aminopeptidases from microbial sources used in the food sector, including flavorzyme used in food industry from *A. oryzae*, neutrase from *A. oryzae* used in dairy industry and fish industry, corolase from *A. sojae* used for hydrolysis of animal and vegetable protein, debitrase from *Lactococcus lactis*, and *A. oryzae* for prevention of bitterness development during natural cheese ripening (Nandan and Nampoothiri 2017).

### 16.4.11 Polysaccharide Lyases

These are basically the carbohydrate modifying enzymes cleaving uronic acid-containing polysaccharides via a β-elimination mechanism and form unsaturated hexenuronic acid residues with a new reducing end (Yip and Withers 2006). These have been known in archaea, eubacteria, fungi, algae, plants, mammals (Sutherland 1995), and even in viruses. These have been characterized from *Thermotoga maritima*, *P. vulgaris*, *Flavobacterium heparinum*, *Erwinia* sp., *Bacillus* sp., *Bacteroides stercoris* (Chakraborty et al. 2017; Kluskens et al. 2003; Pissavin et al. 1998). Another polysaccharide lyases identified as α-1,4-glucan lyases has been characterized from fungi *Morchella costata* and *Morchella vulgaris* (Yu et al. 1997). A novel alginate lyase has also been identified and characterized from the marine bacterium *Vibrio* sp. W13 with broad substrate specificity (Zhu et al. 2015). Other microbial polysaccharide lyases include endopectate lyase from *B. pumilus* BK2, *Erwinia chrysanthemi*, rhamnogalacturonan lyase from *B. licheniformis* DSM13, xanthan lyase from *Microbacterium* sp. XT11, chondroitin AC lyase from *F. heparinum*, chondroitin sulfate ABC lyase from *Bacteroides thetaiotaomicron* (Chakraborty et al. 2017; Klug-Santner et al. 2006; Pissavin et al. 1998; Yang et al. 2014).

Polysaccharide lyases can be produced by both SSF as well as SmF. A high-yielding strain in combination with the optimal carbon and nitrogen sources using low-cost substrates is the foremost requirement of successful fermentation for production of polysaccharide lyases. Rice husk, sawdust, and wheat bran are the most utilized substrates for SSF. Whereas in SmF processes a liquid medium containing the dissolved nutrients such as molasses or corn starch dissolved in water or buffer may be used as nutrient source (Chakraborty et al. 2017). The production of pectate lyase from *Penicillium viridicatum* RFC3 has been carried out both by solid state and SmF. In SSF, a mixture of orange bagasse and wheat bran or orange bagasse, wheat bran, and sugarcane bagasse has been used. In the case of SmF, orange bagasse and wheat bran have been utilized. On comparing both the processes, the production has been found to be higher on solid medium (Chakraborty et al. 2017; Ferreira et al. 2010). Another study evaluated the effect of carbon including fructose, glucose, starch, sucrose, and xanthan as well as a combination of glucose plus xanthan and nitrogen sources, including ammonium sulphate, peptone, tryptone, and yeast extract, as well as combination of corn starch and xanthan on production of a xanthan lyase by *Microbacterium* sp. XT11. Finally, xanthan and yeast extract has been determined to be best among all the carbon and nitrogen sources used for the production (Chakraborty et al. 2017; Yang et al. 2014). Pectate lyases are used in fermentation of tea and coffee. Commercially, pectin lyases are available in numerous enzyme preparations including Rapidase LIQ (GisteBrocades), Pectinex Ultra SP (Novo Nordisk), and Rohapect TF (Rohm), which are utilized in fruit juice-processing industry (Chakraborty et al. 2017; Kashyap et al. 2001a).

### 16.4.12 Phytases

Phytases are a class of phosphatases, which possess the ability to release at least one phosphate from phytate *in vitro*. Among all phytases from microbes are among the most promising ones for biotechnological applications. Extracellular phytate-degrading enzymes have been reported in the molds and yeast, but in bacteria, these enzymes are mainly cell-associated, with the exception of *Bacillus* and *Enterobacter*, periplasmic in *E. coli*, whereas phytate-degrading activity in *Selenomonas ruminantium* and *Mitsuokella multiacidus* has been revealed to be associated with the outer membrane (D'Silva et al. 2000; Greiner et al. 1993; Kim et al. 1998). The cell-bound phytase has been reported from yeast *Pichia anomala* (Kaur and Satyanarayana 2010; Ushasree et al. 2017). A non-cell-bound phytase has been identified from *Pichia kudriavzevii* (Hellström et al. 2015). *Fusarium verticillioides, Penicillium, Rhizopus, Rhizoctonia* sp. have been known to be phytase producers (Marlida et al. 2010; Ushasree et al. 2017).

Phytases have also been reported from thermophilic fungi, including *A. niger, Humicola lanuginose, M. thermophila, Rhizomucor pusillus, Sporotrichum thermophile, T. aurantiacus* (Gupta et al. 2014; Kaur et al. 2017; Kumar et al. 2016, 2017; Mitchell et al. 1997; Ushasree et al. 2017). The phytase activity has been reported from *Y. lipolytica, Candida tropicalis, Zygosaccharomyces bisporus, Williopsis saturnus, Zygosaccharomyces priorionus, Schizosaccharomyces octosporus* (Pable et al. 2014). Phytases are produced both by SSF and SmF.

Further, phytases possess various industrial applications. It has been revealed to be an exceptional bread-making improver (Afinah et al. 2010). It has been known that addition of commercial fungal phytase from *A. niger* in the dough ingredients containing fiber formulation speeds up proofing, improves bread shape, enhances specific volume, adds to softness to the crumb. These improvements in bread quality have been revealed to be associated with an indirect impact of phytase on α-amylase activity (Afinah et al. 2010; Greiner and Carlsson 2006). Phytases have been chiefly used as animal feed additive in diets mainly for swine and poultry, and to some extent for fish. The first commercial phytase products was launched into market in 1991 (Afinah et al. 2010; Greiner and Farouk 2007).

## 16.5 Conclusion and Future Scope

In recent years, the extremozymes from extremophilic microbes have been in high demand in the agriculture, medicine, and food industry, due to their potential applications in diverse processes. In this chapter, we reviewed the biodiversity of extremophiles (acidophile, alkaliphile, halophile, metallotolerant, piezophile, psychrophile, psychrotolerant, radioresistant, thermophile, hyperthermophile, toxitolerant, and xerophile) and extremozymes produced, such as amylase, lipase, pectinase, protease, β-galactosidase, β-glucosidase, cellulase, laccase, chitinase, and xylanase. These have quite diverse potential applications in areas such as agriculture, bioconversion of hemicellulose, biodegradation, bioethanol production, biorefinery, chemical industry, composting, dairy industry, detergent industry, feed supplement, food industry, feed industry, glucose feedstock from cellulose, improving rumen digestion, leather industry, molecular biology, paper and pulp industry, peptide synthesis, pharmaceutical industry, and therapeutic agent. The scope to use extremozymes for biotechnological applications is widening with time. This potential usage in health-related and food related areas is sufficient to search for novel and more efficient extremozymes.

## Acknowledgment

The authors are grateful to Prof. Harcharan Singh Dhaliwal, Vice Chancellor, Eternal University, Baru Sahib for providing infrastructural facilities and constant encouragement.

## References


Afinah, S., Yazid, A., Anis Shobirin, M., and Shuhaimi, M. (2010). Phytase: application in food industry. *Int. Food Res. J.* 17: 13–21.

Aguilar, G. and Huitrón, C. (1990). Constitutive exo-pectinase produced by *Aspergillus* sp. CH-Y-1043 on different carbon source. *Biotechnol. Lett.* 12: 655–660.

Aiyer, P.V. (2005). Amylases and their applications. *Afr. J. Biotechnol.* 4: 1525–1529.

Alam, S., Dube, S., Reddy, G. et al. (2005). Purification and characterisation of extracellular protease produced by *Clostridium* sp. from Schirmacher oasis, Antarctica. *Enzyme Microb. Technol.* 36: 824–831.

Alkorta, I., Garbisu, C., Llama, M.J., and Serra, J.L. (1998). Industrial applications of pectic enzymes: a review. *Process Biochem.* 33: 21–28.

Amoozegar, M., Malekzadeh, F., and Malik, K.A. (2003). Production of amylase by newly isolated moderate halophile, *Halobacillus* sp. strain MA-2. *J. Microbiol. Methods* 52: 353–359.

Andersson, C., Linder, L.E., Sund, M.-L., and Lönnies, H. (1992). Purification and characterization of an aminopeptidase from *Streptococcus mitis* ATCC 903. *Curr. Microbiol.* 25: 261–267.

Ando, S., Ishikawa, K., Ishida, H. et al. (1999). Thermostable aminopeptidase from *Pyrococcus horikoshii*. *FEBS Lett.* 447: 25–28.

Andrade, C.M.C., Aguiar, W.B., and Antranikian, G. (2001). Physiological aspects involved in production of xylanolytic enzymes by deep-sea hyperthermophilic archaeon *Pyrodictium abyssi*. *Appl. Biochem. Biotechnol.* 91: 655–669.

Andronopoulou, E. and Vorgias, C.E. (2003). Purification and characterization of a new hyperthermostable, allosamidin-insensitive and denaturation-resistant chitinase from the hyperthermophilic archaeon *Thermococcus chitonophagus*. *Extremophiles* 7: 43–53.

Anisha, G. (2017). β-galactosidases. In: *Current Developments in Biotechnology and Bioengineering* (eds. A. Pandey, S. Negi and C.R. Soccol), 395–421. Elsevier.

Aravindan, R., Anbumathi, P., and Viruthagiri, T. (2007). Lipase applications in food industry. *Indian J. Biotechnol.* 6 (2): 141–158.

Archana, A. and Satyanarayana, T. (1997). Xylanase production by thermophilic *Bacillus licheniformis* A99 in solid-state fermentation. *Enzyme Microb. Technol.* 21: 12–17.

Arora, G. and Lee, B. (1992). Purification and characterization of aminopeptidase from *Lactobacillus casei* ssp. casei LLG. *J. Dairy Sci.* 75: 700–710.

Asgher, M., Asad, M.J., Rahman, S., and Legge, R. (2007). A thermostable α-amylase from a moderately thermophilic *Bacillus subtilis* strain for starch processing. *J. Food Eng.* 79: 950–955.

Asoodeh, A., Chamani, J., and Lagzian, M. (2010). A novel thermostable, acidophilic α-amylase from a new thermophilic "*Bacillus* sp. *Ferdowsicous*" isolated from Ferdows hot mineral spring in Iran: purification and biochemical characterization. *Int. J. Biol. Macromol.* 46: 289–297.

Atalo, K. and Gashe, B. (1993). Protease production by a thermophilic *Bacillus* species (P-001A) which degrades various kinds of fibrous proteins. *Biotechnol. Lett.* 15: 1151–1156.

Babu, K. and Satyanarayana, T. (1995). α-Amylase production by thermophilic *Bacillus coagulans* in solid state fermentation. *Process Biochem.* 30: 305–309.

Bak, J.S., Ko, J.K., Choi, I.G. et al. (2009). Fungal pretreatment of lignocellulose by *Phanerochaete chrysosporium* to produce ethanol from rice straw. *Biotechnol. Bioeng.* 104: 471–482.

Bakhtiar, S., Andersson, M.M., Gessesse, A. et al. (2003). Stability characteristics of a calcium-independent alkaline protease from *Nesterenkonia* sp. *Enzyme Microb. Technol.* 32: 525–531.

Baracat-Pereira, M.C., Vanetti, M.C.D., de Araujo, E.F., and Silva, D.O. (1993). Partial characterization of *Aspergillus fumigatus* polygalacturonases for the degumming of natural fibers. *J. Ind. Microbiol.* 11: 139–142.

Barnby, F.M., Morpeth, F., and Pyle, D. (1990). Endopolygalacturonase production from *Kluyveromyces marxianus*. I. Resolution, purification, and partial characterisation of the enzyme. *Enzyme Microb. Technol.* 12: 891–897.

Beg, Q., Bhushan, B., Kapoor, M., and Hoondal, G. (2000). Production and characterization of thermostable xylanase and pectinase from *Streptomyces* sp. QG-11-3. *J. Ind. Microbiol. Biotechnol.* 24: 396–402.

Bellamy, W. (1977). Cellulose and lignocellulose digestion by thermophilic actinomyces for single-cell protein production. *Dev. Ind. Microbiol.* 19: 249–254.

Bhardwaj, R., Singh, B., and Bhat, T.K. (2003). Purification and characterization of tannin acyl hydrolase from *Aspergillus niger* MTCC 2425. *J. Basic Microbiol.* 43: 449–461.

Binod, P., Sindhu, R., Madhavan, A. et al. (2017). Recent developments in l-glutaminase production and applications – an overview. *Bioresour. Technol.* 245: 1766–1774.

Brühlmann, F., Leupin, M., Erismann, K.H., and Fiechter, A. (2000). Enzymatic degumming of ramie bast fibers. *J. Biotechnol.* 76: 43–50.

Butt, M.S., Tahir-Nadeem, M., Ahmad, Z., and Sultan, M.T. (2008). Xylanases and their applications in baking industry. *Food Technol. Biotechnol.* 46 (1): 22–31.

Cannio, R., Di Prizito, N., Rossi, M., and Morana, A. (2004). A xylan-degrading strain of *Sulfolobus solfataricus*: isolation and characterization of the xylanase activity. *Extremophiles* 8: 117–124.

Cao, J., Zheng, L., and Chen, S. (1992). Screening of pectinase producer from alkalophilic bacteria and study on its potential application in degumming of ramie. *Enzyme Microb. Technol.* 14: 1013–1016.

Chakraborty, S., Rani, A., Dhillon, A., and Goyal, A. (2017). Polysaccharide Lyases. In: *Current Developments in Biotechnology and Bioengineering*, 527–539. Elsevier.

Chang, P., Tsai, W.-S., Tsai, C.-L., and Tseng, M.-J. (2004). Cloning and characterization of two thermostable xylanases from an alkaliphilic *Bacillus firmus*. *Biochem. Biophys. Res. Commun.* 319: 1017–1025.

Chen, W., Chen, H., Xia, Y. et al. (2008). Production, purification, and characterization of a potential thermostable galactosidase for milk lactose hydrolysis from *Bacillus stearothermophilus*. *J. Dairy Sci.* 91: 1751–1758.

Cheng, C.Y., Yabe, I., and Toda, K. (1989). Predominant growth of α-amylase regulation mutant in continuous culture of *Bacillus caldolyticus*. *J. Ferment. Bioeng.* 67: 176–181.

Chiba, S., Yamada, M., and Isobe, K. (2015). Novel acidophilic β-galactosidase with high activity at extremely acidic pH region from *Teratosphaeria acidotherma* AIU BGA-1. *J. Biosci. Bioeng.* 120: 263–267.

Christopher, L.P., Zambare, V.P., Zambare, A. et al. (2015). A thermo-alkaline lipase from a new thermophile *Geobacillus thermodenitrificans* AV-5 with potential application in biodiesel production. *J. Chem. Technol. Biotechnol.* 90: 2007–2016.

Coker, J.A. and Brenchley, J.E. (2006). Protein engineering of a cold-active β-galactosidase from *Arthrobacter* sp. SB to increase lactose hydrolysis reveals new sites affecting low temperature activity. *Extremophiles* 10: 515–524.

Collins, T., Meuwis, M.-A., Stals, I. et al. (2002). A novel family 8 xylanase, functional and physicochemical characterization. *J. Biol. Chem.* 277: 35133–35139.

Collins, T., Gerday, C., and Feller, G. (2005). Xylanases, xylanase families and extremophilic xylanases. *FEMS Microbiol. Rev.* 29: 3–23.

Cunha, C.C.d.Q.B., Gama, A.R., Cintra, L.C. et al. (2018). Improvement of bread making quality by supplementation with a recombinant xylanase produced by *Pichia pastoris*. *PLoS One* 13: e0192996.

Dashtban, M., Schraft, H., and Qin, W. (2009). Fungal bioconversion of lignocellulosic residues; opportunities & perspectives. *Int. J. Biol. Sci.* 5 (6): 578–595.

Dercova, K., Augustin, J., and Krajčová, D. (1992). Cell growth and α-amylase production characteristics of *Bacillus subtilis*. *Folia Microbiol.* 37: 17–23.

Deutch, C. (2002). Characterization of a salt tolerant extracellular α-amylase from *Bacillus dipsosauri*. *Lett. Appl. Microbiol.* 35: 78–84.

Dos Santos, E.S., Guirardello, R., and Franco, T.T. (2002). Preparative chromatography of xylanase using expanded bed adsorption. *J. Chromatogr. A* 944: 217–224.

D'Silva, C., Bae, H., Yanke, L. et al. (2000). Localization of phytase in *Selenomonas ruminantium* and *Mitsuokella multiacidus* by transmission electron microscopy. *Can. J. Microbiol.* 46: 391–395.

Fang, T.-Y., Hung, X.-G., Shih, T.-Y., and Tseng, W.-C. (2004). Characterization of the trehalosyl dextrin-forming enzyme from the thermophilic archaeon *Sulfolobus solfataricus* ATCC 35092. *Extremophiles* 8: 335–343.

Fang, H., Dong, B., Yan, H. et al. (2010). Characterization of a bacterial strain capable of degrading DDT congeners and its use in bioremediation of contaminated soil. *J. Hazard. Mater.* 184: 281–289.

Ferreira, V., Da Silva, R., Silva, D., and Gomes, E. (2010). Production of pectate lyase by *Penicillium viridicatum* RFC3 in solid-state and submerged fermentation. *Int. J. Microbiol.* https://doi.org/10.1155/2010/276590.

Ferrer, M., Golyshina, O.V., Chernikova, T.N. et al. (2005a). Microbial enzymes mined from the Urania deep-sea hypersaline anoxic basin. *Chem. Biol.* 12: 895–904.

Ferrer, M., Martínez-Abarca, F., and Golyshin, P.N. (2005b). Mining genomes and 'metagenomes' for novel catalysts. *Curr. Opin. Biotechnol.* 16: 588–593.

Fraissinet-Tachet, L. and Fevre, M. (1996). Regulation by galacturonic acid of pectinolytic enzyme production by *Sclerotinia sclerotiorum*. *Curr. Microbiol.* 33: 49–53.

Fuke, Y. and Matsuoka, H. (1993). The purification and characterization of prolyl aminopeptidase from *Penicillium camemberti*. *J. Dairy Sci.* 76: 2478–2484.

Fukushima, T., Mizuki, T., Echigo, A. et al. (2005). Organic solvent tolerance of halophilic α-amylase from a Haloarchaeon, *Haloarcula* sp. strain S-1. *Extremophiles* 9: 85–89.

Gaba, S., Singh, R.N., Abrol, S. et al. (2017). Draft genome sequence of *Halolamina pelagica* CDK2 isolated from natural salterns from Rann of Kutch, Gujarat, India. *Genome Announc.* 5: 1–2.

Ganasen, M., Yaacob, N., Rahman, R.N.Z.R.A. et al. (2016). Cold-adapted organic solvent tolerant alkalophilic family I. 3 lipase from an Antarctic *Pseudomonas*. *Int. J. Biol. Macromol.* 92: 1266–1276.

Gao, J., Bauer, M.W., Shockley, K.R. et al. (2003). Growth of hyperthermophilic archaeon *Pyrococcus furiosus* on chitin involves two family 18 chitinases. *Appl. Environ. Microbiol.* 69: 3119–3128.

Gaur, R., Grover, T., Sharma, R. et al. (2010). Purification and characterization of a solvent stable aminopeptidase from *Pseudomonas aeruginosa*: cloning and analysis of aminopeptidase gene conferring solvent stability. *Process Biochem.* 45: 757–764.

Gaur, R., Hemamalini, R., and Khare, S. (2017). Lipolytic enzymes. In: *Current Developments in Biotechnology and Bioengineering*, 175–198. Elsevier.

Golyshina, O.V., Golyshin, P.N., Timmis, K.N., and Ferrer, M. (2006). The 'pH optimum anomaly'of intracellular enzymes of *Ferroplasma acidiphilum*. *Environ. Microbiol.* 8: 416–425.

Gomes, J., Gomes, I., and Steiner, W. (2000). Thermolabile xylanase of the Antarctic yeast *Cryptococcus adeliae*: production and properties. *Extremophiles* 4: 227–235.

Gomes, I., Gomes, J., and Steiner, W. (2003). Highly thermostable amylase and pullulanase of the extreme thermophilic eubacterium *Rhodothermus marinus*: production and partial characterization. *Bioresour. Technol.* 90: 207–214.

González, M.C., Buenrostro-Figueroa, J., Durán, L.R. et al. (2017). Tannases. In: *Current Developments in Biotechnology and Bioengineering* (eds. A. Pandey, S. Negi and C.R. Soccol), 471–489. Elsevier.

Gopinath, S.C., Anbu, P., Lakshmipriya, T., and Hilda, A. (2013). Strategies to characterize fungal lipases for applications in medicine and dairy industry. *Biomed. Res. Int.* https://doi.org/10.1155/2013/154549.

Greiner, R. and Carlsson, N.-G. (2006). Myo-inositol phosphate isomers generated by the action of a phytate-degrading enzyme from *Klebsiella terrigena* on phytate. *Can. J. Microbiol.* 52: 759–768.

Greiner, R. and Farouk, A.-E. (2007). Purification and characterization of a bacterial phytase whose properties make it exceptionally useful as a feed supplement. *Protein J.* 26 (467): https://doi.org/10.1007/s10930-007-9086-z.

Greiner, R., Konietzny, U., and Jany, K.-D. (1993). Purification and characterization of two phytases from *Escherichia coli*. *Arch. Biochem. Biophys.* 303: 107–113.

Grosová, Z., Rosenberg, M., Gdovin, M. et al. (2009). Production of D-galactose using β-galactosidase and *Saccharomyces cerevisiae* entrapped in poly (vinylalcohol) hydrogel. *Food Chem.* 116: 96–100.

Gummadi, S.N., Kumar, S., and Aneesh, C. (2007). Effect of salts on growth and pectinase production by halotolerant yeast, *Debaryomyces nepalensis* NCYC 3413. *Curr. Microbiol.* 54: 472–476.

Gupta, R., Gangoliya, S., and Singh, N. (2014). Isolation of thermotolerant phytase producing fungi and optimisation of phytase production by *Aspergillus niger* NRF9 in solid state fermentation using response surface methodology. *Biotechnol. Bioprocess Eng.* 19: 996–1004.

Haddar, A., Agrebi, R., Bougatef, A. et al. (2009). Two detergent stable alkaline serine-proteases from *Bacillus mojavensis* A21: purification, characterization and potential application as a laundry detergent additive. *Bioresour. Technol.* 100: 3366–3373.

Hagihara, H., Igarashi, K., Hayashi, Y. et al. (2001). Novel α-amylase that is highly resistant to chelating reagents and chemical oxidants from the Alkaliphilic *Bacillus* isolate KSM-K38. *Appl. Environ. Microbiol.* 67: 1744–1750.

Hamer, R. (1995). Enzymes in the baking industry. In: *Enzymes in Food Processing*, 190–222. Boston, MA: Springer.
Harbak, L. and Thygesen, H. (2002). Safety evaluation of a xylanase expressed in *Bacillus subtilis*. *Food Chem. Toxicol.* 40: 1–8.
Harris, A.D. and Ramalingam, C. (2010). Xylanases and its application in food industry: a review. *J. Exp. Sci.* 1 (7).
Hashim, S.O., Delgado, O., Hatti-Kaul, R. et al. (2004). Starch hydrolysing *Bacillus halodurans* isolates from a Kenyan soda lake. *Biotechnol. Lett.* 26: 823–828.
Hatada, Y., Takeda, N., Hirasawa, K. et al. (2005). Sequence of the gene for a high-alkaline mannanase from an alkaliphilic *Bacillus* sp. strain JAMB-750, its expression in *Bacillus subtilis* and characterization of the recombinant enzyme. *Extremophiles* 9: 497–500.
Hayashi, T., Akiba, T., and Horikoshi, K. (1988). Production and purification of new maltohexaose-forming amylases from alkalophilic *Bacillus* sp. H-167. *Agric. Biol. Chem.* 52: 443–448.
Hayashi, K., Nimura, Y., Ohara, N. et al. (1996). Low-temperature-active cellulase produced by *Acremonium alcalophilum*-JCM 7366-note. *J. Soc. Ferment. Bioeng.* 74: 7–10.
Hayashi, K., Inoue, Y., Shiga, M. et al. (1997). Pectinolytic enzymes from *Pseudomonas marginalis* MAFF 03-01173. *Phytochemistry* 45: 1359–1363.
Hebeda, R., Bowles, L., and Teague, W.M. (1990). Developments in enzymes for retarding staling of baked goods. *Cereal Food World* 35: 453–457.
Hebeda, R., Bowles, L., and Teague, W.M. (1991). Use of intermediate temperature stability enzymes for retarding staling in baked goods. *Cereal Food World (USA)* 36: 619–624.
Hellström, A., Qvirist, L., Svanberg, U. et al. (2015). Secretion of non-cell bound phytase by the yeast *Pichia kudriavzevii* TY13. *J. Appl. Microbiol.* 118: 1126–1136.
Henriksson, G., Akin, D.E., Slomczynski, D., and Eriksson, K.-E.L. (1999). Production of highly efficient enzymes for flax retting by Rhizomucor pusillus. *J. Biotechnol.* 68: 115–123.
Hilhorst, R., Dunnewind, B., Orsel, R. et al. (1999). Baking performance, rheology, and chemical composition of wheat dough and gluten affected by xylanase and oxidative enzymes. *J. Food Sci.* 64: 808–813.
Hill, R. and Needham, J. (1970). *The Chemistry of Life: Eight Lectures on the History of Biochemistry*, 17. London, England: Cambridge University Press.
Hiol, A., Jonzo, M.D., Rugani, N. et al. (2000). Purification and characterization of an extracellular lipase from a thermophilic *Rhizopus oryzae* strain isolated from palm fruit. *Enzyme Microb. Technol.* 26: 421–430.
Hmidet, N., Bayoudh, A., Berrin, J.G. et al. (2008). Purification and biochemical characterization of a novel α-amylase from *Bacillus licheniformis* NH1: cloning, nucleotide sequence and expression of amyN gene in *Escherichia coli*. *Process Biochem.* 43: 499–510.
Hoondal, G., Tiwari, R., Tewari, R. et al. (2002). Microbial alkaline pectinases and their industrial applications: a review. *Appl. Microbiol. Biotechnol.* 59: 409–418.
Hotta, Y., Ezaki, S., Atomi, H., and Imanaka, T. (2002). Extremely stable and versatile carboxylesterase from a hyperthermophilic archaeon. *Appl. Environ. Microbiol.* 68: 3925–3931.
Hu, J.M., Li, H., Cao, L.X. et al. (2007). Molecular cloning and characterization of the gene encoding cold-active β-galactosidase from a psychrotrophic and halotolerant *Planococcus* sp. L4. *J. Agric. Food Chem.* 55: 2217–2224.

Huang, Y., Zhao, L., Jia, B. et al. (2011). Glutaminase dysregulation in HIV-1-infected human microglia mediates neurotoxicity: relevant to HIV-1-associated neurocognitive disorders. *J. Neurosci.* 31: 15195–15204.

Huston, A.L., Krieger Brockett, B.B., and Deming, J.W. (2000). Remarkably low temperature optima for extracellular enzyme activity from Arctic bacteria and sea ice. *Environ. Microbiol.* 2: 383–388.

Hutcheon, G.W., Vasisht, N., and Bolhuis, A. (2005). Characterisation of a highly stable α-amylase from the halophilic archaeon *Haloarcula hispanica*. *Extremophiles* 9: 487–495.

Ito, S. (1997). Alkaline cellulases from alkaliphilic *Bacillus*: enzymatic properties, genetics, and application to detergents. *Extremophiles* 1: 61–66.

Iyer, P. and Singhal, R.S. (2008). Production of glutaminase (EC 3.2. 1.5) from *Zygosaccharomyces rouxii*: statistical optimization using response surface methodology. *Bioresour. Technol.* 99: 4300–4307.

Izidoro, S.C. and Knob, A. (2014). Production of xylanases by an *Aspergillus niger* strain in wastes grain. *Acta Sci. Biol. Sci.* https://doi.org/10.4025/actascibiolsci.v36i3.20567.

Jeon, J.-M., Lee, H.-I., and So, J.-S. (2009). Glutaminase activity of *Lactobacillus reuteri* KCTC3594 and expression of the activity in other *Lactobacillus* spp. by introduction of the glutaminase gene. *Afr. J. Microbiol. Res.* 3: 605–609.

Juturu, V. and Wu, J.C. (2012). Microbial xylanases: engineering, production and industrial applications. *Biotechnol. Adv.* 30: 1219–1227.

Kanekar, P., Nilegaonkar, S., Sarnaik, S., and Kelkar, A. (2002). Optimization of protease activity of alkaliphilic bacteria isolated from an alkaline lake in India. *Bioresour. Technol.* 85: 87–93.

Kang, H.-J., Uegaki, K., Fukada, H., and Ishikawa, K. (2007). Improvement of the enzymatic activity of the hyperthermophilic cellulase from *Pyrococcus horikoshii*. *Extremophiles* 11: 251–256.

Kapoor, M. and Kuhad, R.C. (2002). Improved polygalacturonase production from *Bacillus* sp. MG-cp-2 under submerged (SmF) and solid state (SSF) fermentation. *Lett. Appl. Microbiol.* 34: 317–322.

Kapoor, M., Beg, Q.K., Bhushan, B. et al. (2001). Application of an alkaline and thermostable polygalacturonase from *Bacillus* sp. MG-cp-2 in degumming of ramie (*Boehmeria nivea*) and sunn hemp (*Crotalaria juncea*) bast fibres. *Process Biochem.* 36: 803–807.

Karasova, P., Spiwok, V., Mala, S. et al. (2002). Beta-galactosidase activity in psychrotrophic microorganisms and their potential use in food industry. *Czech J. Food Sci.* 20: 43–47.

Kasana, R.C. and Gulati, A. (2011). Cellulases from psychrophilic microorganisms: a review. *J. Basic Microbiol.* 51: 572–579.

Kasana, R.C., Kaur, B., and Yadav, S.K. (2008). Isolation and identification of a psychrotrophic *Acinetobacter* sp. CR9 and characterization of its alkaline lipase. *J. Basic Microbiol.* 48: 207–212.

Kashyap, D., Vohra, P., Chopra, S., and Tewari, R. (2001a). Applications of pectinases in the commercial sector: a review. *Bioresour. Technol.* 77: 215–227.

Kashyap, D., Vohra, P., Chopra, S., and Tewari, R. (2001b). Biotechnological applications of microbial pectinases. *Bioresour. Technol.* 77: 215–227.

Kaur, P. and Satyanarayana, T. (2010). Improvement in cell-bound phytase activity of *Pichia anomala* by permeabilization and applicability of permeabilized cells in soymilk dephytinization. *J. Appl. Microbiol.* 108: 2041–2049.

Kaur, G., Kumar, S., and Satyanarayana, T. (2004). Production, characterization and application of a thermostable polygalacturonase of a thermophilic mould *Sporotrichum thermophile Apinis*. *Bioresour. Technol.* 94: 239–243.

Kaur, R., Saxena, A., Sangwan, P. et al. (2017). Production and characterization of a neutral phytase of *Penicillium oxalicum* EUFR-3 isolated from Himalayan region. *Nusantara Biosci.* 9: 68–76.

Khan, A.R., Nirasawa, S., Kaneko, S. et al. (2000). Characterization of a solvent resistant and thermostable aminopeptidase from the hyperthermophillic bacterium, Aquifex aeolicus. *Enzyme Microb. Technol.* 27: 83–88.

Kim, T.U., Gu, B.G., Jeong, J.Y. et al. (1995). Purification and characterization of a Maltotetraose-forming alkaline (alpha)-amylase from an Alkalophilic *Bacillus* strain, GM8901. *Appl. Environ. Microbiol.* 61: 3105–3112.

Kim, Y.-O., Kim, H.-K., Bae, K.-S. et al. (1998). Purification and properties of a thermostable phytase from *Bacillus* sp. DS11. *Enzyme Microb. Technol.* 22: 2–7.

Kim, J., Flowers, L., Whiteley, M., and Peeples, T. (2001). Biochemical confirmation and characterization of the family-57-like α-amylase of *Methanococcus jannaschii*. *Folia Microbiol.* 46: 467–473.

Kiran, G.S., Shanmughapriya, S., Jayalakshmi, J. et al. (2008). Optimization of extracellular psychrophilic alkaline lipase produced by marine *Pseudomonas* sp.(MSI057). *Bioprocess Biosyst. Eng.* 31: 483–492.

Klug-Santner, B.G., Schnitzhofer, W., Vršanská, M. et al. (2006). Purification and characterization of a new bioscouring pectate lyase from *Bacillus pumilus* BK2. *J. Biotechnol.* 121: 390–401.

Kluskens, L.D., Van Alebeek, G.-J.W., and Voragen, A.G. (2003). Molecular and biochemical characterization of the thermoactive family 1 pectate lyase from the hyperthermophilic bacterium *Thermotoga maritima*. *Biochem. J.* 370: 651–659.

Kottwitz, B., Upadek, H., and Carrer, G. (1994). Application and benefits of enzymes in detergents. *Chim. Oggi* 12: 21–24.

Kouker, G. and Jaeger, K.-E. (1987). Specific and sensitive plate assay for bacterial lipases. *Appl. Environ. Microbiol.* 53: 211–213.

Kumar, C. (2002). Purification and characterization of a thermostable alkaline protease from alkalophilic *Bacillus pumilus*. *Lett. Appl. Microbiol.* 34: 13–17.

Kumar, B., Trivedi, P., Mishra, A.K. et al. (2004). Microbial diversity of soil from two hot springs in Uttaranchal Himalaya. *Microbiol. Res.* 159: 141–146.

Kumar, S., Kikon, K., Upadhyay, A. et al. (2005). Production, purification, and characterization of lipase from thermophilic and alkaliphilic *Bacillus coagulans* BTS-3. *Protein Expression Purif.* 41: 38–44.

Kumar, L., Singh, B., Adhikari, D.K. et al. (2012). A temperature and salt-tolerant l-glutaminase from Gangotri region of Uttarakhand Himalaya: enzyme purification and characterization. *Appl. Biochem. Biotechnol.* 166: 1723–1735.

Kumar, M., Yadav, A.N., Tiwari, R. et al. (2014). Evaluating the diversity of culturable thermotolerant bacteria from four hot springs of India. *J. Biodivers. Biopros. Dev.* 1: 1–9.

Kumar, V., Yadav, A.N., Saxena, A. et al. (2016). Unravelling rhizospheric diversity and potential of phytase producing microbes. *SM J. Biol.* 2 (1): 1009.

Kumar, V., Yadav, A.N., Verema, P. et al. (2017). β-Propeller phytases: diversity, catalytic attributes, current developments and potential biotechnological applications. *Int. J. Biol. Macromol.* 98: 595–609.

Kumar, A., Wu, G., and Liu, Z. (2018a). Synthesis and characterization of cross linked enzyme aggregates of serine hydroxyl methyltransferase from *Idiomerina leihiensis*. *Int. J. Biol. Macromol.* 117: 683–690.

Kumar, A., Wu, G., Wu, Z. et al. (2018b). Improved catalytic properties of a serine hydroxymethyl transferase from Idiomarina loihiensis by site directed mutagenesis. *Int. J. Biol. Macromol.* 117: 1216–1223.

Lebbink, J.H., Kaper, T., Bron, P. et al. (2000). Improving low-temperature catalysis in the hyperthermostable *Pyrococcus furiosus* β-glucosidase CelB by directed evolution. *Biochemistry* 39: 3656–3665.

Lee, D.-W., Koh, Y.-S., Kim, K.-J. et al. (1999). Isolation and characterization of a thermophilic lipase from *Bacillus thermoleovorans* ID-1. *FEMS Microbiol. Lett.* 179: 393–400.

Lee, D.-W., Kim, H.-W., Lee, K.-W. et al. (2001). Purification and characterization of two distinct thermostable lipases from the gram-positive thermophilic bacterium *Bacillus thermoleovorans* ID-1. *Enzyme Microb. Technol.* 29: 363–371.

Lee, B.-H., Kim, B.-K., Lee, Y.-J. et al. (2010). Industrial scale of optimization for the production of carboxymethylcellulase from rice bran by a marine bacterium, *Bacillus subtilis* subsp. *subtilis* A-53. *Enzyme Microb. Technol.* 46: 38–42.

Lei, F., Zhao, Q., Sun-Waterhouse, D., and Zhao, M. (2017). Characterization of a salt-tolerant aminopeptidase from marine *Bacillus licheniformis* SWJS33 that improves hydrolysis and debittering efficiency for soy protein isolate. *Food Chem.* 214: 347–353.

Li, H. and Zhang, X. (2005). Characterization of thermostable lipase from thermophilic *Geobacillus* sp. TW1. *Protein Expression Purif.* 42: 153–159.

Lin, L.-L., Hsu, W.-H., Wu, C.-P. et al. (2004). A thermostable leucine aminopeptidase from *Bacillus kaustophilus* CCRC 11223. *Extremophiles* 8: 79–87.

Ljungdahl, L.G. (2008). The Cellulase/Hemicellulase system of the anaerobic fungus *Orpinomyces* PC2 and aspects of its applied use. *Ann. N. Y. Acad. Sci.* 1125: 308–321.

Lo, C.-M., Zhang, Q., Callow, N.V., and Ju, L.-K. (2010). Cellulase production by continuous culture of *Trichoderma reesei* rut C30 using acid hydrolysate prepared to retain more oligosaccharides for induction. *Bioresour. Technol.* 101: 717–723.

Loginova, L., Belyakova, L., Guzhova, E. et al. (1983). Thermophilic, cellulose-decomposing *Myceliophthora thermophila* [fungi isolated from cattle rumen]. *Microbiology* 52: 605–608.

Löliger, J. (2000). Function and importance of glutamate for savory foods. *J. Nutr.* 130: 915S–920S.

Lonsane, B. and Ramesh, M. (1990). Production of bacterial thermostable α-amylase by solid-state fermentation: a potential tool for achieving economy in enzyme production and starch hydrolysis. In: *Advances in Applied Microbiology*, vol. 35 (eds. S.L. Neidleman and A.I. Laskin), 1–56. Academic Press.

Lorentzen, M.S., Moe, E., Jouve, H.M., and Willassen, N.P. (2006). Cold adapted features of *Vibrio salmonicida* catalase: characterisation and comparison to the mesophilic counterpart from *Proteus mirabilis*. *Extremophiles* 10: 427.

Luo, Y., Zheng, Y., Jiang, Z. et al. (2006). A novel psychrophilic lipase from *Pseudomonas fluorescens* with unique property in chiral resolution and biodiesel production via transesterification. *Appl. Microbiol. Biotechnol.* 73: 349–355.

Madsen, G., Norman, B., and Slott, S. (1973). A new, heat stable bacterial amylase and its use in high temperature liquefaction. *Starch-Stärke* 25: 304–308.

Maharana, A.K. and Singh, S.M. (2018). Cold active lipases produced by *Cryptococcus* sp. Y-32 and *Rhodococcus erythropolis* N149 isolated from Nella Lake, Antarctica. *Int. J. Curr. Microbiol. Appl. Sci.* 7: 1910–1926.

Maischberger, T., Nguyen, T.-H., Sukyai, P. et al. (2008). Production of lactose-free galacto-oligosaccharide mixtures: comparison of two cellobiose dehydrogenases for the selective oxidation of lactose to lactobionic acid. *Carbohydr. Res.* 343: 2140–2147.

Malathi, V. and Devegowda, G. (2001). In vitro evaluation of nonstarch polysaccharide digestibility of feed ingredients by enzymes. *Poult. Sci.* 80: 302–305.

Marlida, Y., Delfita, R., Adnadi, P., and Ciptaan, G. (2010). Isolation, characterization and production of phytase from endophytic fungus its application for feed. *Pak. J. Nutr.* 9: 471–474.

Marques, N.P., de Cassia Pereira, J., Gomes, E. et al. (2018). Cellulases and xylanases production by endophytic fungi by solid state fermentation using lignocellulosic substrates and enzymatic saccharification of pretreated sugarcane bagasse. *Ind. Crop. Prod.* 122: 66–75.

Matzke, J., Schwermann, B., and Bakker, E.P. (1997). Acidostable and acidophilic proteins: the example of the α-amylase from *Alicyclobacillus acidocaldarius*. *Comp. Biochem. Physiol. A: Physiol.* 118: 475–479.

Medda, S. and Chandra, A. (1980). New strains of *Bacillus licheniformis* and *Bacillus coagulans* producing Thermostable α-amylase active at alkaline pH. *J. Appl. Microbiol.* 48: 47–58.

Méndez, Y., Pérez-Labrada, K., González-Bacerio, J. et al. (2014). Combinatorial multicomponent access to natural-products-inspired peptidomimetics: discovery of selective inhibitors of microbial metallo-aminopeptidases. *ChemMedChem* 9: 2351–2359.

Mijts, B.N. and Patel, B.K. (2002). Cloning, sequencing and expression of an α-amylase gene, amyA, from the thermophilic halophile *Halothermothrix orenii* and purification and biochemical characterization of the recombinant enzymea. *Microbiology* 148: 2343–2349.

Mitchell, D.B., Vogel, K., Weimann, B.J. et al. (1997). The phytase subfamily of histidine acid phosphatases: isolation of genes for two novel phytases from the fungi *Aspergillus terreus* and *Myceliophthora thermophila*. *Microbiology* 143: 245–252.

Mitsuiki, S., Sakai, M., Moriyama, Y. et al. (2002). Purification and some properties of a keratinolytic enzyme from an alkaliphilic *Nocardiopsis* sp. TOA-1. *Biosci. Biotechnol. Biochem.* 66: 164–167.

Miyazaki, K. (2005). A hyperthermophilic laccase from *Thermus thermophilus* HB27. *Extremophiles* 9: 415–425.

Mohagheghi, A., Grohmann, K., Himmel, M. et al. (1986). Isolation and characterization of *Acidothermus cellulolyticus* gen. Nov., sp. nov., a new genus of thermophilic, acidophilic, cellulolytic bacteria. *Int. J. Syst. Evol. Microbiol.* 36: 435–443.

Moh'd, A.S. and Wiegel, J. (2007). Purification and characterization of two highly thermophilic alkaline lipases from *Thermosyntropha lipolytica*. *Appl. Environ. Microbiol.* 73: 7725–7731.

Morita, R.Y. (1975). Psychrophilic bacteria. *Bacteriol. Rev.* 39 (2): 144–167.

Naidu, G. and Panda, T. (1998). Production of pectolytic enzymes–a review. *Bioprocess Eng.* 19: 355–361.

Nakagawa, T., Nagaoka, T., Taniguchi, S. et al. (2004). Isolation and characterization of psychrophilic yeasts producing cold-adapted pectinolytic enzymes. *Lett. Appl. Microbiol.* 38: 383–387.

Nakagawa, T., Ikehata, R., Uchino, M. et al. (2006). Cold-active acid β-galactosidase activity of isolated psychrophilic-basidiomycetous yeast *Guehomyces pullulans*. *Microbiol. Res.* 161: 75–79.

Nandakumar, R., Yoshimune, K., Wakayama, M., and Moriguchi, M. (2003). Microbial glutaminase: biochemistry, molecular approaches and applications in the food industry. *J. Mol. Catal. B Enzym.* 23: 87–100.

Nandan, A. and Nampoothiri, K.M. (2013). Extracellular proline aminopeptidase production by *Streptomyces lavendulae* ATCC14162 under solid-state fermentation. *J. Sci. Ind. Res.* 72: 591–595. https://doi.org/10.1007/s12010-010-9103-0.

Nandan, A. and Nampoothiri, K. (2017). Microbial aminopeptidases. In: *Current Developments in Biotechnology and Bioengineering* (eds. A. Pandey, S. Negi and C.R. Soccol), 491–507. Elsevier.

Nascimento, W.C.A.d. and Martins, M.L.L. (2004). Production and properties of an extracellular protease from thermophilic *Bacillus* sp. *Braz. J. Microbiol.* 35: 91–96.

Nawani, N. and Kaur, J. (2000). Purification, characterization and thermostability of lipase from a thermophilic *Bacillus* sp. J33. *Mol. Cell. Biochem.* 206: 91–96.

Nuyens, F., Verachtert, H., and Michiels, C. (2001. Suppl 1.). Evaluation of a recombinant *Saccharomyces cerevisiae* strain secreting a *Bacillus pumilus* endo-beta-xylanase for use in bread-making. In: *BMC Meeting Abstracts*, 015. BioMed Central Ltd.

Okuda, M., Sumitomo, N., Takimura, Y. et al. (2004). A new subtilisin family: nucleotide and deduced amino acid sequences of new high-molecular-mass alkaline proteases from *Bacillus* spp. *Extremophiles* 8: 229–235.

O'Mahony, M. and Ishii, R. (1986). A comparison of English and Japanese taste languages: taste descriptive methodology, codability and the umami taste. *Br. J. Psychol.* 77: 161–174.

Pable, A., Gujar, P., and Khire, J. (2014). Selection of phytase producing yeast strains for improved mineral mobilization and dephytinization of chickpea flour. *J. Food Biochem.* 38: 18–27.

Palmieri, G., Casbarra, A., Fiume, I. et al. (2006). Identification of the first archaeal oligopeptide-binding protein from the hyperthermophile *Aeropyrum pernix*. *Extremophiles* 10: 393–402.

Palomäki, T. and Saarilahti, H. (1997). Isolation and characterization of new C-terminal substitution mutations affecting secretion of polygalacturonase in *Erwinia carotovora* ssp. *carotovora*. *FEBS Lett.* 400: 122–126.

Pandey, A., Soccol, C.R., and Mitchell, D. (2000). New developments in solid state fermentation: I-bioprocesses and products. *Process Biochem.* 35: 1153–1169.

Panesar, P.S., Kumari, S., and Panesar, R. (2010). Potential applications of immobilized β-galactosidase in food processing industries. *Enzyme Res.* https://doi.org/10.4061/2010/473137.

Panjiar, N., Mishra, S., Yadav, A.N., and Verma, P. (2017). Functional foods from cyanobacteria: an emerging source for functional food products of pharmaceutical importance. In: *Microbial Functional Foods and Nutraceuticals* (eds. V.K. Gupta, H. Treichel, V.O. Shapaval, et al.), 21–37. USA: Wiley https://doi.org/10.1002/9781119048961.ch2.

Pérez-Guzmán, A.E., y Victoria, T.C., Cruz-Camarillo, R., and Hernández-Sánchez, H. (2004). Improvement of fermentation conditions for the production of X-prolyl-dipeptidyl aminopeptidase from *Lactococcus lactis*. *World J. Microbiol. Biotechnol.* 20: 413–417.

Pérez-Pomares, F., Bautista, V., Ferrer, J. et al. (2003). α-Amylase activity from the halophilic archaeon *Haloferax mediterranei*. *Extremophiles* 7: 299–306.

Perombelon, M.C. and Kelman, A. (1980). Ecology of the soft rot erwinias. *Annu. Rev. Phytopathol.* 18: 361–387.

Pirota, R.D.P.B., Tonelotto, M., da Silva Delabona, P. et al. (2013). Enhancing xylanases production by a new Amazon Forest strain of *Aspergillus oryzae* using solid-state fermentation under controlled operation conditions. *Ind. Crop. Prod.* 45: 465–471.

Pissavin, C., Robert-Baudouy, J., and Hugouvieux-Cotte-Pattat, N. (1998). Biochemical characterization of the pectate lyase PelZ of *Erwinia chrysanthemi* 3937. *Biochim. Biophys. Acta Protein Struct. Mol. Enzymol.* 1383: 188–196.

Pivarnik, L.F., Senecal, A.G., and Rand, A.G. (1995). Hydrolytic and transgalactosylic activities of commercial β-galactosidase (lactase) in food processing. In: *Advances in Food and Nutrition Research*, vol. 38 (eds. J.E. Kinsella and S.L. Taylor), 1–102. Academic Press.

Polizeli, M., Rizzatti, A., Monti, R. et al. (2005). Xylanases from fungi: properties and industrial applications. *Appl. Microbiol. Biotechnol.* 67: 577–591.

Prade, R.A. (1996). Xylanases: from biology to biotechnology. *Biotechnol. Genet. Eng. Rev.* 13: 101–132.

Prakasham, R., Rao, C.S., and Sarma, P. (2006). Green gram husk – an inexpensive substrate for alkaline protease production by *Bacillus* sp. in solid-state fermentation. *Bioresour. Technol.* 97: 1449–1454.

Pratuangdejkul, J. and Dharmsthiti, S. (2000). Purification and characterization of lipase from psychrophilic *Acinetobacter calcoaceticus* LP009. *Microbiol. Res.* 155: 95–100.

Pretel, M.T., Lozano, P., Riquelme, F., and Romojaro, F. (1997). Pectic enzymes in fresh fruit processing: optimization of enzymic peeling of oranges. *Process Biochem.* 32: 43–49.

Pretz, M.G., Remigy, H., Swaving, J. et al. (2005). Functional and structural characterization of the minimal sec translocase of the hyperthermophile *Thermotoga maritima*. *Extremophiles* 9: 307–316.

Rahulan, R., Nampoothiri, K.M., Szakacs, G. et al. (2009). Statistical optimization of L-leucine amino peptidase production from *Streptomyces gedanensis* IFO 13427 under submerged fermentation using response surface methodology. *Biochem. Eng. J.* 43: 64–71.

Rai, S.K., Konwarh, R., and Mukherjee, A.K. (2009). Purification, characterization and biotechnological application of an alkaline β-keratinase produced by *Bacillus subtilis* RM-01 in solid-state fermentation using chicken-feather as substrate. *Biochem. Eng. J.* 45: 218–225.

Ramasubbu, N., Paloth, V., Luo, Y. et al. (1996). Structure of human salivary α-amylase at 1.6 Å resolution: implications for its role in the oral cavity. *Acta Crystallogr. D Biol. Crystallogr.* 52: 435–446.

Rastogi, G. (1998). *Vishal's Objective Botany*. Meerut, India: Vishal Publishers.

Ray, A. (2012). Application of lipase in industry. *Asian J. Pharm. Technol.* 2: 33–37.

Ricard, M. and Reid, I.D. (2004). Purified pectinase lowers cationic demand in peroxide-bleached mechanical pulp. *Enzyme Microb. Technol.* 34: 499–504.

Romanowska, I., Polak, J., Janowska, K., and Bielecki, S. (2003). The application of fungal endoxylanase in bread-making. *Commun. Agric. Appl. Biol. Sci.* 68: 317–320.

Rouau, X., Daviet, S., Tahir, T. et al. (2006). Effect of the proteinaceous wheat xylanase inhibitor XIP-I on the performances of an *Aspergillus niger* xylanase in breadmaking. *J. Sci. Food Agric.* 86: 1604–1609.

Ryu, H., Kim, H., Choi, W. et al. (2006). New cold-adapted lipase from Photobacterium *lipolyticum* sp. nov. that is closely related to filamentous fungal lipases. *Appl. Microbiol. Biotechnol.* 70: 321–326.

Sabu, A., Keerthi, T., Kumar, S.R., and Chandrasekaran, M. (2000). L-Glutaminase production by marine *Beauveria* sp. under solid state fermentation. *Process Biochem.* 35: 705–710.

Sabu, A., Kumar, S.R., and Chandrasekaran, M. (2002). Continuous production of extracellular L-glutaminase by Ca-alginate-immobilized marine *Beauveria bassiana* BTMF S-10 in packed-bed reactor. *Appl. Biochem. Biotechnol.* 102: 71–79.

Sahay, H., Singh, S., Kaushik, R. et al. (2011). Characterization of halophilic bacteria from environmental samples from the brackish water of Pulicat Lake, India. *Biologia* 66/5: 741–747. https://doi.org/10.2478/s11756-011-0094-2.

Sahay, H., Mahfooz, S., Singh, A.K. et al. (2012). Exploration and characterization of agriculturally and industrially important haloalkaliphilic bacteria from environmental samples of hypersaline Sambhar lake, India. *World J. Microbiol. Biotechnol.* 28: 3207–3217.

Sahay, H., Yadav, A.N., Singh, A.K. et al. (2017). Hot springs of Indian Himalayas: potential sources of microbial diversity and thermostable hydrolytic enzymes. *3 Biotech* 7: 1–11.

Sakamoto, T., Ihara, H., Kozaki, S., and Kawasaki, H. (2003). A cold-adapted endo-arabinanase from *Penicillium chrysogenum. Biochim. Biophys. Acta (BBA)-Gen. Subj.* 1624: 70–75.

Salazar, L. and Jayasinghe, U. (1999). Fundamentals of purification of plant viruses. In: *Techniques in Plant Virology, CIP., Training Manual, JO, Virus Purification* (eds. L. Salazar and U. Jayasinghe), 1–10. Peru: International Potato Centre.

Sanchez-Porro, C., Mellado, E., Bertoldo, C. et al. (2003). Screening and characterization of the protease CP1 produced by the moderately halophilic bacterium *Pseudoalteromonas* sp. strain CP76. *Extremophiles* 7: 221–228.

Sarada, K. (2013). Production and applications of L-Glutaminase using fermentation technology. *Asia Pac. J. Res.* 1: 1.

Sato, I., Kobayashi, H., Hanya, Y. et al. (1999). *Cryptococcus nodaensis* sp nov, a yeast isolated from soil in Japan that produces a salt-tolerant and thermostable glutaminase. *J. Ind. Microbiol. Biotechnol.* 22: 127–132.

Saxena, A.K., Yadav, A.N., Kaushik, R. et al. (2015). Biotechnological applications of microbes isolated from cold environments in agriculture and allied sectors. In: *International Conference on "Low Temperature Science and Biotechnological Advances"*, 104. Society of low temperature biology https://doi.org/10.13140/RG.2.1.2853.5202.

Saxena, A.K., Yadav, A.N., Rajawat, M. et al. (2016). Microbial diversity of extreme regions: an unseen heritage and wealth. *Indian J. Plant Genet. Resour.* 29: 246–248.

Secades, P., Alvarez, B., and Guijarro, J. (2003). Purification and properties of a new psychrophilic metalloprotease (Fpp2) in the fish pathogen *Flavobacterium psychrophilum. FEMS Microbiol. Lett.* 226: 273–279.

Sharma, R., Chisti, Y., and Banerjee, U.C. (2001). Production, purification, characterization, and applications of lipases. *Biotechnol. Adv.* 19: 627–662.

Sharma, R., Soni, S., Vohra, R. et al. (2002). Purification and characterisation of a thermostable alkaline lipase from a new thermophilic *Bacillus* sp. RSJ-1. *Process Biochem.* 37: 1075–1084.

Sharma, P., Sharma, N., Sharma, P. et al. (2017). Purification and characterization of a Halotolerant and Thermotolerant lipase produced from a novel bacteria "*Brevibacterium halotolerans* PS4| KX671556|" and its application in detergent formulations. *Proc. Indian Natl. Sci. Acad.* 3: 681–687.

Shi, W., Tang, X.-F., Huang, Y. et al. (2006). An extracellular halophilic protease SptA from a halophilic archaeon *Natrinema* sp. J7: gene cloning, expression and characterization. *Extremophiles* 10: 599–606.

Shrestha, P., Khanal, S.K., Pometto, A.L. III, and van Leeuwen, J. (2009). Enzyme production by wood-rot and soft-rot fungi cultivated on corn fiber followed by simultaneous saccharification and fermentation. *J. Agric. Food Chem.* 57: 4156–4161.

Shukla, L., Suman, A., Yadav, A.N. et al. (2016). Syntrophic microbial system for ex-situ degradation of paddy straw at low temperature under controlled and natural environment. *J. Appl. Biol. Biotech.* 4: 30–37.

Sierra, E.M., Pereira, M.R., Maester, T.C. et al. (2017). Halotolerant aminopeptidase M29 from *Mesorhizobium* SEMIA 3007 with biotechnological potential and its impact on biofilm synthesis. *Sci. Rep.* 7 (1): 10684.

Silva, Z., Alarico, S., and Da Costa, M.S. (2005). Trehalose biosynthesis in *Thermus thermophilus* RQ-1: biochemical properties of the trehalose-6-phosphate synthase and trehalose-6-phosphate phosphatase. *Extremophiles* 9: 29–36.

Silverman, R.B. (2002). *The Organic Chemistry of Enzyme-Catalyzed Reactions*, 800. London/San Diego: Academic.

Singh, R.N., Gaba, S., Yadav, A.N. et al. (2016). First, high quality draft genome sequence of a plant growth promoting and cold active enzymes producing psychrotrophic *Arthrobacter agilis* strain L77. *Stand. Genomic Sci.* 11: 54. https://doi.org/10.1186/s40793-016-0176-4.

de Siqueira, F.G., de Siqueira, E.G., Jaramillo, P.M.D. et al. (2010). The potential of agro-industrial residues for production of holocellulase from filamentous fungi. *Int. Biodeterior. Biodegrad.* 64: 20–26.

Sivaramakrishnan, S., Gangadharan, D., Nampoothiri, K.M. et al. (2006). α-amylases from microbial sources–an overview on recent developments. *Food Technol. Biotechnol.* 44: 173–184.

Soden, D., O'callaghan, J., and Dobson, A. (2002). Molecular cloning of a laccase isozyme gene from *Pleurotus sajor-caju* and expression in the heterologous *Pichia pastoris* host. *Microbiology* 148: 4003–4014.

Solingen, P., Meijer, D., Kleij, W.A. et al. (2001). Cloning and expression of an endocellulase gene from a novel streptomycete isolated from an east African soda lake. *Extremophiles* 5: 333–341.

Stutzenberger, F.J. (1979). Degradation of cellulosic substances by *Thermomonospora curvata*. *Biotechnol. Bioeng.* 21: 909–913.

Su, T.-M. and Paulavicius, I. (1975). Paulavicius enzymatic saccharification of cellulose by thermophilic actinomyces. *Appl. Polym. Symp* 28: 221–236.

Sukumaran, R.K., Singhania, R.R., and Pandey, A. (2005). Microbial cellulases—production, applications and challenges. *J. Sci. Ind. Res.* 64: 832–844.

Sukumaran, R.K., Singhania, R.R., Mathew, G.M., and Pandey, A. (2009). Cellulase production using biomass feed stock and its application in lignocellulose saccharification for bio-ethanol production. *Renew. Energy* 34 (12): 421–424.

Suman, A., Verma, P., Yadav, A.N., and Saxena, A.K. (2015). Bioprospecting for extracellular hydrolytic enzymes from culturable thermotolerant bacteria isolated from Manikaran thermal springs. *Res. J. Biotechnol.* 10: 33–42.

Sutherland, I.W. (1995). Polysaccharide lyases. *FEMS Microbiol. Rev.* 16: 323–347.
Sutrisno, A., Ueda, M., Abe, Y. et al. (2004). A chitinase with high activity toward partially N-acetylated chitosan from a new, moderately thermophilic, chitin-degrading bacterium, *Ralstonia* sp. A-471. *Appl. Microbiol. Biotechnol.* 63: 398–406.
Suzuki, T., Nakayama, T., Kurihara, T. et al. (2001). Cold-active lipolytic activity of psychrotrophic *Acinetobacter* sp. strain no. 6. *J. Biosci. Bioeng.* 92: 144–148.
Tanabe, H., Yoshihara, K., Tamura, K. et al. (1987). Pretreatment of pectic wastewater from orange canning process by an alkalophilic *Bacillus* sp. *J. Ferment. Technol.* 65: 243–246.
Tanaka, H., Koike, K., Itakura, S., and Enoki, A. (2009). Degradation of wood and enzyme production by *Ceriporiopsis subvermispora*. *Enzyme Microb. Technol.* 45: 384–390.
Tari, C., Ustok, F.I., and Harsa, S. (2009). Optimization of the associative growth of novel yoghurt cultures in the production of biomass, β-galactosidase and lactic acid using response surface methodology. *Int. Dairy J.* 19: 236–243.
Tatur, J., Hagedoorn, P.-L., Overeijnder, M.L., and Hagen, W.R. (2006). A highly thermostable ferritin from the hyperthermophilic archaeal anaerobe *Pyrococcus furiosus*. *Extremophiles* 10: 139–148.
Taylor, I.N., Brown, R.C., Bycroft, M. et al. (2004). Application of thermophilic enzymes in commercial biotransformation processes. *Biochem. Soc. Trans.* 32: 290–292.
Thirumale, S., Rani, D.S., and Nand, K. (2001). Control of cellulase formation by trehalose in *Clostridium papyrosolvens* CFR-703. *Process Biochem.* 37: 241–245.
Thomas, L., Joseph, A., Singhania, R.R. et al. (2017). Industrial enzymes: xylanases. In: *Current Developments in Biotechnology and Bioengineering* (eds. A. Pandey, S. Negi and C.R. Soccol), 127–148. Elsevier.
Ueno, S., Miyama, M., Ohashi, Y. et al. (1987). Secretory enzyme production and conidiation of *Aspergillus oryzae* in submerged liquid culture. *Appl. Microbiol. Biotechnol.* 26: 273–276.
Ugras, S. (2017). Characterization of a Thermophilic lipase from *Bacillus licheniformis* Ht7 isolated from Hayran thermal springs in Giresun. *Rom. Biotechnol. Lett.* 22: 12297.
Undurraga, D., Markovits, A., and Erazo, S. (2001). Cocoa butter equivalent through enzymic interesterification of palm oil midfraction. *Process Biochem.* 36: 933–939.
Ushasree, M., Vidya, J., and Pandey, A. (2017). Other enzymes: phytases. In: *Current Developments in Biotechnology and Bioengineering* (eds. A. Pandey, S. Negi and C.R. Soccol), 309–333. Elsevier.
Van De Voorde, I., Goiris, K., Syryn, E. et al. (2014). Evaluation of the cold-active *Pseudoalteromonas haloplanktis* β-galactosidase enzyme for lactose hydrolysis in whey permeate as primary step of d-tagatose production. *Process Biochem.* 49: 2134–2140.
Verma, P., Yadav, A.N., Kazy, S.K. et al. (2013). Elucidating the diversity and plant growth promoting attributes of wheat (*Triticum aestivum*) associated acidotolerant bacteria from southern hills zone of India. *Natl. J. Life Sci.* 10: 219–227.
Verma, P., Yadav, A.N., Khannam, K.S. et al. (2015a). Assessment of genetic diversity and plant growth promoting attributes of psychrotolerant bacteria allied with wheat (*Triticum aestivum*) from the northern hills zone of India. *Ann. Microbiol.* 65: 1885–1899.
Verma, P., Yadav, A.N., Shukla, L. et al. (2015b). Hydrolytic enzymes production by thermotolerant *Bacillus altitudinis* IARI-MB-9 and *Gulbenkiania mobilis* IARI-MB-18 isolated from Manikaran hot springs. *Int. J. Adv. Res.* 3: 1241–1250.
Verma, P., Yadav, A.N., Khannam, K.S. et al. (2016a). Molecular diversity and multifarious plant growth promoting attributes of bacilli associated with wheat (*Triticum aestivum* L.) rhizosphere from six diverse agro-ecological zones of India. *J. Basic Microbiol.* 56: 44–58.

Verma, P., Yadav, A.N., Khannam, K.S. et al. (2016b). Appraisal of diversity and functional attributes of thermotolerant wheat associated bacteria from the peninsular zone of India. *Saudi J. Biol. Sci.* https://doi.org/10.1016/j.sjbs.2016.01.042.

Verma, P., Yadav, A.N., Kumar, V. et al. (2017). Beneficial plant-microbes interactions: biodiversity of microbes from diverse extreme environments and its impact for crops improvement. In: *Plant-Microbe Interactions in Agro-Ecological Perspectives* (eds. D.P. Singh, H.B. Singh and R. Prabha), 543–580. Singapore: Springer Nature https://doi.org/10.1007/978-981-10-6593-4_22.

Viikari, L., Tenkanen, M., and Suurnakki, A. (2001). Biotechnology in the pulp and paper industry. In: *Biotechnology*, vol. 10 (ed. H.J. Rehm), 523–546. VCH-Wiley.

Vijayan, N., Swapna, T., Haridas, M., and Sabu, A. (2017). Therapeutic enzymes: l-Glutaminase. In: *Current Developments in Biotechnology and Bioengineering* (eds. A. Pandey, S. Negi and C.R. Soccol), 233–248. Elsevier.

Wainø, M. and Ingvorsen, K. (2003). Production of β-xylanase and β-xylosidase by the extremely halophilic archaeon *Halorhabdus utahensis*. *Extremophiles* 7: 87–93.

Wakayama, M., Yamagata, T., Kamemura, A. et al. (2005). Characterization of salt-tolerant glutaminase from *Stenotrophomonas maltophilia* NYW-81 and its application in Japanese soy sauce fermentation. *J. Ind. Microbiol. Biotechnol.* 32: 383–390.

Wang, C.-Y., Hsieh, Y.-R., Ng, C.-C. et al. (2009). Purification and characterization of a novel halostable cellulase from *Salinivibrio* sp. strain NTU-05. *Enzyme Microb. Technol.* 44: 373–379.

Wang, C.Y., Chan, H., Lin, H.T., and Shyu, Y.T. (2010). Production, purification and characterisation of a novel halostable xylanase from *Bacillus* sp. NTU-06. *Ann. Appl. Biol.* 156: 187–197.

Wang, Y., Ma, R., Li, S. et al. (2018). An alkaline and surfactant-tolerant lipase from *Trichoderma lentiforme* ACCC30425 with high application potential in the detergent industry. *AMB Express* 8: 95. https://doi.org/10.1186/s13568-018-0618-z.

Ward, O.P., Moo-Young, M., and Venkat, K. (1989). Enzymatic degradation of cell wall and related plant polysaccharides. *Crit. Rev. Biotechnol.* 8: 237–274.

Wejse, P.L., Ingvorsen, K., and Mortensen, K.K. (2003). Purification and characterisation of two extremely halotolerant xylanases from a novel halophilic bacterium. *Extremophiles* 7: 423–431.

Wiseman, A. (1985). *Handbook of Enzyme Biotechnology*. Chichester: Horwood.

Wu, Y.T., Zhou, N.D., Zhou, Z.M. et al. (2014). A thermo-stable lysine aminopeptidase from *Pseudomonas aeruginosa*: isolation, purification, characterization, and sequence analysis. *J. Basic Microbiol.* 54: 1110–1119.

Yadav A.N. (2009) Studies of Methylotrophic Community from the Phyllosphere and Rhizosphere of Tropical Crop Plants. M.Sc. Thesis, Bundelkhand University, pp. 66, DOI: 10.13140/2.1.5099.0888.

Yadav A.N. (2015) Bacterial diversity of cold deserts and mining of genes for low temperature tolerance. Ph.D. Thesis, IARI, New Delhi/BIT, Ranchi pp. 234, DOI: 10.13140/RG.2.1.2948.1283/2.

Yadav, A.N. and Saxena, A.K. (2017). Biodiversity and potential biotechnological applications of halophilic microbes for sustainable agriculture. *J. Appl. Biol. Biotechnol.* 5: 1–8.

Yadav, A.N. and Saxena, A.K. (2018). Biodiversity and biotechnological applications of halophilic microbes for sustainable agriculture. *J. Appl. Biol. Biotechnol.* 6: 1–8.

Yadav, A.N., Sharma, D., Gulati, S. et al. (2015a). Haloarchaea endowed with phosphorus solubilization attribute implicated in phosphorus cycle. *Sci. Rep.* 5: https://doi.org/10.1038/srep12293.

Yadav, A.N., Verma, P., Kumar, M. et al. (2015b). Diversity and phylogenetic profiling of niche-specific bacilli from extreme environments of India. *Ann. Microbiol.* 65: 611–629.

Yadav, A.N., Sachan, S.G., Verma, P. et al. (2016). Cold active hydrolytic enzymes production by psychrotrophic bacilli isolated from three sub-glacial lakes of NW Indian Himalayas. *J. Basic Microbiol.* 56: 294–307.

Yadav, A.N., Kumar, R., Kumar, S. et al. (2017a). Beneficial microbiomes: biodiversity and potential biotechnological applications for sustainable agriculture and human health. *J. Appl. Biol. Biotechnol.* 5: 1–13.

Yadav, A.N., Verma, P., Kumar, R. et al. (2017b). Current applications and future prospects of eco-friendly microbes. *EU Voice* 3: 21–22.

Yadav, A.N., Verma, P., Kumar, V. et al. (2017c). Extreme cold environments: a suitable niche for selection of novel psychrotrophic microbes for biotechnological applications. *Adv. Biotechnol. Microbiol.* 2: 1–4.

Yadav, A.N., Verma, P., Sachan, S.G., and Saxena, A.K. (2017d). Biodiversity and biotechnological applications of psychrotrophic microbes isolated from Indian Himalayan regions. *EC Microbiol. ECO* 01: 48–54.

Yadav, A.N., Kour, D., Singh, B. et al. (2018a). Secondary metabolites: from metabolic engineering to synthetic biology/metabolomics for secondary metabolites. In: *New and Future Developments in Microbial Biotechnology and Bioengineering-Metabolites* (ed. P.A. Gupta Vijai Kuamr), 225–265. Amsterdam: Elsevier.

Yadav, A.N., Kumar, V., Prasad, R. et al. (2018b). Microbiome in crops: diversity, distribution and potential role in crops improvements. In: *Crop Improvement through Microbial Biotechnology* (eds. R. Prasad, S.S. Gill and N. Tuteja), 305–332. USA: Elsevier.

Yadav, A.N., Verma, P., Kumar, S. et al. (2018c). Actinobacteria from Rhizosphere: molecular diversity, distributions and potential biotechnological applications. In: *New and Future Developments in Microbial Biotechnology and Bioengineering* (eds. B. Singh, V. Gupta and A. Passari), 13–41. USA: https://doi.org/10.1016/B978-0-444-63994-3.00002-3.

Yadav, A.N., Verma, P., Sachan, S.G. et al. (2018d). Psychrotrophic microbiomes: molecular diversity and beneficial role in plant growth promotion and soil health. In: *Microorganisms for Green Revolution: Volume 2: Microbes for Sustainable Agro-Ecosystem* (eds. D.G. Panpatte, Y.K. Jhala, H.N. Shelat and R.V. Vyas), 197–240. Singapore: Springer Singapore https://doi.org/10.1007/978-981-10-7146-1_11.

Yang, S., Yan, Q., Jiang, Z. et al. (2006). High-level of xylanase production by the thermophilic *Paecilomyces themophila* J18 on wheat straw in solid-state fermentation. *Bioresour. Technol.* 97: 1794–1800.

Yang, F., Yang, L., Guo, X. et al. (2014). Production and purification of a novel xanthan lyase from a xanthan-degrading *Microbacterium* sp. strain XT11. *Sci. World J.* https://doi.org/10.1155/2014/368434.

Yazdi, M., Yazdi, Z., Zarrini, G., and Ghasemian, A. (2008). Purification and characterization of extra-cellular cholesterol oxidase from *Rhodococcus* sp. PTCC 1633. *Biotechnology* 7: 751–756.

Ye, M., Liu, X., and Zhao, L. (2013). Production of a novel salt-tolerant L-glutaminase from *Bacillus amyloliquefaciens* using agro-industrial residues and its application in Chinese soy sauce fermentation. *Biotechnology* 12: 25–35.

Yip, V.L. and Withers, S.G. (2006). Breakdown of oligosaccharides by the process of elimination. *Curr. Opin. Chem. Biol.* 10: 147–155.

Yoshimoto, T., Orawski, A.T., and Simmons, W.H. (1994). Substrate specificity of aminopeptidase P from *Escherichia coli*: comparison with membrane-bound forms from rat and bovine lung. *Arch. Biochem. Biophys.* 311: 28–34.

Yoshimune, K., Galkin, A., Kulakova, L. et al. (2005). Cold-active DnaK of an Antarctic psychrotroph *Shewanella* sp. Ac10 supporting the growth of dnaK-null mutant of *Escherichia coli* at cold temperatures. *Extremophiles* 9: 145–150.

Yu, S., Christensen, T.M., Kragh, K.M. et al. (1997). Efficient purification, characterization and partial amino acid sequencing of two α-1, 4-glucan lyases from fungi. *Biochim. Biophys. Acta Protein Struct. Mol. Enzymol.* 1339: 311–320.

Yu, X., Hao, W., Xie, G. et al. (2015). Clone, purification and characterization of thermostable aminopeptidase ST1737 from *Sulfolobus tokodaii*. *Chem. Res. Chin. Univ.* 31: 98–102.

Yumoto, I., Hirota, K., Sogabe, Y. et al. (2003). *Psychrobacter okhotskensis* sp. nov., a lipase-producing facultative psychrophile isolated from the coast of the Okhotsk Sea. *Int. J. Syst. Evol. Microbiol.* 53: 1985–1989.

Zeikus, J. (1979). Thermophilic bacteria: ecology, physiology and technology. *Enzyme Microb. Technol.* 1: 243–252.

Zeng, R., Zhang, R., Zhao, J., and Lin, N. (2003). Cold-active serine alkaline protease from the psychrophilic bacterium *Pseudomonas* strain DY-A: enzyme purification and characterization. *Extremophiles* 7: 335–337.

Zeng, R., Xiong, P., and Wen, J. (2006). Characterization and gene cloning of a cold-active cellulase from a deep-sea psychrotrophic bacterium *Pseudoalteromonas* sp. DY3. *Extremophiles* 10: 79–82.

Zhai, C., Cao, J., and Wang, Y. (2003). Cloning and expression of a pectate lyase gene from *Bacillus alcalophillus* NTT33. *Enzyme Microb. Technol.* 33: 173–178.

Zhang, J.-W. and Zeng, R.-Y. (2008). Purification and characterization of a cold-adapted α-amylase produced by *Nocardiopsis* sp. 7326 isolated from Prydz Bay, Antarctic. *Mar. Biotechnol.* 10: 75–82.

Zhang, Y., Ji, F., Wang, J. et al. (2018). Purification and characterization of a novel organic solvent-tolerant and cold-adapted lipase from *Psychrobacter* sp. ZY124. *Extremophiles* 13: 1–14.

Zhou, X., Chen, H., and Li, Z. (2004). CMCase activity assay as a method for cellulase adsorption analysis. *Enzyme Microb. Technol.* 35: 455–459.

Zhu, B., Tan, H., Qin, Y. et al. (2015). Characterization of a new endo-type alginate lyase from *Vibrio* sp. W13. *Int. J. Biol. Macromol.* 75: 330–337.

Zimmer, C., Platz, T., Cadez, N. et al. (2006). A cold active (2 R, 3 R)-(−)-di-O-benzoyl-tartrate hydrolyzing esterase from *Rhodotorula mucilaginosa*. *Appl. Microbiol. Biotechnol.* 73: 132–140.

# 17

# Recent Development in Ferulic Acid Esterase for Industrial Production

*Surabhi Singh*[1], *Om Prakash Dwivedi*[1], *and Shashank Mishra*[2]

[1] *Department of Botany, S.V.M. Science and Technology P.G. College, Lalganj, 230 132, India*
[2] *Quality Control and Quality Assurance Laboratory, Biotech Park, Sector G, Jankipuram, Lucknow, 226 021, India*

## 17.1 Introduction

The demand of biofuels, bioenergy, and chemicals has kindled worldwide interest in plant biomass as a renewable source. Millions of tons of agricultural wastes and forest litter are produced per annum; these agro residues can be an abundant source of renewable compounds such as ferulic acid, sinapic acid, caffeic acid, *p*-coumaric acid, 4-vinyl guaiacol, and vanillin. Hydrolysis of highly recalcitrant cellulose and cell wall polysaccharides involves many alkali treatments of agrowaste, which break the ester/ether bonds and releases the products (Nieter et al. 2016), but nature friendly approach demands a microbiological or enzymatic action. The use of enzymatic system in organic synthesis has become a fundamental tool for selective, efficient, and economical transformations, with the goal of using less aggressive reagents, less organic solvents, lower temperatures, and shorter pathways (Parveen et al. 2016). The enzymes including hemicellulases, $\beta$-glucanases, xylanases, pectinases, xyloglucanases, and esterases (Fazary and Ju 2007) are capable of hydrolyzing the ester and ether bond present in plant cell walls releasing bioactive compounds from agroresidues. Ferulic/*p*-coumaric acid esterases represent a subclass of carboxyl esterases that can release the phenolic compounds from plant cell wall. Ferulic acid esterases (FAE; EC 3.1.1.73) are regarded as the key role to release ferulic acid by hydrolysing ferulate esters involved in the cross-links between hemicelluloses and lignin (Uraji et al. 2013).

In recent years, many microbial FAE-related protein sequences have been discerned in growing genome database. Based on substrate utilization, primary amino acid sequence and substrate specificity, four classes of ferulic acid esterases have been characterized and named as type A, B, C, and D. Type A FAEs hydrolyses ester functional groups and release dehydrodiferulic acids from arabinoxylans (Kuhnel et al. 2012). Type B FAEs show poor hydrolytic activity toward acetylated substrates, such as $\alpha$-naphthyl acetate and acetylated xylan (Crepin et al. 2003; Kroon et al. 2000). In addition, Type B FAE releases FA (ferulic acid) that is ester-linked to arabinose of arabinoxylans or pectic arabinans. Type C FAEs show broad substrate specificity and are active toward ferulic, *p*-coumaric, caffeic, and sinapinic acids. Type C FAEs are also active on arabinoxylans as well as pectins and hydrolyze all hydroxycinnamic

*Bioprocessing for Biomolecules Production*, First Edition.
Edited by Gustavo Molina, Vijai Kumar Gupta, Brahma N. Singh, and Nicholas Gathergood.

acid methyl esters, but do not release dehydrodiferulic acids (Kuhnel et al. 2012). Type D FAEs exhibit broad substrate specificity and are active toward ferulic, *p*-coumaric, caffeic, and sinapinic acids and particularly exhibit highest activity toward substrate containing acetyl residues such as 4-nitrophenyl acetate (Kuhnel et al. 2012).

FAEs have a broad range of potential applications in the chemical industry, such as in the pharmaceutical sector (for production of medicinal compounds) and in the leather industry (for liberation of lignin fragments) by hydrolyzing residual xylan (Mathew and Abraham 2005). FAE is also capable of releasing FA from the agroresidues. FA proved to be a renewable source of the biocatalytic or chemical conversion to other useful aromatic chemicals from agricultural byproducts in nature (Mathew and Abraham 2006). One of the most documented activities of FA is its antioxidant property that is due to its phenolic nucleus and extended side chain (Srinivasan et al. 2007). FA also has antifungal and antidiabetic properties (Aragno et al. 2000). FA has been shown used in the production of many flavoring agent, such as vanillin, which has a high demand in pharmaceutical, beverages, and food industries (Hagedorn and Kaphammer 1994). In spite of its widespread application, commercialization of natural FA has been limited by low availability and high cost. A worthwhile goal is to produce the required demand for ferulic acid from agroresidues using FAE.

## 17.2 Microbial Production of Ferulic Acid Esterase

Microbial enzyme productions are often more stable, convenient, facile, and active than their corresponding animal and plant enzymes (Hasan et al. 2006; Huang et al. 2011). FAEs can be produced by both submerged and solid-state fermentation (SSF). In SSF, the productivity per reactor volume is much higher than that attained in submerged culture (Grajek 1987). The operational costs are also lower because of a simpler set up and machinery are required (Roche et al. 1994). Solid culture processes are practical for complex substrates, including agricultural, forestry, food-processing residues and wastes, which are used as inducing carbon sources for the production of FAEs.

*Aspergillus niger* (Panagiotou et al. 2006), several *penicillium* sp. (Donaghy et al. 1998) and the thermophilic fungus *Sporotrichum thermophile* (Topakas et al. 2003) were capable of producing FAEs when grown on agricultural residues such as sugar beet pulp, wheat bran, wheat straw, and brewer's spent grain under SSF. *Penicillium brasilianum* was found to be the most effective FAE producer screened under SSF (Panagiotou et al. 2006).

## 17.3 Microbial Assay for FAE Production

Various quantitative methods have been introduced for the screening of FAE producing microorganisms. They are based on the measurement of the enzyme activity of FAE using high-performance liquid chromatography (HPLC) and UV detection. This prominent method is still used in 2018 (Hegde and Muralikrishna 2009; Chauvigne-Hines et al. 2012). Other methods have also been investigated, i.e., thin layer chromatography and gas chromatography (Borneman et al. 1990), mass spectrometry (Mastihuba et al. 2002) and capillary zone electrophoresis (Donaghy and Mckay 1997). However, these methods were not useful for the screening of large number of microbes. To overcome this, several chromogenic and flourogenic protocols have been introduced, such as phenol red test, which is based on the pH shift of phenol red to yellow as a result of acid formation after the FAE action (Wu et al. 2011). However, the phenol red test is not completely efficient to determine the activity of FAE. For that another accurate, rapid, and easy method i.e. agar plate method (Donaghy et al. 1998) has been introduced, in which FAE producing microorganisms formed a clear halo zone around the culture. Agar screening assay involves the substitution of a carbohydrate component of selective agar with the synthetic phenolic ester ethyl ferulate.

FAEs were first detected in culture of *Streptomyces olivochromogenes* releasing FA from wheat bran (Mackenzie et al. 1987), after which a number of microbes were isolated and characterized with FAE activity such as *Aspergillus awamori* VTT-D-71025, *Aspergillus nidulans, Bacillus subtilis* NCIMB11034, *Clostridium stercorarium* NCIMB 11754 and *P. brasilianum* IBT 20888 (Donaghy et al. 1998, 2000; Kroon et al. 2000; Shin and Chen 2007; Gong et al. 2013) (Table 17.1).

## 17.4 Worldwide Demand and Production of FAE

FAE hydrolyzes the ester bonds between sugar residues and phenolic compounds present in the plant cell wall, therefore facilitating access of hydrolases to the backbone of cell wall polymers. This activity is useful in a diverse array of applications, such as to obtain high-quality fibers for industry and release of free phenolic acids from the ester forms such as FA, *p*-coumaric, caffeic, and sinapic acids. These phenolic compounds have application in the food, health, cosmetic, and pharmaceutical industries (Kroon and Williamson 1999). The next important non-food application of FAE is the production of fuel ethanol from renewable lignocellulosic materials (Tabka et al. 2006) (Figure 17.1).

Currently, industrial enzyme companies sell FAE for a wide variety of applications viz. for pulp and paper process, animal feed additives, and in the production of FA, which is a precursor for natural vanillin. In a biorefinery setting, the production of value-added chemicals besides biofuels and bioenergy derived from a generic feedstock represents an important component of the overall value chain and economics (Bozell 2008). Detergents (37%), textiles (12%), starch (11%), baking (8%), and animal feed (6%) are the main industries; in total, these industries use about 74% of industrially produced FAE. It is also indirectly used in biocatalytic processes involving living or dead and permeabilized microorganisms.

## 17.5 Process Optimization for FAE Production

Finding the most appropriate reaction medium and parameters that favor the modulation of FAE properties toward the synthesis is a key step in the enzymatic acylation of oligosaccharides with phenolic acids. This is due to the reaction media having the ability to alter the reaction equilibrium. Therefore, nonconventional systems with low-water activity are more suited for synthesis reactions.

**Table 17.1** List of ferulic acid esterase (FAE) producing microorganisms (bacteria and fungi).

| Microorganisms | Type of FAE | References |
|---|---|---|
| *Schizophyllum commune* ATCC 38548 | FAE | Johnson et al. 1989 |
| *Fibrobacter succinogenes* S851 | FAE | McDermid et al. 1990 |
| *Aspergillus oryzae* VTTD 85248 | FAE | Tenkanen et al. 1991 |
| *Aspergillus phoenicis* IMI 211395 | FAE | Smith et al. 1991 |
| *Aspergillus terreus* VTT-D-82209 | FAE | Tenkanen et al. 1991 |
| *Neocallimastix* | FAE-I, II | Borneman et al. 1992 |
| *Panicillium pinophilum* CMI 87160ii | FAE | Castanares et al. 1992 |
| *Aspergillus niger* I-1472 | FAE | Faulds and Williamson 1993 |
| *Pseudomonas fluorescens* | FAE-D | Ferreira et al. 1993 |
| *Aspergilus niger* | FAE-I, II | Kroon and Williamson 1996 |
| *Aspergillus tubingensis* | FAE-A | de Vries et al. 1997 |
| *Penicillium expansum* | FAE | Donaghy and McKay 1997 |
| *Aspergillus awamori* IFO4033 | FAE | Koseki et al. 1998 |
| *Bacillus substilis* NCIMB11034 | FAE | Donaghy et al. 1998 |
| *Lactobacillus fermentum* NCFB1751 | FAE | Donaghy et al. 1998 |
| *Clostridium stercorarium* NCIMB 11754 | FAE | Donaghy et al. 2000 |
| *Orpinomyces* sp PC-2 | FAE-A | Blum et al. 2000 |
| *Penicillium funiculosum* IMI-134756 | FAE | Kroon et al. 2000 |
| *Streptomyces olivochromogenes* | FAE-A, B and C | Donaghy et al. 2000 |
| *Aspergillus pullulans* NRRLY 2311–1 | FAE | Rumbold et al. 2003 |
| *Aureobasidium pullulans* | FAE-B | Rumbold et al. 2003 |
| *Fusarium proliferatum* | FAE-B | Rumbold et al. 2003 |
| *Neurospora crassa* | FAE-B and D | Crepin et al. 2003 |
| *Streptococcus thermophile* ATCC 34628 | FAE | Topakas et al. 2003 |
| *Streptomyces avermitilis* CECT 3339 | FAE | Bartolome et al. 2003 |
| *Fusarium oxysporum* | FoFAE-A and B | Nikos and Ioulia 2004 |
| *Fusarium oxysporum* F3 | FAE | Topakas and Christakopoulos 2004 |
| *Neurospora crassa* | FAE-B | Crepin et al. 2004 |
| *Sporotrichum thermophile* | StFAE-A | Topakas and Christakopoulos 2004 |
| *Talaromyces stipitatus* | TsFAE-A and B | Garcia-Conesa et al. 2004 |
| *Aspergillus awamori* VTT-D-71025 | FAE | Mathew and Abraham 2005 |
| *Aspergillus flavipes* | FAE | Mathew and Abraham 2005 |
| *Aspergilus niger* | FAE-A (III) | Faulds et al. 2005 |
| *Sporotrichum thermophile* | StFAE-A, B and C | Vafiadi et al. 2005 |
| *Streptomyces avermitilis* UAH 30 | FAE | Faulds et al. 2005 |
| *Aspergillus niger* NRRL3 | FAE | Shin and Chen 2006 |
| *Fusarium proliferatum* NRRL 26517 | FAE | Shin and Chen 2006 |
| *Fusarium verticillioides* NRRL 26517 JSC-3 | FAE | Shin and Chen 2006 |

*(Continued)*

**Table 17.1** (Continued)

| Microorganisms | Type of FAE | References |
|---|---|---|
| *Aspergillus nidulans* | FAE-B | Shin and Chen 2007 |
| *Streptomyces* sp. S10 | FAE | Mukherjee et al. 2007 |
| *Streptomyces tendae* | FAE | Ferreira et al. 2007 |
| *Aspergillus niger* | FAE | Ou et al. 2011 |
| *Thermobifida* sp. | FAE | Huang et al. 2011 |
| *Chrysosporium lucknowense C1* | FAE | Kuhnel et al. 2012 |
| *Aspergillus flavus* | FAE | Zhang et al. 2013 |
| *Aspergillus usamii* | FAE | Gong et al. 2013 |
| *Pleurotus eryngii* | FAE | Nieter et al. 2014 |
| *Mucor hiemalis* NCIM 837 | FAE | Singh et al. 2015 |

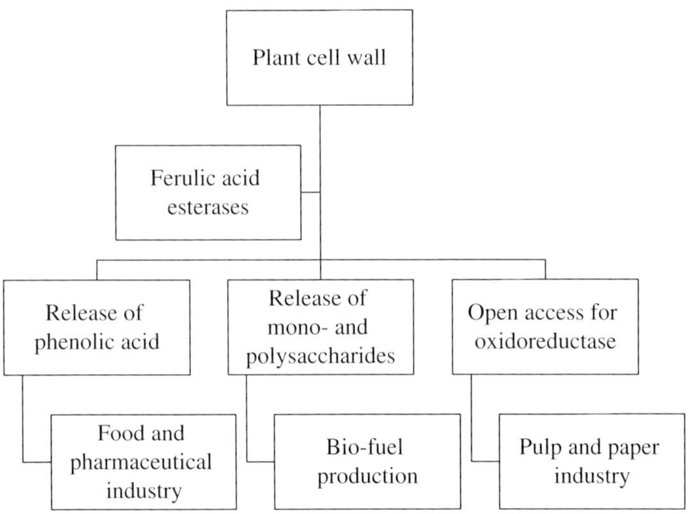


**Figure 17.1** Application of the ferulic acid esterase.

Process parameters like pH and temperature have been observed to alter the FAE production rate. The studies carried out on FAE production show that the optimal pH and temperature for maximum activity ranges from 5.0 to 7.0 and 30 to 60 °C, respectively. *S. olivochromogenes* NRCC 2258 showed 2350 mU $mg^{-1}$ FAE activity at 37 °C (Johnson et al. 1989), whereas *S. thermophile* ATCC 34628 (Topakas et al. 2003) showed 156 mU mg FAE activity at 50 °C. In addition to the parametric conditions, carbon and nitrogen sources are essential media components, required for proper growth and metabolism of the microorganisms. Thus, these are required in appropriate amounts in media for enzyme production. Research has shown that *A. niger* favors FAE production with sugar beet pulp whereas brewer's spent grain is the most suitable natural carbon source for FAE production from *Lactobacillus acidophilus* K1 and *Lactobacillus rhamnosus* (Szwajgier and Jakubczyk 2011). Other agro-residues such as rice bran, corn pectin, and German wheat bran are also used in FAE production (Szwajgier and Jakubczyk 2011). Topakas et al. (2007) reported that complex carbon sources viz. wheat bran, and sugar beet pulp favor FAE production.

Medium optimization or parametric optimization is conducted either by "one variable at a time" or by using different statistical tools such as response surface methodology (RSM). RSM is superior as it analyzes and considers all the possible interactions between variables. Optimization of culture conditions by one factor at a time is not only time consuming but also leads to an incomplete understanding of the system, resulting in confusion and failure of predictive replication. To overcome these issues, RSM is employed more often, and this technique explores the interactions between numerous illustrative variables and one or more response variables.

## 17.6 Recent Development and Genetic Engineering for the Enhancement of FAE Production

Worldwide utilization of FAE suggests that it might have an important role in development and therefore is a target for microbial engineering studies for industrial biomass. Various reports have been published for the enhancement of FAE production using recombinant microbes. AtFAE type A, extracted from maize and triticale bran, can be used as food and fodder for the production of fine chemicals such as FA and *p*-coumaric acid (Zwane et al., 2014). FAE can also be isolated from a wide range of microorganisms (Donaghy and McKay 1997; de Vries et al. 1997) when they are grown on complex substrates, such as xylan, pectin, wheat bran, or sugar beet pulp, as given in Table 17.1. Most of these enzymes are active, with feruloylated oligomers obtained from xylan or pectin, but exhibit less or no activity with the polymeric substrates.

The feruloyl esterase A (FAE-A)-encoding gene (faeA) from *A. niger* has been cloned, and induction of this gene has been studied at the protein level (de Vries et al. 1997). A high level of FAE-A activity in the culture filtrate was detected when an overproducing transformant was grown on a medium containing wheat arabinoxylan. Low levels of FAE-A activity were detected when *A. niger* was grown on sugar beet pectin. Addition of ferulic acid to media containing xylan increased the levels of FAE-A activity in culture filtrates (Faulds et al. 2005). Gong et al. (2013) studied the cloning and heterologous expression of the cDNA gene (AuFAEs) in *Pichia pastoris* GS11S. In further studies, the encoding sequence of putative feruloyl esterase A (AfFAE-A) was cloned from genomic DNA from *Aspergillus flavus* and expressed in *Pichia pastoris*. The recombinant ferulic acid esterases exhibited good pH and thermal tolerance and broad reaction pH, which are important parameter such as for industrial application. Recombinant AfFAE-A expressed in *Puccinia pastoris* would facilitate large-scale fermentation. A putative $\alpha/\beta$ hydrolase fold-encoding gene (locus tag TTE1809) from the genome of *Thermoanaerobacter tengcongensis* was cloned and expressed in *Escherichia coli* as a possible source of thermostable FAE for the production of antioxidant phenolic acids from biomass.

## 17.7 Conclusion

Industries such as pharmaceutical, cosmetic, baking, chocolate, paper, and food have high demand for FAE. Microbes can be a potential tool for FAEs production. Although FAEs have been identified in various plant-cell-wall-degrading microbes, to date fungi are still the main source of FAE producer used in industry. In order to develop microbial production, some features such as optimum parameters for microbial growth and purification protocols have to be taken into consideration and carefully investigated. Improved strategies and new technologies have been developed, and genetically modified microbes are now able to offer different parameters to sustain growth and enzyme production on a large scale together with the optimal bioreactor.

## References

Aragno, M., Parola, S., Tamagno, E. et al. (2000). Oxidative derangement in rat synaptosomes induced by hyperglycaemia: restorative effect of dehydroepiandrosterone treatment. *Biochem. Pharmacol.* 60: 389–395.

Bartolome, B., Gomes-Cordoves, C., Sancho, A.I. et al. (2003). Growth and release of hydroxycinnamic acids from Brewer's spent grain by Streptomyces avermitilis CECT 3339. *Enzyme Microb. Technol.* 32: 140–144.

Blum, D.L., Kataeva, I.A., and Li, X.L. (2000). Feruloyl esterase activity of the *Clostridium thermocellum* cellulosome can be attributed to previously unknown domains of XynY and XynZ. *J. Bacteriol.* 182: 1346–1351.

Borneman, W.S., Hartley, R.D., Morrison, W.H. et al. (1990). Feruloyl and *p*-coumaroyl esterase from anaerobic fungi in relation to plant cell wall degradation. *Appl. Microbiol. Biotechnol.* 33: 345–351.

Borneman, W.S., Ljungdahl, L.G., Hartley, R.D., and Akin, E. (1992). Purification and partial characterization of two feruloyl esterases from the anaerobic fungus *Neocallimastix* strain mc-2. *Appl. Environ. Microbiol.* 58: 3762–3766.

Bozell, J.J. (2008). Feedstocks for the future: biorenewable production of chemicals from renewable carbon. *Clean* 36: 641–647.

Castanares, A., McCrae, S.I., and Wood, T.M. (1992). Purification and properties of a feruloyl/p-coumaroyl esterase from the fungus *Penicillium pinophilum*. *Enzyme Microb. Technol.* 14: 875–884.

Chauvigne-Hines, L.M., Anderson, L.N., Weaver, H.M. et al. (2012). A suite of activity-based probes for cellulose degrading enzymes. *J. Am. Chem. Soc.* 134: 20521–20532.

Crepin, V.F., Faulds, C.B., and Connerton, I.F. (2003). Production and characterization of the *Talaromyces stipitatus* feruloyl esterase FAEc in *Pichia pastoris*: identification of the nucleophilic serine. *Protein Expression Purif.* 29: 176–184.

Crepin, V.F., Faulds, C.B., and Connerton, I.F. (2004). Functional classification of the microbial feruloyl esterases. *Appl. Microbiol. Biotechnol.* 63: 647–652.

Donaghy, J. and McKay, A.M. (1997). Purification and characterization of a feruloyl esterase from the fungus *Penicillium expansum*. *J. Appl. Microbiol.* 83: 718–726.

Donaghy, J., Kelly, P.F., and McKay, A.M. (1998). Detection of ferulic acid esterase production by Bacillus spp. and lactobacilli. *Appl. Microbiol. Biotechnol.* 50: 257–260.

Donaghy, J.A., Bronnenmeier, K., Soto-Kelly, P.F., and McKay, A.M. (2000). Purification and characterization of an extracellular feruloyl esterase from the thermophilic anaerobe *Clostridium stercorarium*. *J. Appl. Microbiol.* 88: 458–466.

Faulds, C.B. and Williamson, G. (1993). Ferulic acid esterase from *Aspergillus niger*: purification and partial characterization of two forms from a commercial source of pectinase. *Biotechnol. Appl. Biochem.* 17: 349–359.

Faulds, C.B., Molina, R., Gonzalez, R. et al. (2005). Probing the determinants of substrate specificity of a feruloyl esterase, AnFAEA, from *Aspergillus niger*. *Eur. J. Biochem.* 272: 4362–4371.

Fazary, A.E. and Ju, Y.H. (2007). Feruloyl esterases as biotechnological tools: current and future perspectives. *Acta Biochim. Biophys. Sin.* 39: 811–828.

Ferreira, L.M., Wood, T.M., Williamson, G. et al. (1993). A modular esterase from *Pseudomonas fluorescens* subsp. cellulosa contains a non- catalytic cellulose-binding domain. *Biochem. J* 29: 4349–4355.

Ferreira, P., Diez, N., Faulds, C.B. et al. (2007). Release of ferulic acid and feruloylated oligosaccharides from sugar beet pulp by *Streptomyces tendae*. *Bioresour. Technol.* 98: 1522–1528.

Garcia-Conesa, M.T., Crepin, V.F., Goldson, A.J. et al. (2004). The feruloyl esterase system of *Talaromyces stipitatus*: production of three discrete feruloyl esterases, including a novel enzyme, TsFAEC, with a broad substrate specificity. *J. Biotechnol.* 108: 227–241.

Gong, Y.Y., Yin, X., Zhang, H.M. et al. (2013). Cloning, expressly of a feruloyl esterase from *Aspergillus usamii* E001 and its applicability in generating ferulic acid from wheat bran. *J. Ind. Microbiol. Biotechnol.* 40: 1433–1441.

Grajek, W. (1987). Comparative studies on the production of cellulases by thermophilic fungi in submerged and solid state fermentation. *Appl. Microbiol. Biotechnol.* 26: 126–129.

Hagedorn, S. and Kaphammer, B. (1994). Microbial biocatalysis in the generation of flavor and fragrance chemicals. *Ann. Rev. Microbiol.* 48: 773–800.

Hasan, F., Shah, A.A., and Hameed, A. (2006). Industrial applications of microbial lipases. *Enzyme Microb. Technol.* 39: 235–251.

Hegde, S. and Muralikrishna, G. (2009). Isolation and partial characterization of alkaline feruloyl esterases from *Aspergillus niger* CFR 1105 grown on wheat bran. *World J. Microbiol. Biotechnol.* 25: 1963–1969.

Huang, Y.C., Chen, Y.F., Chen, C.Y. et al. (2011). Production of ferulic acid from lignocellulolytic agricultural biomass by *Thermobifida fusca* thermostable esterase produced in Yarrowia lipolytica transformant. *Bioresour. Technol.* 102: 8117–8122.

Johnson, K.G., Silva, M.C., MacKenzie, C.R. et al. (1989). Microbial degradation of hemicellulosic materials. *Appl. Biochem. Biotechnol.* 21: 245–258.

Koseki, T., Furuse, S., Iwano, K., and Matsuzawa, H. (1998). Purification and characterization of a feruloyl esterase from *Aspergillus awamori*. *Biosci. Biotechnol. Biochem.* 62: 2032–2034.

Kroon, P.A. and Williamson, G. (1996). Release of ferulic acid from sugar-beet pulp by using arabinanase, arabinofuranosidase and an esterase from *Aspergillus niger*. *Biotechnol. Appl. Biochem.* 23: 263–267.

Kroon, P.A. and Williamson, G. (1999). Hydroxycinnamates in plants and food: current and future perspectives. *J. Sci. Food Agric.* 79: 355–361.

Kroon, P.A., Williamson, G., Fish, N.M. et al. (2000). A modular esterase from *Penicillium funiculosum* which releases ferulic acid from plant cell walls and binds crystalline cellulose contains a carbohydrate binding module. *Eur. J. Biochem.* 267: 6740–6752.

Kuhnel, S., Pouvreau, L., Appeldoorn, M.M. et al. (2012). The ferulic acid esterases of *Chrysosporium lucknowense* C1: purification, characterization and their potential application in biorefinery. *Enzyme Microb. Technol.* 50: 77–85.

Mackenzie, C.R., Bilous, D., Schneider, H., and Johnson, K.G. (1987). Induction of cellulolytic and xylanolytic enzyme systems in *Streptomyces. Appl. Environ. Microbiol.* 53: 2835–2839.

Mastihuba, V., Kremnicky, L., Mastihubova, M. et al. (2002). A spectrophotometric assay for feruloyl esterases. *Anal. Biochem.* 309: 96–101.

Mathew, S. and Abraham, T.E. (2005). Studies on the production of feruloyl esterase from cereal brans and sugar cane bagasse by microbial fermentation. *Enzyme Microb. Technol.* 36: 565–570.

Mathew, S. and Abraham, T.E. (2006). Bioconversions of ferulic acid, a hydroxycinnamic acid. *Crit. Rev. Microbiol.* 32: 115–125.

McDermid, K.P., MacKenzie, C.R., and Forsberg, C.W. (1990). Esterase activities of *Fibrobacter succinogenes* subsp. succinogenes S85. *Appl. Environ. Microbiol.* 56: 127–132.

Mukherjee, G., Singh, R.K., Mitra, A., and Sen, S.K. (2007). Ferulic acid esterase production by *Streptomyces* sp. *Bioresour. Technol.* 98: 211–213.

Nieter, A., Haase-Aschoff, D., Linke, D. et al. (2014). A halotolerant type A feruloyl esterase from *Pleurotus eryngii. Fungal Biol.* 118: 348–357.

Nieter, A., Kelle, S., linke, D., and Berger, R.G. (2016). Feruloyl esterases from *Schizophyllum commune* to treat food industry side streams. *Bioresour. Technol.* 220: 38–46.

Nikos, S.H. and Ioulia, S. (2004). Enantioselectivity and diastereoselectivity in the transesterification of secondary alcohols mediated by feruloyl esterase from *Humicola insolens. Tetrahedron Lett.* 45: 2755–2757.

Ou, S., Zhang, J., Wang, Y., and Zhang, N. (2011). Production of feruloyl esterase from *Aspergillus niger* by solid-state fermentation on different carbon sources. *Enzyme Res.* https://doi.org/10.4061/2011/848939.

Panagiotou, G., Granouillet, P., and Olsson, L. (2006). Production and partial characterization of arabinoxylan-degrading enzymes by *Penicillium brasilianum* under solid-state fermentation. *Appl. Microbiol. Biotechnol.* 72: 1117–1124.

Parveen, I., Donnison, I.S., Insole, J. et al. (2016). An efficient assay for ferulic acid esterase and para-coumari c acid esterases in plants at physiologically relevant activities. *Chem. Biol. Interf.* 6: 300–311.

Roche, N., Desgragnes, C., and Durand, A. (1994). Study on the solid-state production of a thermostable a-L-arabinofuranosidase of *Thermoascus aurantiacus* on sugar beet pulp. *J. Biotechnol.* 38: 43–50.

Rumbold, K., Biely, P., Mastihubova, M. et al. (2003). Purification and properties of a feruloyl esterase involved in lignocellulose degradation by *Aureobasidium pullulans. Appl. Environ. Microbiol.* 69: 5622–5626.

Shin, H.D. and Chen, R.R. (2006). Production and characterization of a type B feruloyl esterase from *Fusarium proliferatum* NRRL 26517. *Enzyme Microb. Technol.* 38: 478–485.

Shin, H.D. and Chen, R.R. (2007). A type b feruloyl esterase from *Aspergillus nidulans* with broad pH applicability. *Appl. Microbiol. Biotechnol.* 73: 1323–1330.

Singh, S., Nigam, V.K., and Sachan, A. (2015). Parametric optimization of ferulic acid esterase production from *Mucor hiemalis NCIM 837*. *Int. J. Pharm. Pharm. Sci.* 7: 230–233.

Smith, D.C., Bhat, K.M., and Wood, T.M. (1991). Xylan-hydrolysing enzymes from thermophilic and mesophilic fungi. *World J. Microbiol. Biotechnol.* 7: 475–484.

Srinivasan, M., Sudheer, A., and Menon, V. (2007). Ferulic acid: therapeutic potential through its antioxidant property. *J. Clin. Biochem. Nutr.* 40: 92–100.

Szwajgier, D. and Jakubczyk, A. (2011). Production of extracellular ferulic acid esterases by Lactobacillus strains using natural and synthetic carbon sources. *Acta Sci. Pol. Technol. Aliment.* 10: 287–302.

Tabka, M.G., Herpoel-Gimbert, I., Monod, F. et al. (2006). Enzymatic saccharification of wheat straw for bioethanol production by a combined cellulase xylanase and feruloyl esteras treatment. *Enzyme Microb. Technol.* 39: 897–902.

Tenkanen, M., Schuseil, J., Puls, J., and Poutanen, K. (1991). Production, purification and characterisation of an esterase liberating phenolic acids from lignocellulosics. *J. Biotechnol.* 18: 69–84.

Topakas, E. and Christakopoulos, P. (2004). Production and partial characterization of alkaline feruloyl esterases by *Fusarium oxysporum* during submerged batch cultivation. *World J. Microbiol. Biotechnol.* 20: 145–150.

Topakas, E., Kalogeris, E., Kekos, D. et al. (2003). Production and partial characterization of feruloyl esterase by *Sporotrichum thermophile* under solid-state fermentation. *Process Biochem.* 38: 1539–1543.

Topakas, E., Vafiadi, C., and Christakopoulos, P. (2007). Microbial production, characterization and applications of feruloyl esterases. *Process Biochem.* 42: 497–509.

Uraji, M., Kimura, M., inoue, Y. et al. (2013). Enzymatic production of ferulic acid from defatted rice bran by using a combination of bacterial enzymes. *Appl. Biochem. Biotechnol.* 171: 1085–1093.

Vafiadi, C., Topakas, E., Wong, K.K.Y. et al. (2005). Mapping the hydrolytic and synthetic selectivity of a type C feruloyl esterase (StFAEC) from *Sporotrichum thermophile* using alkyl ferulates. *Tetrahedron: Asymmetry* 16: 373–379.

de Vries, R.P., Michelsen, B., Poulsen, C.H. et al. (1997). The FAEa genes from *Aspergillus niger* and *Aspergillus tubingensis* encode ferulic acid esterases involved in the degradation of complex cell wall polysaccharides. *Appl. Environ. Microbiol.* 63: 4638–4644.

Wu, M., Abokitse, K., Grosse, S., and Leisch, H. (2011). New feruloyl esterases to access phenolic acids from grass biomass. *Appl. Biochem. Biotechnol.* 168: 129–143.

Zhang, S.B., Zhai, H.C., Wang, L., and Yu, G.H. (2013). Expression, purification and characterization of a feruloyl esterase A from *Aspergillus flavus*. *Protein Expression Purif.* 92: 36–40.

Zwane, E.N., Rose, S.H., van Zyl, W.H. et al. (2014). Overexpression of *Aspergillus tubingensis* faeA in protease-deficient *Aspergillus niger* enables ferulic acid production from plant material. *J. Ind. Microbiol. Biotechnol.* 41: 1027–1034.

# 18

# Research and Production of Second-Generation Biofuels

*H.L. Raghavendra*[1,2,3], *Shashank Mishra*[4], *Shivaleela P. Upashe*[3,5], *and Juliana F. Floriano*[2]

[1] *Department of Biochemistry, School of Medicine, Wollega University, Nekemte, Ethiopia, raghu.biogem@gmail.com*
[2] *Department of Obstetrics and Gynecology, Medical School of Botucatu, São Paulo State University (UNESP), Botucatu, 18603-970, Brazil*
[3] *College of Medical and Health Sciences, Wollega University, PO Box: 395, Nekemte, Oromia, Ethiopia*
[4] *Quality Control and Quality Assurance Laboratory, Lucknow 226 021, India, smishra83@gmail.com*
[5] *College of Nursing Sciences, School of Health Sciences, Dayananda Sagar University, Shavige Malleshwara Hills, Kumaraswamy Layout, Bangalore 560078, Karnataka, India*

## 18.1 Introduction

The biofuels are in high demand in the present scenario as they are potent alternative to nonrenewable energy sources like fossil fuels, which face continuous depletion because of human activities (Guo et al. 2014). The renewable biofuels are an effective substitute. Moreover, the emission of greenhouse gases can be substantially reduced by replacing fossil fuels with biofuels, thereby contributing to a green atmosphere (Luo et al. 2009). The security of the national energy can be enhanced since the use of biofuels induces rural development as well as employment and economy of a developing nation (Domac et al. 2005). Farmers in the rural areas can earn revenue by selling their waste products to biofuel producers. The technologies used for the production of first-generation biofuels have come under tight regulations since confronted with their impact on food and water linked directly to the living system.

First-generation biofuels were commonly synthesized from food crops used as raw materials, which could be converted to sugars that were later fermented (Deenanath et al. 2012). The continuous and excess use of food crops as raw materials would have an adverse effect on humans and animals due to food shortage as well as contamination of water (Naylor et al. 2011). The complete harvest of wheat, sorghum sugar beet, maize, cassava, and sugarcane would be required to replace 85% of the gasoline consumed throughout the world (Rajagopal and Zilberman 2008).

Taking this into account, second-generation biofuels have evolved that replace the food crops with cellulosic biomass. The cellulosic biomass constitutes 60–90% of terrestrial biomass by weight and is naturally available. It is considered as a potent feedstock for the generation of biofuels based on their huge abundance, sustainability, and low cost (Agbor et al. 2011).

*Bioprocessing for Biomolecules Production*, First Edition.
Edited by Gustavo Molina, Vijai Kumar Gupta, Brahma N. Singh, and Nicholas Gathergood.

### 18.1.1 Second-Generation Biofuels

The first, second-generation biofuels were produced from lignocellulose and are still under the development phase. Technologies to produce these biofuels have posed an economic challenge (Justice 2009), so intensive research is required for finding cheap and better viable technologies for mass production (Laser and College 2009). The operation cost of second-generation feedstock is comparatively high, although the first-generation feedstock is higher than that of the second generation (Canilha et al. 2011).

Synthetic fuels or biosynfuels and cellulosic ethanol or bioethanol are two such biofuels obtained from lignocellulose. The bio-feedstocks like jatropha, which are not used as food, are mainly used for the production of biodiesel. Additionally, the biomass is converted to syngas, which contains a substantial amount of carbon monoxide and hydrogen that is converted catalytically into biokerosene and synthetic diesel by Fischer-Tropsch synthesis.

The requirement of thermal energy during the pretreatment and distillation process along with the need for enzymes during cellulosic hydrolysis makes the production cost of bioethanol considerably high (Kazi et al. 2010). In contrast, the capital investment and feedstock cost are very high for the biosynthetic fuels, while the process cost is comparatively less as compared to bioethanol (Haarlemmer et al. 2012).

### 18.1.2 Feedstocks for Biofuels

The two main feedstocks for the production of second-generation biofuels are lignocellulosic and biodiesel feedstocks.

#### 18.1.2.1 Lignocellulose Biomass

The three major polymers present in lignocellulosic materials are 40–60% of cellulose, 20–40% of hemicellulose and 10–25% of lignin (McKendry 2002). The relative constituent of the polymers differs with different lignocellulosic material where the polymers are interconnected with each other, as encapsulated in Table 18.1.

The culture along with the harvesting time should be appropriately chosen for a plant, as the content of cellulose and lignin is dictated by the harvesting time of the crops (Tutt et al. 2013). The plant walls are primarily cellulose, which is fibrous and crystalline and is a potent source to produce biofuels (Raud et al. 2014). The rigidity of the cellulose molecules is due to the presence of the hydrogen bonds between the molecules, which makes it chemically stable. It hinders depolymerization and is water insoluble in nature. It is difficult for the digestion of the lignocellulosic biomass and fermentation to sugars, since the cellulose molecules are covered by a layer of hemicellulose that forms a cellulose-hemicellulose complex (Zabed et al. 2016).

The hemicellulose is another predominant polymer found in the cell walls of plants, which are made of two compounds called mannan and xylan. It is amorphous in nature and has a random structure, lower degree of polymerization, and is chemically heterogeneous, as compared to that of cellulose (Yu et al. 2017). The rate of hydrolysis of the cellulose can be increased by removing the hemicellulose film, which covers the cellulose fibrils.

The lignocellulosic biomass contains another constituent called lignin in small amounts, which consist of three different polymers – namely, 4-propenyl-2,5-dimethoxy

Table 18.1 Constituent polymers in different lignocellulosic material (Saini et al. 2015).

| No. | Feedstock | Cellulose (%) $(C_6H_5O_{10})_n$ | Hemicellulose (%) $(C_5H_8O_4)_n$ | Lignin (%) |
|---|---|---|---|---|
| 1 | Cotton, flax | 80–95 | 5–20 | — |
| 2 | Grass | 25–40 | 25–50 | 10–30 |
| 3 | Hardwood | 45–2 | 30–5 | 20–40 |
| 4 | Barks of hardwood | 22–40 | 20–38 | 30–55 |
| 5 | Softwood | 42–2 | 27–2 | 28–3 |
| 6 | Barks of softwood | 18–38 | 15–33 | 30–60 |
| 7 | Corn stalk | 39–47 | 26–31 | 3–5 |
| 8 | Corn stover | 38–40 | 28 | 7–21 |
| 9 | Sorghum stalks | 27 | 25 | 11 |
| 10 | Sorghum straw | 32 | 24 | 13 |
| 11 | Rice straw | 28–36 | 23–28 | 12–14 |
| 12 | Wheat straw | 33–38 | 26–32 | 17–19 |
| 13 | Barley straw | 31–45 | 27–38 | 14–9 |
| 14 | Baggase | 32–48 | 19–24 | 23–32 |

phenol, 4-propenyl-2-methoxy phenol, and 4-propenyl phenol. It hinders the attack of microbes inside the cell walls because it makes the cell walls impermeable (Hendriks and Zeeman 2009). The lignin contributes to the structural support of the plants and links all the components of the lignocellulosic biomass. Thus, the plant cells cannot degrade easily and are also insoluble in water. Thus, they resist enzymatic hydrolysis and affect biomass digestion. Biomass digestibility can be highly enhanced by removing the lignin content (Chang and Holtzapple 2000).

#### 18.1.2.2 Forest Residues

Forest residues belong to a class of lignocellulosic feedstock. The fuel wood obtained from the forest, the remains obtained from mills after primary and secondary processing of woods, and the logging residues generated from the harvest contribute to forest residues (Perlack et al. 2005). These residues cannot be effectively used based on factors like environmental conditions that encumber the recovery process in the areas of harvest and high transportation cost due to poor accessibility (Richardson 2008).

#### 18.1.2.3 Perennial Forage Crops

The miscanthus and switchgrass belonging to the species of perennial forage crops are potential feedstocks for second-generation biofuels. The major asset of switchgrass is its quality to adapt to low-quality land, its need for water and nutrition, and most significantly, its positive impact on the environment (Keshwani and Cheng 2009). Miscanthus requires a low level of nitrogen and can adapt to cold condition. Based on these characteristics, these crops have become a promising feedstock in Europe. However, its propagation through rhizome cutting requires a time span of two to three years for maturity. Additionally, the establishment cost is high and the plants cannot grow during winter. Also, the genotype is very rare (Lewandowski et al. 2003).

#### 18.1.2.4 Residues from Agriculture

The residue obtained from crops like rice, wheat, barley, sorghum, and sugarcane are mainly employed for the generation of second-generation biofuels. These feedstocks are cost-effective as they do not require extra land for cultivation, thereby enhancing the carbon balance of the fuel (Searchinger et al. 2008). The germination of seeds is induced as the removal of crop residues increases the temperature of the soil and decreases pests and diseases (Andrews 2006). However, the residues of crops are also required for retaining water in the soil and increase its productivity rate. Thus, removal of crops at a high rate may have adverse effects on the soil (Blanco-Canqui and Lal 2009).

#### 18.1.2.5 Energy Crops

Herbaceous and woody trees are considered as energy crops for the second generation, which is used for the production of biogas or gas biomass for anaerobic digester. They can be converted to liquid fuel and hence are also termed as liquid biomass. The major advantage of woody energy crops like eulpt, poplar, and willow are that they are widely distributed in nature, have a high production rate, and minimum input (Smeets et al. 2007). Since a higher fraction of biomass can be utilized, more energy per unit land can be obtained as compared to food crops. However, they should be grown on uncultivated land so as not to affect the land used for growing food crops.

### 18.1.3 Feedstocks for Biodiesel

#### 18.1.3.1 Microalgae

The lipids required for the production of biodiesel can be obtained from microalgae (Li et al. 2008). The algae grow in saline water, and hence coastal areas, non-arable land and waste water can be used for its cultivation. It can be cultivated in salinity affected or arid regions and become a source of economy. Additionally, the amount of oil produced per unit of land is immense (Schenk et al. 2008).

#### 18.1.3.2 Jatropha

Jatropha belongs to the species of oilseeds, which has a great potential for generation of biodiesel. It can be grown in semiarid conditions with minimum input and is currently grown in tropical and subtropical regions like Africa, Latin America, and Asia (Jongschaap et al. 2007). The yields of biodiesel can be enhanced by cultivating jatropha in fertile lands with good input. Moreover, the oil can be either used directly or can be converted into biofuels.

### 18.1.4 Types of Second-Generation Biofuels

#### 18.1.4.1 Biodiesel

Biodiesel is a second-generation biofuel that is much better than diesel on account of biodegradable nature, amount of sulfur present and flash point. The demand for biodiesel in the automobile industry has substantially increased in present time in order to reduce environmental pollution (Demirbas 2007). The biodiesel is composed of methyl eater or ethyl ester obtained from both used or virgin edible as well as inedible oils. It is mainly obtained from seeds of plants like *Jatropha curcas*, *Pongamia pinnata*, and others. The oils extracted from the seeds are trans esterified for obtaining the desired biodiesel. The glycerol and ester are synthesized by the reaction between an alcohol and triglyceride in the presence of a catalyst during the transesterification reaction. The factors responsible for the yield of the synthesized biodiesel are specific gravity, molar ratio, amount of moisture and water as well as the reaction temperature (Sharma et al. 2008). The molecular weight of methyl esters and ethyl esters are one third of that of triglycerides. Thus, the viscosity of the biodiesel is decreased by eight, while the volatility is enhanced slightly. The esters present in biodiesel have 10–11% oxygen by weight, which triggers more combustion as compared to conventional diesel fuels. Additionally, the biodiesel has higher cetane number (48–60), which augments the quality of ignition and flash point, which results in low volatility. It is devoid of organic compounds, has minimum sulfur content, and possesses high lubricity. The biodiesel can be blended with conventional fuels or can be used directly. Moreover, it does not require special storage places. The emission of carbon monoxide and particulate matter is reduced by about 48% by replacing conventional diesel by biodiesel. Moreover, the emission of polyaromatic hydrocarbon can also be limited by its use. However, biodiesel has a higher viscosity, which leads to the formation of gum. The equivalence ratio (leaner fuel/air mixture) of pure biodiesel is low, and the presence of excess oxygen leads to total combustion. These two factors impart high thermal efficiency to the biofuels.

#### 18.1.4.2 Bioethanol

The hydrolysis of household wastes straw and wood converts them into bioethanol in an economical manner. However, the transformation of cellulosic biomass and starch to bioethanol can only take place after the pretreatment process, which is highly expensive. The prehydrolysis of carbohydrates like cellulose and hemicellulose from plants with dilute acids produces sugars. The sugars are converted to alcohol in the presence of microorganisms like yeast by fermentation. Fermentation is a biological anaerobic process that gives ethanol as the final product.

Sucrose undergoes hydrolysis to form glucose and fructose catalyzed by an enzyme called invertase followed by conversion of both glucose and fructose to ethanol catalyzed by another enzyme called zymase (Scheme 18.1).

$$\underset{\text{sucrose}}{C_{12}H_{22}O_{11}} \xrightarrow{\text{Invertase}} \underset{\text{glucose}}{C_6H_{12}O_6} + \underset{\text{fructose}}{C_6H_{12}O_6}$$

$$C_6H_{12}O_6 \xrightarrow{\text{zymase}} \underset{\text{ethanol}}{C_2H_5OH} + 2CO_2$$

**Scheme 18.1** Conversion of sucrose to ethanol.

The system using bioethanol in the pure state should avoid certain polymers and aluminum. The emission of unwanted gases from bioethanol is minimum as compared to other biofuels. However, it is miscible with water and has a low vapor pressure, leading to complications while handling.

#### 18.1.4.3 Biogas

The conventional fuels can be substituted by biogas to produce power and heat. The biogas is used as a fuel in automobiles. Biogas is obtained by anaerobic digestion of bio wastes in digesters which consists of carbon dioxide and methane. The higher version of biogas is *biomethane,* which can be used to replace natural gas in the production of chemicals. The technology of anaerobic digestion method is eco-friendly and energy efficient, as it can limit the emission of greenhouse gases (Nishio and Nakashimada 2007). Digestate is obtained as a byproduct of the anaerobic digestion process, which is a fertilizer of high quality and can substitute for the chemical fertilizers. Many countries have developed the process of biogas production from biowaste and biomass by different technologies (Wagner 2015). The strong bond between cellulose, hemicellulose, and lignin molecules make the lignocellulose a strong and compact structure. Thus, the pretreatment process of the lignocellulosic mass is required to make the anaerobic digestion faster. The production of the biogas can be enhanced by pretreating the lignocellulose and its polymers with microbes, which facilitate the faster degradation of the biomass (Achinas and Euverink 2016). The lignocellulosic mass is converted to sugars, which can be easily decomposed by the microbes during the pretreatment process.

The factors responsible for the economy of biogas production are an efficient process, logistics, and availability of waste products. The lignocellulosic waste of different varieties can be used for commercial generation of biogas at a low cost. The automobile engines must be upgraded for biogas combustion (Banerjee et al. 2010). Most biogas production in European countries is from the sewage sludge, which is used to generate electrical power.

#### 18.1.4.4 Lean Premixed Prevaporized (LPP) Liquid Biofuels

The lean premixed prevaporized (LPP) liquid biofuels can reduce the emission of gases like carbon monoxide and nitrogen oxide, as the equivalence ratio of lean fuel: air mixture is in the range of 0.2–0.5 (Ohkubo et al. 1997). The biodiesel and ethanol are converted to gas that burns with low emission during the process of LPP combustion.

#### 18.1.4.5 Syngas

The biomass constitutes a mixture of different compounds that, on combustion, produce synthetic gas or biosyngas, which is a mixture of hydrogen (22–32%), carbon monoxide (22–36%), along with carbon dioxide (21–30%) in a major amount and methane (8–11%) in a minor amount. The biosyngas can be formed by processes like steam gasification either in the presence or absence of a catalyst. The process of gasification depends on the oxidant and configuration of the reactor. The sugarcane residues are gasified for the generation of syngas, which has a low heating value (Sayigh 1999). The speed of flame propagation of the hydrogen present in biogas is higher than that of other gases.

#### 18.1.4.6 Dimethyl Ether (DME)

The other name of dimethyl ether (DME) is wood ether, which is obtained from biomass. The empirical formula of dimethyl ether is like ethanol, but the structure and properties of both compounds are different. It is a gaseous fuel, and the efficiency of biomass to dimethyl ether is 38% (Zhang 2009). The boiling point of dimethyl ether is very low and hence must be stored in compressed tanks. It is much less toxic than methanol and is prepared by gasification of wood chip in the presence of oxygen and steam to generate synthetic gas, which is converted to methanol and finally dehydrated to produce dimethyl ether (Neilson 1998). The amount of oxygen in the compound is high, which leads to smokeless combustion along with a cetane number of around 55–60. It does not have similarity with diesel as it has a low boiling point and high vapor pressure, while its heat value and the duration of combustion is lesser than the conventional diesel fuel.

Although the emission of gases like nitrogen oxide and hydrocarbons are minimal the emission of carbon monoxide is slightly higher, making it a fuel with low emission and high thermal efficiency (Huang et al. 2009). The compound is devoid of carbon-carbon bond and sulfur with a short carbon chain, which limits the emission of gases and production of soot particles (Lee et al. 2009).

### 18.1.5 Research on Second-Generation Biofuels

The development of second-generation biofuels has been widely researched. Walter and Ensinas (2010) reported two different technologies for the production of second-generation biofuels from cellulosic materials. They reported the generation of biofuels from biomass of sugarcane residue namely trash and bagasse, which is abundant in Brazil. They conducted a comparative study about the conversion of the biomass by hydrolysis and gasification through the Fischer-Tropsch conversion process. They used the fuels for generation of electricity and found that the fuels produced by the hydrolysis unit enhanced the generation of power in the system equipped with backpressure steam cogeneration units. However, a lot of bagasse is required to produce required ethanol. Alternately, they reported the advantage of using the gasification unit, which does not require high process steam to produce fuel. The electricity generated by the gasification unit can be upgraded using an extraction-condensation steam turbine.

The co-processing of the pyrolysis oils acquired from the lignocellulosic biomass is considered as a best candidate (Bridgwater 1994). A complex mixture of oxygen-containing hydrocarbons like phenols, glycols, alcohols, ethers, ketones, aldehydes, and acids form the pyrolysis oils. Additionally, the oil consists of oligomers derived from lignin, along with derivatives of carbohydrates (Samolada et al. 1998). They are chemically unstable, corrosive, highly viscous, and immiscible with a hydrocarbon-containing fuels, and have less heating value based on their oxygen-rich composition (Mohan et al. 2006).

These properties can be modified by diminishing the amount of oxygen that is conducted through co-processing. Fogassy et al. (2010) used a fluid catalytic cracking (FCC) unit for the co-processing of hydrodeoxygenated pyrolysis oils (HDO-oils) with vacuum gas oil (VGO) to generate biofuels more efficiently. HDO-oil must be modified that it can be used in the in-stationary combustion engines. The different reactions that take place in the FCC chamber are simultaneously cracking, reactions, which generate as well as use hydrogen and formation of larger molecules by Diels Alder reaction, which helps in the formation of the carbon–carbon bond. Thus, cracking the HDO-oil in the presence of acid catalyst yields smaller hydrocarbons, coke, carbon dioxide, water, and carbon monoxide. Moreover, the breaking of the carbon–carbon bond takes place prior to the cleavage of the carbon–oxygen bond in the case of unsaturated hydrocarbons present in the HDO-oil leading to the formation of oxygenated hydrocarbons or coke. The major reaction involved in the conversion of HDO-oil is hydrogen consuming reactions which produce water and causes hydrogenation reaction. The main source of hydrogen is the hydrocarbon feedstock. They reported a lesser yield of hydrogen from the cracking of VGO/HDO oil as compared to that of pure VGO oil. The products obtained from the cracking of VGO/HDO-oil feedstock are mainly unsaturated hydrocarbons. The continuous removal of hydrogen leads to the generation of polyaromatic compounds and finally produces coke.

The presence of heavy oxygenated hydrocarbons obtained from lignin derivatives in the VGO/HDO oil feed caused the deposition of coke (Adjaye and Bakhshi 1995). The precursors of the coke are formed on the acid sites due to the strong adsorption of the polar oxygenated compounds on the sites resulting in the production of the larger amount of coke (Samolada et al. 1998). Additionally, a reduction of the concentration of acid sites takes place as the coke produced covers the acid site thereby limiting the rate of diffusion inside the pores (Arandes et al. 2008).

Gupta et al. (2010) reported the application of biofuels for the operation of gas turbines, which will be a promising endeavor in the field of aviation industries. The engine of a gas turbine is a steady-flow engine, which generates a steady flame during stationary operations. Among the different biofuels, they recognized pure hydrogen as advantageous as a clean fuel that produces zero carbon emission and little nitrogen oxide emission. Additionally, the flashback can be eradicated with the use of diffusive burning (Dahl and Suttrop 1998).

Bansal et al. (2011) reported the synthesis of second-generation biofuels by using a strain *Aspergillus niger* NS-2 for hydrolyzing wheat bran. Fungi can generate a mixture of enzymes like amylases, cellulases, pectinases, and hemicellulases, which can degrade the polymers, present in the biomass. The β-glucosidase activity of cellulose obtained from *A. niger* is very high while the level of endoglucanase is low (Kumar et al. 2008). The authors conducted saccharification of a medium consisting of wheat bran by *A. niger* NS-2 to produce pectin, xylan, cellulose, mannan, and other starch substances. The bioconversion of wheat bran was more by solid culture fermentation as compared to solid-state fermentation.

Another report compiles the use of a novel solvent, which is a carboxylate type liquid zwitter ion for dissolving the cellulose (Kosuke Kuroda et al. 2017). The solvent is less toxic in nature and does not affects the microorganisms. Moreover, the process is cheaper for the production of ethanol from second-generation feedstocks. They conducted a comparative study and found that the *Escherichia coli* in the particular solvent produced 21 times more ethanol as compared to that in the ionic solvent. They reported the alteration of plant biomass directly without prior activities like washing to ethanol in a single step in the presence of the zwitter-ionic solvent thus making the process cheap and simple.

Trautmann et al. (2014) reported the generation of biocoals from the biomass wastes by the process of hydrothermal carbonization (HTC) of the wastes. A biocoal having the characteristics of lignite at a temperature range of 453–568 K is generated from the wet biomass by the process of HTC (Cao et al. 2013). The moisture content in the produced biocoal was 2–4% by weight as calculated by proximate and ultimate analysis. Thus, the transport efficiency of the dry coal was enhanced (Koukouzas et al. 2007). Furthermore, the moisture- and ash-free biocoal possessed higher carbon content as compared to wood and lignite thus enhancing the heating value of the coal. The factors dictating the quality of a biocoal are temperature and residence time of HTC as well as the raw materials (Hoekman et al. 2011). The degree of aromaticity and energy of the biocoal can be augmented by carrying out the HTC at high temperature. Thus, the biocoal produced at high temperatures will possess low oxygen and high amount of carbon (Kang et al. 2012). The process of conversion was carried out in the presence of $18Ni/TiO_2$ as a catalyst and the oil obtained from the biocoals could be stored a long time without any alteration of its viscosity.

Chiu and Wu (2013) carried out a study of the effect of second-generation biofuels on water quality. They put forth a mathematical model for calculating the water footprint with the generation of biofuels. They studied three various types of water footprints – namely, green water obtained from rainfall, blue water, which is the ground and surface water polluted due to irrigation and conversion and graywater, which is formed from the discharge of wastewater from two areas in the southeastern United States. The total volume of water polluted due to discharge and water required for reducing the concentration to minimum comprises of the graywater footprints. The amount of water consumed during the production comprises of green and blue water. The authors reported about the pollution caused due to nitrogen. A comparative study of the graywater and blue-water footprints of the woody feedstock indicated that the woody residue had smaller water footprints (ANL 2013). Additionally, the graywater footprint was found to be homogeneous as compared to the green footprints.

Another study by Pourbafrani et al. (2013) revealed the production of biomethane, ethanol, digestate, and limonene from citrus wastes obtained from factories as feedstocks. They conducted a comparative study on configurations of two biorefineries. The large biorefinery study indicated that the substitution of gasoline by ethanol reduced the emission of toxic gases by about 134% while the small refinery studies pointed toward the reduction of toxic gas formation by 77% on replacing natural gas by biomethane for the generation of electricity. The fertilizer can be substituted by digestate while the acetone can be replaced by limonene, which eliminates the emission of hazardous gases. The emission from biomethane dictates the emission of greenhouse gases.

Corrol et al. (2014) described the generation of biogas from the waste coffee pulp by photocatalytic degradation. The group used $Cu/TiO_2$ as a catalyst for the degradation of the complex organic substances present in the waste coffee pulp. The catalyst was semiconducting in nature, exhibiting an absorption band in the visible region, which was used for the photocatalytic degradation of the organic molecules present in the coffee pulp. The degradation of lignin photocatalytically produces small organic molecules, which are feasible for anaerobic digestion. The lignocellulosic mass of the coffee pulp has some refractory compounds whose solubility can be increased in the presence of the calcined $Cu/TiO_2$ catalyst under sunlight. The production cost of the biogas can be greatly reduced using solar radiation as compared to other methods (Ghanbari et al. 2012). The solubility, as well as the biodegradability of the lignocellulosic mass present in the coffee pulp, was enhanced with the catalyst in the presence of sunlight and atmosphere as the oxidizing agent. The anaerobic digestion process of the waste coffee pulp increases the amount of the generated biogas.

### 18.1.6 Production of Second-Generation Biofuels

The production of global biofuels like bioethanol and biodiesel has increased considerably during the last decade. The production of second-generation biofuels commercially is still in the developmental stage. The initial investment cost of the second-generation biofuels is high, which makes the cost of the products higher than that of the fossil fuels. The two main processes for the production of second-generation biofuels are the biochemical process and the thermochemical process.

#### 18.1.6.1 Biochemical Process

The biochemical process includes the enzymatic hydrolysis of the hemicellulose to monomeric and oligomeric sugars, followed by the ethanolic fermentation of the released sugars and finally distillation to collect the bioethanol. The cellulose gets hydrolyzed to glucose (Hamelinck and Faaij 2006). The chemical reaction where organic molecules are broken into simpler molecules in the presence of ferment is termed as alcoholic fermentation. The steps of hydrolysis and fermentation depend on the nature of the raw materials used. The byproducts obtained from lignin are either burned as fuel or raw material for plant chemistry. The disintegration of large organic molecules to smaller units in an anaerobic environment with the help of microorganisms is termed *anaerobic digestion*. The generation of biogas and bionatural gas from the crop residues and manure is carried out through anaerobic digestion (WGBN 2014).

#### 18.1.6.2 Thermochemical Process

The process in which energy products are obtained from the biomass feedstocks by heating is *thermochemical conversion*. The process in which the biomass is decomposed at a very high temperature and pressure at low levels of oxygen is called pyrolysis. The products of pyrolysis are char residues and liquid fuel. Additionally, the biomass is converted to synthetic gas at a high temperature of around 800 °C under controlled environmental conditions by the process of gasification. The synthesis stage in a thermochemical process is dictated by the different types of fuels like dimethyl ether, ethanol, methanol, kerosene, and synthetic diesel. The process of the production of first- and second-generation biofuels based on the thermochemical process is widely different. Additionally, properties of the biodiesel obtained by first- and second-generation thermochemical processes are also not similar. The biodiesel obtained by the second-generation thermochemical process has a low content of aromatic compound, improved cetane number and is devoid of sulfur. The gasification of the biomass and synthesis of the fuel from the generated gas are the two main processes to produce fuel. The yield of the synthesized fuel by either process depends on the nature of the biomass (Figure 18.1).

#### 18.1.6.3 Flexibility of Biofuel Production

The ultimate source of energy required to produce biofuel feedstocks is the process of photosynthesis. The light harvesting complex proteins like LHCI and LHCII are a class of specialized proteins that captures the sunlight. A large family of genes having a high degree of homology encodes the proteins, depending on the intensity of the sunlight (Elrad and Grossman 2004). The carotenoids and the chlorophyll present in the plant are bound to these proteins, which are also engaged in capturing light. The excess

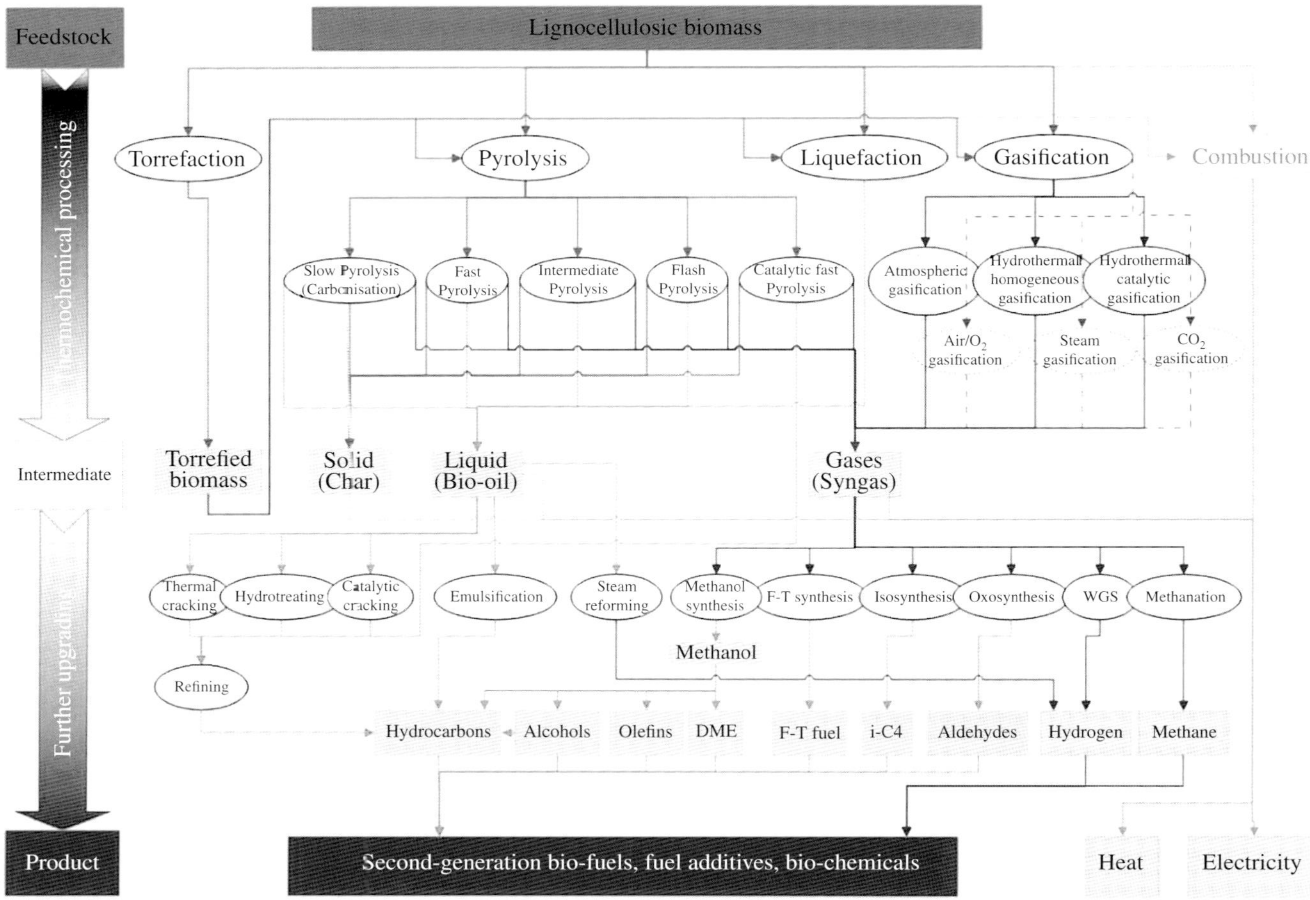


**Figure 18.1** Schematic representation of thermochemical paths for production of second generation biofuels (Zhang 2016).

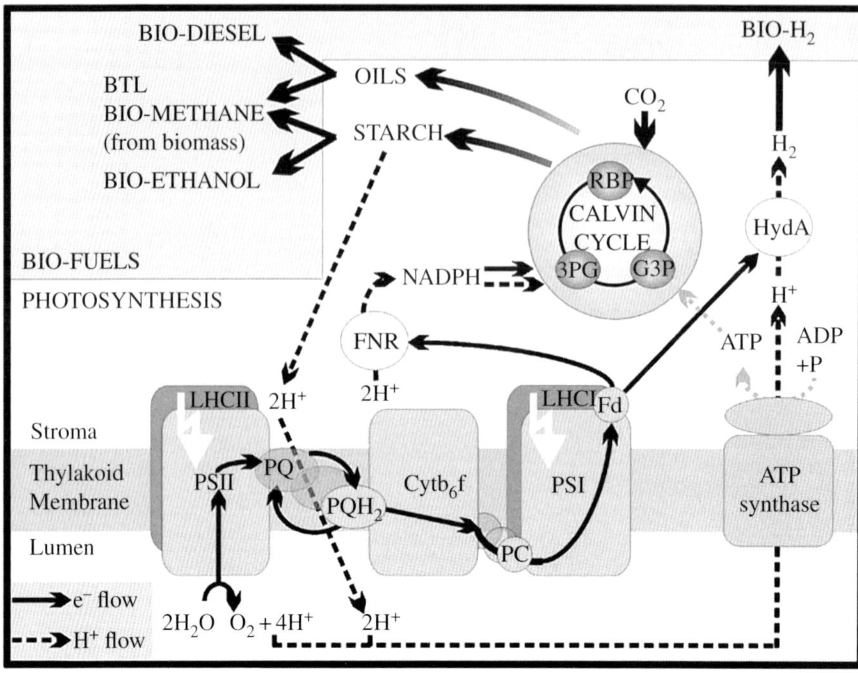


**Figure 18.2** Schematic representation of the conversion of light energy to biofuels (Schenk et al. 2008).

energy is used by these two molecules, thereby preventing the inhibition of the photosynthetic reaction centers (Horton and Ruban 2005). The biochemical pathways utilize ATP and NADPH in the Calvin cycle for the production of biomolecules, starch, sugars, and oils, which are together called *biomass*. The biohydrogen can be directly produced by photochemical reactions where electrons and protons present in starch react with the hydrogenase enzyme through the electron transport chain (Figure 18.2).

#### 18.1.6.4 Area Requirements for the Production of Biofuels

Clearly, solar energy is necessary for the generation of biofuels (WEC 2007). However, even if all the available arable land could be successfully cultivated and covered with abundant sunlight, the world's energy demand cannot be satisfied completely by the generated biofuels. There have been various approaches to using arable land for the cultivation of oil palm and soybean, which are later used to produce biodiesel. Large areas of rainforest in South East Asia and Brazil have been cleared for the cultivation of the oilseed plants, thus reducing green areas.

#### 18.1.6.5 Carbon Balance

The analysis of carbon balance plays a significant role in the production of biofuels. The generation of biofuels from lignocellulosic biomass is a carbon-neutral process, as the carbon present in the biomass is obtained from the atmosphere, which is again released to nature during combustion. Nevertheless, the total emission of carbon dioxide covering the complete process like a refinery, machines used during the cultivation, fertilizer, and transport must all be analyzed (Barbosa et al. 2004).

#### 18.1.6.6 Net Energy Balance

The net energy balance of biofuel is required for calculating the sustainability and value of the production process. The activities like transport, harvesting, farming, and processing are considered.

#### 18.1.6.7 Sequestration of Carbon Dioxide

The coupling of the carbon dioxide sequestration methods with that of the production of biofuels modifies the net balance of carbon dioxide in the biofuels process. The levels of atmospheric carbon dioxide are so high that the generation of biofuels with 0% carbon dioxide emission will keep the level of atmospheric carbon stagnant rather than decreasing it to a normal value (Marris 2006). The process of separating carbon dioxide from the atmospheric gases is difficult in technical terms, thus creating a big hurdle of physical sequestration of carbon dioxide. The plants capture the carbon dioxide through the process of photosynthesis, which is captured in the biomass. There is ample opportunity for the captured carbon dioxide to be stored and later coupled with atmospheric carbon dioxide sequestration and generation of carbon dioxide neutral biofuels. The 70% of the biomass obtained after the production of biodiesel are processed by downstream carbon sequestration process. The slow-burning process of pyrolysis transforms the sequestered carbon to hard C-chips (Bridgwater and Maniatis 2004). The C-chips or the agrichar can be effectively used as fertilizers as it enhances the amount of carbon in the soil, which finally will exhibit a positive effect on the atmosphere as the exchange of gases between the plants and soil can be controlled (Lehmann 2007). The biomass wastes are sterilized during the process of pyrolysis.

### 18.1.7 The Impact on the Environment During the Production of Second-Generation Biofuels

#### 18.1.7.1 Production of Greenhouse Gases

The production pathways of the biofuels decrease the emission of greenhouse gases as compared to that of conventional fossil fuels. However, when land use change is taken into account, either increase or decrease of greenhouse gases may take place (Searchinger et al. 2008). Additionally, it must be stated that indirect emission of greenhouse gases due to land use alteration is less in the case of second-generation biofuels as compared to that of the first generation.

#### 18.1.7.2 Water Footprints

The future production of second-generation biofuels depends immensely on the availability of water. The production process may lead to a scarcity of water because energy crops require a huge amount of water while the demand for agricultural drainage will increase with enhancement in the crop production. The amount of water used by a country is measured by a parameter called water footprint. The volume of fresh water consumed by the people of a country for goods and services is called water footprint (Hoekstra and Chapagain 2007). The total volume of fresh water required to produce a unit of biofuels is called the water footprint of a biofuel (Gerbens-Leenes et al. 2009). The water footprint of sugarcane is lowest while bioethanol obtained from sugar beets and corn is higher. The water footprints of the biofuels depend on climatic condition during the period of cultivation, the yield of crops and the type of feedstocks.

#### 18.1.7.3 Impact on Biodiversity

The increase in the feedstock production for the generation of second-generation biofuels may have either the positive or the negative impact on biodiversity (Anderson et al. 2004). The production of biofuels from forest residues has less impact while the cultivation of the dedicated biomass crops has a strong impact on biodiversity. The intrinsic biodiversity value of biomass crop depends on the type of crop and the different management services. The wildlife will gain from the grass crops while the intensively managed crops will give a negative impact. Similarly, if the intensively managed biomass crop substitutes the grass crops, it will exhibit a negative effect on the effects of landscape scale. The spatial distribution, as well as the location of a biomass crop indicates the effect of the landscape. Thus, land use changes can be promoted by producing huge quantities of feedstocks, which can help to meet the energy demand. The processing facilities should be located near to the areas having the biomass feedstocks, which will decrease the transportation cost rapidly making the biodiversity effect negative. The second-generation biofuels should be produced from eco-friendly crop residues or wood as this biomass being native does not require strong inputs (Groom et al. 2008).

### 18.1.8 Conclusions

The energy, as well as the environmental problems can be solved with the help of second-generation biofuels production. There are different routes to produce the fuels, which makes the production process flexible. The production of the biofuels also involves the supporting energy policies of the government of the respective country. The government must exhibit a strong level of support and contingency to find the most economical route to produce the biofuels.

There will always be a competition for land between the energy crops and the food crops which can be reduced using feedstocks like jatropha, which requires low-quality soil or switchgrass, which uses more biomass per unit land. The utilization of the forest residues and oil from micro algae decreases the competition for agricultural land further. The price of the ethanol obtained from cellulose is thrice as high as that of gasoline based on the cost of feedstock and conversion process. However, the feedstock cost of second-generation biofuels is lower than that of first generation feedstocks.

There are different methods for the transformation of biomass to cleaner fuels. The nature of the oxidant and the configuration of the reactor dictate the gasification products. The biomass of dry low ash can be transformed to around 65% of the organic liquid fuel by means of fast pyrolysis. The use of pure hydrogen is also considered as an efficient biofuel, as it is less polluting with no emission of carbon dioxide.

The major challenge of the second-generation biofuels is on the economic front, which still requires a high cost for production. The policy interventions can enhance the production of the second-generation biofuels at a low cost by inducing the development of good biofuels and reducing the generation of bad biofuels. The preferred pathways should be considered in order to increase productivity. These biofuels can also be produced at low cost by developing the production process of feedstock, which will ultimately decrease the environmental footprints and enhance the energy balance of the biofuels. The process of the collection of raw materials is also expensive in the case of second-generation biofuels. Different incentive mechanisms must be designed for producing biofuels with a variety of advance policy goals. Thus, the policy should address issues like climate change, rural development, and energy provision. Both the regional as well as global issues should be considered in the policy for obtaining maximum benefits. If the policy does not consider the global issues the biofuels would find local application with sustainability issues to be used globally. Incentives should be given as exemptions from tax for producing fuels, which will emit less greenhouse gases. The exemption should increase with higher reduction value of the emission of gases.

## References

Achinas, S. and Euverink, G.J.W. (2016). Consolidated briefing of biochemical ethanol production from lignocellulosic biomass. *Electron. J. Biotechnol.* 23: 44–53.

Adjaye, J.D. and Bakhshi, N.N. (1995). Production of hydrocarbons by catalytic upgrading of a fast pyrolysis bio-oil. Part II: comparative catalyst performance and reaction pathways. *Fuel Process. Technol.* 45: 185–202.

Agbor, V.B., Cicek, N., Sparling, R. et al. (2011). Biomass pretreatment: fundamentals toward application. *Biotechnol. Adv.* 29 (6): 675–685.

Anderson, G., Haskins, L., and Nelson, S. (2004). The effects of bioenergy crops on farmland birds in the UK: a review of current knowledge and future predictions. In: *Biomass and Agriculture: Sustainability, Markets, and Policies* (eds. K. Parris and T. Poincet), 199–218. Paris: OECD.

Andrews, S.S. 2006. Crop Residue Removal for Biomass Energy Production: Effects on Soils and Recommendations. White Paper, USDA-Natural Resource Conservation Service.

Arandes, J.M., Torre, I., Azkoiti, M.J., and Bilbao, J.J. (2008). Effect of atmospheric residue incorporation in the fluidized catalytic cracking (FCC) feed on product stream yields and composition. *Energy Fuels* 22: 2149–2156.

Argonne National Laboratory (2013). *Water Assessment Tool for Energy Resources*. Argonne National Laboratory (http://water.es.anl.gov, accessed 06/152013).

Banerjee, S., Mudliar, S., Sen, R. et al. (2010). Commercializing lignocellulosic bioethanol: technology bottlenecks and possible remedies. *Biofuels, Bioprod. Biorefin.* 4 (1): 77–93.

Bansal, N., Tewari, R., Gupta, J.K. et al. (2011). A novel strain of *Aspergillus niger* producing a cocktail of industrial depolymerising enzymes for the production of second generation biofuels. *BioResources* 6: 552–569.

Barbosa, B., Hadiyanto, M., and Wijffels, R. (2004). Overcoming shear stress of microalgae cultures ın sparged photobioreactors. *Biotechnol. Bioeng.* 85: 78–85.

Blanco-Canqui, H. and Lal, R. (2009). Crop residue removal impacts on soil productivity and environmental quality. *Crit. Rev. Plant Sci.* 28 (3): 139–163.

Bridgwater, A.V. (1994). Catalysis in thermal biomass conversion. *Appl. Catal., A* 116: 5–47.

Bridgwater, A. and Maniatis, K. (2004). The production of biofuels bythermal chemical processing of biomass. In: *Molecular to Global Photosynthesis* (eds. M. Archer and J. Barber). London: Imperial College Press.

Canilha, L., Santos, V.T., Rocha, G.J. et al. (2011). A study on the pretreatment of a sugarcane bagasse sample with dilute sulfuric acid. *J. Ind. Microbiol. Biotechnol.* 38 (9): 1467–1475.

Cao, X., Ro, K.S., Libra, J.A. et al. (2013). Effects of biomass types and carbonization conditions on the chemical characteristics of hydrochars. *J. Agric. Food Chem.* 61: 9401–9411.

Chang, V.S. and Holtzapple, M.T. (2000). Fundamental factors affecting biomass enzymatic reactivity. *Appl. Biochem. Biotechnol.* 84 (1): 5–37.

Chiu, Y.W. and Wu, M. (2013). The water footprint of biofuel produced from forest wood residue via a mixed alcohol gasification process. *Environ. Res. Lett.* 8: 035015.

Corrol, G., Pal, U., and Cebada, S. (2014). Enhanced biogas production from coffee pulp through deligninocellulosic photocatalytic pretreatment. *Energy Sci. Eng.* 2 (4): 177–187.

Dahl, G. and Suttrop, F. (1998). Engine control and low-NOx, combustion for hydrogen fuelled aircraft gas turbines. *Int. J. Hydrogen Energy* 23: 695–704.

Deenanath, E.D., Iyuke, S., and Rumbold, K. (2012). The bioethanol industry in sub-Saharan Africa: history, challenges, and prospects. *J. Biomed. Biotechnol.* 2012: 416491, 11 pages.

Demirbas, A. (2007). Progress and recent trends in bio-fuels. *Prog. Energy Combust. Sci.* 33: 1–18.

Domac, J., Richards, K., and Risovic, S. (2005). Socio-economic drivers in implementing bioenergy projects. *Biomass Bioenergy* 28: 97–106.

Elrad, D. and Grossman, A.R. (2004). A genome's-eye view of the light-harvesting polypeptides of *Chlamydomonas reinhardtii*. *Curr. Genet.* 45: 61–75.

Fogassy, G., Thegarid, N., Toussaint, G. et al. (2010). Biomass derived feedstock co-processing with vacuum gas oil for second-generation fuel production in FCC units. *Appl. Catal., B* 96: 476–485.

Gerbens-Leenes, W., Hoekstra, A.Y., and van der Meer, T.H. (2009). The water footprint of bioenergy. *Proc. Natl. Acad. Sci. U.S.A.* 106 (25): 10219–10223.

Ghanbari, F., Ghoorchi, T., Shawrang, P. et al. (2012). Comparison of electron beam and gamma ray irradiations effects on ruminal crude protein and amino acid degradation kinetics and in vitro digestibility of cottonseed meal. *Radiat. Phys. Chem.* 81: 672–678.

Groom, M., Elizabeth, M., and Townsend, P. (2008). Biofuels and biodiversity: principles for creating better policies for biofuel production. *Conserv. Biol.* 22: 602–609.

Guo, H., Peng, F., Zhang, H., and Xiong, L. (2014). Production of hydrogen rich bio-oil derived syngas from co-gasification of bio-oil and waste engine oil as feedstock for lower alcohols synthesis in two-stage bed reactor. *Int. J. Hydrogen Energy* 39 (17): 9200–9211.

Gupta, K.K., Rehman, A., and Sarviya, R.M. (2010). Bio-fuels for the gas turbine: a review. *Renewable Sustainable Energy Rev.* 14 (9): 2946–2955.

Haarlemmer, G., Boissonnet, G., Imbach, J. et al. (2012). Environmental science second-generation BtL type biofuels – a production cost analysis. *Energy Environ. Sci.* 5: 8445–8456.

Hamelinck, C.N. and Faaij, A. (2006). Outlook for advanced biofuels. *Energy Policy* 34 (17): 3268–3283.

Hendriks, A.T.W.M. and Zeeman, G. (2009). Pretreatments to enhance the digestibility of lignocellulosic biomass. *Bioresour. Technol.* 100 (1): 10–18.

Hoekman, S.K., Broch, A., and Robbins, C. (2011). Hydrothermal carbonization (HTC) of lignocellulosic biomass. *Energy Fuels* 25 (4): 1802–1810.

Hoekstra, A.Y. and Chapagain, A.K. (2007). Water footprints of nations: water use by people as a function of their consumption pattern. *Water Resour. Manag.* 21: 35–48.

Horton, P. and Ruban, A. (2005). Molecular design of the photosystem II light-harvesting antenna: photosynthesis and photoprotection. *J. Exp. Bot.* 56: 365–373.

Huang, Z., Qiao, X., Zhang, W. et al. (2009). Dimethyl ether as alternative fuel for CI engine and vehicle. *Front. Energy Power Eng. China* 3: 99–108.

Jongschaap, R.E.E., Corre, W.J., Bindraban, P.S., and Brandenburg, W.A. (2007). *Claims and Facts on Jatropha curcas L.* Report 158. Wageningen: Plant Research International.

Justice, S. (2009). *Private Financing of Renewable Energy – a Guide for Policy Makers.* UNEP Sustainable Energy Finance Initiative.

Kang, S., Li, X., Fan, J., and Chang, J. (2012). Characterization of hydrochars produced by hydrothermal carbonization of lignin, cellulose, d-xylose, and wood meal. *Ind. Eng. Chem. Res.* 51 (26): 9023–9031.

Kazi, F.K., Fortman, J.A., Anex, R.P. et al. (2010). Techno-economic comparison of process technologies for biochemical ethanol production from corn stover. *Fuel* 89: S20–S28.

Keshwani, D. and Cheng, J. (2009). Switchgrass for bioethanol and other value-added applications: a review. *Bioresour. Technol.* 100: 1515–1523.

Koukouzas, N., Hämäläinen, J., Papanikolaou, D. et al. (2007). Mineralogical and elemental composition of fly ash from pilot scale fluidised bed combustion of lignite, bituminous coal, wood chips and their blends. *Fuel* 86: 2186–2193.

Kumar, R., Singh, S., and Singh, O.V. (2008). Bioconversion of lignocellulosic biomass: biochemical and molecular perspectives. *J. Ind. Microbiol. Biotechnol.* 35: 377–391.

Kuroda, K., Satria, H., Miyamura, K. et al. (2017). Design of Wall-Destructive but Membrane-Compatible Solvents. *J. Am. Chem. Soc.* 139: 16052–16055.

Laser, M. and College, D. (2009). Comparative analysis of efficiency, environmental impact, and process economics for mature biomass refining scenarios. *Biofuels, Bioprod. Biorefin.* 3: 247–270.

Lee, M.C., Seo, S.B., Chung, J.H. et al. (2009). Industrial gas turbine combustion performance test of DME to use as an alternative fuel for power generation. *Fuel* 88: 657–662.

Lehmann, J. (2007). A handful of carbon. *Nature* 447: 143–144.

Lewandowski, I., Scurlock, J., Lindvall, E., and Christou, M. (2003). The development and current status of perennial rhizomatous grasses as energy crops in the US and Europe. *Biomass Bioenergy* 25: 335–361.

Li, Y., Horsman, M., Wu, N. et al. (2008). Biofuels from microalgae. *Biotechnol. Progr.* 24: 815–820.

Luo, L., Van de Voet, E., and Huppes, G. (2009). Life cycle assessment and life cycle costing of bioethanol from sugarcane in Brazil. *Renewable Sustainable Energy Rev.* 13: 1613–1619.

Marris, E. (2006). Black is the new green. *Nature* 442: 624–626.

McKendry, P. (2002). Energy production from biomass (part 1): overview of biomass. *Bioresour. Technol.* 83 (1): 37–46.

Mohan, D., Pittman, C.U. Jr., and Steele, P.H. (2006). Pyrolysis of wood/biomass for bio-oil a critical review. *Energy Fuels* 20: 848 889.

Naylor, R., Liska, A., Burke, M. et al. (2011). The ripple effect: biofuels, food security, and the environment. *Environ. Sci. Policy Sustainable Dev.* 49 (9): 30–43.

Neilson, C.E. (1998). LM 2500 gas turbine modification for biomass fuel operation. *Biomass Bioenergy* 15: 269–273.

Nishio, N. and Nakashimada, Y. (2007). Recent development of anaerobic digestion processes for energy recovery from wastes. *J. Biosci. Bioeng.* 103 (2): 105–112.

Ohkubo, Y., Idota, Y., and Nomura, Y. (1997). Evaporation characteristics of fuel spray and low emissions in a lean premixed-pre-vaporization combustor for a 100 kW automotive ceramic gas turbine. *Energy Convers. Manag.* 38: 1297–1309.

Perlack, R., Wright, L., Turhollow, A. et al. (2005). *Biomass as Feedstock for a Bioenergy and Bioproducts Industry: The Technical Feasibility of a Billion-Ton Annual Supply*. Oak Ridge, TN: Oak Ridge National Laboratory, US Department of Energy /GO-102005-2135. April, 2005.

Pourbafrani, M., McKechnie, J., Mac Lean, H.L., and Saville, B.A. (2013). Life cycle greenhouse gas impacts of ethanol, biomethane and limonene production from citrus waste. *Environ. Res. Lett.* 8: 015007.

Rajagopal, D. and Zilberman, D. (2008). Environmental, economic and policy aspects of biofuels. *Foundations and Trends in Microeconomics* 4 (5): 353–468.

Raud, M., Kesperi, R., Oja, T. et al. (2014). Utilization of urban waste in bioethanol production: potential and technical solutions. *Agron. Res.* 12 (2): 397–406.

Richardson J. 2008. Production of biomass for energy from sustainable forestry systems: Canada and Europe. In: Proceedings of the Short Rotation Crops International Conference.

Saini, J.K., Saini, R., and Tewari, L. (2015). Lignocellulosic agriculture wastes as biomass feedstocks for second-generation bioethanol production: concepts and recent developments. *3 Biotech* 5 (4): 337–353.

Samolada, M.C., Baldauf, W., and Vasalos, A. (1998). Production of a bio-gasoline by upgrading biomass flash pyrolysis liquids via hydrogen processing and catalytic cracking. *Fuel* 77 (14): 1667–1675.

Sayigh, A. (1999). Renewable energy—the way forward. *Appl. Energy* 64: 15–30.

Schenk, P., Thomas-Hall, S.R., Stephens, E. et al. (2008). Second generation biofuels: high-efficiency microalgae for biodiesel production. *Bioenergy Res.* 1: 20–43.

Searchinger, T., Heimlich, R., Houghton, R. et al. (2008). Use of U.S. croplands for biofuels increases greenhouse gases through emissions from land use change. *Science* 319 (5867): 1157–1268.

Sharma, Y.C., Singh, B., and Upadhyay, S.N. (2008). Advancements in development and characterization of bio-diesel: a review. *Fuel* 87: 2355–2373.

Smeets, E.M.W., Faaij, A., Lewandowski, I.M., and Turkenburg, W.C. (2007). A bottom-up assessment and review of global bio-energy potentials for 2050. *Prog. Energy Combust. Sci.* 33: 56–106.

Trautmann, M., Löwe, A., and Traa, Y. (2014). An alternative method for the production of second-generation biofuels. *Green Chem.* 16: 3710.

Tutt, M., Kikas, T., and Olt, J. (2013). Influence of harvesting time on biochemical composition and glucose yield from hemp. *Agron. Res.* 11 (1): 215–220.

Wagner L. 2015. Trends from the use of biogas technology in Germany. In: Proceedings of the VIV Asia Biogas Conference; 2015 Mar 12; Bangkok, Thailand.

Walter, A. and Ensinas, A.V. (2010). Combined production of second-generation biofuels and electricity from sugarcane residues. *Energy* 35: 874–879.

Wisconsin Grasslands Bioenergy Network. Bioenergy Conversion Technologies. [Accessed 20th of November, 2014].

World Energy Council. 2007 Survey of Energy Resources, Executive Summary World Energy Council. Pp 1–24.

Yu, J., Paterson, N., Blamey, J., and Millan, M. (2017). Cellulose, xylan and lignin interactions during pyrolysis of lignocellulosic biomass. *Fuel* 191: 140–149.

Zabed, H., Sahu, J.N., Boyce, A.N., and Faruq, G. (2016). Fuel ethanol production from lignocellulosic biomass: an overview on feedstocks and technological approaches. *Renewable Sustainable Energy Rev.* 66: 751–774.

Zhang, W. (2009). Automotive fuels from biomass via gasification. *Fuel Process. Technol.* 91 (8): 866–876.

Zhang, X. (2016). Essential scientific mapping of the value chain of thermochemically converted second-generation bio-fuels. *Green Chem.* 18: 5086.

# 19

# Research and Production of Third-Generation Biofuels

*Saurabh Singh*[1], *Arthur P.A. Pereira*[2,3], *and Jay Prakash Verma*[1,3]

[1] *Institute of Environment and Sustainable Development, Banaras Hindu University, Varanasi, India*
[2] *Luiz de Queiroz College of Agriculture, University of São Paulo, Piracicaba, Brazil*
[3] *Hawkesbury Institute for the Environment, Hawkesbury Campus, Western Sydney University, Penrith, 2750, Australia*

## 19.1 Introduction

The term *biofuel* has become a buzzword owing to the advantages that it represents in the current world scenario. We are tasked with figuring out solutions for atmospheric pollution and climate impact from greenhouse gas emissions – basically, oxides of nitrogen and carbon due to the burning of fossil fuels. Of major concern, the oxides of nitrogen emitted in the form of NO and oxides of carbon in the form of CO are converted into $N_2O$ and $CO_2$, respectively, which are highly potent greenhouse gases (Rogak et al. 1998; Flessa et al. 2002). Recently, due to the overuse of fossil fuels, the levels of $CO_2$ and $N_2O$ have increased greatly and thus have caused a much higher greenhouse effect than the normal. This abnormal increase in greenhouse gases has led to a phenomenon called global warming. $CO_2$ is one of the two major responsible gases for the greenhouse effect, notably the $CO_2$ concentration has now increased to more than 400 ppm (initially 280 ppm before the preindustrial era) (NOAA 2018). Biofuels can significantly reduce the number of greenhouse gases going into the atmosphere (Figure 19.1).

Biofuels have also been categorized into different categories based on the type of feedstock used for the preparation of the biofuel. To date, there have been four classifications of biofuels. First-generation biofuel, is where the substrate used is competitive in nature with the food crops. Second-generation biofuel is mainly produced from the waste material from farms. Third-generation biofuels are produced from algae. Fourth-generation biofuels employ the technique of carbon capture from the atmosphere for the production of biofuels. Moreover, there is a higher advantage associated with every succeeding generation of biofuel.

With every successive generation of biofuel, the number of net emissions into the atmosphere falls. For example, the third generation of biofuels is called *zero-emission technologies,* whereas the fourth generation of biofuels is called *negative-emission technologies,* owing to their net negative emissions into the atmosphere. Biofuels can be produced from different agricultural substrates, but the production quantities in the algal biofuels are much higher. This makes the technology significantly better in comparison to first and second generation, owing to their land area requirements.

*Bioprocessing for Biomolecules Production,* First Edition.
Edited by Gustavo Molina, Vijai Kumar Gupta, Brahma N. Singh, and Nicholas Gathergood.

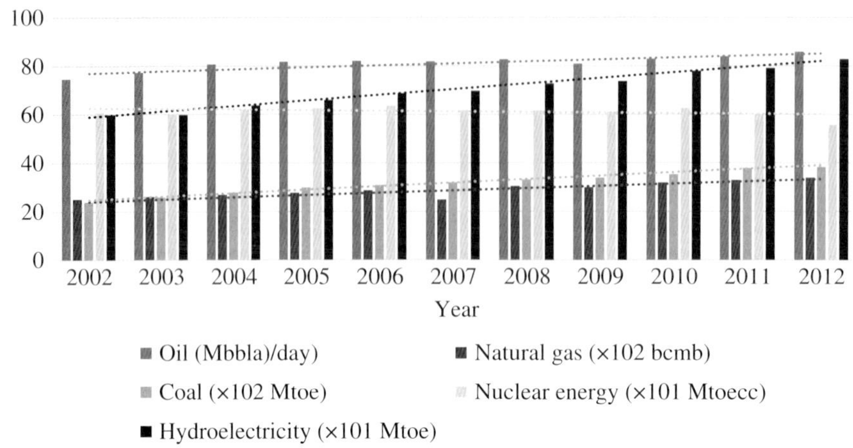


**Figure 19.1** Increasing energy consumption patterns, with their trend lines of different types of energy sources. Source: Adapted and modified from Harun et al. 2014. Mbbl: one thousand barrels; bcm: billion cubic meters; Mtoe: million tons oil equivalent.

In a review by Georgianna and Mayfield, it was found that if only 30 million hectares of land yielding 40,000 liters per hectare of biofuel is used, it could replace almost all the major fuel requirements of the United States with the production of 1,200 billion liters of biofuel (Georgianna and Mayfield 2012). It is predicted that algal biofuels could satisfy all the fuel requirements using just under 4% land area of the United States, which is very small when compared to land required by agricultural crops. In contrast, other crops such as oil palm, jatropha, and soya were calculated to require, 23%, 75%, and 330% of the total land area of United States, respectively (Georgianna and Mayfield 2012). Another major advantage of the use of algae for the production of biofuels is the ease of the process, along with the speed with which it is produced. This makes algal biofuels the best candidate for economically viable biofuel production. Other advantages associated with these are that it can utilize nonarable or the wasteland available for the cultivation of algae and the demand for water can be met with the use of low-quality water, such as wastewater or salt water (Wigmosta et al. 2011).

Third-generation biofuels are produced through the cultivation of algal cells. It is the production of biofuels harnessing the ability of algal cells using sunlight, water, and carbon dioxide (Figure 19.2) (Georgianna and Mayfield 2012). Nonpolar storage lipids, such as triacylglycerides are accumulated by microalgae. Polar lipids are stored in the form of glycolipids, phospholipids, and sterols in the photosynthetic and cellular membranes of the algal cells. Algae are ideal for the production of economically viable crops strains for efficient production of biofuels. Apart from the production of biofuels from the algal cells, these can also be used in the form of lignocellulosic waste for the production of the second generation of biofuel.

## 19.2 Cultivation of Algal Cells

There is basic furcation in the technology employed for the cultivation of the algal biofuel. One is the agricultural cultivation, the other is the industrial cultivation of the

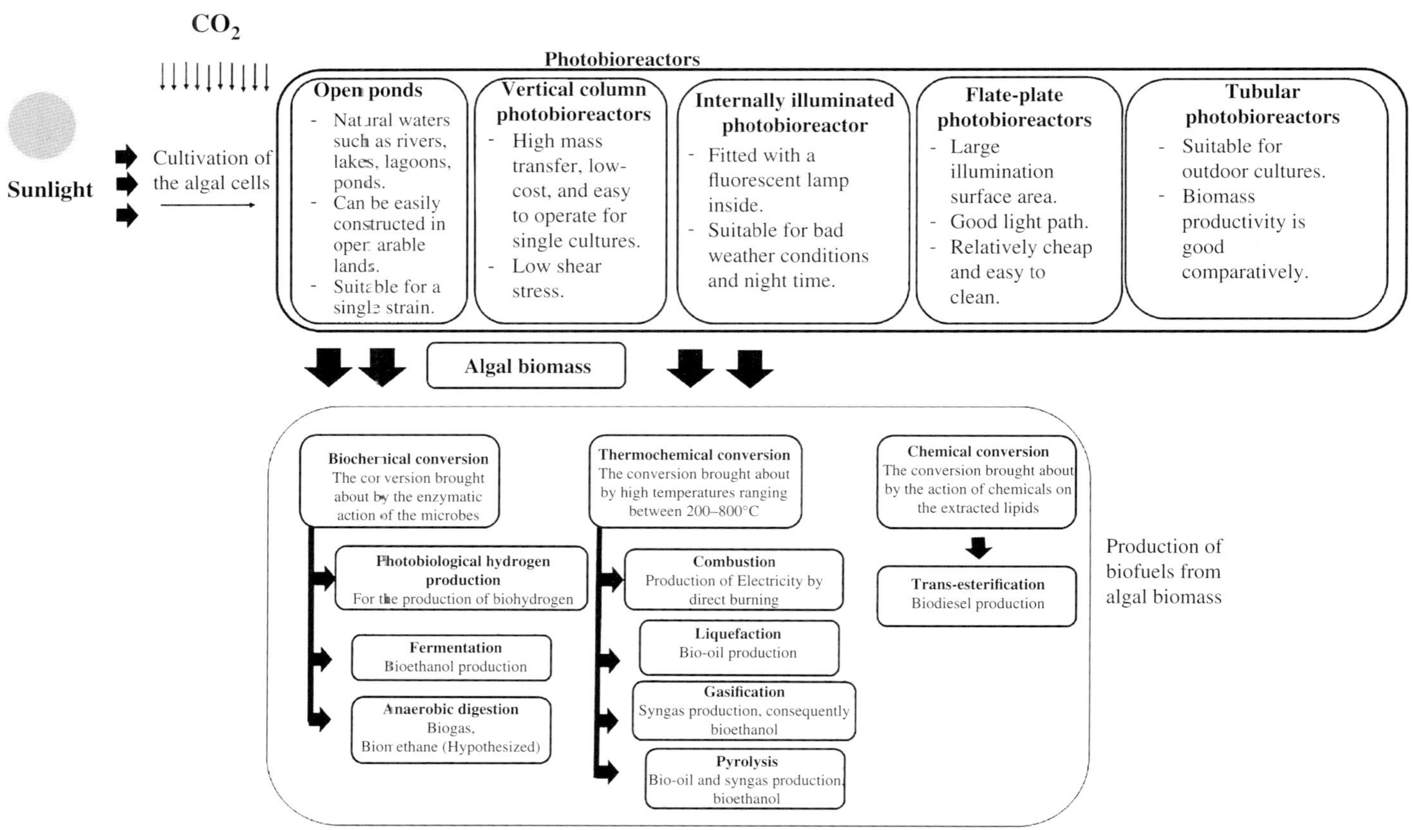


**Figure 19.2** Schematic representation of third-generation biofuels.

algal cells. Algal cells grown in the agricultural cultivation tend to be more resistant in comparison to the ones used in industrial cultivation. Agricultural cultivation of the algal cell is done in nonarable lands, with open spaces, so as to allow the maximum amount of sunlight and carbon capture in the form of $CO_2$. This process is very highly scalable, as the only limiting factors present are the space and capital costs (Georgianna and Mayfield 2012). This method consumes huge amounts of $CO_2$ during the production of the biomass. In the industrial production of algal cells, a similar process is carried out in specialized chambers with reduced carbon sources such as sugars.

Industrial algal biofuel production also uses algal fermentation, which so far is only used for the production of high-value products. But the process of algal fermentation of processed sugars does not reduce the amount of greenhouse-gas emissions when compared to fossil fuel processing, and it is not even cost-effective because sugar prices are high relative to fossil fuels (Darzins et al. 2010). The agricultural cultivation of the algal cells, though, is not fully developed in the form of technology, as the production of algal cells in fully exposed conditions is tedious and needs further development for industrial-level viability. Both the processes of cultivation of cells, i.e., industrial and agricultural production, can be scaled up to economic viability only when the input in the form of energy and capital is greatly reduced and yield is improved.

## 19.3 Strain Selection

Biofuel production from algal cells is greatly supported by the presence of huge diversity of algal strains, which have evolved over billions of years, all around the world (Tirichine and Bowler 2011). Estimates are that there are more types of algae than animals or land plants. But as this diversity has only recently been explored, it should be examined further to study and harness the huge potential that it carries. The algal strains depending on the habitat can be found with various adaptations with respect to temperature, pressure, sunlight, salinity, and various other factors required for its growth. A strain that has varying degree of resistance with respect to its growth factor stresses is considered as a good strain for the production of biofuel. Algal species used in the production of biofuel include green algae, diatoms, and cyanobacteria (Georgianna and Mayfield 2012).

The agricultural cultivation of algal cells, which takes place in open ponds, employs a strain that can tolerate a broad range of temperature, light, salinity, and pathogen load variation. No algal strain used in the agricultural or the industrial cultivation of the cells is the wild type, as the possibility of having all the characters in the broad range is very unlikely. So, the optimization of the strain to several broad-range growth factors is needed to generate adequate amounts of cells and consequently the yield of biofuel.

The algae strains used for industrial cultivation of the cells, which takes place in bioreactors, do not have to face a broad range of conditions, as in the case of agricultural cultivation. Industrial bioreactors can be controlled and hence employ such strains, which can grow quickly, with high lipid production, requiring very low amounts of energy input.

Tolerance against pathogen is still needed in the industrially growing strains as well, to reduce the cost of sterilization. Most of the strains used for algal biofuel production have an optimum temperature of 20–30 °C (Béchet et al. 2010). But in the industrial bioreactors, temperature control is required; otherwise, the temperature may well exceed the optimum range and can lead to cell death. In this case, thermophilic strains of algae can serve the purpose, with an optimum temperature range of 35–50 °C (Miyamoto et al. 1979).

Likewise, in the agricultural systems, where high temperature is not a concern but the low temperature during winter months is, psychrophilic algal strains can serve the purpose (Butterwick et al. 2005). *Chlorella vulgaris* is known to have best production at 10 °C (Bae and Hur 2011). Many studies have been performed regarding the isolation of the best strains for the production of biofuels from it. Different experiments such as growth and lipid production tests have been carried out (Griffiths and Harrison 2009; Rodolfi et al. 2009). The highest biomass productivity is reported in *Porphyridium cruentum* with productivity of 0.21 $gal^{-1}$ $day^{-1}$, with the highest lipid content reported in *Chaetoceros calcitrans* with 39.8% lipid production (Georgianna and Mayfield 2012). Strains of classes chlorophyceae, Eustigmatophyceae, Haptophyceae, and blue-green algae are currently used for the production of biofuels (Larkum et al. 2012).

## 19.4 Types of Micro-Algae Used to Produce Third-Generation Biofuels

Based on the type of carbon substrate and energy assimilation for carbon fixation, microalgae can be divided into different types. Photoautotrophic microalgae depend solely on light for an energy source to convert it into chemical energy through photosynthetic reactions. This type of photosynthesis is also called autotrophic photosynthesis. Heterotrophic microalgae utilize organic compounds such as glucose, acetate, and glycerol, as carbon and energy source. Mixotrophic microalgae depend both on the organic compounds and atmospheric $CO_2$ for the photosynthesis. They are adaptive to the changes in the light intensity of the environment and can grow accordingly. Amphitrophic microalgae live either autotrophically or heterotrophically, depending on the environment to which they are exposed. Photoheterotrophic microalgae or photoorganotrophs use light for the assimilation of the organic compounds as a carbon source. This process is also called photoassimilation or photometabolism (Cuellar-Bermudez et al. 2015).

## 19.5 Biomass Preparation for Third-Generation Biofuel

Third-generation biofuels are the result of increasing concerns over the competitive nature of first- and second-generation biofuel feedstock with food and land, respectively. Also, these are based on improvements in biomass production, as the algae and other feedstock primarily utilized as its energy source are specially engineered through advanced biotechnology (Bonin and Lal 2012; Chisti 2007; Bermudez et al 2015; Rawat et al. 2013). The production of biofuels from algae is advantageous than any other crop (Georgianna and Mayfield 2012). Thus, algal biofuels are preferred over second-generation biofuels.

Algal cells produce oil, which is used for the production of biodiesel, gasoline, and other fuels through different processes such as transesterification, distillation, and cracking (Amin 2009; Luo et al. 2010). The result of such a high efficiency is that the land area requirements do not get compensated as a result of crop production on habitable land. The increasing concern for more and more land for the ever-increasing population raises concerns over land acquirement by the second-generation feed stocks. So, in the midst of such a dilemma, third-generation algal biofuels serve as a welcome alternative. They can grow in a very small space in a limited period of time.

Although average annual yield estimate for algal biofuel production vary between 8000 l $ha^{-1}$ and 140000 l $ha^{-1}$, this suggests that the yields vary considerably with the type of processes employed, which is significantly better than in the case of the second-generation biofuel (US Doe 2010 and USDA-NASS 2012). Also, the amount of land required for the production of algae to meet the whole world's fuel requirement can be calculated to 16% of the total land area of the continental United States, which is far less than the area required for crop production, be it for first-generation or second-generation feedstocks (US Energy 2012). The foremost priority in terms of selection of a good strain for third-generation biofuel is the efficient strain with a high production efficiency, along with high tolerance to pathogens (Georgianna and Mayfield 2012). The presence of such a large diversity of algae, which have evolved over billions of years, presents us with a broad range of choices (Tirichine and Bowler 2011). The algal strains that can be considered ideal for the production of biofuels are those that have high growth rate, high tolerance to pathogens, and high lipid production.

In agricultural systems, qualities such as tolerance to temperature variation, high salinity, and light variation are considered detrimental. Along with this, the strains that are not restricted geographically and seasonally for their growth can be considered as a good strain for biofuel production. There are many external factors involved in the growth of algal strains, such as light, temperature, humidity, and other abiotic factors. These factors need to be clearly understood in order to understand the algal population dynamics. When there is high strength wastewater that contains algal biomass with other forms of microorganisms, then at first dilution only anaerobic fermentative bacteria are present (Benemann et al. 1977). This is due to highly concentrated waste, which has less amount of dissolved oxygen in it.

## 19.6 Photobioreactors

The cultivation of algal cells and production of biofuels takes place in a unit called photobioreactors. They are called photobioreactors because they provide light to the reactions taking place inside the reactor. The algal cultures can grow under artificial as well as solar light. Open ponds with natural illumination are thus suited for growing algal cultures (Hase et al. 2000). There are many other types of bioreactors, too, depending on the type of illumination provided to them.

### 19.6.1 Open Ponds

Open ponds for the cultivation of algae has been categorized into natural waters, such as lakes, lagoons, ponds, and artificial ponds or containers. These systems of algal cultivation include shallow big ponds, tanks, circular ponds, and raceway ponds (Ugwu et al. 2008). The biggest advantage presented by these systems is that they are easy to construct and operate. They can be constructed in nonarable lands considered improper for agricultural production. Open ponds are relatively easier to clean up after cultivation and are therefore considered good for mass cultivation. There are many limitations associated with the open pond type photobioreactor, however. For instance, there is little control on the culture conditions and open ponds are very dependent on the natural conditions for the culture growth. As the ponds are open, there is a high possibility of contamination in the cultures, so cultivation in these systems is restricted to only a few strains of algae that have shown a broad range of resistance against pathogen load. Its production efficiency is low when compared to the artificial photobioreactors and they also cover a large area of land for their setup.

### 19.6.2 Vertical Column Photobioreactors

These are the artificial-type photobioreactors in which there is a high mass transfer. They are compact, low cost, and easy to operate for single culture conditions (Mirón et al. 2002). They present a low shear stress and thus provide good mixing inside the chamber. In addition, they consume low energy with high potentials of scalability. They are easy to sterilize and are good for the immobilization of the algae. Inhibition processes such as photoinhibition and photo-oxidation are reduced. A major disadvantage associated with them is that these photobioreactors have a very small illumination surface area, and their construction requires sophisticated, costly materials (Ugwu et al. 2008).

### 19.6.3 Flat-Plate Photobioreactors

These have a large illumination surface area and are thus more effective in providing light to the algal cultures than vertical-column photobioreactors. They are suitable for outdoor cultures and are good for the immobilization of the algal cultures because they provide a good light path. This, in turn, leads to good production of the biomass. They are relatively cheap and easy to clean, and there is low oxygen accumulation inside the reactor. The cons put forward by the flat-plate bioreactors are the difficulty in controlling the culture temperature and in scaling up of the same, because of the presence of many compartments and support materials for the build-up of the same. Another major drawback in the flat-plate bioreactors is the possibility of a buildup of hydrodynamic stress to algal strains (Ugwu et al. 2008).

### 19.6.4 Tubular Photobioreactors

This type of photobioreactors has a very large illumination surface area and is most suitable for outdoor cultures. They can be divided into horizontal/serpentine type, near horizontal type, vertical type, inclined type, and conical shaped photobioreactors, based on the shape and angle they present (Molina et al. 2001; Tredici and Zittelli 1998; Pirt et al. 1983; Lee and Low 1991; Watanabe and Saiki 1997). The biomass productivities in this are very good, and it is cheaper than other types of photobioreactors (Ugwu et al. 2008). The major cons put forward by this type is the gradients of pH present inside the photobioreactors, and the presence of dissolved oxygen and carbon dioxide along the tubes. One striking disadvantage of this type of photobioreactor is the requirement of large land spaces (Ugwu et al. 2008).

### 19.6.5 Internally Illuminated Photobioreactors

This type of photobioreactors has an internally fitted fluorescent lamp for the illumination of the reactor. They have a well-equipped agitation system that is in the form of impellers for the algal cultures. These bioreactors are designed so they are able to utilize the natural light (solar energy) as well as the artificial light (fluorescent lamp) (Ogbonna et al. 1999). This intelligent design means that whenever the solar light of appropriate intensity is available they utilize it, but as soon as the intensity of the natural light decreases below a certain range, the fluorescent lamps are illuminated. This approach is suitable for bad weather conditions and also enables a continuation of the cultivation process at night (Ugwu et al. 2008).

## 19.7 Production of Biofuels from Algal Cultures

Producing biofuels from algal cells consists of three basic steps. It starts from the harvesting of algal cells in large tanks by harnessing their capability of photosynthesis. They capture the carbon provided to them, be it in the form of atmospheric $CO_2$ or provided organic substrates. The next step is the extraction of the lipid content from the algal cells (Figure 19.3). The algal cells contain polar as well as nonpolar lipids. Currently, fuel production is focused on the use of nonpolar lipids, which constitutes only a small fraction of the total lipids produced by the algal cells. Lipids are then used for further processing and different types of biofuels are produced from these compounds. The process of conversion of algal biomass into biofuel can be divided into three basic types: *biochemical conversion*, *thermochemical conversion*, and *chemical conversion*.

### 19.7.1 Biochemical Conversion

The biochemical conversion can be classified into three basic types, photochemical hydrogen production, fermentation, and anaerobic digestion (Demirbas 2011; Amin 2009). The photochemical bioconversion of the algal cells yields hydrogen as fuel, which is widely considered as the cleanest form of fuel (Wang et al. 1998). Production of hydrogen biologically is performed mainly by microalgae. Biohydrogen production from the microalgae is principally by two processes, direct biophotolysis, and indirect biophotolysis. Direct biophotolysis involves the dissociation of the water molecules using sunlight by the microalgae. In this process, the water molecule in the presence of light is directly converted into hydrogen and oxygen molecule. The microalgal species that produce biohydrogen by direct photolysis method are *Chlamydomonas reinhardii*.

Another method by which biohydrogen production takes place is by indirect biophotolysis. *Cyanobacteria* or the blue-green algae achieves biohydrogen production through indirect photolysis method. The general process in which the indirect biohydrogen production takes place is through the production of glucose as an intermediate product. The glucose is then hydrolyzed in the presence of sunlight to produce hydrogen and carbon monoxide.

Fermentation process involves two basic steps, sugar pathway and syngas pathway. Sugar pathway involves the formation of the monomer sugars before it is fermented into ethanol and other byproducts. This involves the same steps as in the case of

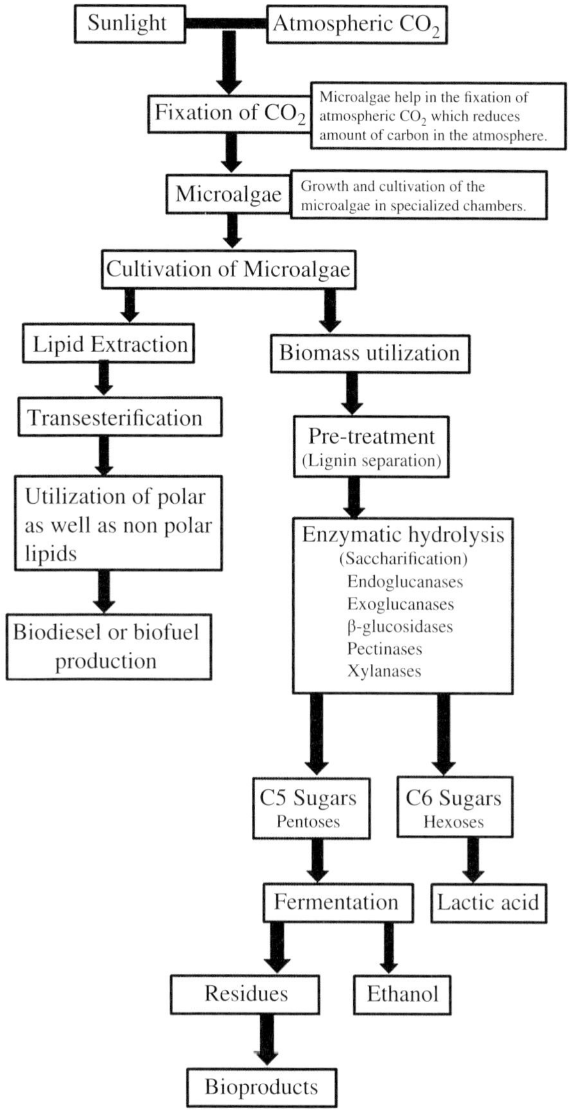


**Figure 19.3** Diagrammatic representation of algal biorefinery concept from $CO_2$ fixation to bioproduct.

second-generation biofuels, i.e., pretreatment, hydrolysis, and fermentation. The pretreatment of algae is not too tedious, as the lignin is absent, which proves to be a rate-limiting factor when using agricultural residues (Harun et al. 2014). The pretreated agricultural residue is then hydrolyzed by α-amylases, as the biomass is mostly starchy in nature, to produce monomer sugars. The monomer sugars are then converted into bioethanol through fermentation. *Syngas pathway* involves the formation of syngas from the algal cells, which is then converted into bioethanol (Harun et al. 2014).

Anaerobic digestion of algal biomass is a process that has long been prevalent and was first reported in 1957 (Golueke et al. 1957). It leads to the formation of biogas. However, the anaerobic digestion of the algal biomass should be performed only after the extraction of the lipids step is completed, as they tend to inhibit the digestion process (Ward et al. 2014). Biomethane production is also predicted by anaerobic digestion, but the actual yield is much less than the theoretical yield proposed (Ward et al. 2014). A comparative study by Chinnasamy et al. predicted a 13% decrease in the theoretical methane potential of the lipid extracted microalgal biomass against nonlipid extracted algal biomass (Chinnasamy et al. 2010). Thus, the 13% decrease in the theoretical methane potential in the study conducted by Chinnasamy et al., confirms the prediction of Ward et al.

### 19.7.2 Thermochemical Conversion

Thermochemical conversion of the microalgal biomass takes place with the help of the following processes: combustion, liquefaction, gasification, and pyrolysis (Ward et al. 2014). Direct combustion is a method of burning the algal biomass in the presence of excess air. The combustion usually takes place in boilers with temperatures around 850 °C (Raheem et al. 2015). This process has a major requirement that the moisture content in the biomass should be less than 50% (McKendry 2002a,b). Gasification involves conversion of organic material into clean fuel or synthetic gases, by reacting them at high temperatures, i.e., 800–1000 °C in the absence of oxygen or steam (Raheem et al. 2015; Brennan and Owende 2010). The product of gasification is syngas, which mainly consists of hydrogen (6–55%) and carbon monoxide (8–53%), with methane as a co-product (2–26%) (Nath and Das 2003; Ramzan et al. 2011). The moisture content of the biomass should be less than 15% to be suitable for gasification (McKendry 2002a,b).

*Pyrolysis* is another thermochemical conversion, which occurs at temperatures of 200–750 °C in the absence of oxygen for the production of mainly bio-oil and biochar (Raheem et al. 2015). The process of pyrolysis can be divided into two types, fast pyrolysis, and slow pyrolysis. Slow pyrolysis leads to the formation of biochar and fast pyrolysis leads to the formation of pyrolysis gas (Suali and Sarbatly 2012). Most of the algal cultures are rich in water content, i.e., around 80–90% moisture (Patil et al. 2008). This reduces the viability of the fuels economically, as an expensive drying step is required to reduce the water content to the required operational level (Raheem et al. 2015). However, utilizing this negative aspect in a positive way, *liquefaction* of the microalgal biomass is a viable option. This is the process of production of liquid biofuels from wet biomass. The temperatures needed for the process to proceed is around 200–500 °C with high-pressure conditions i.e. 5–20 bars, in the presence of a catalyst (Goyal et al. 2008). Liquefaction of the algal biomass can yield 10–73% of bio-oil, 8–20% gaseous mixture, and 0.2–0.5% ash content (Shuping et al. 2010; Ross et al. 2010).

### 19.7.3 Chemical Conversion

Chemical conversion of the algal biomass involves the biodiesel production from the lipid content through the transesterification process. Transesterification is a process in which short-chain alcohols, typically methanol and ethanol, are reacted with lipids to produce biodiesel along with the production of the main byproduct, glycerol. Transesterification processes are of two types, base catalyzed and acid catalyzed. In a base-catalyzed transesterification process (e.g., NaOH), lipids are reacted with short-chain alcohols, producing biodiesel and glycerol. Acid-catalyzed transesterification process (e.g., sulfuric acid (1–2% by weight) (Zullaikah et al. 2005)) is used when the feedstock lipid has a high acid content. Therefore, it is used for the conversion of fatty acids into biodiesel (Figure 19.2).

## 19.8 Factors Governing the Production of Third-Generation Biofuels

As the production of third-generation biofuels is done by the algal cells, production is governed by many factors apart from its growth. For instance, photosynthetic efficiency of the cells, which is dependent on sunlight, temperature, and the internal efficiency of the cells, plays a major role in determining the overall growth of the cells. The major aim of algal biofuel production is to optimize capture of solar energy by the algal cells. This energy capture can be enhanced by increasing the number of cells present in a reactor, either in agricultural or industrial setups, which is directly dependent on the photosynthetic efficiency of the algal cells. In the agricultural setups, the growth is totally dependent on sunlight, temperature, and the strain used for the production. The growing strains have to be adaptable to various environmental variations such as the temperature variation. In cold regions, where in winter the temperatures reach the freezing point of water, the algal strains have to be resistant enough to tolerate these stressful conditions. The pathogen load in open ponds is also a governing factor and significantly affects the overall growth of the cells. For this reason, the used strains should be resistant to pathogens to produce high number of cells. In the industrial setups where the sunlight and the temperature can be controlled, the thing that remains to be explored for the purpose is their internal efficiency, by chloroplasts. In the industrial setup, sunlight and temperature can be easily controlled and hence genomes of the chloroplasts are capable of high rates of homologous recombination. Thus, it becomes facile to explore the high efficiencies of the probable recombinant diversity of the algal strains, as identification of specific genes in these cell organelles is not difficult in most of the species (Georgianna and Mayfield 2012). Apart from this, optimizing the antenna size of certain strains can also enhance the efficiency of the algal cells internally (Ort et al. 2011; Cardol et al. 2008; Georgianna and Mayfield 2012). Rubisco, the primary carbon-fixing enzyme in the photosynthetic pathway also plays a major role in the formation of algal cells. The algal cells' growth can be enhanced by optimizing the enzyme to its working maximum, i.e., overexpression leads to increased growth.

Another major factor for underutilization of algae as a potential source of biofuels is the lipid utilization, i.e., only neutral lipids are used for biofuel production. Algal cells produce both type of lipids, polar (including phospholipids and glycolipids), and nonpolar, of which nonpolar lipids constitute only a small fraction. The utilization of both type of lipids for the production of the biofuels can significantly enhance the production quantities (Georgianna and Mayfield 2012).

## 19.9 Advantages of Third-Generation Biofuel Production

The major advantage of algal biofuel production is that microalgae are capable of fixing $CO_2$ from the atmosphere, which acts as a beneficial sink for the increasing $CO_2$

emissions into the atmosphere. This is in accordance with the strategy of reducing fossil fuel usage, reducing carbon-intensive products, and enhancing the sequestration of atmospheric $CO_2$. Algal cells grow by capturing $CO_2$ from the ambient air, which is in compliance with $CO_2$ sequestration (Figure 19.3). Second, this fuel competes with conventional fossil fuel products and thus helps in reducing the use of carbon-intensive fossil-fuel products (Cuellar-Bermudez et al. 2015).

Carbon sequestration is the process of capturing of $CO_2$ from the ambient air, which would otherwise remain in the atmosphere as a potent greenhouse gas. The industrial carbon capturing techniques are expensive and do not provide benefits to the extent that algal biofuels do. Microalgae can be employed for the same purpose. Apart from capturing atmospheric $CO_2$ and $CO_2$ emissions from power plants and industrial processes, microalgae can also be used to capture $CO_2$ from soluble carbonates (Brennan and Owende 2010; Wang et al. 2008). Thus, they help in the bio-fixation of $CO_2$ removed from industrial flue gases. They also participate in the removal of $NH_4^+$, $NO_3^-$, and $PO_4^{3-}$ because algae assimilate these inorganic ions in the form of nutrients.

After the extraction of the oil is finished, the resulting algal biomass can be processed into different kinds of commercially desirable chemicals such as ethanol, methane, and several other bioproducts. Alternatively, the algal biomass can be used either as a livestock feed or as an organic fertilizer due to high N:P ratio. Depending on the microalgal species, these biomass resources can be used for the production of many other bioproducts such as fats, polyunsaturated fatty acids, oil, natural dyes, sugars, pigments, and other bioactive compounds (Li et al. 2008a,b; Raja et al. 2008).

Algae biofuels can grow in nonarable lands, and thus reduce the stress on the land required for the production of food crops. This is the major advantage of the algal biofuels over first and second generation biofuels. The algal produce can first be used for the production of biofuel and then it can be used in the form of animal feeds such as *Scenedesmus* and *Chlorella,* which have a high nutritional value.

Microalgae are also known for their role in the wastewater treatment studies, since they aid in the removal of contaminants from the wastewater. These contaminants may include chemicals, organic contaminants, heavy metals, and pathogens in the water. Microalgae in general can grow easily in such waters, removing contaminants and ultimately producing biofuels (Brennan and Owende 2010). Depending on the algal strains, they can also remove toxic heavy metals from wastewater, such as chromium, cadmium, and copper. *Spirulina* sp. has been reported to absorb heavy metals and is thus called *biosorbent* (Chojnacka et al. 2005). Toxic compounds are also processed by the microalgae, as they aid the bacterial strains with oxygen produced by them, and consequently help in biodegrading polycyclic aromatic hydrocarbons (PAHs), phenolics, and organic solvents (Munoz and Guieysse 2006).

Another major advantage in support of algal biofuels is the variety of accessible biofuels, including biohydrogen (which is considered the cleanest fuel), bioethanol, biogas, bio-oil, syngas, and biodiesel (Harun et al. 2014).

## 19.10 Conclusions and Future Perspectives

The research and developmental activities in third-generation biofuels mainly depend on the type and efficiency of the photobioreactors. There has been a constant development with the type of photobioreactors with their efficiencies. The use of internally illuminated photobioreactors has taken a major step forward in efficient production of algal cultures for the production of biofuels. Apart from this, the type of algal strain used in the photobioreactor also plays a major role. The use of a well-adapted algal strain that can tolerate a broad range of parameters has enhanced the algal cultivation of biomass significantly. The use of these strains significantly curtails the demerits associated with algal biofuel production. The genetically modified algal strains have also produced significantly enhanced results for the lipid production and consequently biofuel production. Research and development activities with respect to third-generation biofuel are ongoing, because of its potential to become a leader in biofuel production. The fact that third-generation biofuels can be produced in large quantities with comparatively very less land area represents huge potential.

The major challenge that algal biofuel production faces is growth in the highly concentrated waste systems. In a highly concentrated wastewater effluent, the growth of microalgae is suppressed and the anaerobic bacteria dominate. There has been constant development with regard to the same, using new and genetically modified algal strains. Still, to enhance viability, more exploration is required. Genetically modified algal strains could be the solution to this problem. Apart from this, new techniques such as simultaneous fermentation in the photobioreactors and its further development have given some fruitful insights for the research and development of third-generation biofuel production. Further research in this area could be a major boost in tackling the issue of fossil fuel emissions into the atmosphere.

## Acknowledgments

Authors are highly grateful to Council of Scientific and Industrial Research for providing the funds for project to carry out the research work. Authors thank DST and DBT for providing funds for research equipment.

## References

Amin, S. (2009). Review on biofuel oil and gas production processes from microalgae. *Energy Convers. Manage.* 50 (7): 1834–1840.

Bae, J.H. and Hur, S.B. (2011). Selection of suitable species of chlorella, nannochloris, and nannochloropsis in high-end low-temperature seasons for mass culture of the rotifer Brachionus plicatilis. *Fish. Aquat. Sci.* 14 (4): 323–332.

Béchet, Q., Shilton, A., Fringer, O.B. et al. (2010). Mechanistic modeling of broth temperature in outdoor photobioreactors. *Environ. Sci. Technol.* 44 (6): 2197–2203.

Benemann, J.R., Weissman, J.C., Koopman, B.L., and Oswald, W.J. (1977). Energy production by microbial photosynthesis. *Nature* 268 (5615): 19.

Bonin, C. and Lal, R. (2012). Agronomic and ecological implications of biofuels. *Adv. Agron.* 117: 1–50.

Brennan, L. and Owende, P. (2010). Biofuels from microalgae—a review of technologies for production, processing, and extractions of biofuels and co-products. *Renewable Sustainable Energy Rev.* 14 (2): 557–577.

Butterwick, C., Heaney, S.I., and Talling, J.F. (2005). Diversity in the influence of temperature on the growth rates of freshwater algae, and its ecological relevance. *Freshwater Biol.* 50 (2): 291–300.

Cardol, P., Bailleul, B., Rappaport, F. et al. (2008). An original adaptation of photosynthesis in the marine green alga Ostreococcus. *Proc. Natl. Acad. Sci. U.S.A.* 105 (22): 7881–7886.

Chinnasamy, S., Bhatnagar, A., Hunt, R.W., and Das, K.C. (2010). Microalgae cultivation in a wastewater dominated by carpet mill effluents for biofuel applications. *Bioresour. Technol.* 101 (9): 3097–3105.

Chisti, Y. (2007). Biodiesel from microalgae. *Biotechnol. Adv.* 25 (3): 294–306.

Chojnacka, K., Chojnacki, A., and Gorecka, H. (2005). Biosorption of Cr3+, Cd2+ and Cu2+ ions by blue–green algae Spirulina sp.: kinetics, equilibrium and the mechanism of the process. *Chemosphere* 59 (1): 75–84.

Cuellar-Bermudez, S.P., Garcia-Perez, J.S., Rittmann, B.E., and Parra-Saldivar, R. (2015). Photosynthetic bioenergy utilizing $CO_2$: an approach on flue gases utilization for third generation biofuels. *J. Cleaner Prod.* 98: 53–65.

Darzins A., P. Pienkos, and L. Edye. 2010. Current status and potential for algal biofuels production. A report to IEA Bioenergy Task. 39.

Demirbas, M.F. (2011). Biofuels from algae for sustainable development. *Appl. Energy* 88 (10): 3473–3480.

Doe U. S. 2010. National algal biofuels technology roadmap. US Department of Energy, Office of Energy Efficiency and Renewable Energy, Biomass Program.

Flessa, H., Ruser, R., Dörsch, P. et al. (2002). Integrated evaluation of greenhouse gas emissions ($CO_2$, $CH_4$, $N_2O$) from two farming systems in southern Germany. *Agric. Ecosyst. Environ.* 91 (1–3): 175–189.

Georgianna, D.R. and Mayfield, S.P. (2012). Exploiting diversity and synthetic biology for the production of algal biofuels. *Nature* 488 (7411): 329.

Golueke, C.G., Oswald, W.J., and Gotaas, H.B. (1957). Anaerobic digestion of algae. *J. Appl. Microbiol.* 5 (1): 47.

Goyal, H.B., Seal, D., and Saxena, R.C. (2008). Bio-fuels from thermochemical conversion of renewable resources: a review. *Renewable Sustainable Energy Rev.* 12 (2): 504–517.

Griffiths, M.J. and Harrison, S.T. (2009). Lipid productivity as a key characteristic for choosing algal species for biodiesel production. *J. Appl. Phycol.* 21 (5): 493–507.

Harun, R., Yip, J.W., Thiruvenkadam, S. et al. (2014). Algal biomass conversion to bioethanol–a step-by-step assessment. *Biotechnol. J.* 9 (1): 73–86.

Hase, R., Oikawa, H., Sasao, C. et al. (2000). Photosynthetic production of microalgal biomass in a raceway system under greenhouse conditions in Sendai city. *J. Biosci. Bioeng.* 89 (2): 157–163.

Larkum, A.W., Ross, I.L., Kruse, O., and Hankamer, B. (2012). Selection, breeding and engineering of microalgae for bioenergy and biofuel production. *Trends Biotechnol.* 30 (4): 198–205.

Lee, Y.K. and Low, C.S. (1991). Effect of photobioreactor inclination on the biomass productivity of an outdoor algal culture. *Biotechnol. Bioeng.* 38 (9): 995–1000.

Li, Y., Horsman, M., Wang, B. et al. (2008a). Effects of nitrogen sources on cell growth and lipid accumulation of green alga Neochloris oleoabundans. *Appl. Microbiol. Biotechnol.* 81 (4): 629–636.

Li, Y., Horsman, M., Wu, N. et al. (2008b). Biofuels from microalgae. *Biotechnol. Progr.* 24 (4): 815–820.

Luo, Y., Ahmed, I., Kubátová, A. et al. (2010). The thermal cracking of soybean/canola oils and their methyl esters. *Fuel Process. Technol.* 91 (6): 613–617.

McKendry, P. (2002a). Energy production from biomass (part 1): overview of biomass. *Bioresour. Technol.* 83 (1): 37–46.

McKendry, P. (2002b). Energy production from biomass (part 2): conversion technologies. *Bioresour. Technol.* 83 (1): 47–54.

Mirón, A.S., Garcıa, M.C.C., Camacho, F.G. et al. (2002). Growth and biochemical characterization of microalgal biomass produced in bubble column and airlift photobioreactors: studies in fed-batch culture. *Enzyme Microb. Technol.* 31 (7): 1015–1023.

Miyamoto, K., Hallenbeck, P.C., and Benemann, J.R. (1979). Nitrogen fixation by thermophilic blue-green algae (cyanobacteria): temperature characteristics and potential use in biophotolysis. *Appl. Environ. Microbiol.* 37 (3): 454–458.

Molina, E., Fernández, J., Acién, F.G., and Chisti, Y. (2001). Tubular photobioreactor design for algal cultures. *J. Biotechnol.* 92 (2): 113–131.

Munoz, R. and Guieysse, B. (2006). Algal–bacterial processes for the treatment of hazardous contaminants: a review. *Water Res.* 40 (15): 2799–2815.

Nath, K. and Das, D. (2003). Hydrogen from biomass. *Curr. Sci.* 265–271.

NOAA, June 15, 2018. https://www.esrl.noaa.gov/gmd/ccgg/trends/weekly.html.

Ogbonna, J.C., Soejima, T., and Tanaka, H. (1999). An integrated solar and artificial light system for internal illumination of photobioreactors. *J. Biotechnol.* 70 (1–3): 289–297.

Ort, D.R., Zhu, X., and Melis, A. (2011). Optimizing antenna size to maximize photosynthetic efficiency. *Plant Physiol.* 155 (1): 79–85.

Patil, V., Tran, K.Q., and Giselrød, H.R. (2008). Towards sustainable production of biofuels from microalgae. *Int. J. Mol. Sci.* 9 (7): 1188–1195.

Pirt, S.J., Lee, Y.K., Walach, M.R. et al. (1983). A tubular bioreactor for photosynthetic production of biomass from carbon dioxide: design and performance. *J. Chem. Technol. Biotechnol.* 33 (1): 35–58.

Raheem, A., Azlina, W.W., Yap, Y.T. et al. (2015). Thermochemical conversion of microalgal biomass for biofuel production. *Renewable Sustainable Energy Rev.* 49: 990–999.

Raja, R., Hemaiswarya, S., Kumar, N.A. et al. (2008). A perspective on the biotechnological potential of microalgae. *Crit. Rev. Microbiol.* 34 (2): 77–88.

Ramzan, N., Ashraf, A., Naveed, S., and Malik, A. (2011). Simulation of hybrid biomass gasification using Aspen plus: a comparative performance analysis for food, municipal solid and poultry waste. *Biomass Bioenergy* 35 (9): 3962–3969.

Rawat, I., Kumar, R.R., Mutanda, T., and Bux, F. (2013). Biodiesel from microalgae: a critical evaluation from laboratory to large scale production. *Appl. Energy* 103: 444–467.

Rodolfi, L., Chini Zittelli, G., Bassi, N. et al. (2009). Microalgae for oil: strain selection, induction of lipid synthesis and outdoor mass cultivation in a low-cost photobioreactor. *Biotechnol. Bioeng.* 102 (1): 100–112.

Rogak, S.N., Pott, U., Dann, T., and Wang, D. (1998). Gaseous emissions from vehicles in a traffic tunnel in Vancouver, British Columbia. *J. Air Waste Manag. Assoc.* 48 (7): 604–615.

Ross, A.B., Biller, P., Kubacki, M.L. et al. (2010). Hydrothermal processing of microalgae using alkali and organic acids. *Fuel* 89 (9): 2234–2243.

Shuping, Z., Yulong, W., Mingde, Y. et al. (2010). Production and characterization of bio-oil from hydrothermal liquefaction of microalgae Dunaliella tertiolecta cake. *Energy* 35 (12): 5406–5411.

Suali, E. and Sarbatly, R. (2012). Conversion of microalgae to biofuel. *Renewable Sustainable Energy Rev.* 16 (6): 4316–4342.

Tirichine, L. and Bowler, C. (2011). Decoding algal genomes: tracing back the history of photosynthetic life on earth. *Plant J.* 66 (1): 45–57.
Tredici, M.R. and Zittelli, G.C. (1998). Efficiency of sunlight utilization: tubular versus flat photobioreactors. *Biotechnol. Bioeng.* 57 (2): 187–197.
Ugwu, C.U., Aoyagi, H., and Uchiyama, H. (2008). Photobioreactors for mass cultivation of algae. *Bioresour. Technol.* 99 (10): 4021–4028.
US Energy Information Administration. www.eia.gov (US Energy Information Administration, 2012).
USDA-NASS. http://www.nass.usda.gov (2012).
Wang, D., Czernik, S., and Chornet, E. (1998). Production of hydrogen from biomass by catalytic steam reforming of fast pyrolysis oils. *Energy Fuels* 12 (1): 19–24.
Wang, B., Li, Y., Wu, N., and Lan, C.Q. (2008). $CO_2$ bio-mitigation using microalgae. *Appl. Microbiol. Biotechnol.* 79 (5): 707–718.
Ward, A.J., Lewis, D.M., and Green, F.B. (2014). Anaerobic digestion of algae biomass: a review. *Algal Research* 5: 204–214.
Watanabe, Y. and Saiki, H. (1997). Development of a photobioreactor incorporating Chlorella sp. for removal of $CO_2$ in stack gas. *Energy Convers. Manage.* 38: S499–S503.
Wigmosta, M.S., Coleman, A.M., Skaggs, R.J. et al. (2011). National microalgae biofuel production potential and resource demand. *Water Resour. Res.* 47 (3).
Zullaikah, S., Lai, C.C., Vali, S.R., and Ju, Y.H. (2005). A two-step acid-catalyzed process for the production of biodiesel from rice bran oil. *Bioresour. Technol.* 96 (17): 1889–1896.

# 20

# Bioethanol Production from Fruit and Vegetable Wastes

*Meganathan Bhuvaneswari*[1] *and Nallusamy Sivakumar*[2]

[1] *Department of Biotechnology, Sona College of Arts and Science, Salem 636 005, India*
[2] *Department of Biology, College of Science, Sultan Qaboos University, Muscat 123, Sultanate of Oman*

## 20.1 Introduction

Worldwide, researchers are taking the search for sustainable alternate fuel seriously as concerns grow over fossil fuel use. The demand for fossil fuel has increased as urbanization, population, and industrialization expand (Chandel et al. 2007; Hossain et al. 2011). The conventional fossil fuels, which come from dead animals, plants, and microbes, require millions of years to form (Huang et al. 2012). Thus, these fossil fuels such as petrol, oil, coal, and natural gas belong to nonrenewable energy sources. Today, the global consumption of these fossil fuels accounts for 80%. These energy sources often lead to an increase in oil prices, which correspond to economic conflicts in most countries. The growth of many developing countries mainly depends on fossil fuel resources, which are projected to be depleted in another 65 years. Crude oil resources will decline from the current 25 billion barrels to nearly 5 billion barrels in 2050 (Paulova et al. 2015). In 2016, UNO had given agenda for a sustainable environment. One important agenda was to increase the productivity of renewable energy by 2030 to seriously address the effects of fossil fuel, which contributes to environmental and atmospheric pollution, including greenhouse gases, which lead to depletion of the ozone layer. Incomplete combustion of fossil fuels produces toxic gases such as carbon monoxide, carbon (IV) oxide ($CO_2$), methane ($CH_4$), and nitrous and sulfur oxides, etc. (Gosavi et al. 2017). The need for alternate fuel has gained importance due to these major issues.

Moreover, a major drawback of fossil fuel utilization, particularly environmental pollution, has received significant consideration when we search for other resources. The best alternative is the production of biofuel from renewable energy sources for a sustainable environment. Biofuels can be used as a part of the solution and hence can be blended with gasoline, which requires no engine modifications (Guerrero et al. 2018).

Among the biofuels available in the market, bioethanol is indeed a dominant liquid fuel for regular use (Handler et al. 2016). The blending of ethanol with gasoline oxygenate the fuel mixture; consequently, it burns more completely and decreases the polluting emissions. Generally, ethanol is produced from sugar-rich substances like molasses or corn syrup. It is considered the first-generation ethanol production process. Now the base has been shifted to the production of second-generation ethanol

*Bioprocessing for Biomolecules Production*, First Edition.
Edited by Gustavo Molina, Vijai Kumar Gupta, Brahma N. Singh, and Nicholas Gathergood.

from various waste materials of crops, forests, municipal sludge, and solid waste. This second-generation bioethanol production is considered sustainable, and more research has been done to mobilize this production globally. Even though bioethanol has been identified as the dominant renewable source, many improvements must be done to ease the production process as well as to cut short the total cost.

Fruit wastes rich in reducing sugars and vegetable wastes rich in cellulose, hemicellulose, and lignin are the prominent feedstock for the production of second-generation bioethanol. Since large quantities of agro-wastes, including fruit and vegetable wastes, are available throughout the world from different fruit plantations and vegetable growing fields, their disposal has become a major problem. Since they are renewable wastes, utilizing them as a feedstock for bioethanol production would reduce the cost and also solve the issue of waste disposal. This chapter describes the research advances of bioethanol production using fruits and vegetable wastes, which are considered to be decaying waste materials.

## 20.2 Importance of Biofuels

Biofuel is a renewable source of energy and thus can be used as an alternative to conventional fossil fuels. Biofuels offer us many priorities, including sustainability and reduction of greenhouse gas emissions. Increased exploration of biofuels is targeting the increased production of bioethanol and biodiesel. The popular classification for liquid biofuels includes first-generation and second-generation fuels based on edible and nonedible raw materials, respectively. Mainly, the raw materials used for this biofuel production comprise agricultural wastes such as sugarcane, molasses, groundnut shells, rice husks, straw, and corn cobs. These low-cost materials reduce fuel-production costs (Cutzu and Bardi 2017). Biofuels can also be produced from industrial, municipal, and foresting residues (Oniya and Bamgboye 2014). Biofuels are considered to mitigate the effects of climate change (Guerrero and Munoz 2018).

## 20.3 Bioethanol as a Promising Biofuel

Among all the biofuels available in the market, bioethanol is indeed a dominant liquid fuel and can be mixed with gasoline for regular use. Ethyl alcohol (Ethanol) is the most promising liquid biofuel; it can be used as a fuel or gasoline enhancer. The most interesting aspect of ethanol is that it has a higher oxygen content (35%) than other biofuels. This increased percentage of oxygen allows better combustion of hydrocarbons with the reduction in the emission of carbon monoxide and other dreadful hydrocarbons (Gebregergs et al. 2016).

Bioethanol is eco-friendly and has fewer negative environmental impacts than fossil fuels (Umamaheswari et al. 2010). Bioethanol is produced by fermentation of sugar-containing materials using microorganisms. It can be used either directly or mixed with gasoline, thus acting as a substitute for gasoline, with powerful and strategic applications. The mix of bioethanol with gasoline is very appropriate because of its high octane number. Particularly in light-duty vehicles, one of the blends used is called as E85 (comprising 85% bioethanol and 15% gasoline). The year 2015 was remarkable for the production of bioethanol because global production increased by 4%. Brazil (E20), United States (E25), China, India (E5), Australia, Columbia (E10), and Paraguay (E7) use the blended form of bioethanol. Even though bioethanol has some disadvantages, such as a corrosive nature, lesser energy density than gasoline and low vapor pressure (starting problems in cold areas), it is considered as a sustainable energy source because of its many advantages.

## 20.4 Bioethanol from Wastes

Wastes are generated from many sectors, such as industries, agriculture, forestry, corporations, and municipalities. Using agricultural wastes that would otherwise be discarded is an efficient way for the cheaper production of bioethanol (Gosavi et al. 2017). Recently, lignocellulosic biomass received significant consideration as a potential feedstock for sustainable production of chemicals and fuels (Al Azkawi et al. 2018). These waste biomasses are the best feedstock for the production of bioethanol. The production of bioethanol must involve this feedstock rather than the food grains due to limited supplies (Nair et al. 2016). Many countries define bioenergy as energy obtained from biodegradable wastes and residues from agriculture like fruits and vegetables (Panda et al. 2018). Many forms of wastes are used to produce a wide range of biofuels. The waste materials are usually dumped in dumping sites or landfills, which result in the contamination of lands as well as water (Hossain et al. 2011). But when they are used as a feedstock to produce biofuel, we can avoid this unwanted environmental pollution (Tock et al. 2009). Post-harvesting, refrigeration/storage, and inadequate processing are the major problems for fruits and vegetables rotting. The economic value of this wastage is enormous, and cleaning these wastes has become a major challenge (Girish et al., 2014). Developing countries are finding difficulties in disposing of these wastes, which cause excessive environmental damage, including emission of greenhouse gases. Hence, turning these wastes into a resource for energy production solves two issues: It reduces energy costs and provides an environmentally friendly way to dispose of what would otherwise be pollution (Gebregergs et al. 2016).

Novel ideas for using these fruits and vegetable wastes for the production of bioethanol have become the need of the hour. Solid wastes such as fruit and vegetable peels from food-processing industries, juice processing plants, hotels, and restaurants are considered for this production. The food products such as vegetables and fruits are wasted in the various steps from field until they reach the customer (Survase et al. 2013). The production of first-generation bioethanol was primarily considered as a solution for the problems encountered with fossil fuels. But it confronts some negative impacts because of the edible raw materials; the usage of agricultural lands used for crops for biofuels was indeed a great debate. Diverting the food crops into crops for producing biofuels will adversely increase the cost of food and causes socioeconomic problems. To overcome this barrier, researchers have developed second-generation bioethanol from the waste biomass originated from many processes. Waste biomass can have little to no cost, so production cost is automatically reduced. In this chapter, we study the production of bioethanol from various fruits and vegetable wastes easily available, such as waste of banana, citrus, jackfruit, pineapple, dates, mangoes, chestnut, potatoes, tomatoes, and sweet potatoes.

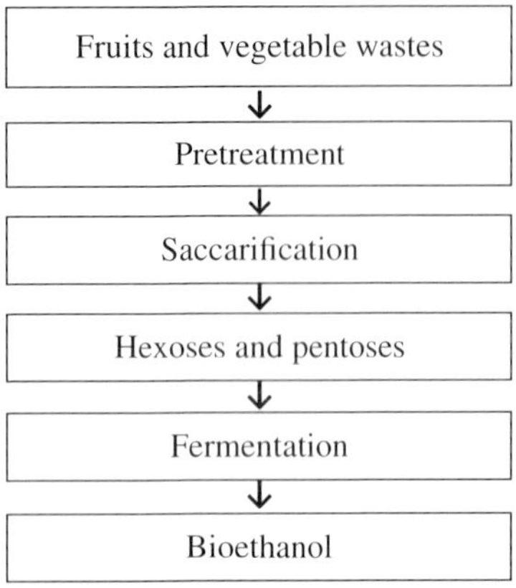


**Figure 20.1** General steps involved in the production of bioethanol from fruits and vegetable wastes.

## 20.5 General Mechanism of Production of Bioethanol

We can produce ethanol from the waste biomass by the following steps: pretreatment of biomass, enzymatic hydrolysis, and sugar fermentation processes. Complex mixtures of carbohydrate polymers such as cellulose, hemicellulose, and lignin are present in the biomass wastes, which are major the components of the plant cell walls. In order to produce sugars from the biomass, the biomass is first pretreated physically and then with acids or alkalies for size reduction of the feedstock and also to open up the lignin barrier in the plant structure (Figure 20.1). Then the homo- and heteropolysaccharide portions are hydrolyzed by enzymes or dilute acids into sugars, which will be the substrate for ethanol fermentation using yeasts (Dudley 2004; Panda et al. 2018).

## 20.6 Ethanol Production Using Fruit Wastes

Fruits and vegetable wastes can be microbially converted into several value-added products as they have a high content of natural sugars. Vast amounts of fruits are consumed worldwide as health supplements and also as functional foods (Sharma 2006). In addition to rich cellulose and hemicellulose content, the fruit wastes have low lignin contents, which make these wastes interesting for bioethanol production. This section describes how various fruit wastes are processed for ethanol production.

### 20.6.1 Bioethanol from Banana Wastes

Banana is the most important fruit crop grown worldwide. Asian and Americans grow it in a high proportion – 56% and 26%, respectively. Australia is also a significant producer of bananas. Approximately, a major proportion of bananas are rejected, wasted, or discarded during the post-harvesting transport and grading process. These rotten, discarded bananas are dumped in the lands which eventually contaminate the environment. These wastes can be used to produce second-generation bioethanol, since these wastes consist of valuable sugars and minerals required for the fermentation process. Thus this low-cost raw material can displace the first generation bioethanol produced using food crops (Guerrero et al. 2018).

Banana wastes (rotten) and banana peels are the feedstock for the ethanol production. *Saccharomyces cerevisiae* is used in the bioethanol production. This fungus is an apt organism for the fermentation process because it can grow well in high sugar concentrations and thus fermenting them into ethanol and $CO_2$. Another important measure for this production process is to hydrolyze lignin, which is a barrier for the conversion of sugars into bioethanol. This can be done using acid hydrolysis of the substrate or the feedstock. Once this pretreatment is done, the yeasts can utilize the sugars, especially glucose, and convert into bioethanol (Hossain et al. 2011). Apart from glucose, the banana wastes also consist of many carbohydrates, which can be mobilized into simple sugars by enzymatic hydrolysis using enzymes such as pectinases, hemicellulases, and cellulases.

For the ethanol production, banana wastes are collected, washed, chopped into small pieces, and then dried under sunlight. The dried pieces are crushed and grounded well. The grounded mixture is then mixed with distilled water. Sometimes the banana wastes can be used without drying. In this case, they must be washed and mashed well. Either the mashed bananas or the powdered mixture has to be autoclaved and the sterile feedstock will be taken for further hydrolysis. The feedstock is mixed with 0.5–2.5% (v/v) of diluted sulfuric acid and kept at 70–110 °C for 10–30 minutes. Enzymes such as pectinases, cellulases, and hemicellulases are added to the hydrolysate around pH 4.5–5.5 (Wu et al. 2016) and incubated at 50 °C for an hour. After this hydrolysis, the sterile hydrolysate is added with 18–24 h grown yeast culture and tightly closed and incubated anaerobically with shaking at 35 °C, pH 5–5.5 for 3–7 days. In between aliquots can be tested for bioethanol concentration. The higher the water content in the feedstock, the higher the bioethanol production, in these optimal conditions. The bioethanol is then collected by distillation (Bello et al. 2014). The element analysis of the bioethanol shows the elements present in it are in very low concentration or nil concentration. Toxic elements such as Pb, Cr, and Mo are showing zero ppm, whereas Mn, P, Mg, and Ca are of higher values (Hossain et al. 2011). Though it shows higher values, these are not toxic to the environment. Thus, the absence of the toxic elements shows bioethanol as a sustainable fuel. Hence, the bioethanol production from the banana wastes shows a promising pathway for the production of second generation bioethanol.

### 20.6.2 Bioethanol from Citrus Fruit Wastes

Like banana, citrus fruits are one of the most important cultivated crops in the world. Moreover, their juices are the most consumed globally. Global production of oranges is around 50.2 million tons (Cypriano et al. 2018). Unless properly stored, transported, and retailed, more amount of fruits are wasted. The production of orange juice also gives out many wastes such as citrus pulp floater, orange peels, internal tissues and fibers, etc. These waste materials are having high levels of sugars, pectins, proteins, hemicelluloses, and celluloses. These waste products when treated in the floatation tank gives out clogging. If it is dumped inland sites can cause microbial contamination and becomes so dreadful to the environment.

This relatively less-expensive citrus fruit waste has abundant sugars that can be used for the fermentative production of bioethanol. The fruit wastes of *Citrus limetta* (Mousambi) and *C. sinensis* (sweet oranges) are having 21 mg and 17 mg $ml^{-1}$ of glucose, respectively (Girish et al. 2014). Moreover, the presence of a very small amount of lignin is an advantage when compared to other lignin-rich fruit wastes for bioethanol production.

Decaying citrus fruits and citrus peels are the raw material for this second-generation bioethanol production process. The raw material is subjected to fermentation with a monoculture of yeasts or by co-culturing yeasts with another strain of yeast called *Candida parapsilosis* NRRLY-12969. This co-culturing is most efficient because the unavailable sugars such as pentoses can be fermented by this new strain and thus bioethanol conversion will be greater.

Citrus wastes and its peels are collected, washed, surface sterilized, and dried in sunlight or ovens. Dried wastes are grounded well and mixed with distilled water. The watery feedstock will be sterilized. Once cooled, the feedstock is mixed with enzymes such as pectinases, cellulases, and hemicellulases. This enzyme hydrolysate is then fermented with a monoculture of yeast at 35 °C for five days aerobically and then from sixth to ninth day under anaerobic conditions (Girish et al. 2014) or by co-culturing *S. cerevisiae* and *C. parapsilosis*. The produced bioethanol can be purified by distillation process and considered for further use.

### 20.6.3 Bioethanol from Pineapple Wastes

Global energy demand is significantly influencing the production technology of the second generation of bioethanol from the wastes. This second-generation bioethanol, as discussed earlier, utilizes potential wastes, especially fruit wastes. One among those is the waste of *Ananas cosmosus* (i.e., pineapple wastes and the potential of these wastes lies in its high content of fermentable sugars and highly hydrolysable cellulose and hemicelluloses). The total pineapple wastes account for 50% (w/w) of its total production worldwide. Post-harvesting damages and inadequate storage facilities are the main reasons for this loss. Waste management by producing bioethanol is one of the leading strategies for a sustainable environment.

Pineapple wastes, including the leafy shoots, are utilized as raw materials for ethanol production. Pretreatments like acid and enzymatic hydrolysis help in the better fermentation of this sugary waste by the yeasts. Hydrolysis, fermentation, distillation, and dehydration are the four main steps involved in bioethanol production. Because of the higher sugar content, these processes can produce more bioethanol (Hossain and Fazliny 2010).

For the ethanol production, pineapple wastes are collected, washed, cut, and oven dried for 48 hours. The dried wastes are powdered and added with distilled water. This mixture is subjected to acid hydrolysis using sulfuric acid and then autoclaved. The autoclaved sample is cooled and enzymes such as celluloses, hemicelluloses, and pectinases are added. This enzyme hydrolysate is then inoculated with the yeast culture and allowed to do the fermentation under aerobic fermentation with proper mixing of the hydrolysate up to 11 days. After that by distillation and dehydration process ethanol produced can be collected (Conesa et al. 2017). Because of the high sugar content of the pineapple wastes (20 mg $ml^{-1}$), more amount of bioethanol can be produced by fermentation (Girish et al. 2014).

### 20.6.4 Bioethanol from Pomegranate

Pomegranate (*Punica granatum*) has become a popular fruit because of its health benefits. Worldwide, people consume it and thus processing industry produces the peel, which is one of the useful raw materials for bioethanol production. As the peels are rich in hexoses and pentoses, co-fermentation method using yeasts and *Pichia stipitis* would be beneficial (Demiray et al. 2018) Acid hydrolysis releases sugars in the available forms and the yeasts convert them efficiently into bioethanol.

### 20.6.5 Bioethanol from Mango Wastes

Mango fruits (*Magnifera indica*) are highly perishable and easily gets damaged from the plucking process until it reaches the consumers. Tropical countries are well known for the cultivation of mangoes. The edible pulp makes up to 85%, whereas the peel and the kernel make up to 24% and 40%, respectively (Buenrostro-Figueroa et al. 2018). Mangoes are having 18–20% of sugar content. Hence, bioethanol production is possible by utilizing the wastes from mangoes. Two types of yeasts are used for the effective fermentation process such as *S. cerevisiae* and *Kluyveromyces marxianus* for the better production.

### 20.6.6 Bioethanol from Jackfruit Wastes

Jackfruit (*Artocarpus heterophyllus*) is one of the sweetest tropical fruits. Fructose, glucose, and sucrose were found to be the major sugars in all parts of this multiple fruit. Moreover, the seeds called jackfruit stones are rich in starch (12.78%), which makes them feasible for ethanol production (Nuriana and Wuryantoro, 2015). The jackfruit wastes and the stones (seeds) are collected, cleaned, washed, and cut into small pieces. The dried pieces are powdered well. The powder is mixed with distilled water and then hydrolyzed using acid and enzyme. The enzyme used for this purpose is α-amylase to mobilize the starch content for fermentation. Then the hydrolysate is fermented using yeast cells. Thus, jackfruit wastes proved to be a potential raw material for bioethanol production (Gosavi et al. 2017).

### 20.6.7 Bioethanol from Date Palm Fruit Wastes

Dates, a huge source of carbohydrates, are wasted in large amounts during picking, transporting, storing, and also in retail outlets. The date-palm waste is rich in biodegradable sugars, and the bioethanol production can be produced by bioconversion of these sugars under anaerobic conditions using yeasts (Boulal et al. 2016). Date-palms are grown in the Gulf countries, such as Egypt, Pakistan, Algeria, and Libya. Dates wastes are washed, immersed in a water bath, and rubbed and rinsed with water. Seeds are separated, grounded, and kept in a water bath at 90–95 °C for extraction of sugars. A series of extractions are carried out with juices of decreasing Brix content. To avoid contamination, the temperature of the date juice is maintained at around 60 °C. After acid hydrolysis, the hydrolysate is added with yeast culture for fermentation around 32 °C. After 72 h the process is stopped and the bioethanol produced can be distilled. Ethanol strength of 90% can be obtained by repeating this process a minimum of four times (Sivakumar 2012).

### 20.6.8 Pistachio-Wastes as Potential Raw Material

Pistachio wastes also account for bioethanol production as equivalent to the earlier discussion with the fruit and vegetable wastes in the process of conversion of biomass as a

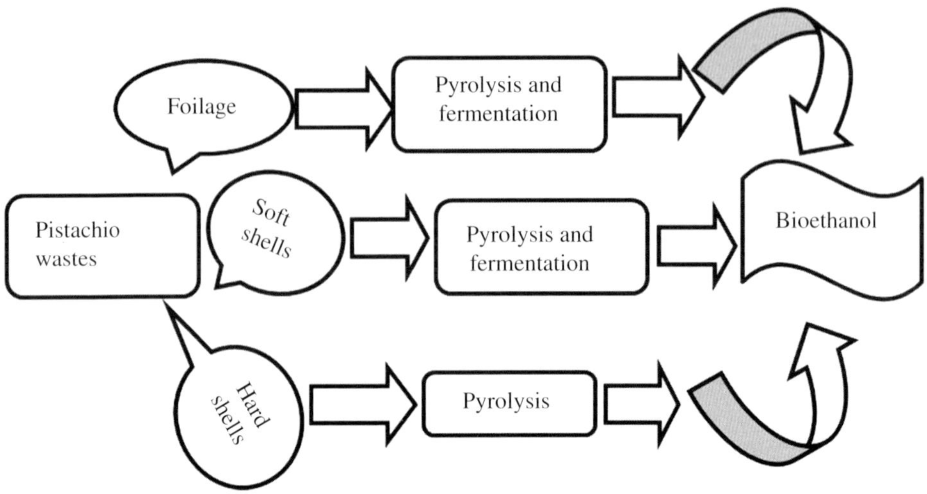


**Figure 20.2** Bioethanol production strategy from pistachios.

renewable energy resource for both eco-friendly and sustainable environment. The pistachio is mostly cultivated in the dry areas of the Middle East, Mediterranean countries, and the United States. Pistachios can be consumed only after the revival of the shells. The waste produced by the pistachios processing industries produces methane that causes environmental hazards. Thus, disposing of the wastes is always done in a reliable, economical, and healthy way. One among the promising way is to use them as a raw material for bioethanol production (Taghizadeh-Alisaraei et al. 2017). The steps involved in this conversion are depicted in Figure 20.2.

### 20.6.9 Bioethanol from Other Fruit Wastes

Apart from the fruit wastes mentioned above, the wastes of apple, papaya (Jahid et al. 2018) and grapes (Janani et al. 2013) can also be used as valid raw materials for ethanol production. Papaya *(Carica papaya)* is a fruit available in every part of the world. This fruit is rich in energy and inverts sugar. Grapes are the fruit of wine (*Vitis vinifera*). Grapes as whole fruit, skin, leaves, and seeds are used in medicine. Thus, the fruit wastes of grapes are also a most promising raw material for bioethanol production. Apples (*Malus pumia*) are grown in the temperate zones and rich in nutrients. These wastes of apples are also a good source for our bioethanol production. These wastes are subjected to the same process of saccharification, fermentation, distillation, and dehydration.

## 20.7 Bioethanol from Vegetable Wastes

Vegetables are very much needed as a source of vital nutrients present in them. However, in the process of harvesting, packing, transporting, storage before they reach the consumers, a lot of vegetable waste is produced. Even at the wholesale and retail outlets, delay in different stages of handling leads to damaged vegetables. The market-refused vegetables sometimes are discolored, spotted, or rotting, which results in a huge loss. These wastes are generally given as feed or kept for composting. Vegetables like tomatoes, cabbage, and cauliflower have heavy postharvest losses, ranges from 30–40%, 25–30%, and 29–35%, respectively. When these wastes are utilized properly, they can be converted into bioethanol. Even coffee and cassava wastes (stems, peels, and leaves) have been studied as possible source materials for bioethanol. These wastes mainly consist of starch, cellulose, and hemicelluloses. Like fruits wastes, pretreatment with acids and enzymes is required prior to fermentations (Singh et al. 2012). The pretreatment thus releases simple sugars so as to ease the fermentation process (Deshpande et al. 2015). To ensure that staple crops are not used as raw materials in biofuels, diversification is required for the feedstock for bioethanol production (Guerrero et al. 2018).

Europe is a significant producer of potatoes and thus disposal of the peels is a major problem in need of environmentally friendly solutions. Potato processing industries are growing exponentially. All these eventually produce a huge amount of wastes. Potato wastes consist of starch, cellulose, hemicelluloses, lignin, and simple sugars in higher concentration, which assures valuable raw material for bioethanol (Arapoglou et al. 2010). The outer covering of water chestnuts and skin peels of sweet potatoes (Chafle et al. 2014) are also a vital source of fermentable sugars. So these wastes can also be used as a potential raw material for bioethanol production.

## 20.8 Conclusion

Production of bioethanol from renewable sources has great economic and ecological importance. Fruit and vegetable wastes are renewable substrates and could serve as a feedstock for bioethanol production. Fruit and vegetable wastes are easily available in large quantities at no cost, thus reducing the production costs of bioethanol. Many studies have reported the successful utilization of fruit and vegetable wastes for bioethanol production. Processing of these wastes for bioethanol production facilitates the recycling of such wastes. However, the challenge is to optimize the high yield of ethanol and build the process at a scale that could meet global market demands.

## References

Al Azkawi, A.S., Sivakumar, N., and Al Bahry, S. (2018). Bioprocessing of cardboard waste for cellulase production. *Biomass Conv. Biorefin.* 8: 597–606.

Arapoglou, D., Varzakas, T., Vlyssides, A., and Israilides, C. (2010). Ethanol production from potato peel waste (PPW). *Waste Manage.* 30: 1898–1902.

Bello, R.H., Linzmeyer, P., Franco, C.M.B. et al. (2014). Pervaporation of ethanol produced from banana waste. *Waste Manage.* 34: 1501–1509.

Boulal, A., Kihal, M., Khelifi, C., and Benali, B. (2016). Bioethanol production from date palm fruit waste fermentation using solar energy. *Afr. J. Agric. Res.* 15 (30): 1621–1627.

Buenrostro-Figueroa, J., Tafolla-Arellano, J.C., Flores-Gallegos, A.C. et al. (2018). Native yeasts for alternative utilization of overripe mango pulp for ethanol production. *Rev. Argent. Microbiol.* 50: 173–177.

Chafle, S., Parmar, V., and Biya, S. (2014). Utilization of vegetable and fruit waste for bioenergy generation. *J. Autom. Control Eng.* 2: 143–145.

Chandel, A.K., Chan, E.S., Rudravanan, R. et al. (2007). Economics and environmental impact of bioethanol production technologies: an appraisal. *Biotechnol. Mol. Biol. Rev.* 2: 14–32.

Conesa, C., Sanchez, L.G., Segui, L. et al. (2017). Ethanol quantification in pineapple waste by an electrochemical impedance spectroscopy- based system and artificialneural networks. *Chemom. Intell. Lab. Syst.* 161: 1–7.

Cutzu, R. and Bardi, L. (2017). Production of bioethanol from agricultural wastes using residual thermal energy of a cogeneration plant in the distillation phase. *Fermentation* 3: 1–8.

Cypriano, D.Z., Silva, L.L.D., and Tasic, L. (2018). High value-added products from the orange juice industry waste. *Waste Manage.* 79: 71–78.

Demiray, E., Karatay, S.E., and Donmez, G. (2018). Evaluation of omegranate peel in ethanol production by *Saccharomyces cerevisiae* and *Pichia stipitis. Energy* 159: 988–994.

Deshpande, D.P., Adwant, A.V., and Hakke, V.S. (2015). Conversion of vegetable waste in to ethanol. *Int. J. Latest Trends Eng. Technol.* 5: 327–333.

Dudley, R. (2004). Ethanol, fruit ripening and the historical origins of human alcoholism in primate frugivory. *Integr. Comp. Biol.* 44: 315–323.

Gebregergs, A., Gebresemati, M., and Sahu, O. (2016). Industrial ethanol from banana peels for developing countries: response surface methodology. *Pac. Sci. Rev. A. Nat. Sci. Eng.* 18: 22–29.

Girish, V., Ravikumar, K., and Girisha, S.T. (2014). Estimation of sugar and bioethanol from different decaying fruits extract. *Adv. Appl. Sci. Res.* 5: 106–110.

Gosavi, P., Chaudhary, Y., and Gupta, A.D. (2017). Production of biofuel from fruits and vegetable wastes. *Eur. J. Biotechnol. Biosci.* 5: 69–73.

Guerrero, A.B. and Munoz, E. (2018). Life cycle assessment of second generation ethanol derived from banana agricultural waste: environmental impacts and energy balance. *J. Clean. Prod.* 174: 710–717.

Guerrero, A.B., Ballesteros, I., and Ballesteros, M. (2018). The potential of agricultural banana waste for bioethanol production. *J. Fuel.* 213: 176–185.

Handler, R.M., Shonnard, D.R., Griffing, E.M. et al. (2016). Life cycle assessments of ethanol production via gas fermentation: anticipated greenhouse gas emissions for cellulosic and waste gas feedstocks. *Ind. Eng. Chem. Res.* 55: 3253–3261.

Hossain, A.B.M.S. and Fazliny, A.R. (2010). Creation of alternative energy by bio-ethanol production from pineapple waste and the usage of its properties for engine. *Afr. J. Microbiol. Res.* 4: 813–819.

Hossain, A.B.M.S., Ahmed, S.A., Alshammari, M.A. et al. (2011). Bioethanol fuel production from rotten banana as on environmental waste management and sustainable energy. *Afr. J. Microbiol. Res.* 5: 586–598.

Huang, D., Zhou, H., and Lin, L. (2012). Biodiesel: an alternative to conventional fuel. *Energy Procedia* 16: 1874–1885.

Jahid, M., Gupta, A., and Sharma, D.K. (2018). Production of bioethanol from fruit wastes (Banana, papaya, pineapple and mango peels) under milder conditions. *J. Bioprocess. Biotech.* 8: 1–11.

Janani, K., Ketzi, M., Megavathi, S. et al. (2013). Comparative studies of ethanol production from different fruit wastes using *Saccharomyces cerevisiae*. *Int. J. Innovative Res. Sci. Eng. Technol.* 2: 7161–7167.

Nair, R.B., Lennartsson, P.R., and Taherzadeh, M.J. (2016). Bio-ethanol production from agricultural and municipal wastes. In: *Current Developments in Biotechnology and Bioengineering. Solid Waste Management* (eds. J. Wong, R.D. Tyagi and A. Pandey), 157–190. USA: Elsevier.

Nuriana, W. and Wuryantoro (2015). Ethanol synthesis from jackfruit (*Artocarpus Heterophyllus* lam). Stone waste as renewable energy source. *Energy Procedia* 65: 372–377.

Oniya, O.O. and Bamgboye, A.I. (2014). Production of biodiesel from groundnut (*Arachis hypogea*,L) oil. *Agric. Eng. Int. CIGR J.* 16: 143–150.

Panda, S.K., Ray, R.C., Mishra, S.S., and Kayitesi, E. (2018). Microbial processing of fruit and vegetable wastes in to potential biocommodities: a review. *Crit. Rev. Biotechnol.* 38: 1–16.

Paulova, L., Patakova, P., Branska, B. et al. (2015). Lignocellulosic ethanol: technology design and its impact on process efficiency. *Biotechnol. Adv.* 33: 1091–1107.

Sharma, D.K. (2006). Bioprospecting for drug research and functional foods for prevention of diseases-role of flavonoids in drug development. *J. Sci. Ind. Res.* 2: 391–401.

Singh, A., Kuila, A., Adak, S. et al. (2012). Utilization of vegetable wastes for bioenergy generation. *Agric. Res.* 1: 213–222.

Sivakumar, N. (2012). Fermentative products using dates as a substrate. In: *Dates: Production, Processing, Food, and Medicinal Values* (eds. A. Manickavasagan, M. Mohamed Essa and E. Sukumar), 305–315. USA: CRC press.

Survase, S.A., Sklavounos, E., Heiningen, A.V., and Granstrom, T. (2013). Market refused vegetables as a supplement for improved acetone-butanol-ethanol production by *Clostridium acetobutylicum* DSM 792. *J. Ind. Crop. Prod.* 45: 349–354.

Taghizadeh-Alisaraei, A., Assar, H.A., Ghobadian, B., and Motevali, A. (2017). Potential of biofuel production from pistachio waste in Iran. *Renewable Sustainable Energy Rev.* 72: 510–522.

Tock, J.Y., Lai, C.L., Lee, K.T. et al. (2009). Banana biomass as potential renewable energy resource: a Malaysian case study. *Renewable Sustainable Energy Rev.* 14: 798–805.

Umamaheswari, M., Jayakumari, M., Maheswari, K. et al. (2010). Bioethanol production from cellulosic materials. *Asian J. Sci. Technol.* 1: 005–011.

Wu, Y., Wang, C., Zheng, M. et al. (2016). Effect of pH on ethanol-type Acidogenic fermentation of fruit and vegetable waste. *Waste Manage.* 60: 158–163.

# 21

# Bioprocessing of Cassava Stem to Bioethanol Using Soaking in Aqueous Ammonia Pretreatment

*Ashokan Anushya*[1], *Moorthi Swathika*[1], *Selvaraju Sivamani*[1], *and Nallusamy Sivakumar*[2]

[1] *Department of Biotechnology, Kumaraguru College of Technology, Coimbatore, India*
[2] *Department of Biology, College of Science, Sultan Qaboos University, Muscat, Sultanate of Oman*

## 21.1 Introduction

In recent years, the production of value-added products from lignocellulosic biomass has gained significant interest due to economic and environmental reasons (Limayem and Ricke 2012). Bioethanol is generally produced by fermenting the sugar components of biomass such as rice straw (Wang et al. 2015; de Assis Castro et al. 2017), wheat straw (Moreno et al. 2016, Nielsen et al. 2017), sugarcane bagasse (Soares et al. 2016; Antunes et al. 2017), sugarcane tops (Sindhu et al. 2014), sugarcane juice (Zabed et al. 2014), cotton stalk (Singh et al. 2017), cotton gin waste (Chandrasekaran and Sivamani 2018), bamboo (Yang et al. 2014; Xin et al. 2015), and lignocellulosic, and other waste biomass as carbon sources (Gupta and Verma 2015; Al Azkawi et al. 2018). Among them, cassava-based materials are a cheaply available lignocellulosic waste in huge quantities in developing countries.

Cassava (*Manihot esculenta Crantz*) belongs to the Euphorbiaceae family, has high starch content, is cultivated in many countries across Africa, Asia, and South America, and is nutritional (Uchechukwu-Agua et al. 2015). Cassava plant contains both edible and non-edible parts. Cassava tuber is the only edible part used to produce starch. The non-edible parts of cassava that produce value-added products are peel (Sivamani and Baskar 2015), bagasse (Escaramboni et al. 2018), stem (Wei et al. 2015, Sivamani and Baskar 2018), rhizome (Ontanee et al. 2017), and leaves (Pooja et al. 2018). The high cellulose content of the cassava stem makes it a preferred source for ethanol production.

The production of bioethanol from lignocellulosic materials involves pretreatment, detoxification, hydrolysis, and fermentation (Figure 21.1). The complex structure of the plant cell wall makes the conversion of lignocellulosic biomass to ethanol more challenging. Hence, pretreatment is a necessary step required to modify or remove the structural barriers of the lignocellulosic biomass to achieve efficient hydrolysis and increased yields of fermentable sugars. Several pretreatment methods such as milling

*Bioprocessing for Biomolecules Production*, First Edition.
Edited by Gustavo Molina, Vijai Kumar Gupta, Brahma N. Singh, and Nicholas Gathergood.

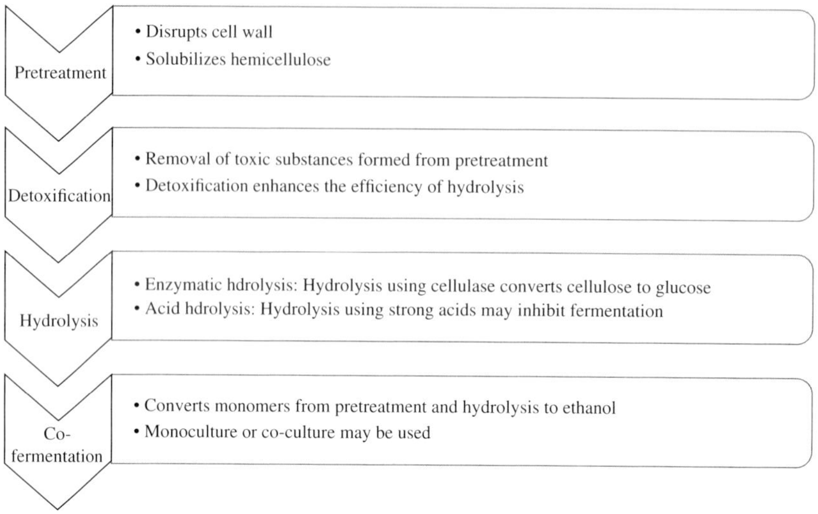


**Figure 21.1** Process for bioethanol production from lignocellulosic biomass.

(Kim et al. 2016), hydrothermolysis (Liggenstoffer et al. 2014), acid (Sivamani and Baskar 2018), alkali (Sindhu et al. 2014), solvents (Amiri and Karimi 2015), ammonia fiber explosion (Bonner et al. 2015), steam explosion (Auxenfans et al. 2017), and biological (Saha et al. 2016) pretreatment techniques have been applied to increase the accessibility of the enzymes to the cellulose available in the lignocellulosic biomass. Pretreatment technologies involving high temperature and/or pressure affect the process economics. Therefore, pretreatment involving moderate temperature and pressure is suitable for the economic production of bioethanol from lignocellulosic materials. One such technique is soaking in aqueous ammonia (SAA). SAA involves immersing milled biomass in dilute ammonia solution for a fixed time at moderate temperature and pressure (Du et al. 2018). After SAA pretreatment, hemicellulose in biomass will be converted to its monomers and retained in the liquid supernatant. Solid residue or water insoluble solids (WIS) left behind contains cellulose and lignin. Lignin acts as an inhibitor for ethanol fermentation. Hence, delignification is important for effective fermentation of cellulose to ethanol.

Detoxification involves the removal of degradation products from the monomers of hemicellulose. Detoxification using lime is widely used (Chandel et al. 2013). Hydrolysis by cellulase converts cellulose in the lignocellulosic material to glucose. As an alternative, strong acids are also used for hydrolysis of cellulose. However, acid hydrolysis leads to the formation of degradation products. So, detoxification is required before co-fermentation. Generally, enzymatic hydrolysis is preferred in bioethanol production. The final step in bioethanol production from biomass is fermentation. Since monomers from pretreatment and hydrolysis are converted to ethanol, co-fermentation increases product yield and process economics. Either monoculture or co-culture is used for the ethanol production. In this chapter, the characterization of cassava stem, pretreatment of cassava stem by SAA, and the bacterial production of ethanol from cassava stem are discussed.

## 21.2 Characterization of Cassava Stem

The biomass of cassava stem constitutes nearly 50% of the root mass and is usually discarded as waste after harvesting the starchy roots. Only 10–20% of the cassava stems are used for propagation, utilized as a substrate for mushroom cultivation, or used as a fuel for cooking and recycled into the soil to maintain the soil fertility. The rest is thrown as a waste or burned in the wild, causing emission-related environmental problems (Zhu et al. 2015). Hence, utilizing this non-food-waste biomass for ethanol production would be a very promising approach.

The cellulose content of cassava stem makes it an appropriate feedstock for ethanol production. To measure the cellulose content, mature cassava stems were collected from tapioca and castor research station, Yethapur, Salem district, Tamil Nadu, India. The collected materials were sun-dried to constant weight and ground well. Finally, the material was stored in sealed plastic bags and placed in an incubator maintained at 4 °C prior to characterization and experimental studies. Cassava stem was quantitatively analyzed for cellulose, hemicellulose, lignin (Goering and Van Soest 1970), holocellulose (Wise 1945), extractives, and ash (Sluiter et al. 2008). The holocellulose content of cassava stem was approximately 60%. This high sugar quantity made the substrate more suitable for ethanol production. Also, the quantity of extractives was low. This made the sample preparation process easier. The ash content was around 5% while the lignin was found to be around 33% (Figure 21.2).

## 21.3 SAA Pretreatment of Cassava Stem

Pretreatment is essential to modify the structural and chemical composition of the lignocellulosic biomass in order to facilitate rapid and efficient hydrolysis of the complex

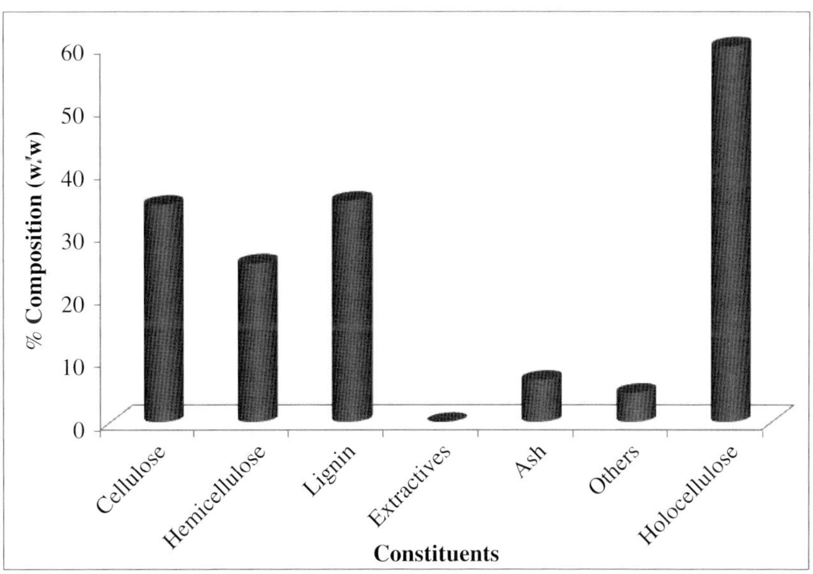


**Figure 21.2** Biochemical composition of cassava stem.

carbohydrates into fermentable sugars. Among the pretreatment methods, SAA pretreatment has generally been used most frequently because it is more efficient for agricultural residues and herbaceous crops (Sun and Cheng 2002). In our lab, we have tried the SAA pretreatment of cassava stem. A known quantity of cassava stem was mixed with a fixed volume of ammonia at a specific concentration. The mixture was allowed to soak for a definite time, and the resultant solution was filtered. The resultant liquid was analyzed for total soluble sugars content, and the solid residue left behind was analyzed for WIS and % delignification.

Total soluble sugars content was determined by the anthrone method (Hedge and Hofreiter 1962). WIS was estimated by drying solid residue at 105 °C in a hot air oven to constant weight. The percentage delignification was calculated from lignin before and after pretreatment.

$$\text{Total soluble sugar content } (\%(\text{w/v})) = \frac{C \times V \times 100}{a \times V_s} \tag{21.1}$$

$$\text{WIS (g)} = \text{Mass of dried solid residue after pretreatment} \tag{21.2}$$

$$\%\text{delignification} = \frac{(\text{Lignin before pretreatment} - \text{Lignin after pretreatment}) \times 100}{\text{Lignin before pretreatment}} \tag{21.3}$$

where C is the total soluble sugars content by referring to the standard curve, $V$ is the total volume of the extracted solution, $a$ is the volume of the sample solution for color development, $V_s$ is the volume of the liquid sample.

The factors affecting SAA pretreatment on cassava stem are temperature, time, solid to liquid ratio (SLR), and ammonia concentration (Jia et al. 2013). The effect of temperature, ammonia concentration, and SLR at various time intervals on total soluble sugars, WIS, and % delignification were studied as follows:

### 21.3.1 Effect of Temperature

The effect of temperature was studied on total soluble sugars content, WIS, and percentage delignification at constant ammonia concentration and SLR at varying times. Total soluble sugars increase with an increase in temperature because monomers in hemicellulose are released. When the temperature is high, monomers are degraded to furfural. WIS increases with temperature due to the fact that it increases when hemicellulose decreases. Delignification increases with increased temperature and becomes saturated with temperature when lignin is completely depolymerized.

One hundred grams of cassava stem were mixed with 1200 ml of ammonia at a concentration of 15% (w/w). The mixture was allowed to soak for 4, 8, and 12 hours at temperatures from 30 to 70 °C and the resultant solution was filtered. The resultant liquid was analyzed for total soluble sugars content, and the solid residue left behind was analyzed for WIS and % delignification. Figure 21.3a describes the effect of temperature on total soluble sugars in the supernatant. For the ethanol production process to be effective, total soluble sugars obtained in the supernatant should be more. From Figure 21.3a, it was clear that the maximum total soluble sugars concentration of 13% (w/v) was achieved at 70 °C for 12 hours with an ammonia concentration of 15% (w/w)

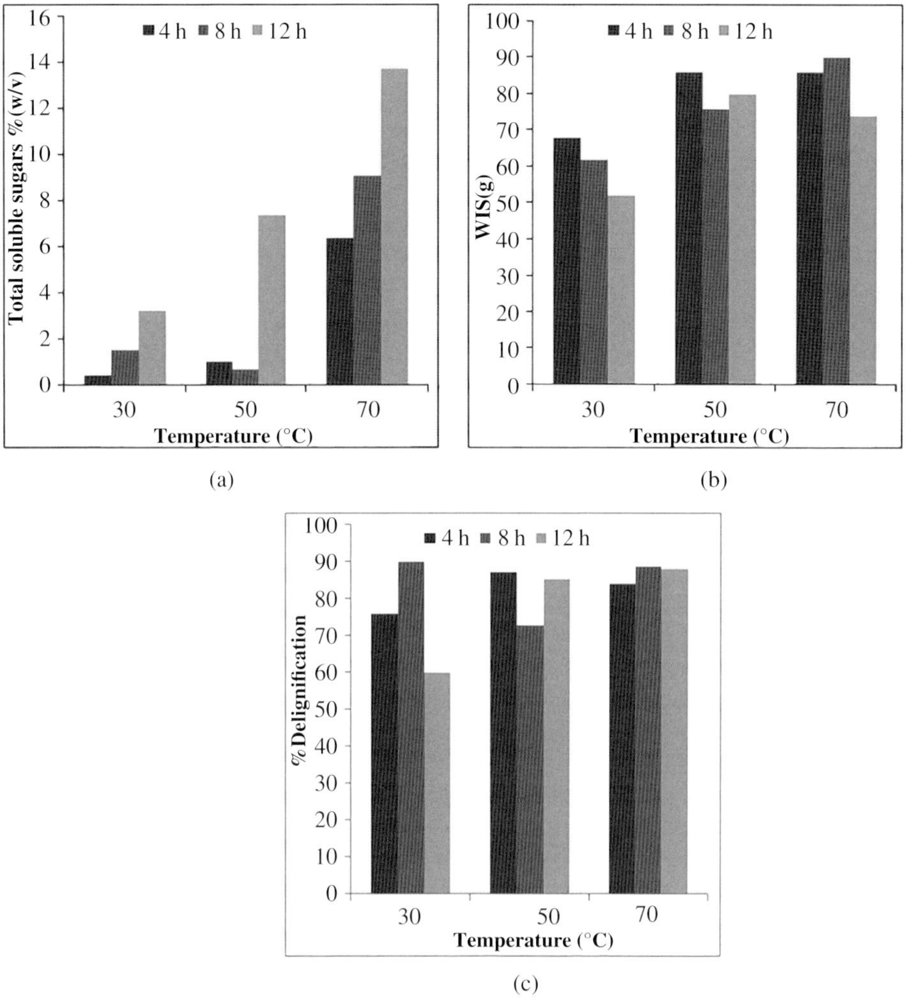


**Figure 21.3** Effect of temperature on (a) total soluble sugars; (b) WIS; (c) delignification during SAA pretreatment of cassava stem.

at the SLR of 1 : 12. The optimum temperature from 40 to 70 °C was reported in the literature (Hedayatkhah et al. 2013; Zhang et al. 2013)

The effect of temperature on WIS in the solid residue is represented in Figure 21.3b. This includes cellulose and lignin (75%). For the ethanol production process to be effective, WIS obtained in the residue should not be more than cellulose and lignin contents together. In this study, WIS obtained was 70% at 70 °C for 12 hours with an ammonia concentration of 15% (w/w) at the SLR of 1 : 12. Zhang et al. 2013 reported 90% glucan retainment at 85 ° C, 11 minutes, 80% for corn stalk. Figure 21.3c shows the effect of temperature on % delignification at constant SLR and ammonia concentration. For the ethanol production process to be effective, delignification in the residue should be more because lignin in cassava stem acts as an inhibitor for microbial fermentation. In this study, the maximum delignification of 90% was achieved at 70 °C for 12 hours with an ammonia concentration of 15% (w/w) at the SLR of 1 : 12. Rémond et al. 2010 reported 38% delignification at room temperature for three days at SLR of 1 : 6 for wheat straw. Li and Kim 2011 obtained 55% delignification at 30 °C for four weeks with ammonia concentration and SLR of 50% (w/w) and 1 : 5 respectively for corn stover. Xin et al. 2015 reported 40% delignification at 70 °C for 72 hours with ammonia concentration and SLR of 25% (w/w) and 1 : 10, respectively, for bamboo fractions.

### 21.3.2 Effect of Ammonia Concentration

The effect of ammonia concentration was studied on total soluble sugars content, WIS, and percentage delignification at constant temperature and SLR at the varying time. Total soluble sugars increase with an increase in ammonia concentration because xylan, arabinan, mannan, and galactan in hemicellulose are converted to their monomers. When the ammonia concentration is high, monomers of hemicellulose convert to their degradation products. WIS increases with ammonia concentration due to the fact that cellulose increases when hemicellulose decreases in lignocellulosic biomass. Delignification increases with an increase in ammonia concentration because lignin is depolymerized by bases.

One hundred grams of cassava stem was mixed with 1200 ml of ammonia at different concentrations from 15 to 25% (w/w). The mixture was allowed to soak for 4, 8, and 12 hours at the temperature of 70 °C and the resultant solution was filtered. The resultant liquid was analyzed for total soluble sugars content, and the solid residue left behind was analyzed for WIS and % delignification.

Figure 21.4a depicts the effect of ammonia concentration on total soluble sugars in the supernatant. It was clear that maximum sugar removal was achieved at 15% (w/w). Garlock et al. 2011 yielded 83% total glucose with an ammonia concentration of 15% (w/w) at 160 °C for 60 minutes at SLR of 1 : 10 for switchgrass. Qin et al. 2013 reported 90.5% glucose and 33.1% xylose yield with an ammonia concentration of 20% (w/w) at 180 °C for 30 minutes for corn stover. Yazici and Isci 2013 obtained 81% xylose and 76% glucose with an ammonia concentration of 30% (w/w) at 50 °C for 45 hours at SLR of 1:6 for wheat straw. Figure 21.4b explains the effect of ammonia concentration on WIS. In this study, 70 g of WIS was obtained at an ammonia concentration of 15% (w/w) for 12 hours at 70 °C and SLR of 1 : 12. Jia et al. 2013 achieved maximum glucan yield of 63% and 96% for wheat straw and corn stover, respectively, with an ammonia concentration of 21% (w/w) at 75 °C for 20 hours at SLR of 1 : 10. Li and Kim 2011 retained 80% xylan with an ammonia concentration of 15% (w/w) at 60 °C for 12 hours at SLR of 1 : 11 for corn stover. Liu et al. 2013 recovered 70% xylan with an ammonia concentration of 30% (w/w) at 150 °C for 15 minutes at SLR of 1:3 for *Miscanthus*.

The effect of ammonia concentration on % delignification at constant SLR and temperature are presented in Figure 21.4c. The maximum % delignification of 60% was achieved at 15% (w/w) ammonia for 12 hours at 70 °C and SLR of 1 : 12. Pryor et al. 2012 attained 57.6% delignification with an ammonia concentration of 15% (w/w) at 40 °C for 24 hours at SLR of 1 : 6 for switchgrass.

### 21.3.3 Effect of SLR

The effect of SLR was studied on total soluble sugars content, WIS, and a percentage delignification at constant ammonia concentration and temperature at the varying

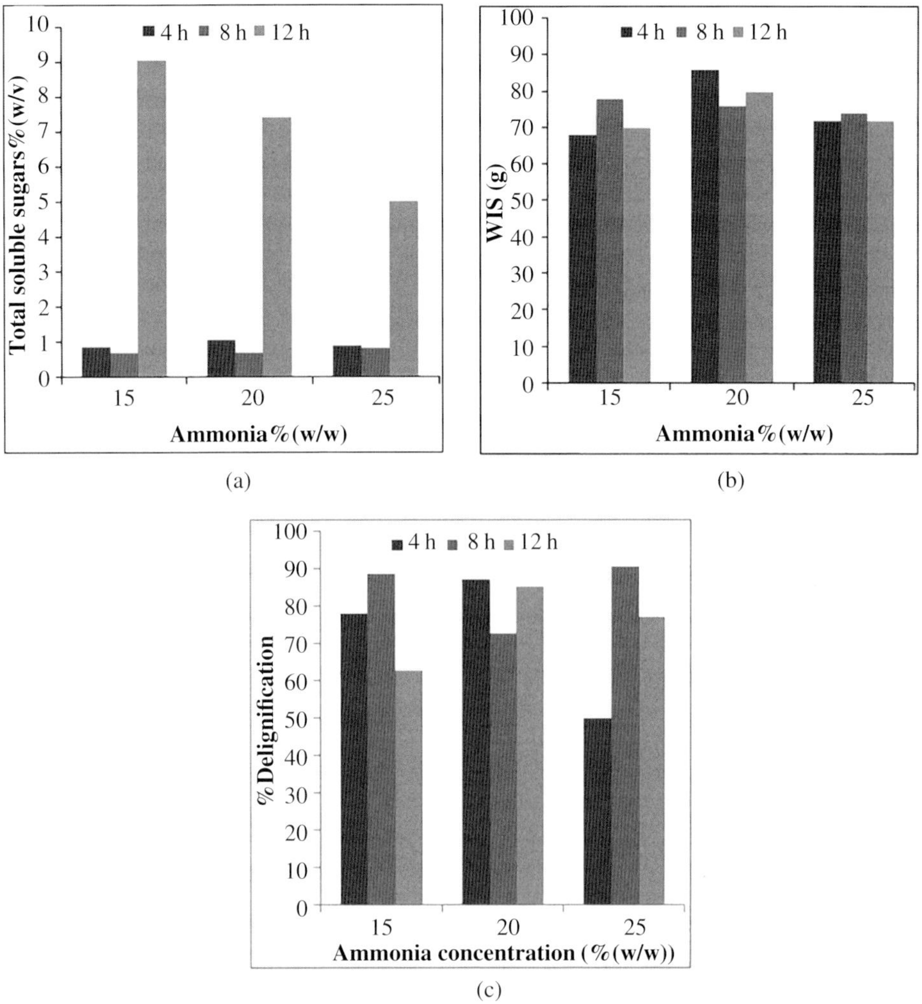


Figure 21.4 Effect of ammonia concentration on (a) total soluble sugars; (b) WIS; (c) delignification during SAA pretreatment of cassava stem.

time. Total soluble sugars increase with SLR because more monomers are released. WIS increases with SLR since there is an increase in cellulose when hemicellulose decreases. Delignification increases with an increase in SLR because of the greater availability of lignin.

One hundred grams of cassava stem was mixed with volumes from 400 to 1200 ml of ammonia at the concentration of 15% (w/w). The mixture was allowed to soak for 4, 8, and 12 hours at 70 °C and the resultant solution was filtered. The resultant liquid was analyzed for total soluble sugars content, and the solid residue left behind was analyzed for WIS and % delignification.

Figure 21.5a illustrates the effect of SLR on total soluble sugars in the supernatant. The maximum total soluble sugars content of 9.5% (w/w) was achieved at 1 : 12 with ammonia concentration of 15% (w/w) at 70 °C for 12 hours. Tao et al. 2013 solubilized

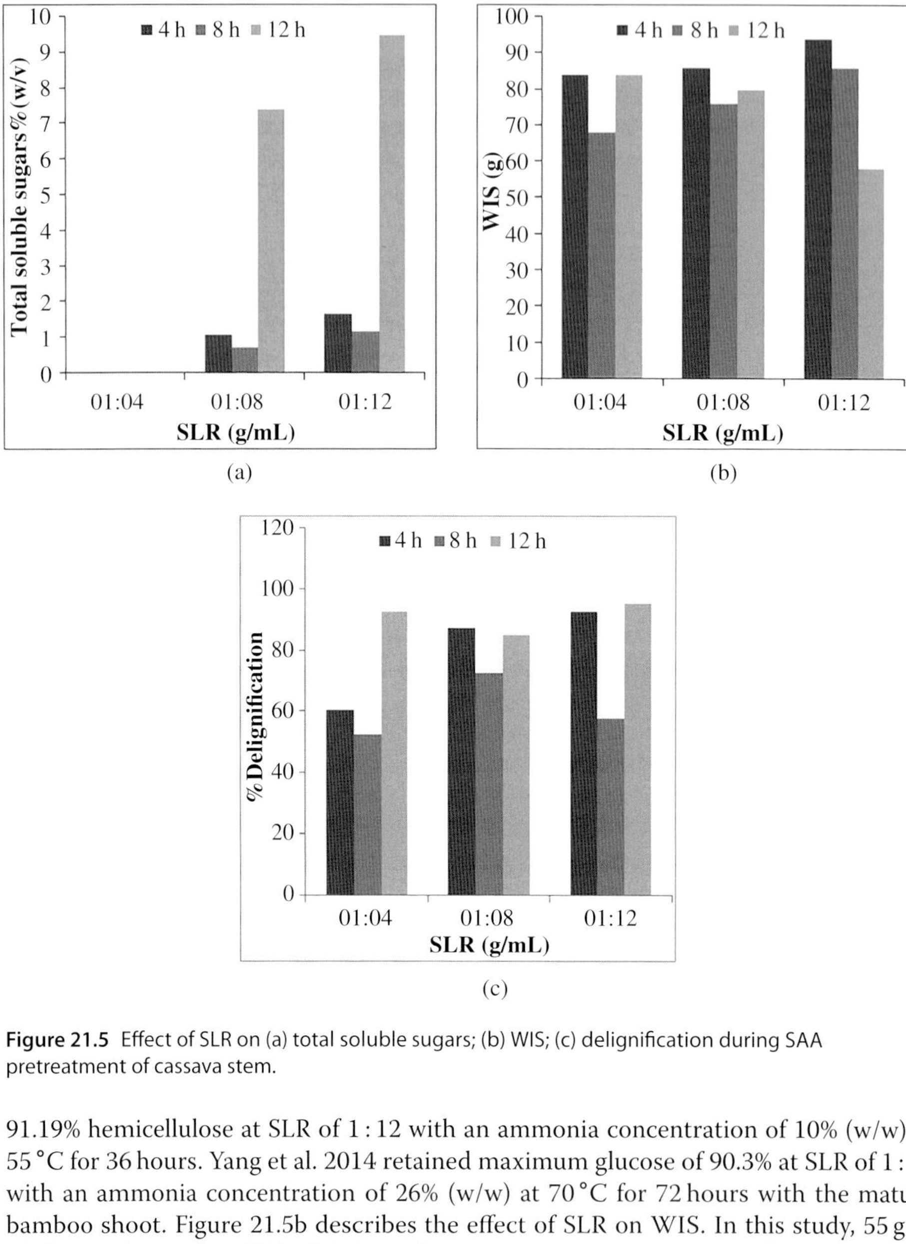


**Figure 21.5** Effect of SLR on (a) total soluble sugars; (b) WIS; (c) delignification during SAA pretreatment of cassava stem.

91.19% hemicellulose at SLR of 1 : 12 with an ammonia concentration of 10% (w/w) at 55 °C for 36 hours. Yang et al. 2014 retained maximum glucose of 90.3% at SLR of 1 : 10 with an ammonia concentration of 26% (w/w) at 70 °C for 72 hours with the mature bamboo shoot. Figure 21.5b describes the effect of SLR on WIS. In this study, 55 g of WIS was obtained at SLR of 1 : 12 with an ammonia concentration of 15% (w/w) for 12 hours at 70 °C. Waeonukul et al. 2012 reported glucan hydrolysis of 91% at SLR of 1:10 with an ammonia concentration of 28% (w/w) at 60 °C for 7 days with rice straw. Azadfar et al. 2015 retained maximum glucan at SLR of 1 : 4 with an ammonia concentration of 20% (w/w) at 160 °C for 4 hours with wheat straw. Figure 21.5c shows the effect of SLR on % delignification at constant temperature and ammonia concentration. In this study, the maximum delignification of 90% was achieved at SLR of 1 : 12 with ammonia concentration of 15% (w/w) at 70 °C for 12 hours. Pallapolu et al. 2011 achieved delignification of 62% at SLR of 1 : 10 and a temperature of 160 °C for 60 minutes with switchgrass.

## 21.4 Ethanol Fermentation

*Zymomonas mobilis* MTCC 92 was purchased from Microbial Type Culture Collection (MTCC) of Institute of Microbial Technology, Chandigarh, India. The medium composition, as recommended by MTCC, was used for the growth of organisms. The suitable medium was sterilized at 121 °C and 15 psi for 15 minutes, added to the culture after cooling, incubated until mid-log phase. The cell suspension was used as inoculum for fermentation.

During fermentation, pentoses and hexoses are converted to ethanol by the ethanologenic organism. The common ethanologenic organism is *Saccharomyces cerevisiae*. Because of its low ethanol tolerance, *Zymomonas* is emerging as an important ethanologenic bacteria. *Zymomonas* metabolizes both pentoses and hexoses simultaneously, which increases ethanol yield. The product yield coefficient $Y_{P/S}$ for ethanol production from sugars is 0.51 g ethanol/g sugar whereas it is 0.57 g ethanol/g polysaccharide for xylans and glucans.

The resultant liquid solution from the pretreatment was subjected to microbial fermentation with *Z. mobilis*. After fermentation, ethanol concentration was determined by the dichromate method (Williams and Reese 1950). The fermentation was carried out at pH 5, inoculum size of 6% (v/v), agitation speed of 25 rpm, and temperature of 30 °C. A maximum ethanol concentration of 13.3 $g\,l^{-1}$ was achieved at a substrate concentration of 1 $g\,l^{-1}$. Table 21.1 summarizes the ethanol yield from various substrates after SAA pretreatment.

## 21.5 Conclusion

Cassava stem was characterized for biochemical traits, which showed that it contains 60% holocellulose, making it suitable for bioethanol production. SAA pretreatment of cassava stem was optimized to obtain elevated total soluble sugars content and delignification. At optimal conditions with ammonia concentration of 15% (w/w) at 70 °C for 12 hours and SLR of 1 : 12, total soluble sugars content, WIS (%) and delignification (%) were obtained at 9.5 (w/w), 55, and 90%, respectively. Then, the pretreated stem was

**Table 21.1** Ethanol yield from substrates after SAA pretreatment.

| Substrate | Result | Reference |
|---|---|---|
| Palm empty fruit bunch | 890 mg ethanol/g substrate | Asli et al. 2013 |
| Oil palm fruits | 65.6% theoretical maximum ethanol yield | Jung et al. 2011 |
| Sugarcane bagasse | 3.83 /L ethanol concentration | Chandel et al. 2013 |
| Napier grass | 91.2% theoretical maximum ethanol yield | Yasuda et al. 2013 |
| Corn stover | 77% theoretical maximum ethanol yield | Kim and Lee 2005 |

subjected to bacterial fermentation with *Z. mobilis*. The maximum bioethanol concentration of 13.3 $g\,l^{-1}$L was obtained at pH of 5, 6% (v/v) inoculum size, agitation speed of 25 rpm, and temperature of 30 °C. Thus, the SAA pretreatment could serve as the promising pretreatment technology for the production of bioethanol from cassava stem.

## References

Al Azkawi, A.S., Sivakumar, N., and Al Bahry, S. (2018). Bioprocessing of cardboard waste for cellulase production. *Biomass Convers. Biorefin.* 8: 597–606.

Amiri, H. and Karimi, K. (2015). Improvement of acetone, butanol, and ethanol production from woody biomass using organosolv pretreatment. *Bioprocess. Biosyst. Eng.* 38: 1959–1972.

Antunes, F.A., Chandel, A.K., dos Santos, J.C. et al. (2017). Hemicellulosic ethanol production in fluidized bed reactor from sugar cane bagasse hydrolysate: interplay among carrier concentration and aeration rate. *ACS Sustain. Chem. Eng.* 5: 8250–8259.

Asli, U.A., Zakaria, Z.A., Sadikin, A.N., and Rasit, R. (2013). Fermentable sugar from palm empty fruit bunch biomass for bioethanol production. *Int. J. Chem. Sci. Eng.* 7: 605–609.

de Assis Castro, R.C., Fonseca, B.G., dos Santos, H.T.L. et al. (2017). Alkaline deacetylation as a strategy to improve sugars recovery and ethanol production from rice straw hemicellulose and cellulose. *Ind. Crops Prod.* 106: 65–73.

Auxenfans, T., Crônier, D., Chabbert, B., and Paës, G. (2017). Understanding the structural and chemical changes of plant biomass following steam explosion pretreatment. *Biotechnol. Biofuels* https://doi.org/10.1186/s13068-017-0718-z.

Azadfar, M., Gao, A.H., and Chen, S. (2015). Structural characterization of lignin: a potential source of antioxidants guaiacol and 4-vinylguaiacol. *Int. J. Biol. Macromol.* 75: 58–66.

Bonner, I.J., Thompson, D.N., Teymouri, F. et al. (2015). Impact of sequential ammonia fiber expansion (AFEX) pretreatment and pelletization on the moisture sorption properties of corn stover. *Drying Technol.* 33: 1768–1778.

Chandel, A.K., Antunes, F.A., Silva, M.B., and da Silva, S.S. (2013). Unraveling the structure of sugarcane bagasse after soaking in concentrated aqueous ammonia (SCAA) and ethanol production by *Scheffersomyces* (*Pichia*) *stipitis*. *Biotechnol. Biofuels* https://doi.org/10.1186/1754-6834-6-102.

Chandrasekaran, A.P. and Sivamani, S. (2018). Statistical modeling and optimization of pretreatment for fermentable sugars production from cotton gin waste. *Energ. Source Part A* 40: 400–405.

Du, R., Su, R., Qi, W., and He, Z. (2018). Enhanced enzymatic hydrolysis of corncob by ultrasound-assisted soaking in aqueous ammonia pretreatment. *3 Biotech* https://doi.org/10.1007/s13205-018-1186-2.

Escaramboni, B., Núñez, E.G.F., Carvalho, A.F.A., and de Oliva Neto, P. (2018). Ethanol biosynthesis by fast hydrolysis of cassava bagasse using fungal amylases produced in optimized conditions. *Ind. Crops Prod.* 112: 368–377.

Garlock, R.J., Balan, V., Dale, B.E. et al. (2011). Comparative material balances around pretreatment technologies for the conversion of switchgrass to soluble sugars. *Bioresour. Technol.* 102: 11063–11071.

Goering, H.K., and P.J. Van Soest. 1970. Forage fiber analysis. USDA Agricultural Research Service. Handbook number 379. US Department of Agriculture. Superintendent of Documents, US Government Printing Office, Washington, DC.

Gupta, A. and Verma, J.P. (2015). Sustainable bio-ethanol production from agro-residues: a review. *Renewable Sustainable Energy Rev.* 41: 550–567.

Hedayatkhah, A., Motamedi, H., Najafzadeh varzi, H. et al. (2013). Improvement of hydrolysis and fermentation of sugarcane bagasse by soaking in aqueous ammonia and methanolic ammonia. *Biosci. Biotechnol. Biochem.* 77: 1379–1383.

Hedge, J.E. and Hofreiter, B.T. (1962). Estimation of carbohydrate. In: *Carbohydrate Chemistry*, vol. 17 (eds. R.L. Whistler and J.N. Be Millee), 17–22. New York: Academic Press.

Jia, L., Sun, Z., Ge, X. et al. (2013). Comparison of the delignifiability and hydrolysability of wheat straw and corn stover in aqueous ammonia pretreatment. *Bioresources* 8: 4505–4517.

Jung, Y.H., Kim, I.J., Kim, J.J. et al. (2011). Ethanol production from oil palm trunks treated with aqueous ammonia and cellulase. *Bioresour. Technol.* 102: 7307–7312.

Kim, T.H. and Lee, Y.Y. (2005). Pretreatment of corn stover by soaking in aqueous ammonia. In: *Twenty-Sixth Symposium on Biotechnology for Fuels and Chemicals*, 121–124:1119–1131.

Kim, S.M., Dien, B.S., Tumbleson, M.E. et al. (2016). Improvement of sugar yields from corn stover using sequential hot water pretreatment and disk milling. *Bioresour. Technol.* 216: 706–713.

Li, X. and Kim, T.H. (2011). Low-liquid pretreatment of corn stover with aqueous ammonia. *Bioresour. Technol.* 102: 4779–4786.

Liggenstoffer, A.S., Youssef, N.H., Wilkins, M.R., and Elshahed, M.S. (2014). Evaluating the utility of hydrothermolysis pretreatment approaches in enhancing lignocellulosic biomass degradation by the anaerobic fungus *Orpinomyces* sp. strain C1A. *J. Microbiol. Methods* 104: 43–48.

Limayem, A. and Ricke, S.C. (2012). Lignocellulosic biomass for bioethanol production: current perspectives, potential issues and future prospects. *Prog. Energ. Combust.* 38: 449–467.

Liu, Z., Padmanabhan, S., Cheng, K. et al. (2013). Aqueous-ammonia delignification of miscanthus followed by enzymatic hydrolysis to sugars. *Bioresour. Technol.* 135: 23–29.

Moreno, A.D., Ibarra, D., Alvira, P. et al. (2016). Exploring laccase and mediators behavior during saccharification and fermentation of steam-exploded wheat straw for bioethanol production. *J. Chem. Technol. Biotechnol.* 91: 1816–1825.

Nielsen, F., Zacchi, G., Galbe, M., and Wallberg, O. (2017). Sequential targeting of xylose and glucose conversion in fed-batch simultaneous saccharification and co-fermentation of steam-pretreated wheat straw for improved xylose conversion to ethanol. *Bioenergy Res.* 10: 800–810.

Ontanee, A., Klinpratoom, B., Subsadsana, M., and Ruangviriyachai, C. (2017). Pptimization of acid hydrolysis of cassava rhizome into fermentable sugars for bioethanol production using response surface methodology. *Orient. J. Chem.* 33: 2507–2517.

Pallapolu, V.R., Lee, Y.Y., Garlock, R.J. et al. (2011). Effects of enzyme loading and $\beta$-glucosidase supplementation on enzymatic hydrolysis of switchgrass processed by leading pretreatment technologies. *Bioresour. Technol.* 102: 11115–11120.

Pooja, N.S., Sajeev, M.S., Jeeva, M.L., and Padmaja, G. (2018). Bioethanol production from microwave-assisted acid or alkali-pretreated agricultural residues of cassava using separate hydrolysis and fermentation (SHF). *3 Biotech* https://doi.org/10.1007/s13205-018-1095-4.

Pryor, S.W., Karki, B., and Nahar, N. (2012). Effect of hemicellulase addition during enzymatic hydrolysis of switchgrass pretreated by soaking in aqueous ammonia. *Bioresour. Technol.* 123: 620–626.

Qin, L., Liu, Z.H., Jin, M. et al. (2013). High temperature aqueous ammonia pretreatment and post-washing enhance the high solids enzymatic hydrolysis of corn stover. *Bioresour. Technol.* 146: 504–511.

Rémond, C., Aubry, N., Crônier, D. et al. (2010). Combination of ammonia and xylanase pretreatments: Impact on enzymatic xylan and cellulose recovery from wheat straw. *Bioresour. Technol.* 101: 6712–6717.

Saha, B.C., Qureshi, N., Kennedy, G.J., and Cotta, M.A. (2016). Biological pretreatment of corn stover with white-rot fungus for improved enzymatic hydrolysis. *Int. Biodeterior. Biodegrad.* 109: 29–35.

Sindhu, R., Kuttiraja, M., Binod, P. et al. (2014). Physico-chemical characterization of alkali pretreated sugarcane tops and optimization of enzymatic saccharification using response surface methodology. *Renew. Energ.* 62: 362–368.

Singh, A., Bajar, S., and Bishnoi, N.R. (2017). Physico-chemical pretreatment and enzymatic hydrolysis of cotton stalk for ethanol production by *Saccharomyces cerevisiae*. *Bioresour. Technol.* 244: 71–77.

Sivamani, S. and Baskar, R. (2015). Optimization of bioethanol production from cassava peel using statistical experimental design. *Environ. Prog. Sustain.* 34: 567–574.

Sivamani, S. and Baskar, R. (2018). Bioconversion of cassava stem to ethanol: oxalic acid pretreatment and co-culture fermentation. *Biofuels* https://doi.org/10.1080/17597269.2018.1432271.

Sluiter, A., B. Hames, R. Ruiz, C. Scarlata, J. Sluiter, and D. Templeton. 2008. Determination of ash in biomass. National renewable energy laboratory, 1–5.

Soares, L.C., Chandel, A.K., Pagnocca, F.C. et al. (2016). Screening of yeasts for selection of potential strains and their utilization for in situ microbial detoxification (ISMD) of sugarcane bagasse hemicellulosic hydrolysate. *Ind. J. Microbiol.* 56: 172–181.

Sun, Y. and Cheng, J. (2002). Hydrolysis of lignocellulosic materials for ethanol production. A review. *Bioresour. Technol.* 83: 1–11.

Tao, L., Ren, J., Liu, H., and Ni, T.R. (2013). Effects of liquid-to-solid ratio and reaction time on ammonia-sulfuric acid pretreatment of *Achnatherum splendens*. *Asian J. Chem.* 25: 3549–3552.

Uchechukwu-Agua, A.D., Caleb, O.J., Manley, M., and Opara, U.L. (2015). Effects of storage conditions and duration on physicochemical and microbial quality of the flour of two cassava cultivars (TME 419 and UMUCASS 36). *CYTA-J. Food* 13: 635–645.

Waeonukul, R., Kosugi, A., Tachaapaikoon, C. et al. (2012). Efficient saccharification of ammonia soaked rice straw by combination of *Clostridium thermocellum* cellulosome and *Thermoanaerobacter brockii* $\beta$-glucosidase. *Bioresour. Technol.* 107: 352–357.

Wang, G., Tan, L., Sun, Z.Y. et al. (2015). Production of bioethanol from rice straw by simultaneous saccharification and fermentation of whole pretreated slurry using *Saccharomyces cerevisiae* KF-7. *Environ. Prog. Sustain.* 34: 582–588.

Wei, M., Zhu, W., Xie, G. et al. (2015). Cassava stem wastes as potential feedstock for fuel ethanol production: A basic parameter study. *Renew. Energ.* 83: 970–978.

Williams, M.B. and Reese, H.D. (1950). Colorimetric determination of ethyl alcohol. *Anal. Chem.* 22: 1556–1561.

Wise, L.E. (1945). Quantitative isolation of hemicelluloses from coniferous woods. Preliminary communication. *Ind. Eng. Chem.* 17: 63–64.

Xin, D., Yang, Z., Liu, F. et al. (2015). Comparison of aqueous ammonia and dilute acid pretreatment of bamboo fractions: Structure properties and enzymatic hydrolysis. *Bioresour. Technol.* 175: 529–536.

Yang, Z., Zhang, M., Xin, D. et al. (2014). Evaluation of aqueous ammonia pretreatment for enzymatic hydrolysis of different fractions of bamboo shoot and mature bamboo. *Bioresour. Technol.* 173: 198–206.

Yasuda, M., Takeo, K., Nagai, H. et al. (2013). Enhancement of ethanol production from napiergrass (*Pennisetum purpureum Schumach*) by a low-moisture anhydrous ammonia pretreatment. *J. Sustain. Bioenerg. Syst.* 3: 179–185.

Yazici, S. and Isci, A. (2013). The effect of aqueous ammonia soaking on enzymatic hydrolysis of wheat straw. *J. Renewable Sustainable Energy* https://doi.org/10.1063/1.4811800.

Zabed, H., Faruq, G., Sahu, J.N. et al. (2014). Bioethanol production from fermentable sugar juice. *Sci. World J.* https://doi.org/10.1155/2014/957102.

Zhang, C., Pang, F., Li, B. et al. (2013). Recycled aqueous ammonia expansion (RAAE) pretreatment to improve enzymatic digestibility of corn stalks. *Bioresour. Technol.* 138: 314–320.

Zhu, W., Lestander, T.A., Örberg, H. et al. (2015). Cassava stems: a new resource to increase food and fuel production. *GCB Bioenergy* 7: 72–83. https://doi.org/10.1111/gcbb.12112.

# 22

# Bioprospecting of Microbes for Biohydrogen Production: Current Status and Future Challenges

*Sunil Kumar[1], Sushma Sharma[1], Sapna Thakur[1], Tanuja Mishra[1], Puneet Negi[1], Shashank Mishra[2], Abd El-Latif Hesham[3], Ali A. Rastegari[4], Neelam Yadav[5], and Ajar Nath Yadav[1]*

[1] *Eternal University, Sirmour, India*
[2] *Biotech Park, Sector G, Jankipuram, Lucknow, India*
[3] *Genetics Department, Faculty of Agriculture, Assiut University, Assiut, Egypt*
[4] *Department of Molecular and Cell Biochemistry, Falavarjan Branch, Islamic Azad University, Isfahan, Iran*
[5] *Gopi Nath P.G. College, Veer Bahadur Singh Purvanchal University, India*

## 22.1 Introduction

A conventional source of energy, i.e. fossils fuel has been the major source for meeting the increasing energy demand globally for many decades. Fossil fuels carry some drawbacks, however. They are not renewable, and are considered a primary cause of global warming and acid rain, which ultimately affects the existing climate and ecosystem. Thus, researchers has been actively involved in exploring sustainable sources of energy that could replace fossils fuel and meet existing energy demand. Hydrogen has emerged as one of the alternative energy sources that can withstand current energy demand and be a future carrier gas. It is considered a clean fuel because there is no $CO_2$ emission. Utilization of hydrogen is not limited to an energy source, but it can also be used as a feedstock to produce various chemicals, hydrogenation of fats and oils in the food industry, and production of methanol. About half of the produced hydrogen is used for the manufacture of ammonia, which is again used for making fertilizer. Huge quantity of hydrogen is also used in oil refineries for purification or upgrading heavier oil to lighter and more valuable product.

The yearly production of hydrogen is 500 b cu m for various purposes, which can produce 6.5 EJ of energy that is equivalent to 1.5 percent of the current energy demand (Chu and Majumdar 2012). According to National Hydrogen Program of the United States, the utilization of hydrogen as an energy source will be 8–10 percent by 2025 (Armor 1999; Kapdan and Kargi 2006). Based on US Department of Energy, all the regions of the United States ensure the availability of hydrogen power and transport fuel by 2040 (Kapdan and Kargi 2006). As the hydrogen has high energy content i.e. 122 kJ $g^{-1}$ which is equivalent to 2.75 times of hydrocarbon fuel (Darvishi Cheshmeh Soltani et al. 2015),

*Bioprocessing for Biomolecules Production*, First Edition.
Edited by Gustavo Molina, Vijai Kumar Gupta, Brahma N. Singh, and Nicholas Gathergood.

vehicle powered with same are three times more efficient than gasoline engine reflect their widespread use in the future. But the major challenge to utilize hydrogen as a future energy source is its low availability, and this could be overcome by inexpensive, renewable source, and efficient technology.

Generally, hydrogen gas production could be achieved by the physicochemical and biological process (Chu and Majumdar 2012). Physicochemical methods include catalytic steam reforming of methane (SRM), steam reforming of hydrocarbon (SRH), naphtha or natural gas, gasification of coal, and electrolysis of water, noncatalytic partial oxidation of fossil fuels (POX) and auto thermal reforming, which combines SRM and POX. Conventional physicochemical methods are neither sustainable nor environmentally friendly because these methods are energy intensive, requiring high temperatures (>850 °C) and fossils fuels used as the substrate. Among other methods developed to improve the existing technologies are the membrane processes, selective oxidation of methane, and oxidative dehydrogenation (Armor 1999).

As aforementioned methods, biohydrogen production from renewable resources is the sustainable energy source and received considerable attention in recent years. Furthermore, hydrogen production from biological means required ambient temperature and pressure hence less energy intensive. Various microorganism produces hydrogen naturally and the application of biotech in the microbial system could enhance the production of clean and renewable energy. Depending on the choice of microorganism, various methods have been adopted to produce hydrogen, which includes photofermentation, photolysis, dark fermentation, and microbial fuel cell. This chapter deals with the various methods adopted to produce biohydrogen along with reactors used for the same. The study also highlights the various parameters that affect productivity.

## 22.2 Biohydrogen Production Process

### 22.2.1 Photofermentation

Understanding the molecular fundamentals of hydrogen production and its biotechnological implications in biological systems is a goal of supreme importance for basic and applied biological research. The production of hydrogen by different groups of microbes (bacteria, fungi, cyanobacteria, and algae), called biohydrogen, has potential as a renewable alternative to current technologies. The states of four different biohydrogen production mechanisms are reviewed, including biophotolysis, indirect biophotolysis, photofermentation, and dark fermentation (Figure 22.1). Photofermentation is the fermentative conversion of organic substrate to biohydrogen manifested by a diverse group of photosynthetic microbes and relative microbes by a series of biochemical reactions involving three steps conversion reactions. Fermentation of renewable materials by microbes may occur in light (photofermentation) or in the absence of light (dark fermentation). However, fermentative biohydrogen has a high production rate, but poor conversion efficiency from the organic substrate to $H_2$.

Biohydrogen production by purple non-sulfur bacteria is mainly due to the presence of nitrogenase under oxygen-deficient conditions using light energy and reduced compounds (organic acids). The reaction is as follows:

$$C_6H_{12}O_6 + 12H_2O + \text{Light energy} \rightarrow 12H_2 + 6CO_2$$

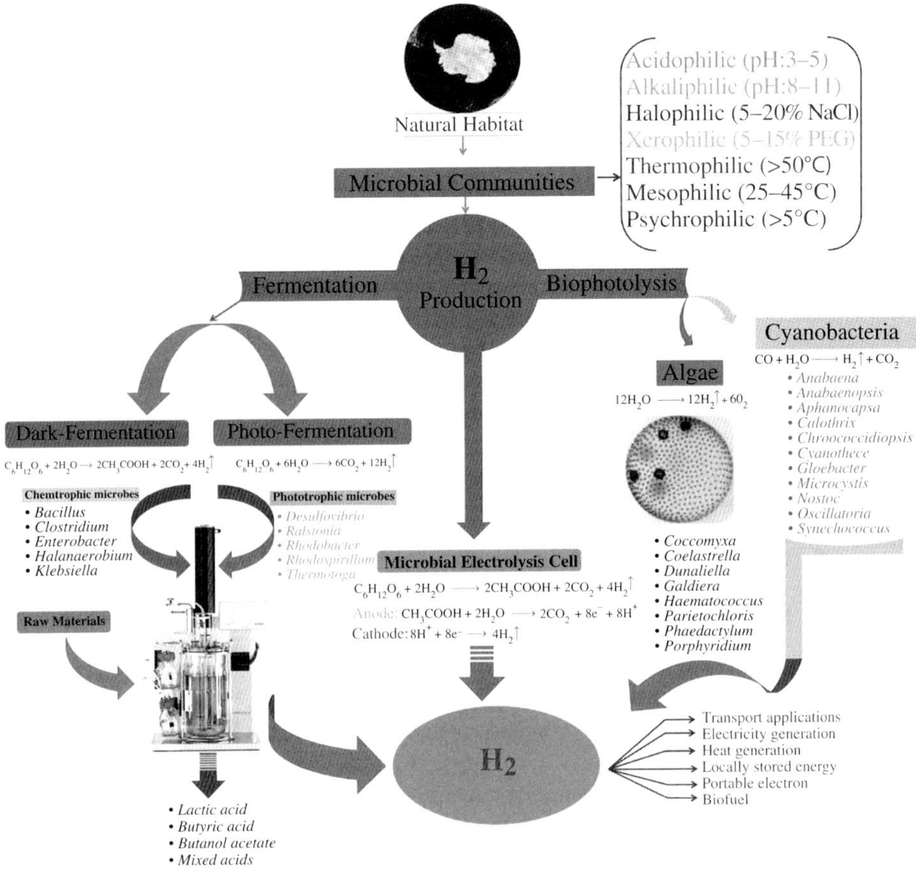


**Figure 22.1** A schematics presentation of hydrogen production through different processes.

Photosynthetic bacteria have long been studied for their capacity to produce significant amounts of $H_2$ (Bolton 1996; Das et al. 2008). The advantage of their use is in the versatile metabolic capabilities of these organisms and the lack of Photo system II (PSII), which automatically eliminates the difficulties associated with $O_2$ inhibition of $H_2$ production. Phototrophic bacteria require an organic or inorganic electron source to drive their photosynthesis. These photoheterotrophic bacteria have been found suitable to convert light energy into $H_2$ using organic wastes as substrate in batch processes, continuous cultures, or immobilized whole cell system using different solid matrices like carrageenan, agar gel, porous glass, and polyurethane foam (Das et al. 2008; Tsygankov and Kosourov 2014; Tsygankov et al. 1994; Vincenzini et al. 1986) (Table 22.1).

In photofermentation, the efficiency of light energy used in producing hydrogen by photosynthesis bacteria is theoretically higher than that of cyanobacteria (Nath and Das 2004a; Yusuf et al. 2016). Currently, molecular $H_2$ is primarily produced from the use of fossil fuels through steam reforming of natural gas or methane ($CH_4$). The worldwide production of $H_2$ currently exceeds 1 billion $m^3$ $day^{-1}$ of which 48% is produced from natural gas, 30% from oil, 18% from coal, and the remaining 4% is produced from $H_2O$-splitting electrolysis (Chandrasekhar et al. 2015). Alternatively, the production of

**Table 22.1** Yields hydrogen production from a different carbon source and industrial wastewaters by microbes through photo-fermentations.

| Microbial strains | Substrate | Maximum $H_2$ yield | References |
|---|---|---|---|
| MC1:PHPBC[a)] | Acetate | 7.8 mmol $H_2$/l | (Lazaro et al. 2012) |
| *Rhodobacter capsulatus* | Acetate | 0.6 mol $H_2$/mol | (Boran et al. 2010) |
| *Rhodobacter sphaeroides* | Acetate | 20 ml $H_2$/l h | (Uyar et al. 2009) |
| *Rhodobacter sphaeroides* | Acetate | 90 ml $H_2$/l h | (Tao et al. 2008) |
| *Rhodopseudomonas faecalis* | Acetate | 2.61 mol $H_2$/mol | (Liu et al. 2009) |
| *Rhodopseudomonas faecalis* | Acetate | 3.17 mol $H_2$/mol | (Ren et al. 2009) |
| *Rhodopseudomonas faecalis* | Acetate | 3.12 mol $H_2$/mol | (Xie et al. 2012) |
| *Rhodopseudomonas faecalis* | Acetate | 2.64 mol $H_2$/mol | (Xie et al. 2013) |
| *Rhodopseudomonas palustris* | Acetate | 2.2 ml $H_2$/l h | (Barbosa et al. 2001) |
| *Rhodopseudomonas* sp. | Acetate | 25.2 ml $H_2$/l h | (Barbosa et al. 2001) |
| *Rhodobacter capsulatus* | Acetate | 3752.7 ml $H_2$ l/l | (Ma et al. 2012) |
| *Rhodobium marinum* A-501 | Acetic acid | 0.2 mmol $H_2$/l | (Ike et al. 1999) |
| MC2:BC1 | Acetic acid | 56.1 mmol $H_2$/l | (Ike et al. 1999) |
| MC1:PHPBC | Butyrate | 9.0 mmol $H_2$/l | (Lazaro et al. 2012) |
| *Rhodobacter sphaeroides* | butyrate | 20 ml $H_2$/l h | (Uyar et al. 2009) |
| *Rhodobacter sphaeroides* | Butyrate | 110 ml $H_2$/l h | (Tao et al. 2008) |
| *Rhodopseudomonas* sp. | Butyrate | 7.6 ml $H_2$/l h | (Barbosa et al. 2001) |
| MC2:BC1 | Cellobiose | 21.7 mmol $H_2$/l | (Ike et al. 1999) |
| MC1:PHPBC | Citrate | 7.9 mmol $H_2$/l | (Lazaro et al. 2012) |
| *Rhodopseudomonas palustris* | Glucose | 0.2 mol $H_2$/mol | (Tian et al. 2010) |
| *Rhodobium marinum* A-501 | Glucose | 21.6 mmol $H_2$/l | (Ike et al. 1999) |
| MC2:BC1 | Glucose | 19.9 mmol $H_2$/l | (Ike et al. 1999) |
| *Rhodobium marinum* A-501 | Glycerol | 8.3 mmol $H_2$/l | (Ike et al. 1999) |
| MC2:BC1 | Glycerol | 15.9 mmol $H_2$/l | (Ike et al. 1999) |
| *Rhodobacter sphaeroides* | Hexose | 8.35 mol $H_2$/mol | (Kim and Kim 2013) |
| MC1:PHPBC | Lactate | 5.6 mmol $H_2$/l | (Lazaro et al. 2012) |
| *Rhodobacter sphaeroides* | Lactate | 20 ml $H_2$/l h | (Uyar et al. 2009) |
| *Rhodopseudomonas palustris* | Lactate | 9.1 ml $H_2$/l h | (Barbosa et al. 2001) |
| *Rhodopseudomonas* sp. | Lactate | 10.7 ml $H_2$/l h | (Barbosa et al. 2001) |
| *Rhodobium marinum* A-501 | Lactic acid | 37.3 mmol $H_2$/l | (Ike et al. 1999) |
| MC2:BC1 | Lactic acid | 82.9 mmol $H_2$/l | (Ike et al. 1999) |
| MC1:PHPBC | Malate | 13.9 mmol $H_2$/l | (Lazaro et al. 2012) |
| *Rhodobacter sphaeroides* | Malate | 24 ml $H_2$/l h | (Uyar et al. 2009) |
| *Rhodobacter sphaeroides* | Malate | 92 ml $H_2$/lh | (Tao et al. 2008) |
| *Rhodopseudomonas palustris* | Malate | 5.8 ml $H_2$/l h | (Barbosa et al. 2001) |
| *Rhodopseudomonas* sp. | Malate | 1.1 ml $H_2$/l h | (Barbosa et al. 2001) |
| *Rhodobium marinum* A-501 | Malate | 13.6 mmol $H_2$/l | (Ike et al. 1999) |
| MC2:BC1 | Malate | 17.6 mmol $H_2$/l | (Ike et al. 1999) |

**Table 22.1** (Continued)

| Microbial strains | Substrate | Maximum $H_2$ yield | References |
|---|---|---|---|
| *Rhodobium marinum* A-501 | Malic acid | 23.4 mmol $H_2$/l | (Ike et al. 1999) |
| MC2:BC1 | Malic acid | 26.4 mmol $H_2$/l | (Ike et al. 1999) |
| *Rhodobacter sphaeroides* | Propionate | 22 ml $H_2$/l h | (Uyar et al. 2009) |
| *Rubrivivax* sp. | Propionate | 10.3 ml $H_2$/l h | (Wu et al. 2010) |
| *Rhodobacter sphaeroides* | Sodium lactate | 2.4 mg/l | (Zhu et al. 2007) |
| *Rhodobium marinum* | Soy sauce | 200 mL H2 | (Anam et al. 2012) |
| MC2:BC1 | Starch | 39.1 mmol $H_2$/l | (Ike et al. 1999) |
| *Rhodobacter sphaeroides* | Succinate | 3.7 mol $H_2$/mol | (Kim et al. 2012) |
| *Rhodobacter sphaeroides* | Succinate | 2.3 mol $H_2$/mol | (Kim et al. 2013) |
| *Rhodobacter sphaeroides* | Succinate | 94 ml $H_2$/l h | (Tao et al. 2008) |
| *Rhodobium marinum* A-501 | Sucrose | 12.3 mmol $H_2$/l | (Ike et al. 1999) |
| MC2:BC1 | Sucrose | 18.3 mmol $H_2$/l | (Ike et al. 1999) |
| *Rhodopseudomonas palustris* | Wastewater | 205 mL $H_2$ L/d | (Lee et al. 2011) |
| *Rhodobacter sphaeroides* | Wastewaters | 2.24 l $H_2$/l medium | (Seifert et al. 2010) |

a) MC: Microbial Consortium, MC1-PHPBC: *Rhodobacter, Rhodospirillum, Rhodopseudomonas,* and *Sulfurospirillum*, MC2-BC1: *Vibriofluvialis, Rhodobium marinum,* and *Proteus vulgaris*.

$H_2$ from biomass through biological pathways is an emerging technology because it is sustainable and eco-friendly.

Biohydrogen is a renewable source of energy and is an alternative to traditional fuels because of its less harmful impact on the environment. Different groups of microbes have been reported to produce biohydrogen under diverse conditions. Archaea are true extremophilic microbes and are ubiquitous in nature habitats such as high- or low-temperature conditions, hypersaline, and in acidic habitats (Yadav et al. 2015c). Chapelle et al. (2002), describe a unique subsurface microbial community in which hydrogen-consuming, methane-producing Archaea far outnumber the bacteria. More than 90% of the 16S ribosomal DNA sequences recovered from hydrothermal waters circulating through deeply buried igneous rocks in Idaho are related to hydrogen-using methanogenic microorganisms. Geochemical characterization indicates that geothermal hydrogen, not organic carbon, is the primary energy source for this methanogen-dominated microbial community. These results demonstrate that hydrogen-based methanogenic communities do occur in Earth's subsurface, providing an analogue for possible subsurface microbial ecosystems on other planets. Among diverse groups of microbes, bacteria are most abundance and are reported as the most dominant groups to produced biohydrogen under photofermentation conditions.

Zürrer and Bachofen (1979) explored continuous photosynthetic production of hydrogen by *Rhodospirillum rubrum* in batch cultures. Hydrogen was observed up to 80 days with the hydrogen donor, pure lactate, or lactic acid-containing wastes, supplied periodically. Hydrogen was produced at an average rate of 6 ml $h^{-1}$ per g (dry weight) of cells with whey as a hydrogen donor. In continuous cultures with glutamate as a growth-limiting nitrogen source and lactate as a hydrogen donor, hydrogen was evolved at a rate of 20 ml $h^{-1}$ per g (dry weight). The composition of the evolved gas remained practically constant (70–75% $H_2$, 25–30% $CO_2$). Photosynthetic bacteria processing specific organic wastes could be an advantage in the large-scale production of hydrogen, together with food protein of high value compared to other biological systems.

Photofermentation is carried out by purple nonsulfur (PNS) photosynthetic bacteria, which can grow as photoheterotrophs, photoautotrophs, or chemoheterotrophs (Basak and Das 2007). These bacteria produce $H_2$ under photoheterotrophic conditions (light, anaerobiosis, organic electron donor) (Redwood et al. 2009). The advantages of this process over photolysis of water using green algae and cyanobacteria are that oxygen does not inhibit the process and that these bacteria can be used in a wide variety of conditions (i.e., batch processes, continuous cultures, and immobilized systems) (Holladay et al. 2009). The hydrogenase and nitrogenase enzymes produced in photosynthesis by green algae and photosynthetic bacteria, respectively, play a crucial role in biohydrogen production. The main PNS bacteria that participate in $H_2$ production are *Rhodospirillum rubrum*, *Rhodopseudomonas palustris*, *Rhodobacter sphaeroides* O.U 001, *R. sphaeroides* RV, *Rhodobacter sulfidophilus* and *Rhodobacter capsulatus*. Kapdan et al. (2009) used three different pure strains of *Rhodobacter sphaeroides* (RV, NRLL, and DSZM) in batch experiments to select the most suitable strain. *R. sphaeroides* RV resulted in the highest cumulative hydrogen gas formation (178 ml), hydrogen yield (1.23 mol $H_2$ $mol^{-1}$ glucose), and specific hydrogen production rate (46 ml $H_2$ $g^{-1}$ biomass $h^{-1}$) at 5 g $l^{-1}$ initial total sugar concentration, among the other pure cultures. Using *Rhodobacter capsulatus* JP91, Keskin and Hallenbeck (2012) compare the photofermentative biohydrogen yield of different feedstocks in a batch culture experiment.

Photofermentative hydrogen ($H_2$) production from glucose with the photosynthetic bacterium *Rhodobacter capsulatus* JP91 ($hup^-$) was examined using a photobioreactor operated in continuous mode (Abo-Hashesh et al. 2013). Stable and high hydrogen yields on glucose were obtained at three different retention times (hydraulic retention time (HRTs); 24, 48, and 72 hours). The $H_2$ production rates vary between 0.57 and 0.81 mmol $h^{-1}$. The highest hydrogen yield, $9.0 \pm 1.2$ mol $H_2$/mol glucose, was obtained at 48 hours HRT. The photosynthetic bacteria have a very versatile metabolic repertoire and have been known for decades to produce hydrogen during photofermentative growth. Hallenbeck and Liu (2016), reported advancements in hydrogen production by phototrophic organisms and projected future directions. Often used as a second stage in two-stage hydrogen production processes, first-stage fermentative sugar to hydrogen and organic acids; second stage, organic acids to hydrogen, recent studies have highlighted their ability to directly convert sugars to hydrogen. Several studies have attempted to optimize a single-stage batch process, and a study with continuous cultures has shown that yields approaching 9 mol $H_2$/mol glucose can be obtained. One of the drawbacks of this system is the dependency on light, necessitating the use of photobioreactors, thus potentially greatly adding to the cost of such a system. In another approach, which avoids the use of light energy, microaerobic fermentation of organic acids to hydrogen, driven by limited oxidative phosphorylation, has been demonstrated in principle.

### 22.2.2 Dark Fermentation

Biohydrogen production through dark fermentation is preferable over other biological processes, owing to its unique features such as rapid cell growth, light independence, and most importantly, economical approach, as it can utilize waste (Hallenbeck and Benemann 2002; Kotay and Das 2008; Nath and Das 2004b). It involves the anaerobic digestion of pyruvate by the action of enzymatic machinery in the microorganisms that carry out fermentation to assimilate carbohydrates into biohydrogen with the help of following enzymes:

- *Pyruvate formate lyase*. It breaks down pyruvate into formic acid releasing acetyl group that shifts to coenzyme A.
- *Pyruvate ferredoxin oxidoreductase*. It reduces the ferredoxin and carbon dioxide is released from pyruvate to activate co-enzyme A.

#### 22.2.2.1 Role of Microbes in Dark Fermentation

The microbes are ubiquitous in nature and have been reported from diverse habitats such as hot springs (Kumar et al. 2014; Sahay et al. 2017; Saxena et al. 2016; Suman et al. 2015), saline environments (Gaba et al. 2017; Yadav et al. 2015c), cold environments (Singh et al. 2016; Yadav 2015; Yadav et al. 2015a,b), acidic/alkaline soil (Biswas et al. 2018; Verma et al. 2013), drought (Verma et al. 2014, 2016b, Verma et al. 2017), and plant associations (Verma et al. 2016a, Verma et al. 2015a, Verma et al. 2015b, Verma et al. 2015c; Yadav and Yadav 2018). The microbiomes from diverse habitats provide indispensable bioresources for agriculture, industry, and allied sector. Various microorganisms that are involved in dark fermentation to produce hydrogen belong to strict anaerobes (e.g. *Clostridia*, rumen bacteria), facultative anaerobes (e.g., *Enterobacter*), and aerobes (e.g. *Bacillus). Clostridium* and *Enterobacter* are the most common bacteria of choice, with numerous reports on different species like *Clostridium beijerinckii* (Zhao et al. 2012), *Clostridium thermocellum* (Magnusson et al. 2008), *Clostridium algidixylanolyticum* (Zieliński et al. 2017), *Enterobacter aerogenes* (Patel et al. 2017), etc. Among aerobes, a few species used for dark fermentation are *Brevibacillus* sp., *Caloranaerobacter* sp. and *Geobacillus* sp. (Khongkliang et al. 2017), *Bacillus cereus, Bacillus megaterium, Bacillus pumilus,* and *Bacillus thuringiensis* (Patel et al. 2017). Microbial species producing hydrogen in psychrophillic conditions are *Rahnella aquatilis, Raoutella terrigena, Carnobacterium maltaromaticum* (Zieliński et al. 2017). However, most of the studies routinely employ mixed culture consortia derived from natural sludge or compost, as they are a cheap source of efficient microorganisms capable of producing hydrogen.

#### 22.2.2.2 Factors Affecting Biohydrogen Production in Dark Fermentation

The main factors that affect the production of biohydrogen in dark fermentation are temperature, pH, and types of the substrate, as well as culture conditions (Wong et al. 2014; Zhao et al. 2012). Temperature plays an important role. Although high temperatures help in substrate solubilization, prolonged exposure can denature the cellular proteins but can restrict interfering hydrogen consumers. So, the optimal temperature of 30–35 °C was reported for efficient biohydrogen production by increasing cell growth and ultimately the product of interest (Cardoso et al. 2014;

Lee et al. 2008). By contrast, other reports suggest that the optimal temperature is above 45 °C as the solubility of hydrogen is low at high temperature (Hallenbeck and Benemann 2002).

Metal ions are involved as co-factors in various enzymatic conversions and, therefore, their optimal concentration in fermentation reactors can efficiently modify the biohydrogen productivity (Sinha and Pandey 2011; Zhao et al. 2012). Cardoso et al. (2014) reported that MgSO4 concentration of 1.2 to 1.6 $g\,l^{-1}L$, resulted in a maximum yield of about 4.84 mol $H_2$/mol lactose. Supplementation with molybdate salts reduces methanogenesis and sulfate reduction and hence enhances biohydrogen production (Ferchichi et al. 2005). Similarly, calcium in the range of 50–150 $mg\,l^{-1}$ (Yuan et al. 2010), nickel (Noyola and Tinajero 2005; Wang and Wan 2008) and iron (18–55 $mg\,l^{-1}$) in media also improve biohydrogen yield by 1.5 times (Kim et al. 2011). (Dong-Jie et al. 2011) reported that a low concentration of Mo (0.0042 $mg\,l^{-1}$) could increase $H_2$ yield by 29%.

During later stages of fermentation, hydrogen yield is observed to reduce due to propionic acid (a metabolite of an $H_2$ consuming pathway) and methanogenic activity (use up $H_2$ in the media). So, it is requisite for the efficient hydrogen production to acidify the media and lower the pH below 4.0 that is inhibitory for the majority of microorganisms interfering with the process. This operational pH has resulted in upregulation of anaerobic biohydrogen production processes (Shin et al. 2004). Many researchers across the globe have proposed that the final pH range of 5.0–6.0 is optimum for maximum hydrogen production (Collet et al. 2004; Kanai et al. 2005; Lay 2001). In another study by (Lee et al. 2008), the optimal pH was observed to be 6.0 for maximum yield of biohydrogen from cassava starch. A recent report describes that under controlled pH environment (pH 6.5), mixed consortia from sewage sludge resulted in better yield as compared to uncontrolled ones, where interference of acid accumulation affected the yield (Penniston and Kana 2018). To achieve acidic pH, HCl has been recommended as most suitable acid for the proliferation of hydrogen producing organisms in the fermentation process. (Lee et al. 2009) investigated that 35% HCl treatment to inoculums sludge effectively enhanced hydrogen production by 2.8-fold.

Any cheap carbohydrate source from agricultural waste or food waste can be used as a substrate for dark fermentation. Numerous reports in literature highlight various readily available substrates such as sugarcane bagasse (Hu et al. 2018), cassava starch (Wang and Wan 2008), food waste (Noblecourt et al. 2018), and molasses (Lay et al. 2010). Substrate specific yield is another concept for biohydrogen production and is evidenced by its higher concentration (1.27 mmol $H_2$/glucose equivalent) by *Clostridium thermocellum* ATCC 27405 (a cellulolytic, thermophillic bacterium) with dried distillers' grain as a substrate as compared to using barley hulls alone as well as those contaminated with *Fusarium* head blight (Magnusson et al. 2008).

The amount of inoculum employed for fermentation is substantial, as optimally 5% inoculum has been reported to yield 123.27 ml hydrogen per gram TOC, i.e., 46% (Argun and Dao 2017). Furthermore, the nature of the substrate is also an important to point of consideration, as the total solid content in the substrate above 15% generally suppresses the substrate breakdown and thus fermentation process due to lactic acid accumulation (Ghimire et al. 2018).

#### 22.2.2.3 Productivity-Enhancing Approaches

Biohydrogen production through dark fermentation can be enhanced either by expediting the hydrogen-producing organisms or by suppressing the interfering hydrogen consuming ones that can only be achieved by targeting the factors contributing to the process. In general, the hydrogen produced is rapidly utilized by hydrogen consuming bacteria to produce methane, acetic acid, or propionic acid, resulting in a low yield of biohydrogen (Mohan 2008). Furthermore, the actual yield gets lower due to hydrogen recycling that is attributed to the presence of uptake hydrogenases that consume the hydrogen (Hallenbeck and Benemann 2002). Different physical as well as chemical pretreatments have been reported to enhance the yield by restricting the growth of hydrogen-utilizing organisms during dark fermentation.

***Physical Treatments*** Physical treatments include heat (Dong et al. 2010; Kawagoshi et al. 2005) that can kill most obnoxious organisms, except for some spore-forming acetogens, lactic acid, or propionic acid producers. These can be restricted by load shock treatment, i.e., to highly enrich the environment with hydrogen producing in ocula so that they dominate the methanogens. This is quite effective for increasing the productivity of biohydrogen (Sompong et al. 2009). Aeration and ultrasonication can also be used effectively to suppress hydrogen consumers (Kotay and Das 2008).

***Chemical Treatments*** Chemical treatments include methanogenic inhibitors like 2-bromoethanesulfonate and iodopropane (Zhu and Béland 2006). Base pretreatment to sewage sludge at pH 11.0 also showed a 1.8-fold increase in hydrogen production and the least methane production (Cai et al. 2004). In a study by (Kim and Shin 2008), alkali pretreatment of food waste at a pH of 12.5 was reported to be more effective than acid pretreatment at a pH 2.0 or carbon dioxide spraying, in regard to stability during 90-day fermentation with a better yield. A study reveals that acid pretreatment as well as heat shock are more effective in comparison to chemical inhibition, aeration, and base pretreatment for complete suppression of methanogenic activity (Zhu and Béland 2006).

Recently, integrating dark fermentation with microbial electrolysis has paved a new path toward a remarkable increase (13-fold) in biohydrogen production (Marone et al. 2017). Similarly, the coupling of microbial electrolysis (with an optimum voltage of 0.6 V) and dark fermentation with *Brevibacillus* sp., *Caloranaerobacter* sp., and *Geobacillus* sp. was reported to enhance the productivity of hydrogen by 2.5-fold (Khongkliang et al. 2017). Another technique to enhance hydrogen production is to promote co-cultures rather than individual cultures. Studies on biohydrogen productivity impart a clear interpretation that mixed consortia perform better in terms of yield. In agreement to this, two cellulolytic, thermophilic, anaerobic bacteria, *Clostridium thermocellum* JN4 and *Thermoanaerobacteriumthermosaccharolyticum* GD17, isolated from rotten wheat straw, were observed to show not only twofold increase in yield but also capability of a vast array substrate utilization like powdered corn stalks and cobs (Liu et al. 2009). Similarly, not only mixed microbial cultures (MMC6—*Bacillus cereus* EGU41, *Bacillus megaterium* (HPC686), *Bacillus pumilus*hpc464, *E. aerogenes* EGU16 and *Proteus mirabilis* strains (EGU21 and EGU30) and MMC4—*Bacillus cereus* EGU43, *Bacillus pumilus*hpc464, *Bacillus sp.* HPC459, *Bacillus thuringiensis* EGU45, *E. aerogenes* EGU16 and *P. mirabilis* EGU21) but also mixed substrates of onion and potato peels resulted in 1.22-fold higher hydrogen production than their individual capacities (Patel et al. 2017).

Ongoing studies employ different microbial consortia to harness the biohydrogen and combat challenges such as lactate accumulation or hydrogen consumers in a reactor so as to provide a sustainable energy source. Interestingly, Cardoso et al. (2014) employed lactose as the major substrate for dark fermentation to produce hydrogen, and they concluded that maximum hydrogen productions was achieved at an optimal temperature of 30–35 °C and $MgSO_4$ concentration of 1.2–1.6 g $l^{-1}$. Similarly, Noblecourt et al. (2018) used microbial consortia from sewage sludge, pretreated with 10 mM 2-bromoethanesulfonate (to inhibit methanogens archea), to produce hydrogen from depackaging wastes and reported a breakdown of different carbon sources including lactate, which may play a role in future directions. Extensive research is still required in this field to achieve the threshold level of biohydrogen production that can easily meet the growing demand.

### 22.2.3 Biophotolysis

Biophotolysis is the action of light on a biological system that results in the dissociation of a substrate, resulting in water to produce hydrogen. $H_2$ is produced under certain growth condition in some microalgae resulting in an overall net dissociation of water. Hydrogen is a nonpolluting source of energy, which is renewable and very abundant. Operational parameters for biophotolysis are pH, temperature, hydraulic retention time, pretreatment, and organic loading rate (ORL). Abilities to split water to hydrogen have been known from 1896 in the case of cyanobacteria and from 1942 in the case of algae. More commercial researches of hydrogen generation by photoautotrophic organisms began in the 1970s after the energy crisis (Akkerman et al. 2002). In biophotolysis, it is necessary to understand appropriate conditions of breeding of organisms and selecting proper substrate preparation, water purity, and flow, with a setting system for suitable sun irradiation.

#### 22.2.3.1 Direct Biophotolysis

This method is like the processes found in plants and algal photosynthesis. In direct biophotolysis, solar energy is directly converted to hydrogen via photosynthetic reactions.

$$2H_2O + \text{light energy} \rightarrow 2H_2 + O_2$$

Algae split water molecules to hydrogen ion and oxygen via photosynthesis (Kapdan and Kargi 2006). Hydrogenase activity has been observed in other green algae, like *Scenedesmus obliquus, Chlorococcum littorale, Platymonas subcordiformis,* and *Chlorella fusca* (Das and Veziroglu 2008). In direct biophotolysis, direct sunlight and water are used for producing hydrogen. Direct biophotolysis is very appealing in principle, but in practice it is severally limited by, among other factors, strong inhibition of $H_2$ production by the simultaneously evolved $O_2$ produced – for example, by respiration using endogenous or exogenous substrates. Disadvantages of direct biophotolysis are its need for a high intensity of light, the dangers of using $O_2$, and lower photochemical efficiency.

#### 22.2.3.2 Indirect Biophotolysis

Hydrogen can be produced from blue-green algae. In this, hydrogen can also be generated from carbohydrates produced by microalgae during normal photosynthesis. Indirect biophotolysis is based on heterocystous cyanobacteria. Cyanobacteria produce hydrogen by breaking down water and organic compounds (Vargas et al. 2018). The heterocysts exclude $O_2$ and reduce $N_2$. Another approach to indirect biophotolysisis is to carry out two reactions, first $O_2$ production (with $CO_2$ fixation) and then $H_2$ production (with $CO_2$ release) in separate stages. Cyanobacteria possess key enzymes (nitrogenase and hydrogenase) that carry out metabolic functions to achieve hydrogen generation (Lindberg et al. 2012). Because of the higher rates of $H_2$ production by Anabaena species and strains, these have been subject to intense study (Levin et al. 2004). In indirect biophotolysis mutant strains of *Anabaena Variabilis* have demonstrated hydrogen production rate of the order of 0.355 mmol $h^{-1}$ per liter (Sveshnikov et al. 1997). Disadvantages of indirect biophotolysis are that uptake hydrogenase enzymes must be removed to stop the degradation of $H_2$ and about 30% $O_2$ is present in the gas mixture.

#### 22.2.3.3 Role of Microbes in Biophotolysis

Green algae and cyanobacteria, previously known as blue-green algae, are microscopic water-borne organisms that carry out plant-type photosynthesis in which water is split by sunlight into $O_2$ and a strong reduction, typically ferrodoxin, is normally used to reduce $CO_2$ to carbohydrates (sugars). They possess chlorophyll a and other pigments to capture sunlight and use photosynthetic systems (PSI and PSII) to carry out plant-like oxygenic photosynthesis. The pigments in PSII (P680) absorbs the photons with a wavelength shorter than 680 nm, generating a strong oxidant capable of splitting water into protons ($H^+$), electrons ($e^-$), and $O_2$. In direct biophotolysis, the photosystem (PS1 and PS II) of algae (*Anabaena* sp and *Chlamydomonas reinhardtii*) convert solar energy into chemical energy, which is required to break down the water molecule to produce biohydrogen.

Cyanobacteria are potential microbial species for hydrogen production via direct biophotolysis (Pinto et al. 2002). By using nitrogenase and/or bidirectional hydrogenase, both heterocystous nitrogen-fixing strains and unicellular non-nitrogen-fixing strains can evolve hydrogen under special conditions. *Anabaena* strains are the representative nitrogen-fixing cyanobacteria in hydrogen. Heterocyst provides an oxygen-free environment to the oxygen-sensitive nitrogenase that reduces molecular nitrogen into $NH_3$ as well as protons into $H_2$.

*Chlamydomonas reinhardtii*, a facultative photoautotrophic and photoheterotrophic microalgae, is the representative of green microalgae for biohydrogen research (White and Melis 2006). Other algal species, such as *Chlorococcum littorale* and *Platymonas subcordiformis*, have also been investigated for hydrogen evolution. Gaffron and Robins first observed that unicellular microalga *S. obliquus* could either use $H_2$ as an electron donor in $CO_2$ fixation or evolve $H_2$ under anaerobic conditions in dark or light (Gaffron and Rubin 1942).

Gfeller and Gibbs grew *Chlamydomonas reinhardtii* F-60 on acetate in air under very low light intensity (0.6 w $m^{-2}$), and the starch content reached 5.5 g glucose/g Chl, or about 15.3% of dry cell mass. In darkness and nitrogen atmosphere, the maximum utilization rate of starch was 2.2 mmol glucose/g Chl/hr and hydrogen evolution

0.95 mmol/g Chl/hr. The hydrogen evolution in dark anaerobic fermentation is about one-sixth of direct biophotolysis. This fact implies that a large portion of hydrogen energy in direct biophotolysis is originated from the light energy absorbed at PSII and/or PSI, which raises the energy level of reducing power or electrons. The fermentative hydrogen yield is quite low, about 0.43 mole of $H_2$ per mole of glucose utilized, and other major fermentative products include 2.07 moles of formate, 1.07 moles of acetate, 0.91 mole of ethanol, and 0.04 mole of glycerol (Gfeller and Gibbs 1984).

Compared with green microalgae, unicellular non-nitrogen-fixing cyanobacteria have attracted more research interest for hydrogen production via indirect biophotolysis. *Spirulina platensis* is a filamentous cyanobacterium cultivated at large commercial scales as a food supplement with high proteins, antioxidants, and other nutrients. *S. platensis* NIES-46 accumulates a high glycogen content (50% dry mass) in nitrogen-limited conditions (Levin et al. 2004) and evolves molecular hydrogen at a moderate rate of 0.11 mmol $g^{-1}$ dry wt/h after induction in dark and nitrogen atmosphere (Aoyama et al. 1997). For 1 mol of hydrogen evolved, the cells also release 1.4 mol of acetate, 0.65 mol of ethanol, 0.4 mol of formate, and 0.1 mol of lactate as electron acceptors. A unicellular non-nitrogen-fixing cyanobacterium *Gloeocapsa alpicola* CALU743 showed a high activity of reversible hydrogenase induced in dark anaerobic conditions. The overall mechanism of hydrogen production in cyanobacteria can be represented by the following reactions:

$$12H_2O + 6CO_2 + \text{'light energy'} \rightarrow C_6H_{12}O_6 + 6O_2$$

$$C_6H_{12}O_6 + 12H_2O + \text{'light energy'} \rightarrow 12H_2 + 6CO_2$$

### 22.2.4 Microbial Electrolysis Cells

In the last few years, emphasis has been on bioelectrochemical or electrohydrogenesis processes for the production of hydrogen gas. A microbial electrolysis cell (MEC) is one of the bioelectrochemical systems in which hydrogen can be produced by combining bacterial metabolism with electrochemistry. In a bioelectrochemical system, oxidation-reduction reactions can be microbially catalyzed. These microorganisms are generally called elctroactive microorganisms, as their metabolic behavior is linked to the electrodes. For example, in MECs, anode-aspiring bacteria or exoelectrogenic bacteria oxidized the organic materials and generated $CO_2$, electrons, and protons. Bacteria extracellularly transfer the electron to the anode in anaerobic condition and protons are released in the solution. Electrons then can travel through a wire to a cathode and combine with the free protons in solution to produce hydrogen. However, this does not occur until an external voltage (>0.2 V) was supplied to the electrodes at neutral pH (Khan et al. 2017; Liu et al. 2005).

#### 22.2.4.1 Advantageous MEC Technology

An MEC is a technology to generate hydrogen from the microbial decomposition of organic compounds by applying an electric current (Khan et al. 2017; Liu et al. 2005; Logan et al. 2008; Meda et al. 2015). Microbial electrolysis cell (MEC) has the potential to accomplish sustainable and clean hydrogen production from biomass and wastewaters. In comparison to water electrolysis and the fermentation process, the rate of $H_2$

production is significantly higher (~100%) in MECs (Khan et al. 2017; Wang et al. 2013). Moreover, the threshold potential is (> 0.2–0.8 V) required in MECs, which is much smaller than the requirements for conventional water electrolysis (>1.6–1.8 V) at neutral pH (Kadier et al. 2016; Khan et al. 2017).

For biohydrogen production, MEC systems are also advantageous over conventional fermentation: Fermentation processes produce 4 mol $H_2$ and 2 mol of acetate from 1 mol of glucose ($C_6H_{12}O_6 + 2H_2O \rightarrow 4H_2 + 2CO_2 + 2C_2H_4O_2$), while MECs can produce 12 mol $H_2$/mol of glucose, as it also utilizes the remaining organic matter (i.e. acetic acid in the present case) (Logan 2004; Parkash 2016). This can be understood as two-step process. The first step is the same as the fermentation process, which produces 4 mol $H_2$/mol of glucose with two acetate molecules. Moreover, in the second step, four hydrogen molecules can be produced by oxidation and reduction processes from each acetate molecule, as follows (Liu et al. 2005):

$$\text{Anode (Oxidation)}: \ CH_3COOH + 2H_2O \rightarrow 2CO_2 + 8e\text{-} + 8H+ \tag{22.1}$$

$$\text{Cathode (Reduction)}: \ 8e\text{-} + 8H+ \rightarrow 4H_2 \tag{22.2}$$

The Gibbs free energy of the reaction (ΔG) must be negative for any reaction to occur spontaneously, and the conversion of most of these organic compounds to hydrogen gives a positive value of ΔG (+54.8 kJ $mol^{-1}$, as in the above reaction for acetate). Hence, this reaction is thermodynamically not favorable and requires an external driving force to take it to the forward direction. Hence, kinematics and thermodynamics also play an important role in the performance of MECs.

#### 22.2.4.2 Possible Designs of MECs and Their Performances

Liu et al. designed the first MEC system, which was inspired by microbial fuel cells (MFCs) (Liu et al. 2005). It was a simple two-chambered reactor consisting of two glass bottles separated by a cation exchange membrane (CEM). The $H_2$ gas was released from the top in the cathode chamber and then collected. Further, optimization of various designs of MECs was elaborated by different ways, for example, increasing membrane size comparative to the electrode-projected surface area, using anodic electrode with larger surface area, decreasing distance between electrodes, designing various membrane less two chamber or single chamber MECs using a MEC-MFC coupled system and dye-sensitized solar cell (DSSC) powered MECs (Han et al. 2010; Harnisch and Schröder 2009; He et al. 2005, 2006; Parkash 2016). The possible design of MECs and their performances are presented in Table 22.2. The two-chamber MEC, constructed with anion exchange membrane, flow-through bioanode and nickel foam cathode, has a maximum hydrogen production rate of 50 l $H_2$/L d and was observed at an applied voltage of 1.0 eV using acetate substrate (Jeremiasse et al. 2010).

#### 22.2.4.3 Limitations in MECs and Their Potential Solution

The research continues to design MECs with better performance for $H_2$ production, and on the basis of the review of previous research, the limitations in MECs and their potential solutions can be elaborated that are presented in Box 22.1 (Cai et al. 2016; He et al. 2006; Jiang et al. 2016; Khan et al. 2017, 2013, 2014; Oliot et al. 2016; Pandey et al. 2016; Rozendal et al. 2007; Wang et al. 2013; Zhang et al. 2009). Various factors affect MEC performance, such as the pH of the solution, solution conductivity, electrode

**Table 22.2** Various types of MEC designs and their performance.

| S.N. | MEC design | Substrate | Applied voltage | Hydrogen production rate | Hydrogen yield | Columbic efficiency | Hydrogen recovery | References |
|---|---|---|---|---|---|---|---|---|
| 1 | MEC was a two-bottle reactor (0.31 L capacity each) with the proton exchange membrane (PEM, NAFION 117) of 3.5 $cm^2$. | Acetate | 0.25 V | — | 2.9 mol $H_2$/mol Acetate | 60% | 90% | (Liu et al. 2005) |
| 2 | Two-chamber MEC was constructed by clamping an anion-exchange membranes (AEM: AMI-7001) between the anode (3 cm in diameter, 2 cm long; 14 ml) and cathode (4 cm long; 28 ml) chambers. | Acetate | 0.6 V | 1.10 L $H_2$/L d | 3.65 mol $H_2$/mol acetate | — — | 91% | (Cheng abd Logan 2007) |
| 3 | Two-chamber MEC was constructed by clamping an anion-exchange membrane (AEM: AMI-7001) between the anode (3 cm in diameter, 2 cm long; 14 ml) and cathode (4 cm long; 28 ml) chambers. | Glucose | | 1.23 L $H_2$/L d | 8.55 mol $H_2$/mol Glucose | — | 71% | (Cheng and Logan 2007) |
| 4 | Single chamber MEC having a brush anode and a flat carbon cathode | Acetate | 0.8 V | 3.12 ± 0.02 L $H_2$/Ld | — | — | 78 ± 1% to 96 ± 1% | (Call and Logan 2008) |
| 5 | Single-chamber membrane free glass tubular MEC using Nickel-Molybdenum cathode | Acetate | 0.6 V | 2.0 L$H_2$/Ld | 2.6 mol $H_{2/}$mol Acetate | 75% | 86% | (Hu et al. 2009) |

| | | | | | | | | |
|---|---|---|---|---|---|---|---|---|
| 6 | Membrane-less continuous flow MEC with a gas-phase cathode. | Acetate | 1.0 V | 6.32 L $H_2$/Ld | 3.9 mol $H_2$/mol Acetate | 80–100% | — | (Tartakovsky et al. 2009) |
| 7 | MEC powered by Dye-sensitized solar cell (DSSC) | Acetate | 602 mV[a)] | — | — | 40 ± 2% | — | (Ajayi et al. 2009) |
| 8 | A two chamber MEC was constructed with anion exchange membrane, flow through bioanode, and Ni foam cathode. | Acetate | 1.0 V | 50.0 L $H_2$/Ld | — | — | 95% | (Jeremiasse et al. 2010) |
| 9 | MEC powered by dark fermentation reactor and microbial fuel cells (MFCs) | Cellulose | 0.435 V[a)] | 0.24 L $H_2$/Ld | 14.3 mmol $H_2$/g cellulose | 57% | — | (Wang et al. 2011) |
| 10 | A double chamber tubular MEC is having components (a platinum-coated titanium mesh cathode, an anion exchange membrane, and a pleated stainless steel felt anode) were arranged in a concentric configuration. | Acetate | 1.0 V | 7.1 L $H_2$/Ld | — | — | 100% | (Guo et al. 2017) |

a) MEC input voltage as MEC was powered by coupled sources (i.e. a series combination of MFCs and $DSSC_S$)

material, substrate, applied voltage, the surface area of electrodes, the distance between the electrodes, electron transfer mechanism, losses within the system like ohmic, and concentration losses.

**Box 22.1**

MEC limitations and their potential solutions (Cai et al. 2016; He et al. 2006; Jiang et al. 2016; Khan et al. 2017, 2013, 2014; Oliot et al. 2016; Pandey et al. 2016; Rozendal et al. 2007; Wang et al. 2013; Zhang et al. 2009) are numerated here.

**Limitations in MEC:**

1) Low energy efficiencies.
2) Ohmic losses, concentration losses, conductivity losses.
3) High internal resistance and overpotential.
4) Microbial communities' abundance.
5) Competing reactions such as methanogenesis: High concentration of hydrogen gas favors the growth of methanogens, which reduces hydrogen gas production and contaminates the gas with methane.
6) Growth of $H_2$ scavenger further reduces the efficiency.

**Potential Solutions:**

1) An improvement in MEC design could increase surface area of electrodes and solution conductivity, decreasing distance between electrodes and leading to good connections in the circuit.
2) Appropriate electrolytes such as $KNO_3$, NaCl, $Na_2SO_4$ will increase the solution conductivity.
3) Electrodes with high porosity will give higher power density, but high porosity may trap $O_2$ within the electrode.
4) The use of electrolytic $O_2$ production can suppress methanogenesis.
5) With suitable exposure of cathode to air or $O_2$, the hydrogenotropic methanogenesis can be suppressed.
6) Adding certain antibiotics such as 2-chloroethane sulfonate, 8-aza-hypoxanthine in MEC can suppress the methanogenesis.

The design and performance analysis of MECs for large-scale applications is still under way. Possibly, in the future, better-performing MECs can be designed that optimize these parameters with improved effects.

## 22.3 Molecular Aspects of Hydrogen Production

Several strains of photosynthetic microorganisms are capable to the production of molecular hydrogen and it is one of the most auspicious tactics for generations of renewable energy (Ghirardi et al. 2009; Kruse et al. 2005; Prince and Kheshgi 2005). *S. obliquus* a green alga was first reported for hydrogen production (Gaffron and Rubin 1942). Then numerous strains of green algae were reported for the production of $H_2$ (Brand et al. 1989). Anaerobic induction and light under ambient temperatures, water, and minimal amounts of macro- and micronutrients necessary to get a high rate of $H_2$ synthesis (Gfeller and Gibbs 1984). $H_2$ synthesis is not sustainable until (i) the different proficiency of light utilization by phototrophs under diverse light intensities (Melis 2009); (ii) the high level of $O_2$ that blights the pathways in cells (Ghirardi et al. 2005); and (iii) the rate of photosynthetic $CO_2$ absorption essential for an effectual accumulation of cell biomass and its further adaptation to biohydrogen is truncated (Allahverdiyeva et al. 2014; Rodionova et al. 2017).

## 22.4 Biotechnological Tools Involved in the Process

Innumerable efforts have been made to develop the biohydrogen synthesis via genetic engineering application. Genetically modified bacteria like *E. coli* (Maeda et al. 2012; Wells et al. 2011; Zhao et al. 2010; Zhou et al. 2015), *Bacillus subtilis* (Jin et al. 2016; Yang et al. 2017; Zhang et al. 2016) *Clostridium* sp. (Olson et al. 2012; Olson et al. 2010; Srivastava et al. 2017; Tripathi et al. 2010) and some species of *Enterobacter* (Cheng et al. 2017; Lin et al. 2016; Maru et al. 2016; Zhang et al. 2011) were productively used for progressive yield of biohydrogen. (Kothari et al. 2017) reported that the perspective *Clostridium acetobutylicum* and *E. coli* are ideal strains due to the ease of use of appropriate genetic tools for gene knockout and gene over expression. The genetic expression in *C. acetobutylicum* to increase the $H_2$ production is regulated by antisense RNA. Bacterial strains of *Clostridium* are found to possess great potential for breaking cellulose in to hydrogen such as *Clostridium cellulolyticum* and *Clostridium populeti*. The belongings of these strains i.e. cellulose degrading pathway can be expressed in *C. acetobutylicum* to attain the highest hydrogen yield. Several early efforts were unproductive to express (Fe—Fe) hydrogenases in *E. coli* by overexpression of hydAfroman organism such as *Clostridium* and have remained unreported. Far along it was observed that to co-express maturation gene hydE, hydF, and hydG that are required for H-cluster maturation, insertion of the organism does not possess these enzyme (Keseler et al. 2005; Maeda et al. 2012). Whereas heterologous expression of hydA is modest and possible without the heterologous expression of the accessory genes if these are encoded by the host genome. Some contemporary works conveyed for the expression hydrogenase gene hydA in *Enterobacter colace* IIT BT08, expressed high hydrogen yield from the strain of *E. aerogens* (ATCC13408), which doubled the hydrogen yield (Zhao et al. 2010).

Thapa et al. (2015) reported genetic engineered bioethanol-producing strain, *E. aerogenes* ATCC 29007, by deleting the D-lactate dehydrogenase gene to block the production of lactic acid. The open-reading frame coding region of ldhA gene was replaced with a kanamycin cassette flanked by FLP recognition target sites by using a one-step method to inactivate chromosomal genes and primers designed to create in-frame deletions upon excision of the resistance cassette. By the combination of gene deletion and over expression, the bioethanol yield was $0.48\,g\,g^{-1}$ when employing $80\,g\,l^{-1}$ glycerol. Hence, a significant enhancement in ethanol production was observed.

Lee et al. (2016), studied enhancement in the production of 2,3-butanediol with the addition of acetate into the culture media using *E. aerogenes* ATCC 29007 and to determine the mechanism of improved 2,3-butanediol production. Song et al. 2016 cloned hydrogenase genes (hoxEFUYH) of *Synechocystis* sp. PCC 6803 and heterologously expressed in *E. aerogenes*, and the hydrogen yield was significantly enhanced using the recombinant strain. A recombinant plasmid containing the gene in-frame with

Glutathione-S-Transferase (GST) gene was transformed into *E. aerogenes* ATCC13408 to produce a GST-fusion protein. During hydrogen fermentation, the recombinant strain produces more acetate and butyrate, but less ethanol.

Recently, Zhao et al. (2017), reported overexpression of the gene of HycE and the gene of HycG in *E. aerogenes* separately by using the pGEX-4T-2-cat vector, to obtain two recombinant strains: *E. aerogenes*/HycE and *E. aerogenes*/HycG. The hydrogen yields were significantly enhanced by the recombinant strains. Cheng et al. (2017) mutated *E. aerogenes* cells using nuclear irradiation of 60 Co $\gamma$-rays. The screened *E. aerogenes* ZJU1 mutant with larger colored circles enhanced the hydrogenase activity from 89.8 of the wild strain to 157.4 ml H2/(g DW h). The hereditary stability of the *E. aerogenes* ZJU1 mutant was certified after over 10 generations of cultivation. The hydrogen yield of 301 ml H2/g glucose with the mutant was higher by 81.8% than that of 166 ml $g^{-1}$ glucose with the wild strain. The mutant produced more acetate and butyrate but less ethanol compared with the wild strain during hydrogen fermentation. Accordingly, the tactic of metabolic engineering facilitates the development of effectual strains to improve the biohydrogen synthesis. Nevertheless, concrete sustainability of these genetically modified strains should be investigated for their long-term sustainability.

## 22.5 Reactors for Biohydrogen Production

The basic function of tubular photobioreactor is to provide a controlled environment in order to achieve optimal growth and maximum yield. A photobioreactor is a four-phase system that includes a solid (microbial cell), a liquid (growth medium), gas, and light radiations (Posten 2009). These reactors generally consist of the transparent wall through which light can easily penetrate and reach the microbial culture system. The performance of the reactor evaluated in terms of photochemical efficiency is defined as the conversion of total incident light to chemically bound energy in the hydrogen produced and the volumetric productivity as hydrogen production per unit illuminated area or per unit land area. While evaluating the efficiency of the process, we must consider not only total input energy of the incident light and operational installation energy but also the cost of construction material, its maintenance, and the resources necessary for operation. Based on structure, two main categories of photobioreactors stand out.

### 22.5.1 Tubular Reactor

The basic function of tubular photobioreactor is to provide a controlled environment to achieve optimal growth and maximum yield. The factor responsible for the performance of the reactor is (i) diameter of the tube, (ii) length of the tube, and (iii) mixing. The choice of the tube diameter and length is the major decision while designing PBR because it affects the surface to volume ratio, biomass concentration, and volumetric productivity. According to Fujita and Kobayashi (1971), the volumetric productivity increases as the tube diameter decreases from 5 to 1.6 cm. Mixing of the culture is necessary because it ensures the uniform distribution of cell and regularly exposed to light. Furthermore, it ensures the uniform nutrient supply and diminishes the nutritional and gaseous gradient surrounding the cell.

### 22.5.2 Flat Panel Reactor

The design of the flat plate reactor is as compact as possible and provides maximum illuminated area in a small land space. The illuminated surface per land space is maximum in a vertical reactor rather than a horizontal reactor. It is observed that the microorganism only settles in the dark, and as soon as the hydrogen is produced, the cells resuspends due to the movement of hydrogen bubbles. The performance of the photobioreactor mainly depends on the light availability for the microorganism. The intensity of light decreases exponentially with culture depth, and this can be managed by the depth of the panel. Since the bacteria can use diffuse light, the intensity of the sunlight can be controlled by the orientation of the photobioreactor.

## 22.6 Scientific Advancements and Major Challenges in Biohydrogen Production Processes

There are numbers of techniques and process advancements for higher-yield production of $H_2$, which includes applications of genetic modification of the microbes, metabolically engineered organisms, and development of improved bioreactor design (Das et al. 2008). The achievement of higher yields is a critical research objective for the sustenance of $H_2$ as the dream fuel of the future. Various strategies have therefore to be applied both at the fermentative and genetic level to improve $H_2$ production rates and yields. Genetic engineering is the key area of focus for the improvement of strains for $H_2$ production.

Major challenges must be overcome for the smooth transition from a fossil fuel-based economy to a hydrogen energy-based economy and may be outlined according to Das et al. (2008) as follows:

- The yield of $H_2$ from any of the processes defined above is low for commercial application. The pathways of $H_2$ production have not been identified and the reaction remains energetically unfavorable.
- Processing some biomass feed stock is too costly. There is a need to develop low-cost methods for growing, harvesting, transporting, and pretreating energy crops and/or biomass waste products.
- There is no clear contender for a robust, industrially capable microorganism that can be metabolically engineered to produce more than 4 mol $H_2$/mol of glucose.
- Several engineering issues must be addressed, including the appropriate bioreactor design for $H_2$ production, difficulty in sustaining steady, continuous $H_2$ production rate in the long term, scale-up, preventing interspecies $H_2$ transfer in nonsterile conditions, and separation/purification of $H_2$.
- The sensitivity of hydrogenase to $O_2$ and $H_2$ partial pressure severely disrupts the efficiency of the processes and adds to lower yields.
- There is insufficient knowledge on the metabolism of $H_2$ producing bacteria and the levels of $H_2$ concentration tolerance of these bacteria.
- There is a lack of understanding on the improvement of the economics of the process by integration of $H_2$ production with other processes.

## 22.7 Conclusions and Future Prospects

Biohydrogen production technology can utilize renewable energy sources such as biomass for the generation of hydrogen, the cleanest form of energy. However, biohydrogen production is the most challenging area of applied microbiology and biotechnology with respect to environmental problems. A challenging problem in establishing biohydrogen as a source of energy is the renewable and environmentally eco-friendly generation of large quantities of $H_2$ gas. The future of biohydrogen production depends not only on research advances, i.e., the improvement in efficiency through genetically engineered microorganisms and/or the development of bioreactors, but also on economic considerations, social espousal, and the development of $H_2$ energy systems. To increase the production of biohydrogen, intensive research work should be carried out on the advancement of processes such as fermentative and biophotolysis for development of genetically modified microbes, the improvement of the bioreactor design, and molecular engineering of the enzyme hydrogenases.

## Acknowledgment

The authors are grateful to Prof. Harcharan Singh Dhaliwal, vice chancellor, Eternal University, Baru Sahib for providing infrastructural facilities and constant encouragement.

## References

Abo-Hashesh, M., Desaunay, N., and Hallenbeck, P.C. (2013). High-yield single stage conversion of glucose to hydrogen by photofermentation with continuous cultures of *Rhodobacter capsulatus* JP91. *Bioresour. Technol.* 128: 513–517.

Ajayi, F.F., Kim, K.Y., Chae, K.J. et al. (2009). Study of hydrogen production in light assisted microbial electrolysis cell operated with dye sensitized solar cell. *Int. J. Hydrogen Energy* 34: 9297–9304.

Akkerman, I., Janssen, M., Rocha, J., and Wijffels, R.H. (2002). Photobiological hydrogen production: photochemical efficiency and bioreactor design. *Int. J. Hydrogen Energy* 27: 1195–1208.

Allahverdiyeva, Y., Aro, E., and Kosourov, S. (2014). Recent developments on cyanobacteria and green algae for biohydrogen photoproduction and its importance in $CO_2$ reduction. In: *Bioenergy Research: Advances and Applications*, 367–387. Elsevier.

Anam, K., Habibi, M.S., Harwati, T.U., and Susilaningsih, D. (2012). Photofermentative hydrogen production using *Rhodobium marinum* from bagasse and soy sauce wastewater. *Int. J. Hydrogen Energy* 37: 15436–15442.

Aoyama, K., Uemura, I., Miyake, J., and Asada, Y. (1997). Fermentative metabolism to produce hydrogen gas and organic compounds in a cyanobacterium, *Spirulina platensis*. *J. Ferment. Bioeng.* 83: 17–20.

Argun, H. and Dao, S. (2017). Bio-hydrogen production from waste peach pulp by dark fermentation: effect of inoculum addition. *Int. J. Hydrogen Energy* 42: 2569–2574.

Armor, J.N. (1999). The multiple roles for catalysis in the production of $H_2$. *Appl. Catal. A* 176: 159–176.

Barbosa, M.J., Rocha, J.M., Tramper, J., and Wijffels, R.H. (2001). Acetate as a carbon source for hydrogen production by photosynthetic bacteria. *J. Biotechnol.* 85: 25–33.

Basak, N. and Das, D. (2007). The prospect of purple non-sulfur (PNS) photosynthetic bacteria for hydrogen production: the present state of the art. *World J. Microbiol. Biotechnol.* 23: 31–42.

Biswas, S., Kundu, D., Mazumdar, S. et al. (2018). Study on the activity and diversity of bacteria in a new Gangetic alluvial soil (Eutrocrept) under rice-wheat-jute cropping system. *J. Environ. Biol.* 39: 379–386.

Bolton, J.R. (1996). Solar photoproduction of hydrogen: a review. *Sol. Energy* 57: 37–50.

Boran, E., Özgür, E., van der Burg, J. et al. (2010). Biological hydrogen production by *Rhodobacter capsulatus* in solar tubular photo bioreactor. *J. Cleaner Prod.* 18: S29–S35.

Brand, J.J., Wright, J.N., and Lien, S. (1989). Hydrogen production by eukaryotic algae. *Biotechnol. Bioeng.* 33: 1482–1488.

Cai, M., Liu, J., and Wei, Y. (2004). Enhanced biohydrogen production from sewage sludge with alkaline pretreatment. *Environ. Sci. Technol.* 38: 3195–3202.

Cai, W., Zhang, Z., Ren, G. et al. (2016). Quorum sensing alters the microbial community of electrode-respiring bacteria and hydrogen scavengers toward improving hydrogen yield in microbial electrolysis cells. *Appl. Energy* 183: 1133–1141.

Call, D. and Logan, B.E. (2008). Hydrogen production in a single chamber microbial electrolysis cell lacking a membrane. *Environ. Sci. Technol.* 42: 3401–3406.

Cardoso, V., Romão, B.B., Silva, F.T. et al. (2014). Hydrogen production by dark fermentation. *Chem. Eng. Trans.* 38: 481–486.

Chandrasekhar, K., Lee, Y.-J., and Lee, D.-W. (2015). Biohydrogen production: strategies to improve process efficiency through microbial routes. *Int. J. Mol. Sci.* 16: 8266–8293.

Chapelle, F.H., O'neill, K., Bradley, P.M. et al. (2002). A hydrogen-based subsurface microbial community dominated by methanogens. *Nature* 415: 312.

Cheng, S. and Logan, B.E. (2007). Sustainable and efficient biohydrogen production via electrohydrogenesis. *Proc. Natl. Acad. Sci. U.S.A.* 104: 18871–18873.

Cheng, J., Liu, M., Song, W. et al. (2017). Enhanced hydrogen production of *Enterobacter aerogenes* mutated by nuclear irradiation. *Bioresour. Technol.* 227: 50–55.

Chu, S. and Majumdar, A. (2012). Opportunities and challenges for a sustainable energy future. *Nature* 488: 294.

Collet, C., Adler, N., Schwitzguébel, J.-P., and Péringer, P. (2004). Hydrogen production by *Clostridium thermolacticum* during continuous fermentation of lactose. *Int. J. Hydrogen Energy* 29: 1479–1485.

Darvishi Cheshmeh Soltani, R., Khataee, A., and Koolivand, A. (2015). Kinetic, isotherm, and thermodynamic studies for removal of direct red 12b using nanostructured biosilica incorporated into calcium alginate matrix. *Environ. Prog. Sustainable Energy* 34: 1435–1443.

Das, D. and Veziroglu, T.N. (2008). Advances in biological hydrogen production processes. *Int. J. Hydrogen Energy* 33: 6046–6057.

Das, D., Khanna, N., and Veziroğlu, N.T. (2008). Recent developments in biological hydrogen production processes. *Chem. Ind. Chem. Eng.* 14: 57–67.

Dong, L., Zhenhong, Y., Yongming, S., and Longlong, M. (2010). Evaluation of pretreatment methods on harvesting hydrogen producing seeds from anaerobic digested organic fraction of municipal solid waste (OFMSW). *Int. J. Hydrogen Energy* 35: 8234–8240.

Dong-Jie, N., Jing-Yuan, W., Bao-Ying, W., and You-Cai, Z. (2011). Effect of Mo-containing additives on biohydrogen fermentation from cassava's stillage. *Int. J. Hydrogen Energy* 36: 5289–5295.

Ferchichi, M., Crabbe, E., Hintz, W. et al. (2005). Influence of culture parameters on biological hydrogen production by *Clostridium saccharoperbutylacetonicum* ATCC 27021. *World J. Microbiol. Biotechnol.* 21: 855.

Fujita, T. and Kobayashi, S. (1971). Experimentally induced granule release in the endocrine cells of dog pyloric antrum. *Cell Tissue Res.* 116: 52–60.

Gaba, S., Singh, R.N., Abrol, S. et al. (2017). Draft genome sequence of *Halolamina pelagica* CDK2 isolated from natural salterns from Rann of Kutch, Gujarat, India. *Genom. Announc.* 5: 1–2.

Gaffron, H. and Rubin, J. (1942). Fermentative and photochemical production of hydrogen in algae. *J. Gen. Physiol.* 26: 219–240.

Gfeller, R.P. and Gibbs, M. (1984). Fermentative metabolism of Chlamydomonas *reinhardtii*: I analysis of fermentative products from starch in dark and light. *Plant Physiol.* 75: 212–218.

Ghimire, A., Trably, E., Frunzo, L. et al. (2018). Effect of total solids content on biohydrogen production and lactic acid accumulation during dark fermentation of organic waste biomass. *Bioresour. Technol.* 248: 180–186.

Ghirardi M.L., P.W. King, M.C. Posewitz, P.C. Maness, A. Fedorov, K. Kim, J. Cohen, K. Schulten, and M. Seibert. 2005. Approaches to Developing Biological $H_2$-Photoproducing Organisms and Processes. International Hydrogenases Conference, Portland Press Limited

Ghirardi, M.L., Dubini, A., Yu, J., and Maness, P.-C. (2009). Photobiological hydrogen-producing systems. *Chem. Soc. Rev.* 38: 52–61.

Guo, K., Prévoteau, A., and Rabaey, K. (2017). A novel tubular microbial electrolysis cell for high rate hydrogen production. *J. Power Sources* 356: 484–490.

Hallenbeck, P.C. and Benemann, J.R. (2002). Biological hydrogen production; fundamentals and limiting processes. *Int. J. Hydrogen Energy* 27: 1185–1193.

Hallenbeck, P.C. and Liu, Y. (2016). Recent advances in hydrogen production by photosynthetic bacteria. *Int. J. Hydrogen Energy* 41: 4446–4454.

Han, Y., Yu, C., and Liu, H. (2010). A microbial fuel cell as power supply for implantable medical devices. *Biosens. Bioelectron.* 25: 2156–2160.

Harnisch, F. and Schröder, U. (2009). Selectivity versus mobility: separation of anode and cathode in microbial bioelectrochemical systems. *ChemSusChem* 2: 921–926.

He, Z., Minteer, S.D., and Angenent, L.T. (2005). Electricity generation from artificial wastewater using an upflow microbial fuel cell. *Environ. Sci. Technol.* 39: 5262–5267.

He, Z., Wagner, N., Minteer, S.D., and Angenent, L.T. (2006). An upflow microbial fuel cell with an interior cathode: assessment of the internal resistance by impedance spectroscopy. *Environ. Sci. Technol.* 40: 5212–5217.

Holladay, J.D., Hu, J., King, D.L., and Wang, Y. (2009). An overview of hydrogen production technologies. *Catal. Today* 139: 244–260.

Hu, H., Fan, Y., and Liu, H. (2009). Hydrogen production in single-chamber tubular microbial electrolysis cells using non-precious-metal catalysts. *Int. J. Hydrogen Energy* 34: 8535–8542.

Hu, B.B., Li, M.-Y., Wang, Y.-T., and Zhu, M.-J. (2018). High-yield biohydrogen production from non-detoxified sugarcane bagasse: fermentation strategy and mechanism. *Chem. Eng. J.* 335: 979–987.

Ike, A., Murakawa, T., Kawaguchi, H. et al. (1999). Photoproduction of hydrogen from raw starch using a halophilic bacterial community. *J. Biosci. Bioeng.* 88: 72–77.

Jeremiasse, A.W., Hamelers, H.V., Saakes, M., and Buisman, C.J. (2010). Ni foam cathode enables high volumetric $H_2$ production in a microbial electrolysis cell. *Int. J. Hydrogen Energy* 35: 12716–12723.

Jiang, Y., Liang, P., Zhang, C. et al. (2016). Periodic polarity reversal for stabilizing the pH in two-chamber microbial electrolysis cells. *Appl. Energy* 165: 670–675.

Jin, P., Kang, Z., Yuan, P. et al. (2016). Production of specific-molecular-weight hyaluronan by metabolically engineered *Bacillus subtilis* 168. *Metab. Eng.* 35: 21–30.

Kadier, A., Simayi, Y., Abdeshahian, P. et al. (2016). A comprehensive review of microbial electrolysis cells (MEC) reactor designs and configurations for sustainable hydrogen gas production. *Alex. Eng. J.* 55: 427–443.

Kanai, T., Imanaka, H., Nakajima, A. et al. (2005). Continuous hydrogen production by the hyperthermophilic archaeon, *Thermococcus kodakaraensis* KOD1. *J. Biotechnol.* 116: 271–282.

Kapdan, I.K. and Kargi, F. (2006). Bio-hydrogen production from waste materials. *Enzyme Microb. Technol.* 38: 569–582.

Kapdan, I.K., Kargi, F., Oztekin, R., and Argun, H. (2009). Bio-hydrogen production from acid hydrolyzed wheat starch by photo-fermentation using different *Rhodobacter* sp. *Int. J. Hydrogen Energy* 34: 2201–2207.

Kawagoshi, Y., Hino, N., Fujimoto, A. et al. (2005). Effect of inoculum conditioning on hydrogen fermentation and pH effect on bacterial community relevant to hydrogen production. *J. Biosci. Bioeng.* 100: 524–530.

Keseler, I.M., Collado-Vides, J., Gama-Castro, S. et al. (2005). EcoCyc: a comprehensive database resource for *Escherichia coli*. *Nucleic Acids Res.* 33: D334–D337.

Keskin, T. and Hallenbeck, P.C. (2012). Hydrogen production from sugar industry wastes using single-stage photofermentation. *Bioresour. Technol.* 112: 131–136.

Khan, M.Z., Sim, Y.L., Lin, Y.J., and Lai, K.M. (2013). Testing biological effects of hand-washing grey water for reuse in irrigation on an urban farm: a case study. *Environ. Technol.* 34: 545–551.

Khan, M.Z., Singh, S., Sreekrishnan, T., and Ahammad, S. (2014). Feasibility study on anaerobic biodegradation of azo dye reactive orange 16. *RSC Adv.* 4: 46851–46859.

Khan, M., Nizami, A., Rehan, M. et al. (2017). Microbial electrolysis cells for hydrogen production and urban wastewater treatment: a case study of Saudi Arabia. *Appl. Energy* 185: 410–420.

Khongkliang, P., Kongjan, P., Utarapichat, B. et al. (2017). Continuous hydrogen production from cassava starch processing wastewater by two-stage thermophilic dark fermentation and microbial electrolysis. *Int. J. Hydrogen Energy* 42: 27584–27592.

Kim, D.-H. and Kim, M.-S. (2013). Development of a novel three-stage fermentation system converting food waste to hydrogen and methane. *Bioresour. Technol.* 127: 267–274.

Kim, S.-H. and Shin, H.-S. (2008). Effects of base-pretreatment on continuous enriched culture for hydrogen production from food waste. *Int. J. Hydrogen Energy* 33: 5266–5274.

Kim, D.-H., Kim, S.-H., Kim, H.-W. et al. (2011). Sewage sludge addition to food waste synergistically enhances hydrogen fermentation performance. *Bioresour. Technol.* 102: 8501–8506.

Kim, D., Cho, S.-Y., Maeng, S.-H. et al. (2012). Diagnosis and constitutional and laboratory features of Korean girls referred for precocious puberty. *Korean J. Pediatr.* 55: 481–486.

Kim, D.-H., Cha, J., Kang, S., and Kim, M.-S. (2013). Continuous photo-fermentative hydrogen production from lactate and lactate-rich acidified food waste. *Int. J. Hydrogen Energy* 38: 6161–6166.

Kotay, S.M. and Das, D. (2008). Biohydrogen as a renewable energy resource—prospects and potentials. *Int. J. Hydrogen Energy* 33: 258–263.

Kothari, R., Kumar, V., Pathak, V.V. et al. (2017). A critical review on factors influencing fermentative hydrogen production. *Front. Biosci.* 22: 1195–1220.

Kruse, O., Rupprecht, J., Mussgnug, J.H. et al. (2005). Photosynthesis: a blueprint for solar energy capture and biohydrogen production technologies. *Photochem. Photobiol. Sci.* 4: 957–970.

Kumar, M., Yadav, A.N., Tiwari, R. et al. (2014). Evaluating the diversity of culturable thermotolerant bacteria from four hot springs of India. *J. Biodivers. Biopros. Dev.* 1: 1–9.

Lay, J.J. (2001). Biohydrogen generation by mesophilic anaerobic fermentation of microcrystalline cellulose. *Biotechnol. Bioeng.* 74: 280–287.

Lay, C.-H., Wu, J.-H., Hsiao, C.-L. et al. (2010). Biohydrogen production from soluble condensed molasses fermentation using anaerobic fermentation. *Int. J. Hydrogen Energy* 35: 13445–13451.

Lazaro, C.Z., Vich, D.V., Hirasawa, J.S., and Varesche, M.B.A. (2012). Hydrogen production and consumption of organic acids by a phototrophic microbial consortium. *Int. J. Hydrogen Energy* 37: 11691–11700.

Lee, K.-S., Hsu, Y.-F., Lo, Y.-C. et al. (2008). Exploring optimal environmental factors for fermentative hydrogen production from starch using mixed anaerobic microflora. *Int. J. Hydrogen Energy* 33: 1565–1572.

Lee, M.-J., Song, J.-H., and Hwang, S.-J. (2009). Effects of acid pre-treatment on bio-hydrogen production and microbial communities during dark fermentation. *Bioresour. Technol.* 100: 1491–1493.

Lee, C.-M., Hung, G.-J., and Yang, C.-F. (2011). Hydrogen production by *Rhodopseudomonas palustris* WP 3-5 in a serial photobioreactor fed with hydrogen fermentation effluent. *Bioresour. Technol.* 102: 8350–8356.

Lee, S.J., Thapa, L.P., Lee, J.H. et al. (2016). Stimulation of 2, 3-butanediol production by upregulation of alsR gene transcription level with acetate addition in *Enterobacter aerogenes* ATCC 29007. *Process Biochem.* 51: 1904–1910.

Levin, D.B., Pitt, L., and Love, M. (2004). Biohydrogen production: prospects and limitations to practical application. *Int. J. Hydrogen Energy* 29: 173–185.

Lin, R., Cheng, J., Ding, L. et al. (2016). Enhanced dark hydrogen fermentation by addition of ferric oxide nanoparticles using *Enterobacter aerogenes*. *Bioresour. Technol.* 207: 213–219.

Lindberg, P., Devine, E., Stensjö, K., and Lindblad, P. (2012). HupW protease specifically required for processing of the catalytic subunit of the uptake hydrogenase in the cyanobacterium *Nostoc* sp. strain PCC 7120. *Appl. Environ. Microbiol.* 78: 273–276.

Liu, H., Grot, S., and Logan, B.E. (2005). Electrochemically assisted microbial production of hydrogen from acetate. *Environ. Sci. Technol.* 39: 4317–4320.

Liu, B.-F., Ren, N.-Q., Ding, J. et al. (2009). Enhanced photo-$H_2$ production of *R. faecalis* RLD-53 by separation of $CO_2$ from reaction system. *Bioresour. Technol.* 100: 1501–1504.

Logan, B. (2004). Biologically extracting energy from wastewater: biohydrogen production and microbial fuel cells. *Environ. Sci. Technol.* 38: 160–167.

Logan, B.E., Call, D., Cheng, S. et al. (2008). Microbial electrolysis cells for high yield hydrogen gas production from organic matter. *Environ. Sci. Technol.* 42: 8630–8640.

Ma, C.X., Wang, L., Guo, X. et al. (2012). Enhanced photo-fermentative hydrogen production by *Rhodobacter capsulatus* with pigment content manipulation. *Bioresour. Technol.* 118: 490–495.

Maeda, T., Sanchez-Torres, V., and Wood, T.K. (2012). Hydrogen production by recombinant *Escherichia coli* strains. *Microb. Biotechnol.* 5: 214–225.

Magnusson, L., Islam, R., Sparling, R. et al. (2008). Direct hydrogen production from cellulosic waste materials with a single-step dark fermentation process. *Int. J. Hydrogen Energy* 33: 5398–5403.

Marone, A., Ayala-Campos, O.R., Trably, E. et al. (2017). Coupling dark fermentation and microbial electrolysis to enhance bio-hydrogen production from agro-industrial wastewaters and by-products in a bio-refinery framework. *Int. J. Hydrogen Energy* 42: 1609–1621.

Maru, B., Lopez, F., Kengen, S. et al. (2016). Dark fermentative hydrogen and ethanol production from biodiesel waste glycerol using a co-culture of *Escherichia coli* and *Enterobacter* sp. *Fuel* 186: 375–384.

Meda, U., Rakesh, S., and Raj, M. (2015). Bio-hydrogen production in microbial electrolysis cell using waste water from sugar industry. *Int. J. Eng. Sci. Res. Technol.* 4: 452–458.

Melis, A. (2009). Solar energy conversion efficiencies in photosynthesis: minimizing the chlorophyll antennae to maximize efficiency. *Plant Sci.* 177: 272–280.

Mohan, S.V. (2008). Fermentative hydrogen production with simultaneous wastewater treatment: influence of pretreatment and system operating conditions. *J. Sci. Ind. Res.* 67: 950–961.

Nath, K. and Das, D. (2004a). Biohydrogen production as a potential energy resource–present state-of-art. *J. Sci. Ind. Res.* 63: 729–738.

Nath, K. and Das, D. (2004b). Improvement of fermentative hydrogen production: various approaches. *Appl. Microbiol. Biotechnol.* 65: 520–529.

Noblecourt, A., Christophe, G., Larroche, C., and Fontanille, P. (2018). Hydrogen production by dark fermentation from pre-fermented depackaging food wastes. *Bioresour. Technol.* 247: 864–870.

Noyola, A. and Tinajero, A. (2005). Effect of biological additives and micronutrients on the anaerobic digestion of physicochemical sludge. *Water Sci. Technol.* 52: 275–281.

Oliot, M., Galier, S., de Balmann, H.R., and Bergel, A. (2016). Ion transport in microbial fuel cells: key roles, theory and critical review. *Appl. Energy* 183: 1682–1704.

Olson, D.G., Tripathi, S.A., Giannone, R.J. et al. (2010). Deletion of the Cel48S cellulase from *Clostridium thermocellum*. *Proc. Natl. Acad. Sci. U.S.A.* 107: 17727–17732.

Olson, D.G., McBride, J.E., Shaw, A.J., and Lynd, L.R. (2012). Recent progress in consolidated bioprocessing. *Curr. Opin. Biotechnol.* 23: 396–405.

Pandey, P., Shinde, V.N., Deopurkar, R.L. et al. (2016). Recent advances in the use of different substrates in microbial fuel cells toward wastewater treatment and simultaneous energy recovery. *Appl. Energy* 168: 706–723.

Parkash, A. (2016). Microbial fuel cells: a source of bioenergy. *J. Microb. Biochem. Technol.* 8: 247–255.

Patel, S.K., Lee, J.K., and Kalia, V.C. (2017). Dark-fermentative biological hydrogen production from mixed biowastes using defined mixed cultures. *Indian J. Microbiol.* 57: 171–176.

Penniston, J. and Kana, E.B.G. (2018). Impact of medium pH regulation on biohydrogen production in dark fermentation process using suspended and immobilized microbial cells. *Biotechnol. Biotechnol. Equip.* 32: 204–212.

Pinto, F.A.L., Troshina, O., and Lindblad, P. (2002). A brief look at three decades of research on cyanobacterial hydrogen evolution. *Int. J. Hydrogen Energy* 27: 1209–1215.

Posten, C. (2009). Design principles of photo-bioreactors for cultivation of microalgae. *Eng. Life Sci.* 9: 165–177.

Prince, R.C. and Kheshgi, H.S. (2005). The photobiological production of hydrogen: potential efficiency and effectiveness as a renewable fuel. *Crit. Rev. Microbiol.* 31: 19–31.

Redwood, M.D., Paterson-Beedle, M., and Macaskie, L.E. (2009). Integrating dark and light bio-hydrogen production strategies: towards the hydrogen economy. *Rev. Environ. Sci. Biotechnol.* 8: 149.

Ren, N.Q., Liu, B.F., Zheng, G.X. et al. (2009). Strategy for enhancing photo-hydrogen production yield by repeated fed-batch cultures. *Int. J. Hydrogen Energy* 34: 7579–7584.

Rodionova, M., Poudyal, R., Tiwari, I. et al. (2017). Biofuel production: challenges and opportunities. *Int. J. Hydrogen Energy* 42: 8450–8461.

Rozendal, R.A., Jeremiasse, A.W., Hamelers, H.V., and Buisman, C.J. (2007). Hydrogen production with a microbial biocathode. *Environ. Sci. Technol.* 42: 629–634.

Sahay, H., Yadav, A.N., Singh, A.K. et al. (2017). Hot springs of Indian Himalayas: potential sources of microbial diversity and thermostable hydrolytic enzymes. *3 Biotech* 7: 1–11.

Saxena, A.K., Yadav, A.N., Rajawat, M. et al. (2016). Microbial diversity of extreme regions: an unseen heritage and wealth. *Indian J. Plant Genet. Resour.* 29: 246–248.

Seifert, K., Waligorska, M., and Laniecki, M. (2010). Brewery wastewaters in photobiological hydrogen generation in presence of *Rhodobacter sphaeroides*. *Int. J. Hydrogen Energy* 35: 4085–4091.

Shin, H.S., Youn, J.H., and Kim, S.H. (2004). Hydrogen production from food waste in anaerobic mesophilic and thermophilic acidogenesis. *Int. J. Hydrogen Energy* 29: 1355–1363.

Singh, R.N., Gaba, S., Yadav, A.N. et al. (2016). First, high quality draft genome sequence of a plant growth promoting and cold active enzymes producing psychrotrophic *Arthrobacter agilis* strain L77. *Stand. Genomic Sci.* 11: 54. https://doi.org/10.1186/s40793-016-0176-4.

Sinha, P. and Pandey, A. (2011). An evaluative report and challenges for fermentative biohydrogen production. *Int. J. Hydrogen Energy* 36: 7460–7478.

Sompong, O., Prasertsan, P., and Birkeland, N.K. (2009). Evaluation of methods for preparing hydrogen-producing seed inocula under thermophilic condition by process performance and microbial community analysis. *Bioresour. Technol.* 100: 909–918.

Song, W., Cheng, J., Zhao, J. et al. (2016). Enhancing hydrogen production of Enterobacter aerogenes by heterologous expression of hydrogenase genes originated from *Synechocystis* sp. *Bioresour. Technol.* 216: 976–980.

Srivastava, N., Srivastava, M., Kushwaha, D. et al. (2017). Efficient dark fermentative hydrogen production from enzyme hydrolyzed rice straw by *Clostridium pasteurianum* (MTCC116). *Bioresour. Technol.* 238: 552–558.

Suman, A., Verma, P., Yadav, A.N., and Saxena, A.K. (2015). Bioprospecting for extracellular hydrolytic enzymes from culturable thermotolerant bacteria isolated from Manikaran thermal springs. *Res. J. Biotechnol.* 10: 33–42.

Sveshnikov, D., Sveshnikova, N., Rao, K., and Hall, D. (1997). Hydrogen metabolism of mutant forms of *Anabaena variabilis* in continuous cultures and under nutritional stress. *FEMS Microbiol.* 147: 297–301.

Tao, Y., He, Y., Wu, Y. et al. (2008). Characteristics of a new photosynthetic bacterial strain for hydrogen production and its application in wastewater treatment. *Int. J. Hydrogen Energy* 33: 963–973.

Tartakovsky, B., Manuel, M.-F., Wang, H., and Guiot, S. (2009). High rate membrane-less microbial electrolysis cell for continuous hydrogen production. *Int. J. Hydrogen Energy* 34: 672–677.

Thapa, L.P., Lee, S.J., Yang, X. et al. (2015). Improved bioethanol production from metabolic engineering of *Enterobacter aerogenes* ATCC 29007. *Process Biochem.* 50: 2051–2060.

Tian, X., Liao, Q., Zhu, X. et al. (2010). Characteristics of a biofilm photobioreactor as applied to photo-hydrogen production. *Bioresour. Technol.* 101: 977–983.

Tripathi, S.A., Olson, D.G., Argyros, D.A. et al. (2010). Development of pyrF-based genetic system for targeted gene deletion in *Clostridium thermocellum* and creation of a pta mutant. *J. Appl. Environ. Microbiol.* 76: 6591–6599.

Tsygankov, A. and Kosourov, S. (2014). Immobilization of photosynthetic microorganisms for efficient hydrogen production. In: *Microbial Bioenergy: Hydrogen Production,* 321–347. Springer.

Tsygankov, A.A., Hirata, Y., Miyake, M. et al. (1994). Photobioreactor with photosynthetic bacteria immobilized on porous glass for hydrogen photoproduction. *J. Ferment. Bioeng.* 77: 575–578.

Uyar, B., Eroglu, I., Yucel, M., and Gunduz, U. (2009). Photofermentative hydrogen production from volatile fatty acids present in dark fermentation effluents. *Int. J. Hydrogen Energy* 34: 4517–4523.

Vargas, S.R., dos Santos, P.V., Zaiat, M., and do Carmo Calijuri, M. (2018). Optimization of biomass and hydrogen production by *Anabaena* sp.(UTEX 1448) in nitrogen-deprived cultures. *Biomass Bioenergy* 111: 70–76.

Verma, P., Yadav, A.N., Kazy, S.K. et al. (2013). Elucidating the diversity and plant growth promoting attributes of wheat (*Triticum aestivum*) associated acidotolerant bacteria from southern hills zone of India. *Natl. J. Life Sci.* 10: 219–227.

Verma, P., Yadav, A.N., Kazy, S.K. et al. (2014). Evaluating the diversity and phylogeny of plant growth promoting bacteria associated with wheat (*Triticum aestivum*) growing in central zone of India. *Int. J. Curr. Microbiol. Appl. Sci.* 3: 432–447.

Verma, P., Yadav, A.N., Khannam, K.S. et al. (2015a). Assessment of genetic diversity and plant growth promoting attributes of psychrotolerant bacteria allied with wheat (*Triticum aestivum*) from the northern hills zone of India. *Ann. Microbiol.* 65: 1885–1899.

Verma, P., Yadav, A.N., Shukla, L. et al. (2015b). Alleviation of cold stress in wheat seedlings by *Bacillus amyloliquefaciens* IARI-HHS2-30, an endophytic psychrotolerant K-solubilizing bacterium from NW Indian Himalayas. *Natl. J. Life Sci.* 12: 105–110.

Verma, P., Yadav, A.N., Shukla, L. et al. (2015c). Hydrolytic enzymes production by thermotolerant *Bacillus altitudinis* IARI-MB-9 and Gulbenkiania mobilis IARI-MB-18 isolated from Manikaran hot springs. *Int. J. Adv. Res.* 3: 1241–1250.

Verma, P., Yadav, A.N., Khannam, K.S. et al. (2016a). Molecular diversity and multifarious plant growth promoting attributes of Bacilli associated with wheat (*Triticum aestivum* L.) rhizosphere from six diverse agro-ecological zones of India. *J. Basic Microbiol.* 56: 44–58.

Verma, P., Yadav, A.N., Khannam, K.S. et al. (2016b). Appraisal of diversity and functional attributes of thermotolerant wheat associated bacteria from the peninsular zone of India. *Saudi J. Biol. Sci.* https://doi.org/10.1016/j.sjbs.2016.01.042.

Verma, P., Yadav, A.N., Kumar, V. et al. (2017). Beneficial plant-microbes interactions: biodiversity of microbes from diverse extreme environments and its impact for crops improvement. In: *Plant-Microbe Interactions in Agro-Ecological Perspectives* (eds. D.P. Singh, H.B. Singh and R. Prabha), 543–580. Singapore: Springer Nature https://doi.org/10.1007/978-981-10-6593-4_22.

Vincenzini, M., Materassi, R., Sili, C., and Florenzano, G. (1986). Hydrogen production by immobilized cells. III—prolonged and stable $H_2$ photoevolution by *Rhodopseudomonas palustris* in light-dark cycles. *Int. J. Hydrogen Energy* 11: 623–626.

Wang, J. and Wan, W. (2008). Influence of Ni2+ concentration on biohydrogen production. *Bioresour. Technol.* 99: 8864–8868.

Wang, A., Sun, D., Cao, G. et al. (2011). Integrated hydrogen production process from cellulose by combining dark fermentation, microbial fuel cells, and a microbial electrolysis cell. *Bioresour. Technol.* 102: 4137–4143.

Wang, J., Xu, S., Xiao, B. et al. (2013). Influence of catalyst and temperature on gasification performance of pig compost for hydrogen-rich gas production. *Int. J. Hydrogen Energy* 38: 14200–14207.

Wells, M.A., Mercer, J., Mott, R.A. et al. (2011). Engineering a non-native hydrogen production pathway into *Escherichia coli* via a cyanobacterial [NiFe] hydrogenase. *Metab. Eng.* 13: 445–453.

White, A.L. and Melis, A. (2006). Biochemistry of hydrogen metabolism in *Chlamydomonas reinhardtii* wild type and a Rubisco-less mutant. *Int. J. Hydrogen Energy* 31: 455–464.

Wong, Y.M., Wu, T.Y., and Juan, J.C. (2014). A review of sustainable hydrogen production using seed sludge via dark fermentation. *Renewable Sustainable Energy Rev.* 34: 471–482.

Wu, X., Wang, X., Yang, H., and Guo, L. (2010). A comparison of hydrogen production among three photosynthetic bacterial strains. *Int. J. Hydrogen Energy* 35: 7194–7199.

Xie, G.J., Liu, B.F., Guo, W.Q. et al. (2012). Feasibility studies on continuous hydrogen production using photo-fermentative sequencing batch reactor. *Int. J. Hydrogen Energy* 37: 13689–13695.

Xie, G.J., Liu, B.F., Wen, H.Q. et al. (2013). Bioflocculation of photo-fermentative bacteria induced by calcium ion for enhancing hydrogen production. *Int. J. Hydrogen Energy* 38: 7780–7788.

Yadav A.N. 2015. Bacterial diversity of cold deserts and mining of genes for low temperature tolerance. PhD Thesis, IARI, New Delhi/BIT, Ranchi pp. 234, DOI: https://doi.org/10.13140/RG.2.1.2948.1283/2

Yadav, N. and Yadav, A. (2018). Biodiversity and biotechnological applications of novel plant growth promoting methylotrophs. *J. Appl. Biotechnol. Bioeng.* 5: 342–344.

Yadav, A.N., Sachan, S.G., Verma, P., and Saxena, A.K. (2015a). Prospecting cold deserts of north western Himalayas for microbial diversity and plant growth promoting attributes. *J. Biosci. Bioeng.* 119: 683–693.

Yadav, A.N., Sachan, S.G., Verma, P. et al. (2015b). Culturable diversity and functional annotation of psychrotrophic bacteria from cold desert of Leh Ladakh (India). *World J. Microbiol. Biotechnol.* 31: 95–108.

Yadav, A.N., Sharma, D., Gulati, S. et al. (2015c). Haloarchaea endowed with phosphorus solubilization attribute implicated in phosphorus cycle. *Sci. Rep.* 5: https://doi.org/10.1038/srep12293.

Yang, S., Du, G., Chen, J., and Kang, Z. (2017). Characterization and application of endogenous phase-dependent promoters in *Bacillus subtilis*. *Appl. Microbiol. Biotechnol.* 101: 4151–4161.

Yuan, Z., Yang, H., Zhi, X., and Shen, J. (2010). Increased performance of continuous stirred tank reactor with calcium supplementation. *Int. J. Hydrogen Energy* 35: 2622–2626.

Yusuf, N., Masdar, M., Isahak, W. et al. (2016). Challenges and prospects of bio-hydrogen production for PEMFC application: a review. *Int. J. Appl. Eng. Res.* 11: 9960–9969.

Zhang, X., Cheng, S., Wang, X. et al. (2009). Separator characteristics for increasing performance of microbial fuel cells. *Environ. Sci. Technol.* 43: 8456–8461.

Zhang, C., Lv, F.X., and Xing, X.H. (2011). Bioengineering of the *Enterobacter aerogenes* strain for biohydrogen production. *Bioresour. Technol.* 102: 8344–8349.

Zhang, L., Huang, H., Wang, H. et al. (2016). Rapid evolution of hyaluronan synthase to improve hyaluronan production and molecular mass in *Bacillus subtilis*. *Biotechnol. Lett* 38: 2103–2108.

Zhao, J.F., Song, W.L., Cheng, J., and Zhang, C.X. (2010). Heterologous expression of a hydrogenase gene in Enterobacter aerogenes to enhance hydrogen gas production. *World J. Microbiol. Biotechnol.* 26: 177.

Zhao, X., Xing, D., Liu, B. et al. (2012). The effects of metal ions and L-cysteine on hydA gene expression and hydrogen production by *Clostridium beijerinckii* RZF-1108. *Int. J. Hydrogen Energy* 37: 13711–13717.

Zhao, J., Song, W., Cheng, J. et al. (2017). Improvement of fermentative hydrogen production using genetically modified *Enterobacter aerogenes*. *Int. J. Hydrogen Energy* 42: 3676–3681.

Zhou, P., Wang, Y., Gao, R. et al. (2015). Transferring [NiFe] hydrogenase gene from *Rhodopseudomonas palustris* into *E. Coli*. BL21 (DE3) for improving hydrogen production. *Int. J. Hydrogen Energy* 40: 4329–4336.

Zhu, H. and Béland, M. (2006). Evaluation of alternative methods of preparing hydrogen producing seeds from digested wastewater sludge. *Int. J. Hydrogen Energy* 31: 1980–1988.

Zhu, H., Fang, H.H., Zhang, T., and Beaudette, L.A. (2007). Effect of ferrous ion on photo heterotrophic hydrogen production by *Rhodobacter sphaeroides*. *Int. J. Hydrogen Energy* 32: 4112–4118.

Zieliński, M., Korzeniewska, E., Filipkowska, Z. et al. (2017). Biohydrogen production at low load of organic matter by psychrophilic bacteria. *Energy* 134: 1132–1139.

Zürrer, H. and Bachofen, R. (1979). Hydrogen production by the photosynthetic bacterium *Rhodospirillum rubrum*. *J. Appl. Environ. Microbiol.* 37: 789–793.

# Index



*Bioprocessing for Biomolecules Production,* First Edition.
Edited by Gustavo Molina, Vijai Kumar Gupta, Brahma N. Singh, and Nicholas Gathergood.

**c**

## *f*

## g

***i***

## j

### k

***m***

## q



## r

## t

***u***



***v***

### y



### z